The HCS12/9S12:

An Introduction to Software and Hardware Interfacing

Han-Way Huang
Minnesota State University · Mankato

THOMSON
DELMAR LEARNING

Australia • Canada • Mexico • Singapore • Spain • United Kingdom • United States

The HCS12/9S12: An Introduction to Software and Hardware Interfacing
Han-Way Huang

Vice President, Technology and Trades SBU:
Alar Elken

Editorial Director:
Sandy Clark

Senior Acquisitions Editor:
Stephen Helba

Senior Development Editor:
Michelle Ruelos Cannistraci

Marketing Director:
Dave Garza

Senior Channel Manager:
Dennis Williams

Marketing Coordinator:
Stacey Wiktorek

Production Director:
Mary Ellen Black

Production Manager:
Andrew Crouth

Production Editor:
Stacy Masucci

Technology Project Manager:
Kevin Smith

Technology Project Specialist:
Linda Verde

Editorial Assistant:
Dawn Daugherty

Library of Congress
Cataloging-in-Publication Data:

Huang, Han-Way.
  The HCS12/9S12: an introduction to software and hardware interfacing/Han-Way Huang.
    p. cm.
  Includes index.
  ISBN 1-4018-9812-2
1. Digital control systems. 2. Electronic controllers. I. Title.
  TJ223.M53H948 2005
  629.8'9—dc22
                                2005007715

ISBN: 1-4018-9812-2

Contents

Chapter 2 HCS12 Assembly Programming 29

Chapter 3 Members and Hardware and Software Development Tools 77

Chapter 5 C Language Programming   181

Chapter 6 Interrupts, Clock Generation, Resets, and Operation Modes   221

Chapter 7 Parallel Ports   255

Chapter 8 Timer Functions 327

Chapter 9 Serial Communication Interface (SCI)   403

Chapter 10 Serial Peripheral Interface (SPI)   433

Chapter 12 Analog-to-Digital Converter   551

Chapter 13 Controller Area Network (CAN)   593

Chapter 14 Internal Memory Configuration and External Memory Expansion   653

Preface

The 16-bit HCS12, also known as 9S12, was designed as an upgrade to the popular 8-bit 68HC11 microcontroller. If the 68HC11 is considered as the first-generation microcontroller from Freescale, then the HCS12 should be considered as the second-generation microcontroller. The design of the HCS12 combines many features from several competitive 8-bit and 16-bit micro-controllers:

1. *Full feature timer system.* The HCS12 timer system provides input-capture, output-compare, pulse-accumulator, pulse-width-modulation (PWM), real-time interrupt (RTI), modulus down-counter, and computer-operate-properly (COP) systems. The enhanced timer system of the HCS12 has the capability to capture two edges before interrupting the CPU. This feature allows the HCS12 to handle high-frequency signals even when the CPU is very busy.

2. *In-system programming* (ISP) *capability.* Most HCS12 members provide on-chip flash memory and allow the software to be upgraded in the system.

3. *Background debug mode* (BDM). The BDM circuit provides a one-line interface for accessing the internal resources of the HCS12 and hence allows a low-cost source-level debugger to be designed.

4. *Multiple serial interfaces.* The HCS12 supports industrial-standard UART, SPI, I²C, BDLC, and the CAN bus. UART allows the user to interface with a PC using the popular RS232 protocol. The SPI and I²C allow the HCS12 to interface with numerous peripheral devices with serial interface, such as an LED driver, LCD, A/D converter, D/A converter, real-time clock, SRAM, EEPROM, and phase-locked loop. The BDLC is mainly used in automotive applications. The CAN bus was proposed as a communication bus for automotive applications. However, this bus is also widely used in process control applications.

5. *Fuzzy logic support.* The HCS12 provides a group of instructions to support fuzzy logic operations. These instructions should facilitate the programming of fuzzy logic applications in assembly language.

With these features, the HCS12 is perfect for those who want to learn modern microcontroller interfacing and applications.

Intended Audience

This book is written for two groups of readers:

1. Students in electrical and computer engineering and technology who are taking an introductory course in microprocessor interfacing and applications. For this group of people, this book provides a broad and systematic introduction to microprocessors and microcontrollers.

2. Senior electrical engineering and computer engineering students and working engineers who want to learn the HCS12 and use it in design projects. For this group of readers, this book provides numerous more complicated examples to explore the functions and applications of the HCS12.

Prerequisites

The writing of this book has assumed that the reader has taken a course on digital logic design and has been exposed to high-level language programming. Knowledge of digital logic will greatly facilitate learning about the HCS12. Knowledge of assembly language programming is not required because one of the goals of this book is to teach the HCS12 assembly language programming.

Approach

During the past few years, we have seen a new trend in teaching microprocessor applications: using the microcontroller along with a high-level language such as C to teach microprocessor interfacing and programming. This book follows the trend. The HCS12 is the latest 16-bit microcontroller developed by Freescale that combines all the useful features from many other 8- and 16-bit microcontrollers. According to the survey posted on the website www.microcontroller.com, more than 80% of the applications are written in C language and 75% are written in assembly language. It is not difficult to tell from these statistics that many applications use both the assembly and C languages. The code generated by a C compiler is still much larger than its equivalent in assembly language. Many time-critical applications are still written in assembly language. Both the assembly and C languages are used in illustrating the programming of the HCS12 in this text.

Organization of the Book

Chapter 1 presents the basic concepts about computer hardware and software, microcontroller applications, the HCS12 addressing modes, and a subset of the HCS12 instructions. Chapter 2 introduces basic assembly programming skills and the HCS12 instructions. Chapter 3 provides a brief summary of the features of the HCS12 members, hardware, and software development tools. A tutorial on using the AsmIDE and the use of the D-Bug12 monitor commands are also provided in this chapter, as are a few tips and examples on software debugging. Chapter 4 covers more advanced assembly programming skills and subroutine calls. A few tips on program debugging involving subroutine calls are provided. Chapter 5 provides a brief tutorial on the C language syntax and the use of the EmbeddedGNU IDE environment. Tips on program debugging in C language are provided. Chapter 6 discusses the concepts and programming of interrupts and resets.

Chapter 7 introduces the basic concepts of parallel I/O. This chapter also covers the interfacing and programming of simple I/O devices, including DIP switches, keypad scanning and debouncing, LEDs, LCDs, D/A controller, and stepper motor control. Electrical compatibility issues for I/O interfacing are also discussed. Chapter 8 explores the operation and applications of the timer system, including input capture, output compare, real-time interrupt, pulse accumulator, and pulse-width modulation. Chapter 9 deals with serial communication interface (SCI). Chapter 10 deals with the SPI interface and the applications of the SPI-compatible peripheral chips. Chapter 11 introduces the I²C protocol and several peripheral chips with I²C interface. Chapter 12 discusses the A/D converter and its applications in temperature, humidity, and barometric pressure measurement. Chapter 13 presents the CAN 2.0 protocol and the HCS12 CAN module. Several examples of the programming of the CAN module are provided. Chapter 14 describes the HCS12 internal SRAM, EEPROM, and flash memory. This chapter also explores issues related to external memory expansion: address space assignment, decoder design, and timing analysis.

Pedagogical Features

Each chapter opens with a list of objectives. Every subject is presented in a step-by-step manner. Background issues are presented before the specifics related to each HCS12 function are discussed. Numerous examples are then presented to demonstrate the use of each HCS12 I/O function. Procedural steps and flowcharts are used to help the reader to understand the program logic in most examples. Each chapter concludes with a summary and numerous exercises and lab assignments.

Demo Boards

A demo board is essential for learning the HCS12. Although general programming can be debugged by using a simulator, the testing of peripheral functions cannot be done without a piece of hardware (demo board). The SSE256 from Shuan-Shizu and the Dragon12 from Wytec are recommended for learning the HCS12 and doing the HCS12-based projects. Both demo boards provide seven-segment displays, LEDs, an LCD kit, RS232 connectors, and numerous peripheral chips to satisfy the needs of laboratory experiments at a reasonable cost. Both demo boards use Freescale MC9S12DP256 as the main controller and the D-Bug12 as their debug monitor. The D-Bug12 monitor allows the user to display and modify the contents of registers and memory locations, set breakpoints, enter short programs, download programs onto the demo boards for execution, and act as the BDM host for a target board.

Information about the SSE256 demo board can be found at the website http://www.evb.com.tw. Information about the Dragon12 demo board can be found at the website http://www.evbplus.com.

Development Software

The AsmIDE and MiniIDE are recommended for the development of assembly programs. Both IDEs combine a text editor, as12 cross assembler, project manager, and terminal program in one package. Tutorials for these two IDEs are provided in Chapter 3.

The GNU C compiler and the EmbeddedGNU IDE are recommended for the development of C programs. The EmbeddedGNU IDE is a front end of the GNU C compiler. It consists of a text editor, project manager, and terminal program. It invokes the GNU C compiler to build the project for the user.

All of these software development tools are freeware and can be downloaded from the Internet. The MiniIDE can be downloaded from the website http://www.mgtek.com\miniide. Both the AsmIDE and the EmbeddedGNU IDE can be downloaded from the website http://www.geocities.com/englere_geo. The user also needs to download and install the GNU C compiler. GNU C compiler can be downloaded from the website http://www.gnu-m68hc11.org.

Complementary CD

This text include a complementary CD that contains the source code of all example programs in the text and also the PDF files of datasheets of the HCS12 microcontroller and peripheral chips discussed in the text. The freeware software AsmIDE, EGNU IDE, and GNU C compiler are included on the complementary CD. Utility programs (in assembly and C languages) on time delay creation, LCD, data conversion, and the SPI function are included.

Supplements

A CD dedicated to instructors who adopt this text is also available from the publisher. This CD contains solutions to all exercise problems and the lecture notes in PowerPoint format. Instructors are encouraged to modify the PowerPoint lecture notes to suit their teaching needs. (ISBN#:1401898114)

Feedback and Update

The website at http://www.microhuang.com and http://www.electronictech.com have been set up for posting errata and additional examples and miscellaneous items. Readers are encouraged to check these websites for additional information. Error reports are welcomed. Please send them to hanwayh@yahoo.com.

Acknowledgments

This book would not have been possible without the help of a number of people. I would like to thank professor Shu-Jen Chen of DeVry University in Tinley Park, Illinois, for his useful suggestions to clarify several issues regarding the HCS12 microcontroller. I would also like to thank Steve Helba, senior acquisition editor, of Thomson Delmar Learning, and Michelle Ruelos Cannistraci, senior development editor, for their enthusiastic support during the preparation of this book. I also appreciate the outstanding work of the production staff led by Stacy Masucci.

I would like to express my thanks for the many comments and suggestions provided by the following colleagues who reviewed this text during the course of its development:

Shu-Jen Chen, DeVry University

Abhiman Hande, Lake Superior State University

I am grateful to my wife, Su-Jane, and my sons, Craig and Derek, for their encouragement and support during the entire preparation of this book.

Han-Way Huang
Mankato, Minnesota

1

Introduction to the HCS12 Microcontroller

1.1 Objectives

After completing this chapter you should be able to:

- Define or explain the following terms: computer, processor, microprocessor, microcontroller, embedded system, hardware, software, cross assembler, cross compiler, RAM, DRAM, SRAM, ROM, PROM, EPROM, EEPROM, flash memory, byte, word, nibble, bus, KB, MB, mnemonic, opcode, and operand.

- Explain the differences among the inherent, immediate, direct, extended, relative, indexed, indirect indexed addressing modes.

- Write a sequence of arithmetic and data transfer instructions to perform simple operations.

1.2 Basic Computer Concepts

A computer is made up of hardware and software. The hardware consists of the following major components:

- *The processor.* The processor is the brain of a computer system. All computation operations are performed in the processor. A computer system may have one or multiple processors. A computer system that consists of multiple processors is called a *multiprocessor* computer system.
- *The memory.* The memory is the place where software programs and data are stored. A memory system can be made of semiconductor chips, magnetic material, and/or optical discs.
- *Input device.* Input devices allow the computer user to enter data and program into the computer so that computation can be performed. Switches, keypads, keyboards, mice, microphones, scanners, and thumb wheels are examples of input devices.
- *Output device.* The results of computation are displayed via output devices so that the user can read them and equipment can be controlled. Examples of output devices include CRT displays, LEDs, seven-segment displays, liquid-crystal displays (LCD), printers, plotters, and so on.

As shown in Figure 1.1, the processor communicates with other components through a *common bus.* A bus is simply a group of conducting wires that allow signals to travel from one point to another. The common bus actually consists of three buses: address bus, data bus, and control bus. Each bus is a group of signals of the same nature.

1.2.1 The Processor

The processor, which is also called the central processing unit (CPU), can be further divided into two major parts (see Figure 1.1):

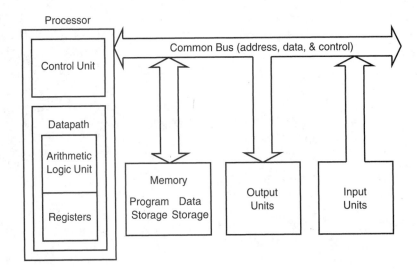

Figure 1.1 ■ Computer organization

Datapath. The datapath consists of a *register file* and an *arithmetic logic unit* (ALU). The register file consists of one or more registers, which is a storage location in the CPU. It is used to hold data or a memory address during the execution of an instruction. Because the register file is small and close to the ALU, accessing data in registers is much faster than accessing data in memory outside the CPU. The register file makes program execution more efficient. The number of registers varies from computer to computer. All of the arithmetic computations and logic evaluations are performed in the ALU. The ALU receives data from main memory and/or the register file, performs a computation, and, if necessary, writes the result back to main memory or registers.

Control unit. The control unit contains the hardware instruction logic. It decodes and monitors the execution of instructions and also acts as an arbiter as various portions of the computer system compete for the resources of the CPU. The system clock synchronizes the activities of the CPU. All CPU activities are measured by clock cycles. At the time of this writing, the highest clock rate of a PC has exceeded 3 GHz, where

1 GHz = 1 billion ticks (or cycles) per second

The control unit maintains a register called the *program counter* (PC), which controls the memory address of the next instruction to be executed. During the execution of an instruction, the presence of overflow, an addition carry, a subtraction borrow, and so forth is flagged by the system and stored in another register called a *status register*. The resultant flags are then used by programmers for program control and decision-making.

WHAT IS A MICROPROCESSOR?

The processor in a very large computer is built from a number of integrated circuits. A *microprocessor* is a processor fabricated on a single integrated circuit. A *microcomputer* is a computer that uses a microprocessor as its CPU. Early microcomputers were quite simple and slow. However, many of today's desktop or notebook computers have become very sophisticated and even faster than many large computers were only a few years ago.

One way to classify microprocessors is to use the number of bits (referred to as *word length*) that a microprocessor can manipulate in one operation. A microprocessor is 8-bit if it can only work on 8 bits of data in one operation. Microprocessors in use today include 4-bit, 8-bit, 16-bit, 32-bit, and 64-bit. The first 4-bit microprocessor was the Intel 4004, which was introduced in 1971. Since then, many companies have joined in the design and manufacturing of microprocessors. Today, 8-bit microprocessors are the most widely used among all microprocessors, whereas 64-bit microprocessors emerged only a few years ago and are mainly used in high-performance workstations and servers.

Many 32-bit and 64-bit microprocessors also incorporate on-chip memory to enhance their performance. Microprocessors must interface to input/output (I/O) devices in many applications. However, the characteristics and speed of the microprocessor and I/O devices are quite different. Peripheral chips are required to interface I/O devices with the microprocessor. For example, the integrated circuit i8255 is designed to interface a parallel device such as a printer or seven-segment display with the Intel 8-bit microprocessor 8085.

Microprocessors have been widely used since their invention. It is not an exaggeration to say that the invention of microprocessors has revolutionized the electronic industry. However, the following limitations of microprocessors have led to the development of microcontrollers:

- A microprocessor requires external memory to execute programs.
- A microprocessor cannot directly interface with I/O devices. Peripheral chips are needed.

- Glue logic (such as address decoders and buffers) is needed to interconnect external memory and peripheral interface chips with the microprocessor.

Because of these limitations, a microprocessor-based design cannot be made as small as might be desired. The development of microcontrollers not only eliminated most of these problems but also simplified the hardware design of microprocessor-based products.

WHAT IS A MICROCONTROLLER?

A microcontroller is a computer implemented on a single very large scale integration (VLSI) chip. A microcontroller contains everything a microprocessor contains plus one or more of the following components:

- *Memory*. Memory is used to store data or programs. A microcontroller may incorporate a certain amount of SRAM, ROM, EEPROM, EPROM, or flash memory.

- *Timer*. The timer function is among the most useful and complicated function in most microcontrollers. The timer function of most 8-bit and 16-bit microcontrollers consists of input capture, output compare, counter, pulse accumulator, and pulse-width modulation (PWM) modules. Timers can be used to measure frequency, period, pulse width, and duty cycle and to create time delay and generate waveforms.

- *Analog-to-digital converter* (ADC). The ADC is often used with a sensor. A *sensor* can convert a certain nonelectric quantity into an electric voltage. The ADC can convert a voltage into a digital value. Therefore, by combining an appropriate sensor and the ADC, a data acquisition system that measures temperature, weight, humidity, pressure, or airflow can be designed.

- *Digital-to-analog converter* (DAC). A DAC can convert a digital value into a voltage output. This function can be used in controlling DC motor speed, adjusting the brightness of a lightbulb, fluid level, and so on.

- *Direct memory access* (DMA) *controller*. DMA is a data transfer method that requires the CPU to perform initial setup but does not require the CPU to execute instructions to control the data transfer. The initial setup for a DMA transfer includes the setup of source and destination addresses and the byte count for the transfer. This mechanism can speed up data transfer by several factors.

- *Parallel I/O interface* (often called parallel *port*). Parallel ports are often used to drive parallel I/O devices such as seven-segment displays, an LCD, and a parallel printer.

- *Asynchronous serial I/O interface*. This transfer method transmits data bits serially without using a clock signal to synchronize the sender and receiver.

- *Synchronous serial I/O interface*. This interface transfers data bits serially and uses a clock signal to synchronize the sender and receiver.

- *Memory component interface circuitry*. All memory devices require certain control signals to operate. This interface generates appropriate control signals to be used by the memory chips.

- *Digital signal processing* (DSP) *feature*. Some microcontrollers (for example, the Motorola HCS12) provide features such as the multiply-accumulate instruction that can execute in one clock cycle to support DSP computation.

The Motorola 68HC12 family microcontrollers were introduced in 1996 as an upgrade for the 68HC11 microcontrollers. The HCS12 family was later designed to improve the performance of the 68HC12 family. The HCS12 family supports the same instruction set and addressing modes but provides much higher external bus speed. The designs of the peripheral functions of the HCS12 family are also different from those of the 68HC12.

Members in the HCS12 family implement the same instruction set and addressing modes but differ in the amount of on-chip memory and peripheral functions. The HCS12 has the following features:

- 16-bit CPU.
- Standard 64-KB address space support.
- Multiplexed (address and data) external bus.
- 0 to 4 KB of on-chip EEPROM.
- 2 KB to 14 KB of on-chip SRAM.
- 10-bit A/D converter.
- 16 KB to 512 KB of on-chip flash memory (or ROM).
- Timer module that includes input capture, output compare, and pulse accumulator functions. This is the most complicated module in the HCS12 microcontroller.
- Pulse-width modulation (PWM).
- Synchronous peripheral interface (SPI).
- Inter-integrated circuit (I²C) interface.
- Asynchronous serial communication interface (SCI).
- Byte data link controller (BDLC).
- Controller area network (CAN).
- Computer operating properly (COP) watchdog timer.
- Single-wire background debug mode (BDM).
- Instructions for supporting fuzzy logic.

A HCS12 member may have from 48 to 144 pins. These signal pins are needed to support the I/O functions.

For the time being, you (if you are new to microcontrollers) may feel overwhelmed by these terms. However, all of them will be explained in detail in appropriate chapters.

1.2.2 Memory

Memory is the place where software programs and data are stored. A computer may contain semiconductor, magnetic, and/or optical memory. Only semiconductor memory will be discussed in this chapter. The semiconductor memory may be further classified into two major types: *random access memory*, or RAM, and *read-only memory*, or ROM.

RANDOM ACCESS MEMORY

Random access memory allows the processor to read from and write to any location on the memory chip. It takes about the same amount of time to perform a read or a write operation to any memory location. Random access memory is *volatile* in the sense that it cannot retain data without power.

There are two types of RAM technology: dynamic RAM (DRAM) and static RAM (SRAM). Dynamic random access memories are memory devices that require a periodic refresh operation in order to maintain the stored information. *Refresh* is the process of restoring binary data stored in a particular memory location. The dynamic RAM uses one transistor and one capacitor to hold 1 bit of information. The information is stored in the capacitor in the form of electric charges. These charges will leak away over time, so a periodic refresh operation is needed to maintain the contents of the DRAM. The time interval over which each memory location of a DRAM chip must be refreshed at least once in order to maintain its contents is called the *refresh period*. Refresh periods typically range from a few milliseconds to over 100 milliseconds for today's high-density DRAMs.

Static random access memories are designed to store binary information without the need of periodic refreshes. Four to six transistors are used to store 1 bit of information. As long as power is stable, the information stored in the SRAM will not be degraded.

RAM is mainly used to stored *dynamic* programs or data. A computer user often wants to run different programs on the same computer, and these programs usually operate on different sets of data. The programs and data must therefore be loaded into RAM from the hard disk or other secondary storage, and for this reason they are called *dynamic*.

READ-ONLY MEMORY (ROM)

ROM is *nonvolatile*. When power is removed from ROM and then reapplied, the original data will still be there. As its name implies, ROM data can only be read, but this is not exactly true. Most ROM technologies require a special algorithm and voltage to write data into the chip. If the special algorithm and voltage are not used, any attempt to write to the ROM memory will not be successful. There are many different kinds of ROM technologies in use today:

Mask-programmed read-only memory (MROM) is a type of ROM that is programmed when it is manufactured. The semiconductor manufacturer places binary data in the memory according to the request of the customer. To be cost-effective, many thousands of MROM memory units, each consisting of a copy of the same data (or program), must be sold. MROM is the major memory technology used to hold microcontroller application programs and constant data. This is the only ROM technology that cannot be written into by the user program. Most people simply refer to MROM as ROM.

Programmable read-only memory (PROM) is a type of read-only memory that can be programmed in the field (often by the end user) using a device called a PROM programmer or PROM *burner*. Once a PROM has been programmed, its contents cannot be changed. Because of this, PROMs are also called one-time programmable ROM (OTP). PROMs are fuse based; i.e., end users program the fuses to configure the contents of the memory.

Erasable programmable read-only memory (EPROM) is a type of read-only memory that can be erased by subjecting it to strong ultraviolet light. The circuit design of EPROM requires the user to erase the contents of a location before writing a new value into it. A quartz window on top of the EPROM integrated circuit permits ultraviolet light to be shone directly on the silicon chip inside. Once the chip is programmed, the window should be covered with a dark tape to prevent gradual erasure of the data. If no window is provided, the EPROM chip becomes one-time programmable (OTP) only. Many microcontrollers incorporate on-chip one-time programmable EPROM to reduce the cost for those users who do not need to reprogram the EPROM. EPROM is often used in prototype computers, where the software may be revised many times until it is perfected. EPROM does not allow erasure of the contents of an individual location. The only way to make changes is to erase the entire EPROM chip and reprogram it. The programming of an EPROM chip is done electrically by using a device called an EPROM programmer. Today, most programmers are universal in the sense that they can program many types of devices including EPROM, EEPROM, flash memory, and *programmable logic devices.*

Electrically erasable programmable read-only memory (EEPROM) is a type of nonvolatile memory that can be erased by electrical signals and reprogrammed. Like EPROM, the circuit design of EEPROM also requires users to erase the contents of a memory location before a new value can be written into it. EEPROM allows each individual location to be erased and reprogrammed. Unlike EPROM, EEPROM can be erased and programmed using the same programmer. However, EEPROM pays the price for being very flexible in its erasability: The cost of an EEPROM chip is much higher than that of an EPROM chip of comparable density.

Flash memory was invented to incorporate the advantages and avoid the drawbacks of both the EPROM and EEPROM technologies. Flash memory can be erased and reprogrammed in the system without using a dedicated programmer. It achieves the density of EPROM, but it does not require a window for erasure. Like EEPROM, flash memory can be programmed and erased electrically. However, it does not allow the erasure of an individual memory location—the user can only erase a block or the entire chip. Today, more and more microcontrollers are incorporating on-chip flash memory for storing programs and static data.

1.3 The Computer's Software

Programs are known as *software*. A *program* is a set of instructions that the computer hardware can execute. A program is stored in the computer's memory in the form of binary numbers called *machine instructions*. For example, the HCS12 machine instruction

0001 1000 0000 0110

adds the contents of accumulator B and accumulator A together and leaves the sum in accumulator A. The machine instruction

0100 0011

decrements the contents of accumulator A by 1. The machine instruction

1000 0110 0000 0110

places the value 6 in accumulator A.

Writing programs in machine language is extremely difficult and inefficient:

- *Program entering.* The programmer will need to memorize the binary pattern of every machine instruction, which can be very challenging because a microprocessor may have several hundred machine instructions, and each machine instruction can have a different length. Constant table lookup will be necessary during the program-entering process. On the other hand, programmers are forced to work on program logic at a very low level because each machine instruction implements only a very primitive operation.
- *Program debugging.* Whenever there are errors, it is extremely difficult to trace the program because it consists only of sequences of 0s and 1s. A programmer will need to identify each machine instruction and then think about what operation is performed by that instruction. This is not an easy job.
- *Program maintenance.* Most programs will need to be maintained in the long run. A programmer who did not write the program will have a hard time reading the program and following the program logic.

Assembly language was invented to simplify the programming job. An *assembly program* consists of assembly instructions. An *assembly instruction* is the mnemonic representation of a machine instruction. For example, in the HCS12:

> aba stands for "add the contents of accumulator B to accumulator A." The corresponding machine instruction is 00011000 00000110.

> deca stands for "decrements the contents of accumulator A by 1." The corresponding machine instruction is 0100 0011.

A programmer no longer needs to scan through 0s and 1s in order to identify what instructions are in the program. This is a significant improvement over machine language programming.

The assembly program that the programmer enters is called *source program* or *source code*. A software program called an *assembler* is then invoked to translate the program written in assembly language into machine instructions. The output of the assembly process is called *object code*. It is a common practice to use a *cross assembler* to translate assembly programs, which is an assembler that runs on one computer but generates machine instructions to be run at a different computer with a totally different instruction set. In contrast, a *native assembler* runs on a computer and generates machine instructions to be executed by machines that have the same instruction set. The freeware **as12** is a cross assembler that runs on an IBM PC and generates machine code that can be downloaded into a HCS12-based computer for execution. The list file generated by an assembler shows both the source code and machine code (encoded in hexadecimal format). An example follows:

Line	Address	Machine Code	Source Code	
1:		=00002000	org	$2000
2:	2000	B6 1000	ldaa	$1000
3:	2003	BB 1001	adda	$1001
4:	2006	BB 1002	adda	$1002
5:	2009	7A 1100	staa	$1100
6:			end	

There are several drawbacks to programming in assembly language:

- The programmer must be very familiar with the hardware organization of the microcontroller on which the program is to be executed.
- A program (especially a long one) written in assembly language is extremely difficult to understand for anyone other than the author.
- Programming productivity is not satisfactory for large programming projects because the programmer needs to work on the program logic at a very low level.

For these reasons, high-level languages such as Fortran, COBOL, PASCAL, C, C++, and JAVA were invented. High-level languages are very close to plain English and hence a program written in high-level language becomes easier to understand. A statement in high-level language often needs to be implemented by tens or even hundreds of assembly instructions. The programmer can now work on the program logic at a much higher level and achieve higher productivity. A program written in a high-level language is also called a *source program*, and it requires a software program called a *compiler* to translate it into machine instructions. A compiler compiles a program into *object code*. Just as there are cross assemblers, there are *cross compilers* that run on one computer but translate programs into machine instructions to be executed on a computer with a different instruction set.

High-level languages have their own problems. One is that the resulting machine code cannot run as fast as its equivalent in assembly language due to the inefficiency of the compilation process. For this reason, many time-critical programs are still written in assembly language.

C language has been used extensively in microcontroller programming in the electronics industry. The website at http://www.microcontroller.com reported that C has been used to program 80% of the embedded systems, whereas assembly language is used in 75% of the embedded systems. Both C and assembly language will be used throughout this text.

1.4 The HCS12 CPU Registers

The HCS12 microcontroller has many registers, which can be classified into two categories: CPU registers and I/O registers. CPU registers are used solely to perform general-purpose operations such as arithmetic, logic, and program flow control. I/O registers are mainly used to configure the operations of peripheral functions, to hold data transferred in and out of the peripheral subsystem, and to record the status of I/O operations. The I/O registers in a microcontroller can further be classified into *data, data direction, control,* and *status registers*. These registers are treated as memory locations when they are accessed. CPU registers do not occupy the HCS12 memory space.

The CPU registers of the HCS12 are shown in Figure 1.2 and are listed next. Some of the registers are 8-bit and others are 16-bit.

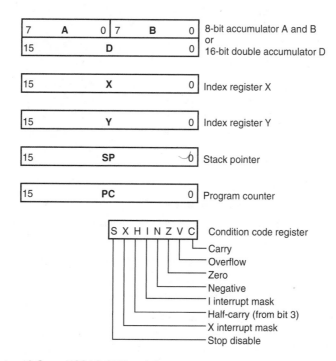

Figure 1.2 ■ HCS12 CPU registers

General-purpose accumulator A and B. Both A and B are 8-bit registers. Most arithmetic functions are performed on these two registers. These two accumulators can also be concatenated to form a single, 16-bit accumulator referred to as the D accumulator.

Index registers X and Y. These two registers are used mainly in forming operand addresses during the instruction execution process. However, they are also used in several arithmetic operations.

Stack pointer (SP). A stack is a *last-in-first-out* data structure. The HCS12 has a 16-bit stack pointer that points to the top byte of the stack (shown in Figure 1.3). The stack grows toward lower addresses. The use of the stack will be discussed in Chapter 4.

Program counter (PC). The 16-bit PC holds the address of the next instruction to be executed. After the execution of an instruction, the PC is incremented by the number of bytes of the executed instruction.

Figure 1.3 ■ HCS12 stack structure

Condition code register (CCR). This 8-bit register is used to keep track of the program execution status, control the execution of conditional instructions, and enable/disable the interrupt handling. The contents of the CCR register are shown in Figure 1.2. The function of each condition code bit will be explained in later sections and chapters.

The HCS12 supports the following types of data:

- Bits
- 5-bit signed integers
- 8-bit signed and unsigned integers
- 8-bit, two-digit binary-coded-decimal (BCD) numbers
- 9-bit signed integers
- 16-bit signed and unsigned integers
- 16-bit effective addresses
- 32-bit signed and unsigned integers

Negative numbers are represented in two's complement format. Five-bit and 9-bit signed integers are formed during addressing mode computations. Sixteen-bit effective addresses are formed during addressing mode computations. Thirty-two-bit integer dividends are used by extended division instructions. Extended multiply and extended multiply-and-accumulate instructions produce 32-bit products.

A multibyte integer (16-bit or 32-bit) is stored in memory from most significant to least significant bytes, starting from low to higher addresses. A number can be represented in binary, octal, decimal, or hexadecimal format. An appropriate prefix (shown in Table 1.1) is added in front of the number to indicate its base:

Base	Prefix	Example
binary	%	%10001010
octal	@	@1234567
decimal		12345678
hexadecimal (shorthand **hex**)	$	$5678

Table 1.1 ■ Prefixes for number bases

1.5 Memory Addressing

Memory consists of a sequence of directly addressable "locations." A memory location is referred to as an *information unit.* A memory location can be used to store data, instruction, and the status of peripheral devices. An information unit has two components: its *address* and its *contents*, as shown in Figure 1.4. Each location in memory has an address that must be supplied before its contents can be accessed. The CPU communicates with memory by first identifying the location's address and then passing this address on the address bus. This is similar to a UPS delivery person needing an address to deliver a parcel. The data are transferred between memory and the CPU along the data bus (see Figure 1.5). The number of bits that can be transferred on the data bus at once is called the *data bus width* of the processor.

Address ⟶ | Contents |

Figure 1.4 ■ The components of a memory location

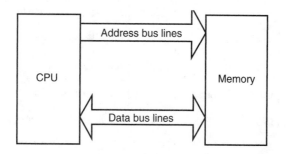

Figure 1.5 ■ Transferring data between CPU and memory

The size of memory is measured in bytes; a byte consists of 8 bits. A 4-bit quantity is called a *nibble*; a 16-bit quantity is called a *word*. To simplify the quantification of memory, the unit *kilobyte* (kB) is often used. k is given by the following formula:

$$k = 2^{10} = 1024$$

Another frequently used unit is *megabyte* (MB), which is given by the following formula:

$$M = k^2 = 2^{20} = 1024 \times 1024 = 1048576$$

The HCS12 has a 16-bit address bus and a 16-bit data bus. The 16-bit data bus allows the HCS12 to access 16-bit data in one operation. The 16-bit address bus enables the HCS12 to address directly up to 2^{16} (65536) different memory locations. Most HCS12 members use paging techniques to allow user programs to access more than 64 kB.

In this book, we will use the notation **m[addr]** and **[reg]** to refer to the contents of a memory location at **addr** and the contents of the register, **reg**, respectively. For example,

m[$20]

refers to the contents of the memory location at $20 and [A] refers to the contents of accumulator A.

1.6 HCS12 Addressing Modes

A HCS12 instruction consists of 1 or 2 bytes of opcode and 0 to 5 bytes of operand addressing information. The opcode byte(s) specifies the operation to be performed and the addressing modes to be used to access the operand(s). The first byte of a 2-byte opcode is $18.

Addressing modes determine how the CPU accesses memory locations to be operated upon. Addressing modes supported by the HCS12 are summarized in Table 1.2.

Addressing Mode	Source Format	Abbrev.	Description
Inherent	**INST** (no externally supplied operands)	INH	Operands (if any) are in CPU registers
Immediate	**INST** #opr8i or **INST** #opr16i	IMM	Operand is included in instruction stream. 8- or 16-bit size implied by context
Direct	**INST** opr8a	DIR	Operand is the lower 8 bits of an address in the range $0000-$00FF
Extended	**INST** opr16a	EXT	Operand is a 16-bit address
Relative	**INST** rel8 or **INST** rel16	REL	An 8-bit or 16-bit relative offset from the current PC is supplied in the instruction
Indexed (5-bit offset)	**INST** oprx5,xysp	IDX	5-bit signed constant offset from x,y,sp, or pc
Indexed (pre-decrement)	**INST** oprx3,-xys	IDX	Auto pre-decrement x, y, or sp by 1 ~ 8
Indexed (pre-increment)	**INST** oprx3,+xys	IDX	Auto pre-increment x, y, or sp by 1 ~ 8
Indexed (post-decrement)	**INST** oprx3,xys-	IDX	Auto post-decrement x, y, or sp by 1 ~ 8

Table 1.2 ■ HCS12 addressing mode summary

Indexed (post-increment)	INST oprx3,xys+	IDX	Auto post-increment x, y, or sp by 1 ~ 8
Indexed (accumulator offset)	INST abd,xysp	IDX	Indexed with 8-bit (A or B) or 16-bit (D) accumulator offset from x, y, sp, or pc
Indexed (9-bit offset)	INST oprx9,xysp	IDX1	9-bit signed constant offset from x, y, sp, or pc (lower 8-bits of offset in one extension byte)
Indexed (16-bit offset)	INST oprx16,xysp	IDX2	16-bit constant offset from x, y, sp, or pc (16-bit offset in two extension bytes)
Indexed-Indirect (16-bit offset)	INST [oprx16,xysp]	[IDX2]	Pointer to operand is found at 16-bit constant offset from (x, y, sp, or pc)
Indexed-Indirect (D accumulator offset)	INST [D,xysp]	[D,IDX]	Pointer to operand is found at x, y, sp, or pc plus the value in D

Note: 1. INST stands for the instruction mnemonic.
        2. opr8i stands for 8-bit immediate value.
        3. opr16i stands for 16-bit immediate value.
        4. opr8a stands for 8-bit address.
        5. opr16a stands for 16-bit address.
        6. oprx3 stands for 3-bit value (amount to be incremented or decremented).

Table 1.2 ■ HCS12 addressing mode summary (continued)

1.6.1 Inherent Mode

Instructions that use this addressing mode either have no operands or all operands are in internal CPU registers. In either case, the CPU does not need to access any memory locations to complete the instruction. For example, the following two instructions use inherent mode:

```
nop            ; this instruction has no operands
inx            ; operand is a CPU register
```

1.6.2 Immediate Mode

Operands for immediate mode instructions are included in the instruction stream. The CPU does not access memory when this type of instruction is executed. An immediate value can be 8-bit or 16-bit depending on the context of the instruction. An immediate value is preceded by a # character in the assembly instruction. For example,

```
ldaa   #$55     ; A ← $55
```

places the hex value $55 in accumulator A when this instruction is executed.

```
ldx    #$2000   ; X ← $2000
```

places the hex value $2000 in index register X when this instruction is executed.

```
ldy    #$44     ; Y ← $0044
```

places the hex value $0044 in index register Y when this instruction is executed. Only an 8-bit value was supplied in this instruction. However, the assembler will generate the 16-bit value $0044 because the CPU expects a 16-bit value when this instruction is executed.

1.6.3 Direct Mode

This addressing mode is sometimes called *zero-page* addressing because it is used to access operands in the address range of $0000–$00FF. Since these addresses begin with $00, only the eight low-order bits of the address need to be included in the instruction, which saves program space and execution time. For example,

    ldaa $20   ; A ← m[$20]

fetches the contents of the memory location at $0020 and puts it in accumulator A.

    ldx $20   ; X_H ← m[$20], X_L ← m[$21]

fetches the contents of memory locations at $0020 and $0021 and places them in the upper and lower bytes (X_H and X_L), respectively, of the index register X.

1.6.4 Extended Mode

In this addressing mode, the full 16-bit address of memory location to be operated on is provided in the instruction. This addressing mode can be used to access any location in the 64-kB memory map. For example,

    ldaa $2000   ; A ← m[$2000]

copies the contents of the memory location at $2000 into accumulator A.

1.6.5 Relative Mode

The relative addressing mode is used only by branch instructions. Short and long conditional branch instructions use the relative addressing mode exclusively. Branching versions of bit manipulation instructions (BRSET and BRCLR) may also use the relative addressing mode to specify the branch target. A short branch instruction consists of an 8-bit opcode and a signed 8-bit offset contained in the byte that follows the opcode. Long branch instructions consist of an 8-bit prebyte, an 8-bit opcode, and a signed 16-bit offset contained in two bytes that follow the opcode.

Each conditional branch instruction tests certain status bits in the condition code register. If the bits are in a specified state, the offset is added to the address of the next memory location after the offset to form an effective address, and execution continues at that address; if the bits are not in the specified state, execution continues with the instruction immediately following the branch instruction.

Both 8-bit and 16-bit offsets are signed two's complement numbers to support branching forward and backward in memory. The numeric range of the short branch-offset values is $80 (−128) to $7F (127). The numeric range of long branch-offset values is $8000 (−32768) to $7FFF (32767). If the offset is zero, the CPU executes the instruction immediately following the branch instruction, regardless of the test result.

Branch offset is often specified using a label rather than a numeric value due to the difficulty of calculating the exact value of the offset. For example, in the following instruction segment:

```
minus .
       .              ; if N (of CCR) = 1
       .              ;         PC ← PC + branch offset
       bmi    minus   ; else
       . . .          ;         PC ← PC
```

The instruction **bmi minus** will cause the HCS12 to execute the instruction with the label **minus** if the N flag of the CCR register is set to 1.

The assembler will calculate the appropriate branch offset when the symbol that represents the branch target is encountered. Using a symbol to specify the branch target makes the programming task easier and the resultant program more readable.

1.6.6 Indexed Modes

There are quite a few variations for the indexed addressing scheme. The indexed addressing scheme uses a postbyte plus zero, one, or two extension bytes after the instruction opcode. The postbyte and extensions implement the following functions:

- Specify which index register is used.
- Determine whether a value in an accumulator is used as an offset.
- Enable automatic pre- or postincrement or pre- or postdecrement.
- Specify the size of increment or decrement.
- Specify the use of 5-, 9-, or 16-bit signed offsets.

The indexed addressing scheme allows

- The stack pointer to be used as an index register in all indexed operations
- The program counter to be used as an index register in all but autoincrement and autodecrement modes
- The value in accumulator A, B, or D to be used as an offset
- Automatic pre- or postincrement or pre- or postdecrement by -8 to $+8$
- A choice of 5-, 9-, or 16-bit signed constant offsets
- Indirect indexing with 16-bit offset or accumulator D as the offset

Table 1.3 is a summary of indexed addressing mode capabilities and a description of postbyte encoding. The postbyte is noted as *xb* in instruction descriptions.

Postbyte Code (*xb*)	Source Code Syntax	Comments rr: 00 = X, 01 = Y, 10 = SP 11 = PC
rr0nnnnn	, r n, r −n, r	**5-bit constant offset** n = −16 to +15 r can be X, Y, SP, or PC
111rr0zs	n, r −n, r	**Constant offset** (9- or 16-bit signed) z- 0 = 9-bit wth sign in LSB of postbyte, (s)　　　−256 < n < 255 　　　1 = 16-bit　　　　　　　　　　　　　　　　0 < n < 65,536 if z = s = 1, 16-bit offset indexed-indirect (see below) r can be X, Y, SP, or PC
111rr011	[n, r]	**16-bit offset indexed-indirect**　　　　　　0 < n < 65,536 rr can be X, Y, SP, or PC
rr1pnnnn	n, −r n, +r n, r− n, r+	**Auto pre-decrement/increment or auto post-decrement/increment;** p = pre-(0) or post-(1), n = −8 to −1, +1 to +8 r can be X, Y, or SP (PC not a valid choice) 　　+8 = 0111 　　... 　　+1 = 0000 　　−1 = 1111 　　... 　　−8 =1000

Table 1.3 ■ Summary of indexed operations

111rr1aa	A, r
	B, r
	D, r
		Accumulator offset (unsigned 8-bit or 16-bit)
		aa- 00 = A
		       01 = B
		       10 = D (16-bit)
		       11 = see accumulator D offset indexed-indirect
		r can be X, Y, SP or PC

111rr111	[D, r]	**Accumulator D offset indexed-indirect** r can be X, Y, SP, or PC

Table 1.3 ■ Summary of indexed operations (continued)

5-BIT CONSTANT OFFSET INDEXED ADDRESSING

This indexed addressing mode adds a 5-bit signed offset that is included in the instruction postbyte to the base index register to form the effective address. The base index register can be X, Y, SP, or PC. The range of the offset is from −16 to +15. Although the range of the offset is short, it is the most often used offset. Using this indexed addressing mode can make the instruction shorter. For example,

    ldaa   0, X

loads the contents of the memory location pointed to by index register X (i.e., X contains the address of the memory location to be accessed) into accumulator A.

    stab   −8, X

stores the contents of accumulator B in the memory location with the address, which is equal to the contents of index register X minus 8.

9-BIT CONSTANT OFFSET INDEXED ADDRESSING

This indexed addressing mode uses a 9-bit signed offset, which is added to the base index register (X, Y, SP, or PC) to form the effective address of the memory location affected by the instruction. This gives a range of −256 through +255 from the value in the base index register. For example,

    ldaa $FF, X

loads the contents of the memory location located at the address, which is equal to 255 plus the value in index register X.

    ldab   −20, Y

loads the contents of the memory location with the address, which is equal to the value in index register Y minus 20.

16-BIT CONSTANT OFFSET INDEXED ADDRESSING

This indexed addressing mode specifies a 16-bit offset to be added to the base index register (X, Y, SP, or PC) to form the effective address of the memory location affected by the instruction. This allows access to any location in the 64-KB address space. Since the address bus and the offset are both 16 bits, it does not matter whether the offset value is considered to be a signed or an unsigned value. The 16-bit offset is provided as two extension bytes after the instruction postbyte in the instruction flow.

16-BIT CONSTANT INDIRECT INDEXED ADDRESSING

This indexed addressing mode adds a 16-bit offset to the base index register to form the address of a memory location that contains a pointer to the memory location affected by the instruction. The square brackets distinguish this addressing mode from 16-bit constant offset indexing; for example,

    ldaa   [10, X]

In this example, index register X holds the base address of a table of pointers. Assume that X has an initial value of $1000, and that $2000 is stored at addresses $100A and $100B. The instruction first adds the value 10 to the value in X to form the address $100A. Next, an address pointer ($2000) is fetched from memory at $100A. Then, the value stored in $2000 is read and loaded into accumulator A.

AUTO PRE/POSTDECREMENT/INCREMENT INDEXED ADDRESSING

This indexed addressing mode provides four ways to automatically change the value in a base index register as a part of instruction execution. The index register can be incremented or decremented by an integer value either before or after indexing takes place. The base index register may be X, Y, or SP.

Predecrement and preincrement versions of the addressing mode adjust the value of the index register before accessing the memory location affected by the instruction—the index register retains the changed value after the instruction executes. Postdecrement and postincrement versions of the addressing mode use the initial value in the index register to access the memory location affected by the instruction, then change the value of the index register.

The HCS12 allows the index register to be incremented or decremented by any integer value in the ranges −8 through −1, or 1 through 8. The value need not be related to the size of the operand for the current instruction. These instructions can be used to incorporate an index adjustment into an existing instruction rather than using an additional instruction and increasing execution time. This addressing mode is also used to perform operations on a series of data structures in memory. For example,

    staa   1, −SP

stores the contents of accumulator A at the memory location with the address equal to the value of stack pointer SP minus 1. After the store operation, the contents of SP are decremented by 1.

    ldx   2, SP+

loads the contents of the memory locations pointed to by the stack pointer SP into index register X and also increments the value of SP by 2 after the instruction execution.

ACCUMULATOR OFFSET INDEXED ADDRESSING

In this indexed addressing mode, the effective address is the sum of the values in the base index register and an unsigned offset in one of the accumulators. The value in the base index register itself is not changed. The base register can be X, Y, SP, or PC and the accumulator can be either of the 8-bit accumulators (A or B) or the 16-bit D accumulator. For example, the instruction

    ldaa   B,X

loads the contents of the memory location into accumulator A with the address equal to the sum of the values of accumulator B and index register X. Neither B nor X is changed after the instruction execution.

ACCUMULATOR D INDIRECT INDEXED ADDRESSING

This indexed addressing mode adds the value in accumulator D to the value in the base index register to form the address of a memory location that contains a pointer to the memory location affected by the instruction. The instruction operand does not point to the memory location to be acted upon, but rather to the location of a pointer to the location to be acted upon. The square brackets distinguish this addressing mode from accumulator D offset indexing. For example, the following instruction sequence implements a computed GOTO statement:

```
          jmp      [D, PC]
GO1       dc.w     target1      ; the keyword dc.w reserves two bytes to hold the
GO2       dc.w     target2      ; value of the symbol that follows
GO3       dc.w     target3      ;        "
          . . .
target1 . . .
          .
          .
          .
target2 . . .
          .
          .
          .
target3 . . .
          .
          .
          .
```

In this instruction segment, the names (also called labels) *target1*, *target2*, and *target3* are labels that represent the addresses of the memory locations that the JMP instruction may jump to. The names GO1, GO2, and GO3 are also labels. They represent the memory locations that hold the values of the labels *target1*, *target2*, and *target3*, respectively.

The values beginning at GO1 are addresses of potential destinations of the jump instructions. At the time the **jmp [D, PC]** instruction is executed, PC points to the address GO1 and D holds one of the values $0000, $0002, or $0004 (determined by the program some time before the JMP). Assume that the value in D is $0002. The jmp instruction adds the values in D and PC to form the address of GO2 and jumps to target2. The locations of target1 through target3 are known at the time of program assembly but the destination of the jmp depends on the value in D computed during program execution.

1.7 Addressing More than 64 KB

The HCS12 devices incorporate hardware that supports addressing a larger memory space than the standard 64 kB. The expanded memory system is accessed by using the bank-switching scheme. The HCS12 treats the 16 kB of memory space from $8000 to $BFFF as a program memory window. The HCS12 has an 8-bit program page register (PPAGE), which allows up to 256 16-kB program memory pages to be switched into and out of the program memory window. This provides up to 4 MB of paged program memory.

(1-26-2012)

⋆1.8 HCS12 Instruction Examples

In this section we will examine several groups of instructions to develop a feel for the HCS12 instruction set and learn some simple applications of these instructions.

1.8.1 The Load and Store Instructions

The load instruction copies the contents of a memory location or places an immediate value into an accumulator or a register. Memory contents are not changed. Store instructions copy the contents of a CPU register into a memory location. The contents of the accumulator or CPU register are not changed. Store instructions automatically update the N and Z flags in the condition code register (CCR). Table 1.4 is a summary of load and store instructions.

Load Instructions		
Mnemonic	Function	Operation
LDAA	Load A	$(M) \rightarrow A$
LDAB	Load B	$(M) \rightarrow B$
LDD	Load D	$(M:M + 1) \rightarrow (A:B)$
LDS	Load SP	$(M:M + 1) \rightarrow SP$
LDX	Load index register X	$(M:M + 1) \rightarrow X$
LDY	Load index register Y	$(M:M + 1) \rightarrow X$
LEAS	Load effective address into SP	Effective address $\rightarrow$ SP
LEAX	Load effective address into X	Effective address $\rightarrow$ X
LEAY	Load effective address into Y	Effective address $\rightarrow$ Y
Store Instructions		
Mnemonic	Function	Operation
STAA	Store A	$(A) \rightarrow M$
STAB	Store B	$(B) \rightarrow M$
STD	Store D	$(A) \rightarrow M, (B) \rightarrow M + 1$
STS	Store SP	$(SP) \rightarrow M, M + 1$
STX	Store X	$(X) \rightarrow M:M + 1$
STY	Store Y	$(Y) \rightarrow M:M + 1$

Table 1.4 ■ Load and store instructions

All except for the relative addressing mode can be used to select the memory location or value to be loaded into an accumulator or a CPU register. All except for the relative and immediate addressing modes can be used to select the memory location to store the contents of a CPU register. For example, the following instruction loads the contents of the memory location pointed to by index register X into accumulator A:

    ldaa   0,X

The following instruction loads the contents of the memory location at $1004 into accumulator B:

    ldab $1004

The following instruction copies the contents of accumulator A into the memory location at $20:

    staa $20

The following instruction stores the contents of index register X in memory locations at $8000 and $8001:

    stx $8000

When dealing with a complex data structure such as a record, we often use an index register or the stack pointer to point to the beginning of the data structure and use the indexed addressing mode to access the elements of the data structure. For example, a record contains the following four fields:

- ID number (unit *none*, size 4 bytes)
- Height (unit *inch*, size 1 byte)
- Weight (unit *pound*, size 2 bytes)
- Age (unit *year*, size 1 byte)

Suppose this record is stored in memory starting at $6000. Then we can use the following instruction sequence to access the weight field:

```
ldx    #$6000   ; set X to point to the beginning of data structure
ldd    5, X     ; copy weight into D
```

1.8.2 Transfer and Exchange Instructions

A summary of transfer and exchange instructions is displayed in Table 1.5. Transfer instructions copy the contents of a register or accumulator into another register or accumulator. Source content is not changed by the operation. TFR is a universal transfer instruction, but other mnemonics are accepted for compatibility with the MC68HC11. The TAB and TBA instructions affect the N, Z, and V condition code bits. The TFR instruction does not affect the condition code bits.

Transfer Instructions		
Mnemonic	Function	Operation
TAB	Transfer A to B	(A) → B
TAP	Transfer A to CCR	(A) → CCR
TBA	Transfer B to A	(B) → A
TFR	Transfer register to register	(A, B, CCR, D, X, Y, or SP) → A, B, CCR, D, X, Y, or SP
TPA	Transfer CCR to A	(CCR) → A
TSX	Transfer SP to X	(SP) → X
TSY	Transfer SP to Y	(SP) → Y
TXS	Transfer X to SP	(X) → SP
TYS	Transfer Y to SP	(Y) → SP
Exchange Instructions		
Mnemonic	Function	Operation
EXG	Exchange register to register	(A, B, CCR, D, X, Y, or SP) ↔ A, B, CCR, D, X, Y, or SP
XGDX	Exchange D with X	(D) ↔ X
XGDY	Exchange D with Y	(D) ↔ Y
Sign Extension Instructions		
Mnemonic	Function	Operation
SEX	Sign extend 8-bit operand	(A, B, CCR) → X, Y, or SP

Table 1.5 ■ Transfer and exchange instructions

It is possible to transfer from a smaller register to a larger one or vice versa. When transfering from a smaller register to a larger one, the upper half of the larger register is cleared to 0, whereas the lower half receives the value of the smaller register. When transfering from a larger register to a smaller one, the smaller register receives the value of the lower half of the larger register. For example,

```
tfr    A,X      ;X ← $00:[A], A is 0 extended to 16 bits before assigning to X
tfr    X,B      ;B ← X[7:0], B receives bits 7 to 0 of X
```

Exchange instructions (eng r1.r2) exchange the contents of pairs of registers or accumulators. For example,

```
exg    A, B
```

exchanges the contents of accumulator A and B.

```
exg    D,X
```

exchanges the contents of double accumulator D and index register X.

The r1 register needs not have the same size as r2. If r1 has larger size than r2 does, then r2 will be 0-extended to 16 bits and loaded into r1 whereas r2 will receive the lower half of r1. For example,

```
exg    X,A      ; X ← $00:[A], A ← X[7:0]
```

If the r2 register is larger in size than r1, then r1 will be 0-extended to 16 bits and loaded into r2 whereas r1 will receive the lower half of r2. For example,

```
exg    A,Y      ; A ← Y[7:0], Y ← $00:[A]
```

The SEX instruction is a special case of the universal transfer instruction that is used to sign-extend 8-bit two's complement numbers so that they can be used in 16-bit operations. The 8-bit number is copied from accumulator A, accumulator B, or the condition code register to accumulator D, index register X, index register Y, or the stack pointer. All the bits in the upper byte of the 16-bit result are given the value of the most significant bit of the 8-bit number. For example,

```
sex    A,X
```

copies the contents of accumulator A to the lower byte of X and duplicates bit 7 of A to every bit of the upper byte of X.

```
sex    B,Y
```

copies the contents of accumulator B to the lower byte of Y and duplicates bit 7 of B to every bit of the upper byte of Y.

Transfer instructions allow operands to be placed in the right register so that the desired operation can be performed. For example, if we want to compute the squared value of accumulator A, we can use the following instruction sequence:

```
tab             ; B ⇐ (A)
mul             ; A:B ⇐ (A) × (B)
```

Applications of other transfer and exchange instructions will be discussed in Chapters 2 and 4.

1.8.3 Move Instructions ✳

A summary of move instructions is listed in Table 1.6. These instructions move data bytes or words from a source ($M \sim M+1_1$) to a destination ($M \sim M+1_2$) in memory. Six combinations of immediate, extended, and indexed addressing are allowed to specify source and destination addresses (IMM ⇒ EXT, IMM ⇒ IDX, EXT ⇒ EXT, EXT ⇒ IDX, IDX ⇒ EXT, IDX ⇒ IDX). Move instructions allow the user to transfer data from memory to memory or from I/O registers to memory and vice versa. For example, the following instruction copies the contents of the memory location at $1000 to the memory location at $2000:

```
movb $1000, $2000
```

Transfer Instructions		
Mnemonic	Function	Operation
MOVB MOVW	Move byte (8-bit) Move word (16-bit)	$(M1) \rightarrow M2$ $(M:M + 1_1) \rightarrow M:M + 1_2$

Table 1.6 ■ Move instructions

The following instruction copies the 16-bit word pointed to by X to the memory location pointed by Y:

    movw 0,X, 0,Y

1.8.4 Add and Subtract Instructions

Add and subtract instructions allow the HCS12 to perform fundamental arithmetic operations. A summary of add and subtract instructions is in Table 1.7.

Add Instructions		
Mnemonic	Function	Operation
ABA	Add B to A	$(A) + (B) \rightarrow A$
ABX	Add B to X	$(B) + (X) \rightarrow X$
ABY	Add B to Y	$(B) + (Y) \rightarrow Y$
ADCA	Add with carry to A	$(A) + (M) + C \rightarrow A$
ADCB	Add with carry to B	$(B) + (M) + C \rightarrow B$
ADDA	Add without carry to A	$(A) + (M) \rightarrow A$
ADDB	Add without carry to B	$(B) + (M) \rightarrow B$
ADDD	Add without carry to D	$(A:B) + (M:M + 1) \rightarrow A:B$
Subtract Instructions		
Mnemonic	Function	Operation
SBA	Subtract B from A	$(A) - (B) \rightarrow A$
SBCA	Subtract with borrow from A	$(A) - (M) - C \rightarrow A$
SBCB	Subtract with borrow from B	$(B) - (M) - C \rightarrow B$
SUBA	Subtract memory from A	$(A) - (M) \rightarrow A$
SUBB	Subtract memory from B	$(B) - (M) \rightarrow B$
SUBD	Subtract memory from D	$(D) - (M:M + 1) \rightarrow D$

Table 1.7 ■ Add and subtract instructions

Example 1.1

Write an instruction sequence to add 3 to the memory locations at $10 and $15.

Solution: A memory location cannot be the destination of an ADD instruction. Therefore, we need to copy the memory content into an accumulator, add 3 to it, and then store the sum back to the same memory location.

```
ldaa   $10      ; copy the contents of $10 into accumulator A
adda   #3       ; add 3 to A
staa   $10      ; store the sum back to $10
ldaa   $15      ; copy the contents of $15 into accumulator A
adda   #3       ; add 3 to A
staa   $15      ; store the sum back to $15
```

▲

Example 1.2
▼

Write an instruction sequence to add the byte pointed to by index register X and the following byte and place the sum at the memory location pointed to by index register Y.

Solution: The byte pointed to by index register X and the following byte can be accessed by using the indexed addressing mode.

```
ldaa   0,X      ; put the byte pointed to by X in A
adda   1,X      ; add the following byte to A
staa   0,Y      ; store the sum at the location pointed to by Y
```

▲

Example 1.3
▼

Write an instruction sequence to add the numbers stored at $1000 and $1001 and store the sum at $1004.

Solution: To add these two numbers, we need to put one of them in an accumulator:

```
ldaa   $1000    ; copy the number in $1000 to accumulator A
adda   $1001    ; add the second number to A
staa   $1004    ; save the sum at $1004
```

▲

1.9 Instruction Queue

The HCS12 uses a three-stage instruction queue to facilitate instruction fetching and increase execution speed. Queue logic prefetches program information and positions it for sequential execution, one instruction at a time. The relationship between bus cycles and execution cycles is straightforward and facilitates tracking and debugging.

There are three 16-bit stages in the instruction queue. Instructions enter the queue at stage 1 and roll out after stage 3. Each byte in the queue is selectable. An opcode-prediction algorithm determines the location of the next opcode in the instruction queue.

Each instruction refills the queue by fetching the same number of bytes that the instruction uses. Program information is fetched in aligned 16-bit words. Each program fetch indicates that two bytes need to be replaced in the instruction queue. Each optional fetch indicates that only one byte needs to be replaced. For example, an instruction composed of five bytes does two program fetches and one optional fetch. If the first byte of the five-byte instruction was even-aligned, the optional fetch is converted into a free cycle. If the first byte was odd-aligned, the optional fetch is executed as a program fetch.

Two external pins, IPIPE[1:0], provide time-multiplexed information about instruction execution and data movement in the queue.

The content of queue stage 1 advances to stage 2, stage 2 advances to stage 3, and stage 1 is loaded with a word of program information from the data bus.

1.10 Instruction Execution Cycle

In order to execute a program, the microprocessor or microcontroller must access memory to fetch instructions or operands. The process of accessing a memory location is called a *read cycle*, the process of storing a value in a memory location is called a *write cycle,* and the process of executing an instruction is called an *instruction execution cycle.* When executing an instruction, the HCS12 performs a combination of the following operations:

- One or multiple read cycles to fetch instruction opcode byte(s) and addressing information
- One or more read cycles to fetch the memory operand(s) (optional)
- The operation specified by the opcode
- One or more write cycles to write back the result to either a register or a memory location (optional)

(start) 1.11 Embedded Systems ﾟ-¹⁴

A microcontroller is designed to be used in a product to provide certain control functions. End users are interested in the features and functions of the product rather than the microcontroller itself. A product that incorporates one or more microcontrollers and has this nature is called an *embedded product* or *embedded system.*

The weather station is one such system and has become a very popular consumer product. With a weather station, one can read the outdoor temperature and humidity without going out of the house, which is very convenient during a harsh winter. A weather station consists of two modules, a block diagram of which is shown in Figure 1.6. The outdoor

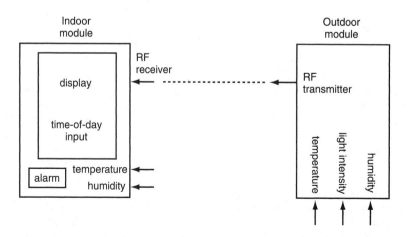

Figure 1.6 ■ Block diagram of a weather station

module is installed outside the house and uses sensors to measure the temperature, humidity, and light intensity, sending their values to the indoor module periodically using the RF transmitter.

The indoor module uses a RF receiver to receive the temperature, humidity, and light intensity readings from the outdoor module periodically. In addition to those readings received from the outdoor module, the indoor module also has its own sensor to measure the indoor temperature and humidity. The indoor module displays the current indoor and outdoor temperature and humidity along with the current time of day, month, and day information on the LCD display. The indoor module can figure out if it is sunny, cloudy, or rainy using the outdoor information. The indoor module allows the user to enter the current time of day, month, and day information and to have the option of using either the 12- or 24-hour format to display the current time information. The user also has the option to enable the alarm and set the alarm time.

Both the indoor and outdoor modules have a microcontroller to control all the operations. The computation performed by the weather station is relatively simple and hence the application program can be stored in the on-chip flash memory of the microcontroller without using any external memory.

A second example of an embedded system is the home security system. A typical home security system consists of a microcontroller along with several sensors and actuators. Among the sensors are the temperature sensors, smoke detectors, humidity sensors, motion detectors, and so on. When the security system detects smoke, it may activate the alarm or even call the fire department. When the temperature is lower than a certain degree, the system may start the furnace to warm the house. When the temperature is higher than a certain degree, it will start the air conditioner. When the owner is away from home, the security system can be armed to call the neighbor or even the police department if the motion detector detects people trying to open the door. A home security system can be as sophisticated as desired.

The cell phone is a very complicated embedded system. It contains a microcontroller that is responsible for making the phone call, accepting incoming calls, accessing Internet websites, displaying the web page, handling all user input and output, keeping track of time, and so on.

Today's automobiles are also good examples of embedded systems. Most new cars incorporate several microcontrollers. One microcontroller is responsible for controlling the instrument display. In order to carry out this function, the display microcontroller needs to collect information using different sensors. These sensors are used to measure temperature, speed, distance, fuel level, and so on. Other microcontrollers are responsible for fuel injection control, cruise control, global positioning, giving warning when the car is too close to other cars or objects, and many other safety features available in high-priced vehicles. These microcontrollers may need to exchange information or even coordinate their operations. The controller area network (CAN) bus is designed for this purpose and will be discussed in detail in Chapter 13.

An example of an embedded product that contains external memory is the network router. A router is connected to two or more networks and is responsible for data packet routing. Whenever a data packet arrives at the router, the router makes sure that there is no error and then looks up the routing table and chooses a route with the lowest cost (in general) and forwards it to the next node (another router or destination). Since the Internet is very complicated, the protocols for handling networking are also very complicated. In addition, the data rate of the Internet can be higher than 10^9 bits per second. Sometimes, the Internet can get congested. The router will need a large amount of memory to buffer the incoming and outgoing data. Having a large external RAM and hard disk is a norm for network routers.

1.12 Summary

The invention of the microprocessor in 1971 resulted in a revolution in the electronics industry. The first microprocessor, the Intel 4004, incorporated a simplified CPU into an integrated circuit (IC). Following the introduction of the 4-bit 4004, Intel introduced the 8-bit 8008, 8080, and 8085 over 3 years. The 8085 was a big success because of its programmability. Through this programmability, many products could be designed and constructed. Other companies also joined in the design and manufacturing of microprocessors, Zilog, Freescale, and Rockwell being among the more successful.

The earliest microprocessors still needed peripheral chips to interface with I/O devices such as seven-segment displays, printers, timers, and so on. Memory chips were also needed to hold the application program and dynamic data. Because of this, the products designed with microprocessors could not be made as small as desired. Then came the introduction of microcontrollers, which incorporated the CPU, some amount of memory, and peripheral functions such as parallel ports, timer, and serial interface functions onto one chip. The development of microcontrollers has had the following impacts:

- I/O interfacing is greatly simplified.
- External memory is no longer needed for many applications.
- System design time is greatly shortened.

A microcontroller is not designed to build a desktop computer. Instead, it is used as the controller of many products. End users of these products do not care what microcontrollers are used in their appliances; they only care about the functionality of the product. A product that uses a certain microcontroller as a controller and has this characteristic is called an *embedded system*. Cell phones, automobiles, cable modems, HDTVs, and home security systems are well-known embedded systems.

Over the last 20 years, we have clearly seen that a microcontroller needs to incorporate some or all of the following peripheral functions in order to be useful:

- Timer module that incorporates input capture, output compare, real-time interrupt, and counting capability
- Pulse-width modulation function for easy waveform generation
- Analog-to-digital converter
- Digital-to-analog converter
- Temperature sensor
- Direct memory access (DMA) controller
- Parallel I/O interface
- Serial I/O interface such as UART, SPI, Microwire, I²C, and CAN
- Memory component interface circuitry

The HCS12 from Freescale incorporates most of these peripheral modules and the CPU in one VLSI chip.

Memory is where software programs and data are stored. Semiconductor memory chips can be classified into two major categories: random access memory (RAM) and read-only memory (ROM). RAM technology includes DRAM and SRAM; MROM, PROM, EPROM, EEPROM, and flash memory are read-only memories.

Programs are known as *software*. A program is a set of instructions that the computer hardware can execute. In the past, system designers mainly used assembly language to write microcontroller application software. The nature of assembly language forces an assembly programmer to work on the program logic at a relatively low level. This hampers programming productivity. In the last 10 years, more and more people have turned to high-level language to improve their programming productivity. C is the most widely used language for embedded system programming.

Although system designers use assembly or high-level language to write their programs, the microcontroller can only execute machine instructions. Programs written in assembly or high-level language must be translated into machine instructions before they can be executed. The program that performs the translation work is called an *assembler* or *compiler*, depending on the language to be translated.

A machine instruction consists of *opcode* and *addressing information* that specifies the operands. Addressing information is also called *addressing mode*. The HCS12 implements a rich instruction set along with many addressing modes for specifying operands. This chapter examines the functions of a few groups of instructions. Examples were used to explore the implementation of simple operations using these instructions.

The execution of an instruction may take several clock cycles. Because the HCS12 does not access memory in every clock cycle, it performs instruction prefetch to speed up the instruction execution. A three-word (16-bit word) instruction prefetch is added to hold the prefetched instructions.

1.13 Exercises

E1.1 What is a processor?

E1.2 What is a microprocessor? What is a microcomputer?

E1.3 What makes a microcontroller different from a microprocessor?

E1.4 How many bits can the HCS12 CPU manipulate in one operation?

E1.5 How many different memory locations can the HCS12 access without the expanded memory?

E1.6 Why must every computer have some amount of nonvolatile memory?

E1.7 Why must every computer have some amount of volatile memory?

E1.8 What is source code? What is object code?

E1.9 Convert 5K, 8K, and 13K to decimal representation.

E1.10 Write an instruction sequence to swap the contents of memory locations at $1000 and $1001.

E1.11 Write an instruction sequence to add 10 to memory locations at $1000 and $1001, respectively.

E1.12 Write an instruction sequence to set the contents of memory locations at $1000, $1010, and $1020 to 10, 11, and 12, respectively.

E1.13 Write an instruction sequence to perform the operations equivalent to those performed by the following high-level language statements:

```
I = 11;
J = 33;
K = I + J − 5;
```

Assume that variables I, J, and K are located at $1000, $1005, and $1010, respectively.

E1.14 Write an instruction sequence to subtract the number stored at $1010 from that stored at $1000 and store the difference at $1005.

E1.15 Write an instruction sequence to add the contents of accumulator B to the 16-bit word stored at memory locations $1000 and $1001. Treat the value stored in B as a signed number.

E1.16 Write an instruction sequence to copy four bytes starting from $1000 to $1100–$1103.

E1.17 Write an instruction sequence to subtract the contents of accumulator B from the 16-bit word at $1000–$1001 and store the difference at $1100–$1101. Treat the value stored in B as a signed value.

E1.18 Write an instruction sequence to swap the 16-bit word stored at $1000–$1001 with the 16-bit word stored at $1100–$1101.

E1.19 Give an instruction that can store the contents of accumulator D at the memory location with an address larger than the contents of X by 8.

E1.20 Give an instruction that can store the contents of index register Y at the memory location with an address smaller than the contents of X by 10.

2

HCS12 Assembly Programming

2.1 Objectives

After completing this chapter you should be able to:

- Use assembler directives to allocate memory blocks, define constants, and create a message to be output.

- Write assembly programs to perform simple arithmetic operations.

- Write program loops to perform repetitive operations.

- Use program loops to create time delays.

- Use Boolean and bit manipulation instructions to perform bit field manipulation.

2.2 Assembly Language Program Structure

An assembly language program consists of a sequence of statements that tells the computer to perform the desired operations. From a global point of view, a HCS12 assembly program consists of three sections. In some cases, these sections can be mixed to provide better algorithm design.

1. *Assembler directives*. Assembler directives instruct the assembler how to process subsequent assembly language instructions. Directives also provide a way to define program constants and reserve space for dynamic variables. Some directives may also set a location counter.

2. *Assembly language instructions*. These instructions are HCS12 instructions. Some instructions are defined with labels.

3. *Comments*. There are two types of comments in an assembly program. The first type is used to explain the function of a single instruction or directive. The second type explains the function of a group of instructions or directives or a whole routine. Adding comments improves the readability of a program.

Each line of a HCS12 assembly program, excluding certain special constructs, is comprised of four distinct fields. Some of the fields may be empty. The order of these fields is

1. Label
2. Operation
3. Operand
4. Comment

2.2.1 The Label Field

Labels are symbols defined by the user to identify memory locations in the programs and data areas of the assembly module. For most instructions and assembler directives, the label is optional. The rules for forming a label are as follows:

- The label field can begin in column one or in any column if it is terminated by a colon (:). It must begin with a letter (A–Z, a–z), and the letter can be followed by letters, digits, or special symbols. Some assemblers permit special symbols to be used. For example, the assembler from IAR Inc. allows a symbol to start with a question mark (?), at character (@), and underscore (_) in addition to letters. Digits and the dollar ($) character can also be used after the first character in the IAR assembler.

- Most assemblers restrict the number of characters in a label name. The as12 assembler reference manual does not mention the limit. The IAR assembler allows a user-defined symbol to have up to 255 characters.

- The **as12** assembler from several companies allows a label to be terminated by ":".

Example 2.1 Valid and Invalid Labels

▼

The following instructions contain valid labels:

```
begin    ldaa    #10         ; label begins in column 1
print:   jsr     hexout      ; label is terminated by a colon
         jmp     begin       ; instruction references the label begin
```

The following instructions contain invalid labels:

```
here is     adda    #5          ; a space is included in the label
   loop     deca                ; label begins at column 2
```

2.2.2 The Operation Field

This field specifies an assembler instruction, a directive, or a macro call. Assembler instructions or directives are case insensitive. The operation field must not start in column one. If a label is present, the opcode or directive must be separated from the label field by at least one space. If there is no label, the operation field must be at least one space from the left margin.

Example 2.2 Examples of Operation Fields

▼

```
            adda    #$02        ; adda is the instruction mnemonic
   true     equ     1           ; the equ directive occupies the operation field
```

2.2.3 The Operand Field

If an *operand field* is present, it follows the *operation field* and is separated from the operation field by at least one space. The operand field is composed of one or more operands separated by commas, followed by optional space or tab characters. The operand field is used to supply arguments to the assembler instruction, directive, or macro that has been used in the operation field. The following instructions include the operand field:

```
   TCNT     equ     $0084       ; the value $0084 is the operand field
   TC0      equ     $0090       ; the value $0090 is the operand field
```

2.2.4 The Comment Field

The comment field is optional and is added mainly for documentation purposes. The comment field is ignored by the assembler. Here are the rules for comments:

- Any line beginning with an * is a comment.
- Any line beginning with a ";" is a comment. In this book, we will use ";" to start a comment.
- You must have a ";" (semicolon) prefixing any comment on a line with mnemonics.

Examples of comments are in the following instruction:

```
; this program computes the square root of N 8-bit integers.
            org     $1000       ; set the location counter to $1000
            dec     lp_cnt      ; decrement the loop count
```

In this chapter, we will use the Freescale Freeware cross assembler as12 as the standard to explain every aspect of assembly programming. The as12 assembler contained in the **MiniIDE** from Mgtek is compatible with the as12 freeware from Freescale.

2.3 Assembler Directives

Assembler directives look just like instructions in an assembly language program, but they tell the assembler to do something other than creating the machine code for an instruction. The available assembler directives vary with the assembler. Interested readers should refer to the user's manual of the specific assembler for details.

We will discuss assembler directives supported by the as12 in detail. In the following discussion, statements enclosed in brackets [] are optional. All directives and assembly instructions can be in either upper- or lowercase.

END

The **end** directive is used to end a program to be processed by the assembler. In general, an assembly program looks like this:

```
(your program)
end
```

The **end** directive indicates the logical end of the source program. Any statement following the end directive is ignored. A warning message will occur if the end directive is missing from the source code; however, the program will still be assembled correctly.

ORG (ORIGIN)

The assembler uses a *location counter* to keep track of the memory location where the next machine code byte should be placed. If the programmer wants to force the program or data array to start from a certain memory location, then this directive can be used. For example, the statement

```
org  $1000
```

forces the location counter to be set to $1000.

The **org** directive is mainly used to force a data table or a segment of instructions to start with a certain address. As a general rule, this directive should be used as infrequently as possible. Using too many **orgs** will make your program less reusable.

DB (DEFINE BYTE), DC.B (DEFINE CONSTANT BYTE), FCB (FORM CONSTANT BYTE)

These three directives define the value of a byte or bytes that will be placed at a given memory location. The **db** (or fcb, or dc.b) directive assigns the value of the expression to the memory location pointed to by the location counter. Then the location counter is incremented. Multiple bytes can be defined at a time by using commas to separate the arguments. For example, the statement

```
array      db      $11,$22,$33,$44,$55
```

initializes five bytes in memory to

```
$11
$22
$33
$44
$55
```

and the assembler will use *array* as the symbolic address of the first byte whose initial value is $11. The program can also force these five bytes to a certain address by adding the *org* directive. For example, the sequence

```
        org     $1000
array   db      $11,$22,$33,$44,$55
```

initializes the contents of memory locations at $1000, $1001, $1002, $1003, and $1004 to $11, $22, $33, $44, and $55, respectively.

DW (DEFINE WORD), DC.W (DEFINE CONSTANT WORD), FDB (FORM DOUBLE BYTES)

These three directives define the value of a word or words that will be placed at a given address. The value can be specified by an integer or an expression. For example, the statement

```
vect_tab    dw    $1234, $5678
```

initializes the two words starting from the current location counter to $1234 and $5678, respectively. After this statement, the location counter will be incremented by 4.

FCC (FORM CONSTANT CHARACTER) *use DC.B instead*

This directive allows us to define a string of characters (a message). The first character in the string is used as the delimiter. The last character must be the same as the first character because it will be used as the delimiter. The delimiter must not appear in the string. The space character cannot be used as the delimiter. Each character is encoded by its corresponding ASCII code. For example, the statement

```
alpha    fcc    "def"   ← delimiter
```

will generate the following values in memory:

```
$64
$65
$66
```

and the assembler will use the label *alpha* to refer to the address of the first letter, which is stored as the byte $64. A character string to be output to the LCD display is often defined using this directive.

FILL (FILL MEMORY)

This directive allows a user to fill a certain number of memory locations with a given value. The syntax of this directive is as follows:

```
        fill    value, count
```

where the number of bytes to be filled is indicated by *count* and the value to be filled is indicated by *value*. For example, the statement

```
spaceLine    fill    $20, 40
```

will fill 40 bytes with the value of $20 starting from the memory location referred to by the label *spaceLine*.

DS (DEFINE STORAGE), RMB (RESERVE MEMORY BYTE), DS.B (DEFINE STORAGE BYTES)

Each of these three directives reserves a number of bytes given as the arguments to the directive. The location counter will be incremented by the number that follows the directive mnemonic. For example, the statement

```
buffer    ds    100
```

reserves 100 bytes in memory starting from the location represented by the label *buffer*. After this directive, the location counter will be incremented by 100. The content(s) of the reserved memory location(s) is not defined.

DS.W (DEFINE STORAGE WORD), RMW (RESERVE MEMORY WORD)

Each of these directives increments the location counter by the value indicated in the number-of-words argument multiplied by 2. In other words, if the *ds.w* expression evaluates to *k*, then the location counter is advanced by 2*k*. These directives are often used with a label. For example, the statement

```
dbuf      ds.w      20
```

reserves 40 bytes starting from the memory location represented by the label *dbuf*. None of these 40 bytes is initialized.

EQU (EQUATE)

This directive assigns a value to a label. Using *equ* to define constants will make a program more readable. For example, the statement

```
loop_cnt   equ      40
```

informs the assembler that whenever the symbol loop_cnt is encountered, it should be replaced with the value of 40.

LOC

This directive increments and produces an internal counter used in conjunction with the backward tick mark ('). By using the loc directive and the ' mark, you can write a program segment like the following example without thinking up new labels:

```
          loc
          ldaa      #2
loop'     deca
          bne       loop'
          loc
loop'     brclr     0,x $55 loop'
```

This code segment will work perfectly fine because the first loop label will be seen as loop001, whereas the second loop label will be seen as loop002. The assembler really sees this:

```
          loc
          ldaa      #2
loop001   deca
          bne       loop001
          loc
loop002   brclr     0,x $55 loop002
```

You can also set the loc directive with a valid expression or number by putting that expression or number in the operand field. The resultant number will be used to increment the suffix to the label.

MACRO, ENDM

A macro is a name assigned to a group of instructions or directives. There are situations in which the same sequence of instructions needs to be included in several places. This sequence of instructions may operate on different parameters. By placing this sequence of instructions in a macro, the sequence need be typed only once. The macro capability not only makes the programmer more productive but also makes the program more readable. The freeware MiniIDE supports macro directives. However, the freeware AsmIDE does not.

The keyword *macro* starts a new macro definition, whereas the keyword *endm* ends the macro definition. For example, a macro may be defined for the HCS12 as follows:

```
sumOf3    macro    arg1,arg2,arg3
          ldaa     arg1
          adda     arg2
          adda     arg3
          endm
```

If you want to add the values of three memory locations at $1000, $1001, and $1002 and leave the sum in accumulator A, you can use the following statement to invoke the previously defined macro:

```
sumOf3 $1000,$1001,$1002
```

When processing this macro call, the assembler will insert the following instructions in your program:

```
ldaa      $1000
adda      $1001
adda      $1002
```

2.4 Software Development Issues

A complete discussion of issues involved in software development is beyond the scope of this text. However, we do need to take a serious look at some software development issues because embedded system designers must spend a significant amount of time on software development.

Software development starts with *problem definition*. The problem presented by the application must be fully understood before any program can be written. At the problem definition stage, the most critical thing is to get the programmer and the end user to agree upon what needs to be done. To achieve this, asking questions is very important. For complex and expensive applications, a formal, written definition of the problem is formulated and agreed upon by all parties.

Once the problem is known, the programmer can begin to lay out an overall plan of how to solve the problem. The plan is also called an *algorithm*. Informally, an algorithm is any well-defined computational procedure that takes some value, or a set of values, as input and produces some value, or a set of values, as output. An algorithm is thus a sequence of computational steps that transforms the input into the output. We can also view an algorithm as a tool for solving a well-specified computational problem. The statement of the problem specifies in general terms the desired input/output relationship. The algorithm describes a specific computational procedure for achieving that input/output relationship.

An algorithm is expressed in *pseudocode* that is very much like C or PASCAL. What separates pseudocode from "real" code is that in pseudocode, we employ whatever expressive method that is most clear and concise to specify a given algorithm. Sometimes, the clearest method is English, so do not be surprised if you come across an English phrase or sentence embedded within a section of "real" code.

An algorithm provides not only the overall plan for solving the problem but also documentation to the software to be developed. In the rest of this book, all algorithms will be presented in the following format:

Step 1

. . . .

Step 2

. . . .

An earlier alternative for providing the overall plan for solving software problems was the use of flowcharts. A flowchart shows the way a program operates. It illustrates the logic flow of the program. Therefore, flowcharts can be a valuable aid in visualizing programs. Flowcharts are not only used in computer programming, they are also used in many other fields, such as business and construction planning.

The flowchart symbols used in this book are shown in Figure 2.1. The *terminal symbol* is used at the beginning and the end of each program. When it is used at the beginning of a program, the word *Start* is written inside it. When it is used at the end of a program, it contains the word *Stop*.

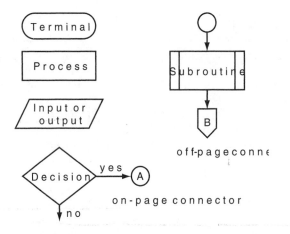

Figure 2.1 ■ Flowchart symbols used in this book

The *process box* indicates what must be done at this point in the program execution. The operation specified by the process box could be shifting the contents of one general-purpose register to a peripheral register, decrementing a loop count, and so on.

The *input/output box* is used to represent data that are either read or displayed by the computer.

The *decision box* contains a question that can be answered either yes or no. A decision box has two exits, also marked yes or no. The computer will take one action if the answer is yes and will take a different action if the answer is no.

The *on-page connector* indicates that the flowchart continues elsewhere on the same page. The place where it is continued will have the same label as the on-page connector.

The *off-page connector* indicates that the flowchart continues on another page. To determine where the flowchart continues, one needs to look at the following pages of the flowchart to find the matching off-page connector.

Normal flow on a flowchart is from top to bottom and from left to right. Any line that does not follow this normal flow should have an arrowhead on it. When the program gets complicated, the flowchart that documents the logic flow of the program also becomes difficult to follow. This is the limitation of the flowchart. In this book, we will mix both the flowchart and the algorithm procedure to describe the solution to a problem.

After the programmer is satisfied with the algorithm or the flowchart, *it is converted to source code in one of the assembly or high-level languages*. Each statement in the algorithm (or each block of the flowchart) will be converted into one or multiple assembly instructions or high-level language statements. If an algorithmic step (or a block in the flowchart) requires many assembly instructions or high-level language statements to implement, then it might be beneficial to either (1) convert this step (or block) into a subroutine and just call the subroutine, or (2) further divide the algorithmic step (or flowchart block) into smaller steps (or blocks) so that it can be coded with just a few assembly instructions or high-level language statements.

The next major step is *testing the program*. Testing a program means testing for anomalies. The first test is for normal inputs that are always expected. If the result is what is expected, then the borderline inputs are tested. The maximum and minimum values of the input are tested. When the program passes this test, then illegal input values are tested. If the algorithm includes several branches, then enough values must be used to exercise all the possible branches. This is to make sure that the program will operate correctly under all possible circumstances.

In the rest of this book, most of the problems are well defined. Therefore, our focus is on how to design the algorithm that solves the specified problem as well as how to convert the algorithm into source code.

2.5 Writing Programs to Do Arithmetic

In this section, we will use small programs that perform simple computations to demonstrate how a program is written.

Example 2.3

Write a program to add the numbers stored at memory locations $1000, $1001, and $1002 and store the sum at memory location $1010.

Solution: This problem can be solved by the following steps:

Step 1
Load the contents of the memory location at $1000 into accumulator A.

Step 2
Add the contents of the memory location at $1001 into accumulator A.

Step 3
Add the contents of the memory location at $1002 into accumulator A.

Step 4
Store the contents of accumulator A at memory location $1010.

These steps can be translated into the as12 assembly program as follows:

```
org     $1500          ; starting address of the program
ldaa    $1000          ; A ⇐ [$1000]
adda    $1001          ; A ⇐ [A] + [$1001]
adda    $1002          ; A ⇐ [A] + [$1002]
staa    $1010          ; $1010 ⇐ [A]
end
```

Example 2.4

Write a program to subtract the contents of the memory location at $1005 from the sum of the memory locations at $1000 and $1002 and store the result at the memory location $1010.

Solution: The logic flow of this program is illustrated in Figure 2.2. The assembly program is as follows:

```
org     $1500          ; starting address of the program
ldaa    $1000          ; A ⇐ [$1000]
adda    $1002          ; A ⇐ [A] + [$1002]
suba    $1005          ; A ⇐ [A] − [$1005]
staa    $1010          ; $1010 ⇐ [A]
end
```

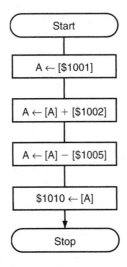

Figure 2.2 ■ Logic flow of program 2.4

Example 2.5

Write a program to subtract 5 from four memory locations at $1000, $1001, $1002, and $1003.

Solution: In the HCS12, a memory location cannot be the destination of an add or sub instruction. Therefore, three steps must be followed to add or subtract a number to or from a memory location:

Step 1
Load the memory contents into an accumulator.

Step 2
Add (or subtract) the number to (from) the accumulator.

Step 3
Store the result at the specified memory location.

The program is as follows:

```
org     $1500
ldaa    $1000        ; A ⇐ [$1000]
suba    #5           ; A ⇐ [A] − 5
staa    $1000        ; $1000 ⇐ [A]
ldaa    $1001        ; A ⇐ [$1001]
suba    #5           ; A ⇐ [A] − 5
staa    $1001        ; $1001 ⇐ [A]
ldaa    $1002        ; A ⇐ [$1002]
suba    #5           ; A ⇐ [A] − 5
staa    $1002        ; $1002 ⇐ [A]
ldaa    $1003        ; A ⇐ [$1003]
suba    #5           ; A ⇐ [A] − 5
staa    $1003        ; $1003 ⇐ [A]
end
```

Example 2.6

Write a program to add two 16-bit numbers that are stored at $1000~$1001 and $1002~$1003 and store the sum at $1010~$1011.

Solution: This program is very straightforward:

```
org     $1500
ldd     $1000        ; place the 16-bit number at $1000~$1001 in D
addd    $1002        ; add the 16-bit number at $1002~$1003 to D
std     $1010        ; save the sum at $1010~$1011
end
```

2.5.1 Carry/Borrow Flag

The HCS12 can add and subtract either 8-bit or 16-bit numbers and place the result in either 8-bit accumulators, A or B, or the double accumulator D. The 8-bit number stored in accumulator B can also be added to index register X or Y. However, programs can also be written to add numbers larger than 16 bits. Arithmetic performed in a 16-bit microprocessor/microcontroller on numbers that are larger than 16 bits is called *multiprecision arithmetic*. Multiprecision arithmetic makes use of the carry flag (C flag) of the condition code register (CCR).

Bit 0 of the CCR register is the C flag. It can be thought of as a temporary 9th bit that is appended to any 8-bit register or 17th bit that is appended to any 16-bit register. The C flag allows us to write programs to add and subtract hex numbers that are larger than 16 bits. For example, consider the following two instructions:

```
ldd     #$8645
addd    #$9978
```

These two instructions add the numbers $8645 and $9978.

```
  $ 8 6 4 5
+ $ 9 9 7 8
  $ 1 1 F B D
```

The result is $11FBD, a 17-bit number, which is too large to fit into the 16-bit double accumulator D. When the HCS12 executes these two instructions, the lower 16 bits of the answer, $1FBD, are placed in double accumulator D. This part of the answer is called *sum*. The leftmost bit is called a *carry*. A carry of 1 following an addition instruction sets the C flag of the CCR register to 1. A carry of 0 following an addition clears the C flag to 0. This applies to both 8-bit and 16-bit additions for the HCS12. For example, execution of the following two instructions:

```
ldd     #$1245
addd    #$4581
```

will clear the C flag to 0 because the carry resulting from this addition is 0. In summary:

- If the addition produces a carry of 1, the carry flag is set to 1.
- If the addition produces a carry of 0, the carry flag is cleared to 0.

2.5.2 Multiprecision Addition

For a 16-bit microcontroller like the HCS12, multiprecision addition is the addition of numbers that are larger than 16 bits. To add the hex number $1A598183 to $76548290, the HCS12 has to perform multiprecision addition:

```
    1    1  1
  $ 1 A 5 9 8 1 8 3
+ $ 7 6 5 4 8 2 9 0
  $ 9 0 A E 0 4 1 3
```

Multiprecision addition is performed one byte at a time, beginning with the least significant byte. The HCS12 does allow us to add 16-bit numbers at a time because it has the addd instruction. The following two instructions can be used to add the least significant 16-bit numbers together:

```
ldd     #$8183
addd    #$8290
```

Since the sum of the most significant digit is greater than 16, it generates a carry that must be added to the next more significant digit, causing the C flag to be set to 1. The contents of double accumulator D must be saved before the higher bytes are added. Let's save these two bytes at $1002–$1003:

```
std     $1002
```

When the second most significant bytes are added, the carry from the lower byte must be added in order to obtain the correct sum. In other words, we need an "add with carry" instruction. There are two versions of this instruction: the adca instruction for accumulator A and the adcb instruction for accumulator B. The instructions for adding the second most significant bytes are

```
ldaa     #$59
adca     #$54
```

We need also to save the *second most significant* byte of the result at $1001 with the following instruction:

```
staa     $1001
```

The most significant bytes can be added using similar instructions, and the complete program with comments appears as follows:

```
ldd      #$8183        ; D ⇐ $8183
addd     #$8290        ; D ⇐ [D] + $8290
std      $1002         ; $1002-$1003 ⇐ [D]
ldaa     #$59          ; A ⇐ $59
adca     #$54          ; A ⇐ [A] + $54 + C
staa     $1001         ; $1001 ⇐ [A]
ldaa     #$1A          ; A ⇐ $1A
adca     #$76          ; A ⇐ [A] + $76 + C
staa     $1000         ; $1000 ⇐ [A]
end
```

Note that the load and store instructions do not affect the value of the C flag (otherwise, the program would not work). The HCS12 does not have a 16-bit instruction with the carry flag as an operand. Whenever the carry needs to be added, we must use the 8-bit instruction adca or adcb. This is shown in the previous program.

Example 2.7

▼

Write a program to add two 4-byte numbers that are stored at $1000~$1003 and $1004~$1007 and store the sum at $1010~$1013.

Solution: The addition should start from the least significant byte and proceed to the most significant byte. The program is as follows:

```
org      $1500         ; starting address of the program
ldd      $1002         ; D ⇐ [$1002~$1003]
addd     $1006         ; D ⇐ [D] + [$1006~$1007]
std      $1012         ; $1012~$1013 ⇐ [D]
ldaa     $1001         ; A ⇐ [$1001]
adca     $1005         ; A ⇐ [A] + [$1005] + C
staa     $1011         ; $1011 ⇐ [A]
ldaa     $1000         ; A ⇐ [$1000]
adca     $1004         ; A ⇐ [A] + [$1004] + C
staa     $1010         ; $1010 ⇐ [A]
end
```

▲

2.5.3 Subtraction and the C Flag

The C flag also enables the HCS12 to borrow from the high byte to the low byte during a multiprecision subtraction. Consider the following subtraction problem:

$$\begin{array}{r} \$39 \\ -\;\$74 \\ \hline \end{array}$$

We are attempting to subtract a larger number from a smaller one. Subtracting $4 from $9 is not a problem:

$$\begin{array}{r} \$39 \\ -\;\$74 \\ \hline 5 \end{array}$$

Now we need to subtract $7 from $3. To do this, we need to borrow from somewhere. The HCS12 borrows from the C flag, thus setting the C flag. When we borrow from the next higher digit of a hex number, the borrow has a value of decimal 16. After the borrow from the C flag, the problem can be completed:

$$\begin{array}{r} \$39 \\ -\;\$74 \\ \hline \$C5 \end{array}$$

When the HCS12 executes a subtract instruction, it always borrows from the C flag. The borrow is either 1 or 0. The C flag operates as follows during a subtraction:

- If the HCS12 borrows a 1 from the C flag during a subtraction, the C flag is set to 1.
- If the HCS12 borrows a 0 from the C flag during a subtraction, the C flag is set to 0.

2.5.4 Multiprecision Subtraction

For a 16-bit microcontroller, multiprecision subtraction is the subtraction of numbers that are larger than 16 bits. To subtract the hex number $16753284 from $98765432, the HCS12 has to perform a multiprecision subtraction:

$$\begin{array}{r} \$98765432 \\ -\;\$16757284 \\ \hline \end{array}$$

Like multiprecision addition, multiprecision subtraction is performed one byte at a time, beginning with the least significant byte. The HCS12 does allow us to subtract two bytes at a time because it has the subd instruction. The following two instructions can be used to subtract the least significant two bytes of the subtrahend from the minuend:

```
ldd     #$5432
subd    #$7284
```

Since a larger number is subtracted from a smaller one, there is a need to borrow from the higher byte, causing the C flag to be set to 1. The contents of double accumulator D should be saved before the higher bytes are subtracted. Let's save these two bytes at $1002~$1003:

```
std     $1002
```

When the second most significant bytes are subtracted, the borrow 1 has to be subtracted from the second most significant byte of the result. In other word, we need a "subtract with borrow" instruction. There is such an instruction, but it is called *subtract with carry*. There are two versions: the sbca instruction for accumulator A and the sbcb instruction for accumulator B. The instructions to subtract the second most significant bytes are

```
ldaa    #$76
sbca    #$75
```

We also need to save the second most significant byte of the result at $1001 with the following instruction:

```
staa      $1001
```

The most significant bytes can be subtracted using similar instructions, and the complete program with comments is as follows:

```
org     $1500        ; starting address of the program
ldd     #$5432       ; D ⇐ $5432
subd    #$7284       ; D ⇐ [D] − $7284
std     $1002        ; $1002~$1003 ⇐ [D]
ldaa    #$76         ; A ⇐ $76
sbca    #$75         ; A ⇐ [A] − $75 − C
staa    $1001        ; $1001 ⇐ [A]
ldaa    #$98         ; A ⇐ $98
sbca    #$16         ; A ⇐ [A] − $16 − C
staa    $1000        ; $1000 ⇐ [A]
end
```

Example 2.8
▼

Write a program to subtract the hex numbers stored at $1004~$1007 from the hex number stored at $1000~$1003 and save the difference at $1010~$1013.

Solution: We will perform the subtraction from the least significant byte toward the most significant byte as follows:

```
org     $1500        ; starting address of the program
ldd     $1002        ; D ⇐ [$1002~$1003]
subd    $1006        ; D ⇐ [D] − [$1006~$1007]
std     $1012        ; $1012~$1013 ⇐ [D]
ldaa    $1001        ; A ⇐ [$1001]
sbca    $1005        ; A ⇐ [A] − [$1005] − C
staa    $1011        ; $1011 ⇐ [A]
ldaa    $1000        ; A ⇐ [$1000]
sbca    $1004        ; A ⇐ [A] − [$1004] − C
staa    $1010        ; $1010 ⇐ [A]
end
```

▲

2.5.5 Binary-Coded-Decimal (BCD) Addition

Although virtually all computers work internally with binary numbers, the input and output equipment generally uses decimal numbers. Since most logic circuits only accept two-valued signals, the decimal numbers must be coded in terms of binary signals. In the simplest form of binary code, each decimal digit is represented by its binary equivalent. For example, 2538 is represented by

0010 0101 0011 1000

This representation is called *binary-coded-decimal* (BCD). If the BCD format is used, it must be preserved during arithmetic processing.

The principal advantage of the BCD encoding method is the simplicity of input/output conversion; its major disadvantage is the complexity of arithmetic processing. The choice between binary and BCD depends on the type of problems the system will be handling.

The HCS12 microcontroller can add only binary numbers, not decimal numbers. The following instruction sequence appears to cause the HCS12 to add the decimal numbers 25 and 31 and store the sum at the memory location $1000:

```
ldaa    #$25
adda    #$31
staa    $1000
```

This instruction sequence performs the following addition:

$$\begin{array}{r} \$25 \\ + \ \$31 \\ \hline \$56 \end{array}$$

When the HCS12 executes this instruction sequence, it adds the numbers according to the rules of binary addition and produces the sum $56. This is the correct BCD answer, because the result represents the decimal sum of 25 + 31 = 56. In this example, the HCS12 gives the appearance of performing decimal addition. However, a problem occurs when the HCS12 adds two BCD digits and generates a sum greater than 9. Then the sum is incorrect in the decimal number system, as the following three examples illustrate:

$$\begin{array}{ccc} \$18 & \$35 & \$19 \\ + \ \$47 & + \ \$47 & + \ \$47 \\ \hline \$5F & \$7C & \$60 \end{array}$$

The answers to the first two problems are obviously wrong in the decimal number system because the hex digits F and C are not between 0 and 9. The answer to the third example appears to contain valid BCD digits, but in the decimal system 19 plus 47 equals 66, not 60; this example involves a carry from the lower nibble to the higher nibble.

In summary, a sum in the BCD format is incorrect if it is greater than $9 or if there is a carry to the next higher nibble. Incorrect BCD sums can be adjusted by adding $6 to them. To correct the examples,

1. Add $6 to every sum digit greater than 9.

2. Add $6 to every sum digit that had a carry of 1 to the next higher digit.

Here are the problems with their sums adjusted:

$$\begin{array}{ccc} \$18 & \$35 & \$19 \\ + \ \$47 & + \ \$47 & + \ \$47 \\ \hline \$5F & \$7C & \$60 \\ + \ \$\ 6 & + \ \$\ 6 & + \ \$\ 6 \\ \hline \$65 & \$82 & \$66 \end{array}$$

The fifth bit of the condition code register is the *half-carry*, or H flag. A carry from the lower nibble to the higher nibble during the addition operation is a half-carry. A half-carry of 1 during addition sets the H flag to 1, and a half-carry of 0 during addition clears it to 0. If there is carry from the high nibble during addition, the C flag is set to 1, which indicates that the high nibble is incorrect. A $6 must be added to the high nibble to adjust it to the correct BCD sum.

Fortunately, we don't need to write instructions to detect the illegal BCD sum following a BCD addition. The HCS12 provides a *decimal adjust accumulator A* instruction, daa, which takes care of all these detailed detection and correction operations. The daa instruction monitors the sums of BCD additions and the C and H flags and automatically adds $6 to any nibble that requires it. The rules for using the daa instruction are

1. The daa instruction can only be used for BCD addition. It does not work for subtraction or hex arithmetic.
2. The daa instruction must be used immediately after one of the three instructions that leave their sum in accumulator A. (These three instructions are adda, adca, aba.)
3. The numbers added must be legal BCD numbers to begin with.

Example 2.9

Write an instruction sequence to add the BCD numbers stored at memory locations $1000 and $1001 and store the sum at $1010.

Solution:

```
ldaa    $1000      ; load the first BCD number in A
adda    $1001      ; perform addition
daa                ; decimal adjust the sum in A
staa    $1010      ; save the sum
```

Multiple-byte BCD numbers can be added and the correct result can be obtained by executing the daa instruction immediately after the addition of each byte.

2.5.6 Multiplication and Division

The HCS12 provides three multiply and five divide instructions. A brief description of these instructions is shown in Table 2.1.

Mnemonic	Function	Operation
emul	unsigned 16 by 16 multiply	(D) × (Y) → Y:D
emuls	signed 16 by 16 multiply	(D) × (Y) → Y:D
mul	unsigned 8 by 8 multiply	(A) × (B) → A:B
ediv	unsigned 32 by 16 divide	(Y:D) ÷ (X) quotient → Y remainder → D
edivs	signed 32 by 16 divide	(Y:D) ÷ (X) quotient → Y remainder → D
fdiv	16 by 16 fractional divide	(D) ÷ (X) → X remainder → D
idiv	unsigned 16 by 16 integer divide	(D) ÷ (X) → X remainder → D
idivs	signed 16 by 16 integer divide	(D) ÷ (X) → X remainder → D

Table 2.1 ■ Summary of HCS12 multiply and divide instructions

The **emul** instruction multiplies the 16-bit unsigned integers stored in accumulator D and index register Y and leaves the product in these two registers. The upper 16 bits of the product are in Y, whereas the lower 16 bits are in D.

The **emuls** instruction multiplies the 16-bit signed integers stored in accumulator D and index register Y and leaves the product in these two registers. The upper 16 bits of the product are in Y, whereas the lower 16 bits are in D.

The **mul** instruction multiplies the 8-bit unsigned integer in accumulator A by the 8-bit unsigned integer in accumulator B to obtain a 16-bit unsigned result in double accumulator D. The upper byte of the product is in accumulator A, whereas the lower byte of the product is in B.

The **ediv** instruction performs an unsigned 32-bit by 16 bits division. The dividend is the register pair Y and D with Y as the upper 16 bits of the dividend. Index register X is the divisor. After division, the quotient and the remainder are placed in Y and D, respectively.

The **edivs** instruction performs a signed 32-bit by 16-bit division using the same operands as the ediv instruction. After division, the quotient and the remainder are placed in Y and D, respectively.

The **fdiv** instruction divides an unsigned 16-bit dividend in double accumulator D by an unsigned 16-bit divisor in index register X, producing an unsigned 16-bit quotient in X and an unsigned 16-bit remainder in D. The dividend must be less than the divisor. The radix point of the quotient is to the left of bit 15. In the case of overflow (the denominator is less than or equal to the nominator) or division-by-zero, the quotient is set to $FFFF and the remainder is indeterminate.

The **idiv** instruction divides an unsigned 16-bit dividend in double accumulator D by the unsigned 16-bit divisor in index register X, producing an unsigned 16-bit quotient in X and an unsigned 16-bit remainder in D. If both the divisor and the dividend are assumed to have radix points in the same positions (to the right of bit 0), the radix point of the quotient is to the right of bit 0. In the case of division by zero, the quotient is set to $FFFF and the remainder is indeterminate.

The **idivs** instruction divides the signed 16-bit dividend in double accumulator D by the signed 16-bit divisor in index register X, producing a signed 16-bit quotient in X and a signed 16-bit remainder in D. If division-by-zero is attempted, the values in D and X are not changed, but the values of the N, Z, and V status bits are undefined.

Example 2.10

▼

Write an instruction sequence to multiply the contents of index register X and double accumulator D and store the product at memory locations $1000~$1003.

Solution: There is no instruction to multiply the contents of double accumulator D and index register X. However, we can transfer the contents of index register X to index register Y and execute the emul instruction. If index register Y holds useful information, then we need to save it before the data transfer.

```
sty     $1010        ; save Y in a temporary location
tfr     x,y          ; transfer the contents of X to Y
emul                 ; perform the multiplication
sty     $1000        ; save the upper 16 bits of the product
std     $1002        ; save the lower 16 bits of the product
ldy     $1010        ; restore the value of Y
```

▲

Example 2.11

▼

Write an instruction sequence to divide the signed 16-bit number stored at memory locations $1005~$1006 by the 16-bit signed number stored at memory locations $1020~$1021 and store the quotient and remainder at $1030~$1031 and $1032~$1033, respectively.

Solution: Before we can perform the division, we need to place the dividend and divisor in D and X, respectively.

```
ldd    $1005    ; place the dividend in D
ldx    $1020    ; place the divisor in X
idivs           ; perform the signed division
stx    $1030    ; save the quotient
std    $1032    ; save the remainder
```

▲

Because most arithmetic operations can be performed only on accumulators, we need to transfer the contents of index register X to D so that further division on the quotient can be performed. The HCS12 provides two exchange instructions in addition to the tfr instruction for this purpose:

- The xgdx instruction exchanges the contents of accumulator D and index register X.
- The xgdy instruction exchanges the contents of accumulator D and index register Y.

The HCS12 provides instructions for performing unsigned 8-bit by 8-bit and both signed and unsigned 16-bit by 16-bit multiplications. Since the HCS12 is a 16-bit microcontroller, we expect that it will be used to perform complicated operations in many sophisticated applications. Performing 32-bit by 32-bit multiplication will be one of them.

Since there is no 32-bit by 32-bit multiplication instruction, we will have to break a 32-bit number into two 16-bit halves and use the 16-bit by 16-bit multiply instruction to synthesize the operation. Assume that M and N are the multiplicand and the multiplier, respectively. These two numbers can be broken down as follows:

$$M = M_H M_L$$
$$N = N_H N_L$$

where M_H and N_H are the upper 16 bits and M_L and N_L are the lower 16 bits of M and N, respectively. Four 16-bit by 16-bit multiplications are performed, and then their partial products are added together, as shown in Figure 2.3 on the next page.

The procedure is as follows:

Step 1
Allocate eight bytes to hold the product. Assume these eight bytes are located at P, P+1, . . ., and P+7.

Step 2
Compute the partial product $M_L N_L$ (in Y:D) and save it at locations P+4 ~ P+7.

Step 3
Compute the partial product $M_H N_H$ (in Y:D) and save it at locations P ~ P+3.

Step 4
Compute the partial product $M_H N_L$ (in Y:D) and add it to memory locations P+2 ~ P+5. The C flag may be set to 1 after this addition.

Step 5

Add the C flag to memory location P+1 using the adca (or adcb) instruction. This addition may also set the C flag to 1. So, again, add the C flag to memory location P.

Step 6

Generate the partial product $M_L N_H$ (in Y:D) and add it to memory locations P+2 ~ P+5. The carry flag may be set to 1, so add the C flag to memory location P+1 and then add it to memory location P.

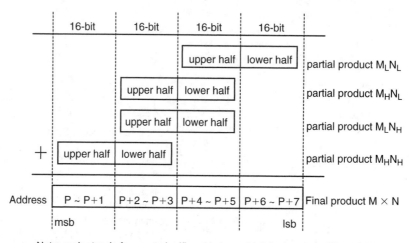

Note: msb stands for most significant byte and lsb for least significant byte.

Figure 2.3 ■ Unsigned 32-bit by 32-bit multiplation

Example 2.12

Write a program to multiply the 32-bit unsigned integers stored at M~M+3 and N~N+3, respectively, and store the product at memory locations P~P+7.

Solution: The following program is a direct translation of the previous multiplication algorithm:

```
          org     $1000
M         rmb     4                ; multiplicand
N         rmb     4                ; multiplier
P         rmb     8                ; product
          org     $1500
          . . . . . . . . . . . .  ; some other instructions
          ldd     M+2              ; place ML in D
          ldy     N+2              ; place NL in Y
          emul                     ; compute MLNL
          sty     P+4              ; save the upper 16 bits of the partial product MLNL
          std     P+6              ; save the lower 16 bits of the partial product MLNL
          ldd     M                ; place MH in D
          ldy     N                ; place NH in Y
          emul                     ; compute MHNH
          sty     P                ; save the upper 16 bits of the partial product MHNH
```

```
        std     P+2          ; save the lower 16 bits of the partial product MₕNₕ
        ldd     M            ; place Mₕ in D
        ldy     N+2          ; place Nₗ in Y
        emul                 ; compute MₕNₗ
; the following seven instructions add MₕNₗ to memory locations P+2~P+5
        addd    P+4          ; add the lower half of MₕNₗ to P+4~P+5
        std     P+4          ;    "
        tfr     y,d          ; transfer Y to D
        adcb    P+3
        stab    P+3
        adca    P+2
        staa    P+2
; the following six instructions propagate carry to the most significant byte
        ldaa    P+1
        adca    #0           ; add C flag to location P+1
        staa    P+1          ;    "
        ldaa    P
        adca    #0           ; add C flag to location P
        staa    P
; the following three instructions compute MₗNₕ
        ldd     M+2          ; place Mₗ in D
        ldy     N            ; place Nₕ in Y
        emul                 ; compute MₗNₕ
; the following seven instructions add MₗNₕ to memory locations P+2~P+5
        addd    P+4          ; add the lower half of MₗNₕ to P+4~P+5
        std     P+4          ;    "
        tfr     y,d          ; transfer Y to D
        adcb    P+3
        stab    P+3
        adca    P+2
        staa    P+2
; the following six instructions propagate carry to the most significant byte
        ldaa    P+1
        adca    #0           ; add C flag to location P+1
        staa    P+1
        ldaa    P
        adca    #0           ; add C flag to location P
        staa    P
        end
```

▲

Example 2.13

▼

Write a program to convert the 16-bit binary number stored at $1000~$1001 to BCD format and store the result at $1010~$1014. Convert each BCD digit into its ASCII code and store it in one byte.

Solution: A binary number can be converted to BCD format using repeated division-by-10. The largest 16-bit binary number corresponds to the five-digit decimal number 65535. The first division by 10 computes the least significant digit and should be stored in the memory location $1014, the second division-by-10 operation computes the ten's digit, and so on. The ASCII code of a BCD digit can be obtained by adding $30 to each BCD digit. The program is as follows:

```
        org    $1000
data    fdb    12345              ; place a number for testing
        org    $1010
result  rmb    5                  ; reserve five bytes to store the result
        org    $1500
        ldd    data               ; make a copy of the number to be converted
        ldy    #result
        ldx    #10                ; divide the number by 10
        idiv                      ;    "
        addb   #$30               ; convert to ASCII code
        stab   4,Y                ; save the least significant digit
        xgdx                      ; swap the quotient to D
        ldx    #10
        idiv
        addb   #$30               ; convert to ASCII code
        stab   3,Y                ; save the second-to-least-significant digit
        xgdx
        ldx    #10
        idiv
        addb   #$30
        stab   2,Y                ; save the middle digit
        xgdx
        ldx    #10
        idiv                      ; separate the most significant and second most
                                  ; significant digits
        addb   #$30
        stab   1,Y                ; save the second most significant digit
        xgdx                      ; swap the most significant digit to B
        addb   #$30               ; convert to ASCII code
        stab   0,Y                ; save the most significant digit
        end
```

2.6 Program Loops

Many applications require repetitive operations. We can write programs to tell the computer to perform the same operation over and over. A *finite loop* is a sequence of instructions that will be executed by the computer for a finite number of times; an *endless loop* is a sequence of instructions that the computer will execute forever.

There are four major loop constructs:

DO STATEMENT S FOREVER

This is an endless loop in which statement S is repeated forever. In some applications, we might add the statement "If C then exit" to leave the infinite loop. An infinite loop is shown in Figure 2.4.

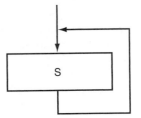

Figure 2.4 ■ An infinite loop

FOR $i = n1$ TO $n2$ DO S OR FOR $i = n2$ DOWNTO $n1$ DO S

Here, the variable i is the *loop counter*, which keeps track of the current iteration of the loop. The loop counter can be incremented (the first case) or decremented (the second case). Statement S is repeated $n2 - n1 + 1$ times. The value of n2 is assumed to be no smaller than n1. If there is concern that the relationship $n1 \le n2$ may not hold, then it must be checked at the beginning of the loop. Four steps are required to implement a FOR loop:

Step 1
Initialize the loop counter and other variables.

Step 2
Compare the loop counter with the limit to see if it is within bounds. If it is, then perform the specified operations. Otherwise, exit the loop.

Step 3
Increment (or decrement) the loop counter.

Step 4
Go to step 2.

A ***For-loop*** is illustrated in Figure 2.5.

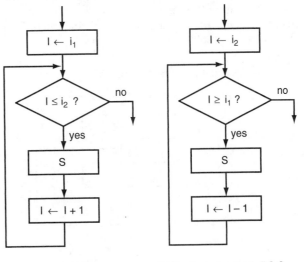

(a) For $I = i_1$ to i_2 DO S   (b) For $I = i_2$ downto i_1 DO S

Figure 2.5 ■ For looping construct

WHILE C DO S

Whenever a While construct is executed, the logical expression C is evaluated first. If it yields a false value, statement S will not be executed. The action of a While construct is illustrated in Figure 2.6. Four steps are required to implement a While loop:

Step 1
Initialize the logical expression C.

Step 2
Evaluate the logical expression C.

Step 3
Perform the specified operations if the logical expression C evaluates to true. Update the logical expression C and go to step 2. (*Note*: The logical expression C may be updated by external conditions or by an interrupt service routine.)

Step 4
Exit the loop.

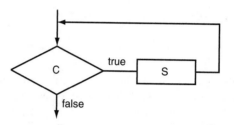

Figure 2.6 ■ The while ... do looping construct

REPEAT S UNTIL C

Statement *S* is first executed then the logical expression *C* is evaluated. If C is false, the next statement will be executed. Otherwise, statement S will be executed again. The action of this construct is illustrated in Figure 2.7. Statement S will be executed at least once. Three steps are required to implement this construct:

Step 1
Initialize the logical expression C.

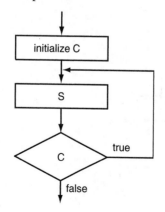

Figure 2.7 ■ The repeat...until looping construct

Step 2
Execute statement S.

Step 3
Go to step 2 if the logical expression C evaluates to true. Otherwise, exit.

To implement one of the looping constructs, we need to use the unconditional branch or one of the conditional instructions. When executing conditional branch instructions, the HCS12 checks the condition flags in the CCR register.

2.6.1 Condition Code Register

The contents of the condition code register are shown in Figure 2.8. The shaded characters are condition flags that reflect the status of an operation. The meanings of these condition flags are as follows:

- C: the carry flag. Whenever a carry is generated as the result of an operation, this flag will be set to 1. Otherwise, it will be cleared to 0.
- V: the overflow flag. Whenever the result of a two's complement arithmetic operation is out of range, this flag will be set to 1. Otherwise, it will be set to 0. The V flag is set to 1 when the carry from the most significant bit and the second most significant bit differ as the result of an arithmetic operation.
- Z: the zero flag. Whenever the result of an operation is zero, this flag will be set to 1. Otherwise, it will be set to 0.
- N: the negative flag. Whenever the most significant bit of the result of an operation is 1, this flag will be set to 1. Otherwise, it will be set to 0. This flag indicates that the result of an operation is negative.
- H: the half-carry flag. Whenever there is a carry from the lower four bits to the upper four bits as the result of an operation, this flag will be set to 1. Otherwise, it will be set to 0.

7	6	5	4	3	2	1	0
S	X	H	I	N	Z	V	C

Figure 2.8 ■ Condition code register

2.6.2 Branch Instructions

Branch instructions cause program flow to change when specific conditions exist. The HCS12 has three kinds of branch instructions, including *short branches, long branches*, and *bit-conditional branches*.

Branch instructions can also be classified by the type of condition that must be satisfied in order for a branch to be taken. Some instructions belong to more than one category.

- *Unary (unconditional) branch* instructions always execute.
- *Simple branches* are taken when a specific bit in the CCR register is in a specific state as a result of a previous operation.
- *Unsigned branches* are taken when a comparison or a test of unsigned quantities results in a specific combination of condition code register bits.
- *Signed branches* are taken when a comparison or a test of signed quantities results in a specific combination of condition code register bits.

When a short branch instruction is executed, a signed 8-bit offset is added to the value in the program counter when a specified condition is met. Program execution continues at the new address. The numeric range of the short branch offset value is $80 (−128) to $7F (127) from the address of the instruction immediately following the branch instruction. A summary of the short branch instructions is in Table 2.2.

Unary Branches		
Mnemonic	Function	Equation or Operation
bra	Branch always	1 = 1
bm	Branch never	1 = 0
Simple Branches		
Mnemonic	Function	Equation or Operation
bcc	Branch if carry clear	C = 0
bcs	Branch if carry set	C = 1
beq	Branch if equal	Z = 1
bmi	Branch if minus	N = 1
bne	Branch if not equal	Z = 0
bpl	Branch if plus	N = 0
bvc	Branch if overflow clear	V = 0
bvs	Branch if overflow set	V = 1
Unsigned Branches		
Mnemonic	Function	Equation or Operation
bhi	Branch if higher	C + Z = 0
bhs	Branch if higher or same	C = 0
blo	Branch if lower	C = 1
bls	Branch if lower or sarne	C + Z = 1
Signed Branches		
Mnemonic	Function	Equation or Operation
bge	Branch if greater than or equal	$N \oplus V = 0$
bgt	Branch if greater than	$Z + (N \oplus V) = 0$
ble	Branch if less than or equal	$Z + (N \oplus V) = 1$
blt	Branch if less than	$N \oplus V = 1$

Table 2.2 ■ Summary of short branch instructions

When a long branch instruction is executed, a signed 16-bit offset is added to the value in the program counter when a specified condition is met. Program execution continues at the new address. Long branch instructions are used when large displacements between decision-making steps are necessary.

The numeric range of long-branch offset values is $8000 (−32768) to $7FFF (32767) from the instruction immediately after the branch instruction. This permits branching from any location in the standard 64-kB address map to any other location in the map. A summary of the long branch instructions is in Table 2.3.

Unary Branches		
Mnemonic	Function	Equation or Operation
lbra	Long branch always	1 = 1
lbrn	Long branch never	1 = 0
Simple Branches		
Mnemonic	Function	Equation or Operation
lbcc	Long branch if carry clear	C = 0
lbcs	Long branch if carry set	C = 1
lbeq	Long branch if equal	Z = 1
lbmi	Long branch if minus	N = 1
lbne	Long branch if not equal	Z = 0
lbpl	Long branch if plus	N = 0
lbvc	Long branch if overflow is clear	V = 0
lbvs	Long branch if overflow set	V = 1
Unsigned Branches		
Mnemonic	Function	Equation or Operation
lbhi	Long branch if higher	C + Z = 0
lbhs	Long branch if higher or same	C = 0
lblo	Long branch if lower	C = 1
lbls	Long branch if lower or same	C + Z = 1
Signed Branches		
Mnemonic	Function	Equation or Operation
lbge	Long branch if greater than or equal	$N \oplus V = 0$
lbgt	Long branch if greater than	$Z + (N \oplus V) = 0$
lble	Long branch if less than or equal	$Z + (N \oplus V) = 1$
lblt	Long branch if less than	$N \oplus V = 1$

Table 2.3 ■ Summary of long branch instructions

Although there are many possibilities in writing a program loop, the following one is a common format:

```
loop:          .

               .

               .
        Bcc (or LBcc) loop
```

where **cc** is one of the condition codes (CC, CS, EQ, MI, NE, PL, VC, VS, HI, HS, LO, LS, GE, GT, LS, and LT). Usually, there will be a comparison or arithmetic instruction to set up the condition code for use by the conditional branch instruction.

2.6.3 Compare and Test Instructions

The HCS12 has a set of compare instructions that are dedicated to the setting of condition flags. The compare and test instructions perform subtraction between a pair of registers or between a register and a memory location or between a register and an immediate value. The result is not stored, but condition codes are set by the operation. In the HCS12, most instructions update condition code flags automatically, so it is often unnecessary to include a separate test or compare instruction. Table 2.4 is a summary of compare and test instructions. The memory location to be compared can be specified by addressing mode, excluding relative mode.

Compare Instructions		
Mnemonic	Function	Operation
cba	Compare A to B	(A) – (B)
cmpa	Compare A to memory	(A) – (M)
cmpb	Compare B to memory	(B) – (M)
cpd	Compare D to memory	(D) – (M:M + 1)
cps	Compare SP to memory	(SP) – (M:M + 1)
cpx	Compare X to memory	(X) – (M:M + 1)
cpy	Compare Y to memory	(Y) – (M:M + 1)
Test instructions		
Mnemonic	Function	Operation
tst	Test memory for zero or minus	(M) – $00
tsta	Test A for zero or minus	(A) – $00
tstb	Test B for zero or minus	(B) – $00

Table 2.4 ■ Summary of compare and test instructions

2.6.4 Loop Primitive Instructions

A lot of the program loops are implemented by incrementing or decrementing a loop count. The branch is taken when either the loop count is equal to zero or not equal to zero, depending on the applications. The HCS12 provides a set of loop primitive instructions for implementing this type of looping mechanism. These instructions test a counter value in a register or accumulator (A, B, D, X, Y, or SP) for zero or nonzero value as a branch condition. There are predecrement, preincrement, and test-only versions of these instructions.

The range of the branch is from $80 (−128) to $7F (127) from the instruction immediately following the loop primitive instruction. Table 2.5 shows a summary of the loop primitive instructions.

Example 2.14

Write a program to add an array of N 8-bit numbers and store the sum at memory locations $1000~$1001. Use the *For i = n1* to *n2 do* looping construct.

Solution: We will use variable **i** as the array index. This variable can also be used to keep track of the current iterations being performed. We will use a 2-byte variable **sum** to hold the sum of array elements. The logic flow of the program is illustrated in Figure 2.9.

Mnemonic	Function	Equation or Operation
dbeq cntr, rel	Decrement counter and branch if = 0 (counter = A, B, D, X, Y, or SP)	counter ← (counter) − 1 If (counter) = 0, then branch else continue to next instruction
dbne, cntr, rel	Decrement counter and branch if ≠ 0 (counter = A, B, D, X, Y, or SP)	counter ← (counter) − 1 If (counter) ≠ 0, then branch else continue to next instruction
ibeq cntr, rel	Increment counter and branch if = 0 (counter = A, B, D, X, Y, or SP)	counter ← (counter) + 1 If (counter) = 0, then branch else continue to next instruction
ibne cntr, rel	Increment counter and branch if ≠ 0 (counter = A, B, D, X, Y, or SP)	counter ← (counter) + 1 If (counter) ≠ 0, then branch else continue to next instruction
tbeq cntr, rel	Test counter and branch if = 0 (counter = A, B, D, X, Y, or SP)	If (counter) = 0, then branch else continue to next instruction
tbne, cntr, rel	Test counter and branch if ≠ 0 (counter = A, B, D, X, Y, or SP)	If (counter) ≠ 0, then branch else continue to next instruction

Note. 1. **cntr** is the loop counter and can be accumulator A, B, or D and register X, Y, or SP.
2. **rel** is the relative branch offset and is usually a label

Table 2.5 ■ Summary of loop primitive instructions

The program is a direct translation of the flowchart in Figure 2.9.

```
N          equ      20                ; array count
           org      $1000             ; starting address of on-chip SRAM
sum        rmb      2                 ; array sum
i          rmb      1                 ; array index
           org      $1500             ; starting address of the program
           ldaa     #0
           staa     i                 ; initialize loop (array) index to 0
           staa     sum               ; initialize sum to 0
           staa     sum+1             ;     "
loop       ldab     i
           cmpb     #N                ; is i = N?
           beq      done              ; if done, then branch
           ldx      #array            ; use index register X as a pointer to the array
           abx                        ; compute the address of array[i]
           ldab     0,x               ; place array[i] in B
           ldy      sum               ; place sum in Y
           aby                        ; compute sum <- sum + array[i]
           sty      sum               ; update sum
           inc      i                 ; increment the loop count by 1
           bra      loop
done       swi                        ; return to D-Bug12 monitor
; the array is defined in the following statement
array      db       1,2,3,4,5,6,7,8,9,10,11,12,13,14,15,16,17,18,19,20
           end
```

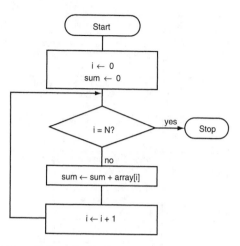

Figure 2.9 ■ Logic flow of Example 2.14

It is a common mistake for an assembly language programmer to forget about updating the variable in memory. For example, we will not get the correct value for sum if we did not add the instruction sty sum in the program of Example 2.14.

Loop primitive instructions are especially suitable for implementing the repeat S until C looping construct, as demonstrated in the following example.

Example 2.15

Write a program to find the maximum element from an array of N 8-bit elements using the *repeat S until C* looping construct.

Solution: We will use the variable **i** as the array index and also as the loop count. The variable max_val will be used to hold the array maximum. The logic flow of the program is shown in Figure 2.10.

The program is as follows:

```
N          equ      20                   ; array count
           org      $1000                ; starting address of on-chip SRAM
max_val    ds.b     1                    ; memory location to hold array max
           org      $1500                ; starting address of program
           ldaa     array                ; set array[0] as the temporary array max
           staa     max_val              ;      "
           ldx      #array+N-1           ; start from the end of the array
           ldab     #N-1                 ; use B to hold variable i and initialize it to N-1
loop       ldaa     max_val
           cmpa     0,x                  ; compare max_val with array[i]
           bge      chk_end              ; no update if max_val is larger
           movb     0,x,max_val          ; update max_val
chk_end    dex                          ; move the array pointer
           dbne     b,loop               ; decrement the loop count, branch if not zero yet.
forever    bra      forever
array      db       1,3,5,6,19,41,53,28,13,42,76,14,20,54,64,74,29,33,41,45
           end
```

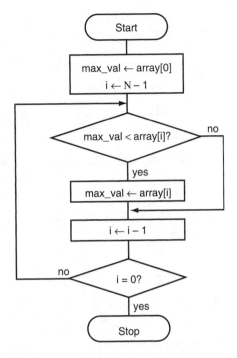

Figure 2.10 ■ Logic flow of Example 2.15

2.6.5 Decrementing and Incrementing Instructions

We often need to add 1 to or subtract 1 from a variable in our program. Although we can use one of the add or sub instructions to achieve this, it would be more efficient to use a single instruction. The HCS12 has a few instructions for us to increment or decrement a variable by 1. A summary of decrement and increment instructions is listed in Table 2.6.

Decrement Instructions		
Mnemonic	Function	Operation
dec	Decrement memory by 1	$M \leftarrow [M] - \$01$
deca	Decrement A by 1	$A \leftarrow [A] - \$01$
decb	Decrement B by 1	$B \leftarrow [B] - \$01$
des	Decrement SP by 1	$SP \leftarrow [SP] - \$01$
dex	Decrement X by 1	$X \leftarrow [X] - \$01$
dey	Decrement Y by 1	$Y \leftarrow [Y] - \$01$

Table 2.6 ■ Summary of compare and test instructions

(continued)

Increment Instructions		
Mnemonic	Function	Operation
inc	Increment memory by 1	M ← [M] + $01
inca	Increment A by 1	A ← [A] + $01
incb	Increment B by 1	B ← [B] + $01
ins	Increment SP by 1	SP ← [SP] + $01
inx	Increment X by 1	X ← [X] + $01
iny	Increment Y by 1	Y ← [Y] + $01

Table 2.6 ■ Summary of compare and test instructions (continued)

Example 2.16

▼

Use an appropriate increment or decrement instruction to replace the following instruction sequence:

```
ldaa   i
adda   #1
staa   i
```

Solution: The three instructions can be replaced by the following instruction:

```
inc    i
```

▲

2.6.6 Bit Condition Branch Instructions

In certain applications, one needs to make branch decisions based on the value of a few bits. The HCS12 provides two special conditional branch instructions for this purpose. The syntax of the first special conditional branch instruction is

brclr opr, msk, rel

where

opr   specifies the memory location to be checked and can be specified using direct, extended, and all indexed addressing modes.

msk   is an 8-bit mask that specifies the bits of the memory location to be checked. The bits to be checked correspond to those bit positions that are 1s in the mask.

rel   is the branch offset and is specified in 8-bit relative mode.

This instruction tells the HCS12 to perform bitwise logical AND on the contents of the specified memory location and the mask supplied with the instruction, then branch if the result is zero. For example, for the instruction sequence

```
here   brclr   $66,$80,here
       ldd     $70
```

the HCS12 will continue to execute the first instruction if the most significant bit of the memory location at $66 is 0. Otherwise, the next instruction will be executed.

The syntax of the second special conditional branch instruction is

brset opr, msk, rel

where

opr	specifies the memory location to be checked and can be specified using direct, extended, and all indexed addressing modes.
msk	is an 8-bit mask that specifies the bits of the memory location to be checked. The bits to be checked correspond to those bit positions that are 1s in the mask.
rel	is the branch offset and is specified in 8-bit relative mode.

This instruction tells the HCS12 to perform the logical AND of the contents of the specified memory location inverted and the mask supplied with the instruction, then branch if the result is zero (this occurs only when all bits corresponding to 1s in the mask byte are 1s in the tested byte). For example, for the following instruction sequence

```
loop    inc     count
        . . .
        brset   $66,$E0,loop
        . . .
```

the branch will be taken if the most significant three bits of the memory location at $66 are all 1s.

Example 2.17
▼

Write a program to count the number of elements that are divisible by 4 in an array of N 8-bit numbers. Use the Repeat S Until C looping construct.

Solution: The lowest two bits of a number divisible by 4 are 00. By checking the lowest two bits of a number, we can determine if a number is divisible by 4. The program is as follows:

```
N           equ     20
            org     $1000
total       rmb     1
            org     $1500           ; starting address of the program
            clr     total           ; initialize total to 0
            ldx     #array          ; use index register X as the array pointer
            ldab    #N              ; use accumulator B as the loop count
loop        brclr   0,x,$03,yes
            bra     chkend
yes         inc     total           ; add 1 to the total
chkend      inx                     ; move the array pointer
            dbne    b,loop
forever     bra     forever
array       db      2,3,4,8,12,13,19,24,33,32,20,18,53,52,80,82,90,94,100,102
            end
```

▲

2.6.7 Instructions for Variable Initialization

We often need to initialize a variable to 0 when writing a program. The HCS12 has three instructions for this purpose. They are

    clr   opr

where **opr** is a memory location specified using the extended mode and all indexed addressing (direct and indirect) modes. The memory location is initialized to 0 by this instruction.

    clra

Accumulator A is cleared to 0 by this instruction.

    clrb

Accumulator B is cleared to 0 by this instruction.

2.7 Shift and Rotate Instructions

Shift and *rotate instructions* are useful for bit field manipulation. They can be used to speed up the integer multiply and divide operations if one of the operands is a power of 2. A shift/rotate instruction shifts/rotates the operand by one bit. The HCS12 has shift instructions that can operate on accumulators A, B, and D or a memory location. A memory operand must be specified using the extended or indexed (direct or indirect) addressing modes. A summary of shift and rotate instructions is shown in Table 2.7.

Logical Shift Instructions		
Mnemonic	Function	Operation
lsl <opr> lsla lslb	Logical shift left memory Logical shift left A Logical shift left B	C b7 b0 ←0
lsld	Logical shift left D	C b7 A b0 b7 B b0 ←0
lsr <opr> lsra lsrb	Logical shift right memory Logical shift right A Logical shift right B	0→ b7 b0 C
lsrd	Logical shift right D	0→ b7 A b0 b7 B b0 C
Arithmetic Shift Instructions		
Mnemonic	Function	Operation
asl <opr> asla aslb	Arithmetic shift left memory Arithmetic shift left A Arithmetic shift left B	C b7 b0 ←0
asld	Arithmetic shift left D	C b7 A b0 b7 B b0 ←0
asr <opr> asra asrb	Arithmetic shift right memory Arithmetic shift right A Arithmetic shift right B	b7 b0 C

Table 2.7 ■ Summary of shift and rotate instructions

Rotate Instructions		
Mnemonic	Function	Operation
rol <opr> rola rolb	Rotate left memory thru carry Rotate left A through carry Rotate left B through carry	C b7 b0
ror <opr> rora rorb	Rotate right memory thru carry Rotate right A through carry Rotate right B through carry	b7 b0 C

Table 2.7 ■ Summary of shift and rotate instructions (continued)

Example 2.18

What are the values of accumulator A and the C flag after executing the **asla** instruction? Assume that originally A contains $95 and the C flag is 1.

Solution: The operation of this instruction is shown in Figure 2.11a.

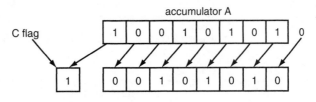

Figure 2.11a ■ Operation of the **asla** instruction

The result is shown in Figure 2.11b:

Original Value	New Value
[A] = 10010101 C = 1	[A] = 00101010 C = 1

Figure 2.11b ■ Execution result of the **asla** instruction

Example 2.19

What are the new values of the memory location at $1000 and the C flag after executing the instruction **asr $1000**? Assume that the memory location $1000 originally contains the value of $ED and the C flag is 0.

Solution: The operation of this instruction is shown in Figure 2.12a.

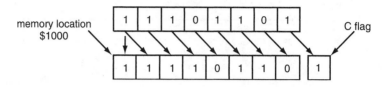

Figure 2.12a ■ Operation of the **asr S1000** instruction

The result is shown in Figure 2.12b.

Original Value	New Value
[$1000] = 11101101 C = 0	[$1000] = 11110110 C = 1

Figure 2.12b ■ Result of the **asr $1000** instruction

Example 2.20

What are the new values of the memory location at $1000 and the C flag after executing the instruction **lsr $1000**? Assume the memory location $1000 originally contains $E7 and the C flag is 1.

Solution: The operation of this instruction is illustrated in Figure 2.13a.

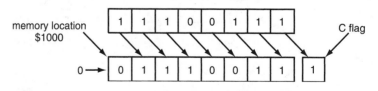

Figure 2.13a ■ Operation of the **lsr S1000** instruction

The result is shown in Figure 2.13b.

Original Value	New Value
[$1000] = 11100111 C = 1	[$1000] = 01110011 C = 1

Figure 2.13b ■ Execution result of **lsr $1000**

Example 2.21

Compute the new values of accumulator B and the C flag after executing the **rolb** instruction. Assume the original value of B is $BD and C flag is 1.

Solution: The operation of this instruction is illustrated in Figure 2.14a.

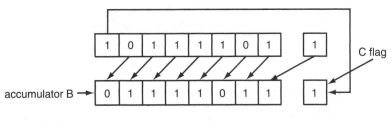

Figure 2.14a ■ Operation of the instruction rolb

The result is shown in Figure 2.14b.

Original Value	New Value
[B] = 10111101 C = 1	[B] = 01111011 C = 1

Figure 2.14b ■ Execution result of **rolb**

Example 2.22

What are the values of accumulator A and the C flag after executing the instruction **rora**? Assume the original value of A is $BE and C = 1.

Solution: The operation of this instruction is illustrated in Figure 2.15a on the next page.

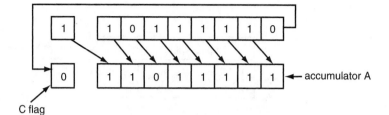

Figure 2.15a ■ Operation of the instruction rora

The result is shown in Figure 2.15b:

Original Value	New Value
[A] = 10111110 C = 1	[A] = 11011111 C = 0

Figure 2.15b ■ Execution result of **rora**

Example 2.23

Write a program to count the number of 0s contained in memory locations $1000~$1001 and save the result at memory location $1005.

Solution: The logical shift right instruction is available for double accumulator D. We can load this 16-bit value into D and shift it to the right 16 times or until it becomes 0. The algorithm of this program is as follows:

Step 1
Initialize the loop count to 16 and the zero count to 0.

Step 2
Place the 16-bit value in D.

Step 3
Shift D to the right one place.

Step 4
If the C flag is 0, increment zero count by 1.

Step 5
Decrement loop count by 1.

Step 6
If loop count is 0, then stop. Otherwise, go to step 3.
The program is as follows:

```
            org     $1000
            dc.w    $2355
zero_cnt    rmb     1
lp_cnt      rmb     1
            org     $1500
```

```
              clr      zero_cnt      ; initialize the zero count to 0
              movb     #16,lp_cnt    ; initialize loop count to 16
              ldd      $1000         ; place the 16-bit number in D
again         lsrd
              bcs      chk_end       ; branch if the lsb is a 1
              inc      zero_cnt
chk_end       dec      lp_cnt
              bne      again         ; have we tested all 16 bits yet?
forever       bra      forever
              end
```

Sometimes one needs to shift a number larger than 16 bits. However, the HCS12 does not have an instruction that does this. Suppose the number has k bytes and the most significant byte is located at **loc**. The remaining $k - 1$ bytes are located at loc + 1, loc + 2, . . ., loc + k − 1, as shown in Figure 2.16.

Figure 2.16 ■ k bytes to be shited

The logical shift-one-bit-to-the-right operation is shown in Figure 2.17.

Figure 2.17 ■ Shift-one-bit-to-the-right operation

As shown in Figure 2.17,

- Bit 7 of each byte will receive bit 0 of the byte on its immediate left with the exception of the most significant byte, which will receive a 0.
- Each byte will be shifted to the right by one bit. Bit 0 of the least significant byte will be shifted out and lost.

The operation can therefore be implemented as follows:

Step 1
Shift the byte at loc to the right one place (using the lsr <opr> instruction).

Step 2
Rotate the byte at loc + 1 to the right one place (using the ror <opr> instruction).

Step 3
Repeat step 2 for the remaining bytes.

By repeating this procedure, the given k-byte number can be shifted to the right as many bits as desired. The operation to shift a multibyte number to the left should start from the least significant byte and rotate the remaining bytes toward the most significant byte.

Example 2.24

Write a program to shift the 32-bit number stored at $1000~$1003 to the right four places.

Solution: The most significant to the least significant bytes are stored at $1000~$1003. The following instruction sequence implements the algorithm that we just described:

```
        ldab    #4          ; set up the loop count
        ldx     #$1000
again   lsr     0,x
        ror     1,x
        ror     2,x
        ror     3,x
        dbne    b,again
```

2.8 Boolean Logic Instructions

When dealing with I/O port pins, we often have the need to change the values of a few bits. For these types of applications, the Boolean logic instructions come in handy. A summary of the HCS12 Boolean logic instructions is described in Table 2.8.

Mnemonic	Function	Operation
anda <opr>	AND A with memory	A ← (A) • (M)
andb <opr>	AND B with memory	B ← (B) • (M)
andcc <opr>	AND CCR with memory (clear CCR bits)	CCR ← (CCR) • (M)
eora <opr>	Exclusive OR A with memory	A ← (A) ⊕ (M)
eorb <opr>	Exclusive OR B with memory	B ← (B) ⊕ (M)
oraa <opr>	OR A with memory	A ← (A) + (M)
orab <opr>	OR B with memory	B ← (B) + (M)
orcc <opr>	OR CCR with memory	CCR ← (CCR) + (M)
clc	Clear C bit in CCR	C ← 0
cli	Clear I bit in CCR	I ← 0
clv	Clear V bit in CCR	V ← 0
com <opr>	One's complement memory	M ← $FF – (M)
coma	One's complement A	A ← $FF – (A)
comb	One's complement B	B ← $FF – (B)
neg <opr>	Two's complement memory	M ← $00 – (M)
nega	Two's complement A	A ← $00 – (A)
negb	Two's complement B	B ← $00 – (B)

Table 2.8 ■ Summary of Boolean logic instructions

The operand **opr** can be specified using all except the relative addressing modes. Usually, one would use the **and** instruction to clear one or a few bits and use the or instruction to set one or a few bits. The **exclusive or** instruction can be used to toggle (change from 0 to 1 and from 1 to 0) one or a few bits. For example, the instruction sequence

```
ldaa    $56
anda    #$0F
staa    $56
```

clears the upper four pins of the I/O port located at $56. The instruction sequence

```
ldaa    $56
oraa    #$01
staa    $56
```

sets the bit 0 of the I/O port at $56. The instruction sequence

```
ldaa    $56
eora    #$0F
staa    $56
```

toggles the lower four bits of the I/O port at $56. The instruction com <opr> can be used if all of the port pins need to be toggled.

2.9 Bit Test and Manipulate Instruction

These instructions use a mask value to test or change the value of individual bits in an accumulator or in a memory location. The instructions bita and bitb provide a convenient

Mnemonic	Function	Operation
bclr <opr>[2], msk8	Clear bits in memory	$M \leftarrow (M) \cdot \overline{(mm)}$
bita <opr>[1]	Bit test A	$(A) \cdot (M)$
bitb <opr>[1]	Bit test B	$(B) \cdot (M)$
bset <opr>[2], msk8	Set bits in memory	$M \leftarrow (M) + (mm)$

Note: 1. <opr> can be specified using all except relative addressing modes for bita and bitb.
2. <opr> can be specified using direct, extended, and indexed (exclude indirect) addressing modes.
3. msk8 is an 8-bit value.

Table 2.9 ■ Summary of Boolean logic instructions

means of testing bits without altering the value of either operand. Table 2.9 shows a summary of bit test and manipulation instructions.
For example, the instruction

```
bclr    0,x,$81
```

clears the most significant and the least significant bits of the memory location pointed to by index register X. The instruction

```
bita    #$44
```

tests bit 6 and bit 2 of accumulator A and updates the Z and N flags of the CCR register accordingly. The V flag in CCR register is cleared. The instruction

```
bitb    #$22
```

tests bit 5 and bit 1 of accumulator B and updates the Z and N flags of CCR register accordingly. The V flag in the CCR register is cleared. The instruction

        bset   0,y,$33

sets bits 5, 4, 1, and 0 of the memory location pointed to by index register Y.

2.10 Program Execution Time

The HCS12 uses the bus clock (we will call it E clock from now on) signal as a timing reference. The generation of the E clock is described in Chapter 6. The execution times of instructions are also measured in E cycles. The execution time of each instruction can be found under the column "Access Detail" in Appendix A. The number of letters in that column indicates the number of E cycles that a specific instruction takes to complete the execution. For example, the Access Detail column of the pula instruction contains three letters, **ufo**, which indicates that the pula instruction takes 3 E cycles to complete.

There are many applications that require the generation of time delays. Program loops are often used to create a certain amount of delay unless the time delay needs to be very accurate. The creation of a time delay involves two steps:

1. Select a sequence of instructions that takes a certain amount of time to execute.

2. Repeat the instruction sequence for the appropriate number of times.

For example, the following instruction sequence takes 40 E clock cycles to execute:

```
loop    psha            ; 2 E cycles
        pula            ; 3 E cycles
        psha
        pula
        psha
        pula
        psha
        pula
        psha
        pula
        psha
        pula
        psha
        pula
        nop             ; 1 E cycle
        nop             ; 1 E cycle
        dbne    x,loop  ; 3 E cycles.
```

Example 2.25

Write an instruction sequence to create a 100-ms time delay for a demo board with a 24-MHz bus clock.

Solution: In order to create a 100-ms time delay, we need to repeat the preceding instruction sequence 60,000 times [100 ms ÷ (40 ÷ 24,000,000) μs = 60,000]. The following instruction sequence will create the desired delay:

```
        ldx    #60000     ; 2 E cycles
loop    psha              ; 2 E cycles
        pula              ; 3 E cycles
        psha              ; 2 E cycles
        pula              ; 3 E cycles
        psha              ; 2 E cycles
        pula              ; 3 E cycles
        psha              ; 2 E cycles
        pula              ; 3 E cycles
        psha              ; 2 E cycles
        pula              ; 3 E cycles
        psha              ; 2 E cycles
        pula              ; 3 E cycles
        psha              ; 2 E cycles
        pula              ; 3 E cycles
        nop               ; 1 E cycle
        nop               ; 1 E cycle
        dbne   x,loop     ; 3 E cycles
```

Example 2.26

Write an instruction sequence to create a delay of 10 sec.

Solution: The instruction sequence in Example 2.25 can only create a delay slightly longer than 100 ms. In order to create a longer time delay, one needs to use a two-layer loop. For example, the following instruction sequence will create a 10-sec delay:

```
              dab    #100       ; 1 E cycle
out_loop      ldx    #60000     ; 2 E cycles
inner_loop    psha
              pula
              psha
              pula
              psha
              pula
              psha
              pula
              psha
              pula
              psha
              pula
              psha
```

(continued)

```
            pula
            psha
            pula
            psha
            pula
            nop
            nop
            dbne    x,inner_loop
            dbne    b,out_loop      ; 3 E cycles
```

▲

The time delay created by using program loops is not accurate. Some overhead is required to set up the loop count. For example, the one-layer loop has a 2-E-cycle overhead while the two-layer loop has much more overhead:

overhead = 1 E cycle (caused by the ldab #100 instruction)
　　　　　+ 100 × 2 E cycles (caused by the out_loop　　　　ldx #60000 instruction)
　　　　　+ 100 × 3 E cycles (caused by the dbne b,out_loop instruction)
　　　　　= 501 E cycles = 20.875 μs (at 24-MHz E clock)

To reduce the overhead, one can adjust the number to be placed in index register X. For example, by placing 59999 in X, one can create a delay of 9.99985 sec with the previous program, which is closer to 10 sec.

2.11 The Multiply and Accumulate Instruction EMACS

The multiply and accumulate (EMACS) instruction multiplies two 16-bit operands stored in memory and adds the 32-bit result to a third memory location (32-bit). EMACS can be used to implement simple digital filters, defuzzification routines, and any operation that involves the evaluation of linear polynomial functions.

When the EMACS instruction is executed, the first source operand is fetched from a location pointed to by index register X, and the second source operand is fetched from a location pointed to by Y. Before the instruction is executed, the X and Y index registers must contain values that point to the most significant bytes of the source operands. The most significant byte of the 32-bit result is specified by an extended address supplied with the instruction.

Example 2.27

▼

Write an assembly program to compute the following expression using the EMACS instruction:

$a \times y + b$

where, a, y, and b are in memory locations $1000, $1002, and $1004, respectively. Leave the result in memory location $1010~1013.

Solution: The value of b should be stored in memory locations $1012~$1013, and index registers X and Y should be set up to point to memory locations $1000 and $1002. We also need to store the constant in the destination. The following program performs the desired computation:

```
            org     $1000
aa          dc.w    10
xx          dc.w    7
bb          dc.w    6
            org     $1010
result      ds.w    2
```

```
        org       $1500
        ldx       #aa            ; set X to point to constant aa
        ldy       #xx            ; set Y to point to variable xx
        movw      bb,result+2    ; store b in the result
        movw      #0,result      ; "
        emacs     result         ; perform the multiplication and accumulation
        swi
        end
```

▲

Many applications (for example, *digital filtering*) involve the evaluation of polynomials. The EMACS instruction can be used to compute the value of a polynomial. Before utilizing this instruction, one must transform the polynomial to a form that can take advantage of this instruction. The following polynomial:

$$a_n x^n + a_{n-1} x^{n-1} + \cdots + a_1 x + a_0$$

can be transformed into the following format:

$$(x(x(\cdots (a_n x + a_{n-1}) + a_{n-2}) + \cdots a_1) + a_0)$$

The expression enclosed by each pair of parentheses in the polynomial requires the computation of a multiplication followed by an addition and hence can be evaluated by using the EMACS instruction. The actual computation is quite straightforward and so will be left as an exercise problem.

2.12 Summary

An assembly language program consists of three major parts: *assembler directives, assembly language instructions,* and *comments.* A statement of an assembly language program consists of four fields: *label, operation code, operand,* and *comment.* Assembly directives supported by the freeware **as12** were all discussed in this chapter.

The HCS12 instructions were explained category by category. Simple program examples were used to demonstrate the applications of different instructions. The HCS12 is a 16-bit microcontroller. Therefore, it can perform 16-bit arithmetic. Numbers greater than 16 bits must be manipulated using multiprecision arithmetic.

Microcontrollers are designed to perform repetitive operations. Repetitive operations are implemented by program loops. There are two types of program loops: *infinite loop* and *finite loop.* There are four major variants of the looping constructs:

- **Do** statement S **forever**
- **For** $i = n_1$ **to** n_2 **do** S or **For** $i = n_2$ **downto** n_1 **do** S
- **While** C **do** S
- **Repeat** S **until** C

In general, the implementation of program loops requires

- The initialization of loop counter (or condition)
- Performing the specified operation
- Comparing the loop count with the loop limit (or evaluating the condition)
- Making a decision regarding whether the program loop should be continued

The HCS12 provides instructions to support the initialization of loop counter, decrementing (or incrementing) the loop counter, and making a decision whether looping should be continued.

The shifting and rotating instructions are useful for bit field operations. Integer multiplication by a power of 2 and division by a power of 2 can be sped up by using the shifting instructions.

The HCS12 also provides many Boolean logical instructions that can be very useful for setting, clearing, and toggling the I/O port pins.

2.13 Exercises

E2.1 Find the valid and invalid labels in the following statements, and explain why the invalid labels are invalid:

```
column 1
   ↓
a. ABC        decb
b.  lp+:      adda    #1
c. too:       mul
d. not_true   nega
e. star+=     ldaa    #10
f.  too_big   dec     count
```

E2.2 Identify the four fields of the following instructions:

```
a.         bne      not_done
b. loop    brclr    0,x,$01,loop    ; wait until the least significant bit is set
c. here:   dec      lp_cnt          ; decrement the variable lp_cnt
```

E2.3 Write a sequence of assembler directives to reserve 10 bytes starting from $1000.

E2.4 Write a sequence of assembler directives to build a table of ASCII codes of lower case letters a–z. The table should start from memory location $2000.

E2.5 Write a sequence of assembler directives to store the message "welcome to the robot demonstration!" starting from the memory location at $2050.

E2.6 Write an instruction sequence to add the two 24-bit numbers stored at $1010~$1012 and $1013~$1015 and save the sum at $1100~$1102.

E2.7 Write an instruction sequence to subtract the 6-byte number stored at $1000~$1005 from the 6-byte number stored at $1010~$1015 and save the result at $1020~$1025.

E2.8 Write a sequence of instructions to add the BCD numbers stored at $1000 and $1001 and store the sum at $1003.

E2.9 Write an instruction sequence to add the 4-digit BCD numbers stored at $1000~$1001 and $1002~$1003 and store the sum at $1010~1011.

E2.10 Write a program to compute the average of an array of N 8-bit numbers and store the result at $1000. The array is stored at memory locations starting from $1010. N is no larger than 255.

E2.11 Write a program to multiply two 3-byte numbers that are stored at $1000~$1002 and $1003~$1005 and save the product at $1010~$1015.

E2.12 Write a program to compute the average of the square of all elements of an array with thirty-two 8-bit unsigned numbers. The array is stored at $1000~$101F. Store the result at $1020~$1021.

E2.13 Write a program to count the number of even elements of an array of N 16-bit elements. The array is stored at memory locations starting from $1010. N is no larger than 255.

E2.14 Write an instruction sequence to shift the 32-bit number to the left four places. The 32-bit number is located at $1000~$1003.

E2.15 Write a program to count the number of elements in an array that are smaller than 16. The array is stored at memory locations starting from $1010. The array has N 8-bit unsigned elements.

E2.16 Write an instruction sequence to swap the upper four bits and the lower four bits of accumulator A (swap bit 7 with bit 3, bit 6 with bit 2, and so on).

E2.17 Write a program to count the number of elements in an array whose bits 3, 4, and 7 are zeroes. The array has N 8-bit elements and is stored in memory locations starting from $1000.

E2.18 Write an instruction sequence to set bits 3, 2, 1, and 0 of memory location at $1000 to 1 and leave the upper four bits of the same location unchanged.

E2.19 Find the values of condition flags N, Z, V, and C in the CCR register after the execution of each of the following instructions, given that [A] = $50 and the condition flags are N = 0, Z = 1, V = 0, and C = 1.

    (a) suba #40   (b) tsta   (c) adda #$50
    (d) lsra   (e) rola   (f) lsla

E2.20 Find the values of condition flags N, Z, V, and C in the CCR register after executing each of the following instructions independently, given that [A] = $00 and the initial condition codes are N = 0, C = 0, Z = 1, and V = 0.

    (a) tsta   (b) adda #$40   (c) suba #$78
    (d) lsla   (e) rola   (f) adda #$CF

E2.21 Write an instruction sequence to toggle the odd-number bits and clear the even-number bits of memory location at $66.

E2.22 Write a program to shift the 8-byte number located at $1000–$1007 to the left four places.

E2.23 Write a program to shift the 6-byte number located at $1010–$1015 to the right three places.

E2.24 Write a program to create a time delay of 100 sec by using program loops, assuming that the frequency of the bus clock is 24 MHz.

E2.25 Write a program to create a time delay of 5 sec using program loops, assuming that the frequency of the bus clock is 24 MHz.

E2.26 Write a program to evaluate the following polynomial:

    $a_2x^2 + a_1x + a_0$

Plug in values to verify that your program is correct.

E2.27 Write a program to evaluate the following polynomial:

    $a_3x^3 + a_2x^2 + a_1x + a_0$

Plug in values to verify that your program is correct.

3

HCS12 Members and
Hardware and Software
Development Tools

3.1 Objectives

After completing this chapter, you should be able to

- Explain the differences among different HCS12 members

- Know the peripheral functions available at different HCS12 members

- Understand the types of hardware and software development tools available

- Explain the functions of a source-level debugger

- Use the D-Bug12 commands to view and change the contents of memory locations and CPU registers

- Use the D-Bug12 commands to set breakpoints and trace program execution on the demo board

- Use the AsmIDE program to enter, assemble, and download programs onto the demo board for execution

- Use the D-Bug12 bootloader mode to upgrade D-Bug12 monitor and program application code into flash memory

- Understand the function of Background Debug Mode

- Use the DragonBDM to download a program to a demo board without a monitor program

3.2 HCS12 Members

The 68HC12 was designed to be an upgrade to the 8-bit 68HC11 microcontroller. However, Motorola discovered that the performance of the 68HC12 microcontroller was not satisfactory after it was introduced to the market. The 68HC12 has the highest bus clock speed of 8 MHz. To be competitive, Motorola revised the design to achieve a bus clock rate of 25 MHz (a few members can run at 33 MHz). The revised 68HC12 was referred to as the Star12 family. It was also named the HCS12 family. The HCS12 MCU has the same instruction set and addressing modes as does the 68HC12. However, many of the internal designs have been changed.

Automotive and process control applications are the two major target markets of the HCS12. This is evidenced by the inclusion of such peripheral functions as *input capture* (IC), *output compare* (OC), *pulse-width modulation* (PWM), *controller area network* (CAN), and *byte data link control* (BDLC). Other peripheral functions such as *serial peripheral interface* (SPI), *serial communication interface* (SCI), and *interintegrated circuit* (I²C) are also included to facilitate interconnection with a wide variety of peripheral chips.

Using flash memory to hold application programs has become the trend of microcontroller design. All HCS12 members incorporate on-chip flash memory to hold programs. Most HCS12 devices also include a certain amount of on-chip SRAM and EEPROM to hold data and/or programs needed in different applications.

Most HCS12 devices have many I/O pins to interface with I/O devices. When on-chip memory is not adequate, external memory can be added. All HCS12 devices adopt the same design for the same peripheral function to facilitate the migration from one device (with less memory or fewer peripheral functions) to another.

The features of all HCS12 devices are shown in Appendix B. All devices with the CAN module are for automotive applications. The numbering system for the HCS12 is shown in Figure 3.1.

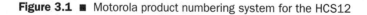

Figure 3.1 ■ Motorola product numbering system for the HCS12

In addition to automotive and control applications, Motorola is also trying to attract users from other application areas. For example, the MC9S12NE64 was designed for applications that need to access the Internet, whereas the MC9S12UF32 was designed for interfacing with the USB bus.

Software debugging support is an important issue for embedded applications. Motorola has implemented the background debug mode (BDM) in each HCS12 member. With this BDM module, the tool developers can design inexpensive software debugging tools for the HCS12. Most of the HCS12 features and peripheral functions will be discussed in this text.

3.3 Development Tools for the HCS12

Development tools are essential for those who want to learn the HCS12 or develop products based on the HCS12. There are software and hardware development tools for learning the HCS12 and developing HCS12-based products.

Function generators, oscilloscopes, in-circuit emulators, logic analyzers, and demo boards are the most important hardware development tools. In this text we will discuss the use of demo boards only. In-circuit emulators and logic analyzers are very useful for hardware development. However, these are expensive debugging tools. The BDM module of the HCS12 microcontroller allows the user to trace instruction execution on the target hardware (a HCS12 microcontroller) from a PC or another HCS12-based demo board running appropriate software. When performing debug activities in the BDM mode, the PC or the host HCS12 communicates with the target HCS12 via the BDM serial interface. This approach allows debug activities to be performed less intrusively.

Software tools include text editors, terminal programs, cross assemblers, cross compilers, simulators, debuggers, and an integrated development environment. Text editors are used to enter and edit the programs. Any text editor can be used; a serious developer, however, will consider using a programmer's editor to improve the program entering productivity because a programmer's editor allows the user to perform operations such as keyword completion, syntax checking, parenthesis matching, keyword highlighting, and all editing functions available in an ordinary text editor.

A *communication program* (also called *terminal* program) allows the PC to communicate with the demo board and download the program onto the demo board for execution. Some demo boards require special software to communicate with the PC. Other demo boards can communicate with the PC using a general-purpose terminal program such as the HyperTerminal bundled with the Windows 98/2000/XP operating system, Kermit, or ProComm.

Cross assemblers and *compilers* translate source programs into executable machine instructions. Cross assemblers and cross compilers generate executable code to be placed in the EPROM, the EEPROM, or the flash memory of a HCS12-based product. There are several freeware cross assemblers for the HCS12. According to a survey posted on the website http://www.microcontroller.com, more than 80% of the embedded applications are written in the C language. Other high-level languages pale by comparison. A tutorial on the C language will be given in Chapter 5. The freeware GNU C compiler will be used to test all the C programs in this book. ✱

A *simulator* allows the user to execute microcontroller programs without having the actual hardware. It uses computer memory to represent microcontroller registers and memory locations. The simulator interprets each microcontroller instruction by performing the operation required by the instruction and then saves the execution results in the computer memory. The simulator also allows one to set the contents of memory locations and registers before the simulation run starts.

The *source-level debugger* is a program that allows one to find problems in one's code at the high-level language (such as C) or assembly-language level. A debugger may have the option to run the program on the demo board or using a simulator. Like a simulator, a debugger can display the contents of registers and memory (internal and external) and program code in separate windows. With a debugger, all debugging activities are done at the source level. One can see the value change of a variable after a statement has been executed. One can also set a break-point at a statement in a high-level language. However, a source-level debugger requires a lot of computation. A debugger may run slowly if it needs to simulate the microcontroller instruction set instead of using the actual hardware to run the program. A source-level debugger needs to communicate with the monitor program on the demo board in order to display the contents of CPU registers and memory locations, set or delete breakpoints, trace program execution, and so on. Since the monitor programs on different evaluation boards may not be the same, a source-level debugger may be used only with one type of demo board. The BDM mode of the HCS12 offers an alternative for implementing the source-level debugger.

Ideally, *integrated development environment* (IDE) software would provide an environment that combines the following software tools so that the user can perform all development activities without needing to exit any program:

- A text editor
- A project manager
- A cross assembler and/or a compiler
- A simulator
- A source-level debugger
- A communication program

A full-blown IDE is certainly very useful but can cost dearly. However, many of today's cross compiler and assembler vendors add a text editor, project manager, and a terminal program into their cross software to make their product more competitive. The ICC12 from ImageCraft is an example. There are also freeware IDEs in this category. Both the AsmIDE (by Eric Engler) and the MiniIDE (from Mgtek) combine an assembler, text editor, and a terminal program. The EmbeddedGNU IDE by Eric Engler contains a text editor, terminal program, and a project manager. Combining the GNU C compiler and the EmbeddedGNU IDE creates a free integrated development environment in C language for the HCS12.

3.4 HCS12 Demo and Evaluation Boards

There are several HCS12DP256-based demo boards that use the D-Bug12 as their debug monitor. One is the Dragon12 from Wytec and the other is SSE256 from Shuan Shizu. Since they have the same debug monitor, most programs will run on both demo boards if they are connected to the same peripheral chips.

3.4.1 The SSE256 Demo Board

The SSE256 from Shuan Shizu has the following features:

1. 24-MHz bus speed (generated from a 16-MHz oscillator)
2. D-Bug12 monitor
3. 20 × 2 LCD kit (8-bit data transfer)
4. 8 LEDs
5. Functional generator with frequencies ranging from 1 Hz up to 16 MHz
6. 8 switches for data input
7. Keypad connector
8. 4 debounced buttons

9.   DS1307 real-time alarm clock chip with I^{2}C interface
10. TC72 digital thermometer with SPI interface
11. 24LC08B serial EEPROM with I^{2}C interface
12. Buzzer for playing songs
13. Potentiometer for testing ATD function
14. CAN transceiver (MCP2551) with a terminating resistor (with a jumper) for accessing CAN network
15. Connectors for accessing HCS12 signals
16. BDM IN and BDM OUT connectors
17. Two RS232 connectors
18. TLV5616 DAC with SPI interface and 12-bit resolution
19. Four seven-segment displays

A photograph of this demo board is shown in Figure 3.2. Product information can be found at the website at http://www.evb.com.tw.

Figure 3.2 ■ The SSE256 demo board (courtesy of Shuan Shizu Enterprise)

3.4.2 The Dragon12 Demo Board

The Dragon12 from Wytec has the following features:

1. 24-MHz bus speed generated from a 4-MHz crystal
2. D-Bug12 monitor
3. 16 × 2 LCD kit (4-bit data transfer)
4. 8 LEDs
5. 8 switches for data input
6. Keypad connector
7. 4 buttons for input
8. 4 seven-segment displays
9. Buzzer for playing songs
10. Potentiometer
11. Infrared transceiver
12. CAN transceiver
13. A small breadboard
14. BDM IN and BDM OUT connectors
15. Two RS232 connectors
16. 24LC16 serial EEPROM with I²C interface
17. LTC1661 DAC with SPI interface and 10-bit resolution

A photograph of this demo board with additional circuits added is shown in Figure 3.3. Product information can be found at the website at http://www.evbplus.com.

Figure 3.3 ■ The Wytec Dragon12 demo board (courtesy of Wytec)

3.5 The D-Bug12 Monitor

The D-Bug12 is a monitor program designed for the HCS12 microcontrollers. Version 4 of the D-Bug12 supports the following devices:

- MC9S12Dx256
- MC9S12A256
- MC9S12Dx128
- MC9S12H256
- MC9S12A128

This monitor is used in several HCS12 demo boards from several companies. It facilitates the writing, evaluation, and debugging of user programs.

Version 4.x.x of the D-Bug12 requires a host terminal program that supports Xon/Xoff software handshaking for proper operation. The HyperTerminal program that comes with Windows 2000 and Windows XP can work with the D-Bug12. In addition, the terminal programs bundled with MiniIDE, AsmIDE, and EmbeddedGNU IDE can all work with the D-Bug12 monitor. The default baud rate out of reset is 9600. However, the baud rate can be set to a higher value if a higher communication speed is desired.

3.5.1 The D-Bug12 Operating Modes

The D-Bug12 monitor has four operating modes. When the D-Bug12 monitor first starts (at power up or reset), it reads the logic levels on the PAD0 and PAD1 pins to enter different operating modes. The operating modes for the four logic level combinations are listed in Table 3.1.

PAD1	PAD0	Operating mode
0	0	D-Bug12;EVB
0	0	Jump to internal EEPROM
1	0	D-Bug12;POD
1	1	SerialBootloader

Table 3.1 ■ D-Bug12 operating modes

3.5.2 EVB Mode

The EVB mode is the most important mode for beginners to learn the HCS12 microcontroller using a demo board. In this mode, the monitor operates as a ROM resident monitor/debugger executing from the on-chip flash memory. Although this mode provides a good environment for testing new algorithms or conducting performance benchmarks, it does have a few limitations. In this mode, the flash memory, 1024 bytes of on-chip SRAM, and one of the SCI serial ports are not available to the developer. The portion of the SRAM that may be used by the user program begins at $1000 and ends at $3BFF. D-Bug12 uses the remainder of the SRAM that begins at $3C00 and ends at $3FFF. User programs can also be loaded onto the on-chip EEPROM for execution. The conceptual operating model for this mode is illustrated in Figure 3.4.

Figure 3.4 ■ EVB model conceptual model

Suppose one has run a terminal program on the PC and selects the EVB mode, then either powering on the demo board or pressing the reset button will bring out the sign-on message shown in Figure 3.5 on the terminal screen. D-Bug12 displays the ASCII greater than character (>), indicating that it is ready to accept a command. When issuing a command that causes a program to run from the internal RAM or EEPROM, D-Bug12 will place the terminal cursor on a blank line, where it will remain until control is returned to D-Bug12. If a running program fails to return to D-Bug12, pressing the EVB's reset button will cause the running program to halt execution and initiate the D-Bug12 initialization sequence. Using this method to regain control of an executing program fails to report any information to the programmer on why or how the program may have failed.

```
D-Bug 12 4.0.0b24
Copyright 1996 - 2002 Motorola Semiconductor
For Commands type "Help"

>
```

Figure 3.5 ■ D-Bug12 EVB mode sign-on message

(Alternatively, if an optional, normally open switch wired to the $\overline{\text{XIRQ}}$ pin is pressed, an $\overline{\text{XIRQ}}$ interrupt it will be generated and cause the running program to halt execution and return control back to D-Bug12, where the CPU register contents are displayed.) Both the Dragon12 and the SSE256 demo boards implement this option.

As with all ROM-based monitors, D-Bug12 utilizes some of the on-chip resources to perform its debugging functions. The D-Bug12's memory maps when running on 256-KB and 128-KB devices are detailed in Table 3.2. For the HCS12 devices with 256-KB flash memory, there are 11 KB of SRAM available for the development of application programs. For devices with 128-KB flash memory, only 7 KB is available for program development. The current implementation of D-Bug12 does not allow any of the on-chip flash memory to be utilized by application code when running in EVB mode.

Note that even though the HCS12Dx256 parts contain 4 KB of EEPROM, only the upper 3 KB are visible, as the lower 1 KB is overlaid with the I/O registers. Table 3.2 shows only the 64-KB memory map. Most of the D-Bug12 code occupies the on-chip paged flash memory beginning on page 38.

Address range	Description
$0000-$03FF	I/O registers
$0400-$0FFF	On-chip EEPROM
$1000-$3BFF	On-chip SRAM (available to user)
$3C00-$3FFF	On-chip SRAM (D-Bug12)
$4000-$EE7F	D-Bug12 code
$EE80-$EEBF	User-accessible function table
$EEC0-$EEFF	Customization data
$EF00-$EF8B	D-Bug12 startup code
$EF8C-$EFFF	Secondary reset/interrupt table
$F000-$FFFF	Bootloader

Table 3.2a ▪ D-Bug12 memory map for HCS12D x 256

Address range	Description
$0000-$03FF	I/O registers
$0400-$0FFF	On-chip EEPROM
$1000-$3BFF	On-chip SRAM (available to user)
$3C00-$3FFF	On-chip SRAM (D-Bug12)
$4000-$EE7F	D-Bug12 code
$EE80-$EEBF	User-accessible function table
$EEC0-$EEFF	Customization data
$EF00-$EF8B	D-Bug12 startup code
$EF8C-$EFFF	Secondary reset/interrupt table
$F000-$FFFF	Bootloader

Table 3.2b ▪ D-Bug12 memory map for HCS12D x 128

D-Bug12 supports a set of commands that can be used for program development on the demo board. A summary of the command set is given in Table 3.3. One can request the D-Bug12 monitor to show the complete command set by typing **help** at the D-Bug12 monitor prompt.

3.6 Using a Demo Board with the D-Bug12 Monitor

We need at least the following software programs in order to develop assembly programs to be downloaded onto a demo board for execution:

1. A text editor
2. A HCS12 cross assembler
3. A terminal program

We prefer using an IDE program in developing assembly programs. Both the **AsmIDE** by Eric Engler and the **MiniIDE** from Mgtek are well-designed IDEs for developing assembly programs for the HCS12 and 68HC11 microcontrollers. These two freeware IDEs allow you to enter, assemble, and download the S-record file onto a demo board for execution without quitting any one of them. AsmIDE can be downloaded from the website at http://www.geocities. com/englere geo; MiniIDE can be downloaded from the website at http://www.mgtek.com/ miniide. AsmIDE does not expire. However, MiniIDE will expire at the end of each year. Users must download it once a year; by doing this, they get an update of MiniIDE. The tutorial for using the AsmIDE is given in the following section. The tutorial for using the MiniIDE is given in Appendix C.

Command	Description
ALTCLK	Specify an alternate BDM communication rate
ASM	Single line assembler/disassembler
BAUD <baudrate> [;t]	Set communications rate for the terminal
BDMBD	Enter the BDM command debugger
BF <startAddress><EndAddress> [<data>]	Fill memory with data
BR [<Address>]	Set/Display breakpoints
BULK	Erase entire on-chip EEPROM contents
CALL [<address>]	Execute a user subroutine; return to D-Bug12 when finished
DEVICE [see description]	Select/define a new target MCU device
EEBASE <Address>	Inform D-Bug12 of the Target's EEPROM base address
FBULK	Erase the target processors on-chip flash EEPROM
FLOAD <AddressOffset>	Program the target processors on-chip Flash EEPROM from S -records
FSERASE	Erase one or more sectors of target flash EEPROM
G [<Address>]	Go-begin execution of user program
GT <Address>	Go-Till-set a temporary breakpoint and begin execution of user program
HELP	Display D-Bug 12 command set and command syntax
LOAD [<AddressOffset>]	Load user program in S-record
MD <StartAddress> [<EndAddress>]	Memory display-display memory contents in hex bytes/ASCII format
MDW <StartAddress> [<EndAddress>]	Memory display word-display memory contents in hex words/ASCII format
MM <Address> [<data>]	Memory modify-interactively examine/change memory contents
MMW <address> [<data>]	Memory modify word--interactively examine/change memory contents
MOVE <StartAddress> <EndAddress> <DestAddress>	Move a block of memory
NOBR [<Address> <Address>]	Remove individual user breakpoints
PCALL	Execute a user subroutine in expanded memory, return to D-Bug12 when finiashed
RD	Register display-display the CPU registers
REGBASE	Inform D-Bug12 of the target 1/0 registers base address
RESET	Reset the target CPU
RM	Register mod ify--interactively examine/change CPU register contents
SO	Step over subroutine calls
STOP	Stop execution of user code on the target processor and place it in background mode
T [<count>]	Trace <count> instructions
TCONFIG	Configure target before erasing or programming
UPLOAD <StartAddress><EndAddress>	S-Record Memory display
USEHBR	Use Hardware Breakpoints
VER	Display the running version of D-Bug12
VERF [<AddressOffset>]	Verify S-records against memory contents
<Register Name><Register Value>	Set register contents
Register Names:	PC, SP, X, Y, A, B, D, PP
CCR Status Bits:	S, XM, H, IM, N, Z, V, C

Table 3.3 ■ D-Bug12 command set summary

3.6.1 Starting the AsmIDE

The AsmIDE can be started by clicking on its icon. The screen should look like Figure 3.6 after the AsmIDE is started. The upper half of the window in Figure 3.6 is used by the user to enter the assembly program. The lower half of the AsmIDE window is used to display messages after an assembly process or as the terminal window to communicate with the D-Bug12 monitor.

Figure 3.6 ■ AsmIDE startup screen

3.6.2 Communicating with the Demo Board

To communicate with the demo board using the D-Bug12 monitor, click on **Terminal** in the lower half of the AsmIDE window. After this, the lower half becomes the terminal window. Press the reset button of the demo board and the screen will change to Figure 3.7.

One can then enter D-Bug12 commands in the terminal window to display registers and memory contents, modify memory and register contents, set breakpoints, trace instruction execution, and download program onto the demo board for execution.

If the D-Bug12 prompt does not appear on the screen, the most likely three causes are as follows:

1. The demo board is not powered up.

2. The RS232 cable connection is wrong.

3. The baud rate of the terminal program does not match that of the demo board.

One should make sure that the demo board has been powered up and also make sure that the RS232 cable is connected to the right connector (if the demo board has two RS232 connectors). If the D-Bug12 prompt still does not appear, then check the baud rate setting. Select **View**

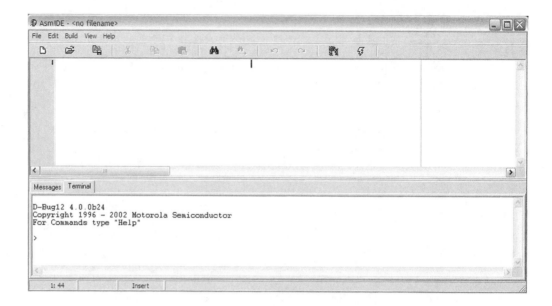

Figure 3.7 ■ Terminal window after pressing the reset button

from the IDE menu and click on **Options.** Then click on **COM** port and the screen should change to that shown in Figure 3.8. To set COM Port options, click on **Set COM Options** and a dialog box will appear. Make sure the settings of the terminal program are identical to those in Figure 3.9. Another option that the user needs to set is *assembler*. Click on **assembler** in the dialog box and make sure the settings are the same as those in Figure 3.10.

3.6.3 Using the D-Bug12 Commands

The D-Bug12 commands are provided to help the program-debugging process. Some of these commands can be used to set and display the contents of memory locations. In this section, character strings entered by the user are in boldface and optional fields are enclosed in brackets [ ]. In the following section, the syntax of a command is presented and then examples are given.

BF <StartAddress> <EndAddress> [<Data>]

The BF command is used to fill a block of memory locations with the same value. For example, the following command clears the internal memory locations from $1000 to $10FF to zero:

>**bf 1000 1FFF 0**

The data field is optional. If one did not specify the data to be filled in, then zero will be entered in the specified memory locations.

MD <StartAddress> [<EndAddress>]

The *MD* command (*memory display*) is used to display memory contents. This command displays memory contents as both hexadecimal bytes and ASCII characters, 16 bytes on each line. The <StartAddress> parameter must be supplied; the <EndAddress> parameter is optional. When the <EndAddress> parameter is not specified, a single line is displayed.

Figure 3.8 ■ AsmIDE options dialog box

The number supplied as the <StartAddress> parameter is rounded down to the next lower multiple of 16, while the number supplied as the <EndAddress> parameter is rounded up to the next higher multiple of 16 minus 1. This causes each line to display memory in the range of $xxx0 to $xxxF. For example, if **$1005** is entered as the start address and **$1020** as the ending address, then the actual memory range displayed would be $1000 through $102F. This command allows the user to examine the program execution result.

Figure 3.9 ■ Appropriate settings for COM port

Example 3.1

```
>md 1000
1000   00 00 00 00 - 00 00 00 00 - 00 00 00 00 - 00 00 00 00      . . . . . . . . . . . .
>md 1005 1020
1000   00 00 00 00 - 00 00 00 00 - 00 00 00 00 - 00 00 00 00      . . . . . . . . . . . .
1010   00 00 00 00 - 00 00 00 00 - 00 00 00 00 - 00 00 00 00      . . . . . . . . . . . .
1020   00 00 00 00 - 00 00 00 00 - 00 00 00 00 - 00 00 00 00      . . . . . . . . . . . .
```

MDW <StartAddress> [<EndAddress>]

This command (memory display words) displays the contents of memory locations as hexadecimal words and ASCII characters, 16 bytes on each line. The <StartAddress> parameter must be supplied; the <EndAddress> parameter is optional. When the <EndAddress> parameter is not supplied, a single line is displayed.

The number supplied as the <StartAddress> parameter is rounded down to the next lower multiple of 16, while the number supplied as the <EndAddress> parameter is rounded up to the next higher multiple of 16 minus 1. This causes each line to display memory in the range of $xxx0 through $xxxF.

Figure 3.10 ■ Appropriate settings for the assembler in AsmIDE

Example 3.2
▼

```
>mdw 1000
1000   FC08 00CD - 0900 CE00 - 0A18 10CB - 306B 44B7          . . . . . . . . . . . . 0kD .
>mdw 1000 1020
1000   FC08 00CD - 0900 CE00 - 0A18 10CB - 306B 44B7          . . . . . . . . . . . . 0kD .
1010   C5CE 000A - 1810 CB30 - 6B43 B7C5 - CE00 0A18          . . . . . . . 0kC . . . . . .
1020   10CB 306B - 42B7 C5CE - 000A 1810 - CB30 6B41          . . 0kB . . . . . . . . 0kA
>
```

▲

MM <Address> [<Data>]
This command (memory modify) allows one to examine and modify the contents of memory locations 1 byte at a time. If the 8-bit data parameter is present on the command line, the

byte at memory location **<Address>** is replaced with **<Data>** and the command is terminated. If no optional data is provided, then D-Bug12 enters the interactive memory modify mode. In the interactive mode, each byte is displayed on a separate line following the address of data. Once the memory modify command has been entered, single-character subcommands are used for the modification and verification of memory contents. These subcommands have the following format:

[<Data>]	<CR>	Optionally update current location and display the next location.
[<Data>]	</> or <=>	Optionally update current location and redisplay the same location.
[<Data>]	<^> or <−>	Optionally update current location and display the previous location.
[<Data>]	<.>	Optionally update current location and exit Memory Modify.

With the exception of the carriage return (CR), the subcommand must be separated from any entered data with at least one space character. If an invalid subcommand character is entered, an appropriate error message will be issued and the contents of the current memory location are redisplayed.

Example 3.3

▼

In this example, each line is terminated with a carriage-return character. However, the carriage-return character is nondisplayable and hence is not shown.

```
>mm 1000
1000 00
1001 00 FF
1002 00 ^
1001 FF
1002 00
1003 00 55 /
1003 55 .
>
```

▲

MMW <Address> [<Data>]

This command (memory modify, word) allows the contents of memory to be examined and/or modified as 16-bit hex data. If the 16-bit data is present on the command line, the word at memory location **<Address>** is replaced with **<Data>** and the command is terminated. If not, D-Bug12 enters the interactive memory modify mode. In the interactive mode, each word is displayed on a separate line following the address of data. Once the memory modify command has been entered, single-character subcommands are used for the modification and verification of memory contents. These subcommands have the following format:

[<Data>]	<CR>	Optionally update current location and display the next location.
[<Data>]	</> or <=>	Optionally update current location and redisplay the current location.
[<Data>]	<^> or <−>	Optionally update current location and display the previous location.
[<Data>]	<.>	Optionally update current location and exit Memory Modify.

With the exception of the carriage return (CR), the subcommand must be separated from any entered data with at least one space character. If an invalid subcommand character is entered, an appropriate error message will be issued and the contents of the current memory location are redisplayed.

Example 3.4

In this example, each line is terminated with a carriage-return character. However, the carriage-return character is nondisplayable and hence is not shown.

```
>mmw 1100
1100 00F0
1102 AA55 0008
1104 0000 ^
1102 0008 aabb
1104 0000
1106 0000 .
>
```

Move <StartAddress> <EndAddress> <DestAddress>

This command (move memory block) is used to move a block of memory from one location to another, 1 byte at a time. Addresses are specified in 16-bit hex values. The number of bytes moved is one more than <EndAddress> − <StartAddress>. The block of memory beginning at the destination address may overlap the memory block defined by <StartAddress> and <EndAddress>. One of the uses of the **move** command might be to copy a program from RAM into the on-chip EEPROM.

Example 3.5

```
>move 1000 10ff 1100
>
```

RD

This command (register display) is used to display the HCS12 CPU registers, including the ppage register.

Example 3.6

```
>rd
PP   PC      SP      X       Y       D = A:B     CCR = SXHI    NZVC
38 1521     3C00    2014    0000      6E:14            1001    0100
xx: 1521    9C42            CPD      $0042
>
```

RM

This command (register modify) is used to examine and/or modify the contents of the CPU12 registers interactively. As each register and its contents are displayed, D-Bug12 allows the user to enter a new value for the register in hex. If modification of the displayed register is not desired, entering a carriage return will cause the next CPU register and its contents to be displayed on the next line. When the last of the CPU registers has been examined and/or modified, the **RM** command displays the first register, giving the user an opportunity to make additional modifications to the CPU register contents. Typing a period as the first nonspace character on the line will exit the interactive mode of the register modify command and return to the D-Bug12 command prompt. The registers are displayed in the following order, one register per line: PC, SP, X, Y, A, B, CCR. The PPAGE register (represented by PP) is not displayed.

Example 3.7

▼

```
>rm
PC=0000 1500
SP=0A00
IX=0000 0100
IY=0000
A=00
B=00 ff
CCR=90 d1
PC=1500 .
>
```

▲

<RegisterName> <RegisterValue>
This command allows one to change the value of any CPU register (PC, SP, X, Y, A, B, D, CCR). Each of the fields in the CCR may be modified by using the bit names shown in Table 3.4.

CCR bit name	Description	Legal Values
S	STOP enable	0 or 1
H	Half carry	0 or 1
N	Negative flag	0 or 1
Z	Zero flag	0 or 1
V	Two's complement over flag	0 or 1
C	Carry flag	0 or 1
IM	IRQ interrupt mask	0 or 1
XM	XIRQ interrupt mask	0 or 1

Table 3.4 ■ Condition code register bits

Examples 3.8

▼

```
>pc 2000
   PC      SP      X       Y      D = A:B    CCR = SXHI    NZVC
  2000    0A00    0100    0000     00:FF          1101     0001
>x 800
   PC      SP      X       Y      D = A:B    CCR = SXHI    NZVC
  2000    0A00    0800    0000     00:FF          1101     0001
>c 0
   PC      SP      X       Y      D = A:B    CCR = SXHI    NZVC
  2000    0A00    0800    0000     00:FF          1101     0000
>z 1
```

PC	SP	X	Y	D = A:B	CCR = SXHI	NZVC
2000	0A00	0800	0000	00:FF	1101	0100

>d 2010

PC	SP	X	Y	D = A:B	CCR = SXHI	NZVC
2000	0A00	0800	0000	20:10	1101	0100

>

▲

ASM <Address>

This command invokes the one-line assembler/disassembler. It allows memory contents to be viewed and altered using assembly language mnemonics. Each entered source line is translated into object code and placed into memory at the time of entry. When displaying memory contents, each instruction is disassembled into its source mnemonic form and displayed along with the hex object code and any instruction operands.

Assembly mnemonics and operands may be entered in any mix of upper- and lowercase letters. Any number of spaces may appear between the assembler prompt and the instruction mnemonic or between the instruction mnemonic and the operand. Numeric values appearing in the operand field are interpreted as *signed* decimal numbers. Placing a $ in front of any number will cause the number to be interpreted as a hex number.

When an instruction is disassembled and displayed, the D-Bug12 prompt is displayed following the disassembled instruction. If a carriage return is the first non-space character entered following the prompt, the next instruction in memory is disassembled and displayed on the next line. If an HCS12 instruction is entered following the prompt, the entered instruction is assembled and placed into memory. The line containing the new entry is erased and the new instruction is disassembled and displayed on the same line. The next instruction location is then disassembled and displayed on the screen.

When entering branch instructions, the number placed in the operand field should be the absolute destination address of the instruction. The assembler calculates the two's complement offset of the branch and places the offset in memory with the instruction. The assembly/disassembly process may be terminated by entering a period as the first nonspace character following the assembler prompt.

The following example displays the assembly instructions from memory location $2000 until $2011. The carriage-return character is entered at the > prompt of each line and the period character is entered at the last line.

```
>asm 2000
2000    FC0800    LDD     $0800     >
2003    CD0900    LDY     #$0900    >
2006    CE000A    LDX     #$000A    >
2009    1810      IDIV              >
200B    CB30      ADDB    #$30      >
200D    6B44      STAB    4,Y       >
200F    B7C5      XGDX              >
2011    CE000A    LDX     #$000A    >.
>
```

The following example enters a short program that consists of three instructions starting from the memory location at $1500:

```
>asm 1500
1500    FC0800    LDD     $0800
1503    F30802    ADDD    $0802
```

1506	7C0900	**STD**	**$0900**	
1509	E78C	TST	12,SP	>.

>

BR [<Address> . . .]

This command (breakpoint set) sets a breakpoint at a specified address or to display any previously set breakpoints. The function of a breakpoint is to halt user program execution when the program reaches the breakpoint address. When a breakpoint address is encountered, D-Bug12 disassembles the instruction at the breakpoint address, prints the CPU register contents, and waits for a D-Bug12 command to be entered by the user.

Breakpoints are set by typing the breakpoint command followed by one or more breakpoint addresses. Entering the breakpoint command without any breakpoint addresses will display all the currently set breakpoints. A maximum of ten user breakpoints may be set at one time. Whenever one's program is not working correctly and one suspects that the instruction at a certain memory location is incorrect, one can set a breakpoint at that location and check the execution result by looking at the contents of CPU registers or memory locations.

Example 3.9

▼

>**br**	**1020**	**1040**	**1050**		; set three breakpoints
Breakpoints:	1020	1040	1050		
>**br**					; display current breakpoints
Breakpoints:	1020	1040	1050		

>

▲

NOBR [<Address> <Address>]

This command removes one or more previously entered breakpoints. If the NOBR command is entered without any argument, all user breakpoints are removed from the breakpoint table.

Examples 3.10

▼

>**br**	**2000**	**2010**	**2020**	**2040**	**2090**
Breakpoints:	2000	2010	2020	2040	2090
>**nobr**	**2000**	**2010**			
Breakpoints:	2020	2040	2090		
>**nobr**					
All Breakpoints Removed					

>

▲

G [<Address>]

This command is used to begin execution of user code in real time. Before beginning execution of user code, any breakpoints that were set with the BR command are placed in memory. Execution of the user program continues until a user breakpoint is encountered, a CPU exception occurs, the STOP or RESET command is entered, or the EVB's reset switch is pressed. When the user code halts for any of these reasons and control is returned to D-Bug12, a message is displayed explaining the reason for user program termination. In addition, D-Bug12 disassembles the instruction at the current PC address, prints the CPU register contents, and waits for the next D-Bug12 command to be entered by the user. If the starting address is not supplied in the command line parameter, program execution will begin at the address defined by the current value of the program counter.

Example 3.11

▼

```
>g    1500
User Bkpt Encountered
PP  PC      SP      X       Y       D = A:B     CCR = SXHI   NZVC
38 150C    3C00    7B48    0000     03:E8             1001   0001
xx: 150C   911E            CMPA    $001E
>
```

▲

GT <Address>

This command (go till) is similar to the **G** command except that a temporary breakpoint is placed at the address supplied at the command line. Any breakpoints that were set by the use of the BR command are **not** placed in the user code before program execution begins. Program execution begins at the address defined by the current value of the program counter. When user code reaches the temporary breakpoint and control is returned to D-Bug12, a message is displayed explaining the reason for user program termination. In addition, D-Bug12 disassembles the instruction at the current PC address, prints the CPU register contents, and waits for a command to be entered by the user.

Example 3.12

▼

```
>pc 1500

PP  PC      SP      X       Y       D = A:B     CCR = SXHI   NZVC
38 1500    3C00    1000    1002     00:00             1001   0101
xx: 1500   CF1500          LDS      #$1500
>gt 1540
Temporary Breakpoint Encountered
PP  PC      SP      X       Y       D = A:B     CCR = SXHI   NZVC
38 1510    1500    1000    1002     1E:00             1001   0000
xx: 1510   3B              PSHD
>
```

▲

T [<count>]

This command (trace) is used to execute one or several user program instructions beginning at the current program counter location. As each program instruction is executed, the CPU register contents and the next instruction to be executed are displayed. A single instruction may be executed by entering the trace command immediately followed by a carriage return.

Because of the method used to execute a single instruction, branch instructions (Bcc, LBcc, BRSET, BRCLR, DBEQ/NE, IBEQ/NE, TBEQ/NE) that contain an offset that branches back to the instruction opcode do not execute. D-Bug12 appears to become stuck at the branch instruction and does not execute the instruction even if the condition for the branch instruction is satisfied. This limitation can be overcome by using the **GT** command to set a temporary breakpoint at the instruction following the branch instruction.

Example 3.13

▼

```
>pc 1500
PP  PC     SP     X      Y       D = A:B    CCR = SXHI   NZVC
38 1500    1500   1000   1002       1E:00          1001   0000
xx: 1500   CF1500              LDS    #$1500
>t

PP  PC     SP     X      Y       D = A:B    CCR = SXHI   NZVC
38 1503    1500   1000   1002       1E:00          1001   0000
xx: 1503   CE1000              LDX    #$1000
>t 2

PP  PC     SP     X      Y       D = A:B    CCR = SXHI   NZVC
38 1506    1500   1000   1002       1E:00          1001   0000
xx: 1506   34                  PSHX

PP  PC     SP     X      Y       D = A:B    CCR = SXHI   NZVC
38 1507    14FE   1000   1002       1E:00          1001   0000
xx: 1507   861E               LDAA   #$1E
>
```

▲

The first command in this example sets the program counter to $1500 so that the user knows where the program execution starts. This command is normally needed when tracing a program.

CALL [<Address>]

This command is used to execute a subroutine and returns to the D-Bug12 monitor program when the final RTS of the subroutine is executed. All CPU registers contain the values at the time the final RTS instruction was executed, with the exception of the program counter. The program counter contains the starting address of the subroutine. If a subroutine address is not supplied on the command line, the current value of the program counter is used as the starting address.

No user breakpoints are placed in memory before execution is transferred to the subroutine. If the called subroutine modifies the value of the stack pointer during its execution, it must restore the stack pointer's original value before executing the RTS instruction. This restriction is required because a return address is placed on the user's stack that returns to D-Bug12 when the final RTS of the subroutine is executed. Obviously, any subroutine must obey this restriction in order to execute properly.

Example 3.14

▼

```
>call 1600
Subroutine Call Returned
pp  PC     SP     X      Y       D = A:B    CCR = SXHI   NZVC
38 1600    0A00   0032   0900       00:31          1001   0000
xx: 1600   FC1000              LDD    $1000

>
```

▲

This command is useful for testing a subroutine without writing a testing program.

3.6.4 Entering an Assembly Program

When a program is first created, one needs to open a new file to hold it. To do that, one presses the **File** menu from the AsmIDE window and selects **New**, as shown in Figure 3.11. After one selects the **New** command from the menu, a popup confirm box as shown in Figure 3.12 will appear to ask the user to confirm the request. The user simply clicks on **Yes** to get rid of the popup confirm box and start to enter a new program.

Figure 3.11 ■ Select NEW from the File menu to create a new file

Figure 3.12 ■ Confirmation dialog for creating a new file

The editing window (upper half of the IDE window) may be too small and should be adjusted. After adjusting the window size, the user enters the program that converts a hex number into BCD digits in Example 2.13. This program performs the repeated divide-by-10 operation to the given number and adds hex number $30 to each remainder to convert to its corresponding ASCII code. The result is shown in Figure 3.13.

```
🔧 AsmIDE - C:\books\hcs12\programs\ch02\eg02_13.asm                        _ □ ✕
File  Edit  Build  View  Help

    1          org     $1000
    2  data    fdb     12345    ; place a number for testing
    3          org     $1010
    4  result  rmb     5        ; reserve five bytes to store the result
    5
    6          org     $1500
    7          ldd     data     ; make a copy of the number to be converted
    8          ldy     #result
    9          ldx     #10      ; divide the number by 10
   10          idiv             ;      "
   11          addb    #$30     ; convert to ASCII code
   12          stab    4,Y      ; save the least significant digit
   13          xgdx             ; swap the quotient to D
   14          ldx     #10
   15          idiv
   16          addb    #$30     ; convert to ASCII code
   17          stab    3,Y      ; save the second-to-least-significant digit
   18          xgdx
   19          ldx     #10
   20          idiv
   21          addb    #$30
   22          stab    2,Y      ; save the middle digit
   23          xgdx
   24          ldx     #10
   25          idiv             ; separate the most-significant and second-to-most-
   26                           ; significant digits
   27          addb    #$30
   28          stab    1,Y      ; save the second-to-most-significant digit
   29          xgdx             ; swap the most significant digit to B
   30          addb    #$30     ; convert to ASCII code
   31          stab    0,Y      ; save the most significant digit
   32          end

Messages  Terminal
```

Figure 3.13 ■ Enter the program to convert hex number to BCD ASCII digits

3.6.5 Assembling the Program

To assemble an assembly program, press the **Build** menu and select **Assemble**, as shown in Figure 3.14. If the program is assembled successfully, the status window will display the corresponding message, as shown in Figure 3.15.

After the program is assembled successfully, one is ready to run and debug the program. The output of the assembly process is an S-record file *eg2_13.s19*. The file name has a suffix **s19**. The S-record format is a common file format defined by Freescale to allow tools from different vendors to work on the same project.

Figure 3.14 ■ Prepare to assemble the program eg2_13.asm

Figure 3.15 ■ Message window shows that previous assembling is successful

3.6.6 Downloading the S-Record File onto the Demo Board for Execution

The D-Bug12 provides the following command for the user to download the S-record file onto the demo board:

LOAD [<AddressOffset>]

This command is used to load S-record objects into memory from an external device. The address offset, if supplied, is added to the load address of each S-record before its data bytes are placed in memory. Providing an address offset other than zero allows object code or data to be loaded into memory at a location other than that for which it was assembled. During the loading process, the S-record data is not echoed to the control console. However, for each ten

S-records that are successfully loaded, an ASCII asterisk character (*) is sent to the control console. When an S-record file has been successfully loaded, the D-Bug12 prompt reappears on the screen.

The **load** command is terminated when D-Bug12 receives an S9 end-of-file record. If the object being loaded does not contain an S9 record, D-Bug12 does not return its prompt and continues to wait for the end-of-file record. Pressing the reset switch returns D-Bug12 to its command line prompt.

Since the lower half of the IDE window was used to display the assembly message, one has to click on **Terminal** to switch it to the terminal window in order to enter the command to download the **eg02_13.s19** file.

Example 3.15
▼

```
>load
*
>
```

▲

This example downloads a file that is equal to or shorter than ten S-records and hence only one * character is displayed on the screen. In addition to entering the **load** command followed by a carriage return, one also needs to go back to the IDE window to specify the file to be downloaded. An example of selecting the **Download** command in AsmIDE is shown in Figure 3.16.

Figure 3.16 ■ Prepare to download a .S19 file

After the **Download** command is selected, a popup dialog box, as shown in Figure 3.17, will appear. This dialog box allows one to specify the file to be downloaded. Click on the **Open** button on the popup window and the file will be transferred to the demo board.

Figure 3.17 ■ Select an S-Record file to be downloaded

3.6.7 Running and Debugging the Program

One needs to go back to the terminal window to run and debug the program. The screen after the program download would look like Figure 3.18.

Figure 3.18 ■ Screen after downloading the program eg2_13.s19 onto the demo board

Before running the program, one should verify that the test data is downloaded correctly into the memory. In this program, the test data is the decimal number 12345, which corresponds to the hex number $3039. One can use the command **md 1000** to verify it. The contents of the memory locations $1000 and $1001 displayed by the D-Bug12 monitor should be $3039. Otherwise, some error might have occurred.

One should also make sure that the program has been downloaded onto the right memory area. This can be verified by using the **asm 1500** command (the eg2_13 program starts at $1500) and pressing the **enter** key several times. The first few lines should look as follows:

```
>asm 1500
xx:1500      FC1000      LDD      $1000      >
xx:1503      CD1010      LDY      #$1010     >
xx:1506      CE000A      LDX      #$000A     >
xx:1509      1810        IDIV                >
xx:150B      CB30        ADDB     #$30       >
xx:150D      6B44        STAB     4,Y        >
```

After making sure that the program has been downloaded correctly, one should execute the program without setting any breakpoint and check the result to see if it is correct. The screen should look like Figure 3.19.

Figure 3.19 ■ Screen after running the eg2_13 program and displaying the result

The second line of the terminal window that contains the statement **User Bkpt Encountered** is caused by the software interrupt (SWI) instruction. The SWI instruction caused the program control to be returned to the D-Bug12 monitor. This instruction is often used as the last instruction of a program to be executed on a demo board with the D-Bug12 monitor. The last line of the terminal window displays the contents of the memory locations $1010 to $101F. The first 5 bytes represent the ASCII codes of 1, 2, 3, 4, and 5. Therefore, the program executed correctly.

If the program does not work correctly, one can set breakpoints at locations that one suspects or even trace the execution of some instructions to find out the error. Suppose the execution result (we take out the first **xgdx** instruction) looks as follows:

```
>md 1010

1010   30 30 35 33 - 35 2D 2D 2D - 2D 2D 2D 2D - 2D 2D 2D 2D   00535---------------
>
```

The first four digits were found to be incorrect. Since this program is short, one can trace through it. One approach is the following.

Step 1
Trace through the first six instructions and check to see if the quotient (in index register X) and the memory contents at $1014 are correct. The last seven lines of the terminal window are

```
PP  PC     SP      X       Y        D = A:B    CCR = SXHI    NZVC
38 150F    3C00    04D2    1010      00:35            1001    0000
xx: 150F   CE000A          LDX       #$000A

>md 1010

1010   30 30 35 33 - 35 2D 2D 2D - 2D 2D 2D 2D - 2D 2D 2D 2D   00535---------------
>
```

These six lines tell us that

- The index register contains hex value $04D2 (equal to decimal 1234) and is the correct quotient.
- The memory location at $1014 contains hex value $35 and is the ASCII code of 5.
- In the next division, the number to be divided by 10 will be 1234.
- The next instruction to be executed is **LDX #$0A.**

Step 2
Trace the next two instructions. Oops! We find one error. The double accumulator D does not contain the value 1234 (it contains $35 instead) before the IDIV instruction is executed. We forgot to swap the value in index register X with double accumulator D before performing the second division!

```
>t
PP  PC     SP      X       Y        D = A:B    CCR = SXHI    NZVC
38 1512    3C00    000A    1010      00:35            1001    0000
xx: 1512   1810            IDIV

>t
PP  PC     SP      X       Y        D = A:B    CCR = SXHI    NZVC
38 1514    3C00    0005    1010      00:03            1001    0000
xx: 1514   CB30    ADDB    #$30

>
```

Step 3
Fix the error by inserting the **xgdx** instruction before the **ldx #$0a** instruction and rerun the program. After rerunning the program, check the contents at $1000 to $1004 again. The last few lines on the screen of the terminal window would be as follows:

```
>load
*
>g 1500
User Bkpt Encountered
PP  PC     SP      X       Y         D = A:B    CCR = SXHI   NZVC
381537  3C00    0032    1010      00:31            1001    0000
xx: 1537  2D2D            BLT     $1566
>md 1010
1010   31 32 33 34 - 35 2D 2D 2D - 2D 2D 2D 2D - 2D 2D 2D 2D   12345--------------
>
```

The memory locations from $1000 to $1004 contain correct values. So, we have fixed the error.

Most debug sessions are similar to this. Of course, longer programs will take a longer time and require more commands to try different things before one can fix the bugs.

3.7 Other D-Bug12 Modes

The other three D-Bug12 modes are not used as often as the EVB mode.

3.7.1 The Pod Mode

The pod mode is intended to use the demo board as a BDM host to control a target board. The arrangement is shown in Figure 3.20. This mode will be discussed in Section 3.9.5.

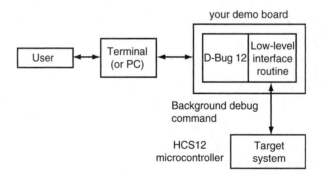

Figure 3.20 ■ D-Bug 12's POD mode conceptual model

3.7.2 The Jump-to-EEPROM Mode

The **Jump-to-EEPROM** mode allows a small program to be executed from the on-chip EEPROM whenever the EVB is powered up or reset. When running on a demo board with an MC9S12Dx256, MC9S12H256, or MC9S12A256 MCU, the bootloader startup code jumps directly to address $0400 without performing any initialization of the CPU registers or peripherals. When running on a demo board with an MC9S12Dx128 or MC9S12A128 MCU, the EEPROM is not visible at its default address of $0000 because it is overlayed by the on-chip SRAM. In this case, the SRAM is relocated to $2000 and the bootloader start code jumps directly to the address $0400 without performing any additional initialization of the CPU registers or peripherals. This mode provides a convenient way to execute a program in a stand-alone manner without having to erase and program the on-chip flash using the bootloader. Program code and data can be programmed into the EEPROM using the D-Bug12's load command.

3.7.3 The Bootloader Mode

The on-chip flash memory includes a boot block area from $F000 to $FFFF containing an S-record bootloader program. The bootloader can be used to erase and reprogram the remainder of on-chip flash memory or erase the on-chip byte-erasable EEPROM. The bootloader utilizes the on-chip SCI for communications and does not require any special programming software on the host computer. The only host software required is a simple terminal program that is capable of communicating at 9600 to 115,200 baud and supports XON/XOFF handshaking.

The bootloader mode is mainly used to program the D-Bug12 monitor into the flash memory or download a user's fully debugged code into the D-Bug12 portion of flash memory. The latter allows the board to be operated in EVB mode or restart user code every time the board is turned on or reset.

When the user code is programmed into the D-Bug12 portion of flash memory, it wipes out the D-Bug12 monitor. One can restore it any time because bootloader is not erased. One cannot erase the bootloader in bootloader mode. The bootloader can only be erased by a BDM kit via the BDM port.

Invoking the bootloader causes the prompt shown in Figure 3.21 to be displayed on the host terminal's screen.

```
HCS912DP256 Bootloader

a) Erase Flash
b) Program Flash
c) Set Baud Rate
d) Erase EEPROM
?
```

Figure 3.21 ■ Serial bootloader prompt

ERASE FLASH COMMAND

This command is selected by typing **b** on the terminal after the **?** character in Figure 3.21. This command causes a bulk erase of all four 64-kB flash memory arrays except for the 4-kB boot block in the upper 64-kB array where the S-record bootloader resides. After the erase operation is completed, a verify operation is performed to ensure that all locations were properly erased. If the erase operation is successful, the bootloader's prompt is redisplayed. If any locations were found to contain a value other than $FF, an error message is displayed on the screen and the bootloader's prompt is redisplayed. If the MC9S12DP256 will not erase after one or two attempts, the device may be damaged.

PROGRAM FLASH COMMAND

The bootloader uses interrupt-driven, buffered serial I/O in conjunction with XON/XOFF software handshaking to control the flow of S-record data from the host computer during the programming process. This allows the bootloader to continue receiving S-record data from the host computer while the data from the previously received S-record is programmed into the flash memory. The terminal program must support XON/XOFF handshaking to properly reprogram the HCS12's flash memory.

Typing a lowercase **b** on the keyboard causes the bootloader to enter the programming mode and wait for S-records to be sent from the host computer. The bootloader will continue to receive and process S-records until it receives an S8 or S9 end-of-file record. If the object file being sent to the bootloader does not contain an S8 or S9 record, the bootloader will not return its prompt and will continue to wait for the end-of-file record. Pressing the reset button on the demo board will cause the bootloader to return to its prompt.

If a flash memory location does not program properly, an error message is displayed on the terminal screen and the bootloader's prompt is redisplayed. If the device does not program correctly after one or two attempts, the device may be damaged or an S-record with a load address outside the range of the available on-chip flash may have been received. The S-record data must have load addresses in the range from $C0000 to $FFFFF. This address represents the upper 256 KB of the 1-MB address space of the MC9S12DP256.

SET BAUD RATE COMMAND

The default baud rate of 9600 is too slow for programming a large S-record file. The user has the option to select a high baud rate in programming the on-chip flash memory. The **Set Baud Rate** command allows the bootloader communication rate to be set to one of four standard baud rates.

Typing a lowercase **c** on the terminal keyboard causes the prompt shown in Figure 3.22 to be displayed on the monitor screen. Entering **1** through **4** on the keyboard will select the associated baud rate and issue a secondary prompt indicating that the terminal baud rate should also be changed. After changing the terminal baud rate, pressing the **Enter** key will return to the main bootloader prompt (in Figure 3.21).

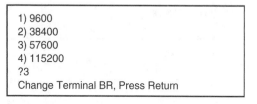

```
1) 9600
2) 38400
3) 57600
4) 115200
?3
Change Terminal BR, Press Return
```

Figure 3.22 ■ Change baud rate (both the demo board and the terminal program)

RELOADING D-BUG12

Motorola adds new features and enhancements to the D-Bug12 monitor frequently. It may be desirable to update to the latest version. A .zip distribution file containing the current version of D-Bug12, including the bootloader and the reference manual, can be downloaded from Motorola's website.

LOADING USER PROGRAMS INTO FLASH MEMORY

A demo board preloaded with the D-Bug12 monitor can also be reprogrammed with the user application to prototype an embedded application. When using the board in this manner, the user-supplied code may occupy all of the on-chip flash memory except the address range from $F000–$FFFF in the fixed flash memory page that begins at $C000. To begin execution of the user's application program, both the PAD0 and PAD1 pins must have jumpers placed in the "0" or "off" position. This will cause the bootloader startup code to jump to the address in the alternate reset vector at $EFFE. When the user code is programmed into flash, an address must be placed in the reset vector position ($EFFE) of the alternate interrupt vector table.

3.8 Tips for Assembly Program Debugging

Assembly program errors can be classified into two categories:

- Syntax errors
- Logical errors

3.8.1 Syntax Errors

Syntax errors are common for beginners. Syntax errors can be divided into following categories:

- Misspelling of instruction mnemonics. This type of error will be highlighted by the assembler and can be fixed easily.
- Starting the assembly program from column one. If a line of the assembly program has a label, then this does not cause any error. However, if a line of the assembly program does not have a label, then the instruction mnemonic is treated as a label by the assembler, whereas the operand (represented in one of the addressing modes) is treated as the instruction mnemonics. This will always cause the undefined mnemonic error!
- Missing operands. Depending on what is missing, the error message will vary. For example, for the instruction:

brclr   PORTA,$04

The resultant error message is "missing target address". For the instruction:

ldaa

The assembler outputs the message "inherent addressing illegal".

As time goes on, one gains experience and also memorizes the instruction mnemonics better; this type of error will reduce and even disappear.

3.8.2 Logical Errors

Beginners make logical errors. The most common ones are as follows:

USING EXTENDED (OR DIRECT) MODE INSTEAD OF IMMEDIATE MODE

This error is very common for beginners. The following assembly program is written to compute the sum of an array of N 8-bit elements:

```
N        equ      20                              ; array count
         org      $1000
array    dc.b     2,4,6,8,10,12,14,16,18,20
         dc.b     22,24,26,28,30,32,34,36,38,40
sum      ds.w     1
         org      $1500
```

```
            ldx      array                  ; place the starting address of array in X
            movw     0,sum                  ; initialize sum to 0
            ldy      N                      ; initialize loop count to N
loop        ldab     1,x+                   ; place one number in B and move array pointer
            sex      B,D                    ; sign-extend the 8-bit number to 16-bit
            addd     sum                    ; add to sum
            std      sum                    ; update the sum
            dbne     y,loop                 ; add all numbers to sum yet?
            swi                             ; return to monitor
            end
```

At first look, this program appears to be fine and should work. After assembling the program, we download it onto the demo board and get the following screen output:

```
>load
....
done
>
```

It is a good idea to make sure that the program has been downloaded onto the demo board by using the **asm** command:

```
>asm 1500
xx:1500     FE1000           LDX      $1000          >
xx:1503     180400001014     MOVW     $0000,$1014    >
xx:1509     DD14             LDY      $0014          >
xx:150B     E630             LDAB     1,X+           >
xx:150D     B714             SEX      B,D            >
xx:150F     F31014           ADDD     $1014          >
xx:1512     7C1014           STD      $1014          >
xx:1515     0436F3           DBNE     Y,$150B        >
xx:1518     3F               SWI                     >
```

Indeed, the program is downloaded correctly. The next thing to check is to make sure that the array data is downloaded correctly by using the **md** command:

```
>md 1000 1010

1000   02 04 06 08 - 0A 0C 0E 10 - 12 14 16 18 - 1A 1C 1E 20   ...............
1010   22 24 26 28 - 00 00 B9 A9 - 2A CA FA DB - AC DA 18 97   "$&(....*.......
>
```

Again, we are convinced that the data array has been downloaded correctly. The array is stored at locations from $1000 to $1013. The variable **sum** occupies the memory locations from $1014 to $1015. Since the program has not been run yet, these two bytes contain zeroes.

The next thing to do is to run the program. This program should run to the SWI instruction and stop. The screen should look as follows:

```
>g 1500
User Bkpt Encountered

PP   PC      SP     X      Y       D = A:B    CCR = SXHI   NZVC
38 1519    3C00   0213   0000       FF:07           1001   1000
xx: 1519   88F4          EORA       #$F4

   >
```

Checking the resultant sum, we can see that it ($FF07) is incorrect. The correct answer should be $1A4 (or 420). The sum can be found by using the **md** command:

```
>md 1010

1010  22 24 26 28 - FF 07 B9 A9 - 2A CA FA DB - AC DA 18 97   "$&(....*.......
>
```

Since this program is short, we can trace through it. To trace this program, one should set the program counter value to the start of the program ($1500) as follows:

```
>pc 1500

PP  PC      SP      X       Y         D = A:B    CCR = SXHI   NZVC
38 1500  3C00     0213    0000        FF:07            1001   1000
xx: 1500  FE1000          LDX       $1000
>
```

Trace one instruction at a time:

```
>t 1

PP  PC      SP      X       Y         D = A:B    CCR = SXHI   NZVC
38 1503  3C00     0204    0000        FF:07            1001   0000
xx: 1503  180400001014   MOVW  $0000,$1014
>
```

The executed instruction is "ldx array". The purpose of this instruction is to place the starting address of the array into X. After the execution of this instruction, the value of X should change to $1000. However, the instruction trace shows otherwise. This is because of the incorrect use of the addressing mode. Change the instruction to "ldx #array", rerun the program, and the sum is still incorrect:

```
>md 1010

1010  22 24 26 28 - FF F0 B9 A9 - 2A CA FA DB - AC DA 18 97   "$&(....*.......
>
```

Trace the program again. This time we trace up to the second instruction and examine the contents of **sum**:

```
>t 2

PP  PC      SP      X       Y         D = A:B    CCR = SXHI   NZVC
38 1503  3C00     1000    0000        FF:F0            1001   0000
xx:1503  180400001014   MOVW  $0000,$1014

PP  PC      SP      X       Y         D = A:B    CCR = SXHI   NZVC
38 1509  3C00     1000    0000        FF:F0            1001   0000
xx:1509  DD14            LDY       $0014
>md 1010

1010  22 24 26 28 - FF 00 B9 A9 - 2A CA FA DB - AC DA 18 97   "$&(....*.......
>
```

At this point, the value of **sum** should be changed to 0 because the second instruction intends to initialize it to 0. Again, the addressing mode is wrong. The correct instruction should be "movw #0,sum". This is not easy to figure out by looking at the program. Rerun the program and display the contents of sum:

```
>load
*
>g 1500
User Bkpt Encountered
```

```
PP  PC     SP     X      Y       D = A:B    CCR = SXHI    NZVC
38 1519   3C00   100F   0000      00:F0          1001     0000
xx: 1519  88F4           EORA    #$F4
>md 1010

1010   22 24 26 28 - 00 F0 B9 A9 - 2A CA FA DB - AC DA 18 97   "$&(....*.......
>
```

The value of sum is still incorrect. Again trace the program up to the third instruction:

```
>pc  1500
PP  PC     SP     X      Y       D = A:B    CCR = SXHI    NZVC
38 1500   3C00   100F   0000      00:F0          1001     0000
xx: 1500  CE1000          LDX    #$1000
>t 3

PP  PC     SP     X      Y       D = A:B    CCR = SXHI    NZVC
38 1503   3C00   1000   0000      00:F0          1001     0000
xx: 1503  180300001014   MOVW   #$0000,$1014

PP  PC     SP     X      Y       D = A:B    CCR = SXHI    NZVC
38 1509   3C00   1000   0000      00:F0          1001     0000
xx: 1509  DD14            LDY    $0014

PP  PC     SP     X      Y       D = A:B    CCR = SXHI    NZVC
38 150B   3C00   1000   000F      00:F0          1001     0000
xx: 150B  E630            LDAB   1,X+
>
```

The third instruction intends to load the array count N into index register Y. Y should receive the value of N, which is 20. The instruction trace shows that Y receives 15 ($F) instead of 20. Again, this is due to the incorrect addressing mode. Change the instruction to "ldy #N", assemble, download, rerun the program, and display the contents of sum:

```
>g 1500
User Bkpt Encountered
PP  PC     SP     X      Y       D = A:B    CCR = SXHI    NZVC
38 151A   3C00   1014   0000      01:A4          1001     0000
xx: 151A  F421BD          ANDB   $21BD
>md 1010

1010   22 24 26 28 - 01 A4 B9 A9 - 2A CA FA DB - AC DA 18 97   "$&(....*.......
>
```

This time the value of sum is $1A4 (420) and is correct.

MISMATCH OF OPERAND SIZE

This is another common mistake made by beginners. For example, some people might write the previous program as follows:

```
N        equ     20                                    ; array count
         org     $1000
array    dc.b    2,4,6,8,10,12,14,16,18,20
         dc.b    22,24,26,28,30,32,34,36,38,40
sum      ds.w    1
         org     $1500
         ldx     #array                                ; place the starting address of array in X
```

```
              movw    #0,sum          ; initialize sum to 0
              ldy     #N              ; initialize loop count to N
      loop    ldd     1,x+            ; place one number in D and move array pointer
              addd    sum             ; add to sum
              std     sum             ; update the sum
              dbne    y,loop          ; add all numbers to sum yet?
              swi                     ; return to monitor
              end
```

At the first run of this program, the sum ($A61F) is also incorrect:

>md 1010

1010 22 24 26 28 - **A6 1F** B9 A9 - 2A CA FA DB - AC DA 18 97   "$&(....*.......
>

Again, the error can be found by tracing. This time the error will not be found until the instruction "loop ldd 1,x+" is traced:

>pc 1500

PP PC	SP	X	Y	D = A:B	CCR = SXHI	NZVC
381500	3C00	1014	0000	A6:1F	1001	1000
xx: 1500	CE1000		LDX	#$1000		

>t

PP PC	SP	X	Y	D = A:B	CCR = SXHI	NZVC
381503	3C00	1000	0000	A6:1F	1001	0000
xx: 1503	180300001014		MOVW	#$0000,$1014		

>t

PP PC	SP	X	Y	D = A:B	CCR = SXHI	NZVC
381509	3C00	1000	0000	A6:1F	1001	0000
xx: 1509	CD0014		LDY	#$0014		

>t

PP PC	SP	X	Y	D = A:B	CCR = SXHI	NZVC
38150C	3C00	1000	0014	A6:1F	1001	0000
xx: 150C	EC30		LDD	1,X+		

>t

PP PC	SP	X	Y	D = A:B	CCR = SXHI	NZVC
38150E	3C00	1001	0014	**02:04**	1001	0000
xx: 150E	F31014		ADDD	$1014		

>

We expect this instruction to place the first array element ($02) in D. Instead, it places $0204 in D. This is obviously wrong! The error is due to the fact that the **ldd** instruction loads a word (2 bytes) instead of 1 byte into D. At this point, we should be able to figure out the right instruction to use and fix the problem.

INAPPROPRIATE USE OF INDEX ADDRESSING MODE

When the index addressing mode is used to step through an array, one needs to increment or decrement the index register to reach the next element. Depending on the size of the element, one needs to increment/decrement the index register by 1, 2, or other values. When such errors occur, the program execution result will be incorrect. This type of error can be discovered by checking the computation result of the first two elements.

STACK FRAME ERRORS
This type of error will be discussed in Chapter 4.

INCORRECT ALGORITHM
An incorrect algorithm can never result in a correct program. This type of problem cannot be fixed by tracing the program. After program tracing fails to fix the problem, one should reexamine the algorithm to see if it is incorrect.

3.9 BDM-Based Source-Level Debugger

In order to carry out debug activities, a source-level debugger must have the capability to access (read and write) the contents of CPU registers and memory locations when the program is running. The Background Debug Mode (BDM) provided by the HCS12 allows the development hardware to access the contents of CPU registers and memory locations via the BKGD pin serially. In addition, the HCS12 also provides the hardware breakpoint and instruction tagging capability to facilitate the setting of breakpoints during the process of software debugging.

3.9.1 Overview of BDM

The block diagram of the BDM module is shown in Figure 3.23. The development host sends commands to the BDM module to perform program debugging through the BKGD pin. Some of the commands are hardware commands, which do not require the CPU to get involved because the BDM module is not part of the CPU. Other commands are firmware commands, which would require the CPU to execute a small program (written in HCS12's instruction). Each firmware command is implemented in a small program. Programs that implement the firmware commands are stored in the firmware ROM and the commands can be utilized to download programs into memory, set breakpoints, execute the user program up to the breakpoint, trace the program execution, display values of the CPU and I/O registers, and display the contents of memory locations. Tool developers can take advantage of these commands to implement low-cost source-level debuggers.

Figure 3.23 ■ BDM block diagram

3.9.2 BDM Commands

A BDM command is either an 8-bit opcode or an 8-bit opcode followed by a 16-bit address or data. BDM commands can be classified into hardware commands and firmware commands. Hardware commands are used to read and write target system memory locations and enter active background debug mode. Target (hardware to be debugged) system memory includes all memory accessible to the CPU, including the on-chip RAM, EEPROM, flash memory, I/O control registers, and all external memory.

Hardware commands are executed with minimal or no CPU intervention and do not require the system to be in active BDM for execution. When executing a hardware command, the BDM subblock waits for a free CPU bus cycle so that the background access does not disturb the running application program. If a free cycle is not found within 128 clock cycles, the CPU is momentarily frozen so that the BDM can steal a cycle. When the BDM finds a free cycle, the operation does not intrude on normal CPU operation, provided that it can be completed in a single cycle. However, if an operation requires multiple cycles, the CPU is frozen until the operation is complete, even though the BDM found a free cycle.

The BDM hardware commands are listed in Table 3.5. The READ_BD and WRITE_BD commands allow access to the BDM register locations. These locations are not normally in the

Command	Opcode	Data	Description
BACKGROUND	$90	None	Enter background mode if firmware is enabled. If enabled, an ACK will be issued when the part enters active background mode.
ACK_ENABLE	$D5	None	Enable handshake. Issues an ACK pulse after the command is executed.
ACK_DISABLE	$D6	None	Disable handshake. This command does not issue an ACK pulse.
READ_BD_BYTE	$E4	16-bit address 16-bit data out	Read from memory with standard BDM firmware lookup table in map. Odd address data on low byte; even address data on high byte.
READ_BD_WORD	$EC	16-bit address 16-bit data out	Read from memory with standard BDM firmware lookup table in map. Must be aligned access.
READ_BYTE	$E0	16-bit address 16-bit data out	Read from memory with standard BDM firmware lookup table out of map. Odd address data on low byte; even address data on high byte.
READ_WORD	$E8	16-bit address 16-bit data out	Read from memory with standard BDM firmware lookup table out of map. Must be aligned access.
WRITE_BD_BYTE	$C4	16-bit address 16-bit data in	Write to memory with standard BDM firmware lookup table in map. Odd address data on low byte; even address data on high byte.
WRITE_BD_WORD	$CC	16-bit address 16-bit data in	Write to memory with standard BDM firmware lookup table in map. Must be aligned access.
WRITE_BYTE	$C0	16-bit address 16-bit data in	Write to memory with standard BDM firmware lookup table out of map. Odd address data on low byte; even address data on high byte.
WRITE_WORD	$C8	16-bit address 16-bit data in	Write to memory with standard BDM firmware lookup table out of map. Must be aligned access.

Note: If enabled, ACK will occur when data is ready for transmission for all BDM READ commands and will occur after the write is complete for all BDM WRITE commands.

Table 3.5 ■ Hardware BDM commands

system memory map, but share addresses with the application in memory. To distinguish between physical memory locations that share the same address space, BDM memory resources are enabled just for the READ_BD and WRITE_BD access cycles. This allows the BDM to access BDM locations unobtrusively, even if the addresses conflict with the application memory map.

Firmware commands are used to access and manipulate CPU registers. The system must be in active BDM mode to execute standard BDM firmware commands. Normal instruction execution is suspended while the CPU executes the firmware located in the standard BDM firmware lookup table. The hardware command BACKGROUND is the usual way to activate BDM.

As the system enters active BDM mode, the standard BDM firmware lookup table and BDM registers become visible in the on-chip memory map at $FF00–$FFFF, and the CPU begins executing the standard BDM firmware. The standard BDM firmware watches for serial commands and executes them as they are received. The firmware commands are shown in Table 3.6.

Command1	Opcode	Data	Description
READ_NEXT	$62	16-bit data out	X = X + 2; then read the word pointed to by X.
READ_PC	$63	16-bit data out	Read program counter.
READ_D	$64	16-bit data out	Read accumulator D.
READ_X	$65	16-bit data out	Read index register X.
READ_Y	$66	16-bit data out	Read index register Y.
READ_SP	$67	16-bit data out	Read stack pointer.
WRITE_NEXT	$42	16-bit data in	X = X + 2; then write word to location pointed to by X.
WRITE_PC	$43	16-bit data in	Write program counter.
WRITE_D	$44	16-bit data in	Write to accumulator D.
WRITE_X	$45	16-bit data in	Write into index register X.
WRITE_Y	$46	16-bit data in	Write into index register Y.
WRITE_SP	$47	16-bit data in	Write to stack pointer.
GO	$08	None	Go to user program. If enabled, ACK will occur when leaving active background mode.
GO_UNTIL[2]	$0C	None	Go to user program. If enabled, ACK will occur upon returning to active background mode.
TRACE1	$10	None	Execute one user instruction then return to BDM. If enabled, ACK will occur upon returning to active background mode.
TAGGO	$18	None	Enable tagging and go to user program. There is no ACK pulse related to this command.

Notes:
1. If enabled, ACK will occur when data is ready for transmission for all BDM READ commands and will occur after the write is complete for all BDM WRITE commands.
2. Both WAIT (with clocks to the S12 CPU core disabled) and STOP disable the ACK function. The GO_UNTIL command will not get an acknowledge if one of these two CPU instructions occurs before the "UNTIL" instruction. This can be a problem for any instruction that uses ACK, but GO_UNTIL is a lot more difficult for the development tool to time out.

Table 3.6 ■ Firmware BDM commands

3.9.3 BDM Serial Interface

The BDM module communicates with external devices via the BKGD pin. During reset, this pin is a mode select input, which selects between normal and special modes of operation. After reset, this pin becomes the dedicated serial interface pin for the BDM.

The timing of the BDM serial interface is controlled by the target clock. The target clock is programmable and can be either the bus clock or the oscillator clock. The BDM serial interface uses a clocking scheme in which the external host generates a falling edge on the BKGD pin to indicate the start of each bit time. This falling edge is sent for every bit whether data is transmitted or received. Data is transferred most significant bit first at 16 target clock cycles per bit. The interface times out if 512 clock cycles occur between falling edges from the development host.

For hardware read commands, the development host must wait for 150 clock cycles after sending the address before attempting to obtain the read data. For hardware write commands, the host must wait for 150 clock cycles after sending data to be written before attempting to send a new command. For firmware read commands, the external host should wait for 44 clock cycles after sending the command opcode and before attempting to obtain the read data. For firmware write commands, the external host must wait for 32 clock cycles after sending data to be written before attempting to send a new command. The external host should wait for 64 clock cycles after a TRACE1 or GO command before starting any new serial command.

The detailed timing of the BDM serial interface is beyond the scope of this text. Interested readers should refer to the *Background Debug Module Guide* published by Freescale.

3.9.4 The BDM-Based Source-Level Debugger

The block diagram of a BDM-based development system is shown in Figure 3.24. The BDM pod is connected to the target board via a BDM cable. To promote the portability of tools that use BDM mode, Freescale defines a 6-pin connector to be installed on the target board, shown in Figure 3.25.

The BDM pod can be connected to the PC or workstation via the serial, parallel, or USB port. There must be a software program running on the PC or workstation to send commands to the POD to perform the appropriate debug operations. This PC-based software should have a good user interface and allow the user to perform at least the following operations:

- Perform basic configuration
- Download program to the target board for execution
- Set breakpoints
- Display and change the values of the CPU registers, I/O registers, and memory locations
- Display the source code
- Set up a watch list to display the values of selected program variables
- Trace program execution to the breakpoint

The PC sends the intended operation of the user to the POD and the POD sends the corresponding BDM commands to the target board to perform the desired operations. A source-level debugger using this approach need not be expensive. A BDM-based source-level debugger can be integrated with an IDE, which allows the user to perform program editing, assembling or compiling, and source-level debugging in the same environment.

Figure 3.24 ■ BDM development system

Figure 3.25 ■ BDM tool connector

3.9.5 The DragonBDM from Wytec

Some designers cannot afford a BDM-based source-level debugger but need to design an embedded product using an HCS12 microcontroller. Starting from scratch, the designer needs at least a programmer that can program the application onto the target chip. The least expensive BDM kit without a source-level debugger is designed with a HCS12DP256, which is programmed with the D-Bug12 monitor and configured in the POD mode. The DragonBDM from Wytec is such a product. It can be used as a programmer to program the application into the HCS12 microcontroller on the target hardware. It also allows the user to use all of the D-Bug12 commands to debug the application program. The user needs only a terminal program (available in MiniIDE, AsmIDE, and EGNU) to communicate with the DragonBDM pod. Combining the MiniIDE (or AsmIDE) with the DragonBDM creates an inexpensive integrated assembly language development system and programmer. Similarly, combining the EGNU and the DragonBDM creates the least expensive integrated C language development environment with the debugging capability available in the D-Bug12 monitor. A photograph of the DragonBDM is shown in Figure 3.26.

3.9.6 Using the Wytec DragonBDM Kit

The DragonBDM kit is connected to the PC through a RS232 cable and connected to the target board via a 6-pin BDM flat cable. The DragonBDM kit receives its power from the target board.

Figure 3.26 ■ Photograph of the DragonBDM (Courtesy of Wytec)

The procedure for starting up the debug session is as follows:

Step 1
Connect the BDM kit to the PC and the target board.

Step 2
Start the terminal program on the PC. One can use an IDE (e.g., MiniIDE or EGNU) to communicate with the DragonBDM kit. Using an IDE allows the user to edit, compile/assemble the source program, and download the program for execution on the target board without quitting any program.

Step 3
Power up the target board (one can use the Dragon12, SSE256, or any HCS12 demo board). After powering up the target board, the following message (if the target board is a Dragon12 or CML12S from Axiom, both demo boards have 4-MHz crystals) will appear:

D-Bug12 4.0.0b24
Copyright 1996–2002 Motorola Semiconductor
For Commands type "Help"

Can't Communicate With Target CPU

1.) Set Target Speed (4000 KHz)
2.) Reset Target
3.) Reattempt Communication
4.) Erase & Unsecure
5.) Enter BDM debugger
?

One must first set the target speed with choice 1. After entering the first choice, one will be prompted to enter the target speed, which is the crystal frequency on the target board, not the bus speed that is boosted up by the on-chip PLL. After a reset, before the PLL is enabled, the target MCU is running from the crystal frequency, not the PLL frequency. The bus speed is one-half of the crystal frequency by default. The PLL function must be initialized by the reset startup routine for a higher bus speed. For instance, if the target is a DRAGON12 or an Axiom CML12S, enter 4000 for the 4-MHz target crystal speed. After the correct speed is entered, the DragonBDM kit will try to communicate with the target board. If it is not successful, enter choice 2 to reset the target board:

Enter Target Crystal Frequency (kHz): 4000

Can't Communicate With Target CPU

1.) Set Target Speed (4000 kHz)
2.) Reset Target
3.) Reattempt Communication
4.) Erase & Unsecure
5.) Enter BDM debugger
? 2
S>

Step 4
Enter D-Bug12 monitor commands to download the program and use all appropriate D-Bug12 commands as described in this chapter to debug the program. For example, the following program increments a counter every 0.2 second and outputs the count value to Port P pins to drive eight LEDs:

```
#include "c:\miniIDE\hcs12.inc"
            org      $1000
cnt         ds.b     1
            org      $1500
            lds      #$1500          ; set up stack pointer
            clr      cnt
            movb     #$FF,DDRP       ; configure Port P for output
forever     ldy      #2
            jsr      DlyBy100ms      ; wait for 0.2 seconds
            inc      cnt
            ldab     cnt
            movb     cnt,PTP
            bra      forever
; *******************************************************************************
; The following subroutine creates a delay that is equal to the value in Y
; multiplied by 100 ms. E clock is 8 MHz.
; *******************************************************************************
DlyBy100ms
            bset     TSCR1,TFFCA     ; enable fast timer flag clear
            movb     #$07,MCCTL      ; enable modulus counter with 1:16 as prescaler
            movw     #50000,MCCNT    ; load the value to be down counted
            brclr    MCFLG,MCZF,*    ; wait until MCCNT down counted to 0
            bclr     MCCTL,MCEN      ; disable modulus down counter
            dbne     y,DlyBy100ms    ; is the desired delay reached?
            rts
```

After assembling the program, enter the load command to download the program to the target board SRAM:

S>load

*

S>

After this, enter g 1500 to run the program and the screen will change to

S>g 1500

R>

The prompt **R>** indicates that the target board is running a program. When the previous program is running, one can see that the LEDs are incrementing from 0 to 255 and repeat. When the target board is running, the BDM kit cannot read the CPU registers for display. If one wants to read the values of target CPU registers, one needs to stop (use the STOP command) or reset (use the reset command) the target board. After entering the STOP command (or the RESET command), the S> prompt reappears.

Either the SSE256 or the Dragon12 demo board configured in POD mode can serve the role of a BDM pod. In the POD mode, these two demo boards have the same functionality as does the DragonBDM kit. Both demo boards have a 6-pin BDM OUT connector. The pin assignment of this connector is shown in Figure 3.27. The PT7 and PT6 pins are used to drive the BKGD and RESET pins of the target board.

Figure 3.27 ■ BDM OUT connector

3.10 Summary

The HCS12 microcontroller was developed as an upgrade for the 8-bit 68HC11 microcontroller family. All members implement the same architecture, i.e., they have the same instruction set and addressing modes, but differ in the amount of peripheral functions that they implement.

All HCS12 members use the same design in each peripheral function, which simplifies the migration from one device to another. All of the HCS12 members implement a 10-bit A/D converter. The HCS12 microcontroller implements a complicated timer system:

- The *input-capture* (IC) function latches the arrival time of a signal edge. This capability allows one to measure the frequency, the period, and the duty cycle of an unknown signal.
- The *output compare* (OC) function allows one to make a copy of the main timer, add a delay to this copy, and store the sum to an output compare register. This capability can be used to create a delay, generate a digital waveform, trigger an action on a signal pin, and so on.

- The *pulse accumulation* function allows one to count the events that arrived within an interval.
- The *pulse-width modulation* module (PWM) is used to generate a digital waveform of a certain frequency with a duty cycle ranging from 0 to 100 percent. This capability is useful in DC motor control.

The HCS12 microcontroller also provides a wide variety of serial interface functions to appeal to different applications:

- The serial communication interface (SCI) function supports the EIA232 standard. A microcontroller demo board would use this interface to communicate with a PC.
- The synchronous peripheral interface (SPI) function is a synchronous interface that requires a clock signal to synchronize the data transfer between two devices. This interface is mainly used to interface with peripheral chips such as shift registers, seven-segment displays and LCD drivers, A/D and D/A converters, SRAM and EEPROM, phase-lock loop chips, and so on. All of these devices must have the SPI interface.
- The interintegrated circuit (I²C) is a serial interface standard proposed by Phillips. This interface standard allows microcontrollers and peripheral devices to exchange data.
- The byte data link communication (BDLC) module was proposed for low-speed data communication in automotive applications. It provides access to an external serial communication multiplexed bus that operates according to the SAE J1850 protocol.
- The controller area network (CAN) was proposed to be used mainly as a vehicle serial data bus to provide reliable operation in the EMI environment and achieve the high bandwidth required in that environment.

Two MC9S12DP256-based demo boards (SSE256 and Dragon12) from Shuan-Shizu and Wytec are recommended for learning the HCS12 microcontrollers.

Software development tools include the text editor, the terminal program, the cross assembler, the cross compiler, the simulator, the source-level debugger, and the integrated development environment (IDE). A sophisticated IDE should contain a text editor, a terminal program, a cross compiler, a cross assembler, and a source-level debugger. It allows the user to perform all the development work without leaving any program.

In this text, we use the freeware AsmIDE by Eric Engler and MiniIDE from Mgtek to enter and test all the assembly programs. A tutorial on how to use AsmIDE and MiniIDE are provided. The debugging activities on a demo board are facilitated by the command set provided by the D-Bug12 monitor.

The D-Bug12 monitor has four modes:

- EVB mode. Most users use this mode to perform development and evaluation work.
- POD mode. This mode allows one to use the demo board to debug another HCS12 target board.
- Jump-to-EEPROM mode. This mode allows the user to run the application programmed into the on-chip EEPROM after power up or reset.
- Bootloader mode. This mode allows the user to update the D-Bug12 monitor or program the application code into the flash memory out of reset or after being powered on.

The only freeware source-level debugger is the limited version of Code Warrior from Metrowerks. The Code Warrior requires a demo board to be programmed with the 2-KB serial monitor from Freescale. Code Warrior is a complicated system with a lot of bells and whistles. Tutorials on programming the Serial Monitor into the Dragon12 demo board can be found on the website at http://www.mecheng.adelaide.edu.au/robotics/WWW_Devs/Dragon12/ Dragon12.htm.

A tutorial on using the Code Warrior will be posted on the website at http://www. microhuang.com in the near future.

3.11 Lab Exercises and Assignment

L3.1 Turn on the PC and start the AsmIDE (or MiniIDE) program to connect to a demo board with the D-Bug12 monitor. Perform the following operations:

    a. Enter a command to set the contents of the memory locations from $1000 to $10FF to 0.

    b. Display the contents of the memory locations from $1000 to $10FF.

    c. Set the contents of the memory locations $1000 to 1, 2, 3, and 4, respectively.

    d. Verify that the contents of memory locations $1003 have been set correctly.

L3.2 Enter monitor commands to display the breakpoints; set new breakpoints at $1520, $1550, and $1500. Delete breakpoints at $1550, and redisplay the breakpoints.

L3.3 Enter commands to place $10 and $0 in accumulators A and B.

L3.4 Use appropriate D-Bug12 commands to perform the following operations:

    a. Set the contents of the memory location at $1000 to 3.

    b. Set the contents of the memory location at $1001 to 4, and redisplay the location $1000.

    c. Set the contents of the memory location at $1002 to 5, and redisplay the same location.

    d. Set the contents of the memory location at $1003 to 6, and return to the D-Bug12 command prompt.

L3.5 Invoke the one-line assembler to enter the following instructions to the demo board, starting from address $1500, trace through the program, and examine the contents of the memory locations at $1000 and $1001:

```
ldd     #$0000
std     $1000
ldab    #$00
incb
ldx     $1000
abx
stx     $1000
cmpb    #$14
bne     $1508
swi
```

L3.6 Use the text editor of AsmIDE (or MiniIDE) to enter the following assembly program as a file with the filename *lab03_6.asm*:

```
        org     $1000
sum     rmb     1
arcnt   rmb     1
        org     $1500
        ldaa    #20
        staa    arcnt
        ldx     #array
        clr     sum
again   ldaa    0,X
        lsra
        bcs     next
        ldaa    sum
        adda    0,X
        staa    sum
next    inx
        dec     arcnt
        bne     again
        swi
        org     $2000
array   fcb     1,3,5,7,2,4,6,8,9,11,13,10,12,14,15,17,19,16,18,20
        end
```

After entering the program, perform the following operations:

    a. Assemble the program.

    b. Download the S-record file (filename **lab03_6.s19**) to the demo board.

    c. Display the contents of memory locations from $1000 to $100F and $2000 to $202F.

    d. Execute the program.

    e. Display the contents of the memory location at $1000.

This program adds all even numbers in the given array and stores the sum at $1000.

L3.7 Write a program to count the number of elements in an array that are divisible by 8. The array has thirty 8-bit elements and is stored immediately after your program. Use the **repeat S until C** looping construct. Leave the result at $1000.

L3.8 Write a program to swap the last element of an array with the first element, second element with the second-to-last element, and so on. The array has thirty 8-bit elements.

L3.9 Write a program to find the greatest common divisor (gcd) of two 16-bit numbers stored at $1000~$1001 and $1002~$1003, and store the result at $1010~$1011. When testing the program, one should enter two 16-bit numbers manually before running the program. The most efficient method for finding the greatest common divisor is the Euclidean method. This algorithm can be found at the website at http://www.fact-index.com/e/eu/euclidean_algorithm.html.

L3.10 (optional) Use the DragonBDM pod to connect to the PC and the demo board to walk through the assignment in L3.6.

4

Advanced Assembly Programming

4.1 Objectives

After completing this chapter, you should be able to

- Access and manipulate arrays, vectors, and strings

- Access parameters stored in the stack and manipulate stack data structure

- Perform binary to ASCII string and ASCII string to binary conversion

- Write subroutines to perform certain functions

- Make subroutine calls

- Perform terminal I/O by calling D-Bug12 library functions

4.2 Introduction

A program consists of two parts: *data structure* and *algorithm*. Data structures deal with how information is organized so that it can be processed efficiently. Algorithms deal with systematic methods for information processing. Common operations applied to data structures include adding and deleting elements, traversing and searching a data structure, balancing and sorting a data structure, and so on. There are many data structures in use today. However, we will deal with only the following data structures:

- *Stacks.* A stack is a data structure with a top and a bottom element. Elements can only be added or deleted from the top of the stack. From this standpoint, a stack is a last-in-first-out (LIFO) data structure.

- *Arrays.* An array is an ordered set of elements of the same type. The elements of the array are arranged so that there is a zeroth, first, second, . . . , and nth element. An array may be one-, two-, or multidimensional. A vector is a one-dimensional array. A matrix is a two-dimensional array.

- *Strings.* A string is a sequence of characters terminated by a special character such as a NULL (ASCII code 0) or EOT (ASCII code 4). A computer system often needs to output a string to inform users of something or obtain some input from the user.

Algorithms are often represented in procedural steps or flowcharts. We will use both approaches in this book.

4.3 Stack

Conceptually, a stack is a list of data items whose elements can be accessed from only one end. A stack element has a top and a bottom. The operation that adds a new item to the top is called *push*. The top element can be removed by performing an operation called *pop* or *pull*. Physically, a stack is a reserved RAM area in main memory where programs agree to perform only push and pull operations. The structure of a stack is shown in Figure 4.1. Depending on the microcontroller, a stack has a pointer that points to the top element or to the memory location above the top element.

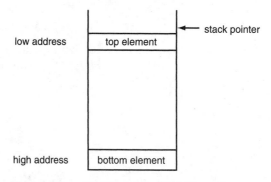

Figure 4.1 ■ Diagram of the HCS12 stack

The HCS12 has a 16-bit stack pointer (SP) to facilitate the implementation of the stack data structure. The stack pointer points to the top byte of the stack. Many companies follow the convention that the stack grows from high addresses toward lower addresses. An area in main memory (must be in RAM) is allocated for use as the stack area. If the stack grows into memory locations with addresses lower than the stack buffer, a *stack overflow* error occurs; if the stack is pulled too many times, a *stack underflow* occurs. We must check the stack overflow and underflow in order to make sure that the program won't crash. Of course, this checking adds overhead to the stack access.

The HCS12 provides instructions for pushing and pulling all CPU registers except the stack pointer, SP. A push instruction writes data from the source to the stack after decrementing the stack pointer. There are six push instructions: *psha, pshb, pshc, pshd, pshx,* and *pshy.* A pull instruction loads data from the top of the stack to a register and then increments the stack pointer. There are also six pull instructions: *pula, pulb, pulc, puld, pulx,* and *puly.* These instructions have equivalent store and load instructions combined with predecrement and postincrement index addressing modes, as shown in Table 4.1.

Mnemonic	Function	Equivalent instruction
psha	push A into the stack	staa 1, –SP
pshb	push B into the stack	stab 1, –SP
pshc	push CCR into the stack	none
pshd	push D into stack	std 2, –SP
pshx	push X into the stack	stx 2, –SP
pshy	push Y into the stack	sty 2, –SP
pula	pull A from the stack	ldaa 1, SP+
pulb	pull B from the stack	ldab 1, SP+
pulc	pull CCR from the stack	none
puld	pull D from the stack	ldd 2, SP+
pulx	pull X from the stack	ldx 2, SP+
puly	pull Y from the stack	ldy 2, SP+

Table 4.1 ■ HCS12 push and pull instructions and their equivalent load and store instructions

Example 4.1

▼

Assuming we have the following instruction sequence to be executed by the HCS12, what would be the contents of the stack after the execution of these instructions?

```
lds    #$1500
ldaa   #$20
staa   1, -SP
ldab   #40
staa   1, -SP
ldx    #0
stx    2, -SP
```

Solution: The first instruction initializes the stack pointer to $1500. The second and the third instructions together push the 8-bit value $20 into the stack. The fourth and fifth instructions push the 8-bit value 40 (hex $28) into the stack. The sixth and seventh instructions push the value 0 to the top two bytes of the stack. The contents of the HCS12 stack are shown in Figure 4.2.

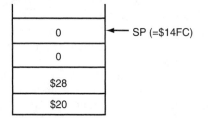

Figure 4.2 ■ The contents of the HCS12 stack

One of the main uses of the stack is saving the return address for a subroutine call. Before the stack can be used, we need to set up the stack pointer. Since the stack can only be implemented in the RAM, one must make sure not to violate this requirement. Because the stack is a last-in-first-out data structure, it can be used to reverse a data structure.

4.4 Indexable Data Structures

Vectors and matrices are indexable data structures. A vector is a sequence of elements in which each element is associated with an index i that can be used to access it. Conceptually, a vector is a one-dimensional data structure. To make address calculation easy, the first element is associated with the index 0 and each successive element with the next integer. However, one can change the index origin of the array to 1 or some other values if one is willing to modify the routines. The elements of a vector or array have the same precision.

The assembler directives *db*, *dc.b*, and *fcb* can be used to define arrays with 8-bit elements. Directives *dw*, *dc.w*, and *fdb* can be used to define arrays with 16-bit elements. Suppose we have an array **vec_x** with six elements: 11, 12, 13, 14, 15, and 16. It can be defined as follows:

```
vec_x    db     11,12,13,14,15,16
```

The assembler directives *ds*, *rmb*, and *ds.b* can be used to reserve memory locations for arrays with 8-bit elements. Directives *ds.w* and *rmw* can be used to reserve memory space for arrays with 16-bit elements.

Searching an array is a common operation for microcontroller applications. If the array is not sorted, the only way to find out whether a value is in the array is by performing the *sequential search*. The sequential search algorithm compares the value to be searched with every element in the array.

Example 4.2
▼

Write a program to find out if the array **vec_x** contains a value **key**. The array has **n** 16-bit elements and is not sorted. Store the address of the element, which is equal to the key. Otherwise, store a −1 in it.

Solution: The algorithm for sequential search is shown in Figure 4.3.

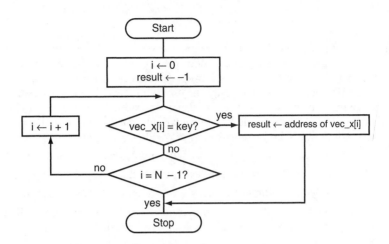

Figure 4.3 ■ Flowchart for sequential search

To implement the sequential search algorithm, we will:
- Use double accumulator D to hold the search key
- Use index register X as a pointer to the array
- Use index register Y to hold the loop count

The program is as follows:

```
N          equ      30              ; array count
notfound   equ      -1
key        equ      190             ; define the searching key
           org      $1000
result     rmw      1               ; reserve a word for result
           org      $1500           ; starting address of the program
           ldy      #N              ; set up loop count
           ldd      #notfound
           std      result          ; initialize the search result
           ldd      #key
           ldx      #vec_x          ; place the starting address of vec_x in X
loop       cpd      2,X+            ; compare the key with array element and
                                    ; increment the array pointer after comparison
           beq      found
           dbne     Y,loop          ; have we gone through the whole array?
           bra      done
found      dex                      ; need to restore the value of X to point to the
           dex                      ; matched element
           stx      result
done       swi
```

```
vec_x      dw      13,15,320,980,42,86,130,319,430,4
           dw      90,20,18,55,30,51,37,89,550,39
           dw      78,1023,897,930,880,810,650,710,300,190
           end
```

▲

If an array (for example, a lookup table) needs to be searched frequently, then sequential search is very inefficient. A better approach is to sort the array and use the *binary search* algorithm. Suppose the sorted array (in *ascending* order) has **n** elements and is stored at memory locations starting at the label **arr**. Let *max* and *min* represent the highest and lowest range of array indices to be searched and the variable *mean* represent the average of *max* and *min*. The idea of a binary search algorithm is to divide the sorted array into three parts:

- The portion of the array with indices ranging from *mean + 1* to *max*
- The element with index equal to *mean* (middle element)
- The portion of the array with indices ranging from *min* to *mean − 1*

The binary search algorithm compares the key with the middle element and takes one of the following actions based on the comparison result:

- If the key equals the middle element, then stop.
- If the key is larger than the middle element, then the key can be found only in the portion of the array with larger indices. The search will be continued in the upper half.
- If the key is smaller than the middle element, then the key can be found only in the portion of the array with smaller indices. The search will be continued in the lower half.

The binary search algorithm can be formulated as follows:

Step 1
Initialize variables *max* and *min* to **n − 1** and **0**, respectively.

Step 2
If *max* < *min*, then stop. No element matches the key.

Step 3
Let *mean* = (*max* + *min*)/2.

Step 4
If *key* equals **arr**[mean], then key is found in the array, exit.

Step 5
If *key* < arr[mean], then set *max* to *mean − 1* and go to step 2.

Step 6
If *key* > arr[mean], then set *min* to *mean + 1* and go to step 2.

The algorithm works for arrays sorted in ascending order but can be modified to work for arrays sorted in descending order.

Example 4.3

Write a program to implement the binary search algorithm and a sequence of instructions to test it. Use an array of n 8-bit elements for implementation.

Solution: We will store a 1 as the result if the key is found in the array. Otherwise, a 0 is stored as the search result. The program is as follows:

```
n           equ    30              ; array count
key         equ    69              ; key to be searched
            org    $1000
max         rmb    1               ; maximum index value for comparison
min         rmb    1               ; minimum index value for comparison
mean        rmb    1               ; the average of max and min
result      rmb    1               ; search result
            org    $1500
            clra
            staa   min             ; initialize min to 0
            staa   result          ; initialize result to 0
            ldaa   #n − 1
            staa   max             ; initialize max to n − 1
            ldx    #arr            ; use X as the pointer to the array
loop        ldab   min
            cmpb   max
            lbhi   notfound        ; if min > max, then not found (unsigned comparison)
            addb   max             ; compute mean
            lsrb                   ;
            stab   mean            ; save mean
            ldaa   b,x             ; get a copy of the element arr[mean]
            cmpa   #key
            beq    found
            bhi    search_lo
            ldaa   mean
            inca
            staa   min             ; place mean+1 in min to continue
            bra    loop
search_lo   ldaa   mean
            deca
            staa   max
            bra    loop
found       ldaa   #1
            staa   result
notfound    swi
arr         db     1,3,6,9,11,20,30,45,48,60
            db     61,63,64,65,67,69,72,74,76,79
            db     80,83,85,88,90,110,113,114,120,123
            end
```

4.5 Strings

A string is a sequence of characters terminated by a NULL (ASCII code 0) or other special character such as EOT (ASCII code 4). In this book, we will use the NULL character to terminate a string. Common operations applied to strings include string concatenation, character and word counting, string matching, substring insertion and deletion, and so on.

Strings are also heavily used in input and output operations. Before a number (in binary format in memory) is output, it must be converted to ASCII code because most output devices only accept ASCII code. The number can be output in BCD or hex format. We have one example in Chapter 2 that converts a 16-bit binary number into BCD format by performing repeated divide-by-10 operations and adding $30 to each remainder (between 0 and 9).

4.5.1 Data Conversion

In this section, we will deal with how data is converted from binary to ASCII and from ASCII to binary.

Example 4.4

Write a program to convert the unsigned 8-bit binary number in accumulator B into BCD digits terminated by a NULL character. Each digit is represented in ASCII code. The resulting string is stored in memory.

Solution: Since an 8-bit number can be from 0 to 255, 4 bytes are needed to hold the result. By using the repeated divide-by-10 method, the conversion can be performed. The first divide-by-10 operation will obtain the least significant digit, whereas the last divide-by-10 operation will extract the most significant digit. If one stores the remainder digit in the buffer and advances the pointer to the buffer for each divide-by-10 operation, then the resultant decimal string is in the reverse order. A simple way to fix the problem is to push the remainder digits into the stack and then pop them out later. The following program implements this idea:

```
test_dat   equ    34
           org    $1000
buf        ds.b   4                 ; to hold the decimal string
temp       ds.b   2                 ;      "
           org    $1500
           ldab   #test_dat
           lds    #$1500            ; initialize stack pointer
           ldy    #buf
           tstb
           bne    normal
           movb   #$30,buf
           clr    buf+1             ; terminate the string
           bra    done
normal     movb   #0,1,-sp          ; push a NULL into the stack
           clra
loop       ldx    #10
           idiv
           addb   #$30              ; convert to ASCII code
           pshb                     ; push into stack
           cpx    #0
```

```
                     beq       reverse
                     xgdx                              ; put quotient back to B
                     bra       loop
        reverse      tst       0,sp
                     beq       done
                     movb      1,sp +, 1,y +
                     bra       reverse
        done         swi
                     end
```

Example 4.5

Write a program to convert the 16-bit signed integer stored in double accumulator D into a string of BCD digits, and save the conversion result in a memory buffer starting with the label *out_buf*. The string must be terminated by a NULL character.

Solution: Because a string must be terminated by a NULL character, a signed 16-bit integer (in the range of −32768 to + 32767) may need 2, 3, 4, 5, 6, or 7 bytes to store all of its digits in BCD format. If the number is negative, we need to add the minus sign as the first character of the converted string. A minus integer must be complemented. After performing these preprocessing steps, a repeated divide-by-10 operation is performed.

The program is as follows:

```
test_dat     equ       −2478
             org       $1000
buf          ds.b      7                   ; to hold the decimal string
             org       $1500
             ldd       #test_dat
             lds       #$1500              ; initialize stack pointer
             ldy       #buf
             cpd       #0
             bmi       negate             ; branch if the value is negative
             bne       normal             ; branch if the value is positive
             movb      #$30,buf
             clr       buf + 1            ; terminate the string
             bra       done
negate       coma                         ; when negative, complement the number
             comb                         ;        "
             addd      #1                 ;        "
             movb      #$2D,1,y +         ; add minus sign and move pointer
normal       movb      #0,1,-sp           ; push a NULL into the stack
loop         ldx       #10
             idiv
             addb      #$30               ; convert to ASCII code
             pshb                         ; push into stack
             cpx       #0                 ; is the quotient zero?
             beq       reverse
             xgdx                         ; put quotient back to B
             bra       loop
```

```
reverse     tst      0,sp
            beq      done
            movb     1,sp + ,1,y +      ; pop one byte and store it in buffer
            bra      reverse
done        swi
            end
```

When the user enters a number from a keyboard, it is often in the form of a NULL-terminated ASCII string. We need to convert this into binary before it can be processed. Supposing *in_ptr* is the pointer to the input ASCII string, then the algorithm to convert an ASCII string into a binary number would be as follows:

Step 1

> sign ← 0
> error ← 0
> number ← 0

Step 2
If the character pointed to by *in_ptr* is the minus sign, then

> sign ← 1
> in_ptr ← in_ptr + 1

Step 3
If the character pointed to by *in_ptr* is the NULL character,

> then go to step 4.

else if the character is not a BCD digit (i.e., m[in_ptr] > $39 or m[in_ptr] < $30), then:

> error ← 1;
> go to step 4;

else:

> number ← number × 10 + m[in_ptr] − $30;
> in_ptr ← in_ptr + 1;
> go to step 3;

Step 4
If sign = 1 and error = 0, then

> number ← two's complement of number;

else

> stop;

The following example converts an ASCII string into a binary number.

Example 4.6

Write a program to convert a BCD ASCII string to a binary number and leave the result in double accumulator D. The ASCII string represents a number in the range of $-2^{15} \sim 2^{15} - 1$.

Solution: The following program is a direct translation of the preceding algorithm:

```
minus     equ     $2D              ; ASCII code of minus sign
          org     $1000
```

```
in_buf      fcc      "9889"           ; input ASCII to be converted
            fcb      0                ; null character
out_buf     rmb      2                ; holds the converted binary value
buf2        rmb      1                ; holds a zero
buf1        rmb      1                ; holds the current digit value
sign        rmb      1                ; holds the sign of the number
error       rmb      1                ; indicates the occurrence of illegal character
            org      $1500
            clr      sign
            clr      error
            clr      out_buf
            clr      out_buf + 1
            clr      buf2
            ldx      #in_buf
            ldaa     0,x
            cmpa     #minus           ; is the first character a minus sign?
            bne      continue         ; branch if not minus
            inc      sign             ; set the sign to 1
            inx                       ; move the pointer
continue    ldaa     1,x +            ; is the current character a NULL character?
            lbeq     done             ; yes, we reach the end of the string
            cmpa     #$30             ; is the character not between 0 to 9?
            lblo     in_error         ; "
            cmpa     #$39             ; "
            lbhi     in_error         ; "
            suba     #$30             ; convert to the BCD digit value
            staa     buf1             ; save the digit temporarily
            ldd      out_buf
            ldy      #10
            emul                      ; perform 16-bit by 16-bit multiplication
            addd     buf2             ; add the current digit value
            std      out_buf
            bra      continue
in_error    ldaa     #1
            staa     error
done        ldaa     sign             ; check to see if the original number is negative
            beq      positive
            ldaa     out_buf          ; if negative, compute its two's complement
            ldab     out_buf + 1      ;          "
            coma                      ;          "
            comb                      ;          "
            addd     #1               ;          "
            std      out_buf
positive    swi
            end
```

4.5.2 Character and Word Counting

People probably will disagree on whether the white-space characters such as space, carriage return, line feed, and form feed should be included in the character count. We will include them in the character count because white-space characters occupy disk space when a string is saved.

The procedure for finding the character count and the word count of a string is shown in Figure 4.4. In Figure 4.4, every non-NULL character causes the character count to be incremented by 1. To look for a new word, one needs to skip the white-space characters during the

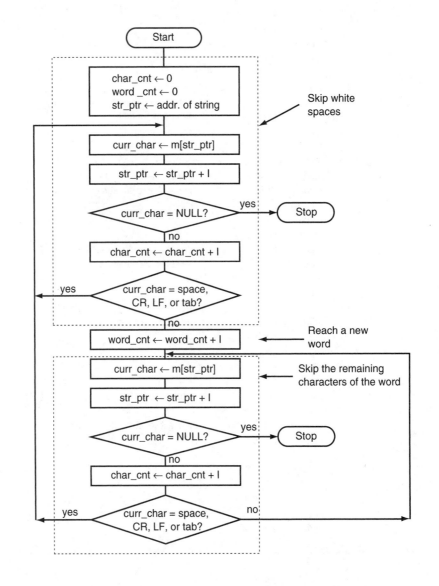

Figure 4.4 ■ Logic flow for counting characters and words in a string

scanning process. A new word is identified by the occurrence of a nonwhite character during the scanning process. Whenever a new word is encountered, one needs to scan through the remaining nonwhite characters until the next white character is encountered.

The variables used in Figure 4.4 are

- char_cnt: character count
- word_cnt: word count
- curr_char: current character
- char_ptr: character pointer

Example 4.7

▼

Write a program to count the number of characters and words contained in a given string.

Solution: The following program is a direct translation of the algorithm in Figure 4.4:

```
tab         equ     $09
sp          equ     $20
cr          equ     $0D
lf          equ     $0A
            org     $1000
char_cnt    rmb     1
word_cnt    rmb     1
string_x    fcc     "this is a strange test string to count chars and words."
            fcb     0
            org     $1500
            ldx     #string_x
            clr     char_cnt
            clr     word_cnt
string_lp   ldab    1,x +        ; get one character and move string pointer
            lbeq    done         ; is this the end of the string?
            inc     char_cnt
; the following eight instructions skip white-space characters between words
            cmpb    #sp
            beq     string_lp
            cmpb    #tab
            beq     string_lp
            cmpb    #cr
            beq     string_lp
            cmpb    #lf
            beq     string_lp
; a nonwhite character is the start of a new word
            inc     word_cnt
wd_loop     ldab    1,x +        ; get one character and move pointer
            beq     done
            inc     char_cnt
; the following eight instructions check the end of a word
            cmpb    #sp
            lbeq    string_lp
```

```
               cmpb      #tab
               lbeq      string_lp
               cmpb      #cr
               lbeq      string_lp
               cmpb      #lf
               lbeq      string_lp
               bra       wd_loop
   done        swi
               end
```

4.5.3 Word Matching

Searching certain words from a string is a common operation. The searching process starts from the beginning of the given string. The algorithm will look for the next new word from the given string and compare it with the given word. The comparison is performed character by character until either two words are found to be matched or unmatched. If two words are not equal, then the remaining characters of the word in the string must be skipped so that the same process can be repeated. The comparison must be performed one character beyond the last character of the given word if the comparison is a match, and there are three possible outcomes:

Case 1

The matched word is not the last word in the string. Comparison of the last characters will yield *"not equal"*.

Case 2

The matched word is the last word in the given string, but there is one or a few whitespace characters between the last word and the NULL character. Comparison of the last characters will again give the result *"not equal"*.

Case 3

The matched word is the last word of the given string, and it is followed by the NULL character. In this case, the comparison result for the last character is *"equal"*.

A detailed flowchart for the word-matching algorithm is shown in Figure 4.5. In our convention, a punctuation mark that follows a word is not considered part of a word. The variables used in Figure 4.5 are

str_ptr:	string pointer
wd_ptr:	word pointer
match_flag:	flag to indicate search result
m[str_ptr]:	the character pointed to by str_ptr
m[wd_ptr]:	the character pointed to by wd_ptr

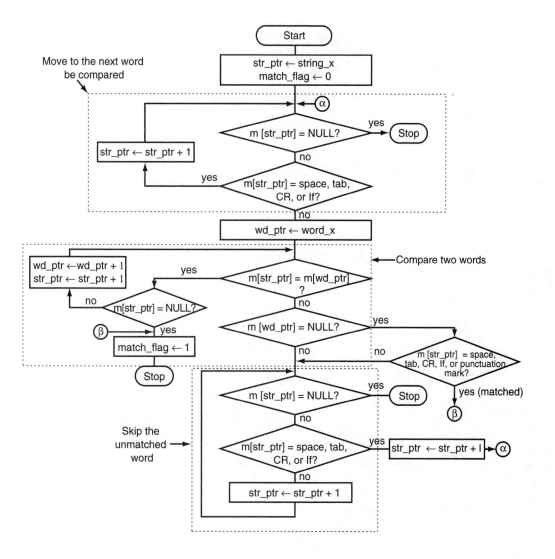

Figure 4.5 ■ Algorithm for searching a word from a string

Example 4.8

Write a program to search a given word from a string.

Solution: The following program is a direct implementation of the preceding algorithm:

```
tab     equ     $09             ; ASCII code of tab
sp      equ     $20             ; ASCII code of space character
cr      equ     $0D             ; ASCII code of carriage return
lf      equ     $0A             ; ASCII code of line feed
```

```
        period      equ     $2E                 ; ASCII code of period
        comma       equ     $2C                 ; ASCII code of comma
        semicolon   equ     $3B                 ; ASCII code of semicolon
        exclamation equ     $21                 ; ASCII code of exclamation
        null        equ     $0                  ; ASCII code of NULL character
                    org     $1000
        match       rmb     1
                    org     $1500
                    clr     match
                    ldx     #string_x
        loop        ldab    1,x+
; the following 10 instructions skip white spaces to look for the next word in string_x
                    tstb
                    beq     done
                    cmpb    #sp
                    beq     loop
                    cmpb    #tab
                    beq     loop
                    cmpb    #cr
                    beq     loop
                    cmpb    #lf
                    beq     loop
; the first nonwhite character is the beginning of a new word to be compared
                    ldy     #word_x
                    ldaa    1,y+
        next_ch     cba
                    bne     end_of_wd
                    cmpa    #null               ; check to see if the end of word is reached
                    beq     matched             ;           "
                    ldaa    1,y +               ; get the next character from the word
                    ldab    1,x +               ; get the next character from the string
                    bra     next_ch
; the following 10 instructions check to see if the end of the given word is reached
        end_of_wd   cmpa    #null
                    bne     next_wd
                    cmpb    #cr
                    beq     matched
                    cmpb    #lf
                    beq     matched
                    cmpb    #tab
                    beq     matched
                    cmpb    #sp
                    beq     matched
                    cmpb    #period
                    beq     matched
                    cmpb    #comma
                    beq     matched
```

```
                    cmpb    #semicolon
                    beq     matched
                    cmpb    #exclamation
                    beq     matched
; the following 11 instructions skip the remaining characters in the unmatched word
next_wd             ldab    1,x +
                    beq     done
                    cmpb    #cr
                    lbeq    loop
                    cmpb    #lf
                    lbeq    loop
                    cmpb    #tab
                    lbeq    loop
                    cmpb    #sp
                    lbeq    loop
                    bra     next_wd
matched             ldab    #1
                    stab    match
done                swi
string_x            fcc     "This string contains certain number of words to be matched."
                    fcb     0
word_x              fcc     "This"
                    fcb     0
                    end
```

4.5.4 String Insertion

Inserting a substring into a string is another common operation. The insertion point will be given for such an operation. To insert a substring into a string, one needs to make room for the substring to be inserted. All of the characters starting from the inserting point until the end of the string must be moved by the distance equal to the length of the substring. The main steps of the algorithm (also illustrated in Figure 4.6) are as follows:

Figure 4.6 ■ Major steps of substring insertion

Step 1
Count the number of characters that need to be moved (the NULL character is included because the resultant new string must be terminated by a NULL character).

Step 2
Count the number of characters in the substring to be inserted (excluding the NULL character).

Step 3
Move the characters in the string starting from the insertion point until the end of the string (move the last character first).

Step 4
Insert the substring (the first character of the substring is inserted first).
The program that implements the substring insertion is shown in Example 4.9.

Example 4.9

▼

Write a program to implement the string insertion algorithm.

Solution: The following program is a direct translation of the preceding algorithm:

```
            org       $1000
ch_moved    rmb       1                    ; number of characters to be moved
char_cnt    rmb       1                    ; number of characters of the substring to be inserted
sub_strg    fcc       "the first and most famous"
            fcb       0
string_x    fcc       "Yellowstone is national park."
            fcb       0
offset      equ       15
ins_pos     equ       string_x + offset    ; insertion point
            org       $1500
; the next seven instructions count the number of characters to be moved
            ldaa      #1
            staa      ch_moved
            ldx       #ins_pos             ; x points to the insertion point
cnt_moved   ldaa      1,x +
            beq       cnt_chars
            inc       ch_moved
            bra       cnt_moved
cnt_chars   dex                            ; subtract 1 from x so it points to the NULL character
            ldy       #sub_strg            ; y points to the substring
            clr       char_cnt
```

```
                   ; the following three instructions count the move distance
         char_loop   ldab     1,y +
                     beq      mov_loop
                     inc      char_cnt
                     bra      char_loop
         mov_loop    tfr      x,y             ; make a copy of x in y
                     ldab     char_cnt
                     aby                      ; compute the copy destination
                     ldab     ch_moved        ; place the number of characters to be moved in B
         again       movb     1,x-,1,y-
                     dbne     b,again         ; make room for insertion
                     ldx      #ins_pos        ; set up pointers to prepare insertion
                     ldy      #sub_strg       ; "
                     ldab     char_cnt
         insert_lp   movb     1,y + ,1,x +
                     dbne     b,insert_lp
                     swi
                     end
```

4.6 Subroutines

Good program design is based on the concept of *modularity*—the partitioning of a large program into subroutines. A main program contains the logical structure of the algorithm, whereas smaller program units (called *subroutines*) execute many of the details.

The principles of program design in high-level language apply even more to the design of assembly language programs. Begin with a simple main program whose steps clearly outline the logical flow of the algorithm, and then assign the execution details to subroutines. Of course, subroutines may call other subroutines themselves. The structure of a modular program is illustrated in Figure 4.7.

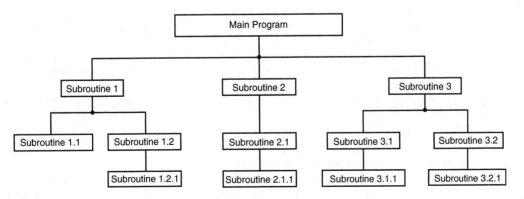

Figure 4.7 ■ A structured program

A *subroutine* is a sequence of instructions that can be called from many different places in a program. A key issue in a subroutine call is to make sure that the program execution returns to the point immediately after the subroutine call (this address is called *return address*) when the subroutine completes its computation. This is normally achieved by saving and retrieving the return address in and from the stack. The program flow change involved in a subroutine call is illustrated in Figure 4.8.

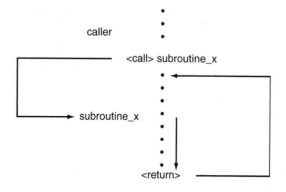

Figure 4.8 ■ Program flow during a subroutine call

A microcontroller has dedicated instructions for making the subroutine call. The subroutine call instruction will save the return address, which is the address of the instruction immediately following the subroutine call instruction, in the system stack. When completing the computation task, the subroutine will return to the instruction immediately following the instruction that makes the subroutine call. This is achieved by executing a *return* instruction, which will retrieve the return address from the stack and transfer CPU control to it.

The HCS12 provides three instructions (BSR, JSR, and CALL) for making subroutine calls and two instructions for returning from a subroutine. The syntax and operations performed by these instructions are listed and described next.

BSR <REL>   ; BRANCH TO SUBROUTINE

This instruction allows one to use only the *relative addressing mode* to specify the subroutine to be called. When this instruction is executed, the stack pointer is decremented by 2 and the return address is saved in the stack; then the offset in the instruction is added to the current PC value and instruction execution is continued from there. In assembly language, the relative address is specified by using a label, and the assembler will figure out the relative offset and place it in the program memory. For example, the following instruction calls the subroutine *bubble*:

```
bsr     bubble
```

JSR <OPR>   ; JUMP TO SUBROUTINE

This instruction allows us to use *direct, extended, indexed,* and *indexed indirect* addressing modes to specify the subroutine to be called. The subroutine can be located anywhere within 64 kB. Like the BSR instruction, the HCS12 first saves the return address in the stack and then jumps to execute the subroutine. Examples of the JSR instruction are as follows:

```
jsr     $ff         ; call the subroutine located at $ff
jsr     sq_root     ; call the subroutine sq_root
jsr     0,x         ; call a subroutine pointed to by index register X
```

CALL <opr>   ; CALL A SUBROUTINE

This instruction is designed to work with expanded memory (larger than 64 kB) supported by some HCS12 members. Members with expanded memory treat the 16 kB memory space from $8000 to $BFFF as a program memory window. An 8-bit program page register (PPAGE) is added to select one of the 256 16-kB program memory pages to be accessed. To support subroutine calls in expanded memory, the *call* instruction pushes the current value of the PPAGE register along with the return address onto the stack and then transfers program control to the subroutine (three bytes are pushed into the stack). The **<opr>** field in the call instruction specifies the page number and the starting address of the subroutine within that page. The new page number will be loaded into the PPAGE register when the call instruction is executed. Extended, indexed, and indexed indirect addressing modes can be used to specify the subroutine address within a page. Writing assembly programs to be run in expanded memory requires an assembler that supports this feature.

RTS   ; RETURN FROM SUBROUTINE

This instruction loads the program counter with a 16-bit value pulled from the stack and increments the stack pointer by 2. Program execution continues at the address restored from the stack.

RTC   ; RETURN FROM CALL

This instruction terminates subroutines in expanded memory invoked by the call instruction. The program page register (PPAGE) and the return address are restored from the stack; program execution continues at the restored address. For code compatibility, CALL and RTC are also executed correctly by the HCS12 members that do not have expanded memory capability.

4.7 Issues in Subroutine Calls

The program unit that makes the subroutine call is referred to as a *caller*, and the subroutine called by the other program unit is referred to as a *callee*. A subroutine is *entered* when it is being executed. There are three issues involved in subroutine calls:

4.7.1 Parameter passing

The caller usually wants the subroutine to perform computation using the parameters passed to it. There are several methods available for passing parameters to the subroutine:

1. *Use registers.* In this method, parameters are placed in CPU registers before the subroutine is called. This method is very convenient when there are only a few parameters to be passed.

2. *Use the stack.* In this method, parameters are pushed into the stack before the subroutine is called. The stack must be cleaned up after the computation is completed. This can be done by either the caller or the callee.

3. *Use the global memory.* Global memory is accessible to both the caller and the callee. As long as the caller places parameters in global memory before it calls the subroutine, the callee will be able to access them.

4.7.2 Result returning.

The result of a computation performed by the subroutine can be returned to the caller using three methods:

1. *Use registers.* This method is most convenient when there are only a few bytes to be returned to the caller.

2. *Use the stack.* The caller creates a hole of a certain size in the stack before making the subroutine call. The callee places the computation result in the hole before returning to the caller.

3. *Use global memory.* The callee simply places the value in the global memory and the caller will be able to access them.

4.7.3 Allocation of local variables

In addition to the parameters passed to it, a subroutine may need memory locations to hold temporary variables and results. Temporary variables are called *local variables* because they only exist when the subroutine is entered. Local variables are always allocated in the stack so that they are not accessible to any other program units.

Although there are several methods for allocating local variables, the most efficient one is using the **leas** instruction. This instruction loads the stack pointer with an effective address specified by the program. The effective address can be any indexed addressing mode except an indirect address. For example, to allocate 10 bytes, one can use the indexed addressing mode with SP as the base register and -10 as the offset:

```
leas      −10,sp       ; allocate ten bytes in the stack
```

This instruction simply subtracts 10 from SP and put the difference back to SP. The general format for allocating space to local variables is

```
leas      −n,sp        ; allocate n bytes in the stack
```

where n is the number of bytes to be allocated.

Before the subroutine returns to the caller, the space allocated to local variables must be deallocated. Deallocation is the reverse of allocation and can be achieved by the following instruction:

```
leas      n,sp         ; deallocate n bytes from the stack
```

4.8 The Stack Frame

The stack is used heavily during a subroutine call: The caller may pass parameters to the callee, and the callee may need to save registers and allocate local variables in the stack. The region in the stack that holds incoming parameters, return address, saved registers, and local variables is referred to as the *stack frame*. Some microprocessors have a dedicated register for managing the stack frame—the register is referred to as the *frame pointer*. The HCS12, however, does not have a register dedicated to the function of the frame pointer. Since the stack frame is created during a subroutine call, it is also called the *activation record* of the subroutine. The stack frame exists as long as the subroutine is not exited. The structure of a stack frame is shown in Figure 4.9.

Figure 4.9 ■ Structure of the HCS12 stack frame

The reason for having a dedicated frame pointer is that the stack pointer may change during the lifetime of a subroutine. Once the stack pointer changes value, there can be problems in accessing the variables stored in the stack frame. The frame pointer is added to point to a fixed location in the stack and can avoid this problem. Since the HCS12 does not have a dedicated frame pointer, we will not use the term *frame pointer* in the following discussion.

Example 4.10
▼

Draw the stack frame for the following program segment after the last **leas −10,sp** instruction is executed:

```
            ldd      #$1234
            pshd
            ldx      #$4000
            pshx
            jsr      sub_xyz
            . . .
sub_xyz     pshd
            pshx
            psy
            leas     −10,sp
            . . .
```

Solution: The caller pushes two 16-bit words into the stack. The subroutine **sub_xyz** saved three 16-bit registers in the stack and allocates 10 bytes in the stack. The resultant stack frame is shown in Figure 4.10.

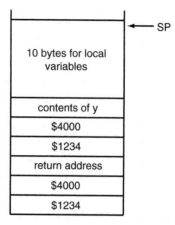

Figure 4.10 ■ Stack frame of Example 4.10

If a subroutine saves registers when it is entered, it must restore them in the reverse order (i.e., *last-in-first-out*) before returning to the caller. For example, if a subroutine has the following instructions to save the CPU registers when it is entered:

```
pshd
pshx
pshy
pshc
```

then it must have the following instructions to restore these registers:

```
pulc
puly
pulx
puld
```

4.9 Examples of Subroutine Calls

The examples in this section illustrate parameter passing, local variable allocation, and result returning.

4.9.1 Finding the Greatest Common Divisor

In this example, we will use CPU registers to pass parameters and return the result to the caller. The algorithm for finding the **gcd** of two integers **m** and **n** is as follows:

Step 1
If m = n then
    gcd ← m;
    return;

Step 2
If n < m then swap m and n.

Step 3
gcd ← 1.
If m = 1 or n = 1 then return.

Step 4
p = n % m;

Step 5
if (p == 0) then m is the gcd.
else
      n ← m;
      m ← p;
goto step 4.

This algorithm was invented by Euclid around 300 BC.

Example 4.11

▼

Write a program to compute the greatest common divisor of two 16-bit unsigned integers.

Solution: The following two local variables are needed for implementing the Euclidean algorithm:

    m_local: used to hold the incoming m value
    n_local: used to hold the incoming n value

Index register X will be used and hence must be saved in the stack. The stack frame is shown in Figure 4.11. The program is as follows:

2 bytes	contents of X	← SP
2 bytes	return address	
2 bytes	n_local	
2 bytes	m_local	

Figure 4.11 ■ Stack frame of Example 4.11

```
m          equ      3575            ; number to be tested
n          equ      2925            ;          "
m_local    equ      6               ; stack offset from SP for local variable m
n_local    equ      4               ; stack offset from SP for local variable n
           org      $1000
gcd        ds.w     1               ; hold the gcd of m and n
           org      $1500
           lds      #$1500          ; set up stack pointer
           ldd      #m              ; pass m in the stack
           pshd                     ;          "
           ldd      #n              ; pass n in the stack
           pshd                     ;          "
           jsr      find_gcd
           std      gcd
           leas     4,sp            ; deallocate space used by parameters
           swi
```

```
;********************************************************************
; The following subroutine uses the iterative division method to compute the gcd of two 16-bit
; integers m and n and returns the result in D.
;********************************************************************
;
find_gcd      pshx
gcd_loop      ldd      n_local,sp                   ; compute p = n % m
              ldx      m_local,sp                   ;         "
              idiv                                  ;         "
              cpd      #0
              beq      done
              movw     m_local,sp,n_local,sp  ; n ← m
              std      m_local,sp             ; m ← p
              bra      gcd_loop
done          ldd      m_local,sp                   ; retrieve gcd
              pulx                                  ; restore X
              rts
              end
```

4.9.2 Multiple Byte Division

The HCS12 provides instructions for 16-bit by 16-bit signed and unsigned divisions and also for 32-bit by 16-bit signed and unsigned divisions. However, the HCS12 does not have instructions for performing higher-precision divisions; say, for example, 32-bit by 32-bit division. We will need to synthesize such operations.

The most popular method for performing high-precision divide operations is the repeated subtraction method. The conceptual hardware for implementing a repeated subtraction algorithm is shown in Figure 4.12 and is detailed in the following steps:

Figure 4.12 ■ Conceptual hardware for implementing the repeated subtraction method

Step 1
Shift the register pair (R, Q) 1 bit left.

Step 2
Subtract the register P from the register R, put the result back to R if the result is non-negative.

Step 3
If the result of step 2 is negative, then set the least significant bit of Q to 0. Otherwise, set the least significant bit of Q to 1.

Perform this procedure **n** times, then the quotient and the remainder will be left in registers Q and R, respectively. Before the division steps are performed, the dividend and divisor must be loaded into register Q and register P, respectively. Register R must be cleared to zero.

Example 4.12

Write a subroutine that implements the division algorithm using the repeated subtraction method for a 32-bit unsigned dividend and divisor. The caller of this subroutine will pass the dividend and divisor in the stack and will allocate space in the stack for this subroutine to return the quotient and remainder. Also, write an instruction sequence to test this subroutine.

Solution: The stack frame of this subroutine is shown in Figure 4.13.

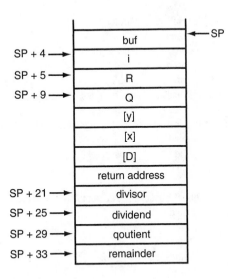

Figure 4.13 ■ Stack frame for Example 4.12

The instruction sequence to test this subroutine and the divide subroutine are as follows:

```
buf          equ       0                      ; distance of buf from the top of the stack
i            equ       4                      ; distance of i from the top of the stack
R            equ       5                      ; distance of R from the top of the stack
Q            equ       9                      ; distance of Q from the top of the stack
divisor      equ       21                     ; distance of divisor from the top of the stack
dividend     equ       25                     ; distance of dividend from the top of the stack
quotient     equ       29                     ; distance of quotient from the top of the stack
remainder    equ       33                     ; distance of remainder from the top of the stack
local        equ       13                     ; number of bytes for local variables
dvdend_hi    equ       $42                    ; dividend to be tested
dvdend_lo    equ       $4c15                  ; "
dvsor_hi     equ       $0                     ; divisor to be tested
dvsor_lo     equ       $64                    ; "

             org       $1000
quo          ds.b      4                      ; memory locations to hold the quotient
rem          ds.b      4                      ; memory locations to hold the remainder
             org       $1500                  ; starting address of the program
             lds       #$1500                 ; initialize stack pointer
             leas      -8,sp                  ; make a hole of 8 bytes to hold the result
             ldd       #dvdend_lo
             pshd
             ldd       #dvdend_hi
             pshd
             ldd       #dvsor_lo
             pshd
             ldd       #dvsor_hi
             pshd
             jsr       div32                  ; call the divide subroutine
; the following instruction deallocates the stack space used by the divisor and dividend
             leas      8,sp
; the following four instructions get the quotient from the stack
             puld
             std       quo
             puld
             std       quo + 2
; the following four instructions get the remainder from the stack
             puld
             std       rem
             puld
             std       rem + 2
             swi
;*************************** ********************************
; The following subroutine divides an unsigned 32-bit integer by another unsigned
; 32-bit integer
;*************************** ********************************
div32        pshd
             pshx
             pshy
```

```
; the following instruction allocates space for local variables
                leas        −local,sp
                ldd         #0
                std         R,sp                    ; initialize register R to 0
                std         R + 2,sp
                ldd         dividend,sp
                std         Q,sp                    ; place dividend in register Q
                ldd         dividend + 2,sp
                std         Q + 2,sp
                ldaa        #32
                staa        i,sp                    ; initialize loop count
loop            lsl         Q + 3,sp                ; shift register pair Q and R to the right
                rol         Q + 2,sp                ; by 1 bit
                rol         Q + 1,sp                ; "
                rol         Q,sp                    ; "
                rol         R + 3,sp                ; "
                rol         R + 2,sp                ; "
                rol         R + 1,sp                ; "
                rol         R,sp                    ; "
; the following eight instructions subtract the divisor from register R
                ldd         R + 2,sp
                subd        divisor + 2,sp
                std         buf + 2,sp
                ldaa        R + 1,sp
                sbca        divisor + 1,sp
                staa        buf + 1,sp
                ldaa        R,sp
                sbca        divisor,sp
                bcs         smaller
; the following six instructions store the difference back to R register
                staa        R,sp
                ldaa        buf + 1,sp
                staa        R + 1,sp
                ldd         buf + 2,sp
                std         R + 2,sp
                bset        Q + 3,sp,$01            ; set the least significant bit of Q register to 1
                bra         looptest
smaller         bclr        Q + 3,sp,$01            ; set the least significant bit of Q register to 0
looptest        dec         i,sp
                lbne        loop
; the following four instructions copy the remainder into the hole
                ldd         R,sp
                std         remainder,sp
                ldd         R + 2,sp
                std         remainder + 2,sp
```

; the following four instructions copy the quotient into the hole

```
        ldd       Q,sp
        std       quotient,sp
        ldd       Q + 2,sp
        std       quotient + 2,sp
        leas      local,sp              ; deallocate local variables
        puly
        pulx
        puld
        rts
        end
```

4.9.3 Bubble Sort

Sorting is among the most common ingredients of programming and many sorting methods are being used. Sorting makes many efficient search methods possible. The *bubble sort* is a simple, but widely known inefficient sorting method. Many other more efficient sorting methods require the use of recursive subroutine calls, which fall outside the scope of this book.

The basic idea underlying the bubble sort is to go through the array or file sequentially several times. Each iteration consists of comparing each element in the array or file with its successor (x[i] with x[i + 1]) and interchanging the two elements if they are not in proper order (either ascending or descending). Consider the following array:

157 13 35 9 98 810 120 54 10 30

Suppose one wants to sort this array in ascending order. The following comparisons are made in the first iteration:

x[0] with x[1] (157 and 13) interchange

x[1] with x[2] (157 with 35) interchange

x[2] with x[3] (157 with 9) interchange

x[3] with x[4] (157 with 98) interchange

x[4] with x[5] (157 with 810) no interchange

x[5] with x[6] (810 with 120) interchange

x[6] with x[7] (810 with 54) interchange

x[7] with x[8] (810 with 10) interchange

x[8] with x[9] (810 with 30) interchange

Thus, after the first iteration, the array is in the order

13 35 9 98 157 120 54 10 30 810

Notice that after this first iteration, the largest element (in this case, 810) is in its proper position within the array. In general, x[n − i] will be in its proper position after iteration i. The method is called the *bubble sort* because each number slowly bubbles up to its proper position. After the second iteration the array is in the order

13 9 35 98 120 54 10 30 157 810

Notice that 157 is now in the second-highest position. Since each iteration places a new element into its proper position, an array or a file of n elements requires no more than $n - 1$ iterations. The complete set of iterations is as follows:

Iteration		
0 (original array)		157 13 35 9 98 810 120 54 10 30
1		13 35 9 98 157 120 54 10 30 810
2		13 9 35 98 120 54 10 30 157 810
3		9 13 35 98 54 10 30 120 157 810
4		9 13 35 54 10 30 98 120 157 810
5		9 13 35 10 30 54 98 120 157 810
6		9 13 10 30 35 54 98 120 157 810
7		9 10 13 30 35 54 98 120 157 810
8		9 10 13 30 35 54 98 120 157 810
9		9 10 13 30 35 54 98 120 157 810

There are some obvious improvements to the foregoing method:

- First, since all elements in positions greater than or equal to $n - i$ are already in proper position after iteration i, they need not be considered in succeeding iterations. Thus, in the first iteration $n - 1$ comparisons are made, on the second iteration $n - 2$ comparisons, and on the $(n - 1)$th iteration, only one comparison is made (between x[0] and x[1]). Therefore, the process is sped up as it proceeds through successive iterations.

- Second, although we have shown that $n - 1$ iterations are sufficient to sort an array or a file of size n, in the preceding sample array of ten elements, the array was sorted after the seventh iteration, making the last two iterations unnecessary. To eliminate unnecessary iterations, we must be able to detect the fact that the array is already sorted. An array is sorted if no swaps are made in an iteration. By keeping a record of whether any swaps are made in a given iteration, it can be determined whether any further iterations are necessary. The logic flow of the bubble sort algorithm is illustrated in Figure 4.14. The following example implements the bubble sort as a subroutine.

Example 4.13
▼

Write a subroutine to implement the bubble sort algorithm and a sequence of instructions along with a set of test data for testing this subroutine. Use an array that consists of n 8-bit unsigned integers for testing purposes.

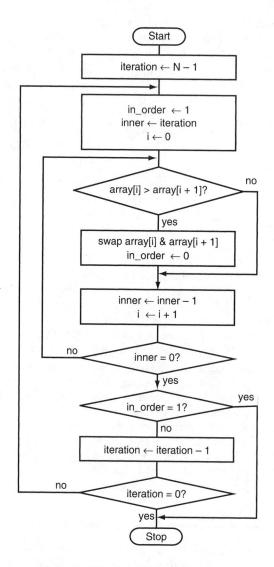

Figure 4.14 ■ Logic flow of bubble sort

Solution: The subroutine has several local variables:
- buf: buffer space for swapping adjacent elements
- in_order: flag to indicate whether the array is in order after an iteration
- inner: loop count for each iteration
- iteration: number of iterations remaining to be performed

The stack frame of this program is shown in Figure 4.15.

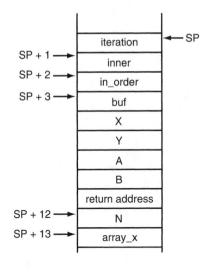

Figure 4.15 ■ Stack frame for bubble sort

```
arr          equ     13           ; distance of the variable arr from stack top
arcnt        equ     12           ; distance of the variable arcnt from stack top
buf          equ     3            ; distance of local variable buf from stack top
in_order     equ     2            ; distance of local variable in_order from stack top
inner        equ     1            ; distance of local variable inner from stack top
iteration    equ     0            ; distance of local variable iteration from stack top
true         equ     1
false        equ     0
n            equ     30           ; array count
local        equ     4            ; number of bytes used by local variables
             org     $1000
array_x      dc.b    3,29,10,98,54,9,100,104,200,92,87,48,27,22,71
             dc.b    1,62,67,83,89,101,190,187,167,134,121,20,31,34,54
             org     $1500
             lds     #$1500       ; initialize stack pointer
             ldx     #array_x
             pshx
             ldaa    #n
             psha
             jsr     bubble
             leas    3,sp         ; deallocate space used by outgoing parameters
             swi                  ; break to D-Bug12 monitor
;*************************** ********************************************
; The following subroutine implements the bubble sort with the array address and array count
; pushed into the stack.
;*************************** ********************************************
bubble       pshd
             pshy
             pshx
```

```
            leas      -local,sp        ; allocate space for local variables
            ldaa      arcnt,sp         ; compute the number of iterations to be performed
            deca                       ; "
            staa      iteration,sp     ; "
ploop       ldaa      #true            ; set array in_order flag to true before any iteration
            staa      in_order,sp      ; "
            ldx       arr,sp           ; use index register X as the array pointer
            ldaa      iteration,sp     ; initialize inner loop count for each iteration
            staa      inner,sp         ; "
cloop       ldaa      0,x              ; compare two adjacent elements
            cmpa      1,x              ; "
            bls       looptest
; the following five instructions swap the two adjacent elements
            staa      buf,sp           ; swap two adjacent elements
            ldaa      1,x              ; "
            staa      0,x              ; "
            ldaa      buf,sp           ; "
            staa      1,x              ; "
            ldaa      #false           ; reset the in-order flag
            staa      in_order,sp      ; "
looptest    inx
            dec       inner,sp
            bne       cloop
            tst       in_order,sp      ; test array in_order flag after each iteration
            bne       done
            dec       iteration,sp
            bne       ploop
; the following instruction deallocates local variables
done        leas      local,sp         ; deallocate local variables
            pulx
            puly
            puld
            rts
            end
```

4.9.4 Finding the Square Root

There are several methods available for finding the square root of a number **q**. One of the methods is based on the following equation:

$$\sum_{i=0}^{n-1} i = \frac{n(n-1)}{2} \qquad (4.1)$$

This equation can be transformed into

$$n^2 = \sum_{i=0}^{n-1} (2i + 1) \qquad (4.2)$$

Equation 4.2 gives us a clue about how to compute the square root of a value. Suppose we want to compute the square root of q, and n is the integer value that is closest to the true square root. One of the following three relationships is satisfied:

$n^2 < q$
$n^2 = q$
$n^2 > q$

The logic flow shown in Figure 4.16 can be used to find a value for n, and this algorithm is translated into the program in Example 4.14.

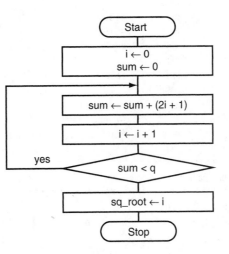

Figure 4.16 ■ Algorithm for finding the square root of integer q

Example 4.14

Write a subroutine to implement the square root algorithm. This subroutine must be able to find the square root of a 32-bit unsigned integer. The parameter q (for which we want to find the square root) is pushed into the stack and the square root is to be returned in double accumulator D.

Solution: To implement the algorithm illustrated in Figure 4.16, we need to use two local variables:

- i: for looping purpose
- sum: for accumulating all of the (2i + 1) terms

The stack frame of this subroutine is shown in Figure 4.17.

Figure 4.17 ■ Stack frame of Example 4.14

The subroutine and the instruction sequence for testing this subroutine are as follows:

```
q_hi        equ     $000F               ; upper word of q
q_lo        equ     $4240               ; lower word of q
i_local     equ     0                   ; distance of local variable i from the top of the stack
sum         equ     2                   ; distance of local variable sum from the top of the stack
q_val       equ     10                  ; distance of parameter q from the top of stack
local       equ     6                   ; number of bytes allocated to local variables
            org     $1000               ; starting address of data area
sq_root     ds.b    2                   ; to hold the square root of q
            org     $1500               ; starting address of program
            lds     #$1500              ; set up stack pointer
            ldd     #q_lo
            pshd
            ldd     #q_hi
            pshd
            jsr     findSqRoot
            std     sq_root
            leas    4,sp                ; deallocate space used by outgoing parameters
            swi
; ***********************************************************
; The following subroutine computes the square root for a 32-bit unsigned integer.
; ***********************************************************
findSqRoot  pshy                        ; save y in the stack
            leas    −local,sp           ; allocate space to local variables
            ldd     #0                  ; initialize local variable i to 0
            std     i_local,sp          ; "
            std     sum,sp              ; initialize local variable sum to 0
            std     sum + 2,sp          ; "
loop        ldd     i_local,sp
            ldy     #2
            emul                        ; compute 2i
            addd    sum + 2,sp          ; add 2i to sum
            std     sum + 2,sp          ;       "
            tfr     y,d                 ;       "
            adcb    sum + 1,sp          ;       "
            stab    sum + 1,sp          ;       "
            adca    sum,sp              ;       "
            staa    sum,sp              ;       "
; add 1 to sum (need to propagate carry to the most significant byte of sum)
            ldd     #1
            addd    sum + 2,sp
            std     sum + 2,sp
            ldaa    sum + 1,sp
            adca    #0
            staa    sum + 1,sp
            ldaa    sum,sp
            adca    #0
            staa    sum,sp
```

```
                ldd        i_local,sp              ; increment i by 1
                addd       #1                      ;      "
                std        i_local,sp              ;      "
; compare sum to q_val by performing subtraction (need consider borrow)
                ldd        sum + 2,sp
                subd       q_val + 2,sp
                ldaa       sum + 1,sp
                sbca       q_val + 1,sp
                ldaa       sum,sp
                sbca       q_val,sp
                lblo       loop
                ldd        i_local,sp              ; place sq_root in D before return
exit            leas       local,sp                ; deallocate space used by local variables
                puly
                rts
                end
```

This subroutine will find the exact square root if the given number has one. If the given number does not have an exact integer square root, then the number returned by the subroutine may not be the closest approximation. The algorithm in Figure 4.16 will stop only when **sum > q**. This may have the undesirable effect that the last **i** value may not be as close to the real square root as is the value **i − 1**. However, this can be fixed easily by comparing the following two expressions:

1. $i^2 -$ q_val
2. q_val $- (i − 1)^2$

If the first expression is smaller, then **i** is a better choice. Otherwise, we should choose **i − 1** as the approximation to the real square root. This will be left as an exercise problem.

4.10 Using the D-Bug12 Functions to Perform I/O Operations

The D-Bug12 monitor provides a few subroutines to support I/O operations. One can utilize these I/O routines to facilitate program developments on a demo board that contains the D-Bug12 monitor. A summary of these routines is in Table 4.2.

These user-accessible routines are written in C. All except the first parameter are passed to the user-callable functions on the stack. Parameters must be pushed onto the stack in the reverse order they are listed in the function declaration (right-to-left). The first parameter is passed to the function in accumulator D. If a function has only a single parameter, then the parameter is passed in accumulator D. Parameters of type *char* must be converted to an *integer* (16-bit). Parameters of type *char* will occupy the lower-order byte (higher address) of a word pushed onto the stack or accumulator B if the parameter is passed in D.

Parameters pushed onto the stack before the function is called remain on the stack when the function returns. It is the responsibility of the *caller* to remove passed parameters from the stack.

All 8- and 16-bit function values are returned in accumulator D. A value of type *char* returned in accumulator D is located in the 8-bit accumulator B. *Boolean* function returns a zero value for false and a nonzero value for true.

None of the CPU register contents, except the stack pointer, are preserved by the called functions. If any of the register values need to be preserved, they should be pushed onto the stack before any of the parameters have been pushed and restored after deallocating the parameters.

Subroutine	Function	Pointer Address
far main ( )	Start of D-Bug12	$EE80
getchar ( )	Get a character from SCI0 or SCI1	$EE84
putchar ( )	Send a character out to SCI0 or SCI1	$EE86
printf ( )	Formatted string output—translates binary values to string	$EE88
far GetCmdLine ( )	Get a line of input from the user	$EE8A
far sscanhex ( )	Convert ASCII hex string to a binary integer	$EE8E
isxdigit ( )	Check if a character (in B) is a hex digit	$EE92
toupper ( )	Convert lower case characters to upper case	$EE94
isalpha ( )	Check if a character is alphabetic	$EE96
strlen ( )	Return the length of a NULL-terminated string	$EE98
strcpy ( )	Copy a NULL-terminated string	$EE9A
far out2hex ( )	Output 8-bit number as two ASCII hex characters	$EE9C
far out4hex ( )	Output a 16-bit number as four ASCII hex characters	$EEAO
SetUserVector ( )	Set up a vector to a user's interrupt service routine	$EEA4
far WriteEEByte ( )	Write a byte to the on-chip EEPROM memory	$EEA6
far EraseEE ( )	Bulk erase the on-chip EEPROM memory	$EEAA
far ReadMem ( )	Read data from the HCS12 memory map	$EEAE
far WriteMem ( )	Write data to the HCS12 memory map	$EEB2

Table 4.2 ■ D-Bug12 monitor (version 4.x.x) routines

4.10.1 Calling D-Bug12 Functions from Assembly Language

Calling the functions from assembly language is a simple matter of pushing the parameters onto the stack in the proper order and loading the first or only function parameter into accumulator D. The function can then be called with a jsr instruction. The code following the jsr instruction should remove any parameters pushed onto the stack. If a single parameter was pushed onto the stack, a simple pulx or puly instruction is one of the most efficient ways to remove the parameter from the stack. If two or more parameters are pushed onto the stack, the leas instruction is the most efficient way to remove the parameters. Any of the CPU registers that were saved on the stack before the function parameters should be restored with a corresponding PULL instruction.

For example, the WriteEEByte() function has two parameters: The first parameter is the address of the memory location to which the data is to be written; the second is the data itself. An example of the instruction sequence to call this function to write the value #$55 into EEPROM is as follows:

```
WriteEEByte     equ         $EEA6
                .
                .
                ldd         #$55              ; write $55 to EEPROM
                pshd
                ldd         EEAddress         ; EEPROM address to write data
                jsr         [WriteEEByte,pcr] ; call the routine
                leas        2,sp              ; remove the parameter from stack
                beq         EEWError          ; zero return value means error
                .
                .
```

The addressing mode used in the **jsr** instruction of this example is a form of indexed indirect addressing that uses the program counter as an index register. The PCR mnemonic used in place of an index register name stands for *Program Counter Relative* addressing. In reality, the HCS12 does not support PCR. Instead, the PCR mnemonic is used to instruct the assembler to calculate an offset to the address specified by the label *WriteEEByte*. The offset is calculated by subtracting the value of the PC at the address of the first object code byte of the next instruction (in this example, **leas 2,sp**) from the address supplied in the indexed offset field (*WriteEEByte*). When the jsr instruction is executed, the opposite occurs. The HCS12 adds the value of the PC at the first object code byte of the next instruction to the offset embedded in the instruction object code. The indirect addressing, indicated by the square brackets, specifies that the address calculated as the sum of the index register (in this case the PC) and the 16-bit offset contains a pointer to the destination of the jsr.

The MiniIDE software supports this syntax. However, if you are using an assembler that does not support program-counter-relative-indexed addressing, the following two-instruction sequence can be used:

```
ldx        WriteEEByte        ; load the address of WriteEEByte()
jsr        0,x                ; call the subroutine
```

If the name of a library function is preceded by the keyword **far**, then it is located in the expanded memory and must be called by using the call instruction. Other library functions can be called by executing the jsr instruction.

4.10.2 Descriptions of Callable Functions

For each of the callable functions, the prototype declaration and the pointer address (where the starting address of the function is stored) are listed.

void far main (void);

Pointer address: $EE80

This function simply restarts the D-Bug12 monitor, which will reinitialize all of D-Bug12's internal tables and variables. Any previously set breakpoints will be lost. This function will not be useful to user application development.

int getchar (void);

Pointer address: $EE84

This function retrieves a single character from the control terminal SCI. If an unread character is not available in the receive data register when this function is called, it will wait until one is received. Because the character is returned as an integer, the 8-bit character is placed in accumulator B.

Adding the following instruction sequence in your program will read a character from the SCI0 port:

```
getchar        equ        $EE84
               . . .
               jsr        [getchar,PCR]
               . . .
```

int putchar(int);

Pointer address: $EE86

This function outputs a single character to the control terminal SCI. If the control SCI's transmit data register is full when the function is called, putchar() will wait until the transmit data register is empty before sending the character. No buffering of characters is provided. Putchar() returns the character that was sent. However, it does not detect any error conditions that may occur in the process and therefore will never return EOF (end-of-file). Adding the following instruction sequence in your program will output the character A to the control SCI (when the program is running on a demo board, the character A will be displayed on the monitor screen):

```
putchar     equ        $EE86
            . . .
            ldd        #'A'
            jsr        [putchar,PCR]
            . . .
```

int printf(char *format, . . .);

Pointer address: $EE88

This function is used to convert, format, and print its arguments on the standard output (the output device could be the monitor screen, printer, LCD, etc.) under the control of the format string pointed to by format. It returns the number of characters that were sent to standard output (sent through serial port SCI0). All except floating-point data types are supported.

The format string can contain two basic types of objects: ASCII characters that are copied directly from the format string to the display device, and conversion specifications that cause succeeding printf() arguments to be converted, formatted, and sent to the display device. Each conversion specification begins with a percent sign (%) and ends with a single conversion character. Optional formatting characters may appear between the percent sign and end with a single conversion character in the following order:

[-] [<FieldWidth>] [.] [<Precision>] [h │ l]

These optional formatting characters are explained in Table 4.3.

Character	Description
– (minus sign)	Left justifies the converted argument.
FieldWidth	Integer number that specifies the minimum field width for the converted argument. The argument will be displayed in a field at least this wide. The displayed argument will be padded on the left or right if necessary.
. (period)	Separates the field width from the precision.
Precision	Integer number that specifies the maximum number of characters to display from a string or the minimum number of digits for an integer.
h	To have an integer displayed as a short.
l (letter ell)	To have an integer displayed as a long.

Table 4.3 ■ Optional formatting characters

The *FieldWidth* or *Precision* field may contain an asterisk (*) character instead of a number. The asterisk will cause the value of the argument in the argument list to be used instead.

The formatting characters supported by the printf() function are listed in Table 4.4. If the conversion character following the percent sign is not one of the formatting characters shown in this table or the characters shown in Table 4.3, the behavior of the printf() function is undefined.

character	Argument Type; Displayed as
d, i	int; signed decimal number
o	int; unsigned octal number (without a leading zero)
x	int; unsigned hex number using abcdef for 10... 15
X	int; unsigned hex number using ABCDEF for 10... 15
u	int; unsigned decimal number
c	int; single character
s	char *; display from the string until a '/0' (NULL)
p	void *; pointer (implementation-dependent representation)
%	no argument is converted; print a %

Table 4.4 ■ Printf() conversion characters

The printf() function can be used to print a message. One example is as follows:

```
CR          equ         $0D
LF          equ         $0A
printf      equ         $EE88
            . . .
            ldd         #prompt
            jsr         [printf,PCR]
            . . .
prompt      db          "Flight simulation",CR,LF,0
```

This instruction sequence will cause the message *Flight simulation* to be displayed and the cursor will be moved to the beginning of the next line.

Suppose labels *m, n,* and *gcd* represent three memory locations and the memory locations (2 bytes) started with the label *gcd* holds the greatest common divisor of two numbers stored at memory locations started with labels *m* and *n*. By adding the following instruction sequence to Example 4.11, the computation result can be displayed on the PC monitor screen when the program is executed on a demo board with D-Bug12 monitor:

```
CR          equ         $0D
LF          equ         $0A
printf      equ         $EE88
            . . .
            ldd         gcd
            pshd
            ldd         n
            pshd
            ldd         m
            pshd
            ldd         #prompt
            jsr         [printf,PCR]
            leas        6,sp
            . . .
prompt      db          "The greatest common divisor of %d and %d is %d",CR,LF,0
            . . .
```

This example is probably not easy to follow. In reality, this example is equivalent to the following C statement:

printf("The greatest common divisor of %d and %d is %d\n", m, n, gcd);

This function call has four parameters: the first parameter is a pointer to the formatting string and should be placed in D (done by the *ldd #prompt* instruction). The other parameters (m, n, and gcd) should be pushed into the stack in the order from right to left. There are three formatting characters corresponding to these three variables to be output.

int far GetCmdLine(char *CmdLineStr, int CmdLineLen);

Pointer address: $EE8A

This function is used to obtain a line of input from the user. GetCmdLine() accepts input from the user one character at a time by calling getchar(). As each character is received, it is echoed back to the user terminal by calling putchar() and placed in the character array pointed to by *CmdLineStr*. A maximum of *CmdLineLen − 1* characters may be entered. Only printable ASCII characters are accepted as input with the exception of the ASCII backspace character ($08) and the ASCII carriage return ($0D). All other nonprintable ASCII characters are ignored by the function.

The ASCII backspace character ($08) is used by the GetCmdLine() function to delete the previously received character from the command line buffer. When GetCmdLine() receives the backspace character, it will echo the backspace to the terminal, print the ASCII space character ($20), and then send a second backspace character to the terminal device. At the same time, the character is deleted from the command line buffer. If a backspace character is received when there are no characters in *CmdLineStr*, the backspace character is ignored.

The reception of an ASCII carriage return character ($0D) terminates the reception of characters from the user. The carriage return is not placed in the command line buffer. Instead, a NULL character ($00) is placed in the next available buffer location.

Before returning, all the entered characters are converted to uppercase. *GetCmdLine()* always returns an error code of *noErr*.

We can use this function to request the user to enter a string from the keyboard when running programs on a demo board with the D-Bug12 monitor. Usually, the program would output a message so that the user knows when to enter the string. The following instruction sequence will ask the user to enter a string from the keyboard:

```
printf        equ    $EE88
GetCmdLine    equ    $EE8A
cmdlinelen    equ    100
CR            equ    $0D
LF            equ    $0A
              . . .
prompt        db     "Please enter a string: ",CR,LF,0
              . . .
inbuf         rmb    100
              . . .
              ldd    #prompt              ; output a prompt to remind the user to
              jsr    [printf,PCR]         ; enter a string
              ldd    #cmdlinelen          ; push the CmdLineLen
              pshd                        ;    "
              ldd    #inbuf
              call   [GetCmdLine,PCR]     ; read a string from the keyboard
              puld                        ; clean up the stack
```

char * far sscanhex(char *HexStr, unsigned int *BinNum);

Pointer address: $EE8E

The sscanhex() function is used to convert a hex string to a binary number. The hex string pointed to by *HexStr* may contain any number of ASCII hex characters. However, the converted value must be no greater than $FFFF. The string must be terminated by either a space or a NULL character.

Suppose one has entered an ASCII string that represents a number, then the following instruction sequence will convert it into a hex number (or binary number):

```
sscanhex    equ     $EE8E
            . . .
HexStr      rmb     10                  ; input buffer to hold a hex string
BinNum      rmb     2                   ; to hold the converted number
            . . .
            ldd     #BinNum
            pshd
            ldd     #HexStr
            call    [sscanhex,PCR]
            leas    2,sp                ; deallocates space used by outgoing parameters
            . . .
```

int isxdigit(int c);

Pointer address: $EE92

The isxdigit() function tests the character passed in c for membership in the character set [0..9, a..f, A..F]. If the character c is in this set, the function returns a nonzero (true) value. Otherwise, a value of zero is returned. The following instruction sequence illustrates the use of this function:

```
isxdigit    equ     $EE92
            . . .
c_buf       rmb     20                  ; buffer that holds data to be validated
            . . .
            clra                        ; clear accumulator A
            ldab    c_buf               ; get one character
            jsr     [isxdigit,PCR]
            . . .
```

int toupper(int c);

Pointer address: $EE94

If c is a lowercase character, toupper() will return the corresponding uppercase letter. If the character is in uppercase, it simply returns c. The following instruction utilizes this function to convert a character contained in B to uppercase:

```
toupper     equ     $EE94
            . . .
c_buf       rmb     20                  ; buffer that contains string to be converted to
            . . .                       ; uppercase
            ldab    c_buf               ; get one character to convert
            clra
            jsr     [toupper,PCR]
            . . .
```

int isalpha(int c);

Pointer address: $EE96

This function tests the character passed in c for membership in the character set [a..z, A..Z]. If the character c is in this set, the function returns a nonzero value. Otherwise, it returns a zero. This function would also be useful for validating input string. The following instruction sequence illustrates the use of this function:

```
isalpha     equ        $EE96
            . . .
c_buf       rmb        20
            . . .
            ldab       c_buf
            clra
            jsr        [isalpha,PCR]        ; check whether the character in B is alphabetic
            . . .
```

unsigned int strlen(const char *cs);

Pointer address: $EE98

The strlen() function returns the length of the string pointed to by cs. The following instruction sequence counts the number of characters contained in the string pointed to by cs:

```
strlen      equ        $ EE98
            . . .
cs          db         "....."
            . . .
            ldd        #cs
            jsr        [strlen,PCR]
            . . .
```

char *strcpy(char *s1, char *s2);

Pointer address: $EE9A

This function copies the string pointed to by s2 into the string pointed to by s1 and returns a pointer to s1. The following instruction sequence copies the string pointed to by s2 to the memory location pointed to by s1:

```
strcpy      equ        $EE9A
            . . .
s2          db         "......"
s1          rmb        . . .
            . . .
            ldd        #s2
            pshd
            ldd        #s1
            jsr        [strcpy,PCR]
            leas       2,sp
            . . .
```

void far out2hex(unsigned int num);

Pointer address: $EE9C

This function displays the lower byte of **num** on the terminal screen as two hex characters. The upper byte of **num** is ignored. The function out2hex () simply calls printf() with a format string of "%2.2X".

The following instruction sequence outputs the number in accumulator B as two hex digits to the terminal screen:

```
out2hex     equ      $EE9C
            . . .
data        rmb      20
            . . .
            ldab     data
            clra
            call     [out2hex,PCR]
            . . .
```

void far out4hex(unsigned int num);

Pointer address: $EEA0

This function displays num on the control terminal as four hex characters. The function out4hex() simply calls printf() with a format string of "%4.4X". The following instruction sequence outputs the 16-bit number stored at memory location *num* as four hex digits:

```
out4hex     equ      $EEA0
            . . .
num         db       . . .
            . . .
            ldd      num
            call     [out4hex,PCR]
            . . .
```

int SetUserVector(int VectNum, Address UserAddress);

Pointer address: $EEA4

Since interrupts will be discussed in Chapter 6, we will cover the use of this function at that time.

Boolean far WriteEEByte (Address EEAddress, Byte EEData);

Pointer address: $EEA6

The WriteEEByte() function provides a mechanism to program individual bytes of the on-chip EEPROM. It does not perform any range checking on the passed *EEAddress*. A user's program can determine the start address and size of the on-chip EEPROM array by examining the data contained in the custom data area fields *CustData.EEBase* and *CustData.EESize*.

A byte erase operation is performed before the programming operation and a verification operation is performed after the programming operation. If the EEPROM data does not match EEData, false (zero value) is returned by the function.

Int far EraseEE(void);

Pointer address: $EEAA

This function performs a bulk erase operation to the on-chip EEPROM without having to manipulate the EEPROM programming control registers. After the bulk erase operation is performed, the memory range described by CustData.EEBase and CustData.EESize is checked for erasure. If any of the bytes does not contain 0xFF, a nonzero error code is returned.

> int far ReadMem(Address StartAddress, Byte *MemDataP, unsigned int NumBytes);
>
> Pointer address: $EEAE

This function is used internally by D-Bug12 for all memory read access.

> int WriteMem(Address StartAddress, Byte *MemDataP, unsigned int NumBytes);
>
> Pointer address: $EEB2

The *WriteMem()* function is used internally by D-Bug12 for all memory write accesses. If a byte is written to the memory range described by CustData.EEBase and CustData.EESize, WriteMem() calls the WriteEEByte() function to program the data into the on-chip EEPROM memory. A nonzero error code is returned if a problem occurs while writing to target memory.

4.10.3 Using the D-Bug12 Functions

A useful program usually consists of several functions. An example is given in this section.

Example 4.15

▼

Write a program that invokes appropriate functions to find the prime number between 1000 and 2000. Output eight prime numbers in one line. To do this, one will need to

1. Write a subroutine to test if an integer is a prime
2. Invoke the printf() function to output the prime number
3. Write a loop to test all the integers between 1000 and 2000

Solution: The logic structure of the program is as follows:

Step 1
Output the message "The prime numbers between 1000 and 2000 are as follows:".

Step 2
For every number between 1000 and 2000, do the following:

1. Call the test_prime() function to see if it is a prime.
2. Output the number (call printf()) if it is a prime.
3. If there are already eight prime numbers in the current line, then also output a carry return.

The algorithm of the *test_prime()* function is as follows:

Step 1
Let *num*, *i*, and *isprime* represent the number to be tested, the loop index, and the flag to indicate if *num* is prime.

Step 2
isprime ← 0;

Step 3
For i = 2 to num/2 do
        if num % i = 0 then
                    return;
isprime ← 1;
return;

The stack frame of the function *test_prime()* is shown in Figure 4.18.

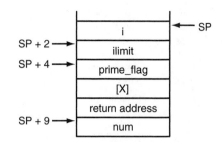

Figure 4.18 ■ Stack frame for the prime test subroutine

The assembly program is as follows:

```
CR          equ     $0D
LF          equ     $0A
i           equ     0           ; distance of variable i from stack top
ilimit      equ     2           ; distance of variable ilimit from stack top
prime_flag  equ     4           ; distance of variable isprime from stack top
num         equ     9           ; distance of variable num from the stack top
plocal      equ     5           ; number of bytes allocated to local variables
upper       equ     2000        ; upper limit for testing prime
lower       equ     1000        ; lower limit for testing prime
printf      equ     $EE88       ; location where the address of printf() is stored

            org     $1000
out_buf     rmb     10
prime_cnt   rmb     1
k           rmb     2
temp        rmb     2
            org     $1500
            ldx     #upper
            stx     temp
            pshx
            ldx     #lower
            stx     k           ; initialize k to 1000 for prime testing
            pshx
            ldd     #format0
            jsr     [printf,PCR]
            leas    4,sp
            clr     prime_cnt
again       ldd     k
            cpd     #upper
            bhi     stop        ; stop when k is greater than upper
            pshd
            jsr     prime_tst   ; test if k is prime
            leas    2,sp        ; deallocate space used by outgoing parameters
            tstb
            beq     next_k      ; test next integer if k is not prime
```

```
                inc      prime_cnt              ; increment the prime count
                ldd      k
                pshd
                ldd      #format1
                jsr      [printf,PCR]           ; output k
                leas     2,sp
                ldaa     prime_cnt
                cmpa     #8                     ; are there 8 prime numbers in the current line?
                blo      next_k
; output a CR, LF if there are already 8 prime numbers in the current line
                ldd      #format2
                jsr      [printf,PCR]
                clr      prime_cnt
                ldx      k
                inx
                stx      k
                bra      again
next_k          ldx      k
                inx
                stx      k
                jmp      again
stop            swi
prime_tst       pshx
                leas     -plocal,sp             ; allocate local variable
                ldaa     #1
                staa     prime_flag,sp          ; set the flag to true
                ldd      num,sp
                cpd      #1
                beq      not_prime              ; 1 is not a prime number
                cpd      #2
                beq      is_prime               ; 2 is a prime number
                lsrd                            ; compute num/2 as the test limit
                std      ilimit,sp              ; "
                ldd      #2
                std      i,sp                   ; set first test number to 2
test_loop       ldx      i,sp
                ldd      num,sp
                idiv
                cpd      #0                     ; if num can be divided by i, then
                beq      not_prime              ; it is not a prime
                ldd      i,sp
                cpd      ilimit,sp
                beq      is_prime               ; num is prime at the end of test loop
                ldx      i,sp
                inx
                stx      i,sp
                bra      test_loop
not_prime       clr      prime_flag,sp
```

```
is_prime      ldab      prime_flag,sp      ; put the result in B
              leas      plocal,sp
              pulx
              rts
format0       db        CR,LF,"The prime numbers between %d and %d are as follows:"
              db        CR,LF,CR,LF,0
format1       db        " %d ",0
format2       db        " ",CR,LF,0
              end
```

The program execution output should look like the following:

```
>load
**
>g 1500
```

The prime numbers between 1000 and 2000 are as follows:

1009	1013	1019	1021	1031	1033	1039	1049
1051	1061	1063	1069	1087	1091	1093	1097
1103	1109	1117	1123	1129	1151	1153	1163
1171	1181	1187	1193	1201	1213	1217	1223
1229	1231	1237	1249	1259	1277	1279	1283
1289	1291	1297	1301	1303	1307	1319	1321
1327	1361	1367	1373	1381	1399	1409	1423
1427	1429	1433	1439	1447	1451	1453	1459
1471	1481	1483	1487	1489	1493	1499	1511
1523	1531	1543	1549	1553	1559	1567	1571
1579	1583	1597	1601	1607	1609	1613	1619
1621	1627	1637	1657	1663	1667	1669	1693
1697	1699	1709	1721	1723	1733	1741	1747
1753	1759	1777	1783	1787	1789	1801	1811
1823	1831	1847	1861	1867	1871	1873	1877
1879	1889	1901	1907	1913	1931	1933	1949
1951	1973	1979	1987	1993	1997	1999	User Bkpt Encountered

```
PP  PC      SP        X        Y        D = A:B      CCR = SXHI  NZVC
30 1560    1500      07D1     15DE        07:D1                 1001 0000
xx: 1560    34                 PSHX

>
```

4.11 Tips for Program Debugging Involving Subroutine Calls

Program debugging becomes much more difficult when a program calls subroutines. To make debugging simpler, one should test each individual subroutine thoroughly to make sure each subroutine works correctly and returns to its caller.

4.11.1 What to Do When the Program Gets Stuck

It is a common problem that the program gets stuck in one of the subroutines and cannot return to the caller. The procedure for debugging in this situation is as follows:

Step 1
Write an instruction sequence to call each subroutine to find out if the program gets stuck in the subroutine. This can be done by setting a breakpoint at the instruction immediately after the **jsr** or **bsr** instruction. One can use the D-Bug12 **asm** command to find out the address of the breakpoint to be set.

Step 2
Find out why the program gets stuck in the subroutine. There are at least the following four causes:

1. Forgetting to restore registers pushed onto the stack before returning to the caller
2. Forgetting to deallocate local variables before returning to the caller
3. Some infinite loops in the subroutine
4. Calling other subroutines that do not return

The first two causes can be identified by simply looking at the program. To determine if the program gets stuck because of forgetting to restore registers, one checks to see if the subroutine has several push instructions at its entrance. If it does, then it should have the same number of pull instructions in the reverse order before it returns (using the RTS instruction). For the second cause, one checks to see if the subroutine has the instruction **leas −k,SP** to allocate space to local variables at the entry point of the subroutine. If it does, then the subroutine should have the instruction **leas k,SP** or its equivalent to deallocate the stack space used by local variables before returning to the caller. The first two causes result in the incorrect return address to be popped out from the stack. Sometimes a subroutine has several returning points. One needs to make sure to restore registers saved in the stack and deallocate local variables before each **rts** instruction.

If the first two causes are not present, it is still possible that the subroutine gets stuck in some loop. To find out if the program gets stuck in a certain loop, one can insert a breakpoint after the last instruction of the suspicious loop. If the breakpoint is never reached, then one knows that the loop has some problems. Once the infinite loop is identified, one should be able to figure out what's wrong.

To make sure that the subroutine does not get stuck after calling other subroutines, one should make sure that the subroutines called by the current routine do not have the first three problems described in this step.

4.11.2 General Debugging Strategy

Subroutines can be classified into two categories: *intermediate* and *leaf* subroutines. An intermediate subroutine may call other subroutines, whereas a leaf subroutine does not call any other subroutines. Making sure that a subroutine returns to its caller does not guarantee that it produces correct results. One needs to use the methods described in Section 3.9 to debug each leaf subroutine to make sure it works correctly. After making sure that each leaf subroutine works correctly, one can start to debug the intermediate subroutines. One needs to make sure that each intermediate subroutine does not get stuck and works correctly using the methods discussed in Sections 4.11.1 and 3.9. After each intermediate subroutine has been debugged, one can perform the top-level program debugging. Again, the method discussed in Section 3.9 can be used.

4.12 Summary

This chapter deals with three commonly used data structures, including the array, the stack, and the string. Conceptually, a stack is a list of data items whose elements can be accessed from only one end. A stack has a top and a bottom. The most common operations that can be applied to a stack are pushing and popping (or pulling). The push operation adds a new item to the top of the stack. The pop operation removes the top element of the stack. Physically, a stack is a reserved area in main memory where programs perform only push and pop operations. The HCS12 has a 16-bit register SP that points to the top byte of the stack. The HCS12 has instructions for pushing and pulling accumulators A, B, and D, the CCR register, and index registers X and Y.

An array is an indexable data structure. It consists of a sequence of elements of the same type, and each element being associated with an index. The index of the array often starts from zero to simplify the address calculation. An array can be one-dimensional or multidimensional. A two-dimensional array is called a *matrix*.

A string is a sequence of characters terminated by a NULL (ASCII code 0) or other special character, such as EOT (ASCII code $04). Common operations applied to strings include concatenation, substring insertion and deletion, character and word counting, word or substring matching, and so on.

Good program design is based on the concept of modularity—the partitioning of program code into subroutines. The main module contains the logical structure of the algorithm, while smaller program units execute many of the details.

The principles of program design in high-level languages apply even more to the design of assembly programs. Begin with a simple main program whose steps clearly outline the logical flow of the algorithm, and then assign the execution details to subroutines. Subroutines may themselves call other subroutines.

The HCS12 provides instructions bsr, jsr, and call for making subroutine calls. The instructions bsr and jsr will save the return address in the stack before jumping to the subroutine. The call instruction will save the contents of the PPAGE register in the stack in addition to the return address. The call instruction is provided to call subroutines located in expanded memory. All subroutines should have rts (rtc for subroutines in expanded memory) as the last instruction. The rts instruction will pop the return address onto the PC register from the stack, and program control will be returned to the point that called the subroutine. The rtc instruction should be used by subroutines that are in expanded memory. The rtc instruction will restore the PPAGE value pushed onto the stack in addition to the return address.

Issues related to subroutine calls include parameter passing, result returning, and local variables allocation.

Parameters can be passed in registers, program memory, the stack, or the global memory. The result computed by the subroutine can be returned in CPU registers, the stack, or the global memory. Local variables must be allocated in the stack so that they are not accessible to the caller and other program units. Local variables come into being only when the subroutine is being executed. The HCS12 provides instructions to facilitate the access to variables in the stack. The **leas** instruction is most effective for local variables allocation and deallocation.

The D-Bug12 monitor provides many functions to support I/O programming on demo boards that include the D-Bug12 monitor.

4.13 Exercises

E4.1 Supposing that we have the following instruction sequence to be executed by the HCS12, what will be the contents of the topmost four bytes of the stack after the execution of these instructions?

```
lds     #$1500
ldaa    #$56
staa    1, -SP
ldab    #22
staa    1, -SP
ldy     #0
sty     2, -SP
```

E4.2 Convert the binary search program into a subroutine so that it can be called by other program units. Let the starting address of the array, array count, and the key to be matched be passed via the stack and the result to be returned in accumulator D.

E4.3 Write a subroutine to create a delay of a multiple of 100 ms. The minimum delay created by this subroutine is 100 ms. The delay to be created is equal to the value in index register Y multiplied by 100 ms, assuming that the E clock frequency is 24 MHz.

E4.4 Convert the program of Example 4.5 into a subroutine. The 16-bit number to be converted and the pointer to the buffer to hold the converted string are passed in D and Y.

E4.5 Convert the program of Example 4.6 into a subroutine. Let the pointer to the ASCII string to be converted be passed in index register X. This subroutine would return the resultant 16-bit hex number in double accumulator D.

E4.6 The label *array_x* is the starting address of an array of 100 8-bit elements. Trace the following code sequence and describe what the subroutine *sub_x* does:

```
        ldx     #array_x
        ldaa    #100
        jsr     sub_x
        . . .
sub_x   deca
        ldab    0,x
        inx
loop    cmpb    0,x
        ble     next
        ldab    0,x
next    inx
        deca
        bne     loop
        rts
```

E4.7 Write a subroutine that can perform a 32-bit by 32-bit multiplication. Both the multiplicand and the multiplier are passed to this subroutine in the stack. The pointer to the buffer to hold the product is passed in index register X.

E4.8 Convert the program in Example 4.7 into a subroutine. The starting address of the string is passed to this subroutine in index register X, and the character and word count are returned to the caller in register D and X, respectively.

E4.9 Convert the program in Example 4.8 into a subroutine. The caller of this subroutine would push the starting addresses of the string and the word into the stack, and this subroutine would return a 1 in B if the word is found in the string. Otherwise, a 0 is returned in B.

E4.10 Draw the stack frame and enter the value of each stack slot (if it is known) at the end of the following instruction sequence:

```
            leas      -2,sp
            clrb
            ldaa      #20
            psha
            ldaa      #$E0
            psha
            ldx       #$7000
            pshx
            jsr       sub_abc
            . . .
sub_abc     pshd
            leas      -12,sp
            . . .
```

E4.11 Draw the stack frame and enter the value of each stack slot (if it is known) at the end of the following instruction sequence:

```
            leas      -8,sp
            ldd       #$1020
            psha
            ldx       #$800
            pshx
            bsr       xyz
            . . .
xyz         pshd
            pshx
            leas      -10,sp
            . . .
```

E4.12 Write a subroutine to convert all the lowercase letters in a string into uppercase. The starting address of the string is passed to this subroutine in index register X.

E4.13 Write a subroutine to convert all the uppercase letters in a string into lowercase. The starting address of the string is passed to this subroutine in index register X.

E4.14 Write a subroutine to generate a 16-bit random number. The result is returned to the caller in D. Pass any appropriate parameters to this subroutine in the stack.

E4.15 Write a subroutine to compute the *least common multiple* of two 16-bit integers. Incoming parameters are passed in the stack and the result should be returned in a (Y,D) pair with the upper 16 bits in Y and the lower 16 bits in D.

E4.16 Write a subroutine to convert the 8-bit signed integer into an ASCII string that represents a decimal number. The 8-bit integer and the pointer to the buffer to hold the ASCII string are passed to this subroutine in accumulator B and index register X, respectively.

E4.17 Write a subroutine that converts a 16-bit unsigned integer into an ASCII string (that represents a decimal number) so that it can be output to the serial communication port. Both the 16-bit integer and the pointer to the buffer to hold the ASCII string are passed to this subroutine in the stack.

E4.18 Give an instruction sequence to call the out4hex() function to output the 16-bit integer stored in memory location $1000-$1001.

E4.19 Give an instruction sequence that outputs a prompt "Please enter a string" and reads the string entered by the user from the keyboard and then echoes this string on the screen again.

E4.20 Write a subroutine that will convert a 32-bit signed integer into a BCD ASCII string so that it can be output to the control terminal (SCI) and appear as a BCD string on the screen. The 32-bit integer to be converted and the pointer to the buffer to hold the resultant string are passed to this subroutine via the stack.

4.14 Lab Exercises and Assignments

L4.1 Enter, assemble, and download the following program for execution on the demo board using MiniIDE:

```
CR          equ     $0D
LF          equ     $0A
printf      equ     $EE88
getcmdline  equ     $EE8A
cmdlinelen  equ     40
            org     $1000
inbuf       rmb     20
err_flag    rmb     1
sign_flag   rmb     1
            org     $1500
            lds     #$1500
            ldd     #prompt1
            jsr     [printf,PCR]        ; output a prompt to remind the user to enter
            ldd     #cmdlinelen         ; an integer
            pshd
            ldd     #inbuf
            call    [getcmdline,PCR]    ; read in a string that represents an integer
            leas    2,sp
            ldd     #prompt3            ; move cursor to the next line
            jsr     [printf,PCR]        ; "
            ldd     #inbuf
            pshd
            ldd     #prompt2
            jsr     [printf,PCR]        ; output the number that you entered
            leas    2,sp
            ldd     #prompt3
            jsr     [printf,PCR]        ; move cursor to the next line
            swi
prompt1     db      "Please enter a number: ",CR,LF,0
prompt2     db      "The entered number is: %s ",0
prompt3     db      " ",CR,LF,0
            end
```

When you see the message **Please enter a number:**, enter an integer followed by a carriage return. The screen output should be similar to what appears here:

```
>g 1500
Please enter a number:

The entered number is: 1234
User Bkpt Encountered

PP  PC        SP        X        Y       D = A:B     CCR = SXHI  NZVC
30 1533      1500      1500     156C      00:03             1001 0000
xx: 1533      50                NEGB

>
```

L4.2 Write a subroutine that will convert the temperature in Fahrenheit to Celsius accurate to one decimal digit. Write a main program that will:

1. Prompt the user to enter a temperature in Fahrenheit by displaying the following message:

   Please enter a temperature in Fahrenheit:

2. Call the GetCmdLine() function to read in the temperature.

3. Call a subroutine to convert the input string (representing a decimal number) into binary number.

4. Call the temperature conversion subroutine to convert it to Celsius.

5. Output the current temperature to the screen in the following format:

   Current temperature:
       xxxx°F   yyyy.y°C

6. Output the next message:

   Want to continue? (y/n)

7. Call the getchar() function to read in one character. If the character entered by the user is **y**, then repeat the process. Otherwise, return to the D-Bug12 monitor by executing the **swi** instruction.

Note: The ASCII code of the degree character ° is 176 (or $BO).

L4.3 Write a program to display the current time of day. This program will

1. Output the message "Please enter current time in the format of hhmmss:"

2. Read in the time entered by the user by calling the GetCmdLine() function

3. Store the current time in memory and display it the following format:

   current time: hh mm ss

4. Call a delay program to delay for one second, update the current time, and also redisplay the time on the screen

5

C Language Programming

5.1 Objectives

After completing this chapter, you should be able to

- Explain the overall structure of a C language program

- Use the appropriate operators to perform desired operations in the C language

- Understand the basic data types and expressions of the C language

- Write program loops in the C language

- Write functions and make function calls in the C language

- Use arrays and pointers for data manipulation

- Perform basic I/O operations in the C language

- Use the EGNU IDE to enter, compile, and debug C programs

5.2 Introduction to C

This chapter is not intended to provide complete coverage of the C language. Instead, it provides only a summary of those C language constructs that will be used in this book. You will be able to deal with the basic HCS12 interfacing programming if you fully understand the contents of this chapter.

The C language is gradually replacing the assembly language in many embedded applications because it has several advantages over the assembly language. The most important one is that it allows the user to work on program logic at a level higher than assembly language, and thus programming productivity is greatly improved.

A C program, whatever its size, consists of functions and variables. A function contains statements that specify the operations to be performed. The types of statements in a function could be a *declaration, assignment, function call, control,* or *null*. A variable stores a value to be used during the computation. The *main( )* function is required in every C program and is the one to which control is passed when the program is executed. A simple C program is as follows:

```
(1)   #include <stdio.h>                          —include information about standard library
(2)   /* this is where program execution begins */
(3)   int main (void)                             —defines a function named main that receives
                                                  —no argument values
(4)   {                                           —statements of main are enclosed in braces
(5)       int a, b, c;                            —defines three variables of type int
(6)       a = 3;                                  —assigns 3 to variable a
(7)       b = 5;                                  —assigns 5 to variable b
(8)       c = a + b;                              —adds a and b together and assigns it to c
(9)       printf("a + b = %d \n", c);            —calls library function printf to print the result
(10)      return 0;                               —returns 0 to the caller of main
(11)  }                                           —the end of main function
```

The first line of the program:

```
#include <stdio.h>
```

causes the file stdio.h to be included in the program. This line appears at the beginning of many C programs. The header file stdio.h contains the prototype declarations of all I/O routines that can be called by the user program and constant declarations that can be used by the user program. The C language requires a function prototype be declared before that function can be called if a function is not defined when it is called. The inclusion of the stdio.h file allows the function printf() be invoked in the program.

The second line is a comment. A comment explains what will be performed and will be ignored by the compiler. A comment in C language starts with /* and ends with */. Everything in between is ignored. Comments provide documentation to the program and enhance readability. Comments affect only the size of the text file and do not increase the size of the executable code. Many commercial C compilers also allow one to use two slashes (//) for commenting out a single line.

The third line main() is where program execution begins. The opening brace on the fourth line marks the start of the main() function's code. Every C program must have one, and only one, main() function. Program execution is also ended with the main function. The fifth line declares three integer variables *a*, *b*, and *c*. In C, all variables must be declared before they can be used.

The sixth line assigns 3 to the variable *a*. The seventh line assigns 5 to the variable *b*. The eighth line computes the sum of variables *a* and *b* and assigns it to the variable *c*. You will see that assignment statements are major components in C programs.

The ninth line calls the library function printf() to print the string **a** + **b** = followed by the value of *c* and move the cursor to the beginning of the next line. The tenth line returns a 0 to the caller of main(). The closing brace in the eleventh line ends the main() function.

5.3 Types, Operators, and Expressions

Variables and constants are the basic objects manipulated in a program. Variables must be declared before they can be used. A variable declaration must include the name and type of the variable and may optionally provide its initial value. A variable name may start with a letter (**A** through **Z** or **a** through **z**) or underscore character followed by zero or more letters, digits, or underscore characters. Variable names cannot contain arithmetic signs, dots, apostrophes, C keywords or special symbols such as @, #, ?, and so on. Adding the underscore character "_" may sometimes improve the readability of long variables. Don't begin variable names with an underscore, however, since library routines often use such names. Upper- and lowercase letters are distinct.

5.3.1 Data Types

There are only a few basic data types in C: void, char, int, float, and double. A variable of type *void* represents nothing. The type *void* is used most commonly with functions. A variable of type *char* can hold a single byte of data. A variable of type *int* is an integer, which is normally the natural size (word length) for a particular machine. The type *float* refers to a 32-bit, single-precision, floating-point number. The type *double* represents a 64-bit, double-precision, floating-point number. In addition, there are a number of qualifiers that can be applied to these basic types. *Short* and *long* apply to integers. These two qualifiers will modify the lengths of integers. An integer variable is 16-bit by default for many C compilers including the ICC12 and GNU C compiler. The modifier *short* does not change the length of an integer. The modifier *long* doubles a 16-bit integer to 32-bit.

5.3.2 Variable Declarations

All variables must be declared before their use. A declaration specifies a type and contains a list of one or more variables of that type, as in

```
int    i, j, k;
char   cx, cy;
```

A variable may also be initialized when it is declared, as in

```
int    i = 0;
char   echo = 'y';          /* the ASCII code of letter y is assigned to variable echo. */
```

5.3.3 Constants

There are four kinds of constants: characters, integers, floating-point numbers, and strings. A character constant is an integer, written as one character within single quotes, such as 'x'. A character constant is represented by the ASCII code of the character. A string constant is a sequence of zero or more characters surrounded by double quotes, as in

```
"HCS12DP256 is a microcontroller made by Freescale"
```

or

```
""        /* an empty string */
```

Each individual character in the string is represented by its ASCII code.

An integer constant such as 3241 is an *int*. A long constant is written with a terminal l (el) or L, as in 44332211L. The following constant characters are predefined in C language:

\a	alert (bell) character	\\	backslash
\b	backspace	\?	question mark
\f	formfeed	\'	single quote
\n	newline	\"	double quote
\r	carriage return	\ooo	octal number
\t	horizontal tab	\xhh	hexadecimal number
\v	vertical tab		

As in assembly language, a number in C can be specified in different bases. The method to specify the base of a number is to add a prefix to the number. The prefixes for different bases are

Base	Prefix	Example	
decimal	none	1357	
octal	0	04723	; preceded by a zero
hexadecimal	0x	0x2A	

5.3.4 Arithmetic Operators

There are seven arithmetic operators:

+	add and unary plus
−	subtract and unary minus
*	multiply
/	divide
%	modulus (or remainder)
++	increment
−−	decrement

The expression

a % b

produces the remainder when a is divided by b. The % operator cannot be applied to float or double. The ++ operator adds 1 to the operand, and the −−operator subtracts 1 from the operand. The / operator performs a division and truncates the quotient to an integer when both operands are integers.

5.3.5 Bitwise Operators

C provides six operators for bit manipulations; these may be applied only to integral operands, that is, *char, short, int,* and *long,* whether is *signed* or *unsigned.*

&	AND
|	OR
^	XOR
~	NOT
>>	right shift
<<	left shift

The & operator is often used to clear one or more bits to zero. For example, the statement:

```
PORTC    = PORTC & 0xBD;      /* PORTC is 8 bits */
```

clears bits 6 and 1 of PORTC to 0. The | operator is often used to set one or more bits to 1. For example, the statement:

```
PORTB    = PORTB | 0x40;      /* PORTB is 8 bits */
```

sets the bit 6 of PORTB to 1. The XOR operator can be used to toggle a bit. For example, the statement:

```
abc    = abc ^ 0xF0;        /* abc is of type char */
```

toggles the upper four bits of the variable *abc*. The >> operator shifts the involved operand to the right for the specified number of places. For example,

```
xyz = abc >> 3;
```

shifts the variable *abc* to the right three places and assigns it to the variable *xyz*. The << operator shifts the involved operand to the left for the specified number of places. For example,

```
xyz = xyz << 4;
```

shifts the variable *xyz* to the left four places. The assignment operator = is often combined with the operator. For example,

```
PTP = PTP & 0xBD;
```

can be rewritten as

```
PTP &= 0xBD;
```

The statement

```
PORTB = PORTB | 0x40;
```

can be rewritten as

```
PORTB |= 0x40;
```

5.3.6 Relational and Logical Operators

Relational operators are used in expressions to compare the values of two operands. If the result of the comparison is true, then the value of the expression is 1. Otherwise, the value of the expression is 0. Here are the relational and logical operators:

==	equal to (two "=" characters)
!=	not equal to
>	greater than
>=	greater than or equal to
<	less than
<=	less than or equal to
&&	and
‖	or
!	not (one's complement)

Here are some examples of relational and logical operators:

```
if (!(ADCTL & 0x80))
        statement₁;          /* if bit 7 is 0, then execute statement₁ */
if (i > 0 && i < 10)
        statement₂;          /* if 0 < i < 10 then execute statement₂ */
if (a1 == a2)
        statement₃;          /* if a1 equals a2 then execute statement₃ */
```

5.3.7 Precedence of Operators

Precedence refers to the order in which operators are processed. The C language maintains a precedence for all operators, shown in Table 5.1. Operators at the same level are evaluated from left to right. A few examples that illustrate the precedence of operators are listed in Table 5.2.

Precedence	Operator	Associativity
Highest	( ) [ ] → .	left to right
	! ~ ++ -- + - * & (type) sizeof	right to left
	* / %	left to right
	+ −	left to right
	<< >>	left to right
	< <= > >=	left to right
	== !=	left to right
	&	left to right
	^	left to right
	|	left to right
	&&	left to right
	||	left to right
	?:	right to left
	= += −= *= /= %= &= ^= |= <<= >>=	right to left
Lowest	,	left to right

Table 5.1 ■ Table of precedence of operators

Expression	Result	Note
15 − 2 * 7	1	* has higher precedence than +
(13 − 4) * 5	45	
(0x20 | 0x01) != 0x01	1	
0x20 | 0x01 != 0x01	0x20	!= has higher precedence than |
1 << 3 + 1	16	+ has higher precedence than <<
(1 << 3) + 1	9	

Table 5.2 ■ Examples of operator precendence

5.4 Control Flow

The control-flow statements specify the order in which computations are performed. In the C language, the semicolon is a statement terminator. Braces { and } are used to group declarations and statements together into a *compound statement,* or *block,* so that they are syntactically equivalent to a single statement.

5.4.1 If Statement

The *if* statement is a conditional statement. The statement associated with the *if* statement is executed based upon the outcome of a condition. If the condition evaluates to nonzero, the statement is executed. Otherwise, it is skipped. The syntax of the *if* statement is as follows:

```
if (expression)
      statement;
```

Here is an example of an *if* statement:

```
if (a > b)
      sum += 2;
```

The value of sum will be incremented by 2 if the variable *a* is greater than the variable *b.*

5.4.2 If-Else Statement

The *if-else statement* handles conditions where a program requires one statement to be executed if a condition is nonzero and a different statement if the condition is zero. The syntax of an *if-else* statement is

```
if (expression)
      statement₁
else
      statement₂
```

The *expression* is evaluated; if it is true (nonzero), $statement_1$ is executed. If it is false, $statement_2$ is executed. Here is an example of the *if-else* statement:

```
if (a != 0)
        r = b;
else
        r = c;
```

The *if-else* statement can be replaced by the **?:** operator. The statement

```
r = (a != 0)? b : c;
```

is equivalent to the previous *if* statement.

5.4.3 Multiway Conditional Statement

A multiway decision can be expressed as a cascaded series of *if-else* statements. Such a series looks like this:

```
if (expression₁)
      statement₁
else if (expression₂)
      statement₂
```

```
else if (expression3)
    statement3
...
else
    statementn
```

Here is an example of a three-way decision:

```
if (abc > 0) return 5;
else if (abc == 0) return 0;
else return -5;
```

5.4.4 Switch Statement

The *switch* statement is a multiway decision based on the value of a control expression. The syntax of the switch statement is

```
switch (expression) {
    case const_expr1:
        statement1;
        break;
    case const_expr2:
        statement2;
        break;
    ...
    default:
        statementn;
}
```

As an example, consider the following program fragment:

```
switch (i) {
    case 1: printf("*");
        break;
    case 2: printf("**");
        break;
    case 3: printf("***");
        break;
    case 4: printf("****");
        break;
    case 5: printf("*****");
    default:
        printf("\n");
}
```

The number of * characters printed is equal to the value of **i**. The break keyword forces the program flow to drop out of the switch statement so that only the statements under the corresponding *case-label* are executed. If any break statement is missing, then all the statements from that case-label until the next break statement within the same switch statement will be executed.

5.4.5 For-Loop Statement

The syntax of a *for-loop statement* is

```
for (expr1; expr2; expr3)
      statement;
```

where *expr1* and *expr3* are assignments or function calls, and *expr2* is a relational expression. For example, the following *for loop* computes the sum of the squares of integers from 1 to 9:

```
sum = 0;
for (i = 1; i < 10; i++)
      sum = sum + i * i;
```

The following *for loop* prints out the first 10 odd integers:

```
for (i = 1; i < 20; i++)
      if (i % 2) printf("%d ", i);
```

5.4.6 While Statement

While an expression is nonzero, the *while* loop repeats a statement or block of code. The value of the expression is checked prior to each execution of the statement. The syntax of a *while* statement is

```
while (expression)
      statement;
```

The *expression* is evaluated. If it is nonzero (true), *statement* is executed and *expression* is reevaluated. This cycle continues until *expression* becomes zero (false), at which point execution resumes after *statement*. The *statement* may be a NULL statement. A NULL statement does nothing and is represented by a semicolon. Consider the following program fragment:

```
intCnt = 5;
while (intCnt);
```

The CPU will do nothing before the variable *intCnt* is decremented to zero. In microprocessor applications, the decrement of *intCnt* is often triggered by external events such as interrupts.

5.4.7 Do-While Statement

The *while* and *for* loops test the termination condition at the beginning. By contrast, the *do-while* statement tests the termination condition at the end of the statement; the body of the statement is executed at least once. The syntax of the statement is

```
do
      statement
while (expression);
```

The following *do-while* statement displays the integers 9 down to 1:

```
int digit = 9;
do
      printf("%d ", digit--);
while (digit >= 1);
```

5.4.8 Goto Statement

Execution of a *goto* statement causes control to be transferred directly to the labeled statement, which must be located in the same function as the *goto* statement. The use of the *goto* statement interrupts the normal sequential flow of a program and thus makes it harder to follow and decipher. For this reason, the use of *goto*s is not considered good programming style, so it is recommended that they not be used.

The syntax of the *goto* statement is

goto   *label*

An example of the use of a *goto* statement is as follows:

```
if (x > 100)
        goto        fatal_error;
. . .
fatal_error:
    printf("Variable x is out of bound!\n");
```

5.5 Input and Output

Input and output facilities are not part of the C language itself. However, input and output are fairly important in applications. The ANSI standard defines a set of library functions that must be included so that they can exist in a compatible form on any system where C exists. Some of the functions deal with file input and output. Others deal with text input and output. In this section we will look at the following four input and output functions:

1. **int *getchar* ( )**. This function returns a character when it is called. The following program fragment returns a character and assigns it to the variable *xch*:

```
char xch;
xch = getchar ();
```

2. **int *putchar* (int)**. This function outputs a character on the standard output device. The following statement outputs the letter *a* from the standard output device:

```
putchar ('a');
```

3. **int *puts* (const char *s)**. This function outputs the string pointed to by **s** on the standard output device. The following statement outputs the string "Learning microcontroller is fun!" from the standard output device:

```
puts ("Learning microcontroller is fun! \n");
```

4. **int *printf* (*formatting string*, arg$_1$, arg$_2$, . . . , arg$_n$)**. This function converts, formats, and prints its arguments on the standard output under control of *formatting string*. arg$_1$, arg$_2$, . . . , arg$_n$ are arguments that represent the individual output data items. The arguments can be written as constants, single variable or array names, or more complex expressions. The formatting string is composed of individual groups of characters, with one character group associated with each output data item. The character group corresponding to a data item must start with %. In its simplest form, an individual character group will consist of the percent sign followed by a *conversion character* indicating the type of the corresponding data item.

Multiple character groups can be contiguous or separated by other characters, including white-space characters. These "other" characters are simply transferred directly to the output device where they are displayed. A subset of the more frequently used conversion characters is listed in Table 5.3. Between the % character and the conversion character there may be, in order,

- A minus sign, which specifies *left adjustment* of the converted argument.
- A number that specifies the minimum *field width*. The converted argument will be printed in a field at least this wide. If necessary, it will be padded on the left (or right, if left adjustment is called for) to make up the field width.
- A period, which separates the field width from precision.
- A number, the precision, that specifies the maximum number of characters to be printed from a string, or the number of digits after the decimal point of a floating-point value, or the minimum number of digits for an integer.
- An *h* if the integer is to be printed as a *short,* or *l* (letter el) if as a *long.*

Several valid printf calls are as follows:

```
printf ("this is a challenging course! \n");      /* outputs only a string */
printf ("%d %d %d", x1, x2, x3);                  /* outputs variables x1, x2, x3 using a minimal number of digits
                                                     with one space separating each value */

printf("Today's temperature is %4.1d \n", temp);  /* display the string Today's temperature is followed by the
                                                     value of temp. Display one fractional digit and use at least four
                                                     digits for the value. */
```

Conversion character	Meaning
c	data item is displayed as a single character
d	data item is displayed as a signed decimal number
e	data item is displayed as a floating-point value with an exponent
f	data item is displayed as a floating-point value without an exponent
g	data item is displayed as a floating-point value using either e-type or f-type conversion, depending on value; trailing zeros, trailing decimal point will not be displayed
i	data item is displayed as a signed decimal integer
o	data item is displayed as an octal integer, without a leading zero
s	data item is displayed as a string
u	data item is displayed as an unsigned decimal integer
x	data item is displayed as a hexadecimal integer, without the leading 0x

Table 5.3 ■ Commonly used conversions characters for data output

5.6 Functions and Program Structure

Every C program consists of one or more functions. If a program consists of multiple functions, their definitions cannot be embedded within another. The same function can be called from several different places within a program. Generally, a function will process information passed to it from the calling portion of the program and return a single value. Information is passed to the function via special identifiers called *arguments* (also called *parameters*) and returned via the *return* statement. Some functions, however, accept information but do not return anything (for example, the library function *printf*).

The syntax of a function definition is as follows:

```
return_type function_name (declarations of arguments)
{
        declarations and statements
}
```

The declaration of an argument in the function definition consists of two parts: the *type* and the *name* of the variable. The return type of a function is *void* if it does not return any value to the caller. An example of a function that converts a lowercase letter to an uppercase letter is as follows:

```
char lower2upper (char cx)
{
    if (cx >= 'a' && cx <= 'z') return (cx − ('a' - 'A'));
    else return cx;
}
```

A character is represented by its ASCII code. A letter is in lowercase if its ASCII code is between 97 (0x61) and 122 (0x7A). To convert a letter from lowercase to uppercase, subtract its ASCII code by the difference of the ASCII codes of letters *a* and *A*.

To call a function, simply put down the name of the function and replace the argument declarations by actual arguments or values and terminate it with a semicolon.

Example 5.1

▼

Write a function to test whether an integer is a prime number.

Solution: The integer 1 is not a prime number. A number is prime if it is indivisible by any integer between 2 and its half.

```
/* this function returns a 1 if k is prime. Otherwise, it returns a 0. */
char test_prime (int k)
{
    int i;
    if (k == 1) return 0;
    for (i = 2; i < k/2; i++)
            if ((k % i) == 0) return 0;
    return 1;
}
```

▲

Example 5.2

▼

Write a program to find out the number of prime numbers between 100 and 1000.

Solution: One can find the number of prime numbers between 100 and 1000 by calling the function in Example 5.1 to find out if a number is prime.

```
#include <stdio.h>
char test_prime (int k);        /* prototype declaration for the function test_prime */

void main (void)
```

```
{
    int i, prime_count;
    prime_count = 0;
    for (i = 100; i <= 1000; i++) {
        if (test_prime(i))
            prime_count ++;
    }
    printf("\n The total prime numbers between 100 and 1000 is %d\n", prime_count);
}
char test_prime (int k)
{
    int i;
    if (k == 1) return 0;
    for (i = 2; i < k/2; i++)
        if ((k % i) == 0) return 0;
    return 1;
}
```

A function cannot be called before it has been defined. This dilemma is solved by using the function prototype statement. The syntax for a function prototype statement is as follows:

```
return_type function_name (declarations of arguments);
```

The statement

```
char test_prime (int a);
```

before *main( )* is a function prototype statement.

5.7 Pointers, Arrays, Structures, and Unions

5.7.1 Pointers and Addresses

A *pointer* is a variable that holds the address of a variable. Pointers are used frequently in C, as they have a number of useful applications. For example, pointers can be used to pass information back and forth between a function and its reference (calling) point. In particular, pointers provide a way to return multiple data items from a function via function arguments. Pointers also permit references to other functions to be specified as arguments to a given function. This has the effect of passing functions as arguments to the given function.

Pointers are also closely associated with arrays and therefore provide an alternative way to access individual array elements. The syntax for declaring a pointer type is

```
type_name *pointer_name;
```

For example,

```
int *ax;
```

declares that the variable *ax* is a pointer to an integer.

```
char *cp;
```

declares that the variable *cp* is a pointer to a character.

To access the value pointed to by a pointer, use the *dereferencing* operator *. For example,

```
int    a, *b;    /* b is a pointer to int */
. . .
a = *b;
```

assigns the value pointed to by *b* to variable *a*.

We can assign the address of a variable to a pointer by using the unary operator &. The following example shows how to declare a pointer and how to use & and *:

```
int x, y;
int *ip;       /* ip is a pointer to an integer */

ip = &x;       /* assigns the address of the variable x to ip */
y = *ip;       /* y gets the value of x */
```

5.7.2 Arrays

Many applications require the processing of multiple data items that have common characteristics (e.g., a set of numerical data, represented by $x_1, x_2, \ldots, x_n$). In such situations it is more convenient to place data items into an *array*, where they will all share the same name. The individual data items can be characters, integers, floating-point numbers, and so on. They must all be of the same type and the same storage class.

Each array element is referred to by specifying the array name followed by one or more *subscripts*, with each subscript enclosed in brackets. Each subscript must be expressed as a nonnegative integer. Thus, the elements of an *n*-element array x are x[0], x[1], . . . , x[n − 1]. The number of subscripts determines the dimensionality of the array. For example, x[i] refers to an element of a one-dimensional array. Similarly, y[i][j] refers to an element of a two-dimensional array. Higher-dimensional arrays can be formed by adding additional subscripts in the same manner. However, higher-dimensional arrays are not used very often in 8- and 16-bit microcontroller applications. In general, a one-dimensional array can be expressed as

```
data-type array_name[expression];
```

A two-dimensional array is defined as

```
data-type array_name[expr1][expr2];
```

An array can be initialized when it is defined. This is a technique used in table lookup, which can speed up the computation process. For example, a data acquisition system that utilizes an analog-to-digital converter can use the table lookup technique to speed up the conversion process (from the digital value back to the original physical quantity).

Example 5.3

▼

Write the bubble sort function to sort an array of integers.

Solution: The algorithm for bubble sort was described in Chapter 4. Here is the C language version:

```
void    swap (int *, int *);
void    bubble (int a[], int n)        /* n is the array count */
{
        int i, j;
        for (i = 0; i < n − 1; i++)
            for (j = 0; j > n - i - 2; j++)
```

```
                    if (a[j] > a[j+1])
                            swap (&a[j], &a[j+1]);
    }
    void swap (int *px, int *py)
    {
        int temp;
        temp = *px;
        *px = *py;
        *py = temp;
    }
```

5.7.3 Pointers and Arrays

In C, there is a strong relationship between pointers and arrays. Any operation that can be achieved by array subscripting can also be done with pointers. The pointer version will in general be faster but somewhat harder to understand. For example,

```
    int ax[20];
```

defines an array *ax* of 20 integral numbers. The notation ax[i] refers to the *i*th element of the array. If *ip* is a pointer to an integer, declared as

```
    int *ip;
```

then the assignment

```
    ip = &ax[0];
```

makes ip contain the address of ax[0]. Now the statement

```
    x = *ip;
```

will copy the contents of ax[0] into x. If *ip* points to ax[0], then *ip+1* points to ax[1], and *ip+i* points to ax[i], etc.

5.7.4 Passing Arrays to a Function

An array name can be used as an argument to a function, thus permitting the entire array to be passed to the function. To pass an array to a function, the array name must appear by itself, without brackets or subscripts, as an actual argument within the function call. When declaring a one-dimensional array as a formal argument, the array name is written with a pair of empty square brackets. The size of the array is not specified within the formal argument declaration. If the array is two-dimensional, then there should be two pairs of empty brackets following the array name.

The following program outline illustrates the passing of an array from the main portion of the program to a function.

```
    int average (int n, int arr[]);
    void main ( )
    {
        int n, avg;              /* variable declaration */
        int arr[50];             /* array definition */
        . . .
        avg = average(n, arr);   /* function call */
        . . .
    }
```

```
int average (int k, int brr[])   /* function definition */
{
    . . .
}
```

Within *main* we see a call to the function *average*. This function call contains two actual arguments—the integer variable *n* and the one-dimensional integer array *arr*. Note that *arr* appears as an ordinary variable within the function call. In the first line of the function definition, we see two formal arguments, *k* and *brr*. The formal argument declarations establish *k* as an integer variable and *brr* as a one-dimensional integer array. Note that the size of *brr* is not defined in the function definition. As formal parameters in a function definition,

```
int brr[];
```

and

```
int *brr;
```

are equivalent.

5.7.5 Initializing Arrays

C allows initialization of arrays. Standard data-type arrays may be initialized in a straightforward manner. The syntax for initializing an array is as follows:

```
array_declarator = { value-list }
```

The following statement shows a five-element integer array initialization:

```
int i[5] = {10, 20, 30, 40, 50};
```

The element i[0] has the value of 10 and the element i[4] has the value of 50.

A string (character array) can be initialized in two ways. One method is to make a list of each individual character:

```
char strgx[5] = {'w', 'x', 'y', 'z', 0};
```

The second method is to use a string constant:

```
char myname [6] = "huang";
```

A null character is automatically appended at the end of "huang". When initializing an entire array, the array size (which is one more than the actual length) must be included:

```
char prompt [24] = "Please enter an integer:";
```

5.7.6 Structures

A structure is a group of related variables that can be accessed through a common name. Each item within a structure has its own data type, which can be different from the other data items. The syntax of a structure declaration is as follows:

```
struct struct_name {      /* struct_name is optional */
    type1   member1;
    type2   member2;
    . . .
};
```

The *struct_name* is optional and if it exists, defines a *structure tag*. A *struct* declaration defines a type. The right brace that terminates the list of members may be followed by a list of variables, just as for any basic type. The following example is for a card catalog in a library:

```
struct catalog_tag {
        char   author [40];
        char   title [40];
        char   pub [40];
        unsigned int date;
        unsigned char rev;
} card;
```

where the variable *card* is of type *catalog_tag*.

A structure definition that is not followed by a list of variables reserves no storage; it merely describes a template or the shape of a structure. If the declaration is tagged (i.e., has a name), however, the tag can be used later in definitions of instances of the structure. For example, suppose we have the following structure declaration:

```
struct point {
    int x;
    int y;
};
```

We can then define a variable *pt* of type *point* as follows:

```
struct point pt;
```

A member of a particular structure is referred to in an expression by a construction of the form:

```
structure-name.member
```

or:

```
structure-pointer→member
```

The structure member operator "." connects the structure name and the member name. As an example, the square of the distance of a point to the origin can be computed as follows:

```
long integer sq_distance;
. . .
sq_distance = pt.x * pt.x + pt.y * pt.y;
```

Structures can be nested. One representation of a circle consists of the center and radius, as shown in Figure 5.1. This circle can be defined as followed:

```
struct circle {
        struct     point center;
        unsigned   int   radius;
};
```

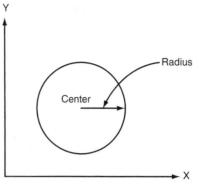

Figure 5.1 ■ A circle

5.7.7 Unions

A *union* is a variable that may hold (at different times) objects of different types and sizes, with the compiler keeping track of size and alignment requirements. Unions provide a way to manipulate different kinds of data in a single area of storage, without embedding any machine-dependent information in the program. The syntax of the union is as follows:

```
union union_name {
        type-name1          element1;
        type-name2          element2;
        . . .
        type-namen          elementn;
};
```

The field *union_name* is optional. When it exists, it is also called *union-tag*. One can declare a union variable at the same time one declares a union type. The union variable name should be placed after the right brace "}". In order to represent the current temperature using both the integer and string, one can use the following declaration:

```
union u_tag {
        int i;
        char c[4];
} temp;
```

Four characters must be allocated to accommodate the larger of the two types. Integer type is good for internal computation, whereas string type is suitable for output. Of course, some conversion may be needed before making a certain kind of interpretation. Using this method, the variable *temp* can be interpreted as an integer or a string, depending on the purpose. Syntactically, members of a union are accessed as

```
union-name.member
```

or

```
union-pointer → member
```

just as for structures.

5.8 Miscellaneous Items

5.8.1 Automatic/External/Static/Volatile

A variable defined inside a function is an *internal variable* of that function. These variables are called *automatic* because they come into existence when the function is entered and disappear when it is left. Internal variables are equivalent to local variables in assembly language. *External variables* are defined outside of any function and are thus potentially available to many functions. Because external variables are globally accessible, they provide an alternative to function arguments and return values for communicating data between functions. Any function may access an external variable by referring to it by name, if the name has been declared somehow. External variables are also useful when two functions must share some data, yet neither calls the other.

The use of *static* with a local variable declaration inside a block or a function causes the variable to maintain its value between entrances to the block or function. Internal *static* variables are local to a particular function just as automatic variables are, but unlike automatic variables, they remain in existence rather than coming and going each time the function is activated. When a variable is declared *static* outside of all functions, its scope is limited to the file that contains the definition. A function can also be declared as static. When a function is declared as static, it becomes invisible outside of the file that defines the function.

A *volatile* variable has a value that can be changed by something other than user code. A typical example is an input port or a timer register. These variables must be declared as *volatile* so the compiler makes no assumptions on their values while performing optimizations. The keyword *volatile* prevents the compiler from removing apparently redundant references through the pointer.

5.8.2 Scope Rules

The functions and external variables that make up a C program need not all be compiled at the same time; the source text of the program may be kept in several files, and previously compiled routines may be loaded from libraries.

The scope of a name is the part of the program within which the name can be used. For a variable declared at the beginning of a function, the scope is the function in which the name is declared. Local (internal) variables of the same name in different functions are unrelated.

The scope of an external variable or a function lasts from the point at which it is declared to the end of the file being compiled. Consider the following program segment:

```
. . .
void f1 (. . .)
{
        . . .
}
int a, b, c;
void f2 (. . .)
{
        . . .
}
```

Variables a, b, and c are accessible to function f2 but not to f1.

When a C program is split into several files, it is convenient to put all global variables into one file so that they can be accessed by functions in different files. Functions residing in different files that need to access global variables must declare them as external variables. In addition, the prototypes of certain functions can be placed in one file so that they can be called by functions in other files. The following example is a skeletal outline of a two-file C program that makes use of external variables:

In file1:

```
extern int xy;
extern long arr[];
main ( )
{

        . . .

}
void foo (int abc) { . . . }
long soo (void) { . . . }
```

In file2:

```
int xy;
long arr[100];
```

5.8.3 Type Casting

Type casting causes the program to treat a variable of one type as though it contains data of another type. The format for type casting is as follows:

```
(type)variable
```

For example, the following expression converts the variable **kk** to a long integer:

```
int kk;
. . .
(long) kk
```

Type casting can avoid many errors caused by size mismatch among operands. For example, in the following program segment:

```
long result;
int     x1, x2;
. . .
result = x1 * x2;
```

if the product of x1 and x2 is larger than $2^{16} - 1$, it will be truncated to 16 bits. Then the variable **result** will receive an incorrect value. To fix the error, use type casting to force x1 and x2 to long integer, as follows:

```
result = ((long) x1) * ((long) x2);
```

This technique is used in several examples in this text.

Another example of the use of type casting is in pointer type. Sometimes one needs to treat the contents of a structure-type variable as a string. The most convenient way to do it is to recast the pointer to a structure-type variable into a pointer to a string (character type). For the declarations

```
struct personal {
    char name    [10];
    char addr [20];
```

```
        char sub1[5];
        char sub2[5];
        char sub3[5];
        char sub4[5];
    }   ptr1;
    char *cp;
```

one can use the following statement to treat the variable ptr1 as a string:

```
        cp = (char *) &ptr1;
```

5.9 Using the C Compiler

There are several C compilers that support the HCS12 microcontroller. The GNU C compiler along with the EmbeddedGNU IDE by Eric Engler (will be called EGNU IDE from now on) will be used to enter, edit, and compile the C programs in this book. The GNU C compiler is a freeware C compiler. The EGNU IDE contains a text editor, a project manager, and a terminal program. The EGNU IDE invokes the GNU C compiler to compile the user C program. Tutorials for using the GNU C compiler and the EGNU IDE will be provided in the following sections.

When accessing peripheral registers, symbolic names are preferred to addresses. All commercial C compilers have header files that provide the symbolic definitions for all peripheral registers so that users can use symbolic names to access the peripheral registers. These header files contain the association of register names and their addresses. The header file hcs12.h that comes with the EGNU (also included in the complementary CD) serves the same purpose for the HCS12DP256. This file should be copied to the directory c:\egnu091\include. One should add the following statement in the C program if one uses the EGNU IDE to develop application programs:

```
#include "c:\egnu091\include\hcs12.h"
```

In addition to the association of register names and addresses, the header file hcs12.h also contains the bit values for many registers. Each bit is associated with a value that is equal to its position weight. For example, the 7th bit of the ATD0CTL2 register is defined as follows:

```
#define     ADPU     0x80
```

With this definition, one can use the following statement to set the ADPU bit of the ATD0CTL2 register:

```
        ATD0CTL2 |= ADPU;      /* set the ADPU bit */
```

To clear the ADPU bit, use the following statement:

```
        ATD0CTL2 &= ~ADPU;  /* clear the ADPU bit */
```

All register names and all bit names are in uppercase. Users can change the case to satisfy their preference. Header files for other HCS12 members can be derived by editing the hcs12.h file.

5.10 Using the GNU C Compiler and EGNU IDE

The GNU-m68hc11 C compiler is a freeware C compiler for the Freescale 68HC11, 68HC12, and HCS12 microcontrollers. It can be downloaded from the website at http://www.gnu-m68hc11.org. The EGNU IDE is a freeware integrated development environment for the GNU C (GCC) compiler to support the Freescale 68HC11, 68HC12, and HCS12 microcontrollers running on Windows platforms. The EGNU IDE, developed by Eric Engler, can be downloaded from the website at http://www.geocities.com/englere_geo. Both need to be downloaded and installed. Assume that GCC is installed at the directory **c:\gnu**, whereas the

EGNU IDE is installed at the directory **c:\egnu091**. To facilitate the use of the EGNU IDE, one should create a shortcut (or icon) for it and place it on the Windows screen.

5.10.1 The GNU C Compiler

The GCC compiler supports three versions of the C standard. The corrected version of the 1990 ISO C standard is fully supported. The support for the latest 1999 C standard is incomplete. GCC also has limited support for the traditional (pre-ISO) C language. However, the use of the pre-ISO C language is not encouraged.

GCC does not provide library support for the HCS12 at the moment. Users will need to deal with the peripheral registers directly to perform the desired function. The EGNU IDE invokes GCC to compile the user programs.

5.10.2 The Procedure for Using the EGNU IDE

The procedure for using the EGNU IDE to enter, compile, and download a C program into an HCS12 demo board with the D-Bug12 monitor for execution is as follows:

Step 1

Invoke the EGNU IDE by double-clicking its icon. The IDE window will appear as shown in Figure 5.2. In Figure 5.2, the block in gray color is used as the *workspace* for program entering and editing. The *project manager* block displays the name of the current active project and the files contained in the project. The pane at the bottom of the window is the *status window*. By default, it is used to display the output of the project build process (referred to as **Make log** in Figure 5.2). The status pane can be used to display compiler output by clicking the keyword "Compiler" above the status pane. The result is shown in Figure 5.3.

The status pane can also be used as the *terminal window* that displays the communications between the PC and the monitor on the demo board. The screen would change to that in Figure 5.4 if the user clicks on the keyword "Terminal" above the status pane and then presses the *Enter* key once or twice when the demo board is powered on.

Figure 5.4 shows the power-on status for a demo board with the D-Bug12 monitor. When the status pane is used as the terminal window, one can enter any D-Bug12 monitor command.

Step 2

Click on the **File** menu and select **New Source File** to create a new file for holding a program. The resultant screen is shown in Figure 5.5. This step must be repeated as many times as the number of files required in the project. As shown in Figure 5.5, one can also press the *Ctrl* and the '*U*' keys to open a new file. To open an existing file, one should select **Reopen File ->** instead of **New Source File**.

Step 3

Type in the C program and save it in an appropriate directory (name it **hcs12_timer.c**). An example is shown in Figure 5.6. The example program performs the following functions:

1. Defines the cnt variable and initializes it to 0.

2. Configures port B for output and outputs the value of cnt to it.

3. Calls the delayby100ms() function to wait for 0.5 seconds and then increments the variable cnt. The argument to this function specifies the number of 100-ms time delays to be created.

4. Output the value of cnt to port B and go to 3.

   Port B is driving eight light-emitting diodes (LEDs). When this program is running, one can see the change of the LED pattern.

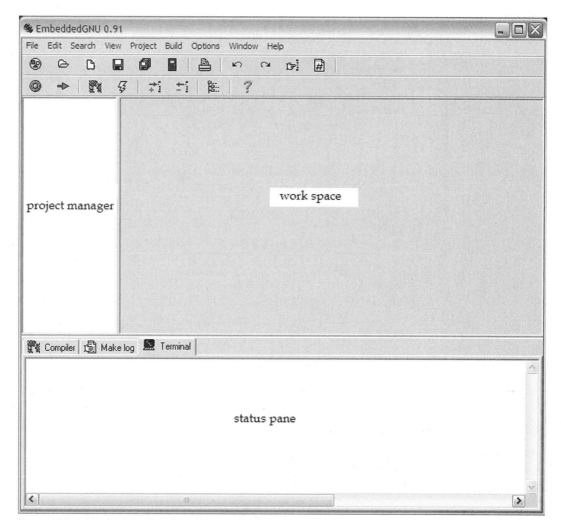

Figure 5.2 ■ Embedded GNU startup window

Step 4

Create a new project by pressing the **File** menu, moving the mouse over **New Project. . .**
and clicking on it. The resultant screen is shown in Figure 5.7. The user will be asked to
enter the project name. In this tutorial, the name "count" is entered. After the project
name is entered, one should also click on OK. As shown in Figure 5.8, the screen will
change to ask the user to select the directory to save the project information. One should
select an appropriate directory and click on the **save** button on the screen. After this,
another popup dialog appears on the screen (shown in Figure 5.9) that will allow the user
to select the Hardware Profile and turn on/off compiler switches. The default compiler
switches are acceptable. This tutorial selects the Dragon12 demo board as its hardware
profile. This tutorial does not use GNU Embedded Library. Therefore, the user should
select No for this choice (in the middle of the dialog box). After the project options are set
properly, click on OK. The screen is changed to that in Figure 5.10.

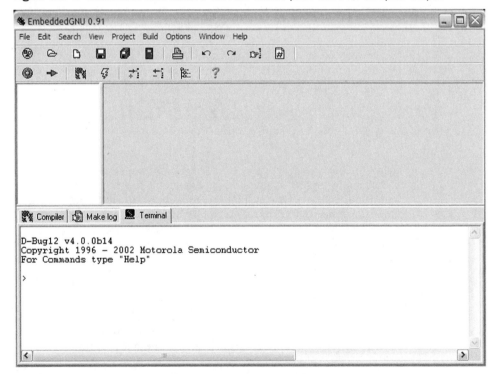

Figure 5.3 ■ Embedded GNU window with the status pane used for compiler output

Figure 5.4 ■ Using the status pane as the terminal window

Figure 5.5 ■ Creating a new source file for entering a program

Figure 5.6 ■ Entering a new program in EGNU IDE

Figure 5.7 ■ Create a new project and name it "count"

Figure 5.8 ■ Embedded GNU screen after entering project name and clicking "OK"

Figure 5.9 ■ Dialog for project options

Figure 5.10 ■ Screen appearance after the project name has been set

Step 5

Add files to the newly created project. Press the **Project** menu (shown in Figure 5.11) and select "**Add to project. . .**". A popup dialog box (shown in Figure 5.12) appears to allow the user to enter the name of the file to be added to the project. Since the project and the files in the project are often stored in the same directory, one can simply click on the file name (hcs12_timer) and then click on **Open** in Figure 5.12 to complete the add-file operation. The screen changes to that in Figure 5.13 after adding *hcs12_timer.c* to the project.

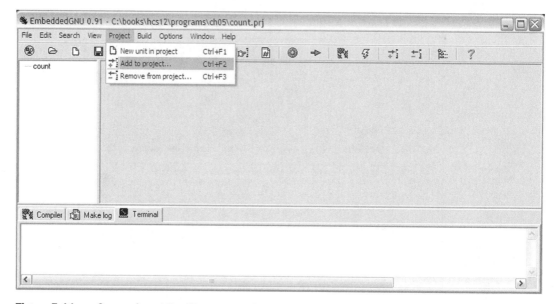

Figure 5.11 ■ Screen for adding files to a project

Figure 5.12 ■ Dialog box for adding a file to the project

```
EmbeddedGNU 0.91 - C:\books\hcs12\programs\ch05\count.prj

File  Edit  Search  View  Project  Build  Options  Window  Help

count
  hcs12_timer.c        hcs12_timer.c

1  #include "c:\egnu091\include\hcs12.h"
2  void delayby100ms(int k);
3  int main (void)
4  {
5      unsigned char cnt;
6      cnt = 0;
7      ddrb = 0xff;
8      portb = cnt;
9      while (1){
10         delayby100ms(5);

Compiler   Make log   Terminal
```

Figure 5.13 ■ Screen after adding files to the project

Step 6

Set up options for the project and environment. Pressing the **Options** menu will reveal that there are two options to be set. The processor options have been set and can be ignored. Select "Environment options" and a popup dialog box will appear, as shown in Figure 5.14, which shows the **Directories** options. The user can enter the search paths for several types of files in this dialog box. If users have created their own library and included files, they may enter them in this dialog box. The setting in Figure 5.14 is acceptable and need not be changed for this tutorial.

After setting the options for directories, one should go on to make sure that the options for **Preferences** are set properly. The dialog box (shown in Figure 5.15) for setting them will appear after the user clicks on the **Preferences** menu. The default directory option is the default place for the EGNU IDE to look for files to be added to the project. Users should change it to match their own setting. The directory "c:\books\hcs12\programs" is where this author stores all the programs in his HCS12 book.

The third option is related to the text editor. Click on **Editor** and the dialog box for setting editor options appears as shown in Figure 5.16. Users can set the available options according to their preference. After the options for the editor have been set properly, click on OK.

The fourth option is related to syntax colors. Click on **Syntax colors** and a dialog box appears as shown in Figure 5.17. EGNU IDE allows the user to set the color of nine different components in a program, including the comment, identifier, keyword, number, space, string, symbol, white space, and directive. The color of any component is indicated by the word "FG" on the color map. The color for comments is shown in Figure 5.17.

The fifth option to be set is related to associations. The default settings are acceptable. The sixth option to be set is related to the COM port. Click on **COM Port** and its associated dialog box will appear as shown in Figure 5.18. The user should select the appropriate COM Port (default is COM 1) and eight data bits, one stop bit, no parity. Environment options need only be set once. EGNU IDE saves the environment options

Figure 5.14 ■ Dialog box for setting environmental options

after they have been set. Therefore, step 6 can be skipped for the next project unless the user wants to change some of them.

Step 7

Build the project. Press the **Build** menu and select **Make**. The EGNU IDE will display the make result as shown in Figure 5.19. This example project has no errors and the resultant machine code has 462 bytes. The machine code is stored in a file with **s19** as its file name extension. Users who are interested in the **s19** file format should refer to the appropriate document from Freescale.

Figure 5.15 ■ Dialog box for setting preferences

Step 8
Download the program onto the demo board for execution. Make the Terminal window active by clicking on Terminal above the status pane of the EGNU IDE window. Type **load** and the **enter** key on the terminal window. Press the **Build** menu and select **Download** (or type function key F10). The S19 file of the current project will be downloaded onto the demo board and the D-Bug12 command prompt will reappear on the screen as follows:

```
>load
. . . . . . . .
done
>
```

Figure 5.16 ■ Setting editor options

Step 9
Execute and debug the program. The user can use all the D-Bug12 commands to debug and execute the downloaded program on the demo board, as illustrated in Chapter 3. The EGNU IDE sets the program starting address to $2000. To run the program that was just downloaded, enter **g 2000**. After this, the example program is running, the user should be able to see the change in the LED patterns. To redisplay the D-Bug12 monitor prompt, press the **reset** button on the demo board and then hit the **enter** key of the keyboard.

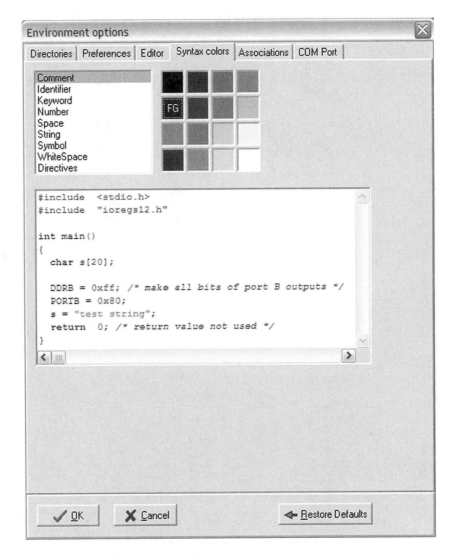

Figure 5.17 ■ Dialog box for setting syntax colors

5.10.3 Inline Assembly Instructions

The GNU C compiler allows the user to add inline assembly instructions in the C program. The syntax for inline assembly is

asm("<string>");

For example, adding the following statement will cause the program control to return to the D-Bug12 monitor when the program is running on a demo board with the D-Bug12 monitor:

asm("swi");

Adding the following statement will enable interrupt globally:

asm("cli");

Figure 5.18 ■ Dialog box for setting COM port options

5.11 C Library Functions for the EGNU IDE

Several groups of functions have been written to create time delays and perform terminal I/O operations and data conversions to facilitate application development using the EGNU IDE. These prototypes and functions are listed in Table 5.4.

The delay functions are contained in delay.c. Data conversion functions are contained in convert.c. Terminal I/O functions are contained in stdio.c. All three files have been placed on the complementary CD for this text. The user needs to include the files that contain these

Figure 5.19 ■ Project make results

Prototype	Function
Category: Time delay	
void delayby10us(int k);	Creates a time delay that is equal to **10 k** us
void delayby50us(int k);	Creates a time delay that is equal to **50 k** us
void delayby1ms(int k);	Creates a time delay that is equal to **k** ms
void delayby10ms(int k);	Creates a time delay that is equal to **10 k** ms
void delayby100ms(int k);	Creates a time delay that is equal to **100 k** ms
void delayby1s(int k);	Creates a time delay that is equal to **k** sec
Category: Terminal I/O	
int putch(char cx);	Outputs the character **cx** to the sci0 port
int puts(const char *cptr);	Outputs the string pointed to by **cptr** to the **sci0** port
int newline (void);	Outputs a carriage return followed by a linefeed
int putsr(char *cptr);	Outputs a string stored in SRAM that is pointed to by **cptr**
char getch (void);	Reads a character from the **sci0** port using polling method
int gets(char *ptr);	Reads a string terminated by a carriage return and saves it in a buffer pointed to by **ptr**
Category: Data conversion	
int int2alpha(int xx, char arr[]);	Converts the 16-bit integer **xx** into a decimal string and stores it in an array **arr[]**
int long2alpha(long xx, char arr[]);	Converts the 32-bit integer **xx** into a decimal string and stores it in an array **arr[]**
int alpha2int(char *ptr);	Converts the ASCII string pointed to by **ptr** into a 16-bit binary number
long alpha2long(char *ptr);	Converts the ASCII string pointed to by **ptr** into a 32-bit binary number
int reverse(char arr[], int k, char sign);	Reverses a string that is stored in the character array **arr[]** with **k** characters and its sign is indicated by **sign**

Table 5.4 ■ Library functions for the EGNU IDE

functions in the program in order to use them. For example, the user needs to add the following statement in order to call the delayby100ms() function:

```
#include "c:\egnu091\include\delay.c"
```

Example 5.4

▼

Write a program to convert a 32-bit integer to a string and then output it to the SCI0 port.

Solution: The program is as follows:

```
#include "c:\egnu091\include\hcs12.h"
#include "c:\egnu091\include\convert.c"
#include "c:\egnu091\include\stdio.c"
const char *msg = "Prepare to convert!\n";
int main(void)
{
    long temp;
    char buf[20];           /* used to hold the converted ASCII string */
    temp = 993450123; /* number to be converted */
    puts(msg);              /* output a message */
    newline();              /* move to the begin of the next line */
    long2alpha(temp,buf);/* convert the 32-bit number to an ASCII string */
    putsr(&buf[0]);         /* output the converted string */
    newline();
    return 0;
}
```

▲

5.12 Tips for C Program Debugging

As in assembly language, C program errors can be divided into two categories:

- Syntax/semantics errors
- Logical errors

Many syntax/semantics errors can be discovered if the programmer follows the rules of C language:

- *Undeclared variables.* A variable must be declared before it can be used. Most errors of this type can be discovered easily.
- *Variable and symbol names are case-sensitive.* Sometimes, the programmer has declared a variable but still has an undeclared variable error with the same variable. This error could occur because one or more characters of the variable name are in the wrong case. For example, in the following program segment

    int   A1, A2, xyClk;

    . . .

    A1 = A1 * xycLk;

    the compiler would output the error message "undefined variable xycLk".

- *Missing parenthesis (or brace, bracket).* A missing parenthesis can cause many other errors. This type of error can be avoided by entering the matching parenthesis whenever the left parenthesis is entered before entering statements between the parentheses.
- *Mismatch of function name in function prototype declaration and function definition.* The EGNU will generate the error message "unknown error—see Make Log". When the user clicks on **Make log** on the Status pane of EGNU, the message "undefined reference to . . ." will appear.

Logical errors are harder to debug than syntax and semantics errors. Whenever the program behavior is not what one expects, the first step is to read the program carefully to find out where the most likely problem spot is. After identifying the potential problem spot, the user can set a breakpoint and examine the program execution result.

If the user has a source-level debugger to debug the program, following actions can be taken to solve the problem:

- Set breakpoints. Breakpoints allow the user to examine program execution results at the suspicious point.
- Set up a watch list. A watch list is used together with the breakpoints. A watch list consists of pairs of program variables of interest and their values at a certain time. It allows the user to find out program execution results quickly at any breakpoint.
- Trace program execution. Users can find out the execution result of a few instructions by tracing instruction execution. Without a source-level debugger, it is very difficult to trace the program execution unless the D-bug12 monitor commands are used.

The freeware EGNU provides very little support for debugging. A commercial source-level debugger for the C language can cost thousands of dollars. Without a source-level debugger, many debugging activities cannot be performed easily.

There are several purposes for setting breakpoints:

- To determine whether a segment of code has ever been executed by the CPU. The programmer can use the embedded assembly instruction *"asm("swi")"* to find out. Using the "swi" instruction allows the programmer to find out where the program execution gets stuck and identify the error.
- To determine whether the execution result is correct up to the breakpoint. Without a source-level debugger, the programmer can output the program execution result to the LCD or terminal monitor to find out if the program executes correctly up to the breakpoint. This also serves as a watch list for the program execution.

The following guidelines can help reduce and identify logic errors:

- Make sure the precedence of operators has been observed.
- Match the size of the source operands and that of the destination variables. Use type casting when appropriate. Type casting has been used in several examples in this text.
- Walk through the program algorithm carefully before converting it into the program code. An incorrect algorithm is often the cause of program bugs.

Use enough data to test the program. The program must be tested with normal inputs, maximum and minimum inputs, and also the illegal inputs to make sure that it operates correctly under all circumstances. When the problem to be solved gets complicated, a structured programming approach should be used to organize the program. The guidelines described in Section 2.4 should be followed to develop the program algorithm and convert the algorithm into program code. Each individual function should be tested thoroughly before the whole program is tested. A comprehensive discussion of structured programming and testing is beyond the scope of this text but can be found in many textbooks on software engineering.

5.13 Summary

A C program consists of one or more functions and variables. The *main ( )* function is required in every C program. It is the entry point of a C program. A function contains statements that specify the operations to be performed. The types of statements in a function could be *declaration, assignment, function call, control,* and *null.*

A *variable* stores a value to be used during the computation. A variable must be declared before it can be used. The declaration of a variable consists of the name and the type of the variable. There are four basic data types in C: int, char, float, and double. Several qualifiers can be added to the variable declarations. They are short, long, signed, and unsigned.

Constants are often needed in forming a statement. There are four types of constants: integers, characters, floating-point numbers, and strings.

There are seven *arithmetic operators:* +, −, ×, /, %, ++, and−−. There are six *bitwise operators:* &, |, ^ , ~, >>, and <<. Bitwise operators can be applied only to integers. *Relational operators* are used in control statements. They are ==, !=, >, >=, <, <=, &&, ||, and !.

The *control-flow statements* specify the order in which computations are performed. Control-flow statements include if-else statement, multiway conditional statement, switch statement, for-loop statement, while statement, and do-while statement.

Every C program consists of one or more functions. If a program consists of multiple functions, their definitions cannot be embedded within another. The same function can be called from several different places within a program. Generally, a function will process information passed to it from the calling portion of the program and return a single value. Information is passed to a function via special identifiers called *arguments* (also called *parameters*) and returned via the *return* statement. Some functions, however, accept information but do not return anything (for example, the library function *printf*).

A *pointer* holds the address of a variable. Pointers can be used to pass information back and forth between a function and its reference (calling) point. In particular, pointers provide a way to return multiple data items from a function via function arguments. Pointers also permit references to other functions to be specified as arguments to a given function. Two operators are related with pointers: * and &. The * operator returns the value of the variable pointed to by the pointer. The & operator returns the address of a variable.

Data items that have common characteristics are placed in an *array*. An array may be one-dimensional or multidimensional. The dimension of an array is specified by the number of square bracket pairs ([]) following the array name. An array name can be used as an argument to a function, thus permitting the entire array to be passed to the function. To pass an array to a function, the array name must appear by itself, without brackets or subscripts. An alternative way to pass arrays to a function is to use pointers.

A variable defined inside a function is an *internal variable* of that function. *External variables* are defined outside of any function and are thus potentially available to many functions. The *scope* of a name is the part of the program within which the name can be used. The scope of an external variable or a function lasts from the point at which it is declared to the end of the file being compiled.

Tutorials for using the EGNU IDE were provided at the end of this chapter. The EGNU IDE is window-driven and is straightforward to use. Programs written for the EGNU IDE in this text can be compiled by other compilers with minor modifications.

Users may also consider commercial C compilers such as the ICC12 from ImageCraft and the CodeWarrior IDE from Metrowerks. The ICC12 has a 4-KB demo version that does not expire. The ICC12 has an integrated text editor, project manager, C compiler, and terminal program. CodeWarrior has a 12-KB special edition that limits the code size to 12 KB. CodeWarrior is a comprehensive IDE that consists of a text editor, cross assembler, C compiler, simulator, and source-level debugger in one package. Code Warrior does not work with the D-Bug12 monitor. However, it can work with a demo board programmed with the Serial Monitor from Freescale and provides source-level debugging capability. Interested users are encouraged to try it out.

5.14 Exercises

E5.1 Write the C function *power(int x, char n)* that computes the value of x^n.

E5.2 Write a C program to find the median and mode of an array of integers. When the array has an even number of elements, the median is defined as the average of the middle two elements. Otherwise, it is defined as the middle element of the array. The mode is the element that occurs most frequently. You need to sort the array in order to find the median.

E5.3 Write a function that tests if a given number is a multiple of 8. A 1 is returned if the given number is a multiple of 8. Otherwise, a 0 is returned. The number to be tested is an integer and is an argument to this function.

E5.4 Write a function that computes the least common multiple (LCM) of two integers m and n.

E5.5 What is a function prototype? What is the difference between a function prototype and function declaration?

E5.6 Write a *switch* statement that will examine the value of an integer variable xx and store one of the following messages in an array of seven characters, depending on the value assigned to xx (terminate the message with a NULL character):

(a) Cold, if xx == 1
(b) Chilly, if xx == 2
(c) Warm, if xx == 3
(d) Hot, if xx == 4

E5.7 Write a C program to clear the screen and then move the cursor to the middle of a line and output the message "Microcontroller is fun to use!" The screen can be cleared by outputting the character "\f" (formfeed character).

E5.8 Write a function that will convert an uppercase letter to lowercase.

E5.9 Write a C program that swaps the first column of a matrix with the last column, swaps the second column of the matrix with the second-to-last column, and so on.

E5.10 Write a loop to compute the sum of the squares of the first 100 odd integers.

E5.11 An *Armstrong number* is a number of n digits that is equal to the sum of each digit raised to the nth power. For example, 153 (which has three digits) equals $1^3 + 5^3 + 3^3$. Write a function to store all three-digit Armstrong numbers in an array.

E5.12 Write a C function to perform a binary search on a sorted array. The binary search algorithm is given in Example 4.3. The starting address, the key, and the array count are parameters to this function. Both the key and array count are integers.

E5.13 Write a program to find the first five numbers that when divided by 2, 3, 4, 5, or 6 leave a remainder of 1, and when divided by 7 have no remainder.

E5.14 Take a four-digit number. Add the first two digits to the last two digits. Now, square the sum. Surprise, you've got the original number again. Of course, not all four-digit numbers have this property. Write a C program to find three numbers that have this special property.

5.15 Lab Exercises and Assignments

L5.1. Enter, compile, and download the following C program onto a demo board for execution using the procedure described in Section 5.10:

```
#include "c:\egnu091\include\hcs12.h"
#include "c:\egnu091\include\delay.c"
#include "c:\egnu091\include\convert.c"
#include "c:\egnu091\include\stdio.c"
int main (void)
{
    const char *ptr = " seconds passed!";
    char buf[20];
    int cnt;
    DDRB = 0xFF; /* configure port B for output */
    cnt = 0;      /* initialize count to 0 */
    while(1) {
        delayby100ms(5);
        int2alpha(cnt,buf);
        putsr(&buf[0]);
        puts(ptr);
        cnt++;
        PORTB = cnt;
        newline();
    }
    return 0;
}
```

L5.2. Write a C program that will generate every third integer, beginning with $i = 2$ and continuing for all integers that are less than 300. Calculate the sum of those integers that are divisible by 5. Store those integers and their sum in an array and as an integer variable, respectively.

L5.3. Write a C function that calculates the *least common multiple* (LCM) of two integers m and n. Integers m and n are parameters to this function. Also write a main program to test this function with several pairs of integers. Use the library functions to convert and output the results.

L5.4. Write a C function to compute the root-mean-square (rms) value of an unknown signal. The rms value will be approximated by **n** (at least 32) samples. The method for computing the square-root value is given in Example 4.14. The input parameters to this function include the starting address of the array and the array count. Also write a main program to test this function. Compile and download the program onto the demo board for execution.

6

Interrupts, Clock Generation, Resets, and Operation Modes

6.1 Objectives

After completing this chapter, you should be able to

- Explain the difference between interrupts and resets

- Describe the handling procedures for interrupts and resets

- Raise one of the HCS12 maskable interrupts to the highest priority

- Enable and disable maskable interrupts

- Use one of the low-power modes to reduce the power consumption

- Use COP watchdog timer reset to detect software failure

- Set up the interrupt vector jump table for demo boards that have the D-Bug12 monitor

- Write interrupt-driven application programs

- Distinguish the HCS12 operation modes

6.2 Fundamental Concepts of Interrupts

Interrupts and resets are among the most useful mechanisms that a computer system provides. With interrupts and resets, I/O operations are performed more efficiently, errors are handled more smoothly, and CPU utilization is improved. This chapter will begin with a general discussion of interrupts and resets and then focus on the specific features of the HCS12 interrupts and resets.

6.2.1 What Is an Interrupt?

An interrupt is an event that requires the CPU to stop normal program execution and perform some service related to the event. An interrupt can be generated internally (inside the chip) or externally (outside the chip). An external interrupt is generated when the external circuitry asserts an interrupt signal to the CPU. An internal interrupt can be generated by the hardware circuitry inside the chip or caused by software errors. Most microcontrollers have timers, I/O interface functions, and the CPU incorporated on the same chip. These subsystems can generate interrupts to the CPU. Abnormal situations that occur during program execution, such as illegal opcodes, overflow, divide-by-zero, and underflow, are called *software interrupts*. The terms *traps* and *exceptions* are also used to refer to software interrupts by some companies.

A good analogy for interrupt is how you act when you are sitting in front of a desk to read this book and the phone rings. You probably will act like this:

1. Remember the page number or place a bookmark on the page that you are reading, close the book, and put it aside.

2. Pick up the phone and say, "Hello, this is so and so."

3. Listen to the voice over the phone to find out who is calling or ask who is calling if the voice is not familiar.

4. Talk to that person.

5. Hang up the phone when you finish talking.

6. Open the book and turn to the page where you placed the bookmark and resume reading this book.

The phone call example spells out a few things that are similar to how the microprocessor handles the interrupt:

1. As a student, you spend most of your time in studying. Answering the phone call happens only occasionally. Similarly, the microprocessor is executing application programs most of the time. Interrupts will only force the microprocessor to stop executing the application program briefly and take some necessary actions.

2. Before picking up the phone, you finish reading the sentence and then place a bookmark to remind you of the page number that you are reading so that you can resume reading after finishing the conversation over the phone. Most microprocessors will finish the instruction they are executing and save the address of the next instruction in memory (usually in the stack) so that they can resume the program execution later.

3. You find out who the person is by listening to the voice over the phone, or you ask questions so that you can decide what to say. Similarly, the microprocessor needs to identify the cause of the interrupt before it can take appropriate actions. This is built into the microprocessor hardware.

4. After identifying the person who called you, you start the phone conversation with that person on some appropriate subjects. Similarly, the microprocessor will take some actions appropriate to the interrupt source.

5. When finishing the phone conversation, you hang up the phone, open the book to the page where you placed the bookmark, and resume reading. Similarly, after taking some actions appropriate to the interrupt, the microprocessor will jump back to the next instruction after the interrupt occurred and resume program execution. This can be achieved easily because the address of the instruction to be resumed was saved in memory (stack). Most microprocessors do this by executing a *return-from-interrupt* instruction.

6.2.2 Why Are Interrupts Used?

Interrupts are useful in many applications, such as the following:

- *Coordinating I/O activities and preventing the CPU from being tied up during the data transfer process.* The CPU needs to know if the I/O device is ready before it can proceed. Without the interrupt capability, the CPU will need to check the status of the I/O device periodically. The interrupt mechanism is often used by the I/O device to inform the CPU that it is ready for data transfer. CPU time can thus be utilized more efficiently because of the interrupt mechanism. Interrupt-driven I/O operations will be explained in more detail in later chapters.

- *Performing time-critical applications.* Many emergent events, such as power failure and process control, require the CPU to take action immediately. The interrupt mechanism provides a way to force the CPU to divert from normal program execution and take immediate actions.

- *Providing a graceful way to exit from an application when a software error occurs.* The service routine for a software interrupt may also output useful information about the error so that it can be corrected.

- *Reminding the CPU to perform routine tasks.* There are many microprocessor applications that require the CPU to perform routine work, such as the following:

  1. *Keeping track of time of day.* Without the timer interrupt, the CPU will need to use program loops in order to update the current time. CPU cannot do anything else without a timer interrupt in this application. The periodic timer interrupts prevent the CPU from being tied up.

  2. *Periodic data acquisition.* Some applications are designed to acquire data periodically.

  3. *Task switching in a multitasking operating system.* In a modern computer system, multiple application programs are resident in the main memory, and the CPU time is divided into many short slots (one slot may be from 10 to 20 ms). A multitasking operating system assigns a program to be executed for one time slot. At the end of a time slot or when a program is waiting for the completion of an I/O operation, the operating system takes over and assigns another program for execution. This technique is called *multitasking*. Multitasking can dramatically improve the CPU utilization and is implemented by using periodic timer interrupts.

6.2.3 Interrupt Maskability

Depending on the situation and application, some interrupts may not be desired or needed and should be prevented from interrupting the CPU. Most microprocessors and microcontrollers have the option of ignoring these interrupts. These types of interrupts are called *maskable interrupts*. There are other types of interrupts that the CPU cannot ignore and must take immediate actions for; these are *nonmaskable interrupts.* A program can request the CPU to service or ignore a maskable interrupt by setting or clearing an *enable bit.* When an interrupt is enabled, the CPU will respond to it. When an interrupt is disabled, the CPU will ignore it. An interrupt is said to be *pending* when it is active but not yet serviced by the CPU. A pending interrupt may or may not be serviced by the CPU, depending on whether or not it is enabled.

To make the interrupt system more flexible, a computer system normally provides a global and local interrupt masking capability. When none of the interrupts are desirable, the processor can disable all the interrupts by clearing the global interrupt enable bit (or setting the global interrupt mask bit for some other processor). In other situations, the processor can selectively enable certain interrupts while at the same time disable other undesirable interrupts. This is achieved by providing each interrupt source an enable bit in addition to the global interrupt mask. Whenever any interrupt is undesirable, it can be disabled while at the same time allowing other interrupt sources to be serviced (attended) by the processor. Today, almost all commercial processors are designed to provide this two-level (or even three-level) interrupt-enabling capability.

6.2.4 Interrupt Priority

If a computer is supporting multiple interrupt sources, then it is possible that several interrupts would be pending at the same time. The CPU has to decide which interrupt should receive service first in this situation. The solution is to prioritize all interrupt sources. An interrupt with higher priority always receives service before interrupts at lower priorities. Many microcontrollers (for example, the HCS12) prioritize interrupts in hardware. For those microcontrollers that do not prioritize interrupts in hardware, the software can be written to handle certain interrupts before others. By doing this, interrupts are essentially prioritized. For most microprocessors and microcontrollers, interrupt priorities are not programmable.

6.2.5 Interrupt Service

The CPU provides service to an interrupt by executing a program called an *interrupt service routine.* After providing service to an interrupt, the CPU must resume normal program execution. How can the CPU stop the execution of a program and resume it later? It achieves this by saving the program counter and the CPU status information before executing the interrupt service routine and then restoring the saved program counter and CPU status before exiting the interrupt service routine. The complete interrupt service cycle involves

1. Saving the program counter value.
2. Saving the CPU status (including the CPU status register and some other registers) in the stack. This step is optional for some microcontrollers and microprocessors.
3. Identifying the source of the interrupt.
4. Resolving the starting address of the corresponding interrupt service routine.
5. Executing the interrupt service routine.
6. Restoring the CPU status from the stack.
7. Restoring the program counter from the stack.
8. Resuming the interrupted program.

For all maskable hardware interrupts, the microprocessor starts to provide service when it completes the execution of the current instruction (the instruction being executed when the interrupt occurred). For some nonmaskable interrupts, the CPU may start the service without completing the current instruction. Many software interrupts are caused by an error in instruction execution that prevents the instruction from being completed. The service to this type of interrupt is simply to output an error message and abort the program.

6.2.6 Interrupt Vector

The term *interrupt vector* refers to the starting address of the interrupt service routine. In general, interrupt vectors are stored in a table called an *interrupt-vector table*. The starting address of each entry (holds one interrupt vector) in the interrupt vector table is called the *vector address*. The interrupt-vector table is fixed for some microprocessors and may be relocated for other microprocessors (for example, the AM29000 family of microprocessors).

The CPU needs to determine the interrupt vector before it can provide service. One of the following methods can be used by a microprocessor/microcontroller to determine the interrupt vector:

1. *Predefined.* In this method, the starting address of the service routine is predefined when the microcontroller is designed. The processor uses a table to store all the interrupt service routines. The Intel 8051 microcontrollers use this approach. Each interrupt is allocated the same number of bytes to hold its service routine. The Intel 8051 allocates eight words to each interrupt service routine. When the service routine requires more than eight words, the solution is to place a jump instruction in the predefined location to jump to the actual service routine.

2. *Fetch the vector from a predefined memory location.* In this approach, the interrupt vector of each interrupt source is stored at a predefined location in the interrupt-vector table, where the microprocessor can get it directly. The Freescale HCS12 and most other Freescale microcontrollers use this approach.

3. *Execute an interrupt acknowledge cycle to fetch a vector number in order to locate the interrupt vector.* During the interrupt acknowledge cycle, the microprocessor performs a read bus cycle, and the external I/O device that requested the interrupt places a number on the data bus to identify itself. This number is called the *interrupt vector number*. The CPU can figure out the starting address of the interrupt service routine by using this number. The CPU needs to perform a read cycle in order to obtain it. The Freescale 68000 and Intel x86 family microprocessors support this method. The Freescale 68000 family of microprocessors also uses the second method. This method is not used by microcontrollers because of the incurred latency.

6.2.7 Interrupt Programming

Interrupt programming deals with how to provide service to the interrupt. There are three steps in interrupt programming:

Step 1

Initialize the interrupt-vector table. (This step is not needed for microprocessors that have predefined interrupt vectors.) This can be done by using the assembler directive **org** (or its equivalent):

```
ORG     $xxxx          ; xxxx is the vector table address
dc.w    service_1      ; store the starting address of interrupt source 1
dc.w    service_2      ;
  .
  .
  .
dc.w    service_n
```

where *service_i* is the starting address of the service routine for interrupt source i. The HCS12 uses this method to set up interrupt vectors. The assembler syntax and the number of bytes needed to store an interrupt vector on your particular microcontroller may be different.

Step 2

Write the interrupt service routine. An interrupt service routine should be as short as possible. For some interrupts, the service routine may only output a message to indicate that something unusual has occurred. A service routine is similar to a subroutine—the only difference is the last instruction. An interrupt service routine uses the *return from interrupt* (or *return from exception*) instruction instead of the *return from subroutine* instruction to return to the interrupted program. The following instruction sequence is an example of an interrupt service routine (in HCS12 instructions):

```
irq_isr    ldx    #msg
           jsr    putst            ; call putst to output a string pointed by X
           rti                     ; return from interrupt
msg        fcc    "This is an error"
```

The service routine may or may not return to the interrupted program, depending on the cause of the interrupt. It makes no sense to return to the interrupted program if the interrupt is caused by a software error such as divide-by-zero or overflow, because the program is unlikely to generate correct results under these circumstances. In such situations, the service routine would return to the monitor program or the operating system instead. Returning to a program other than the interrupted program can be achieved by changing the saved program counter (in the stack) to the desired value. Execution of the *return from interrupt* instruction will then return CPU control to the new address.

Step 3

Enable the interrupts to be serviced. An interrupt can be enabled by clearing the global interrupt mask and setting the local interrupt enable bit in the I/O control register. It is a common mistake to forget enabling interrupts when writing interrupt-driven application programs.

6.2.8 Overhead of Interrupts

Although the interrupt mechanism provides many advantages, it also involves some overhead. The overhead of the HCS12 interrupt includes:

1. Saving the CPU registers, including accumulators (A:B), index registers X and Y, and the condition code register (CCR), and fetching the interrupt vector. This takes at least 9 E cycles.

2. The execution time of RTI instruction. This instruction restores all the CPU registers that have been stored in the stack by the CPU during the interrupt and takes from 8 to 11 E clock cycles to complete for the HCS12.

3. Execution time of instructions of the interrupt service routine. This depends on the type and the number of instructions in the service routine.

The total overhead is thus at least 17 to 20 E clock cycles, which amounts to almost 1 μs for a 24-MHz E clock. One should be aware of the overhead involved in interrupt processing when deciding whether to use the interrupt mechanism.

6.3 Resets

The initial values of some CPU registers, flip-flops, and the control registers in I/O interface chips must be established before the computer can operate properly. Computers provide a reset mechanism to establish initial conditions.

There are at least two types of resets in each microprocessor: the *power-on reset* and the *manual reset*. A power-on reset allows the microprocessor to establish the initial values of registers and flip-flops and to initialize all I/O interface chips when power to the microprocessor is turned on. A manual reset without power-down allows the computer to get out of most error conditions (if hardware hasn't failed) and reestablish the initial conditions. The computer will *reboot* itself after a reset.

The starting address of the reset service routine is either a fixed value or is stored at a fixed location (for HCS12). The reset service routine is stored in the read-only memory of all microprocessors so that it is always ready for execution. At the end of the service routine, control should be transferred to either the monitor program or the operating system.

Like nonmaskable interrupts, resets are also nonmaskable. However, resets are different from the nonmaskable interrupts in that no registers are saved by resets because resets establish the values of registers.

6.4 HCS12 Exceptions

The HCS12 exceptions can be classified into the following categories:

- *Maskable interrupts*: These include the $\overline{\text{IRQ}}$ pin interrupt and all peripheral function interrupts. Since different HCS12 members implement a different number of peripheral functions, they have different numbers of maskable interrupts.

- *Nonmaskable interrupts*: These include the $\overline{\text{XIRQ}}$ pin interrupt, the SWI instruction interrupt, and the unimplemented opcode trap.

- *Resets*: These include the power-on reset, the $\overline{\text{RESET}}$ pin manual reset, the COP (computer operating properly) reset, and the clock monitor reset. For other microcontrollers, the COP reset is also called *Watchdog* reset.

6.4.1 Maskable Interrupts

Since different HCS12 members implement a different number of peripheral functions, they have a different number of maskable interrupts. The I flag in the CCR register is the global mask of all maskable interrupts. Whenever the I flag is 1, all maskable interrupts are disabled. All maskable interrupts have a local enable bit that allows them to be selectively enabled. They are disabled (I flag is set to 1) when the HCS12 gets out of the reset state.

Like any other microcontrollers, all HCS12 exceptions are prioritized. The priorities of resets and nonmaskable interrupts are not programmable. However, one can raise one of the maskable interrupts to the highest priority within the group of maskable interrupts so that it can get quicker attention from the CPU. The relative priorities of the other sources remain the same. The bits 7 to 1 of the HPRIO register select the maskable interrupt at the highest priority within the group of maskable interrupts. The contents of the HPRIO register are shown in Figure 6.1.

7	6	5	4	3	2	1	0	
PSEL7	PSEL6	PSEL5	PSEL4	PSEL3	PSEL2	PSEL1	0	$001F

Figure 6.1 ■ Highest-priority I interrupt register

The priorities and vector addresses of all HCS12 exceptions are listed in Table 6.1. To raise a maskable interrupt source to the highest priority, simply write the low byte of the vector

Vector Address	Interrupt Source	CCR Mask	Local Enable	HPRIO Value to Elevate to Highest 1 Bit
$FFFE	Reset	none	none	—
$FFFC	Clock monitor failure reset	none	PLLCTL(CME, SCME)	—
$FFFA	COP failure reset	none	COP rate select	—
$FFF8	Unimplemented instruction trap	none	none	—
$FFF6	SWI	none	none	—
$FFF4	XIRQ	X bit	none	—
$FFF2	IRQ	I bit	IRQCR(IRQEN)	$F2
$FFF0	Real time interrupt	I bit	CRGINT(RTIE)	$F0
$FFEE	Enhanced capture timer channel 0	I bit	TIE(C0I)	$EE
$FFEC	Enhanced capture timer channel 1	I bit	TIE(C1I)	$EC
$FFEA	Enhanced capture timer channel 2	I bit	TIE(C2I)	$EA
$FFE8	Enhanced capture timer channel 3	I bit	TIE(C3I)	$E8
$FFE6	Enhanced capture timer channel 4	I bit	TIE(C4I)	$E6
$FFE4	Enhanced capture timer channel 5	I bit	TIE(C5I)	$E4
$FFE2	Enhanced capture timer channel 6	I bit	TIE(C6I)	$E2
$FFE0	Enhanced capture timer channel 7	I bit	TIE(C7I)	$E0
$FFDE	Enhanced capture timer overflow	I bit	TSCR2(TOI)	$DE
$FFDC	Pulse accumulator A overflow	I bit	PACTL(PAOVI)	$DC
$FFDA	Pulse accumulator input edge	I bit	PACTL(PAI)	$DA
$FFD8	SPI0	I bit	SP0CR1(SPIE, SPTIE)	$D8
$FFD6	SCI0	I bit	SC0CR2(TIE,TCIE,RIE,ILIE)	$D6
$FFD4	SCI1	I bit	SC1CR2(TIE,TCIE,RIE,ILIE)	$D4
$FFD2	ATD0	I bit	ATD0CTL2(ASCIE)	$D2
$FFD0	ATD1	I bit	ATD1CTL2(ASCIE)	$D0
$FFCE	Port J	I bit	PTJIF(PTJIE)	$CE
$FFCC	Port H	I bit	PTHIF(PTHIE)	$CC
$FFCA	Modulus down counter underflow	I bit	MCCTL(MCZI)	$CA
$FFC8	Pulse accumulator B overflow	I bit	PBCTL(PBOVI)	$C8
$FFC6	CRG PLL lock	I bit	CRGINT(LOCKIE)	$C6
$FFC4	CRG self clock mode	I bit	CRGINT(SCMIE)	$C4
$FFC2	BDLC	I bit	DLCBCR1(IE)	$C2
$FFC0	IIC Bus	I bit	IBCR(IBIE)	$C0
$FFBE	SPI1	I bit	SPICR1(SPIE,SPTIE)	$BE
$FFBC	SPI2	I bit	SP2CR2(SPIE,SPTIE)	$BC
$FFBA	EEPROM	I bit	EECTL(CCIE, CBEIE)	$BA
$FFB8	Flash	I bit	FCTL(CCIE, CBEIE)	$B8
$FFB6	CAN0 wake-up	I bit	CAN0RIER(WUPIE)	$B6
$FFB4	CAN0 errors	I bit	CAN0RIER(CSCIE, OVRIE)	$B4
$FFB2	CAN0 receive	I bit	CAN0RIER(RXFIE)	$B2
$FFB0	CAN0 transmit	I bit	CAN0TIER(TXEIE2-TXEIE0)	$B0
$FFAE	CAN1 wake-up	I bit	CAN1RIER(WUPIE)	$AE
$FFAC	CAN1 errors	I bit	CAN1RIER(CSCIE, OVRIE)	$AC
$FFAA	CAN1 receive	I bit	CAN1RIER(RXFIE)	$AA
$FFA8	CAN1 transmit	I bit	CAN1TIER(TXEIE2-TXEIE0)	$A8
$FFA6	CAN2 wake-up	I bit	CAN2RIER(WUPIE)	$A6
$FFA4	CAN2 errors	I bit	CAN2RIER(CSCIE, OVRIE)	$A4
$FFA2	CAN2 receive	I bit	CAN2RIER(RXFIE)	$A2
$FFA0	CAN2 transmit	I bit	CAN2TIER(TXEIE2-TXEIE0)	$A0
$FF9E	CAN3 wake-up	I bit	CAN3RIER(WUPIE)	$9E
$FF9C	CAN3 errors	I bit	CAN3RIER(CSCIE, OVRIE)	$9C
$FF9A	CAN3 receive	I bit	CAN3RIER(RXFIE)	$9A
$FF98	CAN3 transmit	I bit	CAN3TIER(TXEIE2-TXEIE0)	$98
$FF96	CAN4 wake-up	I bit	CAN4RIER(WUPIE)	$96
$FF94	CAN4 errors	I bit	CAN4RIER(CSCIE, OVRIE)	$94
$FF92	CAN4 receive	I bit	CAN4RIER(RXFIE)	$92
$FF90	CAN4 transmit	I bit	CAN4TIER(TXEIE2-TXEIE0)	$90
$FF8E	Port P interrupt	I bit	PTPIF(PTPIE)	$8E
$FF8C	PWM emergency shutdown	I bit	PWMSDN(PWMIE)	$8C

Table 6.1 ■ Interrupt vector map

address of this interrupt to the HPRIO register. For example, to raise the capture timer channel 7 interrupt to the highest priority, write the value of $E0 to the HPRIO register.

All HCS12 reset and interrupt vectors are stored in a table (shown in Table 6.1) located at $FF8C to $FFFF. In Table 6.1, exceptions that have higher vector addresses are at higher priorities. Not all the exceptions are available in all HCS12 members.

IRQ Pin Interrupt

The IRQ pin is the only external maskable interrupt signal. The IRQ pin interrupt can be edge triggered or level triggered. The triggering method is selected by programming the IRQE bit of the Interrupt Control Register (IRQCR). The IRQ pin interrupt has a local enable bit IRQEN bit, which is bit 6 of the IRQCR. The contents of the IRQCR register are shown in Figure 6.2.

7	6	5	4	3	2	1	0
IRQE	IRQEN	0	0	0	0	0	0

reset: 0   1   0   0   0   0   0   0

IRQE -- IRQ edge sensitive only bit
  IRQE can be written once in normal mode. In special modes, it can be
  written any time, but the first write is ignored.
    1 = IRQ pin responds only to falling edge.
    0 = IRQ pin responds to low level.   ✗ ✓ This method is best
IRQEN -- IRQ enable bit
  IRQEN bit can be written any time in all modes. The IRQ pin has an
  internal pullup.
    1 = IRQ pin interrupt enabled.   ✗
    0 = IRQ pin interrupt disabled.   ✗

Figure 6.2 ■ Interrupt control register (IRQCR)

The advantage of making the IRQ interrupt *level sensitive* (active low) is that it allows multiple external interrupt sources to be tied to this pin. Whenever one of the interrupt sources (that are tied to the IRQ pin) is low, an interrupt request will be detected by the HCS12. The user of this method must make sure that the IRQ signal is de-asserted (goes high) before the HCS12 exits the interrupt service routine if there are no other pending interrupts connected to the IRQ pin.

The major advantage of making the IRQ interrupt *edge sensitive* (falling edge) is that the user does not need to be concerned about the duration of the assertion time of the IRQ signal. However, this approach is not appropriate for a noisy environment. In a noisy environment, any noise spike could generate an undesirable interrupt request on the IRQ pin.

Interrupt Recognition

Once enabled, an interrupt request can be recognized at any time after the I mask bit is cleared. When an interrupt service request is recognized, the CPU responds at the completion of the instruction being executed. Interrupt latency varies according to the number of cycles required to complete the current instruction. The HCS12 has implemented a few instructions to support fuzzy logic rule evaluation. These instructions, which take a much longer time to complete, include fuzzy logic rule evaluation (REV), fuzzy logic rule evaluation weighted (REVW), and weighted average (WAV) instructions. The HCS12 does not wait until the completion of these instructions to service the interrupt request. The instructions will resume at the point when they were interrupted.

Before the HCS12 starts to service an interrupt, it will set the I mask to disable other maskable interrupts. When the CPU begins to service an interrupt, the instruction queue is refilled, a return address is calculated, and then the return address and the contents of all CPU registers (except SP) are saved in the stack in the order shown in Figure 6.3.

[CCR]	← SP
[A]	← SP + 1
[B]	← SP + 2
[X]	← SP + 3
[Y]	← SP + 5
return address	← SP + 7

Figure 6.3 ■ Slack order on entry to interrupts

THE RTI INSTRUCTION

RTI is used to terminate interrupt service routines. RTI is an 8-cycle instruction when no other interrupt is pending and an 11-cycle instruction when another interrupt is pending. In either case, the first 5 cycles are used to restore the CCR, B:A, X, Y, and the return address from the stack. The HCS12 then clears the I mask to enable further maskable interrupts.

If no other interrupt is pending at this point, three program words are fetched to refill the instruction queue from the area of the return address and processing proceeds from there. If another interrupt is pending after registers are restored, a new vector is fetched, and the stack pointer is adjusted to point at the CCR value that was just recovered (SP = SP − 9). This makes it appear that the registers have been stacked again. After the SP is adjusted, three program words are fetched to refill the instruction queue, starting at the address the vector points to. Processing then continues with execution of the instruction that is now at the head of the queue.

6.4.2 Nonmaskable Interrupts

There are three interrupts in this category: $\overline{\text{XIRQ}}$ pin, SWI instruction, and the unimplemented opcode trap.

$\overline{\text{XIRQ}}$ PIN INTERRUPT

The $\overline{\text{XIRQ}}$ pin interrupt is disabled during a system reset and upon entering the interrupt service routine for an $\overline{\text{XIRQ}}$ interrupt.

During reset, both the I and X bits in the CCR register are set. This disables maskable interrupts and interrupt requests made by asserting the $\overline{\text{XIRQ}}$ pin (pulled to low). After minimum system initialization, software can clear the X bit using an instruction such as andcc #$BF. Software cannot reset the X bit from 0 to 1 once it has been cleared, and hence the interrupt requests made via the $\overline{\text{XIRQ}}$ pin become nonmaskable.

When a nonmaskable interrupt is recognized, both the X and I bits are set after CPU registers are saved. The X bit is not affected by maskable interrupts. The execution of an RTI instruction at the end of the $\overline{XIRQ}$ service routine will restore the X and I bits to the preinterrupt request state.

UNIMPLEMENTED OPCODE TRAP

The HCS12 uses up to 16 bits (2 pages) to encode the opcode. All 256 combinations in the page 1 opcode map have been used. However, only 54 of the 256 positions on page 2 of the opcode map are used. If HCS12 attempts to execute one of the 202 unused opcodes on page 2, an unimplemented opcode trap occurs. The 202 unimplemented opcodes are essentially interrupts that share a common interrupt vector address, $FFF8:$FFF9. The HCS12 uses the next address after an unimplemented page 2 opcode as a return address.

SOFTWARE INTERRUPT INSTRUCTION (SWI)

Execution of the SWI instruction causes an interrupt without an interrupt request signal. SWI is not inhibited by the global mask bits in the CCR. So far, we have been using this instruction to jump back to the D-Bug12 monitor.

The SWI instruction is commonly used in the debug monitor to implement *breakpoints* and to transfer control from a user program to the debug monitor. A breakpoint in a user program is a memory location where we want program execution to be stopped and information about instruction execution (in the form of register contents) to be displayed. To implement breakpoints, the debug monitor sets up a breakpoint table. Each entry of the table holds the address of the breakpoint and the opcode byte at the breakpoint. The monitor also replaces the opcode byte at the breakpoint with the opcode of the SWI instruction. When the instruction at the breakpoint is executed, it causes an SWI interrupt. The service routine of the SWI interrupt will look up the breakpoint table and take different actions depending on whether the saved PC value is in the breakpoint table:

Case 1
The saved PC value is not in the breakpoint table. In this case, the service routine will simply replace the saved PC value (in the stack) with the address of the monitor program and return from the interrupt.

Case 2
The saved PC is in the breakpoint table. In this case, the service routine will

1. Replace the SWI opcode with the opcode in the breakpoint table
2. Display the contents of the CPU registers
3. Replace the saved PC value (in the stack) with the address of the monitor program
4. Return from the interrupt (return to the debug monitor because of step 3)

6.4.3 Interrupts in D-Bug12 EVB Mode

Most users would use the EVB mode of the D-Bug12 monitor to develop applications on a demo board having a D-Bug12 monitor. However, the D-bug12 monitor's EVB mode does not allow the use of the on-chip flash memory, which prevents using the default interrupt-vector table to hold interrupt vectors. To allow the user to develop interrupt-driven applications, the D-Bug12 monitor provides an SRAM-based interrupt-vector table. This table starts at $3E00 and has 64 entries of 2 bytes each. The contents of this table are shown in Table 6.2. Initially, all entries in the table contain an address of $0000. Storing a value other than $0000 in any of the RAM interrupt-vector table entries causes execution of the interrupt-service routine pointed to by the address when an associated interrupt occurs.

Interrupt source	RAM vector address	Interrupt source	RAM vector address
Reserved $FF80	$3E00	IIC bus	$3E40
Reserved $FF82	$3E02	DLC	$3E42
Reserved $FF84	$3E04	SCME	$3E44
Reserved $FF86	$3E06	CRG clock	$3E46
Reserved $FF88	$3E08	Pulse accumulator B overflow	$3E48
Reserved $FF8A	$3E0A	Modulus down counter underflow	$3E4A
PWM emergency shutdown	$3E0C	Port H interrupt	$3E4C
Port P interrupt	$3E0E	Port J interrupt	$3E4E
MSCAN 4 transmit	$3E10	ATD1	$3E50
MSCAN 4 receive	$3E12	ATD0	$3E52
MSCAN 4 errors	$3E14	SCI1	$3E54
MSCAN 4 wake-up	$3E16	SCI0	$3E56
MSCAN 3 transmit	$3E18	SPI0	$3E58
MSCAN 3 receive	$3E1A	Pulse accumulator A input edge	$3E5A
MSCAN 3 errors	$3E1C	Pulse accumulator A overflow	$3E5C
MSCAN 3 wake-up	$3E1E	Timer overflow	$3E5E
MSCAN 2 transmit	$3E20	Timer channel 7	$3E60
MSCAN 2 receive	$3E22	Timer channel 6	$3E62
MSCAN 2 errors	$3E24	Timer channel 5	$3E64
MSCAN 2 wake-up	$3E26	Timer channel 4	$3E66
MSCAN 1 transmit	$3E28	Timer channel 3	$3E68
MSCAN 1 receive	$3E2A	Timer channel 2	$3E6A
MSCAN 1 errors	$3E2C	Timer channel 1	$3E6C
MSCAN 1 wake-up	$3E2E	Timer channel 0	$3E6E
MSCAN 0 transmit	$3E30	Real time interrupt	$3E70
MSCAN 0 receive	$3E32	IRQ	$3E72
MSCAN 0 errors	$3E34	XIRQ	$3E74
MSCAN 0 wake-up	$3E36	SWI	$3E76
Flash	$3E38	Unimplemented instruction trap	$3E78
EEPROM	$3E3A	N/A	$3E7A
SPI2	$3E3C	N/A	$3E7C
SPI1	$3E3E	N/A	$3E7E

Table 6.2 ■ D-Bug12 RAM Interrupt vector addresses

If an unmasked interrupt occurs and a table entry contains the default address of $0000, program execution is returned to D-Bug12. The D-bug12 would display a message indicating the source of the interrupt and also display the CPU registers at the point where the program was interrupted. The only exception to this is the SCI0 interrupt. Even though there is an entry for SCI0 in Table 6.2, one cannot use the SCI0 interrupt because it has been used by D-Bug12 for all of its communications. The hcs12.inc (and hcs12.h) file also provides the mnemonic name for each interrupt source so that the user can use the mnemonic name to specify the SRAM vector address. The complete list of these mnemonic names is shown in Table 6.3.

Interrupt Source	RAM Vector Address	Interrupt Source	RAM Vector Address
UserRsrv0x80	$3E00	UserIIC	$3E40
UserRsrv0x82	$3E02	UserDLC	$3E42
UserRsrv0x84	$3E04	UserSCME	$3E44
UserRsrv0x86	$3E06	UserCRG	$3E46
UserRsrv0x88	$3E08	UserPAccBOv	$3E48
UserRsrv0x8a	$3E0A	UserModDwnCtr	$3E4A
UserPWMShDn	$3E0C	UserPortH	$3E4C
UserPortP	$3E0E	UserPortj	$3E4E
UserMSCAN4Tx	$3E10	UserAtoD1	$3E50
UserMSCAN4Rx	$3E12	UserAtoD0	$3E52
UserMSCAN4Errs	$3E14	UserSCI1	$3E54
UserMSCAN4Wake	$3E16	UserSCI0	$3E56
UserMSCAN3Tx	$3E18	UserSPI0	$3E58
UserMSCAN3Rx	$3E1A	UserPAccEdge	$3E5A
UserMSCAN3Errs	$3E1C	UserPAccOvf	$3E5C
UserMSCAN3Wake	$3E1E	UserTimerOvf	$3E5E
UserMSCAN2Tx	$3E20	UserTimerCh7	$3E60
UserMSCAN2Rx	$3E22	UserTimerCh6	$3E62
UserMSCAN2Errs	$3E24	UserTimerCh5	$3E64
UserMSCAN2Wake	$3E26	UserTimerCh4	$3E66
UserMSCAN1Tx	$3E28	UserTimerCh3	$3E68
UserMSCAN1Rx	$3E2A	UserTimerCh2	$3E6A
UserMSCAN1Errs	$3E2C	UserTimerCh1	$3E6C
UserMSCAN1Wake	$3E2E	UserTimerCh0	$3E6E
UserMSCAN0Tx	$3E30	UserRTI	$3E70
UserMSCAN0Rx	$3E32	UserIRQ	$3E72
UserMSCAN0Errs	$3E34	UserXIRQ	$3E74
UserMSCAN0Wake	$3E36	UserSWI	$3E76
UserFlash	$3E38	UserTrap	$3E78
UserEEPROM	$3E3A	N/A	$3E7A
UserSPI2	$3E3C	N/A	$3E7C
UserSPI1	$3E3E	N/A	$3E7E

Table 6.3 ■ Mnemonic names for D-Bug12 RAM interrupt-vector addresses

It is very simple to set up the SRAM interrupt vector. Assuming that the service routine for the $\overline{\text{IRQ}}$ interrupt starts with the label **irq_isr**, then one of the following two methods can be used to establish the SRAM table entry for the $\overline{\text{IRQ}}$ interrupt:

Method 1:

```
org      $3E72          ; one can also use "org UserIRQ"
dc.w     isr_irq
```

Method 2:

```
ldd      #isr_irq
std      $3E72          ; one can also use "std UserIRQ"
```

Example 6.1

▼

Assume that the $\overline{\text{IRQ}}$ pin of the HCS12DP256 is connected to a 1-Hz digital waveform and port B is connected to eight LEDs. Write a program to configure port B for output and enable the $\overline{\text{IRQ}}$ interrupt, and also write the service routine for the $\overline{\text{IRQ}}$ interrupt. The service routine for the $\overline{\text{IRQ}}$ interrupt simply increments a counter and outputs it to port B.

Solution: The program is as follows:

```
#include      "c:\miniide\hcs12.inc"
              org        $1000
count         ds.b       1                    ; reserve 1 byte for count
              org        $1500
              lds        #$1500               ; set up the stack pointer
              movw       #IRQISR, UserIRQ     ; set up interrupt vector in SRAM
              clr        count
              movb       #$FF,DDRB            ; configure port B for output
              movb       count,PORTB          ; display the count value on LEDs
              movb       #$C0,INQCR           ; enable IRQ pin interrupt and select edge-triggering
              cli                             ;    "
forever       nop
              bra        forever              ; wait for IRQ pin interrupt
;**********************************************************************************
; This is the IRQ interrupt service routine.
;**********************************************************************************
IRQISR        inc        count                ; increment count
              movb       count,PORTB          ; and display it
              rti
              end
```

▲

Interrupt programming in C language varies with the C compiler. In the EGNU IDE, one must take the following actions in order to use interrupts:

1. *Use the interrupt attribute of the GCC compiler.* One must include the following statement as one of the first few statements in the program (there are two '_' characters before and after "attribute"):

   #define INTERRUPT__attribute__((interrupt))

2. *Apply the interrupt attribute to the interrupt service routine.* Assuming that rtiisr() and oc5isr() are two interrupt service routines, then one should add the following two prototype declaration statements (for rtiisr() and oc5isr()) to the program:

   void INTERRUPT rtiisr(void);
   void INTERRUPT oc5isr(void);

3. *Include the vectors12.h header file* so that one can use mnemonic names to store the starting addresses of interrupt service routines in the vector table (in SRAM). Include the following statements as one of the first few statements in the program:

   #include "c:\egnu091\include\vectors12.h"

4. *Store interrupt vectors in the SRAM vector table.* This can be done by assignment statements:

```
UserRTI = (unsigned short)&rtiisr;
UserTimerCh5 = (unsigned short)&oc5isr;
```

5. *Write the actual interrupt service routine.*

A C language version of the $\overline{IRQ}$ interrupt program is as follows:

```
#include          "c:\egnu091\include\hcs12.h"
#include          "c:\egnu091\include\vectors12.h"
#define           INTERRUPT__attribute__((interrupt))
void              INTERRUPT IRQISR(void);
unsigned char cnt;

int main(void)
{
    UserIRQ     = (unsigned short)&IRQISR;
    DDRB        = 0xFF;         /* configure Port B for output */
    cnt         = 0;
    DDRJ        | = BIT1;       /* configure PJ1 pin for output */
    PTJ         &= ~BIT1;       /* enable LEDs to light (required for Dragon12 demo board) */
    IRQCR       = 0xC0;         /* enable IRQ interrupt on falling edge */
    asm("cli");                 /* enable interrupt globally */
    while(1);                   /* wait for interrupt forever */
    return 0;
}
void INTERRUPT IRQISR(void)
{
    cnt++;                      /* increment the count value */
    PTB         = cnt;          /* display the count value on LEDs */
}.
```

6.5 Clock and Reset Generation Block (CRG)

This block is responsible for generating the clock signals required by the HCS12 instruction execution and all peripheral operations and providing default values for all on-chip registers. The block diagram of a CRG is shown in Figure 6.4.

A microcontroller needs a clock signal to operate. The clock signal has the waveform of a square wave. Most people use a crystal oscillator to generate the clock signal. However, the output of a crystal oscillator is a sinusoidal waveform that cannot be used to drive digital circuitry directly. Most microcontrollers and microprocessors have an on-chip oscillator to "square up" the incoming sinusoidal waveform so that it can be used as the clock signal. Many crystal oscillators also include the circuitry to square up the sinusoidal waveform. The on-chip oscillator circuitry can be bypassed for this type of crystal oscillator (also called *external oscillators* in Freescale literature).

Figure 6.4 ■ Block diagram of CRG

The phase-locked loop (PLL) is a circuit technique that can accept a low-frequency clock signal and produce a high-frequency clock output. In addition, the PLL has a feedback circuitry to stabilize the frequency of its output signal. The PLL has two operation modes: *acquisition mode* and *tracking mode*. When the PLL is first powered up, the frequency of its output is far from its *target frequency*. During this period, the PLL is in the acquisition mode in which the PLL can make large adjustments to quickly reach the target frequency. After the PLL has attained its target output frequency, it enters tracking mode, in which it makes only small adjustments to not deviate from the target frequency. This mode is useful when the incoming clock signals (square waveform) deviate slightly because the frequency of the crystal oscillator output may change slightly due to a change in temperature, humidity, or barometric pressure.

6.5.1 Modes of CRG Operation

The CRG block can operate in one of the following four modes:

- *Run mode*. All functional parts of the CRG block are running in this mode.
- *Wait mode*. This mode allows the user to disable the system and core clocks by programming the individual bits in the CLKSEL register.
- *Stop mode*. Depending on the setting of the PSTP bit of the CLKSEL register, stop mode can be differentiated between full stop (PSTP = 0) and pseudo stop mode (PSTP = 1). In *full-stop* mode, the oscillator is disabled and thus all system and core clocks are stopped. In *pseudo-stop* mode, the oscillator continues to run and most of the system and core clocks are stopped. If the respective enable bits are set, the Computer Operate Properly (COP) and Real-Time Interrupt (RTI) modules will continue to run.

- *Self-clock mode.* This mode is entered if both the clock monitor enable bit (the CME bit of the PLLCTL register) and the self-clock mode enable bit (the SCME bit of PLLCTL) are set and the clock monitor detects a loss of clock (external oscillator or crystal). As soon as the self-clock mode is entered, the CRG starts to perform a clock quality check. The self-clock mode remains active until the clock check indicates that the required quality of the incoming clock signal is met (frequency and amplitude). The self-clock mode should be used for safety purposes only. It provides reduced functionality to the MCU in case a loss of clock is causing severe system conditions.

6.5.2 CRG Signals

- *VDDPLL, VSSPLL.* The PLL is a critical component for deriving a clock signal with stable frequency. These two pins provide the operating voltage (VDDPLL) and ground (VSSPLL) for the PLL circuitry and allow the supply voltage to the PLL to be independent of the power supply to the rest of the circuit.

- *XFC.* A passive external loop filter must be placed on the XFC pin. The filter is a second-order, low-pass filter to eliminate the VCO input ripple. The value of the external filter network and the reference frequency determines the speed of the corrections and the stability of the PLL. If the PLL usage is not required, the XFC pin must be tied to VDDPLL. The PLL loop filter connection recommended by Freescale is shown in Figure 6.5.

- *EXTAL, XTAL.* These two pins allow the user to connect an external crystal oscillator or a CMOS compatible clock to control the internal clock generator circuitry. The circuit connection for an external crystal oscillator is shown in Figure 6.6; the circuit connection that uses a CMOS compatible external clock is shown in Figure 6.7.

- $\overline{\text{RESET}}$. This pin is an active low, bidirectional reset pin. As an input, this signal initializes the microcontroller to a known state. As an output, it indicates that a system reset (internal to MCU) has been triggered.

- $\overline{\text{XCLKS}}$. This signal is an input that controls whether a crystal in combination with the internal oscillator or an external clock source on the EXTAL pin (oscillator circuitry is bypassed) is used to provide the clock signal required for the MCU operation. The $\overline{\text{XCLKS}}$ signal is sampled on the rising edge of $\overline{\text{RESET}}$. When this signal is high, the crystal and the internal oscillator provide the OSC_CLK signal. When this signal is low, the external clock provides the OSC_CLK signal.

Figure 6.5 ■ PLL loop filter connections

Figure 6.6 ■ Common crystal connections

Figure 6.7 ■ External oscillator connections

6.5.3 The Oscillator Block (OSC)

The OSC block in Figure 6.4 has two external pins, EXTAL and XTAL. The oscillator input pin, EXTAL, is intended to be connected to either a crystal or an external clock source. The selection of the crystal or external clock source depends on the $\overline{\text{XCLKS}}$ signal, which is sampled during the reset. The XTAL pin is an output signal that provides crystal circuit feedback and can be buffered to drive other devices with the same voltage amplitude.

A buffered EXTAL signal, OSCCLK, becomes the internal reference clock. The oscillator is enabled based on the PSTP bit and the STOP condition. The oscillator is disabled when the MCU is in STOP mode except when the pseudo-stop mode is enabled. To improve noise immunity, the oscillator is powered by the VDDPLL and VSSPLL power supply pins.

6.5.4 Phase Locked Loop (PLL)

The PLL is used to run the microcontroller with a clock frequency different from the incoming OSCCLK signal and provides high stability to the system clock. In Figure 6.4, the frequency of the PLLCLK signal is determined by the synthesizer (SYNR) and the reference divide (REFDV) registers using the following equation:

$$\text{PLLCLK} = 2 \times \text{OSCCLK} \times \frac{(\text{SYNR} + 1)}{(\text{REFDV} + 1)} \quad (6.1)$$

The contents of the SYNR register and the REFDV register are shown in Figure 6.8 and 6.9, respectively.

The PLL operates in either the acquisition mode or tracking mode, depending on the difference between the output frequency and the target frequency. The PLL can change between acquisition and tracking modes either automatically or manually. The functional diagram of the PLL is shown in Figure 6.10.

7	6	5	4	3	2	1	0
0	0	SYN5	SYN4	SYN3	SYN2	SYN1	SYN0

reset: 0   0   0   0   0   0   0   0

Figure 6.8 ■ The CRG synthesizer register (SYNR)

7	6	5	4	3	2	1	0
0	0	0	0	REFDV3	REFDV2	REFDV1	REFDV0

reset: 0   0   0   0   0   0   0   0

Figure 6.9 ■ The CRG reference divider register (REFDV)

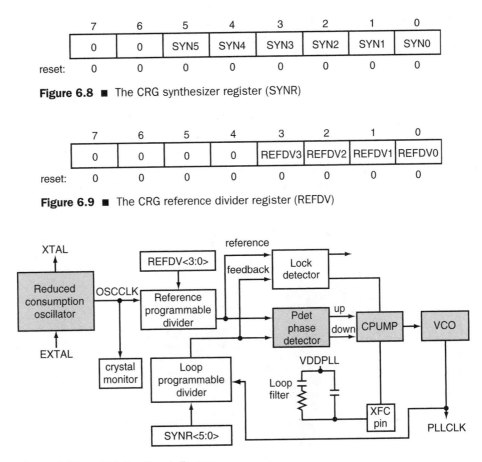

Figure 6.10 ■ PLL functional diagram

PLL OPERATION

In Figure 6.10, the OSCCLK clock is fed through the reference programmable divider and is divided into a range from 1 to 16 (REFDV + 1) to output the reference clock. The VCO output clock (PLLCLK) is fed back through the programmable loop divider and is in a range of 2 to 128 in increments of [2 × (SYNR + 1)] to output the feedback clock.

The phase detector then compares the feedback clock with the reference clock. Correction pulses are generated based on the phase difference between these two signals. The loop filter then slightly alters the DC voltage on the external filter capacitor connected to the XFC pin, based on the width and direction of the correction pulse. The filter can make fast or slow corrections depending on its mode. The values of the external filter network and the reference frequency determine the speed of the corrections and the stability of the PLL.

ACQUISITION AND TRACKING MODES

The lock detector compares the frequencies of the feedback clock and the reference clock. Therefore, the speed of the lock detector is directly proportional to the final reference frequency. The circuit determines the mode of the PLL and the lock condition based on this comparison.

The PLL filter can be manually or automatically configured into one of two possible operating modes: *acquisition* mode and *tracking* mode. In acquisition mode, the filter can make large frequency corrections to the Voltage Controlled Output (VCO). This mode is used at PLL startup or when the PLL has suffered a severe noise hit and the VCO frequency is far off the desired frequency. In tracking mode, the filter makes only small corrections to the frequency of the VCO. PLL jitter is much lower in tracking mode, but the response to noise is also slower. The PLL enters tracking mode when the VCO frequency is nearly correct.

The operation of the PLL is controlled by four registers: CRGINT, CRGFLG, CRGSEL, and PLLCTL. The contents of these four registers are shown in Figures 6.11, 6.12, 6.13, and 6.14.

7	6	5	4	3	2	1	0
RTIE	0	0	LOCKIE	0	0	SCMIE	0
0	0	0	0	0	0	0	0

RTIE: Real-time interrupt enable bit
  0 = Interrupt requests from RTI are disabled.
  1 = Interrupt requests from RTI are enabled.
LOCKIE: Lock interrupt enable bit
  0 = LOCK interrupt requests are disabled.
  1 = LOCK interrupt requests are enabled.
SCMIE: Self-clock mode interrupt enable bit
  0 = SCM interrupt requests are disabled.
  1 = Interrupt will be requested whenever the SCMIF bit is set.

Figure 6.11 ■ The CRG interrupt enable register (CRGINT)

The CRG interrupt register (CRGINT) enables/disables the interrupts associated with the CRG module. The CRG Flag Register (CRGFLG) holds the status flags of the CRG module. The CRGSEL register selects the clock source for the PLL. The PLLCTL register provides the overall control to the PLL module.

The PLL can change the bandwidth or operational mode of the loop filter manually or automatically. In *automatic bandwidth control mode* (the AUTO bit of the PLLCTL register is set to 1), the lock detector automatically switches between acquisition and tracking modes. The automatic bandwidth control mode is also used to determine when the PLL clock (PLLCLK) is safe to use as the source for the system and core clocks. When the LOCK bit of the CRGFLG register is set to 1, the PLLCLK can be used safely. The setting of the LOCK bit can be detected by using the interrupt or polling method. Any change of the LOCK bit will cause the LOCKIF flag bit to be set to 1 and may optionally request an interrupt to the CPU.

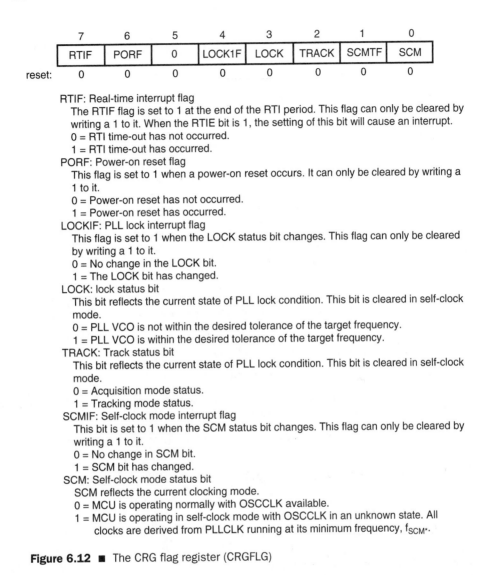

RTIF: Real-time interrupt flag
  The RTIF flag is set to 1 at the end of the RTI period. This flag can only be cleared by writing a 1 to it. When the RTIE bit is 1, the setting of this bit will cause an interrupt.
  0 = RTI time-out has not occurred.
  1 = RTI time-out has occurred.

PORF: Power-on reset flag
  This flag is set to 1 when a power-on reset occurs. It can only be cleared by writing a 1 to it.
  0 = Power-on reset has not occurred.
  1 = Power-on reset has occurred.

LOCKIF: PLL lock interrupt flag
  This flag is set to 1 when the LOCK status bit changes. This flag can only be cleared by writing a 1 to it.
  0 = No change in the LOCK bit.
  1 = The LOCK bit has changed.

LOCK: lock status bit
  This bit reflects the current state of PLL lock condition. This bit is cleared in self-clock mode.
  0 = PLL VCO is not within the desired tolerance of the target frequency.
  1 = PLL VCO is within the desired tolerance of the target frequency.

TRACK: Track status bit
  This bit reflects the current state of PLL lock condition. This bit is cleared in self-clock mode.
  0 = Acquisition mode status.
  1 = Tracking mode status.

SCMIF: Self-clock mode interrupt flag
  This bit is set to 1 when the SCM status bit changes. This flag can only be cleared by writing a 1 to it.
  0 = No change in SCM bit.
  1 = SCM bit has changed.

SCM: Self-clock mode status bit
  SCM reflects the current clocking mode.
  0 = MCU is operating normally with OSCCLK available.
  1 = MCU is operating in self-clock mode with OSCCLK in an unknown state. All clocks are derived from PLLCLK running at its minimum frequency, f_{SCM}.

Figure 6.12 ■ The CRG flag register (CRGFLG)

The PLL can also operate in *manual* mode (AUTO bit = 0). Manual mode is used by systems that do not require an indicator of the lock condition for proper operation. Such systems typically operate well below the maximum system frequency (f_{sys}) and require fast startup. To operate in manual mode, one must observe the following procedure:

1. Assert the ACQ bit (in PLLCTL register) before turning on the PLL manual mode. This configures the filter in acquisition mode.

2. After turning on the PLL, wait for a given time (t_{acq}) before entering tracking mode (clear ACQ to 0). The parameter t_{acq} can be found in the HCS12DP256 block user guide.

3. After entering the tracking mode, wait for t_{al} ns before selecting PLLCLK as the source for system and core clocks (by setting the PLLSEL bit of the PLLCTL register to 1).

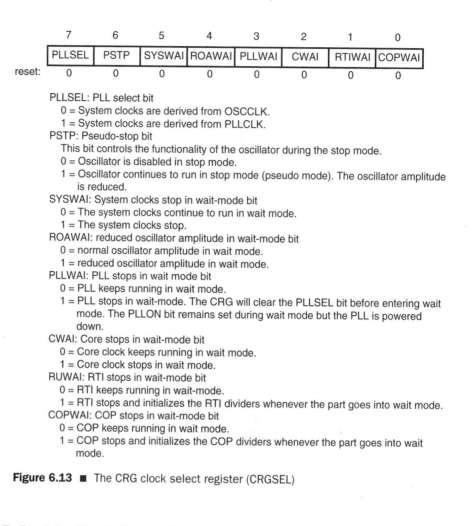

	7	6	5	4	3	2	1	0
	PLLSEL	PSTP	SYSWAI	ROAWAI	PLLWAI	CWAI	RTIWAI	COPWAI
reset:	0	0	0	0	0	0	0	0

PLLSEL: PLL select bit
 0 = System clocks are derived from OSCCLK.
 1 = System clocks are derived from PLLCLK.
PSTP: Pseudo-stop bit
 This bit controls the functionality of the oscillator during the stop mode.
 0 = Oscillator is disabled in stop mode.
 1 = Oscillator continues to run in stop mode (pseudo mode). The oscillator amplitude
   is reduced.
SYSWAI: System clocks stop in wait-mode bit
 0 = The system clocks continue to run in wait mode.
 1 = The system clocks stop.
ROAWAI: reduced oscillator amplitude in wait-mode bit
 0 = normal oscillator amplitude in wait mode.
 1 = reduced oscillator amplitude in wait mode.
PLLWAI: PLL stops in wait mode bit
 0 = PLL keeps running in wait mode.
 1 = PLL stops in wait-mode. The CRG will clear the PLLSEL bit before entering wait
   mode. The PLLON bit remains set during wait mode but the PLL is powered
   down.
CWAI: Core stops in wait-mode bit
 0 = Core clock keeps running in wait mode.
 1 = Core clock stops in wait mode.
RUWAI: RTI stops in wait-mode bit
 0 = RTI keeps running in wait-mode.
 1 = RTI stops and initializes the RTI dividers whenever the part goes into wait mode.
COPWAI: COP stops in wait-mode bit
 0 = COP keeps running in wait mode.
 1 = COP stops and initializes the COP dividers whenever the part goes into wait
   mode.

Figure 6.13 ■ The CRG clock select register (CRGSEL)

6.5.5 System Clock Generation

The clock generator is illustrator in Figure 6.15. This circuit generates the system and core clocks used in the microcontroller. When dealing with external memory or peripheral modules, the *bus clock* (also referred to as the *E clock*) is used. The E clock is derived by dividing the SYSCLK clock by 2. If the PLLCLK is chosen to be the SYSCLK, then the frequency of the E clock is a half that of PLLCLK.

When the MCU enters the self-clock mode, the oscillator clock source is switched to the PLLCLK running at its minimum frequency, f_{SCM}.

Either the oscillator output OSCCLK (when PLLSEL = 0) or the PLL output PLLCLK (PLLSEL = 1) can be selected as SYSCLK. The oscillator can be completely bypassed and turned off by selecting an external clock source instead. The clock monitor, PLL, RTI, COP and all clock signals based on OSCCLK are driven by this external clock instead of the output of the oscillator. As one can see from Figure 6.15, SYSCLK can be a buffered version of the external clock input and the bus clock can be derived by dividing the external clock by 2.

7	6	5	4	3	2	1	0
CME	PLLON	AUTO	ACQ	0	PRE	PCE	SCME

reset: 0   0   0   0   0   0   0   0

CME: Clock monitor enable bit
   0 = Clock monitor is disabled.
   1 = Clock monitor is enabled. Slow or stopped clocks will cause a clock monitor reset
       sequence or self-clock mode.
PLLON: Phase-lock loop on bit
   0 = PLL is turned off.
   1 = PLL is turned on. If AUTO bit is set, the PLL will lock automatically.
AUTO: Automatic bandwidth control bit
   0 = Automatic mode control is disabled and the PLL is under software control, using
       ACQ bit.
   1 = High bandwidth filter is selected.
ACQ: Acquisition bit (if AUTO bit = 1, this bit has no effect)
   0 = Low bandwidth filter is selected.
   1 = High bandwidth filter is selected.
PRE: RTI enable during pseudo-stop bit
   0 = RTI stops running during pseudo-stop mode.
   1 = RTI continues running during pseudo-stop mode.
PCE: COP enable during pseudo-stop bit
   0 = COP stops running during pseudo-stop mode.
   1 = COP continues running during pseudo-stop mode.
SCME: Self-clock mode enable bit
   0 = Detection of crystal clock failure causes clock monitor reset.
   1 = Detection of crystal clock failure forces the MCU in select-clock mode.

Figure 6.14 ■ The CRG PLL control register (PLLCTL)

Example 6.2

▼

    There is a system that derives its bus clock from the PLL circuit, and an external clock of 8 MHz is selected. The desired bus clock is 24 MHz. Write an instruction sequence to perform the desired configuration.

Solution: Since the bus clock frequency is higher, we need to use the PLL circuit. According to Figure 6.15, the frequency of SYSCLK would be 48 MHz. The frequencies of OSCCLK and PLL-CLK are 8 MHz and 48 MHz, respectively. According to equation 6.1,

$$48\text{MHz} = 2 \times 8 \text{ MHz} \times [\text{SYNR} + 1]/[\text{REFDV} + 1]$$

One of the alternatives could be to set SYNR and REFDV to 2 and 0, respectively. The following instruction can achieve the desired configuration:

```
movb    #$02,SYNR        ; set SYNR to 2
movb    #$0,REFDV        ; set REFDV to 0
movb    #$80,CRGSEL      ; enable PLL, keep SYSCLK running in wait mode,
                         ; keep RTI, COP, PLL, and core running in wait mode
movb    #$60,PLLCTL      ; disable clock monitor, enable PLL, set automatic
                         ; bandwidth control, disable RTI and COP in pseudo stop
```

    In addition to this instruction sequence, one also needs to tie the $\overline{\text{XCLKS}}$ pin to ground to bypass the crystal oscillator.

▲

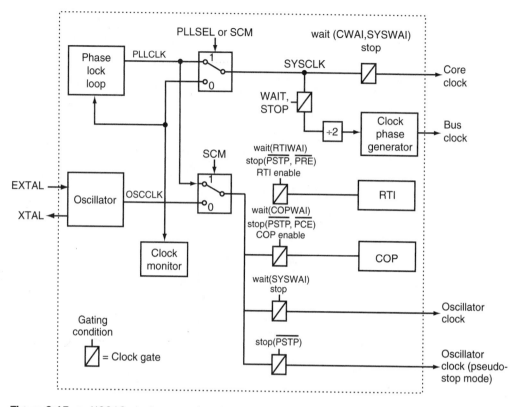

Figure 6.15 ■ HCS12 clock generation circuit

Example 6.3

There is a system that uses a 4-MHz crystal oscillator to derive a 24-MHz bus clock. Write an instruction sequence to perform the required configuration.

Solution: The frequency of SYSCLK would be 48 MHz. The frequencies of OSCCLK and PLL-CLK are 4 MHz and 48 MHz, respectively. According to equation 6.1,

$$48\text{MHz} = 2 \times 4 \text{ MHz} \times [\text{SYNR} + 1] / [\text{REFDV} + 1]$$

One of the alternatives could be to set SYNR and REFDV to 5 and 0, respectively. The following instruction can achieve the desired configuration:

```
movb    #$05,SYNR      ; set SYNR to 5
movb    #$0,REFDV      ; set REFDV to 0
movb    #$80,CRGSEL    ; enable PLL, keep SYSCLK running in wait mode,
                       ; keep RTI, COP, PLL, and core running in wait mode
movb    #$60,PLLCTL    ; disable clock monitor, enable PLL, set automatic
                       ; bandwidth control, disable RTI and COP in pseudo stop
```

One also needs to pull the $\overline{\text{XCLKS}}$ pin to high to select the crystal oscillator to generate the OSCCLK signal.

6.5.6 Clock Monitor

The clock monitor circuit is based on an internal resistor-capacitor (RC) time delay so that it can operate without any MCU clocks. If no OSCCLK edges are detected within this RC time delay, the clock monitor indicates failure, which asserts self-clock mode or generates a system reset depending on the state of the SCME bit. If the clock monitor is disabled or the presence of clocks is detected, no failure is indicated. The clock monitor function is enabled/disabled by the CME control bit.

6.6 Real-Time Interrupt (RTI)

The main function of the RTI circuit is to generate hardware interrupts periodically. If enabled, this interrupt will occur at the rate selected by the RTICTL register. The contents of this register are shown in Figure 6.16. The possible interrupt periods (in number of OSCCLK cycles) are listed in Table 6.4.

7	6	5	4	3	2	1	0
0	RTR6	RTR5	RTR4	RTR3	RTR2	RTR1	RTR0

reset: 0   0   0   0   0   0   0   0

Figure 6.16 ■ CRG RTI control register (RTICTL)

RTR[3:0]	RTR[6:4]							
	000 (off)	001 (2^{10})	010 (2^{11})	011 (2^{12})	100 (2^{13})	101 (2^{14})	110 (2^{15})	111 (2^{16})
0000 (÷1)	off*	2^{10}	2^{11}	2^{12}	2^{13}	2^{14}	2^{15}	2^{16}
0001 (÷2)	off*	2×2^{10}	2×2^{11}	2×2^{12}	2×2^{13}	2×2^{14}	2×2^{15}	2×2^{16}
0010 (÷3)	off*	3×2^{10}	3×2^{11}	3×2^{12}	3×2^{13}	3×2^{14}	3×2^{15}	3×2^{16}
0011 (÷4)	off*	4×2^{10}	4×2^{11}	4×2^{12}	4×2^{13}	4×2^{14}	4×2^{15}	4×2^{16}
0100 (÷5)	off*	5×2^{10}	5×2^{11}	5×2^{12}	5×2^{13}	5×2^{14}	5×2^{15}	5×2^{16}
0101 (÷6)	off*	6×2^{10}	6×2^{11}	6×2^{12}	6×2^{13}	6×2^{14}	6×2^{15}	6×2^{16}
0110 (÷7)	off*	7×2^{10}	7×2^{11}	7×2^{12}	7×2^{13}	7×2^{14}	7×2^{15}	7×2^{16}
0111 (÷8)	off*	8×2^{10}	8×2^{11}	8×2^{12}	8×2^{13}	8×2^{14}	8×2^{15}	8×2^{16}
1000 (÷9)	off*	9×2^{10}	9×2^{11}	9×2^{12}	9×2^{13}	9×2^{14}	9×2^{15}	9×2^{16}
1001 (÷10)	off*	10×2^{10}	10×2^{11}	10×2^{12}	10×2^{13}	10×2^{14}	10×2^{15}	10×2^{16}
1010 (÷11)	off*	11×2^{10}	11×2^{11}	11×2^{12}	11×2^{13}	11×2^{14}	11×2^{15}	11×2^{16}
1011 (÷12)	off*	12×2^{10}	12×2^{11}	12×2^{12}	12×2^{13}	12×2^{14}	12×2^{15}	12×2^{16}
1100 (÷13)	off*	13×2^{10}	13×2^{11}	13×2^{12}	13×2^{13}	13×2^{14}	13×2^{15}	13×2^{16}
1101 (÷14)	off*	14×2^{10}	14×2^{11}	14×2^{12}	14×2^{13}	14×2^{14}	14×2^{15}	14×2^{16}
1110 (÷15)	off*	15×2^{10}	15×2^{11}	15×2^{12}	15×2^{13}	15×2^{14}	15×2^{15}	15×2^{16}
1111 (÷16)	off*	16×2^{10}	16×2^{11}	16×2^{12}	16×2^{13}	16×2^{14}	16×2^{15}	16×2^{16}

*Denotes the default value of reset. This value disables RTI.

Table 6.4 ■ RTI interrupt period (in OSCCLK cycles)

6.7 Computer Operating Properly

The COP (a free-running watchdog timer) enables the user to find out if a user program is running and sequencing properly. When the COP times out, it resets the CPU. The CPU reset by the COP is an indication that the software is no longer being executed in the intended sequence. The software that utilizes the COP function must include an instruction sequence to prevent the COP from timing out.

To prevent the COP module from resetting the MCU, one must write the value $55 followed by $AA into the ARMCOP register. Other instructions can be inserted between the values $55 and $AA, but the sequence ($55, $AA) must be completed prior to the COP timeout period to avoid a COP reset. Writing any value other than $55 or $AA into this register will reset the MCU.

The functioning of the COP module and its timeout period are controlled by the COPCTL register. The contents of this register are shown in Figure 6.17. The COP module is disabled if the CR2:CR0 bits of this register are set to 000. The COP has a windowed option for its operation. The windowed COP operation is enabled by setting the WCOP bit of the COPCTL register. In this mode, writes to the ARMCOP register to clear the COP timer must occur in the last 25% of the selected timeout period. A premature write will immediately reset the MCU.

	7	6	5	4	3	2	1	0
	WCOP	RSBCK	0	0	0	CR2	CR1	CR0
reset:	0	0	0	0	0	0	0	0

WCOP: Windowed COP mode bit
    When set, a write to the ARMCOP register must occur in the last 25% of the selected period. A write during the first 75% of the selected period will reset the MCU.
    0 = Normal COP operation.
    1 = Windowed COP operation.
RSBCK: COP and RTI stop in active BDM mode bit
    0 = Allows the COP and RTI to keep running in active BDM mode.
    1 = Stops the COP and RTI whenever the HCS12 is in active BDM mode.
CRICRO: COP watchdog time rate select (number of OSCCLK cycles)
    000: COP disabled
    001: 2^{14}
    010: 2^{16}
    011: 2^{18}
    100: 2^{20}
    101: 2^{22}
    110: 2^{23}
    111: 2^{24}

Figure 6.17 ■ CRG COP control register (COPCTL)

6.8 Low-Power Modes

When a microcontroller is performing normal operations, power consumption is unavoidable. However, the microcontroller in an embedded system may not always be performing useful operations. Under this situation, it would be ideal for the power consumption to be reduced to the minimum. This issue is especially important for those embedded products powered by batteries. The HCS12 has two low-power modes that can reduce power consumption dramatically: **wait** and **stop** modes.

6.8.1 The WAI Instruction

The wait mode is entered by executing the WAI instruction. The WAI instruction pushes all CPU registers (except the stack pointer) and the return address onto the stack and enters a wait state. During the wait state, CPU clocks are stopped (clock signals that drive the ALU and register file), but other clocks in the microcontroller (clock signals that drive peripheral functions) continue to run.

The CPU leaves the wait state when it senses one or more of the following events:

- Maskable interrupts that are not masked
- Nonmaskable interrupts
- Resets

Upon leaving the wait state, the HCS12 CPU sets the appropriate interrupt mask bit(s), fetches the vector corresponding to the exception sensed, and continues instruction execution at the location the vector points to.

6.8.2 The Stop Instruction

When the S bit in the CCR register is cleared and a stop instruction is executed, the HCS12 saves all CPU registers (except the stack pointer) in the stack, stops all system clocks, and puts the microcontroller in *standby* mode. The standby operation minimizes the system power consumption. The contents of registers and the states of I/O pins remain unchanged.

Asserting the $\overline{RESET}$, $\overline{XIRQ}$, or $\overline{IRQ}$ signal ends standby mode. If it is the $\overline{XIRQ}$ signal that ends the stop mode and the X mask bit is 0, instruction execution resumes with a vector fetch for the $\overline{XIRQ}$ interrupt. If the X mask bit is 1 ($\overline{XIRQ}$ disabled), a two-cycle recovery sequence is used to adjust the instruction queue, and execution continues with the next instruction after the **stop** instruction.

6.9 Resets

There are four possible sources of resets:

- Power-On (POR) and Low-Voltage Detector (LVD) reset
- $\overline{RESET}$ pin
- COP reset
- Clock monitor reset

The COP and clock monitor resets have been discussed earlier. Power-on, Low-Voltage Detector, and the $\overline{RESET}$ pin reset share the same reset vector. The COP reset and the clock monitor reset each have a separate vector.

6.9.1 Power-On Reset

The HCS12 has a circuitry to detect when the V_{DD} supply to the MCU has reached a certain level and asserts reset to the internal circuits. The detector circuit is triggered by the slew rate. As soon as a power-on reset is triggered, the CRG module performs a quality check on the incoming clock signal. Start of the reset sequence is delayed until the clock check indicates a valid clock signal or the clock check was unsuccessful and the CRG module enters self-clock mode.

6.9.2 External Reset

The HCS12 distinguishes between internal and external resets by sensing how quickly the signal on the $\overline{RESET}$ pin rises to logic high after it has been asserted. When the HCS12 senses any of the four reset conditions, internal circuitry drives the $\overline{\text{RESET}}$ pin low for 128 SYSCLK cycles (this number might be increased by 3 to 6 SYSCLK cycles), then releases. Sixty-four SYSCLK cycles later, the CPU samples the state of the signal applied to the $\overline{\text{RESET}}$ pin. If the signal is still low, an external reset has occurred. If the signal is high, the reset has been initiated internally by either the COP system or the clock monitor.

The power supply to an embedded system may drop below the required level. If the microcontroller keeps working under this situation, the contents of the EEPROM might be corrupted. The common solution is to pull the $\overline{\text{RESET}}$ signal to low so that the microcontroller cannot execute instructions. A low-voltage-inhibit (LVI) circuit, such as the Freescale MC34064, can be used to protect against the EEPROM corruption. Figure 6.18 shows an example of the reset circuit with a manual reset and LVI circuit.

Figure 6.18 ■ A typical external reset circuit

6.10 HCS12 Operation Modes

As listed in Table 6.5, the HCS12 can operate in eight different modes. Each mode has a different default memory map and external bus configuration. After reset, most system resources can be mapped to other addresses by writing to the appropriate control registers.

The states of the BKGD, MODB, and MODA pins when the $\overline{\text{RESET}}$ signal is low determine the operating mode after the CPU leaves the reset state. The SMODN, MODB, and MODA bits in the MODE register show the current operation mode and provide limited mode switching during the operation. The states of the BKGD, MODB, and MODA pins are latched into these bits on the rising edge of the $\overline{\text{RESET}}$ signal. During reset, an active pullup (on-chip transistor) is connected to the BKGD pin (as input) and active pulldowns (on-chip transistors) are connected to the MODB and MODA pins. If an open circuit occurs on any of these pins, the device will operate in normal single-chip mode.

The two basic types of operation modes are:

1. Normal modes—Some registers and bits are protected against accidental changes.

2. Special modes—Greater access for special purposes such as testing and emulation to protected control registers and bits is allowed.

The background debug mode (BDM) is a system development and debug feature and is available in all modes. In special single-chip mode, BDM is active immediately after reset.

BKGD	MODB	MODA	Mode	Port A	Port B
0	0	0	Special single chip	General-purpose I/O	General-purpose I/0
0	0	1	Special expanded narrow	ADDR[15:8]DATA[7:0]	ADDR[7:0]
0	1	0	Special peripheral	ADDR/DATA	ADDR/DATA
0	1	1	Special expanded wide	ADDR/DATA	ADDR/DATA
1	0	0	Normal single chip	General-purpose I/O	General-purpose I/0
1	0	1	Normal expanded narrow	ADDR[15:8]DATA[7:0]	ADDR[7:0]
1	1	0	Reserved (forced to peripheral)	—	—
1	1	1	Normal expanded wide	ADDR/DATA	ADDR/DATA

Table 6.5 ■ HCS12 Mode Selection

6.10.1 Normal Operation Modes

These modes provide three operation configurations. Background debugging is available in all three modes, but must first be enabled for some operations by means of a BDM command. BDM can then be made active by another command.

NORMAL EXPANDED WIDE MODE

In this mode, ports A and B are used as the multiplexed 16-bit address and data buses. ADDR[15..8] and DATA[15..8] are multiplexed on port A. ADDR[7..0] and DATA[7..0] are multiplexed on port B.

NORMAL EXPANDED NARROW MODE

The 16-bit external address bus uses port A for the high byte and port B for the low byte. The 8-bit external data bus uses port A. ADDR[15..8] and DATA[7..0] are multiplexed on port A.

NORMAL SINGLE-CHIP MODE

Normal single-chip mode has no external buses. Ports A, B, and E are configured for general-purpose I/O. Port E bits 1 and 0 are input only with internal pullups and the other 22 pins are bidirectional I/O pins that are initially configured as high-impedance inputs. Port E pullups are enabled on reset. Port A and B pullups are disabled on reset.

6.10.2 Special Operation Modes

Special operation modes are commonly used in factory testing and system development.

SPECIAL EXPANDED WIDE MODE

This mode is for emulation of normal expanded wide mode and emulation of normal single-chip mode with a 16-bit bus. The bus-control pins of port E are all configured for their bus-control output functions rather than general-purpose I/O.

SPECIAL EXPANDED NARROW MODE

This mode is for emulation of normal expanded narrow mode. External 16-bit data is handled as two back-to-back bus cycles, one for the high byte followed by one for the low byte. Internal operations continue to use full 16-bit data paths.

SPECIAL SINGLE-CHIP MODE

This mode can be used to force the microcontroller to active BDM mode to allow a system debug through the BKGD pin. The HCS12 CPU does not fetch the reset vector nor execute application code as it would in other modes. Instead, the active background mode is in control of CPU execution, and BDM firmware waits for additional serial commands through the BKGD pin. There are no external address and data buses in this mode. The microcontroller operates as a stand-alone device and all program and data space are on-chip. External port pins can be used for general-purpose I/O.

SPECIAL PERIPHERAL MODE

The HCS12 CPU is not active in this mode. An external master can control on-chip peripherals for testing purposes. It is not possible to change to or from this mode without going through reset. Background debugging should not be used while the microcontroller is in special peripheral mode, as internal bus conflicts between the BDM and the external master can cause improper operation of both modes.

6.10.3 Background Debug Mode

Background debug mode (BDM) is an auxiliary operation mode that is used for system development. This mode is discussed in detail in Chapter 3.

6.11 Summary

Interrupt is a special event that requires the CPU to stop normal program execution and provide certain service to the event. The interrupt mechanism has many applications, including coordinating I/O activities, exiting from software errors, and reminding the CPU to perform routine work, etc.

Some interrupts are *maskable* and can be ignored by the CPU. Other interrupts are *nonmaskable* and cannot be ignored by the CPU. Nonmaskable interrupts are often used to handle emergent events such as power failure.

Multiple interrupts may be pending at the same time. The CPU needs to decide which one to service first. The solution to this issue is to *prioritize* all of the interrupt sources. The pending interrupt with the highest priority will receive the service before other pending interrupts.

The CPU provides service to an interrupt request by executing an *interrupt service routine*. The current program counter value will be saved in the stack before the CPU executes the service routine so that CPU control can be returned to the interrupted program when the interrupt service routine is completed.

In order to provide service to the interrupt, the CPU must have some way to find out the starting address of the interrupt service routine. There are three methods to determine the starting address (called *interrupt vector*) of the interrupt service routine:

1. Each interrupt vector is predefined when the microcontroller is designed. In this method, the CPU simply jumps to the predefined location to execute the service routine.

2. Each interrupt vector is stored in a predefined memory location. When an interrupt occurs, the CPU fetches the interrupt vector from that predefined memory location. The HCS12 uses this approach.

3. The interrupt source provides an interrupt vector number to the CPU so that the CPU can figure out the memory location where the interrupt vector is stored. The CPU needs to perform a read bus cycle to obtain the interrupt vector number.

There are three steps in the interrupt programming:

Step 1
Initialize the interrupt vector table that holds all the interrupt vectors. This step is not needed for those microcontrollers that use the first method to resolve the interrupt vector.

Step 2
Write the interrupt service routine.

Step 3
Enable the interrupt to be serviced.

Users who own a demo board with a resident D-Bug12 monitor would use the EVB mode for application development. The D-Bug12 monitor has occupied the default memory space where the interrupt vectors are to be stored. To enable the user to use interrupts in their applications, D-Bug12 allows vectors to be stored in a table in the SRAM.

Clock signals are critical to the proper operation of the MCU. The HCS12 uses an on-chip PLL circuit to generate the system clock and bus clock. The PLL can generate a high-frequency clock signal using a low-frequency crystal. In addition, the PLL provides stability to the clock signals.

Reset is a mechanism for:

1. Setting up operation mode for the microcontroller

2. Setting up initial values for control registers

3. Exiting from software errors and some hardware errors

All HCS12 microcontrollers have the same number of resets and nonmaskable interrupt sources despite the fact that they may not have the same number of maskable interrupts. The HCS12 has two low-power modes that are triggered by the execution of WAI and STOP instructions. Power consumption will be reduced dramatically in either low-power mode. The HCS12 has a COP timer reset mechanism to detect the software error. A software program that behaves properly will reset the COP timer before it times out and prevent it from resetting the CPU.

The HCS12 has a clock monitor reset mechanism that can detect the slowing down or loss of clock signals. Whenever the clock frequency gets too low, the clock monitor will detect it and reset the CPU.

Two examples are given to illustrate the procedure of interrupt programming using the demo board with the D-Bug12 monitor.

The HCS12 has seven different operation modes divided into two basic categories: normal modes and special modes. Normal modes are used for embedded applications, whereas special modes are used in fabrication testing and development debugging activities.

6.12 Exercises

E6.1 What is the name given to a routine that is executed in response to an interrupt?

E6.2 What are the advantages of using interrupts to handle data inputs and output?

E6.3 What are the requirements for interrupt processing?

E6.4 How do you enable other interrupts when the HCS12 is executing an interrupt service routine?

E6.5 Why would there be a need to promote one of the maskable interrupts to the highest priority among all maskable interrupts?

E6.6 Write the assembler directives to initialize the $\overline{IRQ}$ interrupt vector located at $3000 for the EVB mode of the D-Bug12 monitor.

E6.7 What is the last instruction in most interrupt service routines? What does this instruction do?

E6.8 Suppose that the HCS12 is executing the following instruction segment and the $\overline{IRQ}$ interrupt occurs when the TSY instruction is being executed. What will be the contents of the top ten bytes in the stack?

```
org     $2000
lds     #$2000
clra
ldx     #$0
bset    10,X $48
ldab    #$40
inca
tap
pshb
tsy
adda    #10
```

E6.9 Suppose that the OSCCLK clock frequency is 5 MHz. Compute the COP watchdog timer timeout period for all the possible combinations of the CR2, CR1, and CR0 bits in the COPCTL register.

E6.10 Assume that the starting address of the service routine of the timer overflow interrupt is $3000. Write the assembler directives to initialize its vector table entry on a demo board with D-Bug12 monitor.

E6.11 Write an instruction sequence to clear the X and I bits in the CCR. Write an instruction sequence to set the S, X, and I bits in the CCR.

E6.12 Why does the HCS12 need to be reset when the power supply is too low?

E6.13 Write the instruction sequence to prevent the COP timer from timing out and resetting the microcomputer.

6.13 Lab Exercises and Assignments

L6.1 $\overline{IRQ}$ input interrupt experiment. Use the 555 timer (or take the signal from a function generator) to generate a digital waveform with frequency equal to approximately 1 Hz. The circuit connection of the 555 timer is illustrated in Figure L6.1. Connect the 555 timer output (pin 3) to the $\overline{IRQ}$ pin. The $\overline{IRQ}$ interrupt service routine would output the message "Interrupt k". Enter the program (both assembly and C) to a file, assemble/compile, and download the S-record file onto the demo board for execution. If you are using the SSE256 demo board, take the 1-Hz signal from the on-board function generator output instead of generating it using the 555 timer circuit.

Figure L6.1 ■ HCS12 $\overline{\text{IRQ}}$ pin interrupt circuit

L6.2 *Simple interrupts.* Connect the $\overline{\text{IRQ}}$ pin of the demo board to a debounced switch that can generate a negative-going pulse. Write a main program and an $\overline{\text{IRQ}}$ interrupt service routine. The main program initializes the variable *irq_cnt* to 10, stays in a loop, and keeps checking the value of *irq_cnt*. When *irq_cnt* is decremented to 0, the main program jumps back to monitor. The $\overline{\text{IRQ}}$ service routine simply decrements *irq_cnt* by 1 and returns. The lab procedure is as follows:

Step 1
Connect the $\overline{\text{IRQ}}$ pin of the demo board to a debounced switch that can generate a clean negative-going pulse. A debounced switch is available on the SSE256 demo board.

Step 2
Enter the main program and $\overline{\text{IRQ}}$ service routine, assemble them, and then download them to the single-board computer. Remember to enable the $\overline{\text{IRQ}}$ interrupt in your program.

Step 3
Pulse the switch 10 times. If everything works properly, you should see the D-Bug12 monitor prompt after 10 pulses applied to the $\overline{\text{IRQ}}$ pin.

7

Parallel Ports

7.1 Objectives

After completing this chapter, your should be able to

- Define I/O addressing methods

- Explain the data transfer synchronization methods between the CPU and I/O interface chip

- Explain the data transfer synchronization methods between the I/O interface chip and the I/O device

- Explain input and output handshake protocols

- Verify electrical compatibility when interfacing a microcontroller with peripheral devices

- Input data from simple switches

- Input data from keypads and keyboards

- Output data to LED and LCD displays

- Interface with a D/A converter to generate waveforms

- Explain the operation and programming of a stepper motor

- Explain the operation, application, and programming of key wakeup ports

7.2 Basic Concepts of I/O

I/O devices are also called *peripheral* devices. They are mainly used to exchange data with a computer. Examples of I/O devices include switches, light-emitting diodes (LED), cathode-ray tube (CRT) screens, printers, modems, keyboards, disk drives, and so on. The speed and electrical characteristics of I/O devices are very different from those of the CPU and hence it is not feasible to connect them directly to the CPU. Instead, interface chips are used to resolve the differences between the microprocessor and I/O devices.

The major function of the interface chip is to synchronize data transfer between the CPU and I/O devices. An interface chip consists of control registers, data registers, status registers, data direction registers, and control circuitry. Control registers allow one to set up parameters for the desired I/O operations. The status register reports the progress and status of the I/O operations. The data direction register allows one to select the data transfer direction for each I/O pin. A data register may hold the data to be sent to the output device or the new data placed by the input device.

In an input operation, the input device places data in the data register, which holds the data until they are read by the CPU. In an output operation, the CPU places the data in the data register, which holds the data until they are fetched by the output device.

An interface chip has data pins that are connected to the microprocessor data bus and I/O port pins that are connected to the I/O device, as illustrated in Figure 7.1. In a large embedded system, many I/O devices are attached to the data bus via interface chips. Only one device is allowed to drive data to the data bus at a time. Otherwise, *data bus contention* can result and the system may be damaged. The address decoder in Figure 7.1 makes sure that one and only one device is allowed to drive data to the data bus or accept data from the data bus. An interface chip is allowed to respond to the data transfer request from the microprocessor when its chip enable ($\overline{CE}$) input is low. Otherwise, the interface chip is electrically isolated from the data bus.

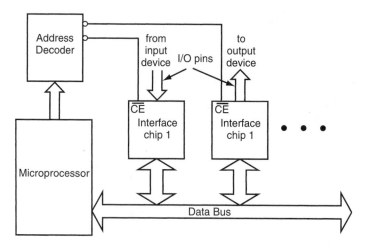

Figure 7.1 ■ Interface chip, I/O devices, and microprocessor

Data transfer between the I/O device and the interface chip can proceed bit-by-bit (serial) or in multiple bits (parallel). Data are transferred serially in low-speed devices such as modems and low-speed printers. Parallel data transfer is mainly used by high-speed I/O devices. Only parallel I/O transfer will be discussed in this chapter.

7.3 I/O Addressing

An interface chip normally has several registers. An address must be assigned to each of these registers so that they can be accessed. Two issues are related to the access of I/O registers:

1. *Address space.* I/O devices can share the same memory space with memory devices (such as SRAM, ROM, and so on) or have their own memory space. In the earlier days of microprocessors, some manufacturers (including Intel and Zilog) used a separate memory space for I/O devices. The current trend among most microprocessor and microcontroller manufacturers is to use the same memory space for memory and I/O devices.

2. *Instruction set and addressing modes.* I/O devices may have their own instruction set and addressing modes or use the same instruction set and addressing modes with memory devices. In the first approach, the microprocessor may use the following instructions for input and output:

   out 3: sends data in accumulator to I/O device at address 3

   in 5: inputs a byte from input device 5 to the accumulator

   In the second approach, the microprocessor uses the same instructions and addressing modes to perform input and output operations. Again, the current trend is the second approach.

Traditionally, Freescale microprocessors and microcontrollers have used the same addressing modes and instruction sets to access I/O and memory devices. Memory and I/O devices share a single memory space.

7.4 I/O Synchronization

The role of an interface chip is shown in Figure 7.2. The microprocessor deals with the interface chip rather than the I/O devices. The electronics in the I/O devices convert electrical signals into mechanical actions or vice versa. The functions of an interface chip and the CPU are incorporated onto the same chip in a microcontroller.

When inputting data, the microprocessor reads data from the interface chip. There must be a mechanism to make sure that data is valid when the microprocessor reads it. When outputting data, the microprocessor writes data into the interface chip. Again, there must be some mechanism to make sure that the output device is ready to accept data when the microprocessor outputs. There are two aspects of I/O synchronization: the synchronization between the microprocessor and the interface chip and the synchronization between the interface chip and the I/O device.

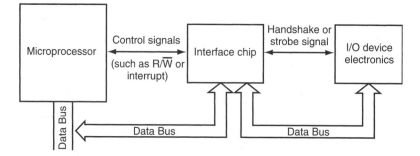

Figure 7.2 ■ The role of an interface chip

7.4.1 Synchronizing the Microprocessor and the Interface Chip

To input valid data from an input device, the microprocessor must make sure that the interface chip has correctly latched the data from the input device. There are two ways to do that.

1. *The polling method.* The interface chip uses a status flag to indicate whether it has valid data (stored in a data register) for the microprocessor. The microprocessor knows that the interface chip has valid data when the status flag is set to 1. The microprocessor keeps checking (reading) the status flag until the status flag is set to 1, and then it reads the data from the data register. When this method is used, the microprocessor is tied up and cannot do anything else when it is waiting for the data. However, this method is very simple to implement and is often used when the microprocessor has nothing else to do when waiting for the completion of the input operation.

2. *Interrupt-driven method.* In this method, the interface chip asserts an interrupt signal to the microprocessor when it has valid data in the data register. The microprocessor then executes the service routine associated with the interrupt to read the data.

To output data successfully, the microprocessor must make sure that the output device is not busy. There are also two ways to do this.

1. *The polling method.* The interface chip has a data register that holds data to be output to an output device. New data should be sent to the interface chip only when the data register of the interface chip is empty. In this method, the interface chip uses a status bit to indicate whether the output data register is empty. The microprocessor keeps checking the status flag until it indicates that the output data register is empty and then writes data into it.

2. *The interrupt-driven method.* The interface chip asserts an interrupt signal to the microprocessor when the output data register is empty and can accept new data. The microprocessor then executes the service routine associated with the interrupt and outputs the data.

The M6821 from Motorola and the i8255 from Intel support both methods. Port C of the 68HC11 also supports both methods. However, the HCS12 supports neither method (there is no data status bit associated with any I/O port). Serial interface functions of the HCS12 still implement polling and interrupt methods as mechanisms for synchronization between the processor and the interface function.

7.4.2 Synchronizing the Interface Chip and I/O Device

The interface chip is responsible for making sure that data are properly transferred to and from I/O devices. The following methods have been used to synchronize data transfer between the interface chip and I/O devices:

1. *The brute-force method*. Nothing special is done in this method. For input, the interface chip returns the voltage levels on the input pins to the microprocessor. For output, the interface chip makes the data written by the microprocessor directly available on output pins. This method is useful in situations in which the timing of data is unimportant. It can be used to test the voltage level of a signal, set the voltage of an output pin to high or low, or drive LEDs. All I/O ports of the HCS12 can perform brute-force I/O.

2. *The strobe method*. This method uses strobe signals to indicate that data are stable on input or output port pins. During input, the input device asserts a strobe signal when the data are stable on the input port pins. The interface chip latches data into the data register using the strobe signal. For output, the interface chip first places data on the output port pins. When the data become stable, the interface chip asserts a strobe signal to inform the output device to latch data on the output port pins. This method can be used if the interface chip and I/O device can keep up with each other. None of the HCS12 parallel ports supports this method.

3. *The handshake method*. The previous two methods cannot guarantee correct data transfer between an interface chip and an I/O device when the timing of data is critical. For example, it takes a much longer time to print a character than it does to send a character to the printer electronics, so data shouldn't be sent to the printer if it is still printing. The solution is to use a handshake protocol between the interface chip and the printer electronics. There are two handshake methods: the *interlocked handshake* and the *pulse-mode handshake*. Whichever handshake protocol is used, two handshake signals are needed—one (call it H1) is asserted by the interface chip and the other (call it H2) is asserted by the I/O device. The handshake signal transactions for input and output are described in the following subsections. Note that the handshake operations of some interface chips may differ slightly from what we describe here. None of the HCS12 parallel ports supports the handshake I/O method.

INPUT HANDSHAKE PROTOCOL

The signal transaction of the input handshake protocol is illustrated in Figure 7.3.

Step 1
The interface chip asserts (or pulses) H1 to indicate its intention to input new data.

Step 2
The input device puts valid data on the data port and also asserts (or pulses) the handshake signal H2.

Step 3
The interface chip latches the data and de-asserts H1. After some delay, the input device also de-asserts H2.

The whole process will be repeated if the interface chip wants more data.

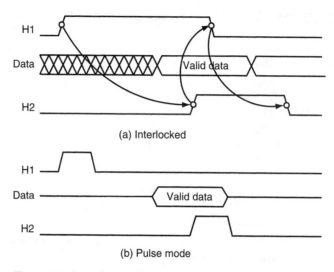

(a) Interlocked

(b) Pulse mode

Figure 7.3 ■ Input handshakes

OUTPUT HANDSHAKE PROTOCOL

The signal transaction of the output handshake protocol is shown in Figure 7.4. It also takes place in three steps:

Step 1

The interface chip places data on the data port and asserts (or pulses) H1 to indicate that it has data to be output.

(a) Interlocked

(b) Pulse mode

Figure 7.4 ■ Output handshakes

Step 2

The output device latches the data and asserts (or pulses) H2 to acknowledge the receipt of data.

Step 3

The interface chip de-asserts H1 following the assertion of H2. The output device then de-asserts H2.

The whole process will be repeated if the microprocessor has more data to be output.

The handshake is a very common synchronization mechanism for older interface chips and I/O devices. However, the microcontrollers (including the HCS12) designed in the last 10 years rarely provide this mechanism. This is probably due to the shift of applications of the parallel I/O ports.

7.5 Overview of HCS12 Parallel Ports

A HCS12 device may have from 48 to 144 pins arranged in 3 to 12 I/O ports and packaged in either a *quad flat pack* (QFP) or a *low-profile quad flat pack* (LQFP) package. The pins of an I/O port are often multiplexed with one or more peripheral functions (e.g., timer or serial peripheral interface) on the device. In general, when a peripheral function is enabled, that pin may not be used as a general-purpose I/O pin. General-purpose I/O pins can be considered the simplest of peripherals. They allow the microcontroller to monitor and control other devices. There are several signal pins that are not part of any ports. They will be covered in appropriate chapters.

I/O ports do not have the same number of pins. The number of pins available in each I/O port for the HCS12 MCU is listed in Table 7.1.

Each I/O port has several associated registers to support its operation and these registers must be dealt with when performing I/O operations by specifying the address of a register in order to read from it or write into it. A better way is to use the **equ** directive (in assembly

Port Name	No. of Pins	Pin Name
A	8	PA7~PA0
B	8	PB7~PB0
E	8	PE7~PE0
H	8	PH7~PH0
J	4	PJ7~PJ0
K	7	PK4~PK0
M	8	PM7~PM0
P	8	PP7~PP0
S	8	PS3~PS0
T	8	PT7~PT0
PAD1, PAD0	16	PAD15~PAD0
L	8	PL7~PL0
U	8	PU7~PU0
V	8	PV7~PV0
W	8	PW7~PW0

Table 7.1 ■ Number of pins available in each parallel port

language) to equate a symbolic name to the address of a given register so that the symbolic name can be used to access it. For example, the following directive:

```
PORTA    equ        $00
```

allows the user to use "PORTA" to refer to the Port A data register in the following instruction:

```
        movb       #$FF,PORTA        ; output $FF to Port A
```

The file **hcs12.inc** that contains the **equ** directive for all peripheral registers has been provided on the complementary CD. (Copy it to the installation directory of either AsmIDE or MiniIDE and include it in the program, then you will be able to use symbolic names to access all peripheral registers.)

7.5.1 Port Registers

All I/O ports (except PAD0 and PAD1) have an associated data direction register and a data register. The name of the data register associated with a specific port is formed by adding the letters "DDR" as the prefix to the port name. For example, DDRA and DDRB are the data direction register for Port A and Port B, respectively. To configure a pin for output, write a 1 to the corresponding bit in the data direction register. To configure a pin for input, write a 0 to the corresponding bit in the data direction register. For example, the following instruction sequence will configure Port A and Port B for output and input, respectively:

```
        movb       #$FF,DDRA        ; configure Port A for output
        movb       #$00,DDRB        ; configure Port B for input
```

The naming of port data registers is not uniform. The names of some of the port data registers are formed by adding "PORT" as the prefix of the port name in Table 7.1, whereas others are formed by adding the prefix "PT" only. A complete list of port data registers is shown in Table 7.2.

Port Name	Data register name
A	PORTA[1]
B	PORTB[1]
E	PORTE[1]
H	PORTK[1]
J	PTH
K	PTJ
M	PTM
P	PTP
S	PTS
T	PTT
PAD1, PAD0	PORTAD0, PORTAD1
L	PTL[2]
U	PTU[2]
V	PTV[2]
W	PTW[2]

[1] PORTA, PORTB, PORTE, and PORTK are also referred to as PTA, PTB, PTE, and PTK in the hcs12.inc and hcs12.h files.

[2] Port L, U, V, and W are available in H-family devices only.

Table 7.2 ■ HCS12 parallel I/O port data register names

One needs to deal with the port data register when outputting data to a port and inputting data from a port. The following instruction sequence configures Port A for input and reads a byte from Port A and places the value in accumulator A:

```
movb    #$00,DDRA        ; configure Port A for input
ldaa    PORTA            ; reads a byte into A from Port A
```

The following instruction sequence configures Port B for output and outputs the value of $A5 to Port B:

```
movb    #$FF,DDRB        ; configure Port B for output
movb    #$A5,PORTB       ; output $A5 to Port B
```

A parallel port may have up to eight associated registers. The number of registers varies with each port and hence will be discussed individually in the following sections.

7.5.2 Port A and Port B

In expanded mode, both Port A and Port B are used as time-multiplexed address/data pins. When configured in single-chip mode, these two ports are used as general-purpose I/O ports. Each Port A or Port B pin can be configured as an input or output pin. When the HCS12 is configured in expanded mode, Port A carries the time-multiplexed upper address and data signals (A15/D15~A8/D8), whereas Port B carries the time-multiplexed lower address and data signals (A7/D7~A0/D0).

7.5.3 Port E

As shown in Figure 7.5, Port E pins are used for bus control and interrupt service request signals. When a pin is not used for one of these specific functions, it can be used as a general-purpose I/O. However, two of the Port E pins (PE[1:0]) can only be used for input, and the states of these pins can be read in the port data register even when they are used for $\overline{IRQ}$ and $\overline{XIRQ}$.

Figure 7.5 ■ Port E pins
and their alternate functions

The PE7 pin serves three functions:

1. A general-purpose I/O pin
2. A signal that indicates the MCU is not accessing external memory
3. A signal that selects between the external clock or crystal oscillator signal as the clock input to the HCS

The PE6 and PE5 pins serve three functions:

1. General-purpose I/O pins (in single-chip mode)
2. Signals that set the operation mode of the HCS12 after reset
3. Instruction queue tracking signals (in expanded mode)

The PE4 pin has two functions:

1. A general-purpose I/O pin (in single-chip mode)
2. The E clock output (in expanded mode)

The PE3 pin has three functions:

1. A general-purpose I/O pin (in single-chip mode)
2. The LCD front-plane segment driver output pin (in H subfamily)
3. The low-byte strobe signal to indicate the type of access (in expanded mode)

The PE2 pin has two functions:

1. A general-purpose I/O pin (in single-chip mode)
2. A signal to indicate whether the current bus cycle is a read cycle or a write cycle (in expanded mode)

The PE1 pin has two functions:

1. A general-purpose I/O pin
2. The $\overline{\text{IRQ}}$ input

The PE0 pin has two functions:

1. A general-purpose I/O pin
2. The $\overline{\text{XIRQ}}$ input

In addition to the DDRE and PORTE registers, Port E also has the following registers:

- Port E assignment register (PEAR)
- Mode register (MODE)
- Pull-up control register (PUCR)
- Reduced drive register (RDRIV)
- External bus interface control register (EBICTL)

PORT E ASSIGNMENT REGISTER (PEAR)

In expanded mode, this register assigns the function of each Port E pin. The contents of the register PEAR are shown in Figure 7.6. The PEAR register is not accessible for reads or writes in peripheral mode.

MODE REGISTER

This register establishes the operation mode and other miscellaneous functions (i.e., internal visibility and emulation of Ports E and K). The contents of this register are shown in Figure 7.7.

PULL-UP CONTROL REGISTER (PUCR)

This register is used to select the pull-up resistors for the pins associated with the core part. The MC9S12DP256 has Ports A, B, E, and K in its core part. This register can be written any time; its contents are shown in Figure 7.8.

7	6	5	4	3	2	1	0
NOACCE	0	PIPOE	NECLK	LSTRE	0	RDWE	0

NOACCE: *No Access output enable*. Can be read/written any time.
    0 = PE7 is used as general-purpose I/O pin.
    1 = PE7 is output and indicates whether the cycle is a CPU free cycle.
PIPOE: Pipe signal output enable.
    In normal modes, write once. Special modes, write anytime except the first
    time. This bit has no effect in single-chip modes.
    0 = PE[6:5] are general-purpose I/O.
    1 = PE[6:5] are outputs and indicate the state of the instruction queue.
NECLK: *No external E clock*. Can be read anytime.
    In expanded modes, writes to this bit have no effect. E clock is required
    for demultiplexing the external address. NECLK can be written once in
    normal single-chip mode and can be written anytime in special single-
    chip mode.
    0 = PE4 is the external E-clock.
    1 = PE4 is a general-purpose I/O pin.
LSTRE: *Low strobe ($\overline{LSTRB}$) enable*. Can be read anytime.
    In normal modes, write once. Special modes, write anytime. This bit has no
    effect in single-chip modes or normal expanded narrow mode.
    0 = PE3 is a general-purpose I/O pin.
    1 = PE3 is configured as the $\overline{LSTRB}$ bus-control output, provided that the
    HCS12 is not in single-chip or normal expanded narrow modes.
RDWE: *Read/write enable*. Can be read anytime.
    In normal modes, write once. Special modes, write anytime except the
    first time. This bit has no effect in single-chip modes.
    0 = PE2 is a general-purpose I/O pin.
    1 = PE2 is configured as the $R/\overline{W}$-pin. In single-chip modes, RDWE has
    no effect and PE2 is a general-purpose I/O pin.
    $R/\overline{W}$ used for external writes. After reset in normal expanded mode, it
    is disabled. If needed, it should be enabled before any external writes.

Figure 7.6 ■ PEAR register

REDUCED DRIVE REGISTER (RDRIV)

This register is used to select reduced drive for the pins associated with the core ports, which gives reduced power consumption and reduced RFI with a slight increase in transition time, a feature used on ports that have a light loading. This register is not in the memory map in expanded mode. The contents of the register are shown in Figure 7.9.

EXTERNAL BUS INTERFACE CONTROL REGISTER (EBICTL)

Only the bit 0 (ESTR) of this register is implemented. It controls the stretching of the external E clock. When the ESTR bit is set to 0, the E clock is free-running and does not stretch (lengthen). When the HCS12 is interfacing with a slower memory device, then the E clock can be lengthened (stretch its high interval) by setting this bit.

7	6	5	4	3	2	1	0
MODC	MODB	MODA	0	IVIS	0	EMK	EME

MODC, MODB, MODA: mode select bits

    000 = special single-chip mode

    001 = emulation narrow mode

    010 = special test mode

    011 = emulation wide mode

    100 = normal single-chip mode

    101 = normal expanded narrow mode (external memory data bus is 8-bit)

    110 = special peripheral mode

    111 = normal expanded wide mode (external memory data bus is 16-bit)

IVIS: internal visibility

    0 = no visibility of internal bus operations on external bus.

    1 = internal bus operations are visible on external bus.

EMK: emulate Port K

    0 = PORTK and DDRK are in memory map and port K can be used in general I/O.

    1 = If in any expanded mode, PORTK and DDRK are removed from memory map.

EME: emulate Port E

    0 = PORTE and DDRE are in the memory map so Port E can be used for general I/O.

    1 = If in any expanded mode or special peripheral mode, PORTE and DDRE are removed from memory map, which allows the user to emulate the function of these registers externally.

Figure 7.7 ■ The MODE register

	7	6	5	4	3	2	1	0
	PUPKE	0	0	PUPEE	0	0	PUPBE	PUPAE
reset:	1	0	0	1	0	0	0	0

PUPKE: Pull-up Port K enable

    0 = pull-up resistors of Port K are disabled.

    1 = pull-up resistors of Port K are enabled.

PUPEE: pull-up Port E enable

    0 = pull-up resistors of Port E input pins 7 and 4-0 are disabled.

    1 = pull-up resistors of Port E input pins 7 and 4-0 are enabled.

PUPBE: Pull-up Port B enable

    0 = pull-up resistors of Port B are disabled.

    1 = pull-up resistors of Port B are enabled.

PUPAE: Pull-up Port A enable

    0 = pull-up resistors of Port A are disabled.

    1 = pull-up resistors of Port A are enabled.

Figure 7.8 ■ Pull-up Control register

	7	6	5	4	3	2	1	0
	RDPK	0	0	RDPE	0	0	RDPB	RDPA
reset:	0	0	0	0	0	0	0	0

RDPK: reduced drive of Port K
  0 = all Port K pins have full drive enabled.
  1 = all Port K pins have reduced drive enabled.
RDPE: reduced drive of Port E
  0 = all Port E pins have full drive enabled.
  1 = all Port E pins have reduced drive enabled.
RDPB: reduced drive of Port B
  0 = all Port B pins have full drive enabled.
  1 = all Port B pins have reduced drive enabled.
RDPA: reduced drive of Port A
  0 = all Port A pins have full drive enabled.
  1 = all Port A pins have reduced drive enabled.

Figure 7.9 ■ Reduced Drive register (RDRIV)

7.5.4 Port K

Port K has a Port K Data register (PORTK) and a Data Direction register (DDRK). In expanded mode, Port K carries the expanded address XADDR14~XADDR19 emulated chip-select ($\overline{ECS}$), and external chip-select ($\overline{XCS}$) signals. The PK6 pin is available only in the H subfamily. The functions of Port K pins are shown in Figure 7.10. At the rising edge of the $\overline{RESET}$ signal, the state of the PK7 pin is latched to the ROMON bit of the MISC register. If this bit is 1, the on-chip flash memory will be enabled in the expanded mode. This bit is forced to 1 in single-chip modes.

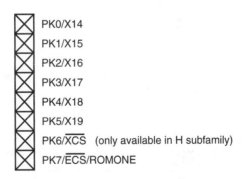

PK0/X14
PK1/X15
PK2/X16
PK3/X17
PK4/X18
PK5/X19
PK6/$\overline{XCS}$ (only available in H subfamily)
PK7/$\overline{ECS}$/ROMONE

Figure 7.10 ■ Port K pins and their alternate functions

7.5.5 Port T

Port T has a Port I/O register (PTT), Port Data Direction register (DDRT), Port Input register (PTIT), Reduced Drive register (RDRT), Pull Device Enable register (PERT), and Port Polarity Select register (PPST).

The PTIT register allows one to read back the status of Port T pins. This register can also be used to detect overload or short circuit conditions on output pins.

The RDRT register configures the drive strength of each port output pin as either full or reduced current. If a port pin is used as input, this bit is ignored. The reduced drive feature for each pin can be enabled individually. The contents of RDRT are shown in Figure 7.11.

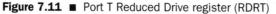

7	6	5	4	3	2	1	0
RDRT7	RDRT6	RDRT5	RDRT4	RDRT3	RDRT2	RDRT1	RDRT0

reset: 0 0 0 0 0 0 0 0

RDRT[7:0]: reduced drive Port T
    0 = full drive strength at output.
    1 = associated pin drives at about 1/3 of the full drive strength.

Figure 7.11 ■ Port T Reduced Drive register (RDRT)

The PERT register configures whether a pull-up or pull-down device is enabled if the port is used as input. Each pin's pull-up or pull-down device can be enabled individually. No pull-up or pull-down device is enabled out of reset. The contents of PERT are shown in Figure 7.12.

7	6	5	4	3	2	1	0
PERT7	PERT6	PERT5	PERT4	PERT3	PERT2	PERT1	PERT0

reset: 0 0 0 0 0 0 0 0

PERT[7:0]: pull device enable Port T
    0 = pull-up or pull-down is disabled.
    1 = either pull-up or pull-down is enabled.

Figure 7.12 ■ Port T Pull Device Enable register (PERT)

The PPST register selects whether a pull-down or a pull-up device is connected to the pin. This register has an effect on input pins only. The contents of PPST are shown in Figure 7.13.

7	6	5	4	3	2	1	0
PPST7	PPST6	PPST5	PPST4	PPST3	PPST2	PPST1	PPST0

reset: 0 0 0 0 0 0 0 0

PPST[7:0]: pull polarity select Port T
    0 = a pull-up device is connected to the associated Port T pin, if
        enabled by the associated bit in register PERT and if the port is used as
        input or as wired-or output.
    1 = a pull-down device is connected to the associated Port T pin, if enabled
        by the associated bit in register PERT and if the port is used as input.

Figure 7.13 ■ Port T Polarity Select register (PPST)

The pin functions of Port T are shown in Figure 7.14. In addition to being used as general I/O pins, Port T pins can also be used as input capture or output compare action pins.

PT0/IOC0
PT1/IOC1
PT2/IOC2
PT3/IOC3
PT4/IOC4
PT5/IOC5
PT6/IOC6
PT7/IOC7

Figure 7.14 ■ Port T pins and their alternate functions

7.5.6 Port S

Port S has a Port S Wired-OR Mode register (WOMS) in addition to all the registers associated with Port T (PTS, DDRS, PTIS, RDRS, PERS, and PPSS). Each bit of this register configures the associated output pin as wired-OR. The contents of this register are shown in Figure 7.15. The contents of PTIS, RDRS, PERS, and PPSS are identical to those of PTIT, RDRT, PERT, and PPST, respectively. As shown in Figure 7.16, Port S pins can be used as general I/O pins or serial interface signals.

7	6	5	4	3	2	1	0
WOMS7	WOMS6	WOMS5	WOMS4	WOMS3	WOMS2	WOMS1	WOMS0

reset: 0   0   0   0   0   0   0   0

WOMS[7:0]: wired-OR mode Port S
   0 = output buffers operate as push-pull outputs.
   1 = output buffers operate as open-drain outputs.

Figure 7.15 ■ Port S wired-Or mode register (WOMS)

PS0/RXD0
PS1/TXD0
PS2/RXD1
PS3/TXD1
PS4/MISO0
PS5/MOSI0
PS6/SCK0
PS7/$\overline{SS0}$

Figure 7.16 ■ Port S pins and their alternate functions

7.5.7 Port M

In addition to having all the equivalent registers (PTM, DDRM, PTIM, RDRM, PERM, PPSM, and WOMM) contained in Port S, Port M also has a Module Routing register (MODRR). This register configures the rerouting of CAN0, CAN4, SPI0, SPI1, and SPI2 on defined port pins. The contents of the MODRR register are shown in Figure 7.17. The functions of Port M pins are illustrated in Figure 7.18.

	7	6	5	4	3	2	1	0
	0	MODRR6	MODRR5	MODRR4	MODRR3	MODRR2	MODRR1	MODRR0
reset:	0	0	0	0	0	0	0	0

CAN0 routing

MODRR1	MODRR0	RXCAN0	TXCAN0
0	0	PM0	PM1
0	1	PM2[1]	PM3[1]
1	0	PM4[2]	PM5[2]
1	1	reserved	

CAN4 routing

MODRR3	MODRR2	RXCAN4	TXCAN4
0	0	PJ6	PJ7
0	1	PM4[3]	PM5[3]
1	0	PM6[4]	PM7[4]
1	1	reserved	

SPI0 routing

MODRR4	MISO0	MOSI0	SCK0	$\overline{SS0}$
0	PS4	PS5	PS6	PS7
1	PM2[5]	PM4[6]	PM5[6]	PM3[5]

SPI1 routing

MODRR5	MISO1	MOSI1	SCK1	$\overline{SS1}$
0	PP0	PP1	PP2	PP3
1	PH0	PH1	PH2	PH3

SPI2 routing

MODRR6	MISO2	MOSI2	SCK2	$\overline{SS2}$
0	PP4	PP5	PP7	PP6
1	PH4	PH5	PH6	PH7

Notes: [1] Routing to this pin takes effect only if CAN1 is disabled.
[2] Routing to this pin takes effect only if CAN2 is disabled.
[3] Routing to this pin takes effect only if CAN2 is disabled, and CAN0 is disabled if routed here.
[4] Routing to this pin takes effect only if CAN3 is disabled.
[5] Routing to this pin takes effect only if CAN1 is disabled, and CAN0 is disabled if routed here.
[6] Routing to this pin takes effect only if CAN2 is disabled, and CAN0 is disabled if routed here and CAN4 is disabled if routed here.

Figure 7.17 ■ Module Routing register (MODRR)

Figure 7.18 ■ Port M pins and their alternate functions

Example 7.1

Give an instruction to configure the MODRR register to achieve the following port routing:

1. CAN0: use pins PM1 and PM0
2. CAN1: use pins PM3 and PM2
3. CAN2: use pins PM5 and PM4
4. CAN3: use pins PM7 and PM6
5. I²C: use PJ7 and PJ6
6. SPI0: use pins PS7~PS4
7. SPI1: use pins PH3~PH0
8. SPI2: use pins PH7~PH4

Solution: For this routing requirement, all we need to do is to prevent CAN4 from using any port pins and keep the default routing after reset. The following instruction will satisfy the requirement:

```
movb    #$60,MODRR      ; CAN4 must be disabled
```

Example 7.2

Give an instruction to configure the MODRR register to achieve the following port routing:

1. CAN0: use pins PM1 and PM0
2. CAN1: use pins PM3 and PM2
3. CAN2: disabled
4. CAN3: disabled
5. I²C: use PJ7 and PJ6
6. SPI0: use pins PS7~PS4

7. SPI1: use pins PP3~PP0
8. SPI2: use pins PH7~PH4

Solution: This routing requirement can be satisfied by the following instruction:

```
movb    #$40,MODRR     ; CAN2~CAN4 must be disabled
```

7.5.8 Ports H, J, and P

These three I/O ports have the same set of registers associated with them. All of the pins associated with these three ports have edge-triggered interrupt capability in the wired-OR fashion. The SPI function pins can be routed to Ports H and P. The rerouting of SPI functions is done by programming the MODRR register. Each of these three ports has eight associated registers:

1. Port I/O register (PTH, PTJ, PTP)
2. Port Input register (PTIH, PTIJ, PTIP)
3. Port Data Direction register (DDRH, DDRJ, DDRP)
4. Port Reduced Drive register (RDRH, RDRJ, RDRP)
5. Port Pull Device Enable register (PERH, PERJ, PERP)
6. Port Polarity Select register (PPSH, PPSJ, PPSP)
7. Port Interrupt Enable register (PIEH, PIEJ, PIEP)
8. Port Interrupt Flag register (PIFH, PIFJ, PIFP)

All except the last two registers have their equivalents in Port T. The contents of the Port H Interrupt Enable register and Port H Interrupt Flag register are shown in Figures 7.19 and 7.20, respectively.

	7	6	5	4	3	2	1	0
	PIEH7	PIEH6	PIEH5	PIEH4	PIEH3	PIEH2	PIEH1	PIEH0
reset:	0	0	0	0	0	0	0	0

PIEH[7:0]: interrupt enable Port H
  0 = interrupt is disabled.
  1 = interrupt is enabled.

Figure 7.19 ■ Port H Interrupt Enable register (PIEH)

	7	6	5	4	3	2	1	0
	PIFH7	PIFH6	PIFH5	PIFH4	PIFH3	PIFH2	PIFH1	PIFH0
reset:	0	0	0	0	0	0	0	0

PIFH[7:0]: interrupt flag Port H
  0 = no active edge pending.
  1 = active edge has occurred (writing a '1' clears the associated flag).

Figure 7.20 ■ Port P Interrupt Flag register (PIFH)

Port H is associated with two SPI modules. Port J is associated with the fifth CAN and the I²C module. Port P is associated with the PWM and two SPI modules. In all modes, Port P pins PP[7:0] can be used for either general-purpose I/O or with the PWM and SPI subsystem. The pins are shared between the PWM channels and the SPI1 and SPI2 modules. If the PWM function is enabled, these pins become PWM output channels, with the exception of pin 7, which can be PWM input or output. If SPI1 or SPI2 are enabled and PWM is disabled, the respective pin configuration is determined by several status bits in the SPI module. Both Port H and Port P have eight pins, whereas Port J has only four pins.

The interrupt enable as well as the sensitivity to rising or falling edges can be individually configured on a per-pin basis. If a pin's pull-down device is enabled, then the interrupt is rising edge-triggered. Otherwise, it is falling edge-triggered. All 8 bits/pins of the port share the same interrupt vector. Interrupts can be used with the pins configured as inputs or outputs.

An interrupt is generated when a bit in the Port Interrupt Flag register and its corresponding port interrupt enable bit are both set. This feature can be used to wake up the CPU when it is in STOP or WAIT mode. Each Port P pin can also be used as an edge-sensitive interrupt source. A digital filter on each pin prevents pulses shorter than a specified time from generating an interrupt. The minimum time varies over process conditions, temperature, and voltage.

The functions of Port H, J, and P pins are shown in Figures 7.21, 7.22, and 7.23, respectively.

PH0/MISO1/KWH0
PH1/MOSI1/KWH1
PH2/SCK1/KWH2
PH3/$\overline{SS1}$/KWH3
PH4/MISO2/KWH4
PH5/MOSI2/KWH5
PH6/SCK2/KWH6
PH7/$\overline{SS2}$/KWH7

Figure 7.21 ■ Port H pins and their alternate functions

PJ0/KWJ0
PJ1/KWJ1
PJ6/KWJ6/RXCAN4/SDA/RXCAN0
PJ7/KWJ7/TXCAN4/SCL/TXCAN0

Figure 7.22 ■ Port J pins and their alternate functions

PP0/MISO1/PWM0/KWP0
PP1/MOSI1/PWM1/KWP1
PP2/SCK1/PWM2/KWP2
PP3/$\overline{SS1}$/PWM3/KWP3
PP4/MISO2/PWM4/KWP4
PP5/MOSI2/PWM5/KWP5
PP6/$\overline{SS2}$/PWM6/KWP6
PP7/SCK2/PWM7/KWP7

Figure 7.23 ■ Port P pins and their alternate functions

7.5.9 Ports AD0 and AD1

Many HCS12 devices have implemented two 8-channel A/D converters (AD1 and AD0). For those devices with a single 8-channel A/D converter, the module is referred to as AD instead of AD0. These two ports are analog input interface to the analog-to-digital subsystem. When analog-to-digital functions are not enabled, these two ports are available for general-purpose I/O. Since these two ports cannot be used as output, there are no data direction registers associated with them.

The ports have no resistive input loads and no reduced drive controls. PTAD0 and PTAD1 are data registers of Port AD0 and AD1, respectively. Each A/D module has an ATD Digital Input Enable register (ATD0DIEN and ATD1DIEN). In order to use an A/D pin as a digital input, its associated bit in this register needs to be set to 1. The contents of this register are shown in Figure 7.24.

	7	6	5	4	3	2	1	0
	IEN7	IEN6	IEN5	IEN4	IEN3	IEN2	IEN1	IEN0
reset:	0	0	0	0	0	0	0	0

IENx: ATD digital input enable on channel x
  0 = disable digital input buffer to PTADx pin.
  1 = enable digital input buffer to PTADx pin.

Figure 7.24 ■ ATD Input Enable register (ATD0DIEN and ATD1DIEN)

7.5.10 Port L

This port is available in the HCS12 H subfamily. In addition to being a general I/O port, Port L is connected to the LCD driver. The HCS12 H subfamily is designed to support LCD interfacing and motor control. Port L has the following registers:

- Port I/O register (PTL)
- Port Data Direction register (DDRL)
- Port Input register (PTIL)
- Reduced Drive register (RDRL)
- Pull Device Enable register (PERL)
- Port Polarity Select register (PPSL)

The contents of these registers are identical to their equivalents in Port T. The functions of Port L pins are shown in Figure 7.25.

PL0/FP16 (LCD frontplane pin 16)
PL1/FP17 (LCD frontplane pin 17)
PL2/FP18 (LCD frontplane pin 18)
PL3/FP19 (LCD frontplane pin 19)
PL4/FP28 (LCD frontplane pin 28)
PL5/FP29 (LCD frontplane pin 29)
PL6/FP30 (LCD frontplane pin 30)
PL7/FP31 (LCD frontplane pin 31)

Figure 7.25 ■ Port L pins and their alternate functions

7.5.11 Ports U, V, and W

These three ports are available in the HCS12's H subfamily. In addition to being general I/O ports, the three ports are provided for motor control. These three ports have the following registers:

- Port I/O register (PTU, PTV, PTW)
- Port Data Direction register (DDRU, DDRV, DDRW)

- Port Input register (PTIU, PTIV, PTIW)
- Pull Device Enable register (PERU, PERV, PERW)
- Port Polarity Select register (PPSU, PPSV, PPSW)
- Port Slew Rate register (SRRU, SRRV, SRRW)

The functions of the first five of these registers are identical to their equivalents in Port L. The contents of the Port Slew Rate register are illustrated in Figure 7.26.

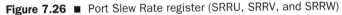

SRRx[7:0]: slew rate port x (x = U, V, or W)
0 = disable slew rate control.
1 = enable slew rate control.

Figure 7.26 ■ Port Slew Rate register (SRRU, SRRV, and SRRW)

The functions of these port pins are shown in Figure 7.27a, 7.27b, and 7.27c, respectively.

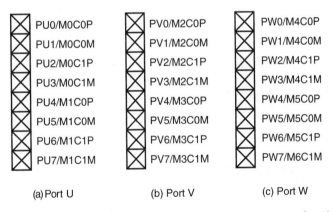

(a) Port U   (b) Port V   (c) Port W

Figure 7.27 ■ Port U, V, and W pins and their alternate functions

7.6 Electrical Characteristic Considerations for I/O Interfacing

Most embedded systems require the use of logic chips—peripheral devices in addition to the microcontroller—to perform their function. Because these chips may use different types of integrated circuit (IC) technologies, there is a concern that the resultant embedded system may not function properly.

The major concern in interfacing IC chips that are made with different technologies is whether they are electrically compatible. There are two issues involved in electrical compatibility:

- *Voltage-level compatibility.* Is the high output level of an IC chip high enough to be considered as a high for the input of another IC chip? Is the low output level of an IC chip low enough to be considered as a low for the input of another IC chip?
- *Current drive capability.* Does the output of an IC chip have enough current to drive its load? Can the output circuit of an IC chip sink the currents of its load?

The signal timing is also an important factor for making sure that the digital circuit functions correctly. The main concern about timing is whether the signal from one chip becomes valid early enough to be used by other chips. This is a *timing compatibility* issue. When the operating frequency becomes very high, the *transmission line effect* and *ground bounce effect* will need to be considered too. However, we are not dealing with high-frequency systems in this text. These two issues will not be discussed.

7.6.1 Voltage-Level Compatibility

There are many IC technologies in use today. Some of them are bipolar; others are unipolar. A bipolar IC technology has both the electron and hole currents in any moment. However, in a unipolar IC technology, the current in any part of the circuit is either the electron current or the hole current. The CMOS technology is unipolar and also the dominant IC technology in use today.

The voltage-level compatibility issue arises because IC technologies differ in the following four voltages:

- Input high voltage (V_{IH}). This is the voltage that will be treated as a logic 1 when applied to the input of a digital circuit.
- Input low voltage (V_{IL}). This is the voltage that will be treated as a logic 0 when applied to the input of a digital circuit.
- Output high voltage (V_{OH}). This is the voltage level when a digital circuit outputs a logic 1.
- Output low voltage (V_{OL}). This is the voltage level when a digital circuit outputs a logic 0.

In order for the digital circuit X to be able to drive circuit Y correctly, the following conditions must be satisfied:

- $V_{OHX} \geq V_{IHY}$ (the output high voltage of circuit X must be higher than the input high voltage of circuit Y). The difference between V_{OH} and V_{IH} of the same technology is referred to as the *noise margin high* (NMH).
- $V_{OLX} \leq V_{ILY}$ (the output low voltage of circuit X must be lower than the input low voltage of circuit Y). The difference between V_{IL} and V_{OL} of the same technology is referred to as the *noise margin low* (NML).

The input and output voltage levels of a few popular logic families are listed in Table 7.3, from which one can draw the following conclusions:

- There is no problem in using CMOS logic chips to drive bipolar logic chips at the same power supply level.
- The HCS12 has no problem in driving the CMOS logic chips and being driven by the CMOS logic chips at the same power supply level.

- The BiCMOS logic is not suitable for driving the HCS12 and other CMOS logic chips at the same power supply level.
- The bipolar logic ICs are not suitable for driving the CMOS logic ICs or the HCS12 microcontroller at the same power supply level.

Logic Family	V_{CC}	V_{IH}	V_{OH}	V_{IL}	V_{OL}
HCS12[3]	5 V	3.25 V	4.2 V	1.75 V	0.8 V
S[4]	5 V	2 V	3.0~3.4 V[1]	0.8 V	0.4~0.5 V[2]
LS[4]	5 V	2 V	3.0~3.4 V[1]	0.8 V	0.4~0.5 V[2]
AS[4]	5 V	2 V	3.0~3.4 V[1]	0.8 V	0.3 V
F[4]	5 V	2 V	3.4 V	0.8 V	0.1 V
HC[3]	5 V	3.5 V	4.9 V	1.5 V	0.1 V
HCT[3]	5 V	3.5 V	4.9 V	1.5 V	0.1 V
ACT[3]	5 V	2 V	4.9 V	0.8 V	0.35 V
ABT[5]	5 V	2 V	3 V	0.8 V	0.55 V
BCT[5]	5 V	2 V	3.3 V	0.8 V	0.42 V
FCT[5]	5 V	2 V2 V	2.4 V	0.8 V	0.55 V

[1] V_{OH} value will get lower when output current is larger.
[2] V_{OL} value will get higher when output current is larger. The V_{OL} values of different logic gates are slightly different.
[3] HCS12, HC, HCT, and ACT are based on the CMOS technology.
[4] S, LS, AS, and F logic families are based on the bipolar technology.
[5] ABT, BCT, and FCT use the BiCMOS technology.

Table 7.3 ■ Input and output voltage levels of common logic families

7.6.2 Current Drive Capability

A microcontroller needs to drive other peripheral I/O devices in an embedded system. The second electrical compatibility issue is whether the microcontroller can source (when the output voltage is high) or sink (when the output voltage is low) the current needed by the I/O devices that it interfaces with. The designer must make sure that the following two requirements are satisfied:

- Each I/O pin can supply and sink the current needed by the I/O devices that it interfaces with.
- The total current required to drive I/O devices does not exceed the maximum current rating of the microcontroller.

Each logic chip has the following four currents that are involved in the current drive calculation:

- *Input high current* (I_{IH}). This is the input current (flowing into the input pin) when the input voltage is high.
- *Input low current* (I_{IL}). This is the input current (flowing out of the input pin) when the input voltage is low.
- *Output high current* (I_{OH}). This is the output current (flowing out of the output pin) when the output voltage is high.

- *Output low current* (I_{OL}). This is the output current (flowing into the output pin) when the output voltage is low.

The current capabilities of several common logic families and the HCS12 are listed in Table 7.4. In the CMOS technology, the gate terminal (one of the three terminals in an N- or P-transistor) draws a significant current only when it is charged up toward V_{CC} or pulled down toward GND level. After that, the gate terminal draws only leakage currents. Bipolar technology differs from CMOS technology in that a DC current always flows into or out of the base terminal of the transistor of a bipolar logic chip.

Logic Family	V_{CC}	I_{IH}	I_{IL}	I_{OH}	I_{OL}
HCS12[23]	5 V	2.5 μA	2.5 μA	25 mA	25 mA
S	5 V	50 μA	1.0 mA	1 mA	20 mA
LS	5 V	20 μA	0.2 mA	15 mA	24 mA
AS	5 V	20 μA	0.5 mA	15 mA	64 mA
F	5 V	20 μA	0.5 mA	1 mA	20 mA
HC[3]	5 V	1 μA	1 μA	25 mA	25 mA
HCT[3]	5 V	1 μA	1 μA	25 mA	25 mA
ACT[3]	5 V	1 μA	1 μA	24 mA	24 mA
ABT[3]	5 V	1 μA	1 μA	32 mA	64 MA
BCT	5 V	20 μA	1 mA	15 mA	64 mA
FCT[3]	5 V	1 μA	1 μA	15 mA	64 mA

[1] Values are based on the 74xx244 of Texas Instrument (xx is the technology name).
[2] The total HCS12 supply current is 65 mA.
[3] The values for I_{IH} and I_{IL} are input leakage currents.

Table 7.4 ■ Current capabilities of common logic families[1]

To determine whether a pin can supply (also called source) and sink currents to all the peripheral pins that it drives directly, the designer needs to check the following two requirements:

1. The I_{OH} of an I/O pin of the microcontroller is equal to or greater than the sum of currents flowing into all peripheral pins that are connected directly to the microcontroller I/O pins.

2. The I_{OL} of an I/O pin of the microcontroller is equal to or greater than the sum of currents flowing out of all peripheral pins that are connected directly to the microcontroller I/O pins.

In addition, the designer must also make sure that the total current needed to drive the peripheral signal pins does not exceed the total current that the microcontroller can supply.

One question that arises here is what should be done if an I/O pin cannot supply (or sink) the current needed to drive the peripheral pins? A simple solution is to add buffer chips (for example, 74ABT244) that can supply enough current between the microcontroller and the peripheral chips. This technique is widely used in microcontroller applications and is illustrated in Example 7.4.

7.6.3 Timing Compatibility

If an I/O pin is driving a peripheral pin that does not contain latches or flip-flops, then timing is not an issue. A latch or flip-flop usually has a control signal or clock signal to control the latching of an input signal. As illustrated in Figure 7.28, the D input to the D flip-flop must be valid t_{su} ns before the rising edge of the CLK signal and remain valid for at least t_{hd} ns after the rising edge of the CLK signal in order for its value to be correctly copied to the output signal Q. The timing parameters t_{su} and t_{hd} are referred to as the *setup* and *hold* time requirements of the D flip-flop. The main timing consideration is that the setup- and hold-time requirements for all latches and flip-flops in a digital system must be satisfied in order for the system to work correctly. A signal may pass through several intermediate chips before it is used by the final latch or flip-flop. The time delays of all intermediate devices must be added when considering the timing analysis. Timing requirement analysis can be very complicated and is best illustrated using examples. Two examples of timing compatibility analysis are given in Section 14.8.

Figure 7.28 ■ D flip-flop and its latching timing requirement

7.7 Interfacing with Simple Input and Output Devices

Many embedded systems only require simple input and output devices such as switches, light-emitting diodes (LED), keypads, and seven-segment displays.

7.7.1 Interfacing with LEDs

The LED is often used to indicate the system operation mode, whether power is turned on, whether system operation is normal, and so forth. An LED can illuminate when it is forward biased and has sufficient current flowing through it. The current required to light an LED may range from a few to more than 10 mA. The forward voltage drop across the LED can range from about 1.6 V to more than 2.2 V.

Figure 7.29 suggests three methods for interfacing with LEDs. Methods (a) and (b) are recommended for use with LEDs that need only 1 or 2 mA to produce enough brightness.

(a) Positive direct drive   (b) Inverse direct drive   (c) Buffered drive

Figure 7.29 ■ An LED connected to a CMOS inverter through a current-limiting resistor.

The circuit (c) is recommended for use with LEDs that need larger current to light. The current-limiting resistor for circuits (a) and (b) should be larger (between 1.5 kΩ to 2 kΩ).

Example 7.3

Use the HCS12 Port B to drive eight LEDs. Light each of them for half a second in turn and repeat, assuming that the HCS12 has a 24-MHz E clock.

Solution: The circuit for driving eight LEDs using Port B is shown in Figure 7.30.

Figure 7.30 ■ Circuit connection for Example 7.3

To turn on the LED driven by pins PB7~PB0 one at a time, one should output the value of $80, $40, . . . , and $01 to Port B in turn. The procedure to achieve the desired LED display pattern is as follows:

Step 1
Place the values $80, $40, . . . , and $01 in a table. Use the index register X to point to the start of this table.

Step 2
Output the value pointed to by X to Port B. Increment the pointer X.

Step 3
Wait for half a second.

Step 4
If X points to the end of the table, reset X to point to the start of the table. Go to step 2.

The assembly program that implements this algorithm is as follows:

```
#include "C:\miniide\hcs12.inc"
org    $1000
org    $1500
movb   #$FF,DDRB     ; configure Port B for output
bset   DDRJ,$02      ; configure PJ1 pin for output
bclr   PTJ,$02       ; enable LEDs to light (Dragon12 demo board)
```

```
forever    ldaa    #16
           ldx     #led_tab        ; Use X as the pointer to LED pattern table
led_lp     movb    1,x+,PORTB      ; turn on one LED
           ldy     #5
           jsr     delayby100ms    ;   "
           dbne    a,led_lp        ; reach the end of the table yet?
           bra     forever         ; start from beginning
led_tab    dc.b    $80,$40,$20,$10,$08,$04,$02,$01
           dc.b    $01,$02,$04,$08,$10,$20,$40,$80
           #include "C:\miniide\delay.asm"
           end
```

The C language version of the program is as follows:

```
#include "c:\egnu091\include\hcs12.h"
#include "c:\egnu091\include\delay.c"
main (void)
{
        char led_tab[8] = {0x80,0x40,0x20,0x10,0x08,0x04,0x02,0x01,
                0x01,0x02,0x04,0x08,0x10,0x20,0x40,0x80};
        char i;
        DDRB    = 0xFF;      /* configure Port B for output */
        DDRJ   |= 0x02;      /* configure PJ1 pin for output (needed for DRAGON12 board only) */
        PTJ    &= 0xFD;      /* enable LEDs to light (needed for DRAGON12 board only)*/
        while (1){
                for (i = 0; i < 16; i++) {
                        PORTB = led_tab[i];
                        delayby100ms(5);
                }
        }
        return 0;
}
```

7.7.2 Interfacing with Seven-Segment Displays

Seven-segment displays are often used when the embedded product needs to display only a few digits. Seven-segment displays are mainly used to display decimal digits and a small subset of letters.

Although a HCS12 device has enough current to drive a seven-segment display, it is not advisable to do so when a HCS12-based embedded product needs to drive many other I/O devices. In Figure 7.31, Port B drives a common-cathode seven-segment display through the buffer chip 74HC244. The V_{OH} (output high voltage) value of the 74HC244 is about 5 V. Adding a 300-Ω resistor will set the display segment current to about 10 mA, which should be sufficient to light an LED segment. The light patterns corresponding to ten BCD (decimal) digits are shown in Table 7.5. The numbers in Table 7.5 require that segments a~g be connected from the second most significant pin to the least significant pin of the output port.

Figure 7.31 ■ Driving a single seven-segment display

BCD Digit	Segments							Corresponding Hex Number
	a	b	c	d	e	f	g	
0	1	1	1	1	1	1	0	$7E
1	0	1	1	0	0	0	0	$30
2	1	1	0	1	1	0	1	$6D
3	1	1	1	1	0	0	1	$79
4	0	1	1	0	0	1	1	$33
5	1	0	1	1	0	1	1	$5B
6	1	0	1	1	1	1	1	$5F
7	1	1	1	0	0	0	0	$70
8	1	1	1	1	1	1	1	$7F
9	1	1	1	1	0	1	1	$7B

Table 7.5 ■ BCD to seven-segment decoder

When an application needs to display multiple BCD digits, the time-multiplexing technique is often used. An example of the circuit that displays six BCD digits is shown in Figure 7.32. In Figure 7.32, the common cathode of a seven-segment display is connected to the collector of an NPN transistor. When a Port K pin voltage is high, the connected NPN transistor will be driven into the saturation region. The common cathode of the display will then be pulled down to low (about 0.1 V), allowing the display to be lighted. By turning the six NPN transistors on and off in turn many times in 1 second, multiple digits can be displayed. A 2N2222 transistor can sink from 100 mA to 300 mA of current. The maximum current that flows into the common cathode is about 70 mA (7 × 10 mA) and hence can be sunk by a 2N2222 transistor. The resistor R should be selected to allow the transistor 2N2222 to be driven into the saturation region. A value of several hundred to 1-kΩ will work.

Figure 7.32 ■ Port B and Pork K together drive six seven-segment displays (MC9S12DP256)

Example 7.4

Write a sequence of instructions to display 4 on the seven-segment display #4 in Figure 7.32.

Solution: To display the digit 4 on the display #4, we need to

1. Output the hex value $33 to Port B
2. Set the PK4 pin to 1
3. Clear pins PK5 and PK3 . . . P0 to 0

The instruction sequence is as follows:

```
        #include  "c:\miniide\hcs12.inc"
four    equ     $33             ; seven-segment pattern of digit 4
        . . .
        movb    #$3F,DDRK       ; configure Port K for output
        movb    #$FF,DDRB       ; configure Port B for output
        bset    PTK,$10         ; turn on seven-segment display #4
        bclr    PTK,$2F         ; turn off seven-segment displays #5, #3. . .#0
        movb    #four,PORTB     ; output the seven-segment pattern to PORTP
        . . .
```

In C language, this can be achieved by the following statements:

```
        DDRK = 0x3F;            /* configure Port K for output */
        DDRB = 0xFF;            /* configure Port B for output */
        PORTK = 0x10;           /* turn on seven-segment display number #4 */
        PORTB = 0x33;           /* output the segment pattern of digit 4 */
```

The circuit in Figure 7.32 can display six digits simultaneously by using the time-multiplexing technique, in which each seven-segment display is lighted in turn briefly and then turned off. When one display is lighted, all other displays are turned off. Within 1 second, each seven-segment display is lighted and then turned off many times. Because of the persistence of vision, the six displays will appear to be lighted simultaneously.

Example 7.5

Write a program to display 123456 on the six seven-segment displays shown in Figure 7.32.

Solution: The digits 1, 2, 3, 4, 5, and 6 are displayed on display #5, #4, . . . , and #0, respectively. The values to be output to Port B and Port K to display one digit at a time are shown in Table 7.6. This table can be created by the following assembler directives:

```
display    dc.b    $30,$20        ; value to display 1 on seven-segment display #5
           dc.b    $6D,$10        ; value to display 2 on seven-segment display #4
           dc.b    $79,$08
           dc.b    $33,$04
           dc.b    $5B,$02
           dc.b    $5F,$01
```

Seven-Segment Display	Displayed BCD Digit	Port B	Port K
#5	1	$30	$20
#4	2	$6D	$10
#3	3	$79	$08
#2	4	$33	$04
#1	5	$5B	$02
#0	6	$5F	$01

Table 7.6 ■ Table of display patterns for Example 7.5

The program logic of this example is shown in Figure 7.33.
The assembly program that implements this algorithm is as follows:

```
           #include    "c:\miniide\hcs12.inc"
pat_port   equ    PORTB              ; Port that drives the segment pattern
pat_dir    equ    DDRB               ; direction register of the segment pattern
sel_port   equ    PORTK              ; Port that selects the digit
sel_dir    equ    DDRK               ; data direction register of the digit select port
           org    $1500
           movb   #$FF,pat_dir       ; configure pattern port for output
           movb   #$3F,sel_dir       ; configure digit select port for output
forever    ldx    #disp_tab          ; use X as the pointer
loop       movb   1,x+,pat_port      ; output the digit pattern and move the pointer
           movb   1,x+,sel_port      ; output the digit select value and move the pointer
           ldy    #1                 ; wait for 1 ms
```

```
              jsr       delayby1ms        ;  "
              cpx       #disp_tab+12      ; reach the end of the table
              bne       loop
              bra       forever
              #include  "c:\miniide\delay.asm"
disp_tab      dc.b      $30,$20           ; seven-segment display table
              dc.b      $6D,$10
              dc.b      $79,$08
              dc.b      $33,$04
              dc.b      $5B,$02
              dc.b      $5F,$01
              end
```

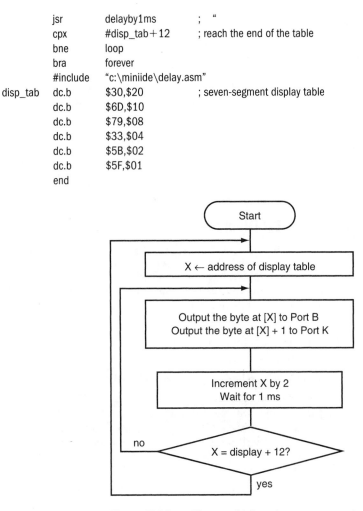

Figure 7.33 ■ Time-multiplexed seven-segment display algorithm

The C language version of this program is as follows:

```
#include "c:\egnu091\include\hcs12.h"
#include "c:\egnu091\include\delay.c"
#define   pat_port   PORTB      /* segment pattern port */
#define   pat_dir    DDRB       /* pattern port data direction register */
#define   sel_port   PORTK      /* digit select port */
#define   sel_dir    DDRK       /* digit select port direction register */
main (void)
{
    char disp_tab[6][2] = {{0x30,0x20},{0x6D,0x10},{0x79,0x08},
                   {0x33,0x04},{0x5B,0x02},{0x5F,0x01}};
```

```
char i;
pat_dir = 0xFF;                  /* configure pat_port for output */
sel_dir = 0x3F;                  /* configure sel_port for output */
while (1) {
        for (i = 0; i < 6; i++) {
                pat_port = disp_tab[i][0];    /* output the segment pattern */
                sel_port = disp_tab[i][1];    /* turn on the display */
                delaybyms(1);                 /* wait for 1 ms */
        }
}
return 0;
}
```

7.7.3 Liquid Crystal Displays (LCDs)

Although seven-segment displays are easy to use, they are bulky and quite limited in the set of characters that they can display. When more than a few letters and digits are to be displayed, seven-segment displays become inadequate. Liquid crystal displays (LCDs) come in handy when the application requires the display of many characters.

A liquid crystal display has the following advantages:

- High contrast
- Low power consumption
- Small footprint
- Ability to display both characters and graphics

The basic construction of an LCD is shown in Figure 7.34. The most common type of LCD allows light to pass through when activated. A segment is activated when a low-frequency bipolar signal in the range of 30 Hz to 1000 Hz is applied to it. The polarity of the voltage must alternate, or else the LCD will not be able to change very quickly.

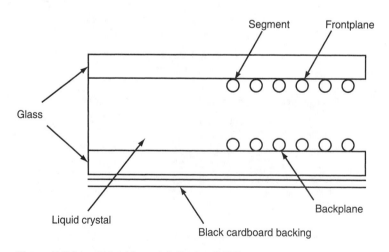

Figure 7.34 ■ A liquid crystal display (LCD)

When a voltage is applied across the segment, an electrostatic field is set up that aligns the crystals in the liquid. This alignment allows the light to pass through the segment. If no voltage is applied across a segment, the crystals appear to be opaque because they are randomly aligned. Random alignment is assured by the AC excitation voltage applied to each segment. In a digital watch, the segments appear darker when they are activated because light passes through the segment to a black cardboard backing that absorbs all light. The area surrounding the activated segment appears to grow brighter in color because the randomly aligned crystals reflect much of the light. In a backlit computer display, the segment appears to grow brighter because of a light placed behind the display; the light is allowed to pass through the segment when it is activated.

In recent years, the price of LCD displays has dropped to such an acceptable level that most PC vendors bundle LCD displays instead of CRT displays with their PC systems. Notebook computers used LCD as displays right from the beginning. Because of the price reduction of LCDs, the prices of notebook computers also have dropped sharply, and more and more computer users have switched from desktop to notebook computers.

Although LCDs can display graphics and characters, only character-based LCDs are discussed in this text. LCDs are often sold in a module that consists of the LCD and its controller. The Hitachi HD44780 (with two slightly different versions: HD44780U and HD44780S) is one of the most popular LCD display controllers in use today. The following section examines the operation and programming of this controller.

7.8 The HD44780U LCD Controller

The block diagram of an LCD kit that incorporates the HD44780U controller is shown in Figure 7.35. The pin assignment shown in Table 7.7 is the industry standard for character-based LCD modules with a maximum of 80 characters. The pin assignment shown in Table 7.8 is the industry standard for character-based LCD modules with more than 80 characters.

The DB7~DB0 pins are used to exchange data with the microcontroller. The E pin is an enable signal to the kit. The R/$\overline{\text{W}}$ signal determines the direction of data transfer. The RS signal selects the register to be accessed. When the RS signal is high, the *data register* is selected. Otherwise, the *instruction register* is selected. The V_{EE} pin is used to control the brightness of the display and is often connected to a potentiometer. The V_{EE} input should not be set to the maximum value (=V_{CC}) for an extended period of time to avoid burning the LCD.

Figure 7.35 ■ Block diagram of a HD44780U-based LCD kit

Pin No.	symbol	I/O	Function
1	V_{SS}	—	Power supply (GND)
2	V_{CC}	—	Power supply (+5 V)
3	V_{EE}	—	Contrast adjust
4	$\overline{RS}$	I	0 = instruction input, 1 = data input
5	$R/\overline{W}$	I	0 = write to LCD, 1 = read from LCD
6	E	I	Enable signal
7	DB0	I/O	Data bus line 0
8	DB1	I/O	Data bus line 1
9	DB2	I/O	Data bus line 2
10	DB3	I/O	Data bus line 3
11	DB4	I/O	Data bus line 4
12	DB5	I/O	Data bus line 5
13	DB6	I/O	Data bus line 6
14	DB7	I/O	Data bus line 7

Table 7.7 ■ Pin assignment for displays with no more than 80 characters

An LCD module can be used as a memory-mapped device and be enabled by an address decoder. The E signal is normally connected to the address decoder output qualified by other control signals to meet the timing requirement. The $R/\overline{W}$ pin can be connected to the same pin of the microcontroller. The RS pin can be connected to the least significant bit of the address pin (A0) from the microcontroller. This approach is applicable only for those microcontrollers that support external memory. The LCD programming in this approach is generally easier and straightforward.

Pin No.	symbol	I/O	Function
1	DB7	I/O	Data bus line 7
2	DB6	I/O	Data bus line 6
3	DB5	I/O	Data bus line 5
4	DB4	I/O	Data bus line 4
5	DB3	I/O	Data bus line 3
6	DB2	I/O	Data bus line 2
7	DB1	I/O	Data bus line 1
8	DB0	I/O	Data bus line 0
9	E1	I	Enable signal row 0 and 1
10	$R/\overline{W}$	I	0 = write to LCD, 1 = read from LCD
11	RS	I	0 = instruction input, 1 = data input
12	V_{EE}	—	Contrast adjust
13	V_{SS}	—	Power supply (GND)
14	V_{CC}	—	Power supply (+5 V)
15	E2	I	Enable signal row 2 and 3
16	N.C	—	

Table 7.8 ■ Pin assignment for displays with more than 80 characters

An LCD module can also be interfaced directly with an I/O port. In this configuration, the designer will need to use I/O pins to control the signals E, R/W, and RS. Programming will be slightly more cumbersome than the memory-mapped approach due to the need to manipulate these three signals.

The HD44780U provides a set of instructions for the user to set up the LCD parameters. The operations performed by these instructions are summarized in Table 7.9. The meanings of certain bits in these instructions are explained in Table 7.10. The HD44780U can be configured to control 1-line, 2-line, and 4-line LCDs. The mappings of the character positions on the LCD screen and the DDRAM addresses are not sequential and are shown in Table 7.11.

Instruction	Code										Description	Execution Time
	RS	R/W	B7	B6	B5	B4	B3	B2	B1	B0		
Clear display	0	0	0	0	0	0	0	0	0	1	Clears display and returns cursor to the home position (address 0).	1.64 ms
Cursor home	0	0	0	0	0	0	0	0	1	*	Returns cursor to home position (address 0). Also returns display being shifted to the original position. DDRAM contents remain unchanged.	1.64 ms
Entry mode set	0	0	0	0	0	0	0	1	I/D	S	Sets cursor move direction (I/D); specifies to shift the display (S). These operations are performed during data read/write.	40 μs
Display on/off control	0	0	0	0	0	1	D	C	B		Sets on/off of all display (D), cursor on/off (C), and blink of cursor position character (B).	40 μs
Cursor /display shift	0	0	0	0	1	S/C	R/L	*	*		Sets cursor-move or display-shift (S/C), shift direction (R/L). DDRAM contents remain unchanged.	40 μs
Function set	0	0	0	1	DL	N	F	*	*		Sets interface data length (DL), number of display line (N), and character font (F).	40 μs
Set CGRAM address	0	0	0	1	CGRAM address						Sets the CGRAM address. CGRAM data are sent and received after this setting.	40 μs
Set DDRAM address	0	0	1	DDRAM address							Sets the DDRAM address. DDRAM data are sent and received after this setting.	40 μs
Read busy flag and address counter	0	1	BF	CGRAM/DDRAM address							Reads busy flag (BF) indicating internal operation being performed and reads CGRAM or DDRAM address counter contents (depending on previous instruction).	0 μs
Write CGRAM or DDRAM	1	0	write data								Writes data to CGRAM or DDRAM.	40 μs
Read from CGRAM or DDRAM	1	1	read data								Reads data from CGRAM or DDRAM.	40 μs

Table 7.9 ■ HD4478U instruction set

7.8.1 Display Data RAM

Display data RAM (DDRAM) stores display data represented in 8-bit character codes. Its extended capacity is 80 × 8 bits, or 80 characters. The area in DDRAM that is not used for display can be used as general data RAM. The relationships between DDRAM addresses and positions on the LCD are shown in Table 7.11.

Bit Name	Settings	
I/D	0 = decrement cursor position.	1 = increment cursor position
S	0 = no display shift.	1 = display shift
D	0 = display off	1 = display on
C	0 = cursor off	1 = cursor on
B	0 = cursor blink off	1 = cursor blink on
S/C	0 = move cursor	1 = shift display
R/L	0 = shift left	1 = shift right
DL	0 = 4-bit interface	1 = 8-bit interface
N	0 = 1/8 or 1/11 duty (1 line)	1 = 1/16 duty (2 lines)
F	0 = 5 × 8 dots	1 = 5 × 10 dots
BF	0 = can accept instruction	1 = internal operation in progress

Table 7.10 ■ LCD instruction bit names

Display Size	Visible	
	Character Positions	DDRAM Addresses
1 × 8	00..07	0x00..0x07
1 × 16	00..15	0x00..0x0F
1 × 20	00..19	0x00..0x13
1 × 24	00..23	0x00..0x17
1 × 32	00..31	0x00..0x1F
1 × 40	00..39	0x00..0x27

Table 7.11a ■ DDRAM address usage for a 1-line LCD

Display Size	Visible	
	Character Positions	DDRAM Addresses
2 × 16	00..15	0x00..0x0F + 0x40..0x4F
2 × 20	00..19	0x00..0x13 + 0x40..0x53
2 × 24	00..23	0x00..0x17 + 0x40..0x57
2 × 32	00..31	0x00..0x1F + 0x40..0x5F
2 × 40	00..39	0x00..0x27 + 0x40..0x67

Table 7.11b ■ DDRAM address usage for a 2-line LCD

Display Size	Visible	
	Character Positions	DDRAM Addresses
4 × 16	00..15	0x00..0x0F + 0x40..0x4F + 0x14..0x23 + 0x54..0x63
4 × 20	00..19	0x00..0x13 + 0x40..0x53 + 0x14..0x27 + 0x54..0x67
4 × 40	00..39 on 1st controller and 00..39 on 2nd controller	0x00..0x27 + 0x40..0x67 on 1st controller and 0x00..0x27 + 0x40..0x67 on 2nd controller

Note: Two LCD controllers are needed to control LCD displays with 4 × 40 characters.

Table 7.11c ■ DDRAM address usage for a 4-line LCD

7.8.2 Character Generator ROM (CGROM)

The character generator ROM generates 5×8 or 5×10 dot character patterns from 8-bit character codes. It can generate 208 5×8 dot character patterns and 32 5×10 dot character patterns.

7.8.3 Character Generator RAM (CGRAM)

The user can rewrite character patterns into the CGRAM by programming. For 5×8 fonts, eight character patterns can be written, and for 5×10 fonts, four character patterns can be written.

7.8.4 Registers

The HD44780U has two 8-bit registers, an *instruction register* (IR) and a *data register* (DR). The IR register stores *instruction codes*, such as "display clear" and "cursor move", and *address information* for display data RAM (DDRAM) and character generator RAM (CGRAM). The microcontroller writes commands into this register to set up the LCD operation parameters. To write data into the DDRAM or CGRAM, the microcontroller writes data into the DR register. Data written into the DR register will be automatically written into DDRAM or CGRAM by an internal operation. The DR register is also used for data storage when reading data from DDRAM or CGRAM. When address information is written into the IR register, data is read and then stored in the DR register from DDRAM or CGRAM by an internal operation. The microcontroller can then read the data from the DR register. After a read operation, data in DDRAM or CGRAM at the next address is sent to the DR register and the microcontroller does not need to send another address. The IR and DR registers are distinguished by the RS signal. The IR register is selected when the RS input is low. The DR register is selected when the RS input is high. Register selection is illustrated in Table 7.12.

RS	R/$\overline{\text{W}}$	Operation
0	0	IR write as an internal operation (display clear, etc.).
0	1	Read busy flag (DB7) and address counter (DB0 to DB6).
1	0	DR write as an internal operation (DR to DDRAM or CGRAM).
1	1	DR read as an internal operation (DDRAM or CGRAM to DR).

Table 7.12 ■ Register selection

Busy Flag (BF)

The HD44780U has a busy flag (BF) to indicate whether the current internal operation is complete. When BF is 1, the HD44780U is still busy with an internal operation. When RS = 0 and R/$\overline{\text{W}}$ is 1, the busy flag is output to the DB7 pin. The microprocessor can read this pin to find out if the HD44780U is still busy.

Address Counter (AC)

The HD44780U uses a 7-bit address counter to keep track of the address of the next DDRAM or CGRAM location to be accessed. When an instruction is written into the IR register, the address information contained in the instruction is transferred to the AC register. The selection of DDRAM or CGRAM is determined by the instruction. After writing into (reading from) DDRAM or CGRAM, the content of the AC register is automatically incremented (decremented) by 1. The content of the AC register is output to the DB6..DB0 pins when the RS signal is low and the R/W̄ signal is high.

7.8.5 Instruction Description

The functions of LCD instructions are discussed in this section.

CLEAR DISPLAY

This instruction writes the space code 0x20 into all DDRAM locations. It then sets 0 into the address counter and returns the display to its original status if it was shifted. In other words, the display disappears and the cursor or blinking goes to the upper left corner of the display. It also sets the I/D bit to 1 (increment mode) in entry mode.

RETURN HOME

This instruction sets DDRAM address 0 into the address counter and returns to its original status if it was shifted. The DDRAM contents are not changed. The cursor or blinking goes to the upper left corner of the display.

ENTRY MODE SET

The **I/D** bit of this instruction controls the incrementing (I/D = 1) or decrementing (I/D = 0) of the DDRAM address. The cursor or blinking will be moved to the right or left depending on whether this bit is set to 1 or 0. The same applies to writing and reading of CGRAM.

The **S** bit of this instruction controls the shifting of the LCD display. The display shifts if S = 1. Otherwise, the display does not shift. If S is 1, it will seem as if the cursor does not move but the display does. The display does not shift when reading from DDRAM. Also, writing into or reading from CGRAM does not shift the display.

DISPLAY ON/OFF CONTROL

This instruction has three bit parameters: **D**, **C**, and **B**. When the D bit is set to 1, the display is turned on. Otherwise, it is turned off. The cursor is turned on when the C bit is set to 1. The character indicated by the cursor will blink when the B bit is set to 1.

CURSOR OR DISPLAY SHIFT

This instruction shifts the cursor position to the right or left without writing or reading display data. The shifting is controlled by two bits, as shown in Table 7.13. This function is used to correct or search the display. When the cursor has moved to the end of a line, it will be moved to the beginning of the next line.

When the displayed data is shifted repeatedly, each line moves only horizontally. The second line of the display does not shift into the first row. The contents of the address counter will not change if the only action performed is a display shift.

S/C	R/L	Operation
0	0	Shifts the cursor position to the left. (AC is decremented by 1)
0	1	Shifts the cursor position to the right. (AC is incremented by 1)
1	0	Shifts the entire display to the left. The cursor follows the display shift.
1	1	Shifts the entire display to the right. The cursor follows the display shift.

Table 7.13 ■ Shift function

FUNCTION SET

This instruction allows the user to set the interface data length, select the number of display lines, and select the character fonts. There are three bit variables in this instruction:

DL: Data is sent or received in 8-bit length (DB7 to DB0) when DL is set to 1, and in 4-bit length (DB7 to DB4) when DL is 0. When the pin count is at a premium for the application, the 4-bit data length should be chosen even though it is cumbersome to perform the programming.

N: This bit sets the number of display lines. When set to 0, one-line display is selected. When set to 1, two-line display is selected.

F: When set to 0, the 5×8 font is selected. When set to 1, the 5×10 font is selected.

SET CGRAM ADDRESS

This instruction contains the CGRAM address to be set into the address counter.

SET DDRAM ADDRESS

This instruction allows the user to set the address of the DDRAM (in address counter).

READ BUSY FLAG AND ADDRESS

This instruction reads the busy flag (BF) and the address counter. The BF flag indicates whether the LCD controller is still executing the previously received instruction.

7.8.6 Interfacing the HD44780U to the HCS12 Microcontroller

The data transfer between the HD44780U and the MCU can be done in 4 bits or 8 bits at a time. When in 4-bit mode, data are carried on the upper four data pins (DB7~DB4). The upper four bits are sent over DB7~DB4 first and followed immediately by the lower four bits.

For those HCS12 members that do not support external memory, the designer must use I/O ports to interface with the LCD module. For those HCS12 members that support external memory, the designer has the choice of using I/O ports to interface with the LCD module or treating the LCD as a memory device. This chapter will treat the LCD only as an I/O device.

The LCD circuit connections for the SSE256 (8-bit data bus) and the Dragon12 (4-bit data bus) are illustrated in Figure 7.36. The R/W̄ signal to the LCD kit in the Dragon12 demo board

Figure 7.36a ■ LCD interface example (8-bit bus, used in SSE256)

Figure 7.36b ■ LCD interface example (4-bit bus, used in Dragon12)

is grounded, which prevents the user from polling the BF flag to determine whether the LCD internal operation has been completed.

Certain timing parameters must be satisfied in order to access the LCD successfully. The read and write timing diagrams are shown in Figures 7.37 and 7.38, respectively. The values of timing parameters depend on the frequency of the operation. HD44780U-based LCDs can operate at either 1 MHz (cycle time of E signal) or 2 MHz. The values of timing parameters at these two frequencies are shown in Tables 7.14 and 7.15, respectively.

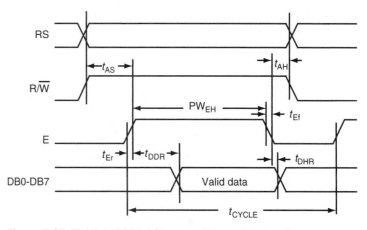

Figure 7.37 ■ HD44780U LCD controller read timing diagram

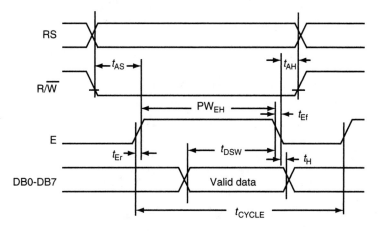

Figure 7.38 ■ HD44780U LCD controller write timing diagram

Symbol	Meaning	Min.	Typ.	Max.	Unit
t_{CYCLE}	Enable cycle time	1000	–	–	ns
PW_{EH}	Enable pulse width (high level)	450	–	–	ns
t_{Er}, t_{Ef}	Enable rise and decay time	–	–	25	ns
t_{AS}	Address setup time, RS, R/$\overline{W}$, E	60	–	–	ns
t_{DDR}	Data delay time	–	–	360	ns
t_{DSW}	Data setup time	195	–	–	ns
t_H	Data hold time (write)	10	–	–	ns
t_{DHR}	Data hold time (read)	5	–	–	ns
t_{AH}	Address hold time	20	–	–	ns

Table 7.14 ■ HD44780U LCD bus timing parameters (1 MHz operation)

Symbol	Meaning	Min.	Typ.	Max.	Unit
t_{CYCLE}	Enable cycle time	500	–	–	ns
PW_{EH}	Enable pulse width (high level)	230	–	–	ns
t_{Er}, t_{Ef}	Enable rise and decay time	–	–	20	ns
t_{AS}	Address setup time, RS, R/$\overline{W}$, E	40	–	–	ns
t_{DDR}	Data delay time	–	–	160	ns
t_{DSW}	Data setup time	80	–	–	ns
t_H	Data hold time (write)	10	–	–	ns
t_{DHR}	Data hold time (read)	5	–	–	ns
t_{AH}	Address hold time	10	–	–	ns

Table 7.15 ■ HD44780U LCD bus timing parameters (2 MHz operation)

Example 7.6

Write a set of LCD routines that perform the following functions for the Dragon12 demo board:

1. cmd2lcd (char cmd). This function sends the command **cmd** to the LCD kit.
2. openlcd (void). This function initializes the LCD.
3. putc2lcd (char cx). This function outputs the character **cx** to the LCD kit.
4. puts2lcd (char *ptr). This function outputs a NULL-terminated string pointed to by **ptr** to the LCD kit.

Solution: The procedure for sending a command to the IR register of the LCD is as follows:

Step 1
Pull the RS and the E signals to low.

Step 2
Pull the R/$\overline{\text{W}}$ signal to low.

Step 3
Pull the E signal to high.

Step 4
Output data to the output port attached to the LCD data bus. One needs to configure Port D for output before writing data to the LCD kit.

Step 5
Pull the E signal to low and make sure that the internal operation is complete.

The procedure for writing a byte to the LCD data register is as follows:

Step 1
Pull the RS signal to high.

Step 2
Pull the R/$\overline{\text{W}}$ signal to low.

Step 3
Pull the E signal to high.

Step 4
Output data to the I/O port attached to the LCD data bus.

Step 5
Pull the E signal to low and make sure that the internal operation is complete.

These procedures need to be repeated once for an LCD kit that has a 4-bit interface.

The following constant definitions will be used in the specified functions:

```
lcd_dat    equ    PTK            ; LCD data pins (PK5~PK2)
lcd_dir    equ    DDRK           ; LCD data direction port
lcd_E      equ    $02            ; E signal pin
lcd_RS     equ    $01            ; RS signal pin
```

All of the LCD commands take a much longer time to complete than an instruction does. There is always a possibility that the LCD is still busy with its internal operation when one intends to send a new command to the LCD. There are two methods for solving this problem:

1. The **cmd2lcd** function calls a subroutine to make sure that the LCD is idle before proceeding with the new command. After performing the desired operation, the

cmd2lcd function simply returns without waiting for the internal LCD operation to complete. This approach cannot work with the Dragon12 demo board.

2. The **cmd2lcd** function performs the desired operation, waits for 40 μs (or slightly longer), and then returns to the caller. By waiting for 40 μs, all except two instructions (clear display and cursor home) will be completed. For these two instructions, the caller needs to call a delay subroutine to wait until the internal LCD operation is completed before proceeding with other operations.

The function that sends a command to the LCD kit using the second approach is as follows:

```
cmd2lcd    psha                             ; save the command in stack
           bclr    lcd_dat,lcd_RS           ; select the instruction register
           bset    lcd_dat,lcd_E            ; pull the E signal high
           anda    #$F0                     ; clear the lower four bits
           lsra                             ; match the upper four bits with the LCD
           lsra                             ; data pins
           oraa    #lcd_E                   ; maintain the E signal value
           staa    lcd_dat                  ; send the command, along with RS, E signals
           nop                              ; extend the duration of the E pulse
           nop                              ;   "
           nop                              ;   "
           bclr    lcd_dat,lcd_E            ; pull the E signal low
           pula                             ; retrieve the LCD command
           anda    #$0F                     ; clear the upper four bits
           lsla                             ; match the lower four bits with the LCD
           lsla                             ; data pins
           bset    lcd_dat,lcd_E            ; pull the E signal high
           oraa    #lcd_E                   ; maintain the E signal value
           staa    lcd_dat                  ; send the lower four bits of command with E and RS
           nop                              ; extend the duration of the E pulse
           nop                              ;   "
           nop                              ;   "
           bclr    lcd_dat,lcd_E            ; clear E signal to complete the write operation
           ldy     #1                       ; adding this delay will complete the internal
           jsr     delayby50us              ; operation for most instructions
           rts
```

Before using the LCD, one must configure it properly. The configuration of the LCD involves at least the following four LCD instructions:

1. Entry mode set. The common setting for this instruction is to move the cursor to the right after reading or writing a character from/to the LCD.

2. Display on/off. The common setting for this instruction is to turn on the display, cursor, and blinking.

3. Function set. This instruction sets the number of rows for display, the font size, and the width of the interface data (4 or 8 bits).

4. Clear display. Before outputting any data, it is always a good idea to clear the LCD screen and move the cursor to the home position (upper left corner).

The following function performs the LCD configuration:

```
openlcd    movb    #$FF,lcd_dir             ; configure Port K for output
           ldy     #10                      ; wait for LCD to be ready
```

```
        jsr     delayby10ms         ;   "
        ldaa    #$28                ; set 4-bit data, 2-line display, 5 × 8 font
        jsr     cmd2lcd             ;   "
        ldaa    #$0F                ; turn on display, cursor, and blinking
        jsr     cmd2lcd             ;   "
        ldaa    #$06                ; move cursor right (entry mode set instruction)
        jsr     cmd2lcd             ;   "
        ldaa    #$01                ; clear display and return to home position
        jsr     cmd2lcd             ;   "
        ldy     #2                  ; wait until "clear display" command is complete
        jsr     delayby1ms          ;   "
        rts
```

The function that outputs a character to the LCD and makes sure that the write operation is complete is as follows:

```
putc2lcd    psha                        ; save a copy of the data
            bset    lcd_dat,lcd_RS      ; select lcd Data register
            bset    lcd_dat,lcd_E       ; pull E to high
            anda    #$F0                ; mask out the lower four bits
            lsra                        ; match the upper four bits with the LCD
            lsra                        ; data pins
            oraa    #$03                ; keep signal E and RS unchanged
            staa    lcd_dat             ; send the upper four bits and E, RS signals
            nop                         ; provide enough duration to the E signal
            nop                         ;   "
            nop                         ;   "
            bclr    lcd_dat,lcd_E       ; pull the E signal low
            pula                        ; retrieve the character from the stack
            anda    #$0F                ; clear the upper four bits
            lsla                        ; match the lower four bits with the LCD
            lsla                        ; data pins
            bset    lcd_dat,lcd_E       ; pull the E signal high
            oraa    #$03                ; keep E and RS unchanged
            staa    lcd_dat
            nop
            nop
            nop
            bclr    lcd_dat,lcd_E       ; pull E low to complete the write cycle
            ldy     #1                  ; wait until the write operation is
            jsr     delayby50us         ; complete
            rts
```

The function that outputs a NULL-terminated string pointed to by the index register X is as follows:

```
puts2lcd    ldaa    1,x+                ; get one character from the string
            beq     done_puts           ; reach NULL character?
            jsr     putc2lcd
            bra     puts2lcd
done_puts   rts
```

Example 7.7

Write an assembly program to test the previous four subroutines by displaying the following messages on two lines:

```
hello world!
I am ready!
```

Solution: The program is as follows:

```
#include        "c:\miniide\hcs12.inc"
lcd_dat    equ    PTK                ; LCD data pins (PK5~PK2)
lcd_dir    equ    DDRK               ; LCD data direction port
lcd_E      equ    $02                ; E signal pin
lcd_RS     equ    $01                ; RS signal pin
           org    $1500
           lds    #$1500             ; set up stack pointer
           jsr    openlcd            ; initialize the LCD
           ldx    #msg1lcd
           jsr    puts2lcd
           ldaa   #$C0               ; move to the second row
           jsr    cmd2lcd            ;    "
           ldx    #msg2lcd
           jsr    puts2lcd
           swi
msg1lcd    fcc    "hello world!"
           dc.b   0
msg2lcd    fcc    "I am ready!"
           dc.b   0
#include        "delay.asm"           ; include delay routines here
;
; include the previous four LCD functions
;
```

Example 7.8

Write the C language versions of the previous four LCD functions and a test program to test them.

Solution: The C functions for the LCD kit on the Dragon12 demo board and their test program are as follows:

```
#include    "c:\egnu091\include\hcs12.h"
#include    "c:\egnu091\include\delay.c"
#define     LCD_DAT  PTK             /* PTK drives LCD data pins, E, and RS */
#define     LCD_DIR  DDRK            /* direction of LCD port */
#define     LCD_E    0x02            /* E signal */
#define     LCD_RS   0x01            /* RS signal */
#define     LCD_E_RS 0x03            /* assert both E and RS signal */
void    cmd2lcd(char cmd);
void    openlcd(void);
```

```
void    putc2lcd(char cx);
void    puts2lcd (char *ptr);
main (void)
{
      char *msg1 = "hello world!";
      char *msg2 = "I am ready!";
      openlcd();
      puts2lcd(msg1);
      cmd2lcd(0xC0);                  /* move cursor to 2nd row, 1st column */
      puts2lcd(msg2);
      return 0;
}
void cmd2lcd(char cmd)
{
      char  temp;
      char  xa, xb, xc;
      temp = cmd;                     /* save a copy of the command */
      cmd &=0xF0;                     /* clear out the lower four bits */
      LCD_DAT    &= (~LCD_RS); /* select LCD instruction register */
      LCD_DAT    |= LCD_E;     /* pull E signal to high */
      cmd >>= 2;                      /* shift to match LCD data pins */
      LCD_DAT    = cmd | LCD_E; /* output upper four bits, E, and RS */
      xa   = 1;                       /* dummy statements to lengthen E */
      xb   = 2;                       /* " */
      xc   = 3;
      LCD_DAT    &= (~LCD_E);  /* pull E signal to low */
      cmd  = temp & 0x0F;             /* extract the lower four bits */
      LCD_DAT    |= LCD_E;     /* pull E to high */
      cmd <<= 2;                      /* shift to match LCD data pins */
      LCD_DAT    = cmd | LCD_E; /* output upper four bits, E, and RS */
      xa   = 1;
      xb   = 2;
      xc   = 3;
      LCD_DAT    &= (~LCD_E);  /* pull E clock to low */
      delayby50us(1);                 /* wait until the command is complete */
}
void openlcd(void)
{
      LCD_DIR    = 0xFF;       /* configure LCD_DAT port for output */
      delayby10ms(10);                /* wait for the LCD to be ready */
      cmd2lcd(0x28);                  /* set 4-bit data, 2-line display, 5 × 7 font */
      cmd2lcd(0x0F);                  /* turn on display, cursor, blinking */
      cmd2lcd(0x06);                  /* move cursor right */
      cmd2lcd(0x01);                  /* clear screen, move cursor to home */
      delayby1ms(2);                  /* wait until "clear display" command is complete */
}
void putc2lcd(char cx)
```

```c
{
        char  temp;
        char  xa, xb, xc;
        temp = cx;
        LCD_DAT    |= LCD_RS;       /* select LCD data register */
        LCD_DAT    |= LCD_E;        /* pull E signal to high */
        cx     &= 0xF0;             /* clear the lower four bits */
        cx     >>= 2;               /* shift to match the LCD data pins */
        LCD_DAT    = cx | LCD_E_RS;/* output upper four bits, E, and RS */
        xa     = 1;                 /* dummy statements to lengthen the width of E pulse */
        xb     = 2;                 /*       "           */
        xc     = 3;                 /*       "           */
        LCD_DAT    &= (~LCD_E);     /* pull E to low */
        cx     = temp & 0x0F;       /* get the lower four bits */
        LCD_DAT    |= LCD_E;        /* pull E to high */
        cx     <<= 2;               /* shift to match the LCD data pins */
        LCD_DAT    = cx | LCD_E_RS;/* output lower four bits, E, and RS */
        xa     = 1;
        xb     = 2;
        xc     = 3;
        LCD_DAT    &= (~LCD_E);     /* pull E to low */
        delayby50us(1);
}
void puts2lcd (char *ptr)
{
        while (*ptr) {
              putc2lcd(*ptr);
              delayby50us(1);
              ptr++;
        }
}
```

The LCD functions for the SSE256 demo board are simpler and are provided in Appendix E. The assembly and C language versions of LCD functions have been grouped into files to be included in the user's program and are provided on the complementary CD.

7.9 Interfacing with DIP Switches

A switch is probably the simplest input device we can find. To make input more efficient, a set of eight switches organized as a dual inline package (DIP) is often used. A DIP package can be connected to any input port with eight pins, such as Port A, Port B, and Port H, as shown in Figure 7.39. When a switch is closed, the associated Port P input is 0. Otherwise, the associated Port A pin has a value of 1. Each Port A pin is pulled up to high via a 10-kΩ resistor when the associated switch is open.

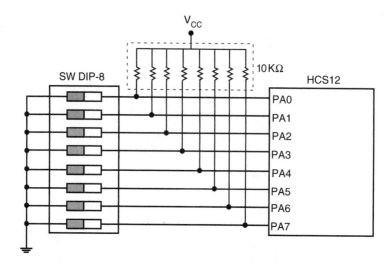

Figure 7.39 ■ Connecting a set of eight DIP switches to port A of the HCS12

Example 7.9

Write a sequence of instructions to read the value from an eight-switch DIP connected to PORTA of the HCS12 into accumulator A.

Solution:

```
#include     "c:\miniide\hcs12.inc"
             movb     #$0,DDRA        ; configure Port A for input
             ldaa     PTA
             . . .
```

In C language, this operation can be achieved as follows:

```
#include <hcs12.h>
void main ()
{
        char    xx;
        . . .
        DDRA = 0;               /* configure Port A for input */
        xx = PTA               /* read a byte from the DIP switch */
        . . .
}
```

When the program needs data input from the DIP switches, it must inform the designer to set up the switches. In addition, the designer must inform the microcontroller to read the data when it is ready. This can be done by using an interrupt signal. A lab assignment is provided to practice this approach.

7.10 Interfacing Parallel Ports to a Keypad

A keypad is another commonly used input device. Like a keyboard, a keypad is arranged as an array of switches, which can be mechanical, membrane, capacitors, or Hall-effect in construction. In mechanical switches, two metal contacts are brought together to complete an electrical circuit. In membrane switches, a plastic or rubber membrane presses one conductor onto another; this type of switch can be made very thin. Capacitive switches internally comprise two plates of a parallel plate capacitor; pressing the key cap effectively increases the capacitance between the two plates. Special circuitry is needed to detect this change in capacitance. In Hall-effect key switches, the motion of the magnetic flux lines of a permanent magnet perpendicular to a crystal is detected as voltage appearing between the two faces of the crystal—it is this voltage that *registers* a switch closure.

Mechanical keypads and keyboards are most popular due to their low cost and strength of construction. However, mechanical switches have a common problem called *contact bounce*. Instead of producing a single, clean pulse output, pressing a mechanical switch generates a series of pulses because the switch contacts do not come to rest immediately. This phenomenon is illustrated in Figure 7.40. When the key is not pressed, the voltage output to the computer is 5 V. In order to detect which key has been pressed, the microprocessor needs to scan every key switch of the keypad. A human being cannot press and release a key switch in less than 20 ms. During this interval, the microprocessor can scan the same key switch closure tens or even hundreds of thousand times, interpreting each low signal as a new input when in fact only one input should be sent.

Figure 7.40 ■ Key switch contact bounce

Because of the contact bounce and the disparity in speed between the microprocessor and human key pressing, a *debouncing* process is needed. A keypad input program can be divided into three stages:

1. *Keypad scanning* to find out which key was pressed
2. *Key switch debouncing* to make sure a key is indeed pressed
3. *Table lookup* to find the ASCII code of the key that was pressed

7.10.1 Keypad Scanning

Keypad scanning is usually performed row-by-row, column-by-column. A 16-key keypad can easily be interfaced with any available I/O port. Figure 7.41 shows a 16-key keypad organized into four rows with each row consisting of four switches. For the keypad input application, the upper four pins (PA7~PA4) of Port A should be configured for output, whereas the lower four pins (PA3~PA0) of Port A should be configured for input.

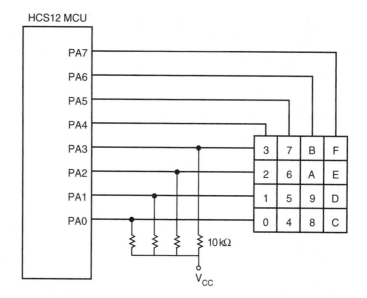

Figure 7.41 ■ Sixteen-key keypad connected to the HCS12

The rows and columns of a keypad are simply conductors. In Figure 7.41, Port A pins PA3~PA0 are pulled up to high by pull-up resistors. Whenever a key switch is pressed, the corresponding row and column are shorted together. In order to distinguish the row being scanned and those not being scanned, the row being scanned is driven low, whereas the other rows are driven high. The row selection of the 16-key keypad is shown in Table 7.16.

PA7	PA6	PA5	PA4	Selected Keys
1	1	1	0	0, 1, 2, and 3
1	1	0	1	4, 5, 6, and 7
1	0	1	1	8, 9, A, and B
0	1	1	1	C, D, E, and F

Table 7.16 ■ Sixteen-key keypad row selections

7.10.2 Keypad Debouncing

Contact bounce is due to the dynamics of a closing contact. The signal falls and rises a few times within a period of about 5 ms as a contact bounces. Since a human being cannot press and release a switch in less than 20 ms, a debouncer will recognize that the switch is closed after the voltage is low for about 10 ms and will recognize that the switch is open after the voltage is high for about 10 ms.

Both hardware and software solutions to the key bounce problem are available. Hardware solutions to contact bounce include an analog circuit that uses a resistor and a capacitor to smooth the voltage and two digital solutions that use set-reset latches or CMOS buffers and double-throw switches. Dedicated scanner chips that perform keypad scanning and debouncing are also available. National Semiconductor 74C922 and 74C923 are two examples.

HARDWARE DEBOUNCING TECHNIQUES

The following are hardware debouncing techniques

- *Set-reset latches.* A key switch can be debounced by using the set-reset latch shown in Figure 7.42a. Before being pressed, the key is touching the set input and the Q voltage is high. When pressed, the key moves toward the reset position. When the key touches the reset position, the Q voltage will go low. When the key is bouncing and touching neither the set nor the reset input, both set and reset inputs are pulled low by the pull-down resistors. Since both set and reset are low, the Q voltage will remain low and the key will be recognized as pressed.

- *Noninverting CMOS buffer with high input impedance.* The CMOS buffer output is identical to its input. When the switch is pressed, the input of the buffer chip 4050 is grounded and hence V_{OUT} is forced to low. When the key switch is bouncing (not touching the input), the resistor R keeps the output voltage low. This is due to the high input impedance of 4050, which causes a negligible voltage drop on the feedback resistor. Thus the output is debounced. This solution is shown in Figure 7.42b.

(a) Set-reset latch

(b) CMOS gate debouncer

(c) Integrating RC circuit debouncer

Figure 7.42 ■ Hardware debouncing techniques

■ *Integrated debouncers.* The RC constant of the integrator determines the rate at which the capacitor charges up toward the supply voltage once the ground connection via the switch has been removed. As long as the capacitor voltage does not exceed the logic 0 threshold value, the V_{OUT} signal will continue to be recognized as a logic 0. This solution is shown in Figure 7.42c.

SOFTWARE DEBOUNCING TECHNIQUES

A simple software debouncing method is *wait-and-see*, i.e., wait for 10 ms and reexamine the same key to see if it is still pressed. A key is considered pressed if its output voltage is still low 10 ms after it is detected low. If the output voltage is high, the program will consider the signal was noise or the key is still bouncing. In either case, the program will continue to perform the scanning. For most applications, this simple wait-and-see method is adequate.

7.10.3 ASCII Code Lookup

For an application that needs a keyboard, the easiest way to find out the ASCII code of the pressed key is to perform a table lookup. However, table lookup is not necessary for the keypad because ASCII code lookup can be embedded in the program that performs scanning and debouncing.

Example 7.10

▼

Write a program to perform keypad scanning and debouncing and return the ASCII code of the pressed key to the caller.

Solution: The following assembly routine will perform scanning and debouncing and return the ASCII code of the pressed key in accumulator A to the caller:

```
          #include "c:\miniide\hcs12.inc"
keyboard  equ    PTA

get_char  movb   #$F0,DDRA        ; configure pins PA7~PA4 for output,PA3~PA0 for input
scan_r0   movb   #$EF,keyboard    ; scan the row containing keys 0123
scan_k0   brclr  keyboard,$01,key0 ; is key 0 pressed?
scan_k1   brclr  keyboard,$02,key1 ; is key 1 pressed?
scan_k2   brclr  keyboard,$04,key2 ; is key 2 pressed?
scan_k3   brclr  keyboard,$08,key3 ; is key 3 pressed?
          bra    scan_r1
key0      jmp    db_key0
key1      jmp    db_key1
key2      jmp    db_key2
key3      jmp    db_key3
scan_r1   movb   #$DF,keyboard    ; scan the row containing keys 4567
scan_k4   brclr  keyboard,$01,key4 ; is key 4 pressed?
scan_k5   brclr  keyboard,$02,key5 ; is key 5 pressed?
scan_k6   brclr  keyboard,$04,key6 ; is key 6 pressed?
scan_k7   brclr  keyboard,$08,key7 ; is key 7 pressed?
          bra    scan_r2
key4      jmp    db_key4
key5      jmp    db_key5
```

```
key6        jmp     db_key6
key7        jmp     db_key7
scan_r2     movb    #$BF,keyboard        ; scan the row containing keys 89AB
scan_k8     brclr   keyboard,$01,key8 ; is key 8 pressed?
scan_k9     brclr   keyboard,$02,key9 ; is key 9 pressed?
scan_kA     brclr   keyboard,$04,keyA ; is key A pressed?
scan_kB     brclr   keyboard,$08,keyB ; is key B pressed?
            bra     scan_r3
key8        jmp     db_key8
key9        jmp     db_key9
keyA        jmp     db_keyA
keyB        jmp     db_keyB
scan_r3     movb    #$7F,keyboard        ; scan the row containing keys CDEF
scan_kC     brclr   keyboard,$01,keyC ; is key C pressed?
scan_kD     brclr   keyboard,$02,keyD ; is key D pressed?
scan_kE     brclr   keyboard,$04,keyE ; is key E pressed?
scan_kF     brclr   keyboard,$08,keyF ; is key F pressed?
            jmp     scan_r0
keyC        jmp     db_keyC
keyD        jmp     db_keyD
keyE        jmp     db_keyE
keyF        jmp     db_keyF
; debounce key 0
db_key0     jsr     delay10ms
            brclr   keyboard,$01,getc0
            jmp     scan_k1
getc0       ldaa    #$30                            ; return the ASCII code of 0
            rts
; debounce key 1
db_key1     jsr     delay10ms
            brclr   keyboard,$02,getc1
            jmp     scan_k2
getc1       ldaa    #$31                            ; return the ASCII code of 1
            rts
db_key2     jsr     delay10ms
            brclr   keyboard,$04,getc2
            jmp     scan_k3
getc2       ldaa    #$32                            ; return the ASCII code of 2
            rts
db_key3     jsr     delay10ms
            brclr   keyboard,$08,getc3
            jmp     scan_r1
getc3       ldaa    #$33                            ; return the ASCII code of 3
            rts
db_key4     jsr     delay10ms
            brclr   keyboard,$01,getc4
            jmp     scan_k5
getc4       ldaa    #$34                            ; return the ASCII code of 4
            rts
```

```
db_key5     jsr     delay10ms
            brclr   keyboard,$02,getc5
            jmp     scan_k6
getc5       ldaa    #$35                    ; return the ASCII code of 5
            rts
db_key6     jsr     delay10ms
            brclr   keyboard,$04,getc6
            jmp     scan_k7
getc6       ldaa    #$36                    ; return the ASCII code of 6
            rts
db_key7     jsr     delay10ms
            brclr   keyboard,$08,getc7
            jmp     scan_r2
getc7       ldaa    #$37                    ; return the ASCII code of 7
            rts
db_key8     jsr     delay10ms
            brclr   keyboard,$01,getc8
            jmp     scan_k9
getc8       ldaa    #$38                    ; return the ASCII code of 8
            rts
db_key9     jsr     delay10ms
            brclr   keyboard,$02,getc9
            jmp     scan_kA
getc9       ldaa    #$39                    ; return the ASCII code of 9
            rts
db_keyA     jsr     delay10ms
            brclr   keyboard,$04,getcA
            jmp     scan_kB
getcA       ldaa    #$41                    ; get the ASCII code of A
            rts
db_keyB     jsr     delay10ms
            brclr   keyboard,$08,getcB
            jmp     scan_r3
getcB       ldaa    #$42                    ; get the ASCII code of B
            rts
db_keyC     jsr     delay10ms
            brclr   keyboard,$01,getcC
            jmp     scan_kD
getcC       ldaa    #$43                    ; get the ASCII code of C
            rts
db_keyD     jsr     delay10ms
            brclr   keyboard,$02,getcD
            jmp     scan_kE
getcD       ldaa    #$44                    ; get the ASCII code of D
            rts
db_keyE     jsr     delay10ms
            brclr   keyboard,$04,getcE
            jmp     scan_kF
```

```
getcE       ldaa    #$45                        ; get the ASCII code of E
            rts
db_keyF     jsr     delay10ms
            brclr   keyboard,$08,getcF
            jmp     scan_r0
getcF       ldaa    #$46                        ; get the ASCII code of F
            rts
;***************************************************************************
; The following subroutine creates a time delay of 10 ms.
;***************************************************************************
delay10ms   movb    #$90,TSCR1                  ; enable TCNT and fast flags clear
            movb    #$06,TSCR2                  ; configure prescale factor to 64
            movb    #$01,TIOS                   ; enable OC0
            ldd     TCNT
            addd    #3750                       ; start an output compare operation
            std     TC0                         ; with 10 ms time delay
wait_lp2    brclr   TFLG1,$01,wait_lp2
            rts
```

▲

The C language version of the function is straightforward and hence will be left as an exercise.

7.11 Interfacing with a D/A Converter

A *digital-to-analog converter* converts a digital code into an analog signal (can be voltage or current). A D/A converter has many applications. Examples are digital gain and offset adjustment, programmable voltage and current sources, programmable attenuators, digital audio, closed-loop positioning, robotics, and so on. Although there are a few microcontrollers incorporating the D/A converter on the chip, most microcontrollers still need to use an off-chip D/A converter to perform the D/A conversion function. The HCS12 is no exception. A D/A converter may use a serial or parallel interface to obtain digital code from the microprocessor or microcontroller.

7.11.1 The AD7302 D/A Converter

The AD7302 is a dual-channel, 8-bit, D/A converter chip from Analog Devices that has a parallel interface with the microcontroller. The AD7302 converts an 8-bit digital value into an analog voltage. The block diagram of the AD7302 is shown in Figure 7.43. The AD7302 is designed to be a memory-mapped device. In order to send data to the AD7302, the $\overline{CS}$ signal must be pulled to low. On the rising edge of the $\overline{WR}$ signal, the values on pins D7–D0 will be latched into the Input register. When the signal $\overline{LDAC}$ is low, the data in the Input register will be transferred to the DAC register and a new D/A conversion is started. The AD7302 needs a reference voltage to perform the D/A conversion. The reference voltage can come from either the external REFIN input or the internal V_{DD}. The A/B signal selects the channel (A or B) to perform the D/A conversion. The PD pin puts the AD7302 in power-down mode and reduces the power consumption to 1 μW.

Figure 7.43 ■ Functional block diagram of the AD7302

The AD7302 operates from a single +2.7- to 5.5-V supply and typically consumes 15 mW at 5 V, making it suitable for battery-powered applications. Each digital sample takes about 2 μs to convert. The output voltage ($V_{OUT}A$ or $V_{OUT}B$) from either DAC is given by

$$V_{OUT} A/B = 2 \times V_{REF} \times (N/256)$$

where

V_{REF} is derived internally from the voltage applied at the REFIN pin or V_{DD}. If the voltage applied to the REFIN pin is within 1 V of the V_{DD}, $V_{DD}/2$ is used as the reference voltage automatically. Otherwise, the voltage applied at the REFIN pin is used as the reference voltage. The range of V_{REF} is from 1 V to $V_{DD}/2$.

N is the decimal equivalent of the code loaded to the DAC register, ranging from 0 to 255.

7.11.2 Interfacing the AD7302 with the HCS12

Interfacing the AD7302 with the HCS12 can be very simple. Both the $\overline{CS}$ and $\overline{LDAC}$ signals can be tied to ground permanently. The value to be converted must be sent to the AD7302 via a parallel port (connect to pins D7~D0). An output pin can be used (as the $\overline{WR}$ signal) to control the transferring of data to the Input register. A typical connection between the HCS12 and the AD7302 is shown in Figure 7.44.

Figure 7.44 ■ Circuit connection between the AD7302 and the HCS12

Example 7.11

Write a program to generate a sawtooth waveform from the $V_{OUT}A$ pin.

Solution: The procedure for generating a sawtooth waveform is as follows:

Step 1
Configure PB7~PB0, PJ0, and PJ1 for output.

Step 2
Output the digital value from 0 to 255 and repeat. For each value, pull the PJ0 to low and then to high so that the value on pins PB7~PB0 can be transferred to the AD7302. Pull the signal PJ1 to low during the process.

The assembly program is as follows:

```
#include    "c:\miniide\hcs12.inc"
            org     $1500
            movb    #$FF,DDRB       ; configure Port B for output
            bset    DDRJ,$03        ; configure PJ1~PJ0 for output
            bclr    PTJ,$02         ; select VOUTA output
loop        inc     PORTB           ; increase the output by one step
            bclr    PTJ,$01         ; generate a rising edge on PJ0 pin
            bset    PTJ,$01         ;    "
            bset    PTJ,$01         ; add 9 more "bset" instructions to provide 2 µs
            bset    PTJ,$01         ; for D/A conversion to complete
            bset    PTJ,$01         ;    "
            bset    PTJ,$01         ;    "
            bset    PTJ,$01         ;    "
            bset    PTJ,$01         ;    "
            bset    PTJ,$01         ;    "
            bset    PTJ,$01         ;    "
            bset    PTJ,$01         ;    "
            bra     loop            ; to complete the D/A conversion
            end
```

The C language version of the program is as follows:

```
#include <hcs12.h>
void main()
{
    DDRB    = 0xFF;        /* configure PORTB for output */
    DDRJ   |= 0x03;        /* configure pins PJ1~PJ0 for output */
    PTJ    &= 0xFD;        /* pull the signal A/B to low too select channel A */
    while (1) {
        PORTB += 1;
        PTJ &= 0xFE;       /* generate a rising edge */
        PTJ |= 0x01;       /*          "            */
        PTJ |= 0x01;       /* use dummy statements to provide 2 µs */
        PTJ |= 0x01;       /* time for D/A conversion to complete */
        PTJ |= 0x01;
        PTJ |= 0x01;
        PTJ |= 0x01;
        PTJ |= 0x01;
```

```
            PTJ │ = 0x01;
            PTJ │ = 0x01;
            PTJ │ = 0x01;
       }
  }
```

▲

The AD7302 can be used to generate many interesting waveforms. Two examples are given as exercise problems.

7.12 Stepper Motor Control

Stepper motors are digital motors. They are convenient for applications where a high degree of positional control is required. Printers, tape drives, disk drives, and robot joints, for example, are typical applications of stepper motors. In its simplest form, a stepper motor has a permanent magnet rotor and a stator consisting of two coils. The rotor aligns with the stator coil that is energized. By changing which coil is energized, as illustrated in the following figures, the rotor is turned.

In Figure 7.45a to 7.45d, the permanent magnet rotor lines up with the coil pair that is energized. The direction of the current determines the polarity of the magnetic field, and thus the angular position of the rotor. Energizing coil pair C3-C4 causes the rotor to rotate 90°. Again, the direction of the current determines the magnetic polarity and thus the angular position of the rotor. In this example, the direction of the current causes the rotor to rotate in a clockwise direction, as shown in Figure 7.45b.

Next, coils C1–C2 are energized again, but with a current opposite to that in step 1. The rotor moves 90° in a clockwise direction, as shown in Figure 7.45c. The last full step moves the rotor another 90° in a clockwise direction. Note that again the coil pair C3–C4 is energized, but with a current opposite to that in step 2.

Figure 7.45a ■ Stepper motor full step 1

Figure 7.45b ■ Stepper motor full step 2

Figure 7.45c ■ Stepper motor full step 3

We can also rotate the stepper motor in the counterclockwise direction. This can be done by reversing the polarities of coils C3 and C4 in Figure 7.45b and 7.45d. Figure 7.46 shows the counterclockwise sequence.

The stepper motor may also be operated with half steps. A *half step* occurs when the rotor (in a four-pole step) is moved to eight discrete positions (45°). To operate the stepper motor in half steps, sometimes both coils may have to be on at the same time. When two coils in close proximity are energized, there is a resultant magnetic field whose center will depend on the relative strengths of the two magnetic fields. Figure 7.47 illustrates the half-stepping sequence.

Figure 7.45d ■ Stepper motor full step 4

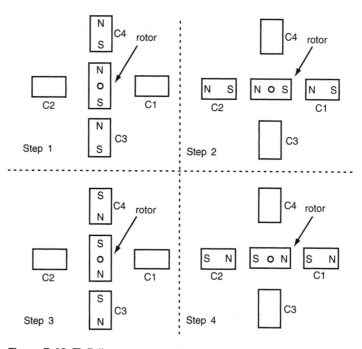

Figure 7.46 ■ Full-step counterclockwise operation of step motor

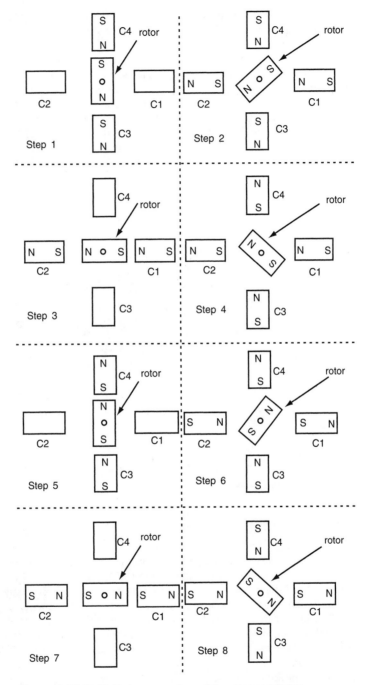

Figure 7.47 ■ Half-step operation of the stepper motor

The step sizes of stepper motors vary from approximately 0.72° to 90°. However, the most common step sizes are 1.8°, 7.5°, and 15°. The steps of 90° or 45° are too crude for many applications.

The actual stator (the stationary electromagnets) of a real motor has more segments on it than previously indicated. One example is shown in Figure 7.48. The rotor is also a little bit different and is also shown in Figure 7.48. In Figure 7.48, the stator has eight individual sections (coils) on it and hence the angle between two adjacent sections is 45°. The rotor has six sections on it and hence there are 60° between two adjacent sections. Using the principle of a Vernier mechanism, the actual movement of the rotor for each step would be 60° − 45°, or 15°. Interested readers should try to figure out how these sections are energized to rotate the motor in the clockwise and counterclockwise directions.

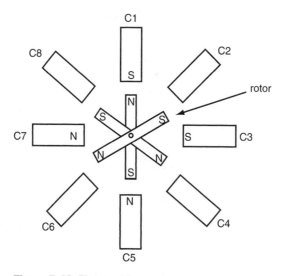

Figure 7.48 ■ Actual internal construction of stepper motor

7.12.1 Stepper Motor Drivers

Driving a stepper motor involves applying a series of voltages to the coils of the motor. A subset of coils is energized at a time to cause the motor to rotate one step. The pattern of coils energized must be followed exactly for the motor to work correctly. The pattern will vary depending on the mode used on the motor. A microcontroller can easily time the duration that the coil is energized and hence control the speed of the stepper motor in a precise manner.

The circuit in Figure 7.49 shows how the transistors are used to switch the current to each of the four coils of the stepper motor. The diodes in Figure 7.49 are called *fly back diodes* and are used to protect the transistors from reverse bias. The transistor loads are the windings in the stepper motor. The windings are inductors, storing energy as a magnetic field. When the current is cut off, the inductor dispenses its stored energy in the form of an electric current. This current attempts to flow through the transistor, reverse biasing its collector-emitter pair. The diodes are placed to prevent this current from going through the transistors.

For higher-torque applications, the normal full-step sequence is used, as shown in Table 7.17. For lower-torque applications, the half-step mode is used and its sequence is shown in

Figure 7.49 ■ Driving a stepper motor

Table 7.18. To control the motor, the microcontroller must output the values in the table in the sequence shown. Tables 7.17 and 7.18 are circular in that after the last step, the next output must be the first step. The values may be output in the order shown to rotate the motor in one direction, or in reverse order to rotate in the reverse direction. It is essential that the order be preserved even if the motor is stopped for a while. The next step to restart the motor must be the next sequential step following the last step used. The mechanical inertia of the motor will require a short delay (usually 5 to 20 ms) between two steps to prevent the motor from missing steps.

Step	Q1 PP0	Q2 PP1	Q3 PP2	Q4 PP3	Value
1	on	off	on	off	1010
2	on	off	off	on	1001
3	off	on	off	on	0101
4	off	on	on	off	0110
1	on	off	on	off	1010

Table 7.17 ■ Full-step sequence for clockwise rotation

Step	Q1 PP0	Q2 PP1	Q3 PP2	Q4 PP3	Value
1	on	off	on	off	1010
2	on	off	off	off	1000
3	on	off	off	on	1001
4	off	off	off	on	0001
5	off	on	off	on	0101
6	off	on	off	off	0100
7	off	on	on	off	0110
8	off	off	on	off	0010
1	on	off	on	off	1010

Table 7.18 ■ Half-step sequence for clockwise rotation

Example 7.12

▼

Suppose that the pins PP3..PP0 are used to drive the four transistors as shown in Figure 7.48. Write a subroutine to rotate the stepper motor clockwise one cycle using the half-step sequence.

Solution: The assembly language subroutine is as follows:

```
#include     "c:\miniide\hcs12.inc"
step1     equ     $0A
step2     equ     $08
step3     equ     $09
step4     equ     $01
step5     equ     $05
step6     equ     $04
step7     equ     $06
step8     equ     $02
half_step movb    #$FF,DDRP       ; configure PTP for output
          movb    #step1,PTP
          bsr     delay10ms
          movb    #step2,PTP
          bsr     delay10ms
          movb    #step3,PTP
          bsr     delay10ms
          movb    #step4,PTP
          bsr     delay10ms
          movb    #step5,PTP
          bsr     delay10ms
          movb    #step6,PTP
          bsr     delay10ms
          movb    #step7,PTP
          bsr     delay10ms
          movb    #step8,PTP
          bsr     delay10ms
          movb    #step1,PTP
          bsr     delay10ms
          rts
```

```
; the following subroutine waits for 10 ms
delay10ms  movb   #$90,TSCR1      ; enable TCNT and fast flags clear
           movb   #$06,TSCR2      ; configure prescale factor to 64
           bset   TIOS,IOS0       ; enable OC0
           ldd    TCNT
           addd   #3750           ; start an output compare operation
           std    TC0             ; with 10 ms time delay 10 ms
wait_lp2   brclr  TFLG1,$01,wait_lp2
           rts
           end
```

The C language version of the routine is straightforward and hence is left as an exercise.

7.13 Key Wakeups

Most embedded products are powered by batteries. In order for batteries to last longer, most microcontrollers have incorporated power-saving modes such as the WAIT mode or the STOP mode in the HCS12. Whenever there is no activity from the end user over a period of time, the application software would put the embedded product in one of the power-saving modes and reduce the power consumption. Whenever the end user wants to use the product, a keystroke would wake up the microcontroller and put the embedded product back to normal operation mode. As we learned in Chapter 6, both the reset and unmasked interrupts can put the microcontroller back to normal operation mode. However, reset is not recommended because it will restart the microcontroller, which would delay the response to the user request. An unmasked interrupt does not have this drawback. All of the HCS12 members have incorporated the key wakeup feature that will issue an interrupt to wake up the MCU when it is in the STOP or WAIT mode. After the appropriate configuration, an active edge on these pins will generate an interrupt to the MCU and force the MCU to exit from the low-power mode.

7.13.1 Key-Wakeup Registers

As described in Section 7.5.8, Ports H, J, and P have the key-wakeup capability. Each port has eight associated registers. The Port Pull Device Enable register and the Port Polarity Select register together allow the user to choose to use either the rising edge or falling edge to wake up the MCU. The Port Interrupt Enable register and the Port Interrupt Flag register together allow the user to use key pressing to wake up the MCU.

7.13.2 Key-Wakeup Initialization

In order to use the HCS12 key-wakeup feature, the user needs to initialize the wakeup port properly. The procedure for using the key-wakeup feature is as follows:

Step 1
Set the direction of the key-wakeup bits to input by writing zeros to the data direction register.

Step 2
Select the rising edge or the falling edge of the wakeup pin to interrupt the MCU by programming the related registers.

Step 3
Write the service routine for the key-wakeup interrupt and initialize the key-wakeup interrupt vector.

Step 4
Clear any flags that have been set in the key-wakeup flag register.

Step 5
Enable the key-wakeup bit by setting the appropriate bits in the wakeup interrupt enable register.

Step 6
Clear the global interrupt mask (the I bit of the CCR register).

7.13.3 Considerations for the Key-Wakeup Application

The main application of the key-wakeup feature is to support the power-saving modes of the HCS12. Application software puts the microcontroller in low-power mode by executing a STOP or a WAI instruction when the inactivity of the end user has exceeded the preset time.

Many applications are designed to be a wait loop that waits for the user to enter a request for service. When a request is entered, the application calls an appropriate routine to provide the service. After the service is done, the routine returns to the wait loop. After providing service to a user request, the application software starts a timer. If the user enters another command before the timer times out, the application software resets the timer and responds to the user request. If the timer times out before the user makes another service request, the application software puts the microcontroller in low-power mode to save power. As long as there is no user request for service, the microcontroller will stay in the low-power mode. The timer timeout interval could be a few minutes or longer depending upon how much power the user wants to save. Either the timer *output compare* or *modulus down counter* function can be used to implement the timeout interval. When the user presses the key, the microcontroller will exit the low-power mode and continue to execute the instruction following the STOP (or WAI) instruction and another cycle of the normal application loop is started.

In order to be used in a key-wakeup application, port pins must be configured for input. Ports H, J, and P can be configured to use the rising or falling edge to wake up the microcontroller. The choice of signal edge will dictate the choice of the pull-up or pull-down resistive device. As shown in Figure 7.50, a rising edge results when a high voltage is applied to a pull-down resistor;

Figure 7.50 ■ (a) Pull-down resistor creates rising edge. (b) Pull-up resistor creates falling edge

a falling edge results when a low voltage is connected to a pull-up resistor. Therefore, the designer should enable the pull-down resistor when a rising edge selected and enable a pull-up resistor when a falling edge is chosen to wake up the microcontroller.

Since the purpose of the key-wakeup feature is to enable the microcontroller to resume normal operation mode, the interrupt service routine need only perform minimal operation. The minimal operation to be performed would be to clear the interrupt flag set by the wakeup interrupt.

The logic flow of an embedded application that incorporates key-wakeup interrupt is illustrated in Figure 7.51.

Figure 7.51 ■ Logic flow of embedded applications that incorporate key-wakeup feature

Example 7.13

▼

Write an instruction sequence to configure the Port P upper four pins for the key-wakeup feature. Program Port P so that pins PP7~PP4 generate an interrupt whenever there is a falling edge applied to any one of these four pins.

Solution:

```
#include     "c:\miniide\hcs12.inc"
     . . .
     bclr    DDRP,$F0     ; configure PP<7:4> pins for input
     bset    PERP,$F0     ; enable PP7~PP4 pins' pull device
     bclr    PPSP,$F0     ; choose pull-up device
     movb    #$FF,PIFP    ; clear the Port P key-wakeup flags
     bset    PIEP,$F0     ; enable Port P interrupt
     cli                  ; enable key-wakeup interrupt globally
```

▲

7.14 Summary

Peripheral devices allow the user to interact with the computer or embedded product. The user uses input devices to provide data needed by the computation and receives the computation result from the output devices. Examples of peripheral devices are switches, keypads, keyboards, LEDs, seven-segment displays, LCDs, printers, and so on. The speed and electrical characteristics of a peripheral device are quite different from those of a microprocessor. Therefore, they are not connected to a microprocessor directly. Instead, interface chips are used to resolve the difference between the microprocessor and peripheral devices.

The major function of an interface chip is to synchronize the data transfer between the microprocessor and an I/O device. An interface chip consists of control registers, status registers, data registers, data direction registers, and control circuitry. There are two aspects of the I/O transfer synchronization: the synchronization between the CPU and the interface chip and the synchronization between the interface chip and the I/O device. There are two methods to synchronize the CPU and the interface chip: the *polling* method and the *interrupt* method. There are three methods for synchronizing the interface chip and the I/O device: the *brute-force*, *strobe*, and *handshake* methods. Because of the shift in microcontroller applications and the improvements in technology, microcontrollers designed in last decade rarely use the strobe or handshake method to synchronize the interface chip and the I/O devices.

Some processors have dedicated instructions and addressing modes for performing I/O operations. In these processors, I/O devices occupy a separate memory space from the main memory. Other processors use the same instructions and addressing modes to access I/O devices. I/O devices and main memory share the same memory space.

An HCS12 member may have from 48 to 144 signal pins. Most of these pins serve multiple purposes and can be used as general I/O pins. These I/O pins are divided among up to 16 ports. Each I/O port has a port data register, a data direction register, and several control registers. All I/O ports except for the Port AD0 and Port AD1 pins can be configured for either input or output.

A set of switches arranged in a DIP package is the simplest input device. Keypads and keyboards are two of the most widely used input devices. When using the switches in a DIP package to input data, one often needs to use an interrupt to inform the MCU to read the data. The computer also needs to output a message to ask the user to set up data with the switches. An input operation is often a two-way interactive process. There are three steps in keyboard and keypad input: key scanning, key debouncing, and ASCII code lookup. The purpose of keyboard scanning is to detect which key has been pressed. The purpose of key debouncing is to make sure that the key has indeed been pressed and not to read the same key press more than once. The software *wait-and-see* method is the most widely used debouncing method.

Light-emitting diodes are the simplest output device. They are often used to indicate whether the computer is working properly or to convey simple information to the user. LEDs organized in seven-segment displays can be used to display decimal digits and a limited number of letters and special characters. Matrix LED displays can display all the letters and special characters. They are organized in 5×7, 5×8, and 8×8 formats. To display multiple digits simultaneously using the seven-segment displays or LED matrix displays, the time-multiplexing technique is often used. The multiplexing of multiple LED matrix displays involves very heavy MCU overhead. A special matrix display driver chip is often used to reduce the workload of the MCU. This will be examined in Chapter 10.

LED-based displays are suitable for displaying a limited number of digits or characters. When many letters and digits need to be displayed, LCD displays are the best choice. An LCD needs a controller to generate the required control signals during the exchange of data between the CPU and the LCD. The most popular character-based LCD controller is the Hitachi HD44780U.

D/A converters are often used to convert digital values into voltages. Sawtooth, triangular, and sinusoidal waveforms can be generated from a D/A converter easily. The digital value to be converted can be represented in 8 bits, 10 bits, or many more bits. The more bits used, the better the resolution gets. A high-fidelity digital audio system uses 24 bits to encode a value to be converted back to audio sound.

A stepper motor is a digital motor in the sense that each step of the rotation rotates a fixed number of degrees. It is most suitable for applications that require a high degree of positional control, such as plotters, disk drives, magnetic tape drives, robot joints, and so on. The resolution of one step of a stepper motor can be as small as 0.72° and as large as 90°. The simplest stepper motor has two pairs of coils.

Driving a stepper motor involves applying a series of voltages to the coils of the motor. A subset of coils is energized at a time to cause the motor to rotate one step. The pattern of coils energized must be followed exactly for the motor to work correctly. The pattern will vary depending on the mode used on the motor. A microcontroller can easily time the duration during which the coil is energized and hence control the speed of the stepper motor in a precise manner.

Saving power is a major concern in most battery-powered embedded products. When the user is not using an embedded system, the microcontroller should be switched to a low-power mode. The HCS12 has two low-power modes: WAIT mode and STOP mode. The HCS12 consumes the least power in the STOP mode. The STOP (WAIT) mode can be entered by executing the STOP (WAI) instruction. To facilitate the exit of the STOP or WAIT mode, the HCS12 provides the key-wakeup feature. Whenever the HCS12 is in one of the low-power modes and a selected signal edge arrives at one of the key-wakeup port pins, an interrupt request will be generated and the HCS12 will be waked up. The service for the key-wakeup interrupt is simply to clear the interrupt flag and resume the execution of the instruction following the WAI or STOP instruction.

The signal edge to wake up the microcontroller could be rising or falling. When the rising edge is selected, the pull-down device should be enabled. When the falling edge is chosen, the pull-up device should be enabled. Ports H, J, and P provide the key-wakeup capability.

7.15 Exercises

E7.1 What is handshake?

E7.2 Write an instruction sequence to output the value $35 to Port P.

E7.3 Write an instruction sequence to input the current signal levels on Port H pins and place the value in accumulator B.

E7.4 Suppose the Port T pins PT4~PT0 are connected to green, yellow, red, blue, and purple LEDs. The circuit connection is similar to that in Figure 7.30. Write a program to light the green, yellow, red, blue, and purple in turn for 0.1, 0.2, 0.3, 0.4, and 0.5 seconds forever. Use both the assembly and C languages.

E7.5 *Traffic Light Controller Simulation.* Use Port B pins PB5~PB0 of the HCS12 and green, yellow, and red LEDs to simulate a traffic-light controller. The traffic-light patterns and durations for the east-west and north-south traffic are given in Table 7E.5. Write an assembly and C program to control the light patterns, and connect the circuit to demonstrate the change of lights.

East-West			North-South			Duration (sec)
Green	Yellow	Red	Green	Yellow	Red	
1	0	0	0	0	1	25
0	1	0	0	0	1	5
0	0	1	1	0	0	20
0	0	1	0	1	0	4

Table 7E.5 ■ Traffic light pattern and duration

E7.6 Assume a demo board uses Port B to drive the seven-segment pattern and uses Port P to drive the digit select signals. The pins PB6~PB0 are connected to segment g~a, whereas the pins PP3~PP0 are connected to digit select 0~3. Write a program to display 1, 2, 3, 4 on displays 3, 2, 1, and 0, respectively.

E7.7 Suppose that Port P pins drive the seven-segment pattern and Port T pins drive six 2N2222 transistors with a connection similar to that in Figure 7.32. Write an assembly and C program to display digits 1, 2, 3, 4, 5, and 6 from the leftmost to the rightmost seven-segment display in turn, each for 1 second repeatedly.

E7.8 Write an assembly and C program to display the following information in two rows in the LCD connected to the demo board:

Date: 10 10 1952

Time: 10:20:10

E7.9 Calculate the period of the sawtooth waveform generated in Example 7.11. What are the voltages that correspond to the digital values 20, 30, 50, 127, and 192?

E7.10 Write a program to generate a sine wave from $V_{OUT}A$ in Figure 7.44. (*Hint:* Use the table lookup method).

E7.11 Write a C program to read a character from the keypad shown in Figure 7.41. This program will perform keypad scanning, key debouncing, and ASCII code lookup.

E7.12 Write a program to generate the waveform shown in Figure 7E.12 from the pin $V_{OUT}A$ in Figure 7.44.

E7.13 Write an assembly and a C program to generate a triangular waveform from the $V_{OUT}A$ pin in Figure 7.44. What is the period of the waveform generated by your program?

E7.14 Write an instruction sequence to rotate the stepper motor shown in Figure 7.49 clockwise for one cycle using the full-step sequence.

E7.15 Write a C program to rotate the stepper motor one cycle in the counterclockwise direction using the half-step sequence. The circuit connection of the stepper motor is shown in Figure 7.49.

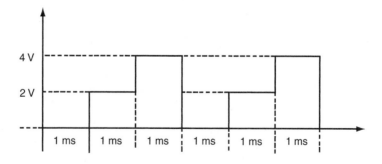

Figure 7E.12 ■ A waveform to be generated

E7.16 Write a C program to rotate the stepper motor clockwise one cycle using the full-step sequence.

E7.17 Write a C function to initialize the Port H key-wakeup function for the HCS12 that interrupts the microcontroller on the falling edge (automatic) of any Port H pin. Also write an interrupt service routine for this interrupt that simply clears the key-wakeup interrupt flags and returns.

E7.18 According to Freescale, we must write a 1 to the key-wakeup interrupt flag in order to clear it. Why shouldn't we use the **bset** instruction with a mask that has 1s at the positions corresponding to the flags that we want to clear?

7.16 Lab Exercises and Assignments

L7.1 Use Port P to drive two 74HC48s. The 74HC48 is a common cathode seven-segment display driver. It accepts a 4-bit binary number and converts it into the corresponding seven-segment pattern. Write a program to display the value from 00 to 99 on these two seven-segment displays three times and then stops. Display each value for a quarter of a second.

L7.2 *Keypad input and LCD output exercises.* Use a 16-key keypad (plugged into the keypad connector) and the LCD kit of the demo board. Write a C program to perform keypad scanning, debouncing, and ASCII code lookup. Output the character read from the keypad to the LCD kit. When reaching the end of a row, output the character to the next row. Stop keypad input whenever an **F** character is entered.

L7.3 *LED flashing.* Write a C (or assembly) program to flash the LEDs on your demo board with the following pattern:

1. Turn on all LEDs for a quarter of a second and turn them off for a quarter of a second.

2. Repeat step 1 for three more times.

3. Turn on one LED at a time from left to right with each LED turned on for a quarter of a second.

4. Repeat step 3 for three more times.

5. Turn on one LED at a time from right to left with each LED turned on for a quarter of a second.

6. Repeat step 5 for three more times.

7. Turn LEDs driven by pins RB4 and RB3 on and off four times. The on-time and off-time are both a quarter of a second.

8. Turn LEDs driven by pins RB5 and RB2 on and off four times. The on-time and off-time are both a quarter of a second.

9. Turn LEDs driven by pins RB6 and RB1 on and off four times. The on-time and off-time are both a quarter of a second.

10. Turn LEDs driven by pins RB7 and RB0 on and off four times. The on-time and off-time are both a quarter of a second.

11. Go to step 1 and repeat.

L7.4 *Interrupt-driven input and output.* When a computer needs the user to input data, the user must be informed to enter the data. On the other hand, the time it takes the user to prepare data is much longer than that needed by a computer to read the data. The computer should read the data only when it is ready. One way to inform the computer that input data is ready is to use the $\overline{IRQ}$ pin interrupt as follows:

Step 1
The computer program clears a flag to indicate that data are not available and outputs a message (on the LCD) to ask the user to enter some data.

Step 2
The user sets up data using the DIP switches. The 8-bit data can be set up to represent a two-digit BCD value. After the value is set up properly, the user presses a button connected to the $\overline{IRQ}$ pin to interrupt the CPU.

Step 3
The user needs to write an $\overline{IRQ}$ service routine that simply sets the flag to 1 and returns.

Step 4
During the time that the user is setting up data using the switches in the DIP package, the computer program keeps checking the flag until it becomes set to 1.

Step 5
The computer reads data from a DIP switch (8 bits) when the flag is 1.

Write a program that uses this method to ask the user to enter the current time (hh:mm:ss) to the computer, updates it once every second, and displays it on the LCD. The program will output three messages to ask the user to enter hour digits, minute digits, and second digits.

L7.5 *Seven-Segment display pattern shifting.* Write a program that uses the four seven-segment displays on the demo boards (either the SSE256 or Dragon12) to display the following pattern forever with each pattern displayed for 1 second:

```
0               ; displayed once
1 0             ;   "
2 1 0           ;   "
3 2 1 0         ; from this row to the last row are repeated forever
4 3 2 1         ;
5 4 3 2         ;
6 5 4 3         ;
7 6 5 4         ;
8 7 6 5         ;
9 8 7 6         ;
0 9 8 7         ;
1 0 9 8         ;
2 1 0 9         ; last row to be repeated
```

8

Timer Functions

8.1 Objectives

After completing this chapter, you should be able to

- Explain the overall structure of the HCS12 timer system
- Use the input-capture function to measure the duration of a pulse or the period of a square wave
- Use the input-capture function to measure the duty cycle of a waveform or the phase difference of two waveforms having the same frequency
- Use the input-capture function to measure the frequency of a signal
- Use the output-compare function to create a time delay or generate a pulse or periodic waveform
- Use the forced output-compare function
- Use the pulse accumulator to measure the frequency of an unknown signal and count the number of events occurring in an interval
- Configure PWM channels to generate waveforms of certain frequencies and duty cycles

8.2 Why Are Timer Functions Important?

There are many applications that require a dedicated timer system, including

- Time delay creation and measurement
- Period and pulse-width measurement
- Frequency measurement
- Event counting
- Arrival time comparison
- Time-of-day tracking
- Waveform generation
- Periodic interrupt generation

These applications will be very difficult to implement without a dedicated timer system. The HCS12 implements a very complicated timer system to support the implementation of these applications.

At the heart of the HCS12 timer system is the 16-bit timer counter (TCNT). This counter can be started or stopped, as you like. One of the timer functions is called *input-capture*. The input-capture function copies the contents of the 16-bit timer to a latch when the specified *event* arrives. An event is represented by a signal edge, which could be a rising or a falling edge. By capturing the timer value, many measurements can be made. Some of them include

- Pulse-width measurement
- Period measurement
- Duty cycle measurement
- Event arrival time recording
- Time reference

Another timer function is called the *output-compare*. The output-compare circuit compares the 16-bit timer value with that of the output-compare register in each clock cycle and performs the following operations when they are equal:

- (optionally) Triggers an action on a pin (set to high, set to low, or toggle its signal level)
- Sets a flag in a register
- (optionally) Generates an interrupt request

The output-compare function is often used to generate a time delay, trigger an action at some future time, and generate a digital waveform. The key to using the output-compare function is to make a copy of the 16-bit timer, add a delay to it, and store the sum in an *output-compare register*. The HCS12 has eight output-compare channels, which share the signal pins and registers with input-capture channels.

The third timer function is the *pulse accumulator*. This circuit is often used to count the events arriving in a certain interval or measure the frequency of an unknown signal.

Certain HCS12 members (for example, A family, C family, and H family) implement a *Standard Timer Module* (TIM) that consists of eight channels of input-capture/output-compare functions and a 16-bit pulse accumulator, PACA. Other HCS members add additional features to the TIM module, and the resultant module is referred to as *Enhanced Captured Timer Module* (ECT). The design goal of the ECT is to support the requirements of automotive, process control, and other applications that need additional timer features.

8.3 Standard Timer Module

The HCS12 standard timer module is illustrated in Figure 8.1. The block diagram in Figure 8.1 shows that the timer module has eight channels of input-capture and output-compare modules. Each channel can be configured to perform input-capture or output-compare, but not both at the same time. The 16-bit counter serves as the base timer for these eight channels. The 16-bit counter needs a clock signal to operate. The clock signal is derived by dividing the E clock by a prescaler.

The timer module has a 16-bit pulse accumulator, which can be used in counting the number of events that have occurred or used to measure the frequency of an unknown signal. The timer module shares the use of Port T pins PT7~PT0. The signal pins IOC7~IOC0 correspond

Figure 8.1 ■ HCS12 Standard Timer (TIM) block diagram

to PT7~PT0. The 16-bit pulse accumulator shares the use of the PT7 pin. The PT7 pin is referred to as the PAI pin when it is used as the pulse accumulator input.

8.4 Timer Counter Register

The timer counter register (TCNT), a 16-bit register, is required for the functioning of all input-capture and output-compare functions. The user must access this register in one access rather than two separate accesses, to its high byte and low byte. Because the TCNT counter does not stop during the access operation, the value accessed in a 16-bit read won't be the same as two separate accesses to its high byte and low byte.

There are three registers related to the operation of the TCNT counter. They are

1. Timer system control register 1 (TSCR1)
2. Timer system control register 2 (TSCR2)
3. Timer interrupt flag 2 register (TFLG2)

8.4.1 Timer System Control Register 1

The contents of the TSCR1 register are shown in Figure 8.2. The timer counter must be enabled before it can count. Setting bit 7 of the TSCR1 register enables the TCNT to count (up). The timer counter can be stopped (by setting bit 6 of TSCR1) during the wait mode to save more power. If the timer function is not needed, it can also be stopped during the freeze mode.

	7	6	5	4	3	2	1	0
value	TEN	TSWAI	TSFRZ	TFFCA	0	0	0	0
after reset	0	0	0	0	0	0	0	0

TEN—timer enable bit.

    0 = disable timer; this can be used to save power consumption.
    1 = allows timer to function normally.

TSWAI—timer stops while in wait mode bit.

    0 = allows timer to continue running during wait mode.
    1 = disables timer when MCU is in wait mode.

TSFRZ—timer and modulus counter stop while in freeze mode.

    0 = allows timer and modulus counter to continue running while in freeze mode.
    1 = disables timer and modulus counter when MCU is in freeze mode.

TFFCA—timer fast flag clear all bit.

    0 = allows timer flag clearing to function normally.
    1 = for TFLG1, a read from an input capture or a write to the output compare channel causes the corresponding channel flag, CnF, to be cleared. For TFLG2, any access to the TCNT register clears the TOF flag. Any access to the PACN3 and PACN2 registers clears the PAOVF and PAIF flags in the PAFLG register. Any access to the PACN1 and PACN0 registers clears the PBOVF flag in the PBFLG register.

Figure 8.2 ■ Timer System Control Register 1 (TSCR1)

All timer interrupt flags can be cleared by writing a 1 to them. However, there is a faster way to clear timer flags. When bit 4 of the TSCR1 register is set to 1, a read from or a write to the appropriate TC register will clear the corresponding flag in the TFLG1 register; any access to the TCNT register will clear the TOF flag, and any access to the PACNT register will clear the PAOVF and PAIF flags in the PAFLG register. This feature reduces the software overhead in a separate clear sequence.

8.4.2 Timer System Control Register 2 (TSCR2)

Timer system control register 2 is another register that controls the operation of the timer counter. Its contents are shown in Figure 8.3. An interrupt will be requested when the TCNT overflows (when TCNT rolls over from $FFFF to $0000) and bit 7 of TSCR2 is set to 1.

The timer counter needs a clock signal to operate. The clock input to the timer counter could be the E clock prescaled by a factor or the PAI pin input prescaled by a factor. The user has the option to choose the prescale factor. When bit 3 and bit 2 of the PACTL register are 00, the clock to the TCNT is the E clock prescaled by a factor. The lowest three bits of the TSCR2 register specify the prescale factor for the E clock, as shown in Table 8.1.

	7	6	5	4	3	2	1	0
value	TOI	0	0	0	TCRE	PR2	PR1	PR0
after reset	0	0	0	0	0	0	0	0

TOI—timer overflow interrupt enable bit
  0 = interrupt inhibited
  1 = interrupt requested when TOF flag is set
TCRE—timer counter reset enable bit
  0 = counter reset inhibited and counter free runs
  1 = counter reset by a successful output compare 7.
  If TC7 = $0000 and TCRE = 1, TCNT stays at $0000
  continuously. If TC7 = $FFFF and TCRE = 1, TOF will never
  be set when TCNT rolls over from $FFFF to $0000.

Figure 8.3 ■ Timer System Control Register 2 (TSCR2)

PR2	PR1	PR0	Prescale Factor
0	0	0	1
0	0	1	2
0	1	0	4
0	1	1	8
1	0	0	16
1	0	1	32
1	1	0	64
1	1	1	128

Table 8.1 ■ Timer counter prescale factor

8.4.3 Timer Interrupt Flag 2 Register (TFLG2)

Only bit 7 (TOF) of this register is implemented. When TCNT rolls over from $FFFF to $0000, bit 7 of this register is set to 1. This flag can be cleared by writing a 1 to it.

8.5 Input-Capture Function

Some applications need to know the arrival time of events. In a computer, *physical time* is represented by the count in a counter, and the occurrence of an event is represented by a signal edge (either the rising or falling edge). The time when an event occurs can be recorded by latching the count value when a signal edge arrives, as illustrated in Figure 8.4.

Figure 8.4 ■ Events represented by signal changes

The HCS12 timer system has eight input-capture channels that implement this operation. Each input-capture channel includes a 16-bit input-capture register, an input pin, input edge-detection logic, and an interrupt generation circuit. In the HCS12, physical time is represented by the count in the timer counter (TCNT).

8.5.1 Input-Capture/Output-Compare Selection

Since input-capture and output-compare functions share signal pins and registers, they cannot be enabled simultaneously. When one is enabled, the other is disabled. The selection is done by the Timer Input-Capture/Output-Compare Select register (TIOS), as shown in Figure 8.5.

IOS[7:0]—Input capture or output compare channel configuration bits
0 = the corresponding channel acts as an input capture.
1 = the corresponding channel acts as an output compare.

Figure 8.5 ■ Timer Input-Capture/Output-Compare Select register (TIOS)

Example 8.1

Write an instruction sequence to enable the output-compare channels 7~4 and input-capture channels 3~0.

Solution: The following instruction sequence will achieve the desired configuration:

```
#include        "c:\miniide\hcs12.inc"
    . . .
movb    #$F0,TIOS
```

8.5.2 Pins for Input-Capture

Port T has eight signal pins (PT7~PT0) that can be used as input-capture/output-compare or general I/O pins. PT3~PT0 can also be used as the *pulse accumulator input* (PA3~PA0). The user must make sure that the PT3~PT0 pins are enabled for one and only one of these three functions (OCn, ICn, and PAn). When these pins are not used for timer functions, they can also be used as general-purpose I/O pins. When being used as general I/O pins, the user must use the DDRT register to configure their direction (input or output). When a Port T pin is used as a general-purpose I/O pin, its value is reflected in the corresponding bit in the PTT register.

8.5.3 Registers Associated with Input-Capture

The user needs to specify what signal edge to capture. The edge selection is done via the Timer Control registers 3 and 4, as shown in Figure 8.6.

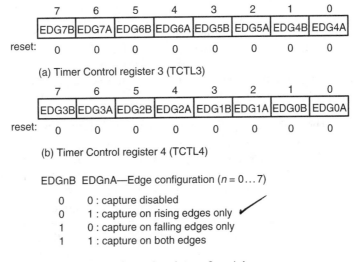

7	6	5	4	3	2	1	0
EDG7B	EDG7A	EDG6B	EDG6A	EDG5B	EDG5A	EDG4B	EDG4A

reset: 0 0 0 0 0 0 0 0

(a) Timer Control register 3 (TCTL3)

7	6	5	4	3	2	1	0
EDG3B	EDG3A	EDG2B	EDG2A	EDG1B	EDG1A	EDG0B	EDG0A

reset: 0 0 0 0 0 0 0 0

(b) Timer Control register 4 (TCTL4)

EDGnB EDGnA—Edge configuration (*n* = 0...7)

```
0    0 : capture disabled
0    1 : capture on rising edges only
1    0 : capture on falling edges only
1    1 : capture on both edges
```

Figure 8.6 ■ Timer Control registers 3 and 4

When an input-capture channel is selected but capture is disabled, the associated pin can be used as a general-purpose I/O pin.

An input-capture channel can optionally generate an interrupt request on the arrival of a selected edge if it is enabled. The enabling of an interrupt is controlled by the Timer Interrupt Enable register (TIE). The enabling of input-capture 7 through input-capture 0 interrupt is controlled by bit 7 through bit 0 of TIE. When a selected edge arrives at the input-capture pin, the corresponding flag in the Timer Flag register 1 (TFLG1) will be set. The contents of TIE and TFLG1 are shown in Figures 8.7 and 8.8, respectively.

7	6	5	4	3	2	1	0
C7I	C6I	C5I	C4I	C3I	C2I	C1I	C0I

reset: 0 0 0 0 0 0 0 0

C7I–C0I: input capture/output compare interrupt enable bits
0 = interrupt disabled
1 = interrupt enabled

Figure 8.7 ■ Timer Interrupt Enable register (TIE)

7	6	5	4	3	2	1	0
C7F	C6F	C5F	C4F	C3F	C2F	C1F	C0F

reset: 0   0   0   0   0   0   0   0

C7F–C0F: input capture/output compare interrupt flag bits
    0 = interrupt condition has not occurred.
    1 = interrupt condition has occurred.

Figure 8.8 ■ Timer Interrupt Flag 1 register (TFLG1)

To clear a flag in the TFLG1 register, write a 1 to it. However, there is a better way to clear the flag that incurs less overhead. Setting bit 4 (TFFCA) of the TSCR1 register allows one to clear a flag by reading the corresponding input-capture register or writing a new value into the output-compare register. This operation is needed for the normal operation of the input-capture/output-compare function.

Each input-capture channel has a 16-bit register (TCx, x = 0 to 7) to hold the count value when the selected signal edge arrives at the pin. This register is also used as the output-compare register when the output-compare function is selected instead.

8.5.4 Input-Capture Applications

There are many applications for the input-capture function. Examples include the following.

- *Event arrival-time recording.* Some applications, for example, swimming competition, need to compare the arrival times of several different swimmers. Input-capture function is very suitable for this application. The number of events that can be compared is limited by the number of input-capture channels.

- *Period measurement.* To measure the period of an unknown signal, the input-capture function should be configured to capture the timer values corresponding to two consecutive rising or falling edges, as illustrated in Figure 8.9.

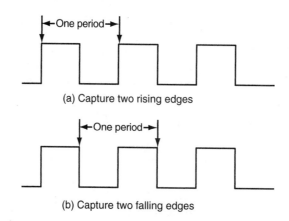

(a) Capture two rising edges

(b) Capture two falling edges

Figure 8.9 ■ Period measurement by capturing two consecutive edges

- *Pulse-width measurement.* To measure the width of a pulse, two adjacent rising and falling edges are captured, as shown in Figure 8.10.

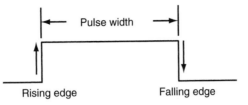

Figure 8.10 ■ Pulse-width measurement using input capture

- *Interrupt generation.* All input-capture pins can serve as edge-sensitive interrupt sources. Once enabled, interrupts will be generated on the selected edge(s).
- *Event counting.* An event can be represented by a signal edge. An input-capture channel can be used in conjunction with an output-compare function to count the number of events that occur during an interval. An event counter can be set up and incremented by the input-capture interrupt service routine. This application is illustrated in Figure 8.11.

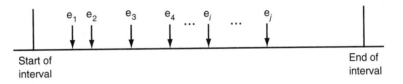

Figure 8.11 ■ Using an input-capture function for event counting

- *Time reference.* In this application, an input-capture function is used in conjunction with an output-compare function. For example, if the designer wants to activate an output signal a certain number of clock cycles after detecting an input event, the input-capture function would be used to record the time at which the edge is detected. A number corresponding to the desired delay would be added to this captured value and stored to an output-compare register. This application is illustrated in Figure 8.12.

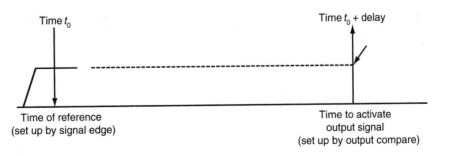

Figure 8.12 ■ A time-reference application

- *Duty cycle measurement.* The *duty cycle* is the percent of time that the signal is high within a period in a periodic digital signal. The measurement of the duty cycle is illustrated in Figure 8.13.

duty cycle = $\dfrac{\Delta T}{T}$ • 100%

Figure 8.13 ■ Definition of duty cycle

- *Phase difference measurement.* The *phase difference* is defined as the difference of arrival times (in percentage of a period) of two signals that have the same frequency but do not coincide in their rising and falling edges. The definition of the phase difference is illustrated in Figure 8.14.

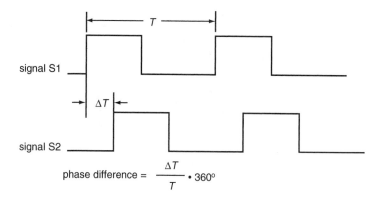

phase difference = $\dfrac{\Delta T}{T}$ • 360°

Figure 8.14 ■ Phase difference definition for two signals

The unit used in most of the measurements is the number of clock cycles. When it is desirable, the unit should be converted into an appropriate unit, such as seconds.

Example 8.2

▼

Period measurement. Use the input-capture channel 0 to measure the period of an unknown signal. The period is known to be shorter than 128 ms. Assume that the E clock frequency is 24 MHz. Use the number of clock cycles as the unit of the period.

Solution: Since the input-capture register is 16-bit, the longest period of the signal that can be measured with the prescaler to TCNT set to 1 is

$2^{16} \div 24\ \text{MHz} = 2.73\ \text{ms}$

To measure a period that is equal to 128 ms, we have two options:

1. Set the prescale factor to 1 and keep track of the number of times that the timer counter overflows.

2. Set the prescale factor to 64 and do not keep track of the number of times that the timer counter overflows.

In this example, we will adopt the second approach to make the programming easier. The result of this measurement will be in number of clock cycles, and the period of each clock cycle is 2.67 μs. The circuit connection for the period measurement is shown in Figure 8.15, and the logic flow for the period measurement is shown in Figure 8.16.

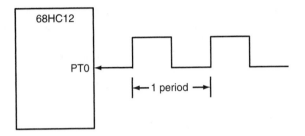

Figure 8.15 ■ Period measurement signal connection

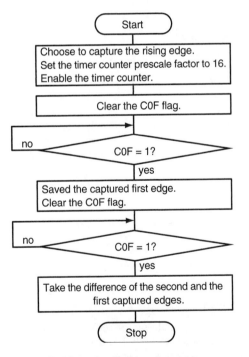

Figure 8.16 ■ Logic flow of period measurement program

The assembly program that measures the period using this method is as follows:

```
#include     "c:\miniide\hcs12.inc"
             org     $1000
edge_1st     rmb     2          ; memory to hold the first edge
period       rmb     2          ; memory to store the period
```

```
           org      $1500
           movb     #$90,TSCR1      ; enable timer counter and enable fast timer flags clear
           bclr     TIOS,IOS0       ; enable input-capture 0
           movb     #$06,TSCR2      ; disable TCNT overflow interrupt, set prescaler to 64
           movb     #$01,TCTL4      ; capture the rising edge of PT0 signal
           movb     #COF,TFLG1      ; clear the COF flag
           brclr    TFLG1,COF,*     ; wait for the arrival of the first rising edge
           ldd      TC0             ; save the first edge and clear the COF flag
           std      edge_1st
           brclr    TFLG1,COF,*     ; wait for the arrival of the second edge
           ldd      TC0
           subd     edge_1st        ; compute the period
           std      period
           swi
           end
```

The C language version of the program is as follows:

```
#include "c:\egnu091\include\hcs12.h"
void main(void)
{
    unsigned int edge1, period;
    TSCR1     = 0x90;               /* enable timer counter, enable fast flag clear*/
    TIOS      &= ~IOS0;             /* enable input-capture 0*/
    TSCR2     = 0x06;               /* disable TCNT overflow interrupt, set prescaler to 64 */
    TCTL4     = 0x01;               /* capture the rising edge of the PT0 pin */
    TFLG1     = COF;                /* clear the COF flag */
    while (!(TFLG1 & COF));         /* wait for the arrival of the first rising edge */
    edge1     = TC0;                /* save the first captured edge and clear COF flag */
    while (!(TFLG1 & COF));         /* wait for the arrival of the second rising edge */
    period    = TC0 − edge1;
    asm ("swi");
}
```

Example 8.3

Write a program to measure the pulse width of a signal connected to the PT0 pin. Assume that the E clock frequency is 24 MHz.

Solution: We will set the prescale factor to 32 and use the clock cycle as the unit of measurement. The period of one clock cycle is 1.33 μs. Since the pulse width could be much longer than 2^{16} clock cycles, we will need to keep track of the number of times that the TCNT counter overflows. Each TCNT overflow adds 2^{16} clock cycles to the pulse width.

Let

$ovcnt$ = TCNT counter overflow count

$diff$ = the difference of two consecutive edges

$edge1$ = the captured time of the first edge

$edge2$ = the captured time of the second edge

The pulse width can be calculated by the following equations:

Case 1

> edge2 $\geq$ edge1
>> pulse width = ovcnt $\times$ 2^{16} + diff

Case 2

> edge2 $<$ edge 1
>> pulse width = (ovcnt $-$ 1) $\times$ 2^{16} + diff

In case 2, the timer overflows at least once even if the pulse width is shorter than 2^{16} $-$ 1 clock cycles. Therefore, we need to subtract 1 from the timer overflow count in order to get the correct result. The pulse width is obtained by appending the difference of the two captured edges to the TCNT overflow count. The logic flow of the program is shown in Figure 8.17.

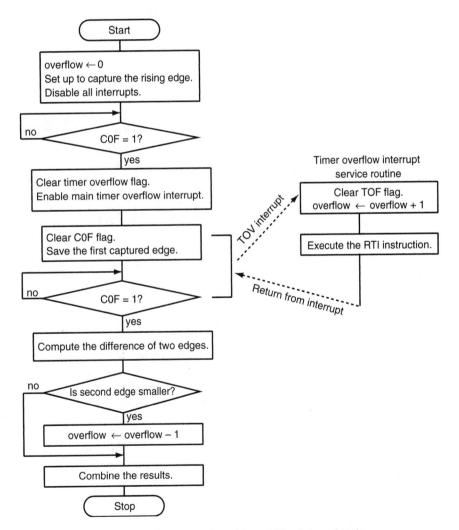

Figure 8.17 ■ Logic flow for measuring pulse width of slow signals

The assembly program that implements this algorithm is as follows:

```
#include      "c:\miniide\hcs12.inc"
              org       $1000
edge1         rmb       2
overflow      rmb       2
pulse_width   rmb       2
              org       $1500
              movw      #tov_isr,UserTimerOvf ; set up TCNT overflow interrupt vector
              lds       #$1500                 ; set up stack pointer
              movw      #0,overflow
              movb      #$90,TSCR1            ; enable TCNT and fast timer flag clear
              movb      #$05,TSCR2            ; disable TCNT interrupt, set prescaler to 32
              bclr      TIOS,IOS0            ; select IC0
              movb      #$01,TCTL4           ; capture rising edge
              movb      #$01,TFLG1           ; clear C0F flag
wait1         brclr     TFLG1,C0F,wait1     ; wait for the first rising edge
              movw      TC0,edge1           ; save the first edge and clear the C0F flag
              movb      #$80,TFLG2          ; clear TOF flag
              bset      TSCR2,$80           ; enable TCNT overflow interrupt
              cli                           ;    "
              movb      #$02,TCTL4          ; capture the falling edge on PT0 pin
wait2         brclr     TFLG1,C0F,wait2     ; wait for the arrival of the falling edge
              ldd       TC0
              subd      edge1
              std       pulse_width
              bcc       next                ; is the second edge smaller?
              ldx       overflow            ; second edge is smaller, so decrement
              dex                           ; overflow count by 1
              stx       overflow            ;    "
next          swi
tov_isr       movb      #$80,TFLG2          ; clear TOF flag
              ldx       overflow
              inx
              stx       overflow
              rti
              end
```

The C language version of this program is as follows:

```
#include    "c:\egnu091\include\hcs12.h"
#include    "c:\egnu091\include\vectors12.h"
#define     INTERRUPT __attribute__((interrupt)) /* there are two '_' characters before and after "attribute" */
unsigned    diff, edge1, overflow;
unsigned    long pulse_width;
void INTERRUPT tovisr(void);

void main(void)
{
            UserTimerOvf = (unsigned short)&tovisr;
            overflow     = 0;
```

```
        TSCR1      = 0x90;                 /* enable timer and fast flag clear */
        TSCR2      = 0x05;                 /* set prescaler to 32, no timer overflow interrupt */
        TIOS       &= ~IOS0;               /* select input-capture 0 */
        TCTL4      = 0x01;                 /* prepare to capture the rising edge */
        TFLG1      = COF;                  /* clear COF flag */
        while(!(TFLG1 & COF));            /* wait for the arrival of the rising edge */
        TFLG2      = 0x80;                 /* clear TOF flag */
        TSCR2      |= 0x80;                /* enable TCNT overflow interrupt */
        asm("cli");
        edge1      = TC0;                  /* save the first edge */
        TCTL4      = 0x02;                 /* prepare to capture the falling edge */
        while (!(TFLG1 & COF));           /* wait for the arrival of the falling edge */
        diff       = TC0 − edge1;
        if (TC0 < edge1)
             overflow -= 1;
        pulse_width = (long) overflow * 65536u + (long) diff;
        asm ("swi");
}
void INTERRUPT tovisr(void)
{
        TFLG2      = TOF;                  /* clear TOF flag */
        overflow = overflow + 1;

}
```

8.6 Output-Compare Function

The HCS12 has eight output-compare channels. Each channel consists of

- A 16-bit comparator
- A 16-bit compare register TCx, x = 0 . . . 7 (also used as input-capture register)
- An output action pin (PTx—can be pulled up to high, pulled down to low, or toggled)
- An interrupt request circuit
- A forced-compared function (CFORCx)
- Control logic

8.6.1 Operation of the Output-Compare Function

One of the major applications of an output-compare function is performing an action at a specific time in the future (when the 16-bit timer counter reaches a specific value). The action might be to toggle a signal, turn on a switch, turn off a valve, and so on. To use an output-compare function, the user.

1. Makes a copy of the current contents of the TCNT register
2. Adds to this copy a value equal to the desired delay
3. Stores the sum into an output-compare register (TCx)

The delay to be added is dependent on the prescaler to TCNT. The prescaler needs to be set before the delay value is chosen. The user has the option of specifying the action to be activated on the selected output-compare pin by programming the TCTL1 and TCTL2 registers. The comparator compares the value of the TCNT and that of the specified output-compare register (TCx) in every clock cycle (the clock input to TCNT). If they are equal, the specified action on the output-compare pin is activated and the associated status bit in TFLG1 will be set to 1. An interrupt request will be generated if it is enabled. The 16-bit output-compare register can be read and written any time.

8.6.2 Registers Related to the Output-Compare Function

The actions that can be activated on an output-compare pin are

- Pull-up to high
- Pull-down to low
- Toggle

The action of an OC pin can be selected by programming the TCTL1 and TCTL2 registers as shown in Figure 8.18. When either OMn or OLn is 1, the pin associated with OCn becomes an output tied to OCn regardless of the state of the associated DDRT bit.

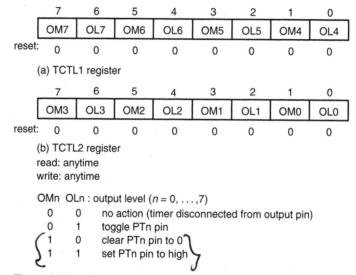

Figure 8.18 ■ Timer Control register 1 and 2 (TCTL1 and TCTL2)

A successful compare will set the corresponding flag bit in the 8-bit TFLG1 register. An interrupt will be generated if it is enabled. An output-compare interrupt is enabled by setting the corresponding bit in the TIE register. The same bit will enable either the input-capture or output-compare interrupt, depending on which one is being selected.

The OC7 channel has the option to reset TCNT when the TC7 register matches the TCNT timer. This option is enabled by setting the TCRE bit of the TSCR2 register. This feature makes the TCNT a programmable modulus counter.

8.6.3 Applications of the Output-Compare Function

An output-compare function can be programmed to perform a variety of functions. Generation of a single pulse, a square wave, and a specific delay are among the most popular applications.

Example 8.4

Generate an active high 1-kHz digital waveform with a 30% duty cycle from the PT0 pin. Use the polling method to check the success of the output-compare operation. The frequency of the E clock is 24 MHz.

Solution: An active high 1-kHz waveform with a 30% duty cycle is shown in Figure 8.19.

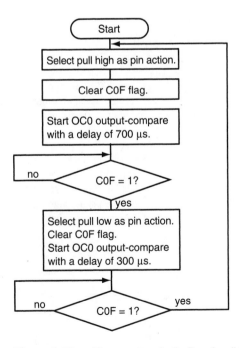

Figure 8.19 ■ 1-kHz 30% duty cycle waveform

The logic flow of this problem is illustrated in Figure 8.20. Suppose that we set the prescale factor to 8 so that the period of the clock input to TCNT is set to $1/3$ μs. Then the intervals of the PT0 signal to be high and low in one period would be 900 and 2100 clock cycles, respectively.

Figure 8.20 ■ The program logic flow for digital waveform generation.

The following assembly program implements this algorithm using the OC0 function:

```
#include     "c:\miniide\hcs12.inc"
hi_time      equ     900
lo_time      equ     2100
             org     $1500
             movb    #$90,TSCR1      ; enable TCNT with fast timer flag clear
             movb    #$03,TSCR2      ; disable TCNT interrupt, set prescaler to 8
             bset    TIOS,OC0        ; enable OC0
             movb    #$03,TCTL2      ; select pull high as pin action
             ldd     TCNT            ; start an OC0 operation with 700 μs as delay
repeat       addd    #lo_time        ;   "
             std     TC0             ;   "
low          brclr   TFLG1,C0F,low   ; wait until OC0 pin goes high
             movb    #$02,TCTL2      ; select pull low as pin action
             ldd     TC0             ; start an OC operation with 300 μs as delay
             addd    #hi_time        ;   "
             std     TC0             ;   "
high         brclr   TFLG1,C0F,high  ; wait until OC0 pin goes low
             movb    #$03,TCTL2      ; select pull high as pin action
             ldd     TC0
             bra     repeat
             end
```

The C language version of the program is as follows:

```
#include "c:\egnu091\include\hcs12.h"
#define hi_time 900
#define lo_time 2100
void main (void)
{
    TSCR1    = 0x90;                  /* enable TCNT and fast timer flag clear */
    TIOS     |= OC0;                  /* enable OC0 function */
    TSCR2    = 0x03;                  /* disable TCNT interrupt, set prescaler to 8 */
    TCTL2    = 0x03;                  /* set OC0 action to be pull high */
    TC0      = TCNT + lo_time;        /* start an OC0 operation */
    while(1) {
        while(!(TFLG1 & C0F));        /* wait for PT0 to go high */
        TCTL2   = 0x02;              /* set OC0 pin action to pull low */
        TC0     += hi_time;          /* start a new OC0 operation */
        while(!(TFLG1 & C0F));        /* wait for PT0 pin to go low */
        TCTL2   = 0x03;              /* set OC0 pin action to pull high */
        TC0     += lo_time;          /* start a new OC0 operation */
    }
}
```

Example 8.5

Write a function to generate a time delay that is a multiple of 1 ms. Assume that the E clock frequency is 24 MHz. The number of milliseconds is passed in Y. Also write an instruction sequence to test this function.

Solution: There are many ways to create a 1-ms time delay using the output-compare function. One method is

- Set the prescaler to TCNT to 64
- Perform the number of output-compare operations (given in Y) with each operation creating a 1-ms time delay. The number to be added to the TC0 register is 375 (375 × 64 ÷ 24000000 = 1 ms).

The corresponding assembly function is as follows:

```
delayby1ms   pshd
             movb    #$90,TSCR1     ; enable TCNT and fast flags clear
             movb    #$06,TSCR2     ; configure prescaler to 64
             bset    TIOS,OC0       ; enable OC0
             ldd     TCNT
again0       addd    #375           ; start an output-compare operation
             std     TC0            ; with 1 ms time delay
wait_lp0     brclr   TFLG1,OC0,wait_lp0
             ldd     TC0
             dbne    y,again0
             puld
             rts
```

The C language version of the function and its test program are as follows:

```c
#include "c:\egnu091\include\hcs12.h"
void delayby1ms(int k);
void main (void)
{
    unsigned char count;
    DDRB = 0xFF;        /* configure port b for output */
    count = 0;
    PORTB = count;
    while(1) {
    delayby1ms(200);
    count++;
    PORTB = count;
    }
}
void delayby1ms(int k)
{
    int ix;
    TSCR1 = 0x90;       /* enable TCNT and fast timer flag clear */
    TSCR2 = 0x06;       /* disable timer interrupt, set prescaler to 64 */
    TIOS |= OC0;        /* enable OC0 */
    TC0  = TCNT + 375;
```

```
            for(ix = 0; ix < k; ix++) {
                    while(!(TFLG1 & OC0));
                    TC0 += 375;
            }
            TIOS &= ~OC0;      /* disable OC0 */
    }
```

▲

Example 8.6

▼

Combining the use of input-capture and output-compare functions. Suppose a signal with unknown frequency is connected to the PT0 pin. Write a program to measure the frequency of this signal.

Solution: One method for measuring the frequency is using one of the output-compare functions to create a 1-s time interval and keep track of the total number of rising (or falling) edges arrived at the PT0 pin. The number of rising edges arrived in 1 s gives the frequency of the signal. By enabling the PT0 interrupt and writing an interrupt service routine for the PT0 interrupt that increases the edge count, the frequency can be measured.

The assembly program that implements this algorithm is as follows:

```
         #include    "c:\MiniIDE\hcs12.inc"
         CR          equ     $0D
         LF          equ     $0A
                     org     $1000
         oc_cnt      rmb     1
         frequency   rmb     2
                     org     $1500
                     movb    #$90,TSCR1    ; enable TCNT and fast timer flags clear
                     movb    #$02,TSCR2    ; set prescale factor to 4
                     movb    #$02,TIOS     ; enable OC1 and IC0
                     movb    #100,oc_cnt   ; prepare to perform 100 OC1 operation, each
                                           ; creates 10-ms delay and total 1 s
                     movw    #0,frequency  ; initialize frequency count to 0
                     movb    #$01,TCTL4    ; prepare to capture the rising edges of PT0
                     movb    #C0F,TFLG1    ; clear the C0F flag
                     bset    TIE,IC0       ; enable IC0 interrupt
                     cli                   ;   "
                     ldd     TCNT          ; start an OC1 operation with 10-ms delay
         continue    addd    #60000        ;   "
                     std     TC1           ;   "
         w_lp        brclr   TFLG1,C1F,w_lp
                     ldd     TC1
                     dec     oc_cnt
                     bne     continue
                     ldd     frequency
                     pshd
                     ldd     #msg
```

```
                jsr         [printf,PCR]
                leas        2,sp
                swi
msg             db          CR,LF,"The frequency is %d",CR,LF,0
TC0_isr         ldd         TC0             ; clear COF flag
                ldx         frequency       ; increment frequency count by 1
                inx                         ;    "
                stx         frequency       ;
                rti
                org         $3E6E           ; set up interrupt vector number
                fdb         TC0_isr         ; for TC0
                end
```

The C language version of the program is as follows:

```c
#include     "c:\egnu091\include\hcs12.h"
#include     "c:\egnu091\include\vectors12.h"
#include     "c:\egnu091\include\convert.c"
#include     "c:\egnu091\include\stdio.c"
#define      INTERRUPT __attribute__((interrupt))
unsigned     int   frequency;
void         INTERRUPT     TC0_isr(void);
void main(void)
{
    char arr[7];
    char *msg = "Signal frequency is ";
    int i, oc_cnt;
    unsigned frequency;
    UserTimerCh0 = (unsigned short)&TC0_isr;
    TSCR1 = 0x90;      /* enable TCNT and fast flag clear */
    TSCR2 = 0x02;      /* set prescale factor to 4 */
    TIOS = 0x02;       /* select OC1 and IC0 */
    oc_cnt = 100;      /* prepare to perform 100 OC1 operations */
    frequency = 0;
    TCTL4 = 0x01;      /* prepare to capture PT0 rising edge */
    TFLG1 = COF;       /* clear COF flag */
    TIE |= IC0;        /* enable IC0 interrupt */
    asm("cli");
    TC1 = TCNT + 60000;
    while (oc_cnt) {
        while(!(TFLG1 & C1F));
        TC1 = TC1 + 60000;
        oc_cnt = oc_cnt - 1;
    }
    int2alpha(frequency, arr);
    puts(msg);
    puts(&arr[0]);
    asm("swi");
```

```
    }
    void INTERRUPT TC0_isr(void)
    {
        TFLG1 = COF; /* clear COF flag */
        frequency ++;
    }
```

8.6.4 Making Sound Using the Output-Compare Function

Using the output-compare function to make sound is easy. A sound can be made by creating a digital waveform of appropriate frequency and using it to drive a speaker or a buzzer. A small speaker of 8-Ω resistance that consumes between 10 mW and 20 mW can produce good-quality sound. The next example illustrates how to use an output-compare channel to generate a siren.

Example 8.7

Describe the circuit for making a sound and write a program that uses an output-compare channel to generate a siren that oscillates between 300 Hz and 1200 Hz.

Solution: A simple 8-mW speaker (or a **buzzer** on a demo board) has two terminals: one terminal is for signal input, whereas the other terminal is for ground connection. The circuit connection for siren generation is shown in Figure 8.21.

The algorithm for generating the siren is as follows:

Step 1
Enable an appropriate OC channel (OC5 in this example) to drive the speaker circuit.

Step 2
Start an output-compare operation and enable its interrupt with a delay count equal to half the period of the siren.

Step 3
Wait for certain amount of time (say half a second). During the waiting period, interrupts will be requested by the output-compare match many times. The interrupt service routine simply starts the next output-compare operation and then returns.

Step 4
At the end of the delay, choose a different delay count for the output-compare operation so the siren sound with a different frequency can be generated.

Step 5
Wait for the same amount of time as in step 3. Again, the interrupt caused by the output-compare match will be requested many times. At the end of the delay, switch back to the delay count used in step 2.

Step 6
Go to step 2.

Figure 8.21 ■ Circuit connection for siren generation

The assembly program that implements this algorithm for a HCS12DP256 running with a 24-MHz E clock is as follows:

```
#include        "c:\miniide\hcs12.inc"
hi_freq         equ     1250            ; delay count for 1200 Hz (with 1:8 prescaler)
lo_freq         equ     5000            ; delay count for 300 Hz (with 1:8 prescaler)
toggle          equ     $04             ; value to toggle the TC5 pin
                org     $1000
delay           ds.w    1               ; store the delay for output-compare operation
                org     $2000
                lds     #$2000
                movw    #oc5_isr,UserTimerCh5 ; initialize the interrupt vector entry
                movb    #$90,TSCR1      ; enable TCNT, fast timer flag clear
                movb    #$03,TSCR2      ; set main timer prescaler to 8
                bset    TIOS,OC5        ; enable OC5
                movb    #toggle,TCTL1   ; select toggle for OC5 pin action
                ldd     #hi_freq
                std     delay           ; use high-frequency delay count first
                ldd     TCNT            ; start the high-frequency sound
                addd    delay           ;   "
                std     TC5             ;   "
                bset    TIE,OC5         ; enable OC5 interrupt
                cli                     ;   "
forever         ldy     #50             ; wait for half a second
                jsr     delayby10ms     ;   "
                movw    #lo_freq,delay  ; switch to low-frequency delay count
                ldy     #50
                jsr     delayby10ms
                movw    #hi_freq,delay  ; switch to high-frequency delay count
                bra     forever
oc5_isr         ldd     TC5
                addd    delay
                std     TC5
                rti
;****************************************************************
; The following subroutine creates a time delay that is equal to [Y] times
; 10 ms. The timer prescaler is 1:8.
;****************************************************************
```

```
delayby10ms    pshd                              ; save accumulator D
               bset    TIOS,OC0                  ; enable OC0
               ldd     TCNT
again1         addd    #30000                    ; start an output-compare operation
               std     TC0                       ; with 10 ms time delay
wait_lp1       brclr   TFLG1,C0F,wait_lp1
               ldd     TC0
               dbne    y,again1
               bclr    TIOS,OC0                  ; disable OC0
               puld                              ; restore accumulator D
               rts
               end
```

In this program, both the OC0 and OC5 channels rely on the TCNT for timing control. One should use the same prescaler throughout all the timing operations.

The C language version of the program to generate the siren is as follows:

```c
#include    "c:\egnu091\include\hcs12.h"
#include    "c:\egnu091\include\delay.c"
#include    "c:\egnu091\include\vectors12.h"
#define     INTERRUPT __attribute__((interrupt))
#define     HiFreq    1250
#define     LoFreq    5000
int    delay;               /* delay count for OC5 operation */
void INTERRUPT oc5ISR(void);
int main(void)
{
    UserTimerCh5  = (unsigned short)&oc5ISR;
    TSCR1         = 0x90;            /* enable TCNT and fast timer flag clear */
    TSCR2         = 0x03;            /* set prescaler to TCNT to 1:8 */
    TIOS         |= BIT5;            /* enable OC5 */
    TCTL1         = 0x04;            /* select toggle for OC5 pin action */
    delay         = HiFreq;         /* use high frequency delay count first */
    TC5           = TCNT + delay;    /* start an OC5 operation */
    TIE          |= BIT5;           /* enable TC5 interrupt */
    asm("cli");
    while(1) {
      delayby100ms(5);              /* wait for half a second */
      delay = LoFreq;               /* switch to low frequency tone */
      delayby100ms(5);              /* wait for half a second */
      delay = HiFreq;               /* switch to high frequency tone */
    }
    return 0;
}
void INTERRUPT oc5ISR(void)
{
    TC5 += delay;
}
```

A siren can be considered as a song with only two notes. A song can be created by modifying the siren program. To make the switching of frequency easier, the whole score should be stored in a table. A note has two components: one is the frequency and the other is the duration. The duration of a quarter note is about 0.4 s. The durations of other notes can be derived proportionally. The complete list of the music notes and their frequencies is in Appendix F.

To play a song from the speaker, one places the frequencies and durations of all the notes in a music score in a table. For every note, the user program uses the output-compare function to generate the digital waveform with the specified frequency and duration. The following example illustrates this idea.

Example 8.8

Use the circuit shown in Figure 8.21 and write a program to play "The Star-Spangled Banner". Assume that the demo board is running with a 24-MHz E clock.

Solution: The following assembly program plays the national anthem:

```
#include     "c:\miniide\hcs12.inc"
G3      equ     7653        ; delay count to generate G3 note (with 1:8 prescaler)
B3      equ     6074        ; delay count to generate B3 note (with 1:8 prescaler)
C4      equ     5733        ; delay count to generate C4 note (with 1:8 prescaler)
C4S     equ     5412        ; delay count to generate C4S (sharp) note
D4      equ     5108        ; delay count to generate D4 note (with 1:8 prescaler)
E4      equ     4551        ; delay count to generate E4 note (with 1:8 prescaler)
F4      equ     4295        ; delay count to generate F4 note (with 1:8 prescaler)
F4S     equ     4054        ; delay count to generate F4S note (with 1:8 prescaler)
G4      equ     3827        ; delay count to generate G4 note (with 1:8 prescaler)
A4      equ     3409        ; delay count to generate A4 note (with 1:8 prescaler)
B4F     equ     3218        ; delay count to generate B4F note (with 1:8 prescaler)
B4      equ     3037        ; delay count to generate B4 note (with 1:8 prescaler)
C5      equ     2867        ; delay count to generate C5 note (with 1:8 prescaler)
D5      equ     2554        ; delay count to generate D5 note (with 1:8 prescaler)
E5      equ     2275        ; delay count to generate E5 note (with 1:8 prescaler)
F5      equ     2148        ; delay count to generate F5 note (with 1:8 prescaler)
notes   equ     101
toggle  equ     $04         ; value to toggle the TC5 pin
        org     $1000
delay   ds.w    1           ; store the delay for output-compare operation
rep_cnt ds.b    1           ; repeat the song this many times
ip      ds.b    1           ; remaining notes to be played
        org     $1500
        lds     #$1500
; establish the SRAM vector address for OC5
        movw    #oc5_isr,UserTimerCh5
        movb    #$90,TSCR1      ; enable TCNT, fast timer flag clear
        movb    #$03,TSCR2      ; set main timer prescaler to 8
        bset    TIOS,OC5        ; enable OC5
        movb    #toggle,TCTL1   ; select toggle for OC5 pin action
        ldx     #score          ; use as a pointer to score table
        ldy     #duration       ; points to duration table
```

```
                movb    #2,rep_cnt        ; play the song twice
                movb    #notes,ip
                movw    2,x+,delay        ; start with zeroth note
                ldd     TCNT              ; play the first note
                addd    delay             ;   "
                std     TC5               ;   "
                bset    TIE,C5I           ; enable OC5 interrupt
                cli                       ;   "
forever         pshy                      ; save duration table pointer in stack
                ldy     0,y               ; get the duration of the current note
                jsr     delayby10ms       ;   "
                puly                      ; get the duration pointer from stack
                iny                       ; move the duration pointer
                iny                       ;   "
                ldd     2,x+              ; get the next note, move pointer
                std     delay             ;   "
                dec     ip
                bne     forever
                dec     rep_cnt
                beq     done              ; if not finished playing, reestablish
                ldx     #score            ; pointers and loop count
                ldy     #duration         ;   "
                movb    #notes,ip         ;   "
                movw    2,x+,delay        ; get the first note delay count
                ldd     TCNT              ; play the first note
                addd    #delay            ;   "
                std     TC5
                bra     forever
done            swi
;***************************************************************
; The OC5 interrupt service routine simply starts a new OC5 operation.
;***************************************************************
oc5_isr         ldd     TC5
                addd    delay
                std     TC5
                rti
;***************************************************************
; The following subroutine creates a time delay which is equal to [Y] times
; 10 ms. The timer prescaler is 1:8.
;***************************************************************
delayby10ms bset    TIOS,OC0              ; enable OC0
                ldd     TCNT
again1          addd    #30000            ; start an output-compare operation
                std     TC0               ; with 10-ms time delay
wait_lp1        brclr   TFLG1,C0F,wait_lp1
                ldd     TC0
                dbne    y,again1
                bclr    TIOS,OC0          ; disable OC0
                rts
```

```
; *****************************************************************
; This table determines the frequency of each note.
; *****************************************************************
;
score       dw      D4,B3,G3,B3,D4,G4,B4,A4,G4,B3,C4S
            dw      D4,D4,D4,B4,A4,G4,F4S,E4,F4S,G4,G4,D4,B3,G3
            dw      D4,B3,G3,B3,D4,G4,B4,A4,G4,B3,C4S,D4,D4,D4
            dw      B4,A4,G4,F4S,E4,F4S,G4,G4,D4,B3,G3,B4,B4
            dw      B4,C5,D5,D5,C5,B4,A4,B4,C5,C5,C5,B4,A4,G4
            dw      F4S,E4,F4S,G4,B3,C4S,D4,D4,G4,G4,G4,F4S
            dw      E4,E4,E4,A4,C5,B4,A4,G4,G4,F4S,D4,D4
            dw      G4,A4,B4,C5,D5,G4,A4,B4,C5,A4,G4
; *****************************************************************
; Each of the following entries multiplied by 10 ms gives the duration of a note.
; *****************************************************************
;
duration    dw      30,10,40,40,40,80,30,10,40,40,40
            dw      80,20,20,60,20,40,80,20,20,40,40,40,40,40
            dw      30,10,40,40,40,80,30,10,40,40,40,80,20,20
            dw      60,20,40,80,20,20,40,40,40,40,40,20,20
            dw      40,40,40,80,20,20,40,40,40,80,40,60,20,40
            dw      80,20,20,40,40,40,80,40,40,40,20,20
            dw      40,40,40,40,20,20,20,20,40,40,20,20
            dw      60,20,20,20,80,20,20,60,20,40,80
            end
```

The C language version of the program is as follows:

```
#include    "c:\egnu091\include\hcs12.h"
#include    "c:\egnu091\include\vectors.h"
#define     G3      7653
#define     B3      6074
#define     C4      5733
#define     C4S     5412
#define     D4      5108
#define     E4      4551
#define     F4      4295
#define     F4S     4054
#define     G4      3827
#define     A4      3409
#define     B4F     3218
#define     B4      3037
#define     C5      2867
#define     D5      2554
#define     E5      2275
#define     F5      2148
#define     notes   101
#define     toggle  0x04
#define     INTERRUPT __attribute__((interrupt))
int   delay;
void delayby10ms(int kk);
void INTERRUPT oc5isr(void);
```

```
unsigned int score [101] = {D4,B3,G3,B3,D4,G4,B4,A4,G4,B3,C4S,
                D4,D4,D4,B4,A4,G4,F4S,E4,F4S,G4,G4,D4,B3,G3,
                D4,B3,G3,B3,D4,G4,B4,A4,G4,B3,C4S,D4,D4,D4,
                B4,A4,G4,F4S,E4,F4S,G4,G4,D4,B3,G3,B4,B4,
                B4,C5,D5,D5,C5,B4,A4,B4,C5,C5,C5,B4,A4,G4,
                F4S,E4,F4S,G4,B3,C4S,D4,D4,G4,G4,G4,F4S,
                E4,E4,E4,A4,C5,B4,A4,G4,G4,F4S,D4,D4,
                G4,A4,B4,C5,D5,G4,A4,B4,C5,A4,G4};
unsigned int dur [101] = {30,10,40,40,40,80,30,10,40,40,40,
                80,20,20,60,20,40,80,20,20,40,40,40,40,40,
                30,10,40,40,40,80,30,10,40,40,40,80,20,20,
                60,20,40,80,20,20,40,40,40,40,40,20,20,
                40,40,40,80,20,20,40,40,40,80,40,60,20,40,
                80,20,20,40,40,40,80,40,40,40,20,20,
                40,40,40,40,20,20,20,20,40,40,20,20,
                60,20,20,20,80,20,20,60,20,40,80};
main (void)
{
     int j;
     UserTimerCh5 = (unsigned short)&oc5isr;
     TSCR1     = 0x90;                    /* enable TCNT, fast timer flag clear */
     TSCR2     = 0x03;                    /* set TCNT prescaler to 8 */
     TFLG1     = 0xFF;                    /* clear all TxF flags */
     TIE       |= C5I;                    /* enable TC5 interrupt */
     TIOS      |= OC5;                    /* enable OC5 function */
     TCTL1     = toggle;                  /* select toggle as OC5 pin action */
     asm(" cli ");                        /* enable TC5 interrupt */
     j = 0;
     delay     = score[0];
     TC5       = TCNT + delay;            /* play the first note */
     while (j < notes) {
          delay = score[j];              /* play the jth note */
          delayby10ms(dur[j]);
          j++;
     }
     TIOS &= ~OC5;                        /* stop playing the song */
     asm ("swi");                         /* return to D-Bug12 monitor */
}
void delayby10ms(int kk)
{
     int i;
     TIOS      |= OC0;                    /* enable OC0 */
     TC0       = TCNT + 30000;            /* start one OC0 operation */
     for (i = 0; i < kk; i++) {
          while(!(TFLG1 & C0F));
          TC0 += 30000;
     }
     TIOS &= ~OC0;
}
```

```
void INTERRUPT oc5isr(void)
{
    TC5 + = delay;
}
```

There are at least three drawbacks in using this method to play a song:

1. Contiguous identical notes become one note. This problem can be solved by inserting a short period of silence between two contiguous identical notes.

2. A periodic square waveform is only an approximation to an actual note. Those who have learned communication theory know that a periodic square waveform also contains harmonics with higher frequencies. The solution to this problem is beyond the scope of this text.

3. There is no loudness control. Using the PWM function rather than the output-compare function and adjusting the duty cycle of each note is one possible solution to this problem.

8.6.5 Using OC7 to Control Multiple Output-Compare Functions

The output-compare function OC7 is special because it can control up to eight output-compare functions at the same time. The register OC7M specifies the output-compare channels to be controlled by OC7. The value that any PTx (x = 0, . . . , 7 pin to assume when the value of TC7 equals that of TCNT is specified by the OC7D register. To control an output-compare pin using OC7, the user sets the corresponding bit in OC7M. When a successful OC7 compare is made, each affected pin assumes the value of the corresponding bit of OC7D. The contents of the OC7M and OC7D registers are shown in Figures 8.22 and 8.23.

reset   0   0   0   0   0   0   0   0

OC7Mn *n* = 0...7

     0 = PTn pin is not affected by OC7 function.
     1 = a successful OC7 action will override a successful OC6–OC0 compare action during the same cycle and the OCn action taken will depend on the corresponding OC7D bit.

Figure 8.22 ■ Output-Compare 7 Mask register (OC7M)

reset   0   0   0   0   0   0   0   0

Figure 8.23 ■ Output-Compare 7 Data register (OC7D)

For the OC7 pin, the HCS12 document does not specify what will happen when both the TCTL1 register and the OC7M:OC7D register pair specify the OC7 pin action on a successful OC7 compare. Avoid using this combination.

For OC0–OC6, when the OC7Mn $(n = 0...6)$ bit is set, a successful OC7 action will override a successful OC6–OC0 compare action during the same cycle; therefore, the OCn action taken will depend on the corresponding OC7D bit. This feature allows a PT_n pin to be controlled by two output-compare functions simultaneously (OC7 and OCn $(n = 0...6)$).

Example 8.9

▼

What value should be written into OC7M and OC7D if one wants pins PT2, PT3, and PT4 to assume the values of 1, 0, and 1, respectively, when OC7 compare succeeds?

Solution: Bits 4, 3, and 2 of OC7M must be set to 1, and bits 4, 3, 2, of OC7D should be set to 1, 0, and 1, respectively. The following instruction sequence will set up these values:

```
movb    #$1C,OC7M
movb    #$14,OC7D
```

The following C statements will achieve the same goal:

```
OC7M = 0x1C;
OC7D = 0x14;
```

▲

The ability to control multiple output-compare pins is very useful in applications that require multiple actions to be triggered simultaneously in the near future.

Example 8.10

▼

Suppose that there is an application that requires the following operations to be triggered 50 ms later:

- Turn off the light controlled by the TC5 pin
- Turn on the temperature sensor controlled by the TC4 pin
- Turn off the heater controlled by the TC3 pin
- Turn on the music controlled by the TC2 pin

Write an instruction sequence to perform the desired operation.

Solution: The values to be written into OC7M and OC7D are $3C and $14, respectively. Set the prescaler to TCNT to 64. Then the value to be added to create a 50-ms delay can be computed as follows:

Delay = (24000000 ÷ 64) × 0.05 = 18,750

The following instruction sequence can perform the desired operations 50 ms later:

```
movb    #$90,TSCR1      ; enable TCNT and fast timer flag clear
movb    #$06,TSCR2      ; set prescaler to 64
movb    #$3C,OC7M
movb    #$14,OC7D
ldd     TCNT
addd    #18750
std     TC7             ; start the OC7 operation
```

▲

Two output-compare functions can control the same pin simultaneously. Thus, OC7 can be used in conjunction with one or more other output-compare functions to achieve even more timing flexibility. We can generate a digital waveform with a given duty cycle by using the OC7 and any other output-compare function.

Example 8.11
▼

Use OC7 and OC0 together to generate a 2-kHz digital waveform with a 40% duty cycle on the PT0 pin. Assume that the E clock frequency is 24 MHz.

Solution: We will set the prescale factor to 8. The period of a 2-kHz waveform is 0.005 s. The period of this signal corresponds to $0.005 \times (24000000 \div 8) = 15{,}000$ cycles. The high interval in one period of the waveform is 6,000 clock cycles, whereas the low interval is 9,000 clock cycles.

The idea in using two OC functions in generating the waveform is as follows:

- Use OC0 (or OC7) to pull the PT0 pin to high every 15,000 clock cycles.
- Use OC7 (or OC0) to pull the PT0 to low after it is high for 6,000 clock cycles.

We will use interrupt-driven approach to implement this algorithm. After starting the first action on OC7 and OC0, the program will stay in a wait loop to wait for interrupts to be requested from OC7 and OC0. The interrupt service routine will simply start the next output-compare operation with 15,000 as the delay count. The assembly program that implements this idea is as follows:

```
        #include "c:\miniide\hcs12.inc"
high_cnt  equ    6000                  ; interval during which pulse is high
delay     equ    15000                 ; a delay corresponds to the period of 2 kHz
          org    $1500
          lds    #$1500
          movw   #oc7_isr,UserTimerCh7 ; set up OC7 interrupt vector
          movw   #oc0_isr,UserTimerCh0 ; set up OC0 interrupt vector
          movb   #$90,TSCR1            ; enable TCNT and fast timer flags clear
          movb   #$03,TSCR2            ; set prescaler to 8
          movb   #$81,TIOS             ; enable OC7 and OC0
          movb   #$01,OC7M             ; allow OC7 to control OC0 pin
          movb   #$01,OC7D             ; OC7 action on PT0 pin is to pull high
          movb   #$02,TCTL2            ; select pull low as the OC0 action
          movb   #$81,TIE             ; enable OC7 and OC0 to interrupt
          ldd    TCNT
          addd   #delay
          std    TC7                  ; start an OC7 operation
          addd   #high_cnt
          std    TC0                  ; start an OC0 operation
          cli                          ; enable interrupt
loop      bra    loop                  ; infinite loop to wait for interrupt
          swi
oc7_isr   ldd    TC7                  ; start the next OC7 action with 15000
          addd   #delay                ; clock cycles delay, also clear the C7F
          std    TC7                  ; flag
          rti
```

```
oc0_isr    ldd     TC0              ; start the next OC0 action with 15000
           addd    #delay           ; clock cycles delay, also clear the COF
           std     TC0              ; flag
           rti
           end
```

The C language version of this program is very straightforward and hence will be left as an exercise.

8.6.6 Forced Output-Compare

There may be applications in which the user requires an output-compare pin action to occur immediately instead of waiting for a match between TCNT and the proper output-compare register. This situation arises in the spark-timing control in some automotive engine control applications. To use the forced output-compare mechanism, the user would write to the CFORC register with 1s in the bit positions corresponding to the output-compare channels to be forced. At the next timer count after the write to CFORC, the forced channels will trigger their programmed pin actions to occur.

The forced actions are synchronized to the timer counter clock input. The forced output-compare signal causes pin action but does not affect the timer flag or generate interrupt. Normally, the force mechanism would not be used in conjunction with the automatic pin action that toggles the corresponding output-compare pin. The contents of CFORC are shown in Figure 8.24. CFORC always reads as all zeroes.

Figure 8.24 ■ Contents of the CFORC register

Example 8.12

Suppose that the contents of the TCTL1 and TCTL2 registers are $D6 and $6E, respectively. The contents of the TFLG1 are $00. What would occur on pins PT7 to PT0 on the next clock cycle if the value $7F is written into the CFORC register?

Solution: The TCTL1 and TCTL2 configure the output-compare actions as shown in Table 8.2. Since the contents of the TFLG1 are $0, none of the started output-compare operations have succeeded yet.

Because the CFORC register specifies that the output-compare channels 6 to 0 are to be forced immediately, the actions specified in the fourth column in Table 8.2 will occur immediately.

Register	Bit Positions	Value	Action to Be Triggered
TCTL1	7 6	1 1	Set the PT7 pin to high
	5 4	0 1	Toggle the PT6 pin
	3 2	0 1	Toggle the PT5 pin
	1 0	1 0	Pull the PT4 pin to low
TCTL2	7 6	0 1	Toggle the PT3 pin
	5 4	1 0	Pull the PT2 pin to low
	3 2	1 1	Set the PT1 pin to high
	1 0	1 0	Pull the PT0 pin to low

Table 8.2 ■ Pin actions on PT7-PT0 pins

8.7 Pulse Accumulator

The HCS12 standard timer system has a 16-bit pulse accumulator, PACA, whereas the Enhanced Captured Timer system has four 8-bit pulse accumulators (PAC3~PAC0). PAC3 and PAC2 can be concatenated into the 16-bit pulse accumulator A (PACA), whereas PAC1 and PAC0 can be concatenated into the 16-bit pulse accumulator B (PACB). The block diagrams of four 8-bit pulse accumulators and two 16-bit pulse accumulators are shown in Figures 8.25 and 8.26, respectively.

There are four possible configurations for these pulse accumulators:

- Two 16-bit pulse accumulators PACA and PACB
- One 16-bit pulse accumulator PACA and two 8-bit pulse accumulators PAC1 and PAC0
- One 16-bit pulse accumulator PACB and two 8-bit pulse accumulators PAC3 and PAC2
- Four 8-bit accumulators PAC3~PAC0

8.7.1 Signal Pins

Four 8-bit pulse accumulators PAC3~PAC0 are sharing the signal pins PT3~PT0 with the lower four input-capture modules IC3~IC0, one pin per pulse accumulator. However, when concatenated into a 16-bit pulse accumulator, PACA and PACB use signal pins PT7 and PT0, respectively.

8.7.2 Operation Modes

The 16-bit pulse accumulator PACA has two operation modes:

1. *Event counting mode.* In this mode, the PACA counter increments on the active edge of the PT7 pin. Many microcontroller applications involve counting things. These things are called *events*, but in real applications they might be anything: pieces on an assembly line, cycles of an incoming signal, or units of time. To be counted by the accumulator, these things must be translated into rising or falling edges on the PT7 pin. A trivial example of an event might be to count pieces on the assembly line: a

light emitter/detector pair could be placed across a conveyor so that as each piece passes the sensor, the light beam is interrupted and a logic-level signal that can be connected to the PT7 pin is produced.

2. *Gated time accumulation mode.* The 16-bit PACA counter is clocked by a free-running E ÷ 64 signal, subject to the PT7 signal being active. One common use of this mode is to measure the duration of a single pulse. The counter is set to 0 before the pulse starts, and the resultant pulse time is read directly when the pulse is finished.

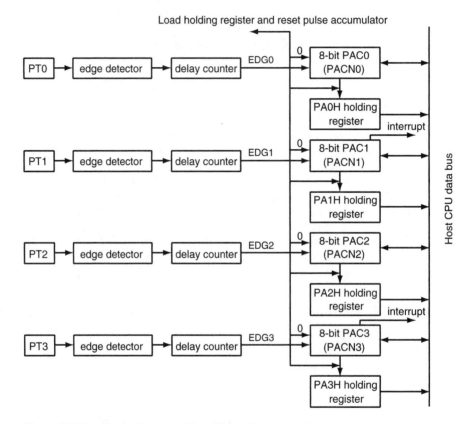

Figure 8.25 ■ Block diagram of four 8-bit pulse accumulators

The 16-bit pulse accumulator B and all four 8-bit pulse accumulators have only one mode: event-counting mode. That is, their counters are incremented on the active edge of their associated pins. The active edges of PACB and PAC3~PAC0 are identical to those of timer capture functions IC0 (PT0 pin) and IC3~IC0 (PT3~PT0 pins), respectively. The active edges of the PT3~PT0 pins are defined by the TCTL4 register.

8.7.3 Interrupt Sources

The 16-bit PACA has two interrupt sources:

1. PT7-edge interrupt
2. PACA counter overflow

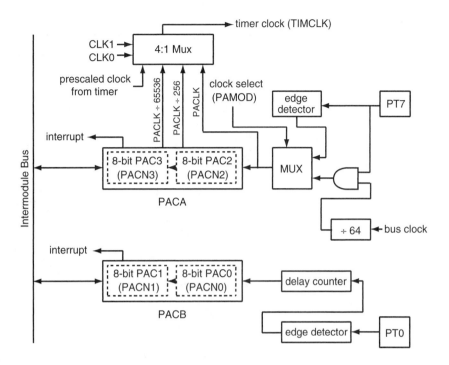

Figure 8.26 ■ 16-bit pulse accumulator block diagram

Only two (PAC3 and PAC1) of the four 8-bit pulse accumulators may request interrupt to the MCU. Interrupts are requested whenever PAC3 or PAC1 rolls over from $FF to $00. The 16-bit PACB may request interrupt to the MCU whenever its upper eight bits counter (PAC1) rolls over from $FF to $00.

8.7.4 Registers Related to Pulse Accumulator

The operation of the16-bit PACA is controlled by the PACTL register. The status of the PACA is recorded in the PAFLG register. The PACA has a 16-bit counter. Since PACA is formed by concatenating PAC3 and PAC2, one can use the instruction ldd PAC3 to copy the contents of the 16-bit counter. To be backward compatible with the 68HC12, the register name PACNT is added to the header file so that one can also use the instruction ldd PACNT to copy the contents of the 16-bit counter. The contents of the PACTL and PAFLG registers are shown in Figures 8.27 and 8.28, respectively.

The PACTL register also controls the clock source for the timer counter (TCNT). When the CLK1 and CLK0 bits are not 00, the PACLK signal (from the PT7 pin) is prescaled by 1, 256, or 65536 and used as the clock input to the timer counter.

Bits 1 and 0 of the PAFLG register keep track of the status of the operation of the PACA, as shown in Figure 8.28. Any access to the PAC3 or PAC2 register will clear all the flag bits in the PAFLG register if the TFFCA bit in the TSCR1 register is set to 1.

The PACB pulse accumulator is controlled by the PBCTL register, and the PBFLG register records its status. The contents of these two registers are shown in Figure 8.29.

PAEN—Pulse accumulator system enable bit
   0 = PACA is disabled (PACN3 and PACN2 can be enabled).
   1 = PACA is enabled (PACN3 and PACN2 cannot be enabled).
PAMOD—Pulse accumulator mode bit
   0 = event counter mode
   1 = gated time accumulation mode
PEDGE—Pulse accumulator edge control bit
   For PAMOD = 0 (event counter mode)
     0 = falling edges on the PAI pin cause the count to increment.
     1 = rising edges on the PAI pin cause the count to decrement.
   For PAMOD = 1 (gated time accumulation mode)
     0 = PAI pin high enables $E \div 64$ clock to pulse accumulator and
        the trailing falling edge on the PAI pin sets the PAIF flag.
     1 = PAI pin low enables $E \div 64$ clock to pulse accumulator and
        the trailing rising edge on the PAI pin sets the PAIF flag.
CLK1 and CLK0—Clock select bits
   00 = use timer prescaler clock as timer counter clock.
   01 = use PACLK as input to timer counter (TCNT) clock.
   10 = use PACLK/256 as timer counter clock.
   11 = use PACLK/65536 as timer counter clock.
PAOVI—Pulse accumulator overflow interrupt enable bit
   0 = disable
   1 = enable
PAI—PAI pin interrupt enable bit
   0 = disabled
   1 = enabled

Figure 8.27 ■ Pulse Accumulator Control register (PACTL)

PAOVF—pulse accumulator overflow flag
   This flag is set when PACNT overflows from $FFFF to $0000 and can
   be cleared by writing a 1 to it.
PAIF—PT7 pin edge flag
   When in event-counter mode, this bit is set when the selected edge on the
   PT7 pin is detected.
   When in gated-accumulation mode, the selected trailing edge sets this
   flag.

Figure 8.28 ■ Pulse Accumulator Flag register (PAFLG)

PBEN—pulse accumulator B system enable bit
 0 = 16-bit pulse accumulator disabled. Eight-bit PAC1 and PAC0 can
   be enabled when their related enable bits in ICPACR are set.
 1 = pulse accumulator B system enabled.
PBOVI—pulse accumulator B overflow interrupt enable bit
 0 = interrupt inhibited.
 1 = interrupt requested if PBOVF is set.

(a) Pulse Accumulator B Control register (PBCTL)

PBOVF—pulse accumulator B overflow flag
This bit is set when the 16-bit pulse accumulator B overflows from
$FFFF to $0000 or when 8-bit accumulator 1 (PAC1) overflows from
$FF to $00. It is cleared by writing 1 to it or by accessing PACN1
and PACN0 when the TFFCA bit in the TSCR1 register is set.

(b) Pulse accumulator B Flag register (PBFLG)

Figure 8.29 ■ Pulse Accumulator B Control and Flag register

Each of the 8-bit pulse accumulators can be enabled if its associated 16-bit pulse accumulator is disabled. The enabling of an 8-bit pulse accumulator is done by programming the ICPAR register. The contents of ICPAR are shown in Figure 8.30.

PAxEN—8-bit pulse accumulator x enable bit
 0 = pulse accumulator x disabled
 1 = pulse accumulator x enabled

Figure 8.30 ■ Input Control Pulse Accumulator Control register (ICPAR)

Each of the 8-bit pulse accumulator also has an 8-bit holding register (PA3H~PA0H).

8.7.5 Operations of the Enhanced Pulse Accumulators

The 16-bit PACA has two operation modes: event-counting mode and gated time accumulation mode. When in the *even-counting* mode, the PACA counter increases on the selected edge of the PT7 signal. The PAIF flag of the PAFLG register is set to 1 whenever the selected signal edge is detected on the PT7 pin. The setting of this flag may request an interrupt to the MCU if the PAI bit of the PACTL register is set to 1. When in *gated time accumulation* mode, the 16-bit counter is enabled to increment by the E/64 clock signal if the selected signal level is applied on the PT7 pin.

All other pulse accumulators (PACB, PAC3~PAC0) count the number of active edges at their associated pins. Whenever the PAC3 or PAC1 rolls over from $FF to $00, its associated flag (PAOVF or PBOVF) will be set to 1 and may optionally request an interrupt to the MCU. Pulse accumulators PAC2 and PAC0 do not have the interrupt capability. The user can prevent 8-bit pulse accumulators counting further than $FF by setting the PACMX bit in the ICSYS register. In this case, a value of $FF means that 255 counts or more have occurred. The contents of the ICSYS register are illustrated in Figure 8.31.

	7	6	5	4	3	2	1	0
	SH37	SH26	SH15	SH04	TFMOD	PACMX	BUFEN	LATQ
reset:	0	0	0	0	0	0	0	0

SHxy—share input action of input capture channel x and y bits
  0 = normal operation
  1 = the channel input x causes the same action on the channel y.
TFMOD—timer flag-setting mode bit
  0 = the timer flags C3F–C0F in TFLG1 are set when a valid input
      capture transition on the corresponding port pin occurs.
  1 = if in the queue mode (BUFEN = 1 and LATQ = 0), the timer flags
      C3F–C0F in TFLG1 are set only when a latch on the
      corresponding holding register occurs. If the queue mode is not
      engaged, the timer flags C3F–C0F are set the same way as for
      TFMOD = 0.
PACMX—8-bit pulse accumulator maximum count bit
  0 = normal operation. When the 8-bit pulse accumulator has reached
      the value $FF, with the next active edge, it will be incremented
      to $00.
  1 = when the 8-bit pulse accumulator has reached the value $FF, it
      will not be incremented further. The value $FF indicates a count
      of 255 or more.
BUFEN—IC buffer enable bit
  0 = input capture and pulse accumulator holding registers are
      disabled.
  1 = input capture and pulse accumulator holding registers are
      enabled.
LATQ—input capture latch or queue mode select bit
  The BUFEN bit should be set to enable IC and the pulse accumulator's
  holding registers. Otherwise, the LATQ latching mode is disabled.
  0 = the queue mode of input capture is enabled.
  1 = the latch mode is enabled. Latching function occurs when modulus
      down-counter reaches 0 or a 0 is written into the count register
      MCCNT. With a latching event, the contents of IC registers and
      8-bit pulse accumulators are transferred to their holding registers.
      The 8-bit pulse accumulators are cleared.

Figure 8.31 ■ Input Control System Control register (ICSYS)

8.7.6 Pulse Accumulator Applications

The pulse accumulator has a few interesting applications, such as interrupting after N events, frequency measurement, and pulse duration measurement.

Example 8.13

Suppose that certain events are converted into pulses and connected to the PT7 pin. Write a program so that the pulse accumulator generates an interrupt to the HCS12 when N events have occurred. Assume that N is less than 65536.

Solution: By writing the two's complement of N into PACNT, it will overflow after N events and generate an interrupt. The assembly program is as follows:

```
#include      "c:\miniide\hcs12.inc"
N             equ     1350
              org     $1500
              lds     #$1500              ; set up stack pointer
              movw    #paov_isr,UserPAccOvf ; set up PAOV interrupt vector
              ldd     #N                  ; place the 2's complement in PACNT
              coma                        ;   "
              comb                        ;   "
              addd    #1                  ;   "
              std     PACNT               ;   "
              movb    #$52,PACTL          ; enable PACA, event counting mode, active
                                          ; edge is rising
              cli                         ; enable PAOV interrupt
;             ...
              swi
paov_isr      movb    #PAOVF,PAFLG        ; clear the PAOVF flag
              end
```

The C language version of the program is as follows:

```
#include      "c:\egnu091\include\hcs12.h"
#include      "c:\egnu091\include\vectors12.h"
#define    N        1252
#define    INTERRUPT           __attribute__((interrupt))
void   INTERRUPT paov_isr (void);
void main(void)
{
    UserPAccOvf = (unsigned short)&paov_isr;
    PACNT     = ~N + 1;        /* load the 2's complement of N into PACNT */
    PACTL     = 0x52;
    asm("cli");                /* enable PAOV interrupt */
    ....
}
void INTERRUPT paov_isr(void)
{
    PAFLG = PAOVF;
}
```

The input-capture function was used to measure the frequency of a signal in Example 8.6. The drawback in that approach is that the interrupt overhead of every input signal edge sets the upper limit of the frequency that can be measured. Using the pulse accumulator system to measure frequency dramatically reduces the incurred interrupt overhead.

The procedure for measuring the frequency of a signal using the pulse accumulator system is as follows:

Step 1
Connect the unknown signal to the PT7 pin.

Step 2
Set up the PACA to operate in event-counting mode.

Step 3
Use one of the output-compare functions to create a 1-s time interval.

Step 4
Use a memory location to keep track of the number of pulse accumulator counter overflow interrupts.

Step 5
Enable the PAOV interrupt.

Step 6
Disable the PAOV interrupt at the end of 1 s.

Example 8.14
▼

Write a program to measure the frequency of a signal connected to the PT7 pin using the algorithm described previously.

Solution: We will use the OC0 function to create a 1-s delay. Fifty OC0 operations will be performed, with each OC0 operation creating a 20-ms delay. The service routine for the PACNT overflow interrupt will increate the overflow count by 1. Let *paov_cnt* represents the PACNT overflow count. At the end of 1 s, the frequency is equal to the following expression:

$$\text{Frequency} = \text{paov\_cnt} \times 2^{16} + \text{PACNT}$$

The assembly program that implements this procedure is as follows:

```
#include        "c:\miniide\hcs12.inc"
                org     $1000
oc_cnt          rmb     1
paov_cnt        rmb     2                       ; PACNT overflow count
frequency       rmb     4                       ; signal frequency
                org     $1500
                lds     #$1500
                movw    #paov_isr,UserPAccOvf   ; set up PAOV interrupt vector
                movb    #50,oc_cnt              ; prepare to perform 50 OC0 actions
                ldd     #0
                std     PACNT                   ; let PACNT count up from 0
                std     paov_cnt                ; initialize PACNT overflow count to 0
                std     frequency               ; initialize frequency to 0
                std     frequency+2             ;    "
                movb    #$90,TSCR1              ; enable TCNT and fast timer flag clear
```

```
                    bset      TIOS,OC0                ; select OC0 function
                    movb      #$03,TSCR2              ; set prescaler to TCNT to 8
                    bclr      DDRT,$80                ; configure PT7 for input
        ; configure PA function: enable PA, select event-counting mode, rising edge
        ; of PAI signal increments the PACNT counter, enable PAOV interrupt
                    movb      #$52,PACTL
                    cli                               ; enable PAOV interrupt
                    ldd       TCNT
sec_loop            addd      #60000
                    std       TC0
                    brclr     TFLG1,C0F,*             ; wait for 20 ms here
                    ldd       TC0
                    dec       oc_cnt
                    bne       sec_loop
                    movb      #0,PACTL                ; disable PA function
                    sei                               ; disable interrupt
                    ldd       PACNT
                    std       frequency+2
                    ldd       paov_cnt
                    std       frequency
                    swi
paov_isr            movb      #PAOVF,PAFLG            ; clear the PAOVF flag
                    ldx       paov_cnt               ; increment PACNT overflow
                    inx                               ; count by 1
                    stx       paov_cnt               ;    "
                    end
```

The C language version of the program is as follows:

```
#include        "c:\egnu091\include\hcs12.h"
#include        "c:\egnu091\include\vectors12.h"
#define         INTERRUPT    __attribute__((interrupt))
unsigned        long   int   frequency;
unsigned        int    paov_cnt;
void            INTERRUPT paov_isr (void);
void main (void)
{
    int oc_cnt;

    UserPAccOvf = (unsigned short)&paov_isr;
    PACNT       = 0;
    frequency   = 0;
    paov_cnt    = 0;
    TSCR1       = 0x90;        /* enable TCNT and fast flag clear */
    TIOS        = OC0;         /* select OC0 function */
    TSCR2       = 0x03;        /* set prescale factor to 8 */
    PACTL       = 0x52;        /* enable PA function, enable PAOV interrupt */
    DDRT        &= 0x7F;       /* configure the PT7 pin for input */
    asm("cli");               /* enable interrupt globally */
    oc_cnt      = 50;
    TC0         = TCNT + 60000u;
```

```
            while (oc_cnt) {
                while(!(TFLG1 & COF));
                TC0 = TC0 + 60000u;
                oc_cnt--;
            }
            PACTL      = 0×00;       /* disable PA function */
            asm("sei");                /* disable interrupt */
            frequency  = (long)paov_cnt * 65536L + (long)PACNT;
            asm("swi");
    }
    void INTERRUPT paov_isr (void)
    {
        PAFLG      = PAOVF;      /* clear PAOVF flag */
        paov_cnt   = paov_cnt + 1;
    }
```

The pulse accumulator module can be set up to measure the duration of a pulse using the gated time accumulation mode. When the active level is applied to the PT7 pin, the PACNT can count and will stop counting on the trailing edge of the PT7 signal. The clock input to the pulse accumulator is E ÷ 64. The procedure for measuring the duration of a pulse is as follows:

Step 1
Set up the pulse accumulator system to operate in the gated time accumulation mode, and initialize PACNT to 0.

Step 2
Select the falling edge as the active edge (for measuring positive pulse). In this setting, the pulse accumulator counter will increment when the signal connected to the PAI pin is high and generate an interrupt to the HCS12 on the falling edge.

Step 3
Enable the PAI active edge interrupt and wait for the arrival of the active edge of PAI.

Step 4
Stop the pulse accumulator counter when the interrupt arrives.

Without keeping track of the PACNT overflows, the longest pulse width (E = 24 MHz) that can be measured is

$$\text{pulse\_width} = 2^{16} \times 64T_E = 2^{16} \times 64 \times 1/24 \ \mu s = 174.763 \ ms$$

To measure a longer pulse width, one will need to keep track of the number of times that the PACNT counter overflows in the duration of the pulse. Let *paov_cnt* be the overflow count of the PACNT counter, then

$$\text{pulse\_width} = [(2^{16} \times \text{paov\_cnt}) + \text{PACNT}] \times 64T_E$$

Example 8.15

▼

Write a program to measure the duration of an unknown signal connected to the PAI pin.

Solution: The assembly program that implements the previous algorithm is as follows:

```
#include        "c:\miniide\hcs12.inc"
                org      $1000
paov_cnt        ds.b     1              ; use to keep track of the PACNT overflow count
pulse_width     ds.b     3              ; hold the pulse width
                org      $1500
                movw     #paov_isr,UserPAccOvf ; set up PAOV interrupt vector
                ldd      #0
                std      PACNT          ; let PACNT count up from 0
                clr      paov_cnt       ; initialize PACNT overflow count to 0
                movb     #$0,TSCR2      ; set TCNT timer prescaler to 1
; configure PA function: enable PA, select gated time accumulation mode, high level
; of PAI signal enables PACNT counter, enable PAOV interrupt
                movb     #$62,PACTL
                bclr     DDRT,$80       ; configure PAI pin for input
                cli                     ; enable PAOV interrupt
                brclr    PAFLG,PAIF,*   ; wait for the arrival of the falling edge of PAI
                movb     #0,PACTL       ; disable PA function
                sei                     ; disable interrupt
                ldd      PACNT
                std      pulse_width+1
                ldaa     paov_cnt
                staa     pulse_width
                swi
paov_isr        movb     #PAOVF,PAFLG   ; clear the PAOVF flag
                inc      paov_cnt       ; increment PACNT overflow count by 1
                end
```

The C language version of the program is as follows:

```
#include        "c:\egnu091\include\hcs12.h"
#include        "c:\egnu091\include\vectors12.h"
#define         INTERRUPT    __attribute__((interrupt))
unsigned     int   paov_cnt;
long     unsigned   int pulse_width;
void     INTERRUPT paov_isr (void);

void main ( )
{
        UserPAccOvf = (unsigned short)&paov_isr;
        PACNT = 0;              /* let PACNT counts up from 0 */
        TSCR1 = 0x90;           /* enable TCNT and fast timer flag clear */
        paov_cnt = 0;
        pulse_width = 0;
        TSCR2 = 0x00;           /* set TCNT prescaler to 1 */
        DDRT &= 0x7F;           /* configure the PT7 pin for input */
/* configure PA function: enable PA, select gated time accumulator mode, high level
   of PAI enables PACNT to count, enable PAOV interrupt */
        PACTL = 0x62;
        asm("cli");
```

```
            while(!(PAFLG & PAIF));    /* wait for the arrival of the PAI falling edge */
            PACTL = 0x00;              /* disable PA system */
            asm("sei");
            pulse_width = (long)paov_cnt * 65536l + (long)PACNT;
            asm ("swi");
    }
    void INTERRUPT paov_isr (void)
    {
            PAFLG     = PAOVF;     /* clear PAOVF flag */
            paov_cnt  = paov_cnt + 1;
    }
```

Example 8.16

▼

Write an instruction sequence to enable the 8-bit pulse accumulator PAC1 and PAC3 and let them increase on the rising edge of their associated pins. Disable their overflow interrupts.

Solution: To enable PAC1 and PAC3, the PAEN and PBEN bits must be cleared and the PA3EN and PA1EN bits set. To select the rising edge as their active edge, the value $44 needs to be written into the TCTL4 register and also the PT3 and PT1 pins must be configured for input-capture. To disable overflow interrupt, clear the PAOVI and PBOVI bits.

The following instruction sequence will perform the required configuration:

```
bclr    PACTL,$42       ; disable 16-bit PACA, disable overflow interrupt
bclr    PBCTL,$42       ; disable 16-bit PACB, disable overflow interrupt
movb    #$44,TCTL4      ; select the rising edges as active edge
bset    ICPAR,$0A       ; enable PAC3 and PAC1
bclr    DDRT,$0A        ; configure PT3 and PT1 for input
```

▲

8.8 Modulus Down-Counter

The HCS12 timer system contains a modulus down-counter that can be used as a time base to generate periodic interrupts. It can also be used to latch the value of the IC registers and the pulse accumulators to their holding registers. The action of latching can be periodic or only once.

The modulus down-counter has a prescaler, which divides the bus clock and uses its output as the clock input to the down-counter. The prescaler can be 1, 4, 8, and 16. The operation of the modulus down-counter is controlled by the MCCTL register. The MCFLG register records the status of the modulus down-counter. The contents of the MCCTL and MCFLG registers are shown in Figures 8.32 and 8.33, respectively.

8.8.1 Using the Modulus Down-Counter to Generate Periodic Interrupts

Periodic interrupts can be generated by selecting the modulus mode and enabling its interrupt. The interrupt period is equal to the value written into the load register multiplied by the prescaler. When writing a value into MCCNT, the same value is also written into the load register. In the modulus mode, when the MCCNT is decremented to zero, the value in the load register will be reloaded into the MCCNT register and another cycle of down-counting is started.

7	6	5	4	3	2	1	0
MCZI	MODMC	RDMCL	ICLAT	FLMC	MCEN	MCPR1	MCPR0

reset: 0   0   0   0   0   0   0   0

MCZI—modulus counter underflow interrupt enable bit
   0 = modulus counter underflow interrupt is disabled.
   1 = modulus counter underflow interrupt is enabled.
MODMC—modulus mode enable bit
   0 = the counter counts once from the value written to it and will stop at $0000.
   1 = modulus mode is enabled. When the counter reaches $0000, the counter is
        loaded with the latest value written into to the modulus count register.
RDMCL—read modulus down-counter load bit.
   0 = reads of the modulus count register will return the present value of
        the count register.
   1 = reads of the modulus count register will return the contents of the
        load register (i.e., the reload value is returned).
ICLAT—input capture force latch action bit
   This bit has effect only when both the LATQ and BUFEN bits in ICSYS are set.
   0 = no effect
   1 = forces the contents of the input capture registers TC0 to TC3 and their
        corresponding 8-bit pulse accumulators to be latched into the associated holding
        registers. The pulse accumulators will be cleared when the latch action occurs.
FLMC—force load register into the modulus counter count register bit
   This bit has effect only when MCEN = 1.
   0 = no effect
   1 = loads the load register into the modulus counter count register. This also resets
        the modulus counter prescaler.
MCEN—modulus down-counter enable bit
   0 = modulus counter is disabled and preset to $FFFF.
   1 = modulus counter is enabled.
MCPR1 and MCPR0—modulus counter prescaler select bits
   0 0 = prescale rate is 1.
   0 1 = prescale rate is 4.
   1 0 = prescale rate is 8.
   1 1 = prescale rate is 16.

Figure 8.32 ■ Modulus Down-Counter Control register (MCCTL)

7	6	5	4	3	2	1	0
MCZF	0	0	0	POLF3	POLF2	POLF1	POLF0

reset: 0   0   0   0   0   0   0   0

MCZF—modulus counter underflow interrupt flag
   This flag is set when the modulus down-counter reaches 0. Writing
   1 to this bit clears the flag.
POLF3–POLF0—first input capture polarity status bits
   These are read-only bits. Writing to these bits has no effect. Each
   status bit gives the polarity of the first edge that has caused an input
   capture to occur after capture latch has been read.
   0 = the first input capture has been caused by a falling edge.
   1 = the first input capture has been caused by a rising edge.

Figure 8.33 ■ Modulus Down-Counter Flag register (MCFLG)

Example 8.17

Write an instruction sequence to generate periodic interrupt every 10 ms.

Solution: One possible value to be written into the MCCTL register is $C0, which will

- Enable MCCNT
- Enable MCCNT interrupt
- Enable modulus mode
- Set the prescaler to 16

The instruction sequence to achieve the desired setting is as follows:

```
movb    #$C7,MCCTL
movw    #15000,MCCNT    ; place the value that will be decremented
                        ; to 0 in 10 ms
cli                     ; enable interrupt
```

8.8.2 Using the Modulus Down-Counter to Generate Time Delays

The method for generating a time delay by using the modulus down-counter is similar to that for generating periodic interrupts. However, one can choose to use either the modulus mode or nonmodulus mode. The following example use the nonmodulus mode to generate a time delay that is equal to 10 ms multiplied by the value in register Y.

Example 8.18

Write a subroutine to generate a time delay that is equal to 10 ms multiplied by the value in register Y.

Solution: The assembly program that can generate a delay that is equal to 10 ms multiplied by the value in register Y is as follows:

```
delay10ms   bset    TSCR1,TFFCA       ; enable timer fast flag clear
            movb    #$07,MCCTL        ; enable modulus down-counter with 1:16 as prescaler
            movw    #15000,MCCNT      ; load the value to be down-counted
            brclr   MCFLG,MCZF,*
            bclr    MCCTL,$04         ; disable modulus down counter
            dbne    y,delay10ms
            rts
```

The C language version of the function is straightforward and hence will be left as an exercise problem.

By storing the previous delay function in the MiniIDE directory (or other appropriate directory), one can include it in the program. The following program calls the previous subroutine to create a delay of 200 ms, increments a count value, and then outputs this count value to port B:

```
#include     "c:\MiniIDE\hcs12.inc"
             org     $1000
count        ds.b    1
             org     $1500
             movb    #$FF,DDRB
```

```
             clr       count
             movb      count,PORTB
loops        ldy       #20              ; create 200 ms delay
             jsr       delayby10ms      ;    "
             inc       count            ; increment count by 1
             movb      count,PORTB      ; update LED displays
             bra       loops
#include      "c:\MiniIDE\delay.asm"
             end
```

Delay functions that create time delays by a multiple of other delay times can be created by modifying the previous delay function. The file **delay.asm** contains the following delay functions and is included in the complementary CD:

1. delayby10μs
2. delayby50μs
3. delayby1ms
4. delayby10ms
5. delayby100ms
6. delayby1s

There are also C language versions of these functions. They are contained in the **delay.c** file and should be copied to the **c:\egnu091\include** directory so that it can be included in the user's program.

8.9 Enhanced Capture Timer (ECT) Module

Some of the HCS12 members (e.g., MC9S12DP256) implement an enhanced capture timer (ECT) module that has the features of the standard timer (TIM) module enhanced by additional features in order to enlarge the field of applications. These additional features are

- One 16-bit buffer register for each of the four input-capture (IC) channels.
- Four 8-bit pulse accumulators. Each of these 8-bit pulse accumulators has an associated 8-bit buffer. Two of these 8-bit pulse accumulators can be concatenated into a single 16-bit pulse accumulator.
- A 16-bit modulus down-counter with 4-bit prescaler.
- Four user-selectable delay counters for increasing input noise immunity.

8.9.1 Enhanced Capture Timer Modes of Operation

The enhanced capture timer has eight input-capture/output-compare (IC/OC) channels, the same as on the HCS12 standard timer module. Four IC channels (IC7~IC4) are the same as the standard timer with one capture register that memorizes the timer value captured by an action on the associated input pin. Four other IC channels (IC3~IC0), in addition to the capture register, also have one buffer, called the *holding register*. This permits the register to memorize two different timer values without generating any interrupt. This feature can reduce software overhead in applications that require capturing two edges in order to perform further computation. In addition, the ECT module provides the option of preventing a captured value from being overwritten before it was read or transferred to the holding register. This option is controlled by the ICOVW register. The contents of this register are shown in Figure 8.34. This capability will be useful when external events occur at a rate that the CPU cannot read quickly enough.

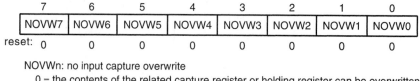

7	6	5	4	3	2	1	0
NOVW7	NOVW6	NOVW5	NOVW4	NOVW3	NOVW2	NOVW1	NOVW0

reset: 0   0   0   0   0   0   0   0

NOVWn: no input capture overwrite

    0 = the contents of the related capture register or holding register can be overwritten
when a new input capture or latch occurs.

    1 = the related capture register or holding register cannot be written by an event
unless it is empty. This will prevent the captured value to be overwritten
until it is read or latched in the holding register.

Figure 8.34 ■ Input Control Overwrite register (ICOVW)

Four 8-bit pulse accumulators are associated with the four IC buffered channels. Each pulse accumulator has a holding register to memorize its value by an action on its external input. Each pair of pulse accumulators can be used as a 16-bit pulse accumulator.

8.9.2 Why the Enhanced Capture Timer Module?

There are applications that require the capture of two consecutive edges (could be both rising and falling, or one rising and the other falling) at very high frequencies. In Example 8.2, the following instructions (faster than movw TCx,edge_ist by one E cycle) are executed before we have time to wait for the arrival of the second edge (after we detect the first edge):

```
ldd    TCx
std    . . . .
```

It takes five E clock cycles to execute these two instructions, which sets the upper limit on the signal frequency that can be dealt with. By providing the capability of setting the interrupt flag (or interrupting the CPU) after two signal edges have been captured, the upper limit of the signal frequency that can be handled can be significantly improved.

The input-capture function of the original standard timer module allows the newly captured value to overwrite the old one even if the CPU has not read the old value yet. This can cause a problem when the event frequency is very high. The enhanced input-capture function allows the user to prevent the overwriting of captured values by enabling the nonoverwrite feature.

8.9.3 The Operation of the Enhanced Input-Capture Function

The enhanced input-capture function has all the registers in the input-capture function of the standard timer module. New registers are added to implement the additional features. An input-capture register is called *empty* if its value has been read or latched into its associated holding register. A holding register is *empty* if its value has been read. An enhanced input-capture channel can be configured to operate in either *latch* mode or *queue* mode. The block diagrams of latch mode and queue mode are shown in Figures 8.35 and 8.36.

Figures 8.35 and 8.36 illustrate the registers related to the operation of the latch mode and queue mode, respectively. In both diagrams, channels IC0 to IC3 are identical and IC4 to IC7 are identical. Only one channel in each group is shown in the figures. The latch mode and queue mode are selected by setting and clearing the LATQ bit of the ICSYS register (shown in Figure 8.31).

In the *latch mode*, the latching function occurs when the modulus down-counter reaches 0 or a 0 is written into the count register MCCNT. With a latching event, the contents of the IC registers and 8-bit pulse accumulators are transferred to their holding registers. After this,

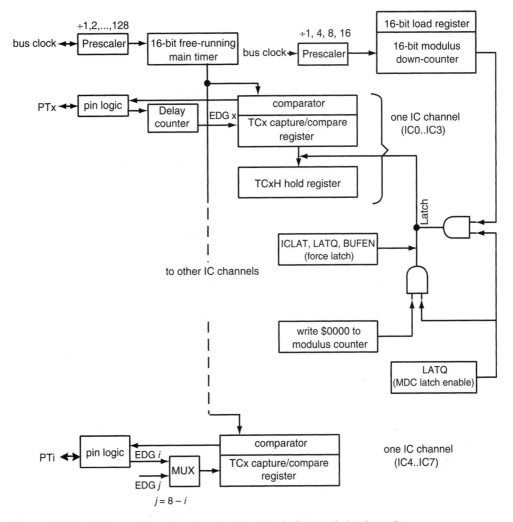

Figure 8.35 ■ Enhanced input capture function block diagram in latch mode

the 8-bit pulse accumulators are cleared. In the *queue mode,* the main timer value is copied into the IC register (TCx register) by a valid input pin transition. With a new occurrence of a capture, the value of the IC register will be transferred to its holding register and the IC register holds the new timer value. The queue mode and latch mode can be entered only when the holding registers are enabled, which can be done by setting the BUFEN bit of the ICSYS register.

The HCS12 is often used in noisy environments such as automotive applications. Due to the noise, false signal edges often occur in input-capture pins. To distinguish between true edge and false edge (usually very short), a delay counter is added to the enhanced input-capture function. Any detected edge with a duration shorter than the preprogrammed value is ignored. The duration of the delay is controlled by the DLYCT register. The contents of the DLYCT register are shown in Figure 8.37. If enabled, after detection of a valid edge on the input-capture pin, the

Figure 8.36 ■ Enhanced input capture function block diagram in queue mode (channels IC0...IC3 block diagram)

DLYx—delay counter select bits
0 0—disabled (bypassed)
0 1—256 P clock cycles
1 0—512 P clock cycles
1 1—1024 P clock cycles

Figure 8.37 ■ Delay Counter Control register (DLYCT)

delay counter counts the preselected number of E clock cycles; then it will generate a pulse to latch the TCNT value into the input-capture register (TCx). The pulse will be generated only if the level of input signal, after the preset delay, is the opposite of the level before the transition. This will avoid reaction to narrow pulses caused by noise. After counting, the counter will be cleared automatically.

Example 8.19

▼

Modify the program in Example 8.2 to take advantage of the queue mode of the enhanced input-capture function.

Solution: The modified program is as follows:

```
#include "c:\miniide\hcs12.inc"
            org     $1000
period      ds.w    1                   ; memory to store the period
            org     $1500
            movb    #$90,TSCR           ; enable timer counter and fast timer flag clear
            bclr    TIOS,IOS0           ; select input-capture 0
            movb    #$04,TSCR2          ; disable TCNT overflow interrupt, set prescale
                                        ; factor to 16
            movb    #$01,TCTL4          ; choose to capture the rising edge of PT0 pin
            movb    #$0A,ICSYS          ; enable timer flag-setting, IC buffer, and queue mode
            clr     DLYCT               ; disable delay counter
            bset    ICOVW,NOVW0         ; no input-capture overwrite for IC0
            ldd     TC0                 ; empty the input-capture register TC0
            ldd     TC0H                ; empty the holding register TC0H
            brclr   TFLG1,$FE,*         ; wait for the arrival of the second rising edge
            ldd     TC0
            subd    TC0H                ; subtract the first edge from the second edge
            std     period
            swi
            end
```

The C language version of the program is as follows:

```
#include "c:\egnu091\include\hcs12.h"
void main(void)
{
    unsigned    int         period;
    TSCR1       = 0x90;         /* enable timer counter, enable fast timer flag clear*/
    TIOS        &= ~IOS0;       /* enable input-capture 0 */
    TSCR2       = 0x04;         /* set prescale factor to 16 */
    TCTL4       = 0x01;         /* capture the rising edge of PT0 pin */
/* enable timer flag-setting mode, IC buffer, and queue mode */
    ICSYS       = 0x0A;
    DLYCT       = 0x00;         /* disable delay counter */
    ICOVW       |= NOVW0;       /* disable input-capture overwrite */
    period      = TC0;         /* empty TC0 and clear the COF flag */
    period      = TC0H;        /* empty the TC0H register */
/* wait for the arrival of the second rising edge */
    while (!(TFLG1 & COF));
    period      = TC0 - TC0H;
    asm ("swi");
}
```

▲

Example 8.20

Suppose one wants to measure the pulse width of a signal connected to the PT0 pin in a noisy environment. Write a program to perform the operation. Ignore any noise pulse shorter than 256 P clock cycles.

Solution: Since the range of the pulse width is unknown, we will need to consider the timer overflow. The program in Example 8.3 is modified to perform the measurement as follows:

```
#include        "c:\miniide\hcs12.inc"
                org     $1000
edge1           rmb     2
overflow        rmb     2
pulse_width     rmb     2
                org     $1500
                movw    #tov_isr,UserTimerOvf ; set up timer overflow interrupt vector
                ldd     #0
                std     overflow
                movb    #$90,TSCR1      ; enable TCNT and fast timer flag clear
                movb    #$04,TSCR2      ; set prescaler to TCNT to 16
                bclr    TIOS,IOS0       ; enable input-capture 0
                movb    #$01,DLYCT      ; select delay count to 256 E cycles
                movb    #$01,ICOVW      ; prohibit overwrite to TC0 register
                movb    #$0,ICSYS       ; disable queue mode
                movb    #$01,TCTL4      ; capture the rising edge on PT0 pin
                movb    #COF,TFLG1      ; clear COF flag
                brclr   TFLG1,COF,*     ; wait for the arrival of the first rising edge
                movb    #$80,TFLG2      ; clear the TOF flag
                bset    TSCR2,TOI       ; enable TCNT overflow interrupt
                cli                     ;  "
                movw    TC0,edge1       ; clear COF flag and save the captured first edge
                movb    #$02,TCTL4      ; capture the falling edge on PT0 pin
                brclr   TFLG1,COF,*     ; wait for the arrival of the falling edge
                ldd     TC0
                subd    edge1
                std     pulse_width
                bcc     next
; second edge is smaller, so decrement overflow count by 1
                ldx     overflow
                dex
                stx     overflow
next            swi
tov_isr         movb    #TOF,TFLG2      ; clear TOF flag
                ldx     overflow        ; increment TCNT overflow count
                inx                     ;  "
                stx     overflow        ;  "
                rti
                end
```

The modification to the C program in Example 8.3 is also minor and hence will be left as an exercise problem.

8.10 Pulse-Width Modulation (PWM) Function

There are many applications that require the generation of digital waveforms. The output-compare function has been used to generate digital waveforms with any duty cycle in Section 8.6. However, the generation of waveforms using the output-compare function requires frequent attention from the MCU. Most microcontrollers designed in the last few years have incorporated the Pulse-Width Modulation (PWM) function to simplify the task of waveform generation.

The MC9S12DP256 and many other HCS12 members implement an 8-channel, 8-bit PWM function. As shown in Figure 8.38, each channel has a period register, a duty cycle register, a control register, and a dedicated counter to support the waveform generation. The clock signal is critical to the setting of the frequency of the generated waveform. The clock source of the counter is programmable through a two-stage circuitry.

8.10.1 PWM Clock Select

There are four possible clock sources for the PWM function: clock A, clock B, clock SA (scaled A), and clock SB (scaled B). These four clocks are derived from the bus clock (E clock). Clocks A and B are derived by dividing the E clock by a factor of 1, 2, 4, 8, 16, 32, 64, or 128. Clock SA (SB) is derived by dividing clock A (B) by an 8-bit reloadable counter.

Each PWM channel has the option of selecting one of two clocks, either the prescaled clock (clock A or B) or the scaled clock (clock SA or SB). Figure 8.39 illustrates the block diagram of the four different clocks and how the scaled clocks are created.

PRESCALE

The input clock to the PWM prescaler is the bus (E) clock. It can be disabled whenever the MCU is in freeze mode by setting the PFRZ bit in the PWMCTL register. If this bit is set, whenever the MCU is in freeze mode the active clock to the prescaler is disabled. This is useful for emulation in order to freeze the PWM. The input clock can also be disabled when all eight PWM channels are disabled (PWME7-0 = 0). This is useful for reducing power consumption by disabling the prescale counter.

Clock A and clock B are scaled values of the input clock. The possible values for clock A and clock B is derived by dividing the E clock by 1, 2, 4, 8, 16, 32, 64, and 128. The value selected for clock A and clock B is determined by the PCKA2~PCKA0 and PCKB2~PCKB0 bits in the PWMPRCLK register, respectively. The contents of the PWMCTL and PWMPRCLK registers are shown in Figure 8.40 and Figure 8.41, respectively.

CLOCK SCALE

The SA clock takes clock A as one of its inputs and divides it further with a user-programmable value (from 1 to 256) and then divides it by 2. The SB clock is derived similarly, but with clock B as its input.

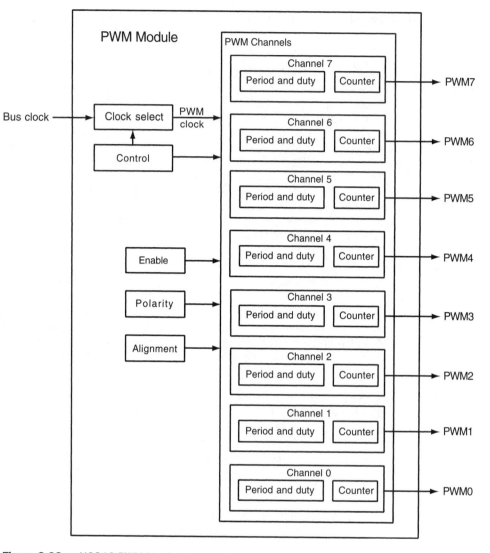

Figure 8.38 ■ HCS12 PWM block diagram

In Figure 8.39, clock A is an input to an 8-bit down-counter. This down-counter loads a user-programmable scale value from the scale register (PWMSCLA). When the down-counter is decremented to 1, two things happen: A pulse is output and the 8-bit counter is reloaded. The output signal from this circuit is further divided by 2. In other words, the clock SA is derived by the following equation:

Clock SA = clock A / (2 × PWMSCLA)

When PWMSCLA equals $00, the PWMSCLA value is considered a full-scale value of 256. Similarly,

Clock SB = clock B /(2 × PWMSCLB)

Figure 8.39 ■ PWM clock select block diagram

	7	6	5	4	3	2	1	0
	CON67	CON45	CON23	CON01	PSWAI	PFRZ	0	0
reset:	0	0	0	0	0	0	0	0

CONjk: concatenate channels j and k (j = 0, 2, 4, or 6; k = j + 1)
   0 = channel j and k are separate 8-bit PWMs.
   1 = channels j and k are concatenated to create one 16-bit PWM channel. Channel j
      becomes the high-order byte and channel k becomes the low-order byte. Channel
      k output pin is used as the output for this 16-bit PWM. Channel k clock select bit
      determines the clock source, channel k polarity bit determines the polarity,
      channel k enable bit enables the output, and channel k center-aligned enable bit
      determines the output mode.
PSWAI: PWM stops in wait mode
   0 = allow the clock to the prescaler to continue while in wait mode.
   1 = stop the input clock to the prescaler whenever the MCU is in wait mode.
PFRZ: PWM counters stop in freeze mode
   0 = allow PWM to continue while in freeze mode.
   1 = disable PWM input clock to the prescaler whenever the part is in freeze mode.

Figure 8.40 ■ PWM Control register (PWMCTL)

	7	6	5	4	3	2	1	0
	0	PCKB2	PCKB1	PCKB0	0	PCKA2	PCKA1	PCKA0
reset:	0	0	0	0	0	0	0	0

Figure 8.41 ■ PWM Prescale Clock Select register (PWMPRCLK)

PCKB2	PCKB1	PCKB0	Value of Clock B
0	0	0	E clock
0	0	1	E clock/2
0	1	0	E clock/4
0	1	1	E clock/8
1	0	0	E clock/16
1	0	1	E clock/32
1	1	0	E clock/64
1	1	1	E clock/128

Table 8.3 ■ Clock B prescaler selects

PCKA2	PCKA1	PCKA0	Value of Clock A
0	0	0	E clock
0	0	1	E clock/2
0	1	0	E clock/4
0	1	1	E clock/8
1	0	0	E clock/16
1	0	1	E clock/32
1	1	0	E clock/64
1	1	1	E clock/128

Table 8.4 ■ OClock A prescaler selects

CLOCK SELECT

Each PWM channel has a choice of two clock signals to use as the clock source for that channel. The clock source selection is done by the PWMCLK register. The contents of this register are shown in Figure 8.42.

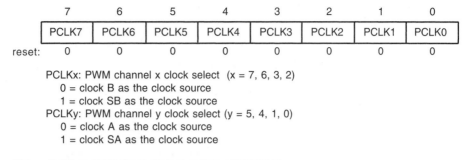

7	6	5	4	3	2	1	0
PCLK7	PCLK6	PCLK5	PCLK4	PCLK3	PCLK2	PCLK1	PCLK0

reset: 0   0   0   0   0   0   0   0

PCLKx: PWM channel x clock select (x = 7, 6, 3, 2)
    0 = clock B as the clock source
    1 = clock SB as the clock source
PCLKy: PWM channel y clock select (y = 5, 4, 1, 0)
    0 = clock A as the clock source
    1 = clock SA as the clock source

Figure 8.42 ■ PWM Clock Select register (PWMCLK)

8.10.2 PWM Channel Timers

The main part of the PWM module consists of the actual timers. Each of the PWM channels has an 8-bit counter, an 8-bit period register, and an 8-bit duty cycle register. The waveform output period is controlled by a match between the period register and the value in the counter. The duty cycle is controlled by a match between the duty cycle register and the counter value that causes the state of the output to change during the period. The starting polarity of the output is selectable on a per channel basis is selected by programming the PWMPOL register. The contents of the PWMPOL register are shown in Figure 8.43.

7	6	5	4	3	2	1	0
PPOL7	PPOL6	PPOL5	PPOL4	PPOL3	PPOL2	PPOL1	PPOL0

reset: 0   0   0   0   0   0   0   0

PPOLx: PWM channel x polarity
    0 = PWM channel x output is low at the start of a period, then goes high when the
        duty count is reached.
    1 = PWM channel x output is high at the start of a period, then goes low when the
        duty count is reached.

Figure 8.43 ■ PWM Polarity register (PWMPOL)

The block diagram of a PWM channel is shown in Figure 8.44. A PWM channel must be enabled to work. A PWM channel is enabled by setting a bit in the PWME register. The contents of the PWME register are shown in Figure 8.45. There is an edge-synchronizing circuit (labeled as GATE in Figure 8.44) to guarantee that the clock will only be enabled or disabled at an edge.

PWM Polarity

Each PWM channel has a polarity bit to allow starting a waveform cycle with a high or low level. This is shown in Figure 8.44 as a MUX select of either the Q or $\overline{Q}$ output of the PWM output flip-flop. When the PPOLx bit is 0, the Q output is selected and the PWM output will start with a low. Otherwise, the output $\overline{Q}$ is selected and the PWM output starts with a high.

Figure 8.44 ■ PWM channel block diagram

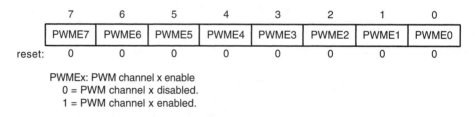

7	6	5	4	3	2	1	0
PWME7	PWME6	PWME5	PWME4	PWME3	PWME2	PWME1	PWME0

reset: 0   0   0   0   0   0   0   0

PWMEx: PWM channel x enable
0 = PWM channel x disabled.
1 = PWM channel x enabled.

Figure 8.45 ■ PWM Enable register (PWME)

PWM Period and Duty

Dedicated period and duty registers exist for each channel and are double buffered so that if they change while the channel is enabled, the change will not take effect until one of the following occurs:

- The effective period ends.
- The counter is written (counter is reset to $00).
- The channel is disabled.

In this way, the output will always be either the old waveform or the new waveform, not some variation in between. If the channel is not enabled, then writes to the period and duty registers will go directly to the latches and the buffer. A change in duty or period can be forced into effect immediately by writing the new values to the duty and/or period registers and then writing to the counter. This forces the counter to reset and the new duty and/or period values to be latched.

PWM COUNTERS

Each channel has a dedicated 8-bit up/down-counter that runs at the rate of the selected clock source. The counter is compared to the duty and period registers in each clock cycle. When the counter matches the duty register, the output flip-flop changes state, causing the PWM waveform to also change state. A match between the PWM counter and the period register behaves differently depending on what output mode is selected.

Any value written to the counter causes the counter to reset to $00 and start to count up, both the duty and period registers to be loaded with values from their buffers, and the output to change according to the polarity bit. When the channel is disabled, the counter stops.

LEFT-ALIGNED OUTPUT

The PWM timer provides the choice of two types of outputs, left-aligned or center-aligned outputs. They are selected with the CAEx bits in the PWMCAE register. The contents of the PWMCAE register are shown in Figure 8.46. If the CAEx bit is 0, the corresponding PWM output will be left-aligned.

In left-aligned output mode, the 8-bit counter is configured as an up-counter only. When the PWM counter equals the duty register, the output flip-flop changes state, which causes the PWM waveform to also change state. A match between the PWM counter and the period register resets the counter and the output flip-flop. The counter counts from 0 to the value in the period register − 1.

    CAEx: center-aligned enable bit for channel x
        0 = PWM channel x output left aligned.
        1 = PWM channel x output center aligned.

Figure 8.46 ■ PWM Center Align Enable register (PWMCAE)

The waveform of the left-aligned mode is shown in Figure 8.47. The frequency of the PWM output is given by the following equation:

PWMx frequency = clock (A, B, SA, or SB) ÷ PWMPERx

The duty cycle of the waveform depends on the selected polarity:

Polarity = 0, PWMx duty cycle = [(PWMPERx − PWMDTYx) ÷ PWMPERx] × 100%

Polarity = 1, PWMx duty cycle = [PWMDTYx ÷ PWMPERx] × 100%

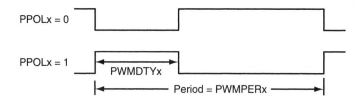

Figure 8.47 ■ PWM left-aligned output waveform

CENTER-ALIGNED MODE

In this mode, the 8-bit PWM counter operates as an up/down-counter and is set to count up whenever the counter is equal to $00. The counter compares with two registers, a duty register and a period register, in each clock cycle. When the counter matches the duty register, the output flip-flop changes state, causing the PWM waveform to also change state. A match between the PWM counter and the period register changes the counter direction from an up-count to a down-count. When the PWM counter decrements and matches the duty register again, the output flip-flop changes state, causing the PWM output to also change state. When the PWM counter decrements and reaches zero, the counter direction changes from a down-count back to an up-count and the period and duty registers are reloaded from their buffers. Since the PWM counter counts from 0 up to the value in the period register and then back down to 0, the effective period is PWMPERx × 2.

The output waveform of the center-aligned mode is shown in Figure 8.48. The frequency of the center-aligned PWM output can be calculated using the following expression:

PWMx frequency = clock(A, B, SA, or SB) ÷ (2 × PWMPERx)

The duty cycle of the waveform depends on the selected polarity:

Polarity = 0, PWMx duty cycle = [(PWMPERx − PWMDTYx) ÷ PWMPERx] × 100%

Polarity = 1, PWMx duty cycle = [PWMDTYx ÷ PWMPERx] × 100%

Figure 8.48 ■ PWM center-aligned output waveform

PWM 16-BIT FUNCTIONS

Two 8-bit PWM modules can be concatenated into one 16-bit PWM module. The concatenation of the PWM channels is controlled by the CON bits of the PWMCTL register. The 16-bit mode PWM system is illustrated in Figure 8.49.

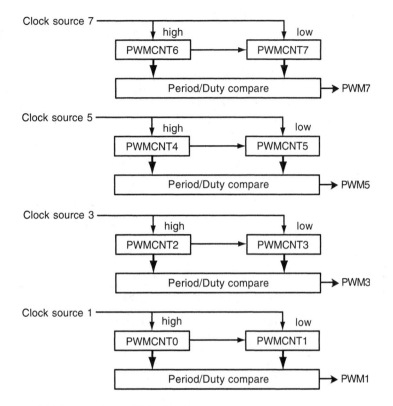

Figure 8.49 ■ PWM 16-bit mode

When PWM channel k and k + 1 are concatenated, channel k is the high-order channel, whereas the channel k + 1 is the low-order channel. When using the 16-bit concatenated mode, the clock source is determined by the low-order 8-bit channel clock select control bits. That is, it is determined by channel k + 1 (k = 0, 2, 4, or 6) when channel k and k + 1 are concatenated. The resulting PWM is output to the pin of the corresponding low-order 8-bit channel. The polarity of the resulting PWM output is controlled by the PPOLx bit of the corresponding low-order 8-bit channel as well.

Once concatenated mode is enabled, then enabling/disabling the corresponding 16-bit PWM channel is controlled by the low-order PWMEx bit. In this case, the PWMEx bits of the high-order bytes have no effect and their corresponding PWM outputs are disabled. In concatenated mode, writes to the 16-bit counter by using a 16-bit access or writes to either the low- or high-order byte of the counter will reset the 16-bit counter. Reads of the 16-bit counter must be made by a 16-bit access to maintain data coherency. Either left-aligned or center-aligned output mode can be used in concatenated mode and is controlled by the low-order CAEx bit. The high-order CAEx bit has no effect.

PWM BOUNDARY CASES

Table 8.5 summarizes the boundary conditions for the PWM regardless of the output mode and 8-bit or 16-bit.

EMERGENCY PWM SHUTDOWN

The PWM system can be shut down under emergency conditions. The emergency shutdown is controlled by the PWMSDN register. The contents of this register are shown in Figure 8.50.

PWMDTYx	PWMPERx	PPOLx	PWMx Output
$00 (indicates no duty)	>$00	1	Always low
$00 (indicates no duty)	>$00	0	Always high
xx	$00[1] (indicates no period)	1	Always high
xx	$00[1] (indicates no period)	0	Always low
>= PWMPERx	xx	1	Always high
>= PWMPERx	xx	0	Always low

[1] Counter = $00 and does not count.

Table 8.5 ■ PWM boundary cases

7	6	5	4	3	2	1	0
PWMIF	PWMIE	PWMRSTRT	PWMLVL	0	PWM7IN	PWM7INL	PWM7ENA

reset: 0 | 0 | 0 | 0 | 0 | 0 | 0 | 0

PWMIF: PWM interrupt flag
    0 = no change on PWM7IN input
    1 = change on PWM7IN input
PWMIE: PWM interrupt enable
    0 = PWM interrupt is disabled.
    1 = PWM interrupt is enabled.
PWMRSTRT: PWM restart
    The PWM can only be restarted if the PWM channel input 7 is de-asserted. After writing a logic 1 to this bit, the PWM channels start running after the corresponding counter passes the next "counter == 0" phase.
PWMLVL: PWM shutdown output level
    0 = PWM outputs are forced to 0.
    1 = PWM outputs are forced to 1.
PWM7IN: PWM channel 7 input status
    This bit reflects the current status of the PWM7 pin.
PWM7INL: PWM shutdown active input level for channel 7
    0 = active level is low.
    1 = active level is high.
PWM7ENA: PWM emergency shutdown enable
    If this bit is logic 1, the pin associated with channel 7 is forced to input and the emergency shutdown feature is enabled. All the other bits in this register are meaningful only if PWM7ENA = 1.
    0 = PWM emergency feature is disabled.
    1 = PWM emergency feature is enabled.

Figure 8.50 ■ PWM Shutdown register (PWMSDN)

Example 8.21

Write an instruction sequence to program the PWM channel 0 to output a waveform with 50% duty cycle and 100-kHz frequency. Assume that the E clock is 24 MHz.

Solution: To achieve the 50% duty cycle, 100-kHz PWM output, we will use the following parameters:

- Clock source prescale factor: set to 2.
- Select clock A as the clock input to PWM channel 0.
- Left-aligned mode.
- Write the value 120 into the PWMPER0 register (frequency is 24MHz ÷ 120 ÷ 2 = 100 kHz).
- Write the value 60 into the PWMDTY0 register.

The following instruction sequence will perform the configuration:

```
#include "c:\miniide\hcs12.inc"
    . . .
        movb    #0,PWMCLK       ; select clock A as the clock source for PWM0
        movb    #1,PWMPRCLK     ; set clock A prescaler to 2
        movb    #1,PWMPOL       ; channel 0 output high at the start of the period
        movb    #0,PWMCAE       ; select left-aligned mode
        movb    #$0C,PWMCTL     ; 8-bit mode, stop PWM in wait and freeze mode
        movb    #120,PWMPER0    ; set period value
        movb    #60,PWMDTY0     ; set duty value
        movb    #0,PWMCNT0      ; reset the PWM0 counter
        bset    PWMEN,PWME0     ; enable PWM channel 0
```

Example 8.22

Assume that the E clock frequency is 24 MHz. Write an instruction sequence to generate a square wave with a period of 20 μs and 60% duty cycle using PWM channel 0. Use center-aligned mode.

Solution: Select clock A with prescale factor 2 as the clock source to PWM0, then:

20 μs = 2/24 μs × PWMPER0 × 2 ⇒ PWMPER0 = 120

Select the PWM0 waveform to start with high level. Then the value to be loaded into the PWMDTY0 is computed as follows:

60% = (PWMDTY0 ÷ PWMPER0) × 100% ⇒ PWMDTY0 = 72

We will need to choose the following parameters for this waveform:

- Clock source prescale factor: set to 2.
- Select clock A as the clock input to PWM channel 0.
- Center-aligned mode with PPOL0 equals 1.
- Write the value 120 into the PWPER0 register.
- Write the value 72 into the PWDTY0 register.

The following instruction sequence will configure the PWM0 properly:

```
movb    #0,PWMCLK        ; select clock A as the clock source
movb    #1,PWMPOL        ; set PWM0 output to start with high level
movb    #1,PWMPRCLK      ; set the PWM0 prescaler to clock A to 2
movb    #1,PWMCAE        ; select PWM0 center-aligned mode
movb    #$0C,PWMCTL      ; select 8-bit mode, stop PWM in wait mode
movb    #120,PWMPER0     ; set period value
movb    #72,PWMDTY0      ; set duty value
bset    PWME,PWME0       ; enable PWM channel 0
```

The following C language statements will configure PWM0 properly:

```
PWMCLK   = 0;             /* select clock A as the clock source to PWM0 */
PWMPOL   = 1;             /* PWM0 output start with high level */
PWMCTL   = 0x0C;          /* select 8-bit PWM0 */
PWMCAE   = 1;             /* PWM0 center-aligned */
PWMPRCLK = 1;             /* set clock A prescaler to 2 */
PWPER0   = 120;
PWDTY0   = 72;
PWEN    |= PWME0;         /* enable PWM0 */
```

▲

Example 8.23

▼

Assume that the E clock frequency is 24 MHz. Write an instruction sequence to generate a 50-Hz digital waveform with 80% duty cycle using the 16-bit mode from the PWM1 pin output.

Solution: One of the possible settings is as follows:

- Select clock A as the clock source.
- Set the prescaler to 16.
- Select left-aligned mode.
- Select polarity of the waveform to be 1.
- The value to be loaded into the PWMPER0:PWMPER1 is computed to be 30000.
- The value to be loaded into the PWMDTY0:PWMDTY1 is computed to be 24000.

The following instruction sequence will achieve the desired setting:

```
movb    #0,PWMCLK        ; select clock A as the clock source
movb    #2,PWMPOL        ; set PWM0:PWM1 output to start with high level
movb    #4,PWMPRCLK      ; set prescaler to 16
movb    #$1C,PWMCTL      ; concatenate PWM0:PWM1, stop PWM in wait mode
movb    #0,PWMCAE        ; select left align mode
movw    #30000,PWMPER0   ; set period to 30000
movw    #24000,PWMDTY0   ; set duty to 24000
bset    PWME,PWME1       ; enable PWM0:PWM1
```

▲

The PWM function can be used in many applications that require the average value of output voltages. Some of the examples are the lamp dimmers and DC motor speed control.

Example 8.24

Using PWM in dimming the light. Suppose we are using the PWM0 of the HCS12 to control the brightness of a light bulb. The circuit connection is shown in Figure 8.51. Write a program so that the light is turned down to 10% brightness gradually in 5 s.

Figure 8.51 ■ Using PWM0 function to dim the light

Solution: We will dim the light in the following manner. Use the PWM0 output to control the brightness of the light bulb. Set the duty cycle to 100% at the beginning and then dim the brightness by 10% in the first second, and then 20% per second in the following four seconds. Use 100 as the initial duty and period values.

Since the brightness of the light is proportional to the average current that flows through the bulb, we can dim the light bulb by reducing the duty cycle of the PWM output from 100% down to 10% in five seconds. We will reduce the duty cycle in steps. In one second, we will reduce the duty value 10 times. Therefore, we will reduce the duty value by 1 every 100 ms in the first second and reduce the duty value by 2 every 100 ms in the following four seconds.

The assembly program that implements this idea is as follows:

```
#include     "c:\miniide\hcs12.inc"
             org     $1000
dim_cnt      rmb     1
             org     $1500
             movb    #0,PWMCLK        ; select clock A as the clock source
             movb    #1,PWMPOL        ; make waveform to start with high level
             movb    #$0C,PWMCTL      ; select 8-bit mode
             movb    #2,PWMPRCLK      ; set clock prescaler to 4
             movb    #0,PWMCAE        ; select left-aligned mode
             movb    #100,PWMPER0     ; set period value (PWM frequency doesn't matter)
             movb    #100,PWMDTY0     ; set duty value
             bset    PWME,PWME0       ; enable PWM channel 0
; The following instruction segment reduces duty count by 1 per 100 ms
             movb    #10,dim_cnt
```

```
loop1        ldy        #1
             jsr        delayby100ms
             dec        PWMDTY0              ; decrement duty by 1
             dec        dim_cnt
             bne        loop1
; The following instruction segment reduces duty count by 2 per 100 ms in four seconds
             movb       #40,dim_cnt
loop2        ldy        #1                   ; wait for 100 ms
             jsr        delayby100ms         ;     "
             dec        PWMDTY0              ; decrement duty cycle by 2%
             dec        PWMDTY0              ; per 100 ms
             dec        dim_cnt
             bne        loop2
             swi
#include      "c:\miniide\delay.asm"         ; include delayby100ms here
             end
```

The C language version of this program is as follows:

```
#include        "c:\egnu091\include\hcs12.h"
#include        "c:\egnu091\include\delay.c"
void main ()
{
    int    dim_cnt;
    PWMCLK   = 0;              /* select clock A as the clock source */
    PWMPOL   = 1;              /* make waveform to start with high level */
    PWMCTL   = 0x0C;           /* select 8-bit mode */
    PWMPRCLK = 2;              /* set clock A prescaler to 4 */
    PWMCAE   = 0;              /* select left-aligned mode */
    PWMPER0  = 100;            /* set period of PWM0 to 0.1 ms */
    PWMDTY0  = 100;            /* set duty cycle to 100% */
    PWME    |= 0x01;           /* enable PWM0 channel */
                              /* reduce duty cycle 1% per 100 ms in the first second */
    for (dim_cnt = 0; dim_cnt < 10 ; dim_cnt ++) {
        delayby100ms(1);
        PWMDTY0--;
    }
                              /* reduce duty cycle 2% per 100 ms in the next 4 seconds */
    for (dim_cnt = 0; dim_cnt < 40; dim_cnt ++) {
        delayby100ms(1);
        PWMDTY0 -= 2;
    }
    asm ("swi");
}
```

8.11 DC Motor Control

DC motors are used extensively in control systems as positional devices because their speeds and their torques can be precisely controlled over a wide range. The DC motor has a permanent magnetic field and its armature is a coil. When a voltage and a subsequent current flow are applied to the armature, the motor begins to spin. The voltage level applied across the armature determines the speed of rotation.

The microcontroller can digitally control the angular velocity of a DC motor by monitoring the feedback lines and driving the output lines. Almost every application that uses a DC motor requires it to reverse its direction of rotation or vary its speed. Reversing the direction is done simply by changing the polarity of the voltage applied to the motor. Changing the speed requires varying the voltage level of the input to the motor, and that means changing the input level to the motor driver. In a digitally controlled system, the analog signal to the driver must come from some form of D/A converter. However, adding a D/A converter to the circuit increases the chip count, which means increasing the system cost and power consumption. The other alternative is to vary the pulse width of a digital signal input to the motor. By varying the pulse width, the average voltage delivered to the motor changes and so does the speed of the motor. The HCS12 PWM subsystem can be used to control the DC motor.

The HCS12 can interface with a DC motor through a driver, as shown in Figure 8.52. This circuit takes up only three I/O pins. The pin that controls the direction can be an ordinary I/O pin but the pin that controls the speed must be a PWM pin. The pin that receives the feedback must be an input capture pin.

Although some DC motors can operate at 5 V or less, the HCS12 cannot supply the necessary current to drive a motor directly. The minimum current required by any practical DC motor is much higher than any microcontroller can supply. Depending on the size and rating of the motor, a suitable driver must be selected to take control signals from the HCS12 and deliver the necessary voltage and current to the motor.

Figure 8.52 ■ Simplified circuit for DC motor control

8.11.1 Drivers

Standard motor drivers are available in many current and voltage ratings. Examples are the L292 and L293 made by SGS Thompson Inc. The L293 has four channels and can output up to 1 A of current per channel with a supply of 36 V. It has a separate logic supply and takes a logic input (0 or 1) to enable or disable each channel. The L293D also includes clamping diodes needed to protect the driver from the back electromagnetic frequency (EMF) generated during the motor reversal. The pin assignment and block diagram of the L293 are shown in Figure 8.53. There are two supply voltages: V_{SS} and V_S. V_{SS} is the logic supply voltage, which can be from 4.5 to 36 V (normally 5.0 V). V_S is the analog supply voltage and can be as high as 36 V.

(a) Pin assignment   (b) Motor connection

Figure 8.53 ■ Motor driver L293 pin assignment and motor connection

8.11.2 Feedback

The DC motor controller needs information to adjust the voltage output to the motor driver circuit. The most important information is the speed of the motor, which must be fed back from the motor by a sensing device. The sensing device may be an optical encoder, infrared detector, Hall-effect sensor, and so on. Whatever the means of sensing is, the result is a signal, which is fed back to the microcontroller. The microcontroller can use the feedback to determine the speed and position of the motor. Then it can make adjustments to increase or decrease the speed, reverse the direction, or stop the motor.

Assume that a Hall-effect transistor is mounted on the armature (stator) of a DC motor and two magnets are mounted on the shaft (rotor). As shown in Figure 8.54, every time the magnet passes by the Hall-effect transistor, a pulse is generated. The input capture function of the HCS12 can capture the passing time of the pulse. The time between two captures is half of a revolution. Thus the motor speed can be calculated. By storing the value of the capture registers each time, and comparing it with its previous value, the controller can constantly measure and adjust the speed of the motor. Using this method, a motor can be run at a precise speed or synchronized with another event.

The schematic of a motor-control system is illustrated in Figure 8.55. The PWM output from the PWM3 pin is connected to one end of the motor, whereas the PP7 pin is connected to the other end of the motor. The circuit is connected so that the motor will rotate clockwise when the voltage of the PP7 pin is 0 while the PWM output is nonzero (positive). The direction of motor rotation is illustrated in Figure 8.56. By applying appropriate voltages on PP7 and PP3 (PWM3), the motor can rotate clockwise, counterclockwise, or even stop. Input capture channel 0 is used to capture the feedback from the Hall-effect transistor.

Figure 8.54 ■ The output waveform of the Hall-effect transistor

All diodes are the same and could be any one of the 1N4000 series.

Hall-effect switch

Figure 8.55 ■ Schematic of a HCS12-based motor-control system

When a motor is first turned on, it cannot reach a steady speed immediately. A certain amount of startup time should be allowed for the motor to get to speed. A smaller motor can reach steady speed faster than a larger one. It is desirable for the motor speed to be constant for many applications. However, when a load is applied to the motor, it will be slowed down. To keep the speed constant, the duty cycle of the voltage applied to the motor should be increased. When the load gets lighter, the motor will accelerate and run faster than desired. To slow down the motor, the duty cycle of the applied voltage should be reduced.

The response time will be slow if the change to the duty cycle is small. However, a large variation in duty cycle tends to cause the motor to overreact and causes oscillation. There are control algorithms for motors that you can find in textbooks on control. A DC motor cannot

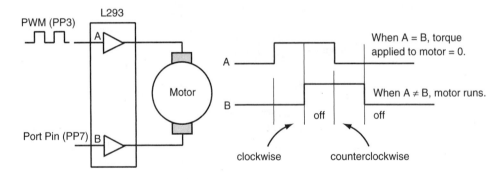

Figure 8.56 ■ The L293 motor driver

respond to the change of duty cycle instantaneously because of its inertia. A certain amount of time should be allowed for the motor to speed up or slow down before the effect of the change of duty cycle is measured.

Example 8.25

▼

Write a subroutine in C language to measure the motor speed (in rpm) assuming that the HCS12 is running with a 16-MHz E clock.

Solution: To measure the motor speed, one needs to capture two consecutive rising edges. Let the difference of two consecutive edges be ***diff*** and the period of the timer be set to 1 μs, then the motor speed (rpm) is

Speed = $60 \times 10^6 \div (2 \times$ diff)

The C function that measures the motor speed is as follows:

```
#include "c:\egnu091\include\hcs12.h"
unsigned int motor_speed (void)
{
        unsigned int edge1, diff, rpm;
        long int temp;

        TSCR1      = 0x90;              /* enable TCNT and fast flag clear */
        TIOS       &= IOS0;            /* select IC0 function */
        TSCR2      = 4;                 /* set TCNT prescale factor to 16 */
        TCTL4      = 0x01;             /* select to capture the rising edge of PT0 */
        TFLG1      = C0F;              /* cleared C0F flag */
        while (!(TFLG1 & C0F));         /* wait for the first edge */
        edge1      = TC0;
        while (!(TFLG1 & C0F));         /* wait for the second edge */
        diff       = TC0 − edge1;
        temp       = 1000000l / (long)(2 * diff);
        rpm        = temp * 60;
        return rpm;
}
```

8.11.3 Electrical Braking

Once a DC motor is running, it picks up speed. Turing off the voltage to the motor does not make it stop immediately because the momentum will keep it rotating. After the voltage is turned off, the momentum will gradually wear out because of friction. If the application does not require an abrupt stop, then the motor can be brought to a gradual stop by removing the driving voltage.

An abrupt stop may be required by certain applications in which the motor must run a few turns and stop quickly at a predetermined point. This could be achieved by electrical braking. Electrical braking is done by reversing the voltage applied to the motor. The length of time that the reversing voltage is applied must be precisely calculated to ensure a quick stop while not starting the motor in the reverse direction. A discussion of good motor braking algorithms is outside the scope of this textbook. In a closed-loop system, the feedback can be used to determine where or when to start and stop braking and when to discontinue. In Figure 8.56, the motor can be braked by (1) reducing the PWM duty count to 0 and (2) setting port pin PP7 output to high for an appropriate amount of time.

8.12 Summary

Many applications require a dedicated timer. Without a timer the following applications will become very difficult or even impossible to implement:

- The measurement of pulse width, frequency, period, duty cycle, and phase difference
- The detection of certain events
- The creation of time delays
- The generation of waveforms
- The generation of a siren and playing of songs

Some HCS12 members implement an Enhanced Capture Timer (ECT) module, which has the features of the Standard Timer Module (TIM) enhanced by additional functions. The heart of the timer system is the 16-bit main timer, TCNT. This timer must be enabled in order to run. Its clock signal is derived by dividing the E clock by a prescaler. The prescale factor can be from 1 to 128 for the standard timer module and the enhanced capture timer module.

The input-capture function can be programmed to latch the main timer value (TCNT) into the input-capture register on the arrival of an active edge and optionally generate an interrupt. The input-capture function is often used to measure the period, pulse width, duty cycle, and phase shift. It can also be used as a time reference and count the events that occur within an interval. There can be up to eight channels (IC0 to IC7) of input-capture function. The input-capture function of TIM has several limitations that make it unsuitable for higher frequency applications and a noisy environment. The ECT module adds a holding register to each of the four input-capture channels (IC0 to IC3). This enhancement adds the following capabilities:

- *Interrupt the CPU after two edges have been captured instead of each edge.* This capability reduces the software overhead.
- *Ignore the short pulse.* This capability enables the user to capture event arrival time and measure pulse width or period in a noisy environment.
- *Selective no-overwrite.* This capability allows the user to do measurements in very high frequency and will not miss the true events because the CPU is busy with other chores.

There can be up to eight output-compare channels (OC0 to OC7). Each output-compare channel has a 16-bit register, a 16-bit comparator, and an output-compare action pin. The output-compare function is often used to create a time delay, to generate a waveform, to activate an operation at a predetermined future time, and so on. To use the output-compare function, one makes a copy of the main timer, adds a delay to it, and stores the sum into an output-compare register. The 16-bit comparator compares the contents of the main timer with that of the output-compare register. When they are equal, the corresponding timer flag will be set and an optional action on the associated signal pin will be triggered: pull to high, pull to low, or toggle. An output-compare operation can be forced to take effect immediately by writing a 1 into the corresponding bit in the FORC register. This action will not set the timer flag and will not generate an interrupt either.

The output-compare channel 7 can control up to eight output-compare channels at the same time. This capability allows the user to use two output-compare channels to control the same signal pin.

The TIM module has a 16-bit pulse accumulator module (PACA). The PACA function has two operation modes: *event counting* and *gated time accumulation*. This function has been used to generate an interrupt after **n** events have occurred, measure the frequency of a signal, count events, measure the pulse width, and so on.

The ECT module adds four buffered 8-bit pulse accumulators. Each pulse accumulator has an 8-bit counter that will increment when an active edge arrives. Each pair of the pulse accumulator can be concatenated into a 16-bit pulse accumulator.

A set of delay functions are created and stored in the files *delay.asm* and *delay.c* and can be called by the end user. These functions utilize the counter mode of the Modulus Down-Counter to create the time delay.

Using the output-compare function to generate digital waveforms requires frequent attention from the CPU. This reduces the CPU time available for other applications. The pulse-width modulation (PWM) is designed to reduce the CPU load from waveform generation. The HCS12 provides eight channels of 8-bit PWM modules. The clock source, prescale factor, the polarity, the waveform alignment, the duty cycle, and the period of the waveform are all programmable. Two adjacent 8-bit PWM channels can be concatenated into one 16-bit PWM channel. The PWM function is used in motor control, light dimming, and any application that requires the control of the average voltage level.

8.13 Exercises

Assume that the E clock frequency of the HCS12 is 24 MHz for the following questions unless it is specified otherwise.

E8.1 Write a program to configure all Port T pins to be used in capturing event arrival times. Use the interrupt-driven approach. Stay in a wait loop after completing the configuration. Exit the wait loop when all eight channels have arrived. Store the arrival times in memory locations starting from $1000.

E8.2 Use input-capture channel 1 to measure the duty cycle of a signal. Write an assembly and a C program (in a subroutine format) to do the measurement.

E8.3 Assume that two signals having the same frequency are connected to the pins PT1 and PT0. Write an assembly and a C program to measure their phase difference.

E8.4 Write an assembly and a C program to generate a 2-kHz, 70% duty cycle waveform from the PT6 pin.

E8.5 Write an assembly and a C program to generate a 4-kHz, 80% duty cycle waveform from the PT5 pin.

E8.6 Write a subroutine that can generate a time delay from 1 to 100 s using the modulus down-counter. The number of seconds is passed to the subroutine in accumulator A.

E8.7 What would be the output frequency of the PT0 signal generated by the following program segment?

```
clr      PTT
movb     #$01,DDRT
bset     TIOS,$81
movb     #$01,OC7M
movb     #$01,OC7D
movb     #$90,TSCR1
movb     #$02,TCTL2
movb     #$0,TIE
movb     #$08,TSCR2
ldd      #$0
std      TC0
ldd      #$1
std      TC7
```

E8.8 Write a program to generate a 25-Hz digital waveform with a 50% duty cycle on the PT0 pin as long as the voltage level on the PT0 pin is high. Your program should consist of two parts:

1. Entry test. As long as the PP7 pin is low, it stays in this loop.

2. Waveform generation body. This part generates a pulse with 20 ms high time and 20 ms low time, and at the end of a period tests the PP7 signal. If PP7 is still high, it generates the next pulse. Otherwise, it jumps to entry test.

E8.9 Write a program to wait for an event (rising edge) to arrive at the PT0 pin. After that, the program will wait for 100 ms and trigger a pulse 20 ms wide on the PT6 pin.

E8.10 In Example 8.2, we used the polling method to check for the arrival of edges. Write a program that uses the interrupt-driven approach to measure the period of an unknown signal. There will be two interrupts related to the PT0 active edges and zero or more TCNT overflow interrupts to be dealt with.

E8.11 Write a program to generate ten pulses from the PT6 pin. Each pulse has 60 μs high time and 40 μs low time.

E8.12 Write a program to generate an interrupt to the HCS12 20 ms after the rising edge on the pin PT2 has been detected.

E8.13 Suppose that the contents of the TCTL1 and TCTL2 registers are $79 and $9B, respectively. The contents of the TFLG1 are $00. What would occur on pins PT7 to PT0 on the next clock cycle if the value $7F is written into the CFORC register?

E8.14 Write an instruction sequence to configure the modulus down-counter so that it generates periodic interrupts to the microcontroller every 40 ms.

E8.15 Modify the programs in Example 8.8 to avoid the drawback in the generated song.

E8.16 Find the score of the song "Home, Sweet Home" and modify the program in Example 8.8 to play it.

E8.17 Write an instruction sequence to generate a 160-kHz digital waveform with 80% duty cycle from the PWM2 pin output. Use left-aligned mode.

E8.18 Write an instruction sequence to generate a 120-kHz digital waveform with 40% duty cycle from the PWM2 pin output. Use center-aligned mode.

E8.19 Write an instruction sequence to generate a 10-kHz digital waveform with 60% duty cycle from the PWM2 pin output. Use left-aligned mode.

E8.20 Write an instruction sequence to generate a 5-kHz digital waveform with 70% duty cycle from the PWM2 pin output. Use center-aligned mode.

E8.21 Write an instruction sequence to generate a 20-Hz digital waveform with 50% duty cycle using the 16-bit mode from the PWM1 pin output. Use left-aligned mode.

E8.22 Write an instruction sequence to generate a 10-Hz digital waveform with 60% duty cycle using the 16-bit mode from the PWM3 pin output. Use center-aligned mode.

E8.23 What is the slowest clock signal that can be generated from the PWM output?

8.14 Lab Exercises and Assignments

L8.1 *Frequency measurement.* Use the pulse accumulator function to measure the frequency of an unknown signal. The procedure is as follows:

Step 1
Set the function generator output to square wave and adjust the output to between 0 and 5 V. Connect the signal to the PAI (PT7) pin. If your demo board (e.g., SSE256) already has function generator outputs, then you don't need this step.

Step 2
Also connect the signal to an oscilloscope or a frequency counter. This is for verification purposes.

Step 3
Output the message "Do you want to continue to measure the frequency? (y/n)".

Step 4
Set up the frequency of the signal to be measured and enter y or n to inform the microcontroller if you want to continue the measurement.

Step 5
Your program would read in the answer from the user. If the character read in is n, then stop. If the answer is y, then repeat the measurement. If the character is something else, then repeat the same question.

Step 6
Perform the measurement and display the frequency in Hz in decimal format on the screen and go back to step 3. Use as many digits as necessary. The output format should look like:

The signal frequency is xxxxxx Hz.

Crank up the frequency until the measurement becomes inaccurate. What is the highest frequency that you can measure?

L8.2 *Pulse-width measurement.* Use the input-capture function to measure the pulse width. The procedure is as follows:

Step 1
Set the function generator output to be square wave and adjust the output to between 0 and 5 V (or take the square wave from your demo board). Connect the signal to the PAI (PT7) pin.

Step 2
Connect the signal to an oscilloscope or a frequency counter to verify your measurement.

Step 3
Output the message "Do you want to continue to measure the pulse width? (y/n)".

Step 4
Set up the appropriate period (frequency) of the signal to be measured and enter y or n to inform the microcontroller if you want to continue the measurement.

Step 5
Your program would read in the answer from the user. If the character read in is n, then stop. If the answer is y, then repeat the measurement. If the character is something else, then repeat the same question.

Step 6
Perform the measurement and display the period in μs in decimal format on the screen and go back to step 3. Use as many digits as necessary. The output format should look like:

The signal period is xxxxx microseconds.

Crank up the frequency until the measurement becomes inaccurate. What is the shortest period that you can measure?

L8.3 *Playing the National Anthem.* Modify the national anthem in Example 8.8 to solve the contiguous identical note problem.

L8.4 *Waveform generation.* Use the PWM function to generate a digital waveform with a certain frequency and duty cycle. The procedure is as follows:

Step 1
Connect the selected PWM channel pin to the oscilloscope.

Step 2
Output a message to ask the user to enter the frequency of the waveform to be generated. Select the range first, and then select the step. The frequency should be as follows:

        10 to 90 Hz (nine steps) (use 16-bit PWM and take the output from PWM1)

        100 to 900 Hz (nine steps)

        1000 to 9000 Hz (nine steps)

        10000 to 90000 Hz (nine steps)

        100000 to 900000 Hz (nine steps)

If an illegal value is entered, ask the user to reenter.

Step 3
Output a message to request the user to enter the duty cycle of the waveform. The duty cycle could be from 10% to 90% (9 steps only). If an illegal value is entered, ask the user to reenter.

Step 4
Read in the value and start the waveform.

9

Serial Communication Interface (SCI)

9.1 Objectives

After completing this chapter, you should be able to

- Explain the four aspects of the EIA-232 standard
- Explain the errors that occur in data transmission
- Establish null-modem connection
- Explain the operation of the SCI subsystem
- Wire the SCI pins to the EIA-232 connector
- Program the SCI subsystem to perform data transmission and reception

9.2 Fundamental Concept of Serial Communications

The need to exchange data between the MCU and peripheral devices can be satisfied by using parallel data transfer (multiple bits in one transfer operation). However, there are a few drawbacks:

- Parallel data transfer requires many I/O pins. This requirement prevents the microcontroller from interfacing with as many devices as desired in the application.

- Many I/O devices do not have a high enough data rate to justify the use of parallel data transfer.

- Data synchronization for parallel transfer is difficult to achieve over a long distance. This requirement is one of the reasons that data communications always use serial transfer.

- Higher cost.

The *serial communication interface* (SCI) was designed to transfer data in asynchronous mode that utilizes the industrial standard EIA-232 protocol. The EIA-232 was originally called RS-232 because it was a *recommended standard*. You have been using this interface to communicate with and download programs onto the demo board for execution. Only two wires are used by the SCI function.

9.3 The RS-232 Standard

The RS-232 standard was established in 1960 by the Electronic Industry Association (EIA) for interfacing between a computer and a modem. It has experienced several revisions since then. The latest revision, EIA-232E, was published in July 1991. In this revision, EIA replaced the prefix "RS" with the prefix "EIA." This change represents no change in the standard, but was made to allow users to identify the source of the standard. We refer to the standard as both RS-232 and EIA-232 throughout this chapter. In data communication terms, both computers and terminals are called *data terminal equipment* (DTE), whereas modems, bridges, and routers are referred to as *data communication equipment* (DCE).

There are four aspects to the EIA-232 standard:

1. *Electrical specifications*—specify the voltage level, rise time and fall time of each signal, achievable data rate, and the distance of communication.

2. *Functional specifications*—specify the function of each signal.

3. *Mechanical specifications*—specify the number of pins and the shape and dimensions of the connectors.

4. *Procedural specifications*—specify the sequence of events for transmitting data, based on the functional specifications of the interface.

9.3.1 EIA-232E Electrical Specification

The following electrical specifications of the EIA-232E are of interest to us:

1. *Data rates.* The EIA-232 standard is applicable to data rates of up to 20,000 bits per second (the usual upper limit is 19,200 baud). Fixed baud rates are not set by the EIA-232E standard. However, the commonly used values are 300, 1200, 2400, 9600, and 19200 baud. Other accepted values that are not often used are 110 (mechanical teletype machines), 600, and 4800 baud.

2. *Signal state voltage assignments.* Voltages of −3 V to −25 V with respect to signal ground are considered logic 1 (the *mark* condition), whereas voltages of +3 V to +25 V are considered logic 0 (the *space* condition). The range of voltages between −3 V and +3 V is considered a transition region for which a signal state is not assigned.

3. *Signal transfer distance.* The signal should be able to transfer correctly within 15 m. Greater distance can be achieved with good design.

9.3.2 EIA-232E Functional Specification

The EIA-232E standard specifies 22 signals. A summary of these signals is given in Table 9.1. These signals can be divided into six categories:

1. *Signal ground and shield.*

2. *Primary communications channel.* This is used for data interchange, and includes flow control signals.

Pin No.	Circuit	Description
1	—	Shield
2	BA	Transmitted data
3	BB	Received data
4	CA/CJ	Request to send/ready for receiving[1]
5	CB	Clear to send
6	CC	DCE ready
7	AB	Signal common
8	CF	Received line signal detector
9	—	(reserved for testing)
10	—	(reserved for testing)
11	—	unassigned[3]
12	SCF/CI	Secondary received line signal detection/data rate selector (DCE source)[2]
13	SCB	Secondary clear to send
14	SBA	Secondary transmitted data
15	DB	Transmitter signal element timing (DCE source)
16	SBB	Secondary received data
17	DD	Receiver signal element timing
18	LL	Local loopback
19	SCA	Secondary request to send
20	CD	DTE ready
21	RL/CG	Remote loopback/signal quality detector
22	CE	Ring indicator
23	CH/CI	Data signal rate selector (DTE/DCE source)[2]
24	DA	Transmitter signal element timing (DTE source)
25	TM	Test mode

1. When hardware flow control is required, circuit CA may take on the functionality of circuit CJ. This is one change from the former EIA-232.
2. For designs using interchange circuit SCF, interchange circuits CH and CI are assigned to pin 23. If SCF is not used, CI is assigned to pin 12.
3. Pin 11 is unassigned. It will not be assigned in future versions of EIA-232. However, in international standard ISO 2110, this pin is assigned to select transmit frequency.

Table 9.1 ■ Functions of EIA-232E signals

3. *Secondary communications channel*. When implemented, this is used for control of the remote modem, requests for retransmission when errors occur, and governance over the setup of the primary channel.

4. *Modem status and control signals*. These signals indicate modem status and provide intermediate checkpoints as the telephone voice channel is established.

5. *Transmitter and receiver timing signals*. If a synchronous protocol is used, these signals provide timing information for the transmitter and receiver, which may operate at different baud rates.

6. *Channel test signals*. Before data is exchanged, the channel may be tested for its integrity and the baud rate automatically adjusted to the maximum rate that the channel could support.

SIGNAL GROUND

Pin 7 and 1 and the shell are included in this category. Cables provide separate paths for each, but internal wiring often connects pin 1 and the cable shell/shield to the signal ground on pin 7. All signals are referenced to a common ground as defined by the voltage on pin 7. This conductor may or may not be connected to protective ground inside the DCE device.

PRIMARY COMMUNICATION CHANNEL

Pin 2 carries the *transmit data* (TxD) signal, which is active when data is transmitted from the DTE device to DCE device. When no data is transmitted, the signal is held in the mark condition (logic 1, negative voltage).

Pin 3 carries the *received data* (RxD), which is active when the DTE device receives data from the DCE device. When no data is received, the signal is held in the mark condition.

Pin 4 carries the *request to send* (RTS) signal, which will be asserted (logic 0, positive voltage) to prepare the DCE device for accepting transmitted data from the DTE device. Such preparation might include enabling the receive circuits or setting up the channel direction in half-duplex applications. When the DCE is ready, it acknowledges by asserting the CTS signal.

Pin 5 carries the *clear to send* (CTS) signal, which will be asserted (logic 0) by the DCE device to inform the DTE device that transmission may begin. RTS and CTS are commonly used as handshaking signals to moderate the flow of data into the DCE device.

SECONDARY COMMUNICATION CHANNEL

Pin 14 is the *secondary transmitted data* (STxD). Pin 16 is the *secondary received data* (SRxD). Pin 19 carries the *secondary request to send* (SRTS) signal. Pin 13 carries the *secondary clear to send* (SCTS) signal. These signals are equivalent to the corresponding signals in the primary communications channel. The baud rate, however, is typically much slower in the secondary channel, for increased reliability.

MODEM STATUS AND CONTROL SIGNALS

This group includes the following signals:

Pin 6—DCE ready (DSR). When originating from a modem, this signal is asserted (logic 0) when all the following three conditions are satisfied:
1. The modem is connected to an active telephone line that is *off-hook*.
2. The modem is in data mode, not voice or dialing mode.
3. The modem has completed dialing or call setup functions and is generating an answer tone.

If the line goes off-hook, a fault condition is detected, or a voice connection is established, the DCE ready signal is de-asserted (logic 1).

Pin 20—DTE ready (DTR). This signal is asserted (logic 0) by the DTE device when it wishes to open a communications channel. If the DCE device is a modem, the assertion of DTR prepares the modem to be connected to the telephone circuit and, once connected, maintains the connection. When DTR is de-asserted, the modem is switched to *on-hook* to terminate the connection (same as placing the phone back on the telephone socket).

Pin 8—Received line signal detector (CD). Also called *carrier detect*, this signal is relevant when the DCE device is a modem. It is asserted (logic 0) by the modem when the telephone line is off-hook, a connection has been established, and an answer tone is being received from the remote modem. The signal is de-asserted when no answer tone is being received or when the answer tone is of inadequate quality to meet the local modem's requirements.

Pin 12—Secondary received line signal detector (SCD). This signal is equivalent to the CD (pin 8) signal but refers to the secondary channel.

Pin 22—Ring indicator (RI). This signal is relevant when the DCE device is a modem and is asserted (logic 0) when a ringing signal is being received from the telephone line. The assertion time of this signal will approximately equal the duration of the ring signal, and it will be de-asserted between rings or when no ringing is present.

Pin 23—Data signal rate selector. This signal may originate in either the DTE or DCE devices (but not both) and is used to select one of two prearranged baud rates. The assertion condition (logic 0) selects the higher baud rate.

TRANSMITTER AND RECEIVER TIMING SIGNALS

This group consists of the following signals:

Pin 15—Transmitter signal element timing (TC). Also called *transmitter clock*, this signal is relevant only when the DCE device is a modem and is operating with a synchronous protocol. The modem generates this clock signal to control exactly the rate at which data is sent on TxD (pin 2) from the DTE device to the DCE device. The logic 1 to logic 0 (negative to positive transition) transition on this line causes a corresponding transition to the next data element on the TxD line. The modem generates this signal continuously, except when it is performing internal diagnostic functions.

Pin 17—Receiver signal element timing (RC). Also called *receiver clock*, this signal is similar to TC, except that it provides timing information for the DTE receiver.

Pin 24—Transmitter signal element timing (ETC). Also called *external transmitter clock*, with timing signals provided by the DTE device for use by a modem, this signal is used only when TC and RC (pins 15 and 17) are not in use. The logic 1 to logic 0 transition indicates the time center of the data element. Timing signals will be provided whenever the DTE is turned on regardless of other signal conditions.

CHANNEL TEST SIGNALS

This group consists of the following signals:

Pin 18—Local loopback (LL). This signal is generated by the DTE device and is used to place the modem into a test state. When LL is asserted (logic 0, positive voltage), the modem redirects its modulated output signal, which is normally fed into the telephone line, back into its receive circuitry. This enables data generated by the DTE to be echoed back through the local modem to check the condition of the modem circuitry. The modem asserts its test mode signal on pin 25 to acknowledge that it has been placed in LL condition.

Pin 21—Remote loopback (RL). This signal is generated by the DTE device and is used to place the remote modem into a test state. When RL is asserted (logic 0), the remote modem redirects its received data back to its transmitted data input, thereby remodulating the received data and returning it to its source. When the DTE initiates such a test, transmitted data is passed through the local modem, the telephone line, the remote modem, and back, to exercise the channel and confirm its integrity. The remote modem signals the local modem to assert test mode on pin 25 when the remote loopback test is underway.

Pin 25—Test mode (TM). This signal is relevant only when the DCE device is a modem. When asserted (logic 0), it indicates that the modem is in a LL or RL condition. Other internal self-test conditions may also cause the TM signal to be asserted, depending on the modem and the network to which it is attached.

9.3.3 EIA-232E Mechanical Specification

The EIA-232E uses a 25-pin D-type connector, as shown in Figure 9.1a. Since only a small subset of the 25 signals is actually used, a 9-pin connector (DB9) is used in most PCs. The signal assignment of DB9 is shown in Figure 9.1b. The DB9 is not part of the EIA-232E standard.

Signal Direction	Signal Name		Signal Name	Signal Direction
to DCE	Secondary transmitted data	14 — 1	Protective ground	Both
to DTE	Transmit clock	15 — 2	Transmitted data	to DCE
to DTE	Secondary received data	16 — 3	Received data	to DTE
to DTE	Receiver clock	17 — 4	Request to send	to DCE
	Unassigned	18 — 5	Clear to send	to DTE
to DCE	Secondary request to send	19 — 6	Data set ready	to DTE
to DCE	Data terminal ready	20 — 7	Signal ground	Both
to DTE	Signal quality detect	21 — 8	Carrier detect	to DTE
to DTE	Ring indicator	22 — 9	Reserved	
Both	Data rate select	23 — 10	Reserved	
to DCE	Transmit clock	24 — 11	Unassigned	
	Unassigned	25 — 12	Secondary carrier detect	to DTE
		13	Secondary clear to send	to DTE

Figure 9.1a ■ EIA-232E DB25 connector and pin assignment

Figure 9.1b ■ EIA-232E DB9 connector and signal assignment

9.3.4 EIA-232E Procedural Specification

The sequence of events that occurs during data transmission using the EIA-232E is easier to understand by studying examples. Two examples are used to explain the procedure.

In the first example, two DTEs are connected with a point-to-point link using a modem. The modem requires only the following circuits to operate:

- Signal ground (AB)
- Transmitted data (BA)
- Received data (BB)
- Request to send (CA)
- Clear to send (CB)
- Data set ready (CC)
- Carrier detect (CF)

Before the DTE can transmit data, the DSR circuit must be asserted to indicate that the modem is ready to operate. This signal should be asserted before the DTE attempts to make a request to send data. The DSR pin can simply be connected to the power supply of the DCE to indicate that it is switched on and ready to operate. When a DTE is ready to send data, it asserts the RTS signal. The modem responds, when ready, with the CTS signal asserted, indicating that data may be transmitted over circuit BA. If the arrangement is half-duplex, then the assertion of the RTS signal also inhibits the receive mode. The DTE sends data to the local modem bit serially. The local modem modulates the data into the carrier signal and transmits the resultant signal over the dedicated communication lines. Before sending out modulated data, the local modem sends out a carrier signal to the remote modem so that it is ready to receive the data. The remote modem detects the carrier and asserts the CD signal. The assertion of the CD signal tells the remote DTE that the local modem is transmitting. The remote modem receives the modulated signal, demodulates it to recover the data, and sends it to the remote DTE over the RxD pin. The circuit connections are illustrated in Figure 9.2.

Figure 9.2 ■ Point-to-point asynchronous connection

The next example involves two computers exchanging data through a public telephone line. One of the computers (initiator) must dial the phone (automatically or manually) to establish the connection, just like people talking over the phone. Two additional leads are required for this application:

- Data terminal ready (DTR)
- Ring indicator (RI)

The data transmission in this setting can be divided into three phases:

Phase 1

Establishing the connection. The following events occur in this phase:

1. The transmitting computer asserts the DTR signal to indicate to the local modem that it is ready to make a call.
2. The local modem opens the phone line and dials the destination number. The number can be stored in the modem or transmitted to the modem by the computer via the TxD pin.
3. The remote modem detects a ring on the phone line and asserts the RI signal to inform the remote computer that a call has arrived.
4. The remote computer asserts the DTR signal to accept the call.
5. The remote modem answers the call by sending a carrier signal to the local modem via the phone line. It also asserts the DSR signal to inform the remote computer that it is ready for data transmission.
6. The local modem asserts both the DSR and CD signals to indicate that the connection is established and it is ready for data communication.
7. For full-duplex data communication, the local modem also sends a carrier signal to the remote modem. The remote modem then asserts the CD signal.

Phase 2

Data transmission. The following events occur during this phase:

1. The local computer asserts the RTS signal when it is ready to send data.
2. The local modem responds by asserting the CTS signal.
3. The local computer sends data bits serially to the local modem over the TxD pin. The local modem then modulates its carrier signal to transmit the data to the remote modem.
4. The remote modem receives the modulated signal from the local modem, demodulates it to recover the data, and sends it to the remote computer over the RxD pin.

Phase 3

Disconnection. Disconnection requires only two steps:

1. When the local computer has finished the data transmission, it drops the RTS signal.
2. The local modem then de-asserts the CTS signal and drops the carrier (equivalent to hanging up the phone).

The circuit connection for this example is shown in Figure 9.3. A timing signal is not required in an asynchronous transmission.

Figure 9.3 ■ Asynchronous connection over public phone line

9.3.5 Data Format

In asynchronous data transfer, data is transferred character by character. Each character is preceded by a start bit (a low), followed by eight or nine data bits, and terminated by a stop bit. The data format of a character is shown in Figure 9.4.

Figure 9.4 ■ The format of a character

As shown in Figure 9.4, the least significant bit is transmitted first, and the most significant bit is transmitted last. The stop bit is high. The start bit and stop bit identify the start and end of a character.

Since there is no clock information in the asynchronous format, the receiver uses a clock signal with a frequency that is a multiple (usually 16) of the data rate to sample the incoming data in order to detect the arrival of the start bit and determine the logical value of each data bit. A clock, with a frequency that is 16 times the data rate, can tolerate a frequency difference in the clocks slightly over 3% at the transmitter and receiver.

The method for detecting the arrival of a start bit is similar among all microcontrollers. When the RxD pin is *idle* (*high*) for at least three sampling times and then is followed by a low voltage, the SCI will look at the third, fifth, and seventh samples after the first low sample (these are called verification samples) to determine if a valid start bit has arrived. This process is illustrated in Figure 9.5. If the majority of these three samples are low, then a valid start bit is detected. Otherwise, the SCI will restart the process. After detecting a valid start bit, the SCI will start to shift in the data bits.

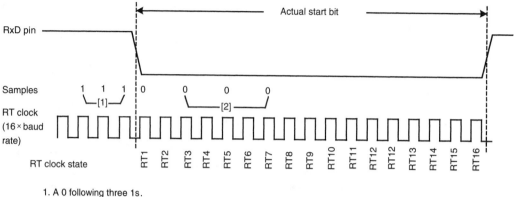

1. A 0 following three 1s.
2. Majority of samples 3, 5, and 7 are 0s.

Figure 9.5 ■ Detection of a start bit (ideal case)

The method for determining the data bit value is also similar for most microcontrollers.

The SCI uses a clock with a frequency about 16 times (most often) that of the data rate to sample the RxD signal. If the majority of the eighth, ninth, and tenth samples are 1s, then the data bit is determined to be 1. Otherwise, the data bit is determined to be 0.

The stop bit is high. Using this format, it is possible to transfer data character by character without any gap.

The term *baud rate* is defined as the number of bit changes per second. Since the RS-232 standard uses a *non-return-to-zero* (NRZ) encoding method, baud rate is identical to bit rate. In a NRZ code, a logic 1 bit is sent as a high value and a logic 0 bit is sent as a low value. When a logic 1 follows another logic 1 during a data transfer, the voltage does not drop to zero before it goes high.

Example 9.1

Sketch the output of the letter g when it is transmitted using the format of one start bit, eight data bits, and one stop bit.

Solution: Letters are represented in ASCII code. The ASCII code of letter g is $67 (= 01100111). Since the least significant bit goes out first in the EIA-232 protocol, the format of the output of letter g is as shown in Figure 9.6.

(a) Output waveform on microcontroller interface

(b) Output waveform on EIA-232E interface

Figure 9.6 ■ Data format for letter g

9.3.6 Data Transmission Errors

The following errors may occur during the data transfer process using asynchronous serial transmission:

- *Framing error.* A framing error occurs when a received character is improperly framed by the start and stop bits; it is detected by the absence of the stop bit. This error indicates a synchronization problem, faulty transmission, or a break condition. A *break* is defined as the transmission or reception of a logic 0 for one frame or more.

- *Receiver overrun.* One or more characters in the data stream were received but were not read from the buffer before subsequent characters were received.

- *Parity error.* A parity error occurs when an odd number of bits change value. It can be detected by a parity error detecting circuit.

9.3.7 Null Modem Connection

When two DTE devices are located side by side and use the EIA-232E interface to exchange data, there is really no reason to use two modems to connect them. However, the EIA-232E standard does not allow the direct connection of two DTEs. In order to make this scheme work, a *null modem* is needed. The null modem interconnects leads in such a way as to fool both DTEs into thinking that they are connected to modems. The null modem connection is shown in Table 9.2.

In Table 9.2, the signals of DTE1 and DTE2 that are to be wired together are listed in the same row. The transmitter timing and receiver timing signals are not needed in asynchronous data transmission. Ring indicator is not needed either because the transmission is not through a public phone line.

Signal Name	DTE 1		DTE 2		Signal Name
	DB25 pin	DB9 pin	DB9 pin	DB25 pin	
FG (frame ground)	1	—	—	1	FG
TD (transmit data)	2	3	2	3	RD
RD (receive data)	3	2	3	2	TD
RTS (request to send)	4	7	8	5	CTS
CTS (clear to send)	5	8	7	4	RTS
SG (signal ground)	7	5	5	7	SG
DSR (data set ready)	6	6	4	20	DTR
CD (carrier detect)	8	1	4	20	DTR
DTR (data terminal ready)	20	4	1	8	CD
DTR (data terminal ready)	20	4	6	6	DSR

Table 9.2 ■ Null modem connection

9.4 The HCS12 Serial Communication Interface

An HCS12 device may have one or two serial communication interfaces (SCI). These two SCI modules are referred to as SCI0 and SCI1, respectively. The SCI0 module shares the use of Port S pins PS1 (TxD0) and PS0 (RxD0); the SCI1 shares the use of Port S pins PS3 (TxD1) and PS2 (RxD1). The block diagram of the HCS12 SCI module is shown in Figure 9.7.

The operation of an SCI module involves the following registers:

- Two baud rate registers: SCIxBDH and SCIxBDL (x = 0 or 1)
- Two control registers: SCIxCR1 and SCIxCR2 (x = 0 or 1)
- Two status registers: SCIxSR1 and SCIxSR2 (x = 0 or 1)
- Two data registers: SCIxDRH and SCIxDRL (x = 0 or 1)

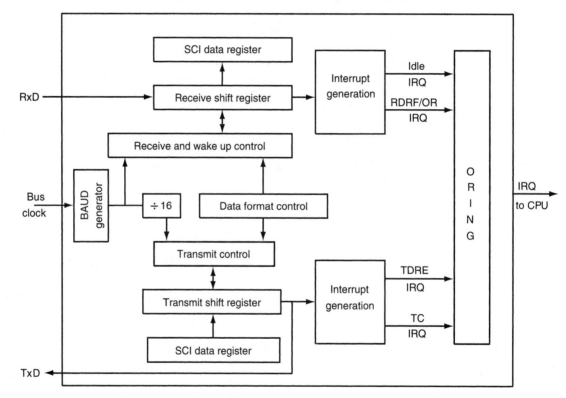

Figure 9.7 ■ HCS12 SCI block diagram

The HCS12 SCI interface uses a data format of one start bit, eight or nine data bits, and one stop bit. When the SCI is configured to use nine data bits, one of the bits can be used as the parity bit. The collection of the start bit, data bits, and the stop bit is called a *frame*. The SCI function has the capability to send break to attract the attention of the other party of the data communication. The SCI function supports hardware parity for transmission and reception. When enabled, parity bit is generated in hardware for transmitted data and received data. Received parity errors are flagged in hardware. The SCI module supports two wakeup methods, including *idle line* wakeup and *address mark* wakeup, which allows the HCS12 to operate in a *multiple-node* environment.

9.5 SCI Baud Rate Generation

The HCS12 SCI modules use a clock signal that is 16 times the data rate to detect the arrival of the start bit and determine the logic value of data bits. The HCS12 SCI modules use a 13-bit counter to generate this clock signal. This circuit is called *baud rate generator*. To set the baud rate to a certain value, one needs to write an appropriate value to the SCIxBDH:SCIxBDL register pair. The upper three bits of the SCIxBDH register are tied to zeros. The contents of these two registers are shown in Figure 9.8.

7	6	5	4	3	2	1	0
0	0	0	SBR12	SBR11	SBR10	SBR9	SBR8

reset: 0   0   0   0   0   0   0   0

(a) SCI baud rate control register high (SC0BDH/SC1BDH)

7	6	5	4	3	2	1	0
SBR7	SBR6	SBR5	SBR4	SBR3	SBR2	SBR1	SBR0

reset: 0   0   0   0   0   1   0   0

(b) SCI baud rate control register low (SC0BDL/SC1BDL)

Figure 9.8 ■ SCI baud rate control register

This baud rate generator divides down the bus clock to derive the clock signal for reception and transmission. The value (referred to as SBR) to be written into the SCIxBDH:SCIxBDL register pair can be derived by rounding the following expression to an integer value:

SBR = $f_E \div 16 \div$ baud rate

The divide factors for the baud rate generator for the 16-MHz and 24-MHz bus clocks are listed in Table 9.3.

Desired SCI Baud Rate	Baud Rate Divisor for f_E = 16 MHz	Baud Rate Divisor for f_E = 24 MHz
300	3333	5000
600	1667	2500
1200	833	1250
2400	417	625
4800	208	313
9600	104	156
14,400	69	104
19,200	52	78
38,400	26	39

Table 9.3 ■ Baud rate generation

9.6 The SCI Operation

The operation of the SCI module is controlled by two control registers: SCI0CR1 (SCI1CR1) and SCI0CR2 (SCI1CR2). Their contents are shown in Figures 9.9 and 9.10, respectively. The SCI module allows full duplex, asynchronous, non-return-to-zero (NRZ) serial communication between the CPU and remote devices, including other CPUs. The SCI transmitter and receiver operate independently, although they use the same baud rate generator. The CPU monitors the status of the SCI, writes the data to be transmitted, and processes received data.

7	6	5	4	3	2	1	0
LOOPS	SCISWAI	RSRC	M	WAKE	ILT	PE	T

reset: 0   0   0   0   0   0   0   0

LOOPS: loop select bit
    0 = loop operation disabled.
    1 = loop operation enabled.
SCISWAI: SCI stop in wait mode
    0 = SCI enabled in wait mode.
    1 = SCI disabled in wait mode.
RSRC: receiver source bit
    When LOOPS = 1, the RSRC bit determines the source for the receiver shift register
    0 = receiver input connected to the transmitter internally (not TxD pin).
    1 = receiver input connected externally to the transmitter (TxD pin).
M: data format mode bit
    0 = one start bit, eight data bits, one stop bit.
    1 = one start bit, nine data bits, one stop bit.
WAKE: wakeup condition bit
    0 = idle line wakeup.
    1 = address mark wakeup (last data bit set).
ILT: idle line type bit
    0 = idle character bit count begins after start bit.
    1 = idle character bit count begins after stop bit.
PE: parity enable bit
    0 = parity disabled.
    1 = parity enabled.
PT: parity type bit (for both transmit and receive)
    0 = even parity selected.
    1 = odd parity selected.

Figure 9.9 ■ SCI control register 1 (SCOCR1/SC1CR1)

	7	6	5	4	3	2	1	0
value	TIE	TCIE	RIE	ILIE	TE	RE	RWU	SBK
after reset	0	0	0	0	0	0	0	0

TIE: transmit interrupt enable bit

  0 = TDRE interrupt disabled.

  1 = TDRE interrupt enabled.

TCIE: transmit complete interrupt enable bit

  0 = TC interrupt disabled.

  1 = TC interrupt enabled.

RIE: receiver full interrupt enable bit

  0 = RDRF and OR interrupt disabled.

  1 = RDRF and OR interrupt enabled.

ILIE: idle line interrupt enable bit

  0 = IDLE interrupt disabled.

  1 = IDLE interrupt enabled.

TE: transmitter enable bit

  0 = transmitter disabled.

  1 = transmitter enabled.

RE: receiver enable

  0 = receiver disabled.

  1 = receiver enabled.

RWU: receiver wakeup bit

  0 = normal SCI receiver.

  1 = enables the wakeup function and inhibits further receiver

     interrupts. Normally, hardware wakes up the receiver by

     automatically clearing this bit.

SBK: send break bit

  0 = no break characters.

  1 = generate a break code, at least 10 or 11 contiguous 0s. As long

     as SBK remains set, the transmitter sends 0s.

Figure 9.10 ■ SCI control register 2 (SC0CR2/SC1CR2)

9.6.1 Character Transmission

The block diagram of the SCI transmitter is shown in Figure 9.11. To transmit data, the MCU writes the data bits to the SCI data registers (SCIxDRH/SCIxDRL), which in turn are transferred to the transmit shift register. The transmit shift register then shifts a frame out through the TxD pin, after it has prefaced the bits with a start bit and appended a stop bit. The SCI data registers are the write-only buffers between the internal data bus and the transmit shift register.

The SCI also sets a flag, the transmit data register empty flag (TDRE), every time it transfers data from the buffer (SCIxDRH/L) to the transmit shift register. The transmit driver routine may respond to this flag by writing another byte to the transmit buffer, while the shift register is still shifting out the first byte.

Figure 9.11 ■ SCI transmitter block diagram

There are three major steps in the SCI transmission:

1. *Configure the SCI transmission*:
   a. Select a baud rate. The user must write an appropriate value to the SCI baud registers to set the baud rate. Writing to the SCIxBDH register has no effect without also writing to the SCIxBDL register.
   b. Write to the SCIxCR1 register to configure the word length, parity, and other configuration bits (LOOPS, RSRC, M, WAKE, ILT, PR, PT).
   c. Enable the transmitter, interrupt, receive, and wakeup as required by writing to the SCIxCR2 register bits (TIE, TCIE, RIE, ILIE, TE, RE, RWU, and SK). A preamble will now be shifted out of the transmitter shift register.

2. *Transmit procedure for each character*:
   a. Poll the TDRE flag by reading the SCIxSR1 register or responding to the TDRE interrupt.
   b. If the TDRE flag is set, write the data to be transmitted to SCIxDRH/L, where the ninth bit is written to the T8 bit in the SCIxDRH register if the SCI is in 9-bit data format. A new transmission will not result until the TDRE flag has been cleared.

3. *Repeat step 2 for each subsequent transmission*

The contents of the SCIxSR1 and SCIxSR2 registers are shown in Figure 9.12 and 9.13, respectively.

Setting the TE bit from 0 to 1 automatically loads the transmit shift register with a preamble of 10 logic 1s (if $M = 0$) or 11 logic 1s (if $M = 1$). After the preamble shifts out, control logic transfers the data from the SCI data register into the transmit shift register. A logic 0 start bit automatically goes into the most significant bit position. Hardware supports odd or even parity. When parity is enabled, the most significant bit of the data character is the parity bit. The transmit data register empty flag, TDRE, in the SCIxSR1 register becomes set when the SCI data register transfers a byte to the transmit shift register. The TDRE flag indicates that the SCI data register can accept new data from the internal data bus. If the transmit interrupt enable bit, TIE, in the SCIxCR2 register is also set, the TDRE flag generates a transmit interrupt request.

When the transmit shift register is not transmitting a frame, the TxD output signal goes to the idle state, logic 1. If at any time software clears the TE bit in the SCIxCR2 register, the TxD signal goes idle. If software clears the TE bit while a transmission is in progress, the frame in

	7	6	5	4	3	2	1	0
	TDRE	TC	RDRF	IDLE	OR	NF	FE	PF
reset:	1	1	0	0	0	0	0	0

TDRE: transmit data register empty flag

    0 = no byte was transferred to the transmit shift register

    1 = transmit data register is empty

TC: transmit complete flag

    0 = transmission in progress

    1 = no transmission in progress

RDRF: receiver data register full flag

    0 = SCIxDR empty

    1 = SCIxDR full

IDLE: idle line detected flag

    0 = RxD line active

    1 = RxD line becomes idle

OR: overrun error flag

    0 = no overrun

    1 = overrun detected

NF: noise error flag

    Set during the same cycle as the RDRF bit but not set in the case

    of an overrun (OR).

    0 = no noise

    1 = noise

FE: framing error flag

    Set when a 0 is detected where a stop bit was expected

    0 = no framing error

    1 = framing error

PF: parity error flag

    0 = parity correct

    1 = incorrect parity detected

Figure 9.12 ■ SCI status register 1 (SCI0SR1/SCI1SR1)

the transmit shift register continues to shift out. To avoid accidentally cutting off the last frame in a message, always wait for TDRE to go high after the last frame before clearing the TE bit.

Use the following procedure if it is desirable to separate messages with preambles with minimum idle line time:

1. Write the last byte of the first message to the transmit data register.
2. Wait for the TDRE flag to go high, indicating the transfer of the last frame to the transmit shift register.
3. Queue a preamble by clearing and then setting the TE bit.
4. Write the first byte of the second message to the transmit data register.

The TDRE bit in the SCIxSR1 register is cleared by reading the SCIxSR1 register and followed by writing a byte into the SCIxDRL register. All status flags related to reception are cleared by reading the SCI status register followed by reading the SCIxDRL register.

7	6	5	4	3	2	1	0
0	0	0	0	0	BK13	TXDIR	RAF

reset:

| 0 | 0 | 0 | 0 | 0 | 0 | 0 | 0 |

BK13: break transmit character length
  0 = break character is 10- or 11-bit long.
  1 = break character is 13- or 14-bit long.
TXDIR: transmit pin data direction in single-wire mode
  0 = TxD pin to be used as an input in single-wire mode.
  1 = TxD pin to be used as an output in single-wire mode.
RAF: receiver active flag
  RAF is set when the receiver detects a logic 0 during the RT1 time period of the start bit search. RAF is cleared when the receiver detects an idle character.
  0 = no reception in progress.
  1 = reception in progress.

Figure 9.13 ■ SCI status register 2 (SCI0SR2/SCI1SR2)

9.6.2 Break Characters

Whenever one party in the data communications discovers an error, it can send break characters to discontinue the communication and start over again. To send a break character, the user sets the SBK bit in the SCIxCR1 register to 1. As long as the SBK bit is 1, transmitter logic continuously loads break characters into the transmit shift register. After software clears the SBK bit, the shift register finishes transmitting the last break character and then transmits at least one logic 1. The automatic logic 1 at the end of a break character guarantees the recognition of the start bit of the next frame.

The SCI module recognizes a break character when a start bit is followed by eight or nine logic 0 data bits and a logic 0 where the stop bit should be. Receiving a break character has these effects on SCI registers:

- Sets the framing error flag FE
- Sets the receive data register full flag RDRF

- Clears the SCI data registers (SCIxDRH/L)
- May set the overrun flag OR, noise flag NF, parity error flag PE, or receiver active flag RAF

9.6.3 Idle Characters

An idle character contains all 1s and has no start, stop, or parity bit. The length of the idle character depends on the M bit in the SCIxCR1 register. The preamble is a synchronizing idle character that begins the first transmission initiated after setting the TE bit from 0 to 1.

If the TE bit is cleared during a transmission, the TxD signal becomes idle after the completion of the transmission in progress. Clearing and then setting the TE bit during a transmission queues an idle character to be sent after the frame currently being transmitted.

9.6.4 Character Reception

The block diagram of the SCI receiver is shown in Figure 9.14. The SCI receiver can accommodate either 8-bit or 9-bit data characters. The state of the M bit in the SCI control register 1 determines the length of data characters. When receiving 9-bit data, the R8 bit of the SCIxDRH register holds the ninth bit.

Figure 9.14 ■ SCI receiver block diagram

During an SCI reception, the receive shift register shifts in a frame from the RxD pin. The SCI data register is the read-only buffer between the internal bus and the receive shift register. After a complete frame is shifted into the receive shift register, the data portion of the frame is transferred to the SCI data register. The receive data register full flag, RDRF, in the SCIxSR1 register becomes set, indicating that the receive byte can be read. If the receive interrupt enable bit RIE in the SCIxCR2 register is also set, then an interrupt is requested to the MCU.

The receiver uses the method illustrated in Figure 9.5 to detect the arrival of the start bit and uses the majority function of the samples RT8, RT9, and RT10 to determine the logic value of a bit.

9.6.5 Receiver Wakeup

The SCI module supports the HCS12 to operate in a multiple-receiver system. When a message is not intended for this MCU, the SCI module will put itself in a standby state to ignore the rest of the message. This is done by setting the RWU bit of the SCIxCR2 register. In the standby state, the SCI module will still load the receive data into the SCIxDRH/L registers, but it will not set the RDRF flag.

The transmitting device can address messages to selected receivers by including addressing information in the initial frame or frames of each message. One can choose to use idle line or address mark to wake up the receiver to compare the address information of the incoming frame.

IDLE LINE WAKEUP

When the idle line wakeup method is chosen, an idle condition on the RxD pin clears the RWU bit in the SCIxCR2 register and wakes up the SCI. The initial frame or frames of every message contain addressing information. All receivers evaluate the addressing information, and receivers for which the message is addressed process the frames that follow. Any receiver for which a message is not addressed can set its RWU bit and return to the standby state. The RWU bit remains set and the receiver remains on standby until another idle character appears on the RxD signal.

Idle line wakeup requires that messages be separated by at least one idle character and that no message contains idle characters. The idle character that wakes a receiver does not set the receiver idle bit, IDLE, or the receive data register full flag, RDRF.

The idle-line-type bit, ILT, determines whether the receiver begins counting logic 1s as idle character bits after the start bit or after the stop bit. The ILT bit is in the SCIxCR1 register.

ADDRESS MARK WAKEUP

In this wakeup method, a logic 1 in the most significant bit (msb) position of a frame clears the RWU bit and wakes up the SCI. The logic 1 in the msb position marks a frame as an address frame that contains addressing information. All receivers evaluate the addressing information, and the receivers for which the message is addressed process the frames that follow. Any receiver for which a message is not addressed can set its RWU bit and return to the standby state.

The logic 1 of an address frame clears the receiver's RWU bit before the stop bit is received and sets the RDRF flag. Address mark wakeup allows messages to contain idle characters but requires that the msb be reserved for use in address frames.

9.6.6 Single-Wire Operation

Normally, the SCI uses two pins for transmitting and receiving. In single-wire operation, the RxD pin is disconnected from the SCI module. The SCI module uses the TxD pin for both receiving and transmitting, as illustrated in Figure 9.15.

Single-wire operation is enabled by setting the LOOPS and the RSRC bits in the SCIxCR1 register. Setting the LOOPS bit disables the path from the RxD pin to the receiver. Setting the RSRC bit connects the receiver input to the output of the TxD pin driver. Both the transmitter and receiver must be enabled. The TXDIR bit determines whether the TxD pin is going to be used as an input (TXDIR = 0) or an output (TXDIR = 1) in this mode of operation.

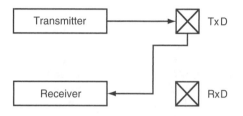

Figure 9.15 ■ Single-wire operation

9.6.7 Loop Operation

In this operation, the transmitter output goes to the receiver input. As shown in Figure 9.16, the RxD signal is disconnected from the SCI. The loop operation is enabled by setting the LOOPS bit and clearing the RSRC bit in the SCIxCR1 register. Setting the LOOPS bit disables the path from the RxD pin to the receiver. Clearing the RSRC bit connects the transmitter output to the receiver input. Both the transmitter and receiver must be enabled.

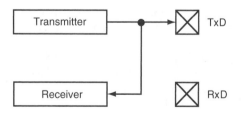

Figure 9.16 ■ Loop operation

9.7 Operation of the SCI in Different MCU Modes

When the MCU is in normal run mode, the SCI module performs normal operation. When the MCU enters the WAIT mode, the SCI's operation depends on the state of the SCISWAI bit in the SCIxCR1 register:

1. If the SCISWAI bit is 0, the SCI operates normally when the CPU is in WAIT mode.

2. If the SCISWAI bit is 1, the SCI clock generation ceases and the SCI module enters a power-conservation state when the CPU is in WAIT mode. Setting the SCISWAI bit does not affect the state of the receiver enable bit (RE) or the transmitter enable bit (TE).

If the SCISWAI bit is set, any transmission or reception in progress stops at WAIT mode entry. The transmission or reception resumes when either an internal or external interrupt brings the CPU out of WAIT mode. Exiting WAIT mode by reset aborts any transmission or reception in progress and resets the SCI module.

When the MCU enters the STOP mode, the SCI becomes inactive. The STOP instruction does not affect the SCI register states, but the SCI module clock will be disabled. The SCI operation resumes from where it left off after an external interrupt brings the CPU out of STOP mode. Exiting STOP mode by reset aborts any transmission or reception in progress and resets the SCI.

9.8 Flow Control of USART in Asynchronous Mode

The SCI module will transmit data as fast as the baud rate allows. In some circumstances, the software that is responsible for reading the data from the SCIxDRH/L register may not be able to do so as fast as the data is being received. In this case, there is a need for the HCS12 MCU to tell the transmitting device to suspend transmission of data temporarily. Similarly, the HCS12 MCU may need to be told to suspend transmission temporarily. This is done by means of flow control. There are two common methods of flow control, XON/XOFF and hardware. Here, **X** stands for **transmission**.

The XON/XOFF flow control can be implemented completely in software with no external hardware, but full-duplex communications is required. When incoming data need to be suspended, an XOFF byte is transmitted back to the other device that is transmitting the data being received. To start the other device transmitting again, an XON byte is transmitted. XON and XOFF are standard ASCII control characters. This means that when sending raw data instead of ASCII text, care must be taken to ensure that XON and XOFF characters are not accidentally sent with the data. The ASCII codes of XON and XOFF are 0x11 and 0x13, respectively. The XON character is called Device Control 1 (DC1) character; the XOFF character is called Device Control 3 (DC3) character.

Hardware flow control uses extra signals to control the flow of data. To implement hardware flow control on a HCS12 device, extra I/O pins must be used. Generally, an output pin is controlled by the receiving device to indicate that the transmitting device should suspend or resume transmissions. The transmitting device tests an input pin before a transmission to determine whether data can be sent.

Example 9.2
▼

Write an instruction sequence to configure the SCI0 to operate with the following parameters:

- 9600 baud (E clock is 24 MHz)
- 1 start bit, 8 data bits, and 1 stop bit format
- No interrupt
- Address mark wakeup
- Disable wakeup initially
- Long idle line mode
- Enable receive and transmit
- No loop back
- Disable parity
- SCI stops in WAIT mode

Solution: The following instruction sequence will achieve the desired configuration:

```
movb    #$00,SCI0BDH    ; set up baud rate to 9600
movb    #156,SCI0BDL    ;            "
movb    #$4C,SCI0CR1    ; select 8 data bits, address mark wakeup
movb    #$0C,SCI0CR2    ; enable transmitter and receiver
```

The equivalent C statements are as follows:

```
SCI0BDH    = 0x00;
SCI0BDL    = 0x9C;
SCI0CR1    = 0x4C;
SCI0CR2    = 0x0C;
```

▲

9.9 Interfacing SCI with EIA-232

Because the SCI circuit uses 0 V and 5 V (or other lower voltages) to represent logic 0 and 1, respectively, it cannot be connected to the EIA-232 interface circuit directly. A voltage translation circuit, which is called *EIA-232 transceiver*, is needed to translate the voltage levels of the SCI signals (RxD and TxD) to and from those of the corresponding EIA-232 signals.

EIA-232 transceiver chips are available from many vendors. LT1080/1081 from Linear Technology, ST232 from SGS Thompson, ICL232 from Intersil, MAX232 from MAXIM, and DS14C232 from National Semiconductor are EIA-232 transceiver chips that can operate with a single 5-V power supply and generate EIA-232 compatible outputs. These chips are also pin-compatible. In this section, we will discuss the use of DS14C232 from National Semiconductor to perform the voltage translation. The pin assignment and the use of each pin are shown in Figure 9.17.

Figure 9.17 ■ Pin assignments and connection of the DS14C232

Adding an EIA-232 transceiver chip will then allow the HCS12 to use the SCI interface to communicate with the EIA-232 interface circuit. An example of such a circuit is shown in Figure 9.18. A null-modem wiring is followed in this circuit.

Figure 9.18 ■ Diagram of SC10 and EIA-232 DB9 connector wiring in SSE256 demo board

Example 9.3

▼

Write a subroutine to send a break to the communication port controlled by the SCI0 interface. The duration of the break is approximately 24,000 E clock cycles, or 1 ms at 24 MHz.

Solution: A break character is represented by 10 or 11 consecutive 0s and can be sent out by setting the bit 0 of the SCI0CR2 register. As long as the bit 0 of the SCI0CR2 register remains set, the SCI will keep sending out the break character. The following assembly subroutine will perform the desired operation:

```
#include    "c:\miniidel\hcs12.inc"
sendbrk     bset    SCI0CR2,SBK        ; turn on send break
            ldy     #1
            jsr     delayby1ms
            bclr    SCI0CR2,SBK        ; turn off send break
            rts
#include    "c:\miniiide\delay.asm"
```

The C language version of the function is as follows:

```
#include     "c:\egnu091\include\hcs12.h"
#include     "c:\egnu091\include\delay.c"
void send_break (void)
{
    SCI0CR2 |= SBK;    /* start to send break /
    delayby1ms(1);
    SCI0CR2 &= ~SBK;   /* stop sending break */
}
```

Example 9.4

▼

Write a subroutine to output the character in accumulator A to the SCI0 module using the polling method.

Solution: A new character should be sent out only when the transmit data register is empty. When the polling method is used, the subroutine will wait until bit 7 of the SCI0SR1 register is set before sending out the character in accumulator A. The assembly language function is as follows:

```
#include     "c:\miniide\hcs12.inc"
putcSCI0     brclr    SCI0SR1,TDRE,*    ; wait for TDRE to be set
             staa     SCI0DRL           ; output the character
             rts
```

The C language version of the function is as follows:

```
void putcSCI0 (char cx)
{
    while (!(SCI0SR1 & TDRE));
    SCI0DRL = cx;
}
```

Example 9.5

▼

Write a subroutine to read a character from the SCI0 using the polling method. The character will be returned in accumulator A.

Solution: Using the polling method, the subroutine will wait until the RDRF bit (bit 5) of the SCI0SR1 register becomes set and then read the character held in the SCI0DRL register. The subroutine that reads a character from SCI0 is as follows:

```
#include     "c:\miniide\hcs12.inc"
getcSCI0     brclr    SCI0SR1,RDRF,*    ; wait until RDRF bit is set
             ldaa     SCI0DRL           ; read the character
             rts
```

The C language version of the function is as follows:

```c
char getcSCI0 (void)
{
    while(!(SCI0SR1 & RDRF));
    return (SCI0DRL);
}
```

▲

Example 9.6

▼

Write a subroutine that outputs a NULL-terminated string pointed to by index register X to the SCI0 using the polling method.

Solution: The subroutine will call putcSCI0() repeatedly until all characters of the string have been output.

```
putsSCI0    ldaa    1,x+        ; get a character and move the pointer
            beq     done        ; is this the end of the string?
            jsr     putcSCI0
            bra     putsSCI0
done        rts
```

The C language version of the program is as follows:

```c
void    putsSCI0 (char *cx)
{
    while (!(*cx)) {
        putcSCI0(*cx);
        cx++;
    }
}
```

▲

Example 9.7

▼

Write a subroutine that inputs a string from the SCI0 module. The string is terminated by the carriage return character and must be stored in a buffer pointed to by index register X.

Solution: The subroutine will call getcSCI0() repeatedly until the carriage return character is input.

```
CR          equ     $0D
getsSCI0    jsr     getcSCI0
            cmpa    #CR         ; is the character a carriage return?
            beq     exit
            staa    1,x+        ; save the character in the buffer pointed to by X
            bra     getsSCI0    ; continue
exit        clr     0,x         ; terminate the string with a NULL character
            rts
```

The C language version of the function is as follows:

```
void getsSCI0 (char *buf)
{
    while ((*buf++ = getcSCI0()) != CR);
    *buf = 0;   /* terminate the string with a NULL character */
}
```

▲

9.10 Summary

When high-speed data transfer is not necessary, using serial data transfer enables the chip designer to make the most use of the limited number of I/O pins. Serial data transfer can be performed asynchronously or synchronously. The SCI interface is an asynchronous interface that is designed to be compatible with the EIA-232 standard. The EIA-232 standard has four aspects: electrical, mechanical, procedural, and functional. For a short distance, modems are not needed for two computers to communicate. A connection called *null modem* is used instead. This is achieved by connecting signals in such a way as to fool two computers into thinking that they are connected through modems.

Since there is no common clock signal for data transfer synchronization, devices participating in communication must agree on a common data rate before data exchange is started. Users of the EIA-232 standard follow a common data format: Each character is framed with a start bit and a stop bit. A character can have eight or nine data bits. The ninth bit is often used as a parity bit for error checking or an address mark to alert the receiver to perform address comparison. The receiver uses a clock signal with a frequency that is 16 times the data rate to sample the RxD pin to detect the start bit and determine the logic values of data bits.

Errors could happen during the data transmission process. The most common errors include framing, receiver overrun, and parity errors. A framing error occurs when the start and stop bits improperly frame a received character. A framing error is detected by a missing stop bit. A receiver-overrun error occurs when one or multiple characters are received but not read by the processor. A parity error occurs when an odd number of bits change value.

Some HCS12 members have two identical serial communication interface (SCI) subsystems. Since the EIA-232 standard uses a voltage level different from those to represent logic 1 and 0, a transceiver is required to do the voltage translation so that the SCI subsystem can interface with the EIA-232 circuit. Due to the widespread use of the EIA-232 standard, transceiver chips are available from many vendors. The DS14C232 from National Semiconductor is used as an example to illustrate the SCI hardware interfacing.

9.11 Exercises

E9.1 Sketch the output of the letter k and p at the TxD pin when they are transmitted using the format of 1 start bit, 8 data bits, and 1 stop bit.

E9.2 Write an instruction sequence to configure the SCI1 to operate with the following parameters:

- 19200 baud (E clock is 24 MHz)
- 1 start bit, 8 data bits, and 1 stop bit format
- Enable both transmit and receive interrupts

- Idle line wakeup
- Disable wakeup initially
- Long idle line mode
- Enable receive and transmit
- No loop back
- Enable parity

E9.3 Write a subroutine to send a break to the communication port controlled by SCI1. The duration of the break must be approximately 480,000 E clock cycles.

E9.4 Write a subroutine to output the contents of accumulator A as two hex digits to channel SCI0.

E9.5 Modify Example 9.4 so that the *putcSCI0* routine will expand the CR character into the CR/LF pair and expand the LF character into the LF/CR pair. You can add a flag to indicate whether the expansion should be performed.

E9.6 Add an echo flag (1 byte) and modify Example 9.5 so that the received character will be echoed back to the SCI0 when the flag is 1. Otherwise, no echo will be performed.

E9.7 Write a subroutine to input two hex digits from the SCI0 module and echo them back to SCI0. When this routine is run on a demo board, it will allow you to input two hex digits from the keyboard and echo them on the terminal screen.

E9.8 Modify the *getcSCI0* routine in Example 9.5 so that it will handle the backspace character like this:

- Check if the input character is the backspace character (ASCII code $08).

- If the entered character is the backspace character, then echo it, output a space character, and output another backspace character. Why would you do this?

9.12 Lab Exercises and Assignment

L9.1 Write a program to be run on your demo board. This program and the user interact as follows:

1. The program outputs the message *Please enter your age:* and waits for the user to enter his or her age.

2. The user enters his or her age (followed by a carriage return) and the program reads it.

3. The program outputs the message *Please enter your height in inches:* and waits for the user to enter his or her height.

4. The user enters his or her height in inches (followed by a carriage return) and the program reads it.

5. The program outputs the message *Please enter your weight in lbs:* and waits for the user to enter his or her weight.

6. The user enters his or her weight in lbs (followed by a carriage return) and the program reads it.

7. The program outputs the following messages on the terminal screen and exits:

    You are *xxx* years old.
    You are *kk* ft *mm* inches tall.
    You weigh *zzz* lbs.

L9.2 If your demo board has two EIA-232 connectors, then perform this experiment.

1. Connect one of the EIA-232 connectors to the PC so that you can download your program onto the demo board.

2. Connect the second EIA-232 connector to another demo board.

3. Write a program to be run on two demo boards. The program outputs one character per second to the SCI port that is connected to another demo board and then reads in characters from the same SCI port.

4. The program displays the SCI activities on the LCD screen as follows:

   The first row displays what has been sent out: *snd:*xxxxx. . .

   The second row displays what has been received: *rvd:*yyyyy. . .

   Only the last 16 characters sent and received will be displayed on the LCD screen if a 16 × 2 LCD is used.

10

Serial Peripheral
Interface (SPI)

10.1 Objectives

After completing this chapter, you should be able to

- Describe the HCS12 SPI module
- Configure the SPI operation parameters
- Interface with peripheral devices with SPI interface
- Use the SPI function to interface with the shift register 74HC595
- Use SPI to interface with the digital temperature sensor TC72
- Use SPI to interface with the 10-bit D/A converter TLV5616
- Use SPI to interface with the matrix LED display driver MAX6952

10.2 Introduction to the SPI Function

The serial peripheral interface (SPI) allows the HCS12 to communicate synchronously with peripheral devices and other microcontrollers. The SPI system in the HCS12 can operate as a master or as a slave. When the SPI module is configured as a master, it is responsible for generating the clock signal (SCK) during an SPI data transfer. The SPI subsystem is mainly used in interfacing with peripherals such as TTL shift registers, LED/LCD display drivers, phase-locked loop (PLL) chips, memory components with serial interface, or A/D and D/A converter chips that do not need a very high data rate.

The SPI must be enabled to operate. When the SPI module is enabled, the four associated SPI port pins are dedicated to the SPI function as

- Slave select ($\overline{SS}$)
- Serial clock (SCK)
- Master out/slave in (MOSI)
- Master in/slave out (MISO)

The main element of the SPI system is the SPI data register. The 8-bit data register in the master and the 8-bit data register in the slave are linked by the MOSI and the MISO pins to form a distributed 16-bit register. When a data transfer is performed, this 16-bit register is serially shifted eight bit positions by the SCK clock from the master; data is exchanged between the master and the slave. Data written to the master SPI data register becomes the output data for the slave, and data read from the master SPI data register after a transfer operation is the input data from the slave.

A write to the SPI data register puts data into the transmit buffer if the previous transmission was complete. When a transfer is complete, received data is moved into a receiver data register. Data may be read from this double-buffered system any time before the next transfer is completed. This 8-bit register acts as the SPI receive data register for reads and as the SPI transmit data register for writes. A single SPI register address is used for reading data from the read data buffer and for writing data to the shifter.

There are four possible clock formats to choose from. The designer selects one of these four clock formats for data transfer by programming the CPOL and CPHA bits of the SPI control register 1. The CPOL bit simply selects a noninverted (idle low) or inverted clock (idle high). The CPHA bit is used to accommodate two fundamentally different protocols by shifting the clock by a half-cycle or not shifting the clock.

The SPI function on the HCS12 has been modified slightly from that in the 68HC11 with the expectation of improving its applicability. An HCS12 device may have from one to three identical SPI modules (SPI0, SPI1, and SPI2).

10.2.1 SPI Signal Pins

The SPI0 function shares the use of the upper four Port S pins: MISO0 (PS4), MOSI0 (PS5), SCK0 (PS6), and $\overline{SS0}$ (PS7). Out of reset, the SPI1 and SPI2 share the use of the lower four Port P pins and upper four Port P pins, respectively. However, the SPI1 and SPI2 pins can also be rerouted to the lower four Port H pins and upper four Port H pins by programming the MODRR register. The assignment of SPI1 and SPI2 signal pins is described in Section 7.5.8. There is no need to configure the pin directions when the SPI function is enabled.

- MISOx(x = 0, 1, or 2): *master in slave out* (serial data input). This pin is used to transmit data out of the SPI module when it is configured as a slave and receive data when it is configured as a master.

- MOSIx: *master out slave in* (serial data output). This pin is used to transmit data out of the SPI module when it is configured as a master and receive data when it is configured as a slave.
- SCKx: *serial clock*. This pin is used to carry the clock signal that synchronizes SPI data transfer. It is an output if the SPI is configured as a master but an input if the SPI is configured as a slave.
- $\overline{\text{SSx}}$: *slave select*. When configured as a slave, this pin must be pulled low for the SPI module to operate.

10.3 Registers Related to the SPI Subsystem

Most of the SPI operational parameters are set by two SPI control registers: SPIxCR1 (x = 0, 1, or 2) and SPIxCR2 (x = 0, 1, or 2). Their contents are shown in Figure 10.1 and 10.2, respectively. The SPI must be enabled before it can start data transfer. Setting bit 6 of the SPIxCR1

	7	6	5	4	3	2	1	0
	SPIE	SPE	SPTIE	MSTR	CPOL	CPHA	SSOE	LSBFE
reset:	0	0	0	0	0	1	0	0

SPIE: SPI interrupt enable bit
    0 = SPI interrupts are disabled.
    1 = SPI interrupts are enabled.
SPE: SPI system enable bit
    0 = SPI disabled.
    1 = SPI enabled and pins PS4–PS7 are dedicated to SPI function.
SPTIE: SPI transmit interrupt enable
    0 = SPTEF interrupt disabled.
    1 = SPTEF interrupt enabled.
MSTR: SPI master/slave mode select bit
    0 = slave mode
    1 = master mode
CPOL: SPI clock polarity bit
    0 = active high clocks selected; SCK idle low
    1 = active low clocks selected; SCK idle high
CPHA: SPI clock phase bit
    0 = The first SCK edge is issued one-half cycle into the 8-cycle transfer operation.
    1 = The SCK edge is issued at the beginning of the 8-cycle transfer operation.
SSOE: slave select output enable bit
    The $\overline{\text{SS}}$ output feature is enabled only in master mode by asserting the
    SSOE bit and the MODFEN bit of the SPIxCR2 register.
LSBF: SPI least significant bit first enable bit
    0 = data is transferred most significant bit first.
    1 = data is transferred least significant bit first.

Figure 10.1 ■ SPI control register 1 (SPIxCR1, x = 0, 1, or 2) register will enable the SPI subsystem. Bit 1 (SSOE) of the SPIxCR1 register allows the user to use the $\overline{SS}$ pin to select the slave device for data transfer automatically. However, this feature is not useful if one wants to use the SPI to interface with multiple slave devices at the same time. The selection of $\overline{SS}$ for input and output is shown in Table 10.1. Some slave devices may transfer data with the least significant bit first. Bit 0 (LSBF) of the SPIxCR1 register allows the user to have this flexibility.

MODFEN	SSOE	Master Mode	Slave Mode
0	0	$\overline{SS}$ not used by SPI	$\overline{SS}$ input
0	1	$\overline{SS}$ not used by SPI	$\overline{SS}$ input
1	0	$\overline{SS}$ input with MODF feature	$\overline{SS}$ input
1	1	$\overline{SS}$ output	$\overline{SS}$ input

Table 10.1 ■ ISS input/output selection

7	6	5	4	3	2	1	0
0	0	0	MODFEN	BIDIROE	0	SPSWAI	SPC0

reset: 0   0   0   0   1   0   0   0

MODFEN: mode fault enable bit
    0 = disable the MODF error.
    1 = enable setting the MODF error.
BIDIROE: output enable in the bidirectional mode of operation
    0 = output buffer disabled
    1 = output buffer enabled
SPSWAI: SPI stop in wait mode
    0 = SPI clock operates normally in stop mode.
    1 = stop SPI clock generation in wait mode.
SPC0: serial pin control bit 0
    With the MSTR bit in the SPIxCR1 register, this bit enables bidirectional pin
    configuration, as shown in Table 10.2.

Figure 10.2 ■ SPI conrol register 2 (SPIxCR2, x = 0, 1, or 2)

Pin mode		SPC0	MSTR	MISO[1]	MOSI[2]	SCK[3]	$\overline{SS}$[4]
A	normal	0	0	slave out	slave in	SCK in	SS in
B			1	master in	master out	SCK out	SS I/O
C	bidirectional	1	0	slave I/O	—	SCK in	SS in
D			1	—	master I/O	SCK out	SS I/O

[1] Slave output is enabled if BIDIROE bit = 1, $\overline{SS}$ = 0, and MSTR = 0, (C).
[2] Master output is enabled if BIDIROE bit = 1 and MSTR = 1, (D).
[3] SCK output is enabled if MSTR = 1 (B,D).
[4] $\overline{SS}$ output is enabled if MODFEN = 1, SSOE = 1, and MSTR = 1, (B,D).

Table 10.2 ■ Bidirectional pin configurations

The SPI data shift rate (also called *baud rate*) is programmable. The baud rate is set by programming the SPIxBR register to appropriate values. The contents of the SPIxBR register are shown in Figure 10.3. The method for computing the baud rate is also given in Figure 10.3.

7	6	5	4	3	2	1	0
0	SPPR2	SPPR1	SPPR0	0	SPR2	SPR1	SPR0

reset: 0   0   0   0   0   0   0   0

SPPR2~SPPR0: SPI baud rate preselection bits
SPR2~SPR0: SPI baud rate selection bits

$$BaudRateDivisor = (SPPR + 1) \times 2^{(SPR + 1)}$$
$$Baud\ Rate = Bus\ Clock \div BaudRateDivisor$$

Figure 10.3 ■ SPI baud rate register (SPIxBR, x = 0, 1, or 2)

Example 10.1
▼

Give a value to be loaded into the SPIxBR register to set the baud rate to 2 MHz for a 24-MHz bus clock.

Solution: 24 MHz ÷ 2 MHz = 12. By setting SPPR2~SPPR0 and SPR2~SPR0 to 010 and 001, respectively, one can set the baud rate to 2 MHz. The value to be loaded to the SPIxBR register is $21.

▲

Example 10.2
▼

What is the highest possible baud rate for the SPI with 24-MHz bus clock?

Solution: The highest SPI baud rate occurs when both SPPR2~SPPR0 and SPR2~SPR0 are set to 000. Under this condition, BaudRateDivisor is 2, and hence the baud rate is 24 MHz / 2 = 12 MHz.

▲

The SPI has a status register that records the progress of data transfer and errors, as shown in Figure 10.4. The application program can check bit 7 of the SPIxSR register or wait for the SPI interrupt to find out if the SPI transfer has completed. When transferring data in higher frequency using the SPI format, using the polling method is more efficient due to the overhead involved in interrupt handling.

The setting of the SPIF flag may request an interrupt to the CPU if the SPIE bit of the SPIxCR1 register is also set to 1.

The SPTEF bit is set when there is room in the transmit data buffer. It is cleared by reading the SPIxSR register with SPTEF set, followed by writing a data value into the SPI data (SPIxDR) register. The SPTEF bit is set whenever the byte in the SPIxDR register is transferred to the transmit shift register. The setting of the SPTEF flag may request an interrupt to the CPU if the SPTIE bit of the SPIxCR1 register is also set to 1.

The MODF bit is set if the $\overline{SS}$ input becomes low while the SPI is configured as a master. The flag is cleared automatically by a read of the SPIxSR register followed by a write to the SPIxCR1 register. The MODF bit is set only if the MODFEN bit of the SPIxCR2 register is set.

7	6	5	4	3	2	1	0
SPIF	0	SPTEF	MODF	0	0	0	0

reset: 0   0   1   0   0   0   0   0

SPIF: SPI interrupt request bit
    SPIF is set after the eight SCK cycles in a data transfer, and it is
    cleared by reading the SP0SR register (with SPIF set) followed by
    a read access to the SPI data register.
    0 = transfer not yet complete
    1 = new data copied to SPIxDR
SPTEF: SPI data register empty interrupt flag
    0 = SPI data register not empty
    1 = SPI data register empty
MODF: mode error interrupt status flag
    0 = mode fault has not occurred.
    1 = mode fault has occurred.

Figure 10.4 ■ SPI status register (SPIxSR)

The 8-bit SPIxDR register is both the input and output register for SPI data. A write to this register allows a byte to be queued and transmitted. For an SPI configured as a master, a queued data byte is transmitted immediately after the previous transmission has completed. *Do not write to the SPIxDR register unless the SPTEF bit is 1.*

10.4 SPI Operation

Only a master SPI module can initiate transmission. A transmission begins by writing to the master SPI data register. Data is transmitted and received simultaneously. The serial clock (SCK) synchronizes shifting and sampling of the information on the two serial data lines. The $\overline{SS}$ line allows selection of an individual slave SPI device; slave devices that are not selected do not interfere with SPI bus activities. Optionally, on a master SPI device, the $\overline{SS}$ signal can be used to indicate multiple-master bus contention.

10.4.1 Transmission Formats

The CPHA and CPOL bits in the SPIxCR1 register allow the user to select one of the four combinations of serial clock phase and polarity. The clock phase control bit (CPHA) selects one of two fundamentally different transmission formats. Clock phase and polarity should be identical for the master SPI device and the communicating slave device.

When the CPHA bit is set to 0, the first edge on the SCK line is used to clock the first data bit of the slave into the master and the first data bit of the master into the slave. In some peripheral devices, the first bit of the slave's data is available at the slave data out pin as soon as the slave is selected. In this format, the first SCK edge is not issued until a half-cycle into the 8-cycle transfer operation.

The SCK output from the master remains in the inactive state for a half SCK period before the first edge appears. A half SCK cycle later, the second edge appears on the SCK pin. When this second edge appears, the value previously latched from the serial data input is shifted into the least significant bit of the shifter. After this second edge, the next bit of the SPI transfer data is transmitted out of the MOSI pin of the master to the serial data input pin of the slave device. This process continues for a total of 16 edges on the SCK pin, with data being latched on odd-numbered edges and shifted (to the shift register) on even-numbered edges. Data reception is double-buffered. Data is shifted serially into the SPI shift register during the transfer and is transferred to the parallel SPI data register after the last bit is shifted in.

After the 16th (last) SCK edge, the SPIF flag in the SPIxSR register is set, indicating that the transfer is complete. The timing diagram for this transfer format (CPHA bit = 0) is shown in Figure 10.5.

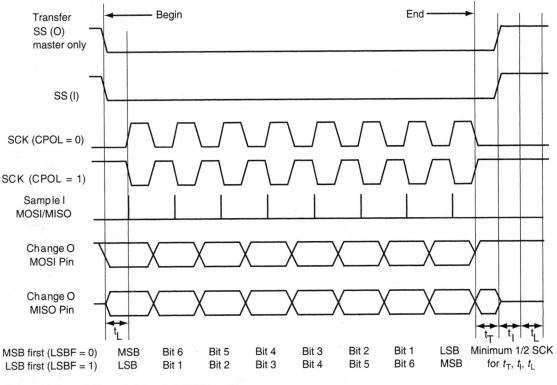

Figure 10.5 ■ SPI clock format 0 (CPHA = 0)

Some peripheral devices require the first SCK edge to appear before the first data bit becomes available at the data out pin; the second edge clocks data into the system. In this format, the first SCK edge is issued by setting the CPHA bit at the beginning of the 8-cycle transfer operation. The first edge of SCK occurs immediately after the half SCK clock cycle synchronization delay. This first edge commands the slave to transfer its most significant data bit to the serial data input pin of the master. A half SCK cycle later, the second edge appears on the SCK pin. This is the latching edge for both the master and slave. When the third edge occurs, the value previously latched from the serial data input pin is shifted into the least significant bit of the SPI shifter. After this edge, the next bit of the master data is shifted out of the serial data output pin of the master to the serial input pins on the slave.

This process continues for a total of 16 edges on the SCK line with data being latched on even-numbered edges and shifting taking place on odd-numbered edges. Again, the SPIF flag is set after the 16th SCK edge. The timing diagram of this transmission format is shown in Figure 10.6.

Figure 10.6 ■ SPI clock format 1 (CPHA = 1)

10.4.2 Bidirectional Mode (MOMI or SISO)

The SPI subsystem can be used in bidirectional mode. In this mode, the SPI uses only one serial data pin for the interface with external device(s). The MSTR bit of the SPIxCR1 register decides which pin is to be used. The MOSI pin becomes a serial data I/O (MOMI) pin for the master mode, and the MISO pin becomes a serial data I/O (SISO) pin for slave mode. The MISO pin in the master mode and the MOSI pin in the slave mode are not used by the SPI in bidirectional mode. The possible combinations are shown in Figure 10.7.

The direction of each serial I/O pin depends on the BIDIROE bit. If the pin is configured as an output, serial data from the shift register is driven out on the pin. The same pin is also the serial input to the shift register. If one wants to read data from a peripheral device, then the BIDIROE bit should be cleared to 0. The use of bidirectional mode is illustrated in exercise problem 10.8.

When SPE = 1	Master Mode MSTR = 1	Slave Mode MSTR = 0
Normal Mode SPC0 = 0	SWOM enables open-drain output	SWOM enables open-drain output
Bidirectional mode SPC0 = 1		

Figure 10.7 ■ Normal mode and bidirectional mode

10.4.3 Mode Fault Error

If the $\overline{SSx}$ input becomes low while the SPIx is configured as a master, it indicates a system error in which more than one master may be trying to drive the MOSIx and SCKx lines simultaneously. This condition is not permitted in normal operation. The MODF bit in the SPIxSR register is set automatically, provided that the MODFEN bit in the SPIxCR2 is set.

In the special case where the MODFEN bit is cleared, the $\overline{SSx}$ pin is a general-purpose I/O pin for the SPI system configured in master mode. In this special case, the mode fault function is prohibited and the MODF flag remains cleared. In case the SPI system is configured as a slave, the $\overline{SSx}$ pin is a dedicated input pin. Mode fault error does not occur in slave mode.

When a mode fault error occurs, the MSTR bit in the SPIxCR1 is cleared, the MODF bit in the SPIxSR register is set, and the output enable for the SCKx, MISOx, and MOSIx pins are de-asserted. If the mode fault error occurs in the bidirectional mode for an SPI system configured in master mode, output enable of the MOMI pin is cleared if it was set but SISO is not affected. No mode fault error occurs in the bidirectional mode for the SPI system configured in slave mode.

10.4.4 Low Power Mode Options

The HCS12 has two low power modes and the SPI module behaves differently in these two modes.

SPI in Wait Mode

SPI operation in wait mode depends upon the state of the SPISWAI bit in the SPIxCR2 register.

- If the SPISWAI bit is cleared, the SPI operates normally when the CPU is in wait mode.
- If the SPISWAI bit is set, the SPI clock generation ceases and the SPI module enters a power conservation state when the CPU is in the wait mode. If the SPI

is configured as a master, any transmission and reception in progress stops at wait mode entry. The transmission and reception resumes when the SPI exits wait mode. If the SPI is configured as a slave, any transmission and reception in progress continues if the SCK continues to be driven from the master. This keeps the slave synchronized to the master and the SCK.

SPI IN STOP MODE

The stop mode is dependent on the system. The SPI enters stop mode when the module clock is disabled. If the SPI is in master mode and exchanging data when the CPU enters stop mode, the transmission is frozen until the CPU exits stop mode. After exiting the stop mode, data to and from the external SPI device is exchanged correctly. In slave mode, the SPI will stay synchronized with the master. The stop mode is equivalent to the wait mode with the SPISWAI bit set except that the stop mode is dependent on the system and cannot be controlled with the SPISWAI bit.

10.5 SPI Circuit Connection

In a system that uses the SPI subsystem, one device (normally a microcontroller) is configured as the master and the other devices are configured as slaves. Either a peripheral chip or a HCS12 can be configured as a slave device. The master SPI device controls the data transfer and can control one or more SPI slave devices simultaneously.

In a single-slave configuration, the circuit would be connected as shown in Figure 10.8. If the SSx output enable feature is set, then the SSx pin will automatically go low to enable the slave device before the data transfer is started. This feature is not used in Figure 10.8.

Figure 10.8 ■ Master/slave transfer block diagram

There could be several connection methods in a multislave SPI environment. One possibility is shown in Figure 10.9. In this connection method, the HCS12 can choose any peripheral device for data transfer. In Figure 10.9, Port P pins are used to drive the $\overline{SS}$ inputs of peripheral devices. Any other unused general-purpose I/O pins can be used for this purpose.

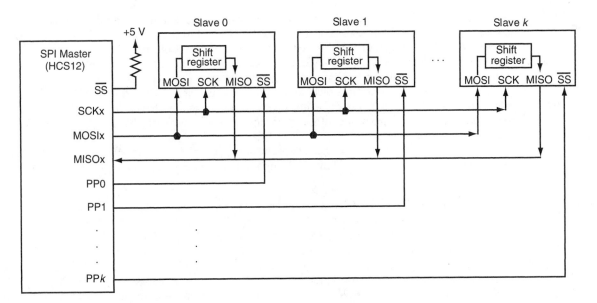

Figure 10.9 ■ Single-master and multiple-slave device connection (method 1)

If one doesn't need the capability of selecting an individual peripheral device for data transfer, then the connection shown in Figure 10.10 can be used, which will save quite a few I/O pins. Figure 10.10 differs from Figure 10.9 in the following ways:

1. The MISO pin of each slave is wired to the MOSI pin of the slave device to its right. The MOSI pins of the master and slave 0 are still wired together.

2. The MISO pin of the master is wired to the same pin of the last slave device.

3. The $\overline{SS}$ inputs of all slaves are tied to ground to enable all slaves.

Figure 10.10 ■ Single-master and multiple-slave device connection (method 2)

Thus the shift registers of the SPI master and slaves become a ring. The data of slave k are shifted to the master SPI, the data of the master are shifted to the slave 0, the data of slave 0 are shifted to slave 1, and so on. In this configuration, a minimal number of pins controls a large number of peripheral devices. However, the master does not have the freedom to select an arbitrary slave device for data transfer without going through other slave devices.

This type of configuration is often used to extend the capability of the SPI slave. For example, suppose there is an SPI-compatible seven-segment display driver/decoder that can drive only four digits. By using this configuration, up to $4 \times k$ digits can be displayed when k driver/decoders are cascaded together.

Depending on the capability and role of the slave device, either the MISO or MOSI pin may not be used in the data transfer. Many SPI-compatible peripheral chips do not have the MISO pin.

10.6 Configuration of and Data Transfer in SPI

Most peripheral chips with the SPI interface have either a chip-select (CS) or chip-enable (CE) signal to enable data exchange with the chip. One needs to assert this signal in order to perform data exchange with the device. Before data transfer can be started, the SPI function must be properly configured. The designer needs to deal with four registers: the WOMS, SPIxBR, SPIxCR1, and SPIxCR2 registers. In addition, the Port S pin pull-up must be enabled so that no external pull-up device is needed.

Example 10.3

▼

Configure the SPI0 module to operate with the following settings assuming that the bus clock is 24 MHz:

- Baud rate at 6 MHz
- No interrupt
- Enable SPI
- SCK idle low
- Data to be latched on the rising edge of SCK
- Master mode
- Transfer data the most significant bit first
- Mode fault and $\overline{SS}$ output disabled
- Normal SPI operation
- SPI stopped in wait mode

Solution: Since 24 MHz / 6 MHz = 4, one needs to set SPPR2~SPPR0 and SPR2~SPR0 to 001 and 000, respectively. Therefore, the value to be written into the SPI0BR register would be $10. For other parameters, the designer needs to write $50 and $02 into SPI0CR1 and SPI0CR, respectively. The WOMS register must also be cleared to enable the Port S pull-up.

The following instruction sequence will configure the SPI as desired:

```
movb  #$10,SPI0BR      ; set baud rate to 6 MHz
movb  #$50,SPI0CR1     ; disable interrupt, enable SPI, SCK idle low,
                       ; data latched on rising edge, data transferred msb first
```

```
movb  #$02,SPI0CR2      ; disable bidirectional mode, stop SPI in wait mode
movb  #0,WOMS           ; enable Port S pull-up
```

The same setting can be achieved by the following C statements:

```
SPI0BR  = 0x20;   /* set baud rate to 4 MHz */
SPI0CR1 = 0x50;   /* enable SPI, data shift on SCK's rising edge, master mode */
SPI0CR2 = 0x02;   /* disable bidirectional mode, SPI stops in wait mode */
WOMS    = 0;      /* enabled Port S pull up */
```

▲

Sending and reading data using the SPI interface is very straightforward. The following four functions are used most often:

1. *putcSPIx* (x = 0, 1, or 2). This function sends out a character through the SPI interface. In assembly language, the character to be output is passed in accumulator A.

2. *putsSPIx* (x = 0, 1, or 2). This function outputs a NULL-terminated string through the SPI interface. In assembly language, the string is pointed to by index register X.

3. *getcSPIx* (x = 0, 1, or 2). This function reads in a character from the SPI interface. In assembly language, the character is returned in accumulator A.

4. *getsSPIx* (x = 0, 1, or 2). This function reads in a string from the SPI interface. The buffer to hold the string is pointed to by index register X, whereas the number of characters to read is specified in accumulator B.

Example 10.4

▼

Write the four common SPI data transfer functions in the assembly and C languages.

Solution: Before writing a byte to the SPI data register, one must make sure that the SPTEF flag is 1. To make sure that the SPI transfer is completed, the program must wait until the SPIF flag is set to 1.

The following assembly function will output the character in accumulator A to SPI0:

```
putcspi0  brclr  SPI0SR,SPTEF,*  ; wait until write operation is permissible
          staa   SPI0DR          ; output the character to SPI0
          brclr  SPI0SR,SPIF,*   ; wait until the byte is shifted out
          ldaa   SPI0DR          ; clear the SPIF flag
          rts
```

The following assembly function outputs a NULL-terminated string pointed to by index register X:

```
putsspi0  ldaa  1,x+       ; get 1 byte to be output to SPI port
          beq   doneps0    ; reach the end of the string?
          jsr   putcspi0   ; call subroutine to output the byte
          bra   putsspi0   ; continue to output
doneps0   rts
```

To read a byte from the SPI interface, a byte must be written into the SPIxDR register to trigger eight clock pulses to be sent out from the SCK pin. The following assembly function reads a character from the SPI0 interface and returns the character in accumulator A:

```
getcspi0  brclr  SPI0SR,SPTEF,*  ; wait until write operation is permissible
          staa   SPI0DR          ; trigger eight clock pulses for SPI transfer
```

```
        brclr  SPI0SR,SPIF,*    ; wait until a byte has been shifted in
        ldaa   SPI0DR           ; return the byte in A and clear the SPIF flag
        rts
```

The following assembly function reads a string from the SPI0 interface:

```
getsspi0  tstb                  ; check the byte count
          beq   donegs0         ; return when byte count is zero
          jsr   getcspi0        ; call subroutine to read a byte
          staa  1,x+            ; save the returned byte in the buffer
          decb                  ; decrement the byte count
          bra   getsspi0
donegs0   clr   0,x             ; terminate the string with a NULL character
          rts
```

These four assembly functions are stored in the file *spi0util.asm* and are included on the complementary CD.

The C language versions of the previous four functions are as follows:

```c
void putcspi0 (char cx)
{
      char      temp;
      while(!(SPI0SR & SPTEF));      /* wait until write is permissible */
      SPI0DR = cx;                   /* output the byte to the SPI */
      while(!(SPI0SR & SPIF));       /* wait until write operation is complete */
      temp = SPI0DR;                 /* clear the SPIF flag */
}
void putsspi0(char *ptr)
{
      while(*ptr) {          /* continue until all characters have been output */
              putcspi0(*ptr);
              ptr++;
      }
}

char getcspi0(void)
{
      while(!(SPI0SR & SPTEF));      /* wait until write is permissible */
      SPI0DR = 0x00;                 /* trigger eight SCK pulses to shift in data */
      while(!(SPI0SR & SPIF));       /* wait until a byte has been shifted in */
      return SPI0DR;                 /* return the character */
}
void getsspi0(char *ptr, char count)
{
      while(count) {         /* continue while byte count is nonzero */
              *ptr++ = getcspi0(); /* get a byte and save it in buffer */
              count--;
      }
      *ptr = 0;`             /* terminate the string with a NULL */
}
```

These four C functions are stored in the file *spi0util.c* and are included on the complementary CD.

▲

10.7 SPI-Compatible Chips

The SPI is a protocol proposed by Freescale to interface peripheral devices to a microcontroller. As long as a peripheral device supports the SPI interface protocol, it can be used with any microcontroller that implements the SPI subsystem. Many semiconductor manufacturers are producing SPI-compatible peripheral chips. The Freescale SPI protocol is compatible with the National Semiconductor Microwire protocol. Therefore, any peripheral device that is compatible with the SPI can also be interfaced with the Microwire protocol.

10.8 The 74HC595 Shift Register

As shown in Figure 10.11, the 74HC595 consists of an 8-bit shift register and an 8-bit D-type latch with three-state parallel outputs. The shift register accepts serial data and provides a serial output. The shift register also provides parallel data to the 8-bit latch. The shift register and the latch have different clock sources. This device also has an asynchronous reset input. The frequency of the shift clock can be as high as 100 MHz. The functions of the pins in Figure 10.11 are as follows:

- DS: *Serial data input.* The data on this pin is shifted into the 8-bit shift register.
- SC: *Shift clock.* A low-to-high transition on this signal causes the data at the serial input pin to be shifted into the 8-bit shift register.
- $\overline{\text{Reset}}$. A low on this pin resets the shift register portion of this device only. The 8-bit latch is not affected.

VCC = Pin 16
GND = Pin 8

Figure 10.11 ■ The 74HC595 block diagram and pin assignment

- LC: *Latch clock.* The rising edge of this signal loads the contents of the shift register into the output latch.
- $\overline{OE}$: *Output enable.* A low on this pin allows the data from the latch to be presented at the output pins Q_A. . .Q_H.
- Q_A to Q_H: Noninverted, tri-state latch outputs.
- SQ_H: *Serial data output.* This is the output of the eighth stage of the 8-bit shift register. This output does not have tri-state capability.

The 74HC595 is designed to shift in 8-bit data serially and then transfer it to the latch to be used as parallel data. The 74HC595 can be used to add parallel output ports to the HCS12. Both the connection methods shown in Figures 10.9 and 10.10 can be used in appropriate applications.

Example 10.5

▼

Describe how to use two 74HC595 to drive eight common-cathode seven-segment displays, assuming that the bus clock frequency of the HCS12 is 24 MHz.

Solution: Two 74HC595s can be cascaded using the method shown in Figure 10.10. One 74HC595 is used to hold the seven-segment pattern, whereas the other 74HC595 is used to carry digit-select signals. The circuit connection is shown in Figure 10.12. Since there are only seven segments, the Q_H bit of the segment-control 74HC595 is not needed. The PK7 pin is used to control the LC input of the 74HC595. The time-multiplexing technique illustrated in Example 7.5 will be used to display multiple digits in Figure 10.12. To light the digit on display

Figure 10.12 ■ Two 74HC595s together drive eight seven-segment displays

#7, the voltage at Q_H of the digit-select 74HC595 must be driven to high. To light the digit on display #6, the voltage at Q_G of the digit-select 74HC595 must be driven to high, and so on.

There are four parts in the program to be written:

Part 1. Display table setup

The segment patterns and digit-select values are stored in a table so that the table lookup method can be used to display the desired pattern.

Part 2. SPI module initialization

Configure the SPI module to shift data at 12 MHz on the rising clock edge and most significant data bit first.

Part 3. Timer delay function

Include the library function *delayby1ms* to multiplex the digit per millisecond.

Part 4. SPI data transfer

Include the library function *putcspi0.asm* (or putcspi0.c) to output the display patterns and digit-select data.

The following program will display 87654321 on display #7 through display #0:

```
#include     "c:\miniide\hcs12.inc"
             org     $1000
icnt         ds.b    1                 ; loop count
             org     $1500
             lds     #$1500            ; set up stack pointer
             bset    DDRK,$80          ; configure the PK7 pin for output
             jsr     openspi0          ; configure SPI0
forever      ldx     #disp_tab         ; use X as a pointer to the table
             movb    #8,icnt           ; set loop count to 8
loop         ldaa    1,x+              ; send the digit-select byte to the 74HC595
             jsr     putcspi0          ;          "
             ldaa    1,x+              ; send segment pattern to 74HC595
             jsr     putcspi0          ;          "
             bclr    PTK,BIT7          ; transfer data from shift register to output
             bset    PTK,BIT7          ; latch
             ldy     #1                ; display the digit for 1 ms
             jsr     delayby1ms        ;          "
             dec     icnt              ;
             bne     loop              ; if not reach digit 1, then next
             bra     forever           ; start from the start of the table
;*********************************************************************************
; The following function configures the SPI module properly.
;*********************************************************************************
openspi0     movb    #0,SPI0BR         ; set baud rate to 12 MHz
             movb    #$50,SPI0CR1      ; disable interrupt, enable SPI, SCK idle low,
                                       ; latch data on rising edge, transfer data msb first
             movb    #$02,SPI0CR2      ; disable bidirectional mode, stop SPI in wait mode
             movb    #0,WOMS           ; enable Port S pull-up
             rts
#include     "c:\miniide\delay.asm"
#include     "c:\miniide\spi0util.asm"
```

```
; ****************************************************************************
; Each digit consists of two bytes of data. The first byte is
; digit select, the second byte is the digit pattern.
; ****************************************************************************
disp_tab    dc.b    $80,$7F,$40,$70,$20,$5F,$10,$5B
            dc.b    $08,$33,$04,$79,$02,$6D,$01,$30
            end
```

The C language version of the program is as follows:

```c
#include    "c:\egnu091\include\hcs12.h"
#include    "c:\egnu091\include\spi0util.c"
#include    "c:\egnu091\include\delay.c"
void openspi0(void);
void main (void)
{
    unsigned char disp_tab[8][2] = {{0x80,0x7F},{0x40,0x70},{0x20,0x5F},{0x10,0x5B},
                            {0x08,0x33},{0x04,0x79},{0x02,0x6D},{0x01,0x30}};
    char i;
    openspi0();             /* configure the SPI0 module */
    DDRK |= BIT7;           /* configure pin PK7 as output */
    while(1) {
        for (i = 0; i < 8; i++) {
            putcspi0(disp_tab[i][0]);       /* send out digit select value */
            putcspi0(disp_tab[i][1]);       /* send out segment pattern */
            PTK &= ~BIT7;                   /* transfer values to latches of 74HC595s */
            PTK |= BIT7;                    /* " */
            delayby1ms(1);                  /* display a digit for 1 ms */
        }
    }
}
void openspi0(void)
{
    SPI0BR  = 0;        /* set baud rate to 12 MHz */
    SPI0CR1 = 0x50;    /* disable SPI interrupt, enable SPI, SCK idle low, shift
                                data on rising edge, shift data msb first */
    SPI0CR2 = 0x02;    /* disable bidirectional mode, disable SPI in wait mode */
    WOMS    = 0;        /* enable Port S pull up */
}
```

This example illustrates one of the applications of the shift register 74HC595. The main drawback of the example program is that the CPU spends all its time on the task of multiplexing the displays. The program can be modified to be interrupt-driven to free the CPU from idle waiting. Several manufacturers produce LED display drivers to eliminate the display multiplexing task altogether. The MC14489 from Freescale and the MAX7221 from Maxim are two examples.

10.9 The TC72 Digital Thermometer

The TC72 from Microchip is a digital temperature sensor with the SPI interface. The TC72 has a 10-bit resolution, i.e., it uses 10 bits to represent the ambient temperature. The pin assignment and functional block diagram are shown in Figure 10.13.

Figure 10.13 ■ TC72 pin assignment and functional block diagram

10.9.1 Functioning of TC72

The TC72 is capable of reading temperatures from −55°C to +125°C. The TC72 can be used either in the *continuous temperature conversion mode* or the *one-shot conversion mode*. The continuous conversion mode measures temperature approximately every 150 ms and stores the data in temperature registers. The TC72 has an internal clock generator that controls the automatic temperature conversion sequence. In contrast, the one-shot mode performs a single temperature measurement and returns to the power-saving shutdown mode.

10.9.2 Temperature Data Format

Temperature data is represented by a 10-bit two's complement word with a resolution of 0.25°C per least significant bit. The analog-to-digital converter is scaled from −128°C to +127°C with 0°C represented as 0x0000, but the operating range of the TC72 is specified from −55°C to +125°C. The 10-bit temperature value is stored in two 8-bit registers. Because the TC72 A/D converter is scaled from −128°C to +127°C and the conversion is represented in 10 bits, the temperature value is equal to the A/C conversion result divided by 4. The lowest two bits of the temperature are stored in the highest two bits of the lower byte of the temperature register. The temperature value is represented in the two's complement format. Whenever the most significant bit of the upper byte of the temperature register is 1, the temperature is negative. The magnitude of a negative temperature can be found by taking the two's complement of the temperature reading. After this is done, the upper eight bits become the integer part of the temperature, whereas the least significant two bits become the fractional part of the temperature. A sample of the temperature readings and their corresponding temperature values are shown in Table 10.3.

Binary High Byte/Low Byte	Hex	Temperature
0010 0001/0100 0000	2140	33.25°C
0100 1010/1000 0000	4A80	74.5°C
0001 1010/1100 0000	1AC0	26.75°C
0000 0001/1000 0000	0180	1.5°C
0000 0000/0000 0000	0000	0°C
1111 1111/1000 0000	FF80	−0.5°C
1111 0010/1100 0000	F2C0	−13.25°C
1110 0111/0000 0000	E700	−24°C
1100 1001/0100 0000	C900	−55°C

Table 10.3 ■ TC72 Temperature output data

10.9.3 Serial Bus Interface

The serial interface consists of chip enable (CE), serial clock (SCK), serial data input (SDI), and serial data output (SDO). The CE input is used to select TC72 when there are multiple SPI slaves connected to the microcontroller. TC72 can operate as an SPI slave only.

The SDI input writes data into the TC72's control register, while the SDO output pin reads the temperature data from the temperature register and the status of the *shutdown bit* of the control register. The TC72 can shift data in/out using either the rising or the falling edge of the SCK input. The CE signal is active high. The SCK idle state is detected when the CE signal goes high. As shown in Figure 10.14, the clock polarity (CP) of SCK determines whether data is shifted on the rising or the falling edge. The highest SCK frequency (f_{SCK}) is 7.5 MHz.

(a) Data shift on rising edge   (b) Data shift on falling edge

Figure 10.14 ■ Serial clock polarity

Data transfer to and from the TC72 consists of one address byte followed by one or multiple data (2 to 4) bytes. The most significant bit (A7) of the address byte determines whether a read (A7 = 0) or a write (A7 = 1) operation will occur. A multiple-byte read operation will start from a high address toward lower addresses. The user need send in only the temperature result high byte address and read out the temperature result high byte, low byte, and the control register. The timing diagrams for the single-data-byte write, single byte read, and three-data-byte read are shown in Figure 10.15a, 10.15b, and 10.15c, respectively.

The procedure for reading the temperature result is as follows:

Step 1
Pull the CE pin to high to enable SPI transfer.

Step 2
Send the temperature result high-byte-read address (0x02) to the TC72. Wait until the SPI transfer is complete.

Step 3
Read the temperature result high byte. The user needs to write a dummy byte into the SSPBUF register to trigger eight pulses to be sent out from the SCK pin so that the temperature result high byte can be shifted in.

Figure 10.15a ■ Single-data-byte write operation

Figure 10.15b ■ Single-data-byte read operation

Figure 10.15c ■ SPI multiple-data-byte transfer

Step 4
Read the temperature result low byte. Likewise, the user also needs to write a dummy
byte into the SSPBUF register to shift in the temperature low byte.

Step 5
Pull CE pin to low so that a new transfer can be started.

10.9.4 Internal Register Structure

The TC72 has four internal registers: control register, LSB temperature, MSB temperature,
and manufacturer ID. Their contents are shown in Table 10.4.

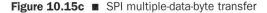

Register	Read address	Write address	Bit 7	Bit 6	Bit 5	Bit 4	Bit 3	Bit 2	Bit 1	Bit 0	Value on POR/BOR
Control	0x00	0x80	0	0	0	OS[1]	0	0	0	SHDN[2]	0x05
LSB temperature	0x01	N/A	T1	T0	0	0	0	0	0	0	0x00
MSB temperature	0x02	N/A	T9	T8	T7	T6	T5	T4	T3	T2	0x00
Manufacturer ID	0x03	N/A	0	1	0	1	0	1	0	0	0x54

[1] OS is One-Shot.
[2] SHDN is Shutdown.

Table 10.4 ■ Register for TC72

CONTROL REGISTER

The control register is read/writable and is used to select the shutdown, continuous, or one-shot conversion operating mode. The temperature conversion mode selection logic is shown in Table 10.5.

Operation Mode	One-Shot Bit	Shutdown Bit
Continuous temperature conversion	0	0
Shutdown	0	1
Continuous temperature conversion	1	0
One-shot	1	1

Table 10.5 ■ Control register temperature conversion mode selection

At power-up, the SHDN bit is 1. Thus, the TC72 is in the shutdown mode at startup. The shutdown mode disables the temperature conversion circuitry; however, the serial I/O communication port remains active. If the SHDN bit is 0, the TC72 will perform a temperature conversion approximately every 150 ms. In normal operation, a temperature conversion will be initialized by a write operation to the control register to select either the continuous temperature conversion or the one-shot operation mode. The temperature data will be available in the upper byte and lower byte of the temperature register approximately 150 ms after the control register is written into.

The one-shot mode performs a single temperature measurement and returns to the power-saving mode. After completion of the temperature conversion, the one-shot bit is reset to 0. The user must set the one-shot bit to 1 to initiate another temperature conversion.

TEMPERATURE REGISTER

The temperature register is a read-only register and contains a 10-bit two's complement representation of the temperature measurement. Bit 0 through bit 5 are always read as 0. After reset, the temperature register is reset to 0.

MANUFACTURING ID REGISTER

This register is read-only and is used to identify the temperature sensor as a Microchip component.

Example 10.6

▼

Describe the circuit connection between the HCS12 MCU and the TC72 for digital temperature reading, and write a C function to read the temperature every 200 ms. Convert the temperature value into a string so that it can be displayed in an appropriate output device. A pointer to the buffer to hold the string will be passed to this function. The bus clock frequency of the demo board is assumed to be 24 MHz.

Solution: A possible circuit connection is shown in Figure 10.16. A 0.1-μF to 1.0-μF capacitor should be added between the V_{DD} and the GND pins to filter out power noise.

Figure 10.16 ■ Circuit connection between the TC72 and the HCS12

The C function (read_temp( )) that starts a temperature measurement and converts the temperature reading into a string is as follows:

```
#include "c:\egnu091\include\hcs12.h"
#include "c:\egnu091\include\spi0util.c"
#include "c:\egnu091\include\delay.c"
#include "c:\egnu091\include\convert.c"
void read_temp (char *ptr);
void openspi0(void);
char buf[10];
void main (void)
{
        DDRM |= BIT1;        /* configure the PM1 pin for output */
        openspi0();          /* configure SPI0 module */
        read_temp(&buf[0]);
}
void openspi0(void)
{
        SPI0BR  = 0x10;                   /* set baud rate to 6 MHz */
        SPI0CR1 = 0x50;
        SPI0CR2 = 0x02;
        WOMS    = 0x00;                   /* enable Port S pull-up */
}
void read_temp (char *ptr)
{
        char hi_byte, lo_byte, temp, *bptr;
        unsigned int result;
        bptr  = ptr;
        PTM |= BIT1;                      /* enable TC72 data transfer */
        putcspi0(0x80);                   /* send out TC72 control register write address */
        putcspi0(0x11);                   /* perform one-shot conversion */
        PTM &= ~BIT1;                     /* disable TC72 data transfer */
        delayby100ms(2);                  /* wait until temperature conversion is complete */
```

```
PTM | = BIT1;                          /* enable TC72 data transfer */
putcspi0(0x02);                        /* send MSB temperature read address */
hi_byte = getcspi0();                  /* read the temperature high byte */
lo_byte = getcspi0();                  /* save temperature low byte and clear SPIF */
PTM &= ~BIT1;                          /* disable TC72 data transfer */

lo_byte &= 0xC0;                       /* make sure the lower six bits are 0s */
result = (int) hi_byte * 256 + (int) lo_byte;
if (hi_byte & 0x80) {                  /* temperature is negative */
        result = ~result + 1;          /* take the two's complement of result */
        result >>= 6;
        temp = result & 0x0003;        /* place the lowest two bits in temp */
        result >>= 2;                  /* get rid of fractional part */
        *ptr++ = 0x2D;                 /* store the minus sign */
        int2alpha(result, ptr);
}
else {       /* temperature is positive */
        result >>= 6;
        temp      = result & 0x0003;   /* save fractional part */
        result >>= 2;                  /* get rid of fractional part */
        int2alpha(result, ptr);        /* convert to ASCII string */
}
while(*bptr){ /* search the end of the string */
        bptr++;
};
        switch (temp){                 /* add fractional digits to the temperature */
        case 0:
                break;
        case 1:                        /* fractional part is .25 */
                *bptr++ = 0x2E;        /* add decimal point */
                *bptr++ = 0x32;
                *bptr++ = 0x35;
                *bptr = '\0';
                break;
        case 2:                        /* fractional part is .5 */
                *bptr++ = 0x2E;        /* add decimal point */
                *bptr++ = 0x35;
                *bptr = '\0';
                break;
        case 3:                        /* fractional part is .75 */
                *bptr++ = 0x2E;        /* add decimal point */
                *bptr++ = 0x37;
                *bptr++ = 0x35;
                *bptr = '\0';
                break;
        default:
                break;
        }
}
```

10.10 The D/A Converter TLV5616

The TLV5616 from Texas Instrument is a 12-bit voltage output digital-to-analog converter (DAC) with a flexible four-wire serial interface. The four-wire serial interface allows glueless interface to SPI, QSPI, and Microwire serial ports.

The TLV5616 has an output settling time of 3 μs in fast mode and 9 μs in slow mode. A D/A conversion operation is started by writing a 16-bit serial string that contains 4 control and 12 data bits to the TLV5616. Designed for a wide range of supply voltages, the TLV5616 can operate from 2.7 V to 5.5 V. The output settling time is programmable to allow the designer to optimize speed versus power consumption. The settling time is chosen by the control bits within the 16-bit serial input string. A high-impedance buffer is integrated on the REFIN terminal to reduce the need for a low source impedance drive to the terminal.

10.10.1 Signal Pins

The pin assignment and functional block of the TLV5616 are shown in Figure 10.17. The functions of the TLV5616 signals are as follows:

AGND: analog ground.

$\overline{\text{CS}}$: chip select. This signal enables/disables inputs.

DIN: serial digital data input.

FS: frame sync. A data transfer (in SPI format) must be preceded by a pulse on this pin.

OUT: DAC analog output.

REFIN: reference analog input voltage. The reference voltage should not exceed $V_{DD}/2$.

SCLK: serial clock input.

V_{DD}: positive power supply.

(a) Pin assignment   (b) Functional block diagram

Figure 10.17 ■ The TLV5616 DAC pins and block diagram

10.10.2 Data Format

The microprocessor or microcontroller sends in a 16-bit frame that consists of 4 bits of command and 12 bits of data to the TLV5616 to start a D/A conversion. The contents of the 16-bit data frame are shown in Figure 10.18. The control bit SPD allows the user to choose between the fast and slow conversion times. The control bit PWR allows the user to shut down the TLV5616 to save power when it is not needed.

15	14	13	12	11	10	9	8	7	6	5	4	3	2	1	0
x	SPD	PWR	x	D11	D10	D9	D8	D7	D6	D5	D4	D3	D2	D1	D0

SPD: speed control bit
   0 = slow mode
   1 = fast mode
PWR: power control bit
   0 = normal operation
   1 = power down
D11~D0: new DAC value

Figure 10.18 ■ Input data format for the TLV5616 DAC

10.10.3 TLV5616 Output Voltage

The output voltage of the TLV5616 is given by the following expression:

$$V_{OUT} = 2 \times REF \times code \div 2^n$$

This equation would generate an output voltage greater than V_{DD} when the digital code to be converted is equal to or larger than 2^{n-1} if one uses V_{DD} as the reference voltage. Due to the characteristics of the electronic circuit, any output greater than V_{DD} will be clipped off. To prevent output voltage from being clipped off, one should use a reference voltage no larger than $V_{DD}/2$.

10.10.4 Data Shifting Timing

The timing diagram of the data shifting is shown in Figure 10.19. As shown in Figure 10.19, a FS pulse must be generated before data shifting can start. The requirements for all timing

Figure 10.19 ■ TLV5616 data shifting timing diagram

parameters must be satisfied in order for data shifting to be successful. The requirements for timing parameters are listed in Table 10.5.

Name	Description	Min Max Nom	Unit
$t_{su(CS-FS)}$	Setup time, $\overline{CS}$ low before FS falls	10	ns
$t_{su(FS-CK)}$	Setup time, FS low before the first SCLK falling edge	8	ns
$t_{su(C16-FS)}$	Setup time, the time difference from the 16th falling edge of SCLK and the rising edge of FS signal	10	ns
$t_{su(C16-CS)}$	Set up time, from the 16th rising edge of SCLK until the rising edge of $\overline{CS}$ signal	10	ns
t_{wH}	Pulse duration, SCLK high	25	ns
t_{wL}	Pulse duration, SCLK low	25	ns
$t_{su(D)}$	Setup time, data ready before SCLK falling edge	8	ns
$t_{h(D)}$	Hold time, data held valid after SCLK falling edge	5	ns
$t_{wH(FS)}$	Pulse duration, FS high	20	ns

Table 10.6 ■ Timing parameter requirements

10.10.5 Interfacing the TLV5616 with the HCS12

A typical circuit connection for interfacing the HCS12 with the TLV5616 is shown in Figure 10.20. Two unused I/O port pins must be used to drive the $\overline{CS}$ and FS pins.

Figure 10.20 ■ TLV5616 to HCS12 interface

Example 10.7

▼

Verify that the timing requirements for the TLV5616 are all satisfied assuming that the bus clock frequency for the HCS12 is 24 MHz.

Solution: The highest baud rate for the SPI is 12 MHz at 24 MHz bus clock. Both the high and low pulse widths for the SCK0 signal are equal to 40.5 ns. Since the SCLK signal is driven by the SCK0 signal, the timing requirement for t_{wL} and t_{wH} is satisfied. Since the data input is latched at the center of a bit time, the data setup time ($t_{su(D)}$) and hold time ($t_{h(D)}$) are also satisfied.

The triggering sequence of signals $\overline{CS}$ and FS is as follows:

1. Pull FS to high.
2. Pull $\overline{CS}$ to low.
3. Pull FS to low.
4. Send out control and data bits using SPI transfer.
5. Wait until all 16 bits have been shifted out, pull FS to high.
6. Pull $\overline{CS}$ to high.

One would use the *bset* instruction to pull a signal to high and use the *bclr* instruction to pull a signal to low. Both instructions take four E clock cycles to execute. The resultant time delay from the moment that $\overline{CS}$ is pulled low until FS is pulled low would be four E clock cycles (164 ns) and is longer than $t_{su(CS\text{-}FS)}$ (10 ns). The actual pulse width for the FS signal is eight E clock cycles (328 ns) and is much longer than $t_{wH(FS)}$ (10 ns).

We will call the function *putcspi0* to send data to the TLV5616. An *ldaa* instruction must be executed to place the value to be written to the SPI0DR register in A. The following instructions must be executed from the moment that the signal FS is pulled low until the falling edge of the first SCK signal:

```
ldaa    zz              ; 1 or 3 E cycles depending on the operand
jsr     putcspi0        ; 4 E cycles
brclr   SPI0SR, SPTEF,* ; 5 E cycles (first instruction of the putcspi0 subroutine)
staa    SPI0DR          ; 3 E cycles (second instruction of the putcspi0
                        ; subroutine)
```

The resultant value for the parameter $t_{su(FS\text{-}CK)}$ is 13 to 15 E cycles and is much longer than the required value of 8 ns.

The subroutine *putcspi0* executes the brclr SPI0SR,SPIF,* instruction to wait for the SPIF to be set (which occurs at the end of the 16th SCLK cycle). The following instructions are executed from the moment that the 16th SCLK signal ends until the FS signal is pulled high:

```
rts                 ; 5 E cycles
bset    PTM,BIT7    ; 4 E cycles
```

The resultant value for the parameter $t_{su(C16\text{-}FS)}$ is 9.5 E cycles and is much longer that the required value of 10 ns. An additional instruction bset PTM,BIT6 needs to be executed to pull the signal $\overline{CS}$ to high. Therefore, the resultant value (13.5 E cycles) for the parameter $t_{su(C16\text{-}CK)}$ is also much longer than the requirement (8 ns).

All timing requirements are satisfied from the previous analysis.

▲

Example 10.8

Configure the SPI properly so that it can work with the TLV5616.

Solution: One can set the baud rate to 12 MHz. This requires a zero to be loaded to the SPI0BR register.

The setting for the SPI0CR1 register is as follows:

- Disable all interrupts.
- Enable SPI0.
- Select master mode.
- Set the SCLK clock to idle low.
- Disable $\overline{SS0}$ output.
- Shift data most significant bit first.

The setting for the SPI0CR2 register is as follows:

- Disable mode fault.
- Disable bidirectional mode.
- Stop SPI0 in wait mode.

The instruction sequence to configure the SPI0 module is as follows:

```
movb    #0,SPI0BR      ; set baud rate to 12 MHz
movb    #$54,SPI0CR1
movb    #$02,SPI0CR2
movb    #0,WOMS        ; enable Port S pull-up
```

Example 10.9

Write a program to generate a waveform that is a repetition of the waveform shown in Figure 10.21 using the circuit shown in Figure 10.20.

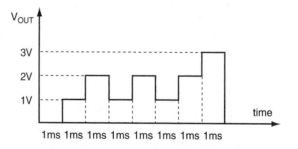

Figure 10.21 ■ Waveform to be generated

Solution: The values to be sent to the TLV5616 to generate 0-V, 1-V, 2-V, and 3-V output are as follows:

$$val\ (0) = 0$$
$$val\ (1) = 2^{12}/4 = 1024$$
$$val\ (2) = 2 \times 2^{12}/4 = 2048$$
$$val\ (3) = 3 \times 2^{12}/4 = 3072$$

We will use fast mode to perform the D/A conversion. The 16-bit values to be written into the TLV5616 are as follows:

$$code\ (0) = \$4000$$
$$code\ (1) = \$4400$$
$$code\ (2) = \$4800$$
$$code\ (3) = \$4C00$$

The assembly program for generating the desired waveform is as follows:

```
#include      "c:\miniide\hcs12.inc"
PM7           equ    BIT7
PM6           equ    BIT6
prolog        macro
              bset   PTM,PM7       ; pull FS to high
              bclr   PTM,PM6       ; pull CS to low
              bclr   PTM,PM7       ; pull FS to low
              endm
epilog        macro
              bset   PTM,PM7       ; pull FS to high
              bset   PTM,PM6       ; pull CS to high
              endm
              org    $1500
              lds    #$1500        ; set up the stack pointer
              bset   DDRM,$C0      ; configure the PM6 and PM7 pins for output
              jsr    openspi0      ; configure SPI0 properly
forever       ldx    #D2ATab       ; Use X to point to the table
              ldab   #8            ; entry count set to 8
iloop         prolog               ; activate an FS pulse and pull CS to low
              ldaa   1,x+          ; output 0 V from OUT pin
              jsr    putcspi0      ;  "
              ldaa   1,x+          ;  "
              jsr    putcspi0      ;  "
              epilog               ; pull PM7 and PM6 to high
              ldy    #1            ; wait for 1 ms
              jsr    delayby1ms    ;  "
              dbne   b,iloop       ; reach the end of the table?
              bra    forever       ; yes, start from the beginning
D2ATab        dc.b   $40,$00,$44,$00,$48,$00,$44,$00
              dc.b   $48,$00,$44,$00,$48,$00,$4C,$00
```

```
openspi0        movb    #0,SPI0BR        ; set baud rate to 12 MHz
                movb    #$54,SPI0CR1
                movb    #$02,SPI0CR2
                movb    #0,WOMS          ; enable Port S pull-up
                rts
#include         "c:\miniide\delay.asm"
#include         "c:\miniide\spi0util.asm"
                end
```

This program can be modified to be interrupt-driven so that the MCU can still perform other operations. The C language version of the program is straightforward and hence will be left as an exercise problem.

10.11 Matrix LED Displays

Many organizations have the need to display important information at the entrance or some corners of their buildings. The information need not be displayed all at once but can be rotated. Temperature, date, humidity, and other data in turn are displayed at many crossroads. Schools display their upcoming games, the result of games yesterday, or other important events using large display panels. Many of these display panels use matrix LED displays because of their brightness and versatility.

10.11.1 The Organization of Matrix LED Displays

Matrix LED displays are denoted by the number of their columns and rows. The most popular matrix LED display has five columns and seven rows (5 × 7). Other configurations, such as 5 × 8 and 8 × 8, are also available. It is obvious that the more rows and columns of LEDs are used, the better the resolution will be.

10.11.2 Colors of Matrix LED Displays

Like seven-segment displays, there are green, yellow, red, and bicolor (red/green) matrix LED displays.

10.11.3 Connection Method

Matrix LED displays can be organized as **cathode-row** (anode column) or **anode-row** (cathode column). In a cathode-row organization, the LEDs in a row have a common cathode. In an anode-row organization, the LEDs in a row have a common anode. The cathode-row organization and anode-row organizations are shown in Figures 10.22 and 10.23, respectively.

10.11.4 Dimension of Matrix LED Displays

The dimension of a matrix LED display is often indicated by its height. Common heights for matrix LED displays include 0.7″, 1.2″, 1.7″, and 2.3″. Larger matrix LED displays are also available.

10.11.5 Method of Driving Matrix LED Displays

Because of the internal connection of a matrix LED display, two parallel ports are needed to drive it. The matrix LED displays need to be scanned one row at a time, from top to bottom

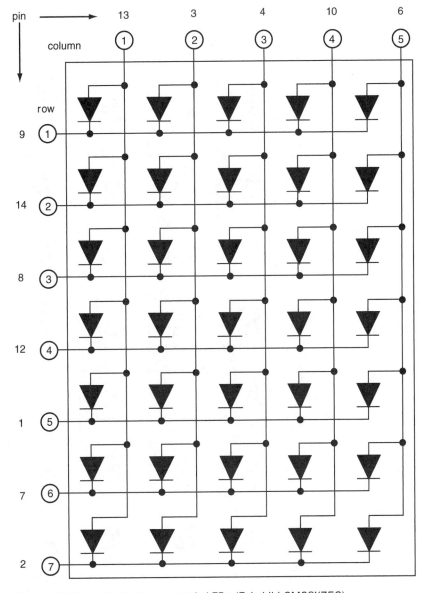

Figure 10.22 ■ Cathode-row matrix LEDs (Fairchild GMC8X75C)

(or bottom to top). Usually, multiple matrix LED displays are needed in the application, so a time-multiplexing technique needs to be used, which is quite demanding on CPU time.

Because of the popularity of matrix LED displays, a few companies have produced driver chips for them. Among them, Maxim provides driver chips with an SPI (MAX6952) or I²C (MAX6953) interface for cathode-row matrix LEDs.

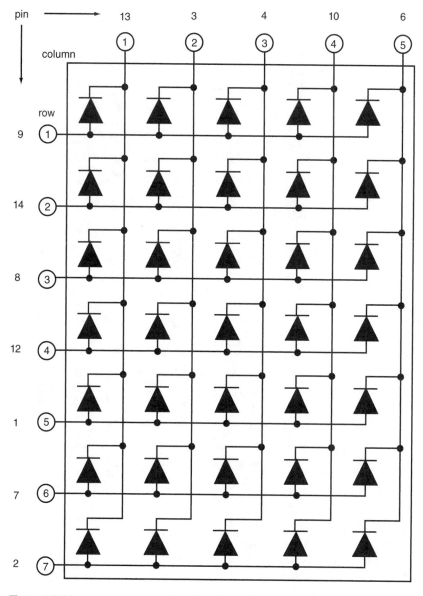

Figure 10.23 ■ Anode-row matrix LEDS (Fairchild GMA8X75C)

10.12 The MAX6952 Matrix LED Display Driver

The MAX6952 is designed to drive cathode-row matrix LED displays with a 5 × 7 organization. This chip can operate with a power supply from 2.7 V to 5.5 V and can drive four monocolor or two bicolor cathode-row matrix LED displays. It has a built-in ASCII 104-character font and 24 user-definable characters. The built-in characters follow the Arial font, with the addition of the following common symbols: £, € , ¥, °, μ, ±, ↑, and ↓. The 24 user-definable characters are uploaded by the user into on-chip RAM through the serial interface and are lost when the device is powered down. It allows automatic blinking control for each segment and provides 16-step digital brightness control. Both the 36-pin SSOP and 40-pin packages are available. The device includes a low-power shutdown mode, segment blinking (synchronized across multiple drivers, if desired), and a test mode that forces all LEDs on.

10.12.1 Pin Functions

The pin functions are described in Table 10.7. The connection for four monocolor digits and two bicolor digits are shown in Tables 10.8 and 10.9, respectively. The typical circuit connection that drives four monocolor digits using the MAX6952 is shown in Figure 10.24.

Name	Pin		Function
	SSOP	PDIP	
O0 to O13	1, 2, 3, 6–14, 23, 24	1, 2, 3, 7–15, 26, 27	LED cathode drivers. O0 to O13 output sink current from the displays's cathode rows.
GND	4, 5, 6	4, 5, 6, 18	Ground
ISET	15	17	Segment current setting. Connect ISET to GND through series resistor R_{SET} to set the peak current.
BLINK	17	19	Blink clock output. Output is open-drain.
DIN	18	20	Serial data input. Data is loaded into the internal 16-bit shift register on the rising edge of the CLK.
CLK	19	21	Serial-clock input. On the rising edge of CLK, data is shifted into the internal shift register. On the falling edge of CLK, data is clocked out of DOUT. CLK input is active only when $\overline{CS}$ is low.
DOUT	20	22	Serial data output. Data clocked into DIN is output to DOUT 15.5 clock cycles later. Data is clocked out on the falling edge of CLK. Output is push-pull.
$\overline{CS}$	21	23	Chip-select input. Serial data is loaded into the shift register while $\overline{CS}$ is low. The last 16 bits of serial data are latched on $\overline{CS}$'s rising edge.
OSC	22	24	Multiplex clock input. To use the internal oscillator, connect capacitor CSET from OSC to GND. To use the external clock, drive OSC with a 1-MHz to 8-MHz CMOS clock.
O14 to O23	25–31, 34, 35, 36	28–34, 38, 39, 40	LED anode drivers. O14 to O23 output source current to the display's anode columns.
V+	32, 33	35, 36, 37	Positive supply voltage. Bypass V+ to GND with a 47-μF bulk capacitor and a 0.1-μF ceramic capacitor.

Table 10.7 ■ MAX6952 4-digit matrix LED display driver pin functions

Digit	O0~O6	O7~O13	O14~O18	O19~O23
1	Digit 0 rows (cathodes) R1 to R7 Digit 1 rows (cathodes) R1 to R7	—	Digit 0 columns (anodes) C1 to C5	Digit 1 columns (anodes) C6 to C10
2	—	Digit 2 rows (cathodes) R1 to R7 Digit 3 rows (cathodes) R1 to R7	Digit 2 columns (anodes) C1 to C5	Digit 3 columns (anodes) C6 to C10

Table 10.8 ■ Connection scheme for four monocolor digits

Digit	O0~O6	O7~O13	O14~O18	O19~O23
1	Digit 0 rows (cathodes) R1 to R14	—	Digit 0 columns (anodes) C1 to C10	
			The 5 green anodes	The 5 red anodes
2	—	Digit 1 rows (cathodes) R1 to R14	Digit 1 columns (anodes) C1 to C10	
			The 5 green anodes	The 5 red anodes

Table 10.9 ■ Connection scheme for two bicolor digits

The MAX6952 uses four pins to interface with the MCU: CLK, $\overline{CS}$, DIN, and DOUT. The $\overline{CS}$ signal must be low to clock data in and out of the device, and the DIN signal is shifted in on the rising edge of CLK. The MAX6952 can shift data at 26 MHz. When the MAX6952 is not being accessed, DOUT is not in high impedance, contrary to the SPI standard.

Multiple MAX6952 can be daisy-chained by connecting the DOUT pin of one device to the DIN pin of the next and driving the CLK and $\overline{CS}$ lines in parallel (Figure 10.25). Data at DIN propagates through the internal shift registers and appears at DOUT 15.5 clock cycles later, clocked out on the falling edge of CLK. When sending commands to daisy-chained MAX6952s, all devices are accessed at the same time. An access requires $16 \times n$ clock cycles, where n is the number of MAX6952s connected together. To update just one device in a daisy chain, the user can send the no-op command (0x00) to the others.

10.12.2 Internal Registers

The block diagram of the MAX6952 is shown in Figure 10.26. The MAX6952 contains a 16-bit shift register (in the serial interface block in Figure 10.26) into which DIN data are shifted on the rising edge of the CLK signal, when the $\overline{CS}$ signal is low. The 16 bits in the shift register are loaded into a latch on the rising edge of the $\overline{CS}$ signal. The 16 bits in the latch are then decoded and executed. The CLK input must be idle low. This signal must be taken low before data transfer is started.

Figure 10.24 ■ MAX6952 driving four matrix LED diplays

Figure 10.25 ■ MAX6952 daisy-chain connection

Figure 10.26 ■ MAX6952 functional diagram

The upper eight bits of the shift register select the destination register to which the lower eight bits of the shift register are to be transferred. The address map of the MAX6952 is shown in Table 10.10.

The procedure for writing the MAX6952 is as follows:

Step 1
Pull the CLK signal to low.

Step 2
Pull the $\overline{CS}$ signal to low to enable the internal 16-bit shift register.

Step 3
Shift in 16 bits of data from the DIN pin with the most significant bit first. The most significant bit (D15) must be low for a write operation.

Step 4
Pull the $\overline{CS}$ signal to high.

Step 5
Pull the CLK signal to low.

Register	Address (command byte)								Hex Code
	D15	D14	D13	D12	D11	D10	D9	D8	
No op	R/W̄	0	0	0	0	0	0	0	0x00
Intensity10	R/W̄	0	0	0	0	0	0	1	0x01
Intensity32	R/W̄	0	0	0	0	0	1	0	0x02
Scan limit	R/W̄	0	0	0	0	0	1	1	0x03
Configuration	R/W̄	0	0	0	0	1	0	0	0x04
User-defined fonts	R/W̄	0	0	0	0	1	0	1	0x05
Factory reserved (do not write into)	R/W̄	0	0	0	0	1	1	0	0x06
Display test	R/W̄	0	0	0	0	1	1	1	0x07
Digit 0 plane P0	R/W̄	0	1	0	0	0	0	0	0x20
Digit 1 plane P0	R/W̄	0	1	0	0	0	0	1	0x21
Digit 2 plane P0	R/W̄	0	1	0	0	0	1	0	0x22
Digit 3 plane P0	R/W̄	0	1	0	0	0	1	1	0x23
Digit 0 plane P1	R/W̄	1	0	0	0	0	0	0	0x40
Digit 1 plane P1	R/W̄	1	0	0	0	0	0	1	0x41
Digit 2 plane P1	R/W̄	1	0	0	0	0	1	0	0x42
Digit 3 plane P1	R/W̄	1	0	0	0	0	1	1	0x43
Write digit 0 plane P0 and plane P1 with same data (reads as 0x00)	R/W̄	1	0	0	0	0	0	0	0x60
Write digit 1 plane P0 and plane P1 with same data (reads as 0x00)	R/W̄	1	1	0	0	0	0	1	0x61
Write digit 2 plane P0 and plane P1 with same data (reads as 0x00)	R/W̄	1	1	0	0	0	1	0	0x62
Write digit 3 plane P0 and plane P1 with same data (reads as 0x00)	R/W̄	1	1	0	0	0	1	1	0x63

Table 10.10 ■ MAX6952 register address map

Any register data within the MAX6952 may be read by setting the D15 bit to 1. The procedure to read a register is as follows:

Step 1
Pull CLK to low.

Step 2
Pull the $\overline{CS}$ signal to low to enable the internal shift register.

Step 3
Clock 16 bits of data into the DIN pin with bit 15 first. Bit 15 must be 1. Bit 14 to bit 8 contain the address of the register to be read. Bits 7 through 0 contain dummy data.

Step 4
Pull the $\overline{CS}$ signal to high. Bits 7 to 0 of the serial shift register will be loaded with the data in the register addressed by bits 15 through 8.

Step 5
Pull CLK to low.

Step 6
Issue another read command (which can be a no-op), and examine the bit stream at the DOUT pin. The second eight bits are the contents of the register addressed by bits 14 to 8 in step 3.

DIGIT REGISTERS

The MAX6952 uses eight digit registers to store the characters that the user wishes to display on the four 5 × 7 LED digits. These digit registers are implemented with two planes of 4 bytes, called P0 and P1. Each LED digit is represented by 2 bytes of memory, one byte in plane P0 and the other in plane P1. The digit registers are mapped so that a digit's data can be updated in plane P0, or plane P1, or both at the same time, as shown in Table 10.10.

If the blink function is disabled through the blink enable bit E in the configuration register, then the digit register data in plane P0 is used to multiplex the display. The digit register data in P1 is not used. If the blink function is enabled, then the digit register data in both plane P0 and P1 are alternately used to multiplex the display. Blinking is achieved by multiplexing the LED display using data planes P0 and P1 on alternate phases of the blink clock (shown in Table 10.11).

The data in the digit registers does not control the digit segments directly. Instead, the register data is used to address a character generator, which stores the data of a 128-character font. The lower seven bits of the digit data (D6 to D0) select the character font. The most significant bit of the register data (D7) selects whether the font data is used directly (D7 = 0) or whether the font is inverted (D7 = 1). The inversion feature can be used to enhance the appearance of bicolor displays by displaying, for example, a red character on a green background.

Segment's Bit Setting in Plane P1	Segment's Bit Setting in Plane P0	Segment Behavior
0	0	Segment off
0	1	Segment on only during the first half of each blink period
1	0	Segment on only during the second half of each blink period
1	1	Segment on

Table 10.11 ■ Digit register mapping with blink globally enabled

CONFIGURATION REGISTER

The configuration register is used to enter and exit shutdown, select the blink rate, globally enable and disable the blink function, globally clear the digit data, and reset the blink timing. The contents of the configuration register are shown in Figure 10.27.

INTENSITY REGISTERS

Display brightness is controlled by four pulse-width modulators, one for each display digit. Each digit is controlled by a nibble of one of the two intensity registers, *Intensity10* and *Intensity32*. The upper nibble of the Intensity10 register controls the intensity of the matrix display 1, whereas the lower nibble of the same register controls the intensity of the matrix display 0. Matrix displays 3 and 2 are controlled by the upper and lower nibbles of the Intensity32 register, respectively. The modulator scales the average segment current in 16 steps from a maximum of 15/16 down to 1/16 of the peak current. The minimum interdigit blanking time is, therefore, 1/16 of a cycle. The maximum duty cycle is 15/16.

P: blink phase read back select
   0 = P1 blink phase
   1 = P0 blink phase
R: global clear digit data
   0 = digit data on both plane P0 and P1 are not affected.
   1 = clear digit data on both planes P0 and P1.
T: global blink timing synchronization
   0 = blink timing counter are unaffected.
   1 = blink timing counters are reset on the rising edge of $\overline{CS}$.
E: global blink enable/disable
   0 = blink function is disabled.
   1 = blink function is enabled.
B: blink rate selection
   0 = select slow blinking (refreshed for 1 s by plane P0, then 1 s
       by P1 at 4MHz).
   1 = select fast blinking.
S: shutdown mode
   0 = shutdown mode
   1 = normal operation

Figure 10.27 ■ The MAX6952 configuration register

SCAN-LIMIT REGISTER

The scan limit register sets how many monocolor digits are displayed, either two or four. A bicolor digit is connected as two monocolor digits. The multiplexing scheme drives digits 0 and 1 at the same time, then digits 2 and 3 at the same time. To increase the effective brightness of the displays, the MAX6952 drives only two digits instead of four. By doing this, the average segment current doubles, but this also doubles the number of MAX6952s required for driving a given number of digits.

The contents of the scan-limit register are shown in Figure 10.28. This register has only one bit implemented. When this bit is 0, only digits 0 and 1 are displayed. Otherwise, all four digits are displayed.

7	6	5	4	3	2	1	0
X	X	X	X	X	X	X	2or4

2or4: scan two digits (0 and 1) or all four digits.
   0 = display digits 0 and 1 only.
   1 = display digits 0, 1, 2, and 3.

Figure 10.28 ■ The MAX6952 scan-limit register

DISPLAY TEST REGISTER

The display test register switches the drivers between one of two modes: normal and display test. Display-test mode turns on all LEDs by overriding, but not altering, all control and digit registers (including the shutdown register). In display-test mode, eight digits are scanned and the duty cycle is 7/16 (half power). The contents of the display test register are shown in Figure 10.29. Only bit zero of this register is implemented.

7	6	5	4	3	2	1	0
X	X	X	X	X	X	X	test

Test: test bit
   0 = normal operation
   1 = display test

Figure 10.29 ■ The MAX6952 display test register

CHARACTER-GENERATOR FONT MAPPING

The character font is a 5 × 7 matrix. The character generator comprises 104 characters in ROM and 24 user-definable characters. The selection from a total of 128 characters is represented by the lower seven bits of the 8-bit digit registers. The character map follows the Arial font for 96 characters in the range from 0x28 to 0x7F. The first 32 characters map the 24 user-defined positions (RAM00 to RAM23), plus 8 extra common characters in ROM. When the most significant bit is 0, the device will display the font normally. Otherwise, the chip will display the font inversely.

USER-DEFINED FONT REGISTER

The 24 user-definable characters are represented by 120 entries of 7-bit data, five entries per character, and are stored in the MAX6952's internal RAM. The 120 user-definable font data are written and read through a single register at the address 0x05. An autoincrementing font address register pointer in the MAX6952 indirectly accesses the font data. The font address pointer can be written, setting one of 120 addresses between 0x00 and 0xF7, but cannot be read back. The font data is written to and read from the MAX6952 indirectly, using this font address pointer. Unused font locations can be used as general-purpose scratch RAM. Font registers are only 7 bits wide.

To define new fonts, the user first needs to set the font address pointer. This is done by placing the address in the font address pointer register and setting bit 7 to 1. After this, one can write the font data in the lower seven bits (to the font address pointer position) and clear bit 7.

The font address pointer autoincrements after a valid access to the user-definable font data. Autoincrementing allows the 120 font data entries to be written and read back very quickly because the font pointer address needs to be set only once. When the last data location, 0xF7, is written into, the font address pointer increments to 0x80 automatically. If the font address pointer is set to an out-of-range address by writing data in the range from 0xF8 to 0xFF, then the address is set to 0x80 instead.

The memory mapping of user-defined font register 0x05 is detailed in Table 10.12. The behavior of the font pointer address is shown in Table 10.13. To display the font defined by the user, one must send in the RAM address from 0x00 through 0x17, corresponding to the font address pointer value that is 5 × RAM address (one character needs five bytes).

Address Code (hex)	Register Data	SPI Read or Write	Function
0x85	0x00–0x7F	Read	Read 7-bit user-definable font data entry from current font address. MSB of the register data is clear. Font address pointer is incremented after the read.
0x05	0x00–0x7F	Write	Write 7-bit user-definable font data entry to current font address. Font address pointer is incremented after the write.
0x05	0x80–0xFF	Write	Write font address pointer with the register data.

Table 10.12 ■ Memory mapping of user-defined font register 0x05

Font Pointer Address	Action
0x80–0xF6	Valid range to set the font address pointer. Pointer autoincrements after a font data read or write, while pointer address remains in this range.
0xF7	Font address resets to 0x80 after a font data read or write to this pointer address.
0xF8 to 0xFF	Invalid range to set the font address pointer. Pointer is set to 0x80.

Table 10.13 ■ Font pointer address behavior

10.12.3　Blinking Operation

The display blinking facility, when enabled, makes the LED drivers flip automatically between displaying the digit register data in planes P0 and P1. If the digit register data for any digit is different in two planes, then that digit appears to flip between two characters. To make a character appear to blink on or off, write the character to one plane and use the blank character (0x20) for the other plane. Once blinking has been configured, it continues automatically without further intervention. Blinking is enabled by setting the E bit of the configuration register.

The blink speed can be programmed to be fast or slow and is determined by the frequency of the multiplex clock, OSC, and by setting the B bit of the configuration. The blink rate selection bit B of the configuration register sets either fast or slow blink speed for the whole display.

Blink Synchronization

When multiple digits are displayed, one can choose to synchronize the blinking operation of these digits. Internally, blink synchronization is achieved by resetting the display multiplexing sequence. As long as all MAX6952s are daisy-chained with one device's DOUT connected to the DIN of the next device, global synchronization is achieved by toggling the $\overline{CS}$ pin for each device, either together or in quick succession.

BLINK OUTPUT

The blink output (the BLINK pin) indicates the blink phase and is high during the P0 period and low during the P1 period. Blink phase status can be read back as the P bit in the configuration register. Typical uses for this output are

- To provide an interrupt to the processor so that segment data can be changed synchronously to the blinking. For example, a clock application may have colon segments blinking every second between hours and minute digits, and the minute display is best changed in step with the colon segments. Also, if the rising edge of blink is detected, there is half a blink period to change the P1 data (P0 data drives the displays during this interval). Similarly, if the falling edge of blink is detected, there is half a blink period to change the P0 digit data.

- If OSC is driven with an accurate frequency, blink can be used as a seconds counter.

10.12.4 Choosing Values for R_{SET} and C_{SET}

The RC oscillator uses an external resistor, R_{SET}, and an external capacitor, C_{SET}, to set the oscillator frequency, f_{OSC}. The allowed range of f_{OSC} is 1 MHz to 8 MHz. R_{SET} also sets the peak segment current. The recommended values for R_{SET} and C_{SET} set the oscillator to 4 MHz, which sets the slow and fast blink frequencies to 0.5 Hz and 1 Hz. The recommended value of R_{SET} also sets the peak current to 40 mA, which makes the segment current adjustable from 2.5 mA to 37.5 mA in 2.5-mA steps:

$$I_{SEG} = K_I / R_{SET} \text{ mA}$$
$$f_{OSC} = K_F / (R_{SET} \times C_{SET} + C_{STRAY}) \text{ MHz}$$

where

$K_I = 2144$

$K_F = 6000$

R_{SET} = external resistor in kΩ

C_{SET} = external capacitor in pF

C_{STRAY} = stray capacitance from OSC pin to GND in pF, typically 2 pF

The recommended value for R_{SET} is 53.6 KΩ and the recommended value for C_{SET} is 26 pF. The recommended value for R_{SET} is the minimum allowed value since it sets the display driver to the maximum allowed segment current. R_{SET} can be set to a higher value to set the segment current to a lower peak value whenever it is desirable. The effective value of C_{SET} includes not only the actual external capacitor used but also the stray capacitance from OSC to GND.

Example 10.10

▼

Daisy-chain two MAX6952s to drive eight of the cathode-row, monocolor, matrix LED displays GMC8975C made by Fairchild, assuming that the PM5 pin is used to drive the $\overline{CS}$ input of two MAX6952s. The connection of two MAX6952s with the HCS12 (24-MHz bus clock) is shown in Figure 10.30. Each MAX6952 is driving four matrix LED displays. Write a program to configure the SPI module to operate in master mode, and transfer data at 12 MHz, shift data in/out on the rising edge of the SCK clock. Display "MSU ECET" on the matrix LED displays without blinking.

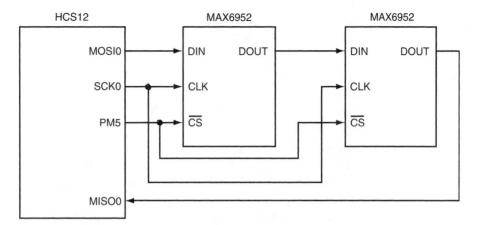

Figure 10.30 ■ HCS12 driving two MAX6952S

Solution: The SPI0 module should be configured with the following features:
- 12-MHz baud rate
- Master mode with interrupts disabled
- Shift data on the rising edge with clock idle low
- Shift data out most significant bit first
- Disable mode fault
- Stop SPI0 on wait mode

Two MAX6952 chips should be configured as follows:

Intensity10 register
We will set the intensity of both MAX6952s to maximum by sending out the following value:

0x01, 0xFF, 0x01, 0xFF

Sixteen bits need to be written to each MAX6952, of which the upper eight bits are the address to select the Intensity10 register.

Intensity32 register
Set the 2nd and 3rd displays to maximum intensity by sending the same data as to Intensity10:

0x02, 0xFF, 0x02, 0xFF

Scan limit register
Configure the MAX6952 to drive four monocolor displays by writing the following value to the MAX6952:

0x03, 0x01, 0x03, 0x01   ; drive four displays

Configuration register
Configure this register to
- Select P1 blink phase
- Not to clear digit data on both plane P0 and P1
- Reset blink timing counter on the rising edge of $\overline{CS}$
- Disable blink function

- - Select slow blinking (don't care)
  - Select normal mode

Send the following values to the MAX6952s:

0x04, 0x11, 0x04, 0x11

Display test register
Disable display test as the start by writing the following values to the MAX6952:

0x07, 0x00, 0x07, 0x00

Digit 0 (rightmost digit) plane P0
We will display **space** character and letter **T**, respectively, on display 0 of the first and second MAX6952 by sending the following data:

0x20, 0x54, 0x20, 0x20

Digit 1 (second rightmost digit) plane P0
We will display letter **U** and letter **E**, respectively, on display 1 of the 1st and 2nd MAX6952 by sending the following data:

0x21, 0x45, 0x21, 0x55

Digit 2 (second leftmost digit) plane P0
We will display letter **S** and letter **C**, respectively, on display 2 of the 1st and 2nd MAX6952 by sending the following data:

0x22, 0x43, 0x22, 0x53

Digit 3 (leftmost digit) plane P0
We will display letter **M** and letter **E**, respectively, on display 3 of the 1st and 2nd MAX6952 by sending the following data:

0x23, 0x45, 0x23, 0x4D

The following C program will configure the SPI and MAX6952 properly:

```
#include "c:\egnu091\include\hcs12.h"
#include "c:\egnu091\include\spiOutil.c"
void sendtomax(char x1, char x2, char x3, char x4);
void openspi0(void);
void main (void)
{
        openspi0();
        DDRM |= BIT5; // configure PM5 pin for output
        sendtomax(0x01, 0xFF, 0x01, 0xFF);     // set intensity for digits 0 and 1
        sendtomax(0x02, 0xFF, 0x02, 0xFF);     // set intensity for digits 2 and 3
        sendtomax(0x03, 0x01, 0x03, 0x01);     // set scan limit to drive four digits
        sendtomax(0x04, 0x11, 0x04, 0x11);     // set configuration register
        sendtomax(0x07, 0x00, 0x07, 0x00);     // disable test
        sendtomax(0x20, 0x54, 0x20, 0x20);     // value for digit 0
        sendtomax(0x21, 0x45, 0x21, 0x55);     // value for digit 1
        sendtomax(0x22, 0x43, 0x22, 0x53);     // value for digit 2
        sendtomax(0x23, 0x45, 0x23, 0x4D);     // value for digit 3
}
void sendtomax (char c1, char c2, char c3, char c4)
{
```

```
        char      temp;
        PTM &= ~BIT5;     /* enable SPI transfer to MAX6952 */
        putcspi0(c1);     /* send c1 to MAX6952 */
        putcspi0(c2);     /* send c2 to MAX6952 */
        putcspi0(c3);     /* send c3 to MAX6952 */
        putcspi0(c4);     /* send c4 to MAX6952 */
        PTM |= BIT5;      /* load data from shift register to latch */
}
void openspi0(void)
{
        SPI0BR   = 0x00;  /* set baud rate to 12 MHz */
        SPI0CR1  = 0x50;  /* disable interrupt, set master mode, shift data on
                             rising edge, clock idle low */
        SPI0CR2  = 0x02;  /* disable mode fault, disable SPI in wait mode */
        WOMS     = 0;     /* enable Port S pull-up */
}
```

Example 10.11

Modify the previous example to blink the display at a slow rate.

Solution: We need to change the setting of the configuration register and also send the space character (0x20) to the four digits in plane P1. New data to be sent to the configuration registers are as follows:

0x04, 0x19, 0x04, 0x19

The main program should be modified to as follows:

```
void main (void)
{
        openspi0();
        DDRM |= BIT5;            // configure PM5 pin for output
        sendtomax(0x01, 0xFF, 0x01, 0xFF);
        sendtomax(0x02, 0xFF, 0x02, 0xFF);
        sendtomax(0x03, 0x01, 0x03, 0x01);
        sendtomax(0x04, 0x19, 0x04, 0x19);   // configuration register, blink at phase P1
        sendtomax(0x07, 0x00, 0x07, 0x00);   // disable test
        sendtomax(0x20, 0x54, 0x20, 0x20);   // value for digit 0 on plane P0
        sendtomax(0x21, 0x45, 0x21, 0x55);   // value for digit 1
        sendtomax(0x22, 0x43, 0x22, 0x53);   // value for digit 2
        sendtomax(0x23, 0x45, 0x23, 0x4D);   // value for digit 3
        sendtomax(0x40, 0x20, 0x40, 0x20);   // value for digit 0 on plane P1 (space)
        sendtomax(0x41, 0x20, 0x41, 0x20);   // value for digit 1    "
        sendtomax(0x42, 0x20, 0x42, 0x20);   // value for digit 2    "
        sendtomax(0x43, 0x20, 0x43, 0x20);   // value for digit 3    "
}
```

Example 10.12

▼

For the circuit shown in Figure 10.30, write a program to display the following message and shift the information from right-to-left every half a second and enable blinking:

08:30:40 Wednesday, 72°F, humidity: 60%

Solution: We will use plane P0 to shift the message once every half a second. The message in plane P0 will be used to multiplex the display in half a second and the message sent to plane P1 will be used to multiplex the display in the next half of a second. We will display the space character in each matrix display using plane P1 and display the normal characters using plane P0. A delay function will be invoked once every second to shift the message to plane P0. In this manner, 16 characters are displayed in 1 second. The program is as follows:

```
#include "c:\egnu091\include\hcs12.h"
#include "c:\egnu091\include\spi0util.c"
#include "c:\egnu091\include\delay.c"
void    send2max (char x1, char x2, char x3, char x4);
void    openspi0 (void);
char    msgP0[41] = "08:30:40 Wednesday, 72°F, humidity: 60% "; /* there are 40 characters */

void main (void)
{
        char i1, i2, i3, i4;
        char j1, j2, j3, j4;
        char k;
        openspi0();
        DDRM |= BIT5; // configure PM5 pin for output
        send2max(0x01, 0xFF, 0x01, 0xFF);
        send2max(0x02, 0xFF, 0x02, 0xFF);
        send2max(0x03, 0x01, 0x03, 0x01);
        send2max(0x04, 0x1D, 0x04, 0x1D); // configuration register
        send2max(0x40, 0x20, 0x40, 0x20); /* send space character to plane P1 */
        send2max(0x41, 0x20, 0x41, 0x20);
        send2max(0x42, 0x20, 0x42, 0x20);
        send2max(0x43, 0x20, 0x43, 0x20);
        k = 0;
        while (1) {
            i1 = k;
            i2 = (k+1)%40;
            i3 = (k+2)%40;
            i4 = (k+3)%40;
            j1 = (k+4)%40;
            j2 = (k+5)%40;
            j3 = (k+6)%40;
            j4 = (k+7)%40;
            sendtomax(0x20, msgP0[i1], 0x20, msgP0[j1]);
            sendtomax(0x21, msgP0[i2], 0x21, msgP0[j2]);
            sendtomax(0x22, msgP0[i3], 0x22, msgP0[j3]);
            sendtomax(0x23, msgP0[i4], 0x23, msgP0[j4]);
```

```
                        delayby100ms(10); /* wait for 1 s */
                        k = (k+1)%40;
                }
        }

        void sendtomax (char c1, char c2, char c3, char c4)
        {
                PTM &= ~BIT5;      /* enable SPI transfer to MAX6952 */
                putcspi0(c1);      /* send c1 to MAX6952 */
                putcspi0(c2);      /* send c2 to MAX6952 */
                putcspi0(c3);      /* send c3 to MAX6952 */
                putcspi0(c4);      /* send c4 to MAX6952 */
                PTM |= BIT5;       /* load data from shift register to latch */
        }
        void openspi0(void)
        {
                SPI0BR   = 0x00;   /* set baud rate to 12 MHz */
                SPI0CR1  = 0x50;   /* disable interrupt, set master mode, shift data on
                                        rising edge, clock idle low */
                SPI0CR2  = 0x02;   /* disable mode fault, disable SPI in wait mode */
                WOMS     = 0;      /* enable Port S pull-up */
        }
```

▲

Example 10.13
▼

Write a program to define fonts for three special characters as shown in Figure 10.31. Store the font of these three special characters at locations from 0x00 to 0x0E of the MAX6952.

Figure 10.31 ■ User-definable font example

Solution: The program that sets fonts is very straightforward. The following program will set up the fonts as specified:

```
#include "c:\egnu091\include\hcs12.h"
#include "c:\egnu091\include\spiOutil.c"
char fonts [15] = {0x70,0x40,0x7F,0x40,0x70,0x48,0x38,0x0F,0x38,0x48,0x49,
                   0x49,0x7F,0x49,0x49};
void send_font (char xc);
void openspi0 (void);
void main (void)
{
      char i;
      DDRM | = BIT5;      /* configure PM5 pin for output */
      openspi0();         /* configure SPI module properly */
      send_font(0x80);    /* set font address pointer address to 0x00 */
      for (i = 0; i < 15; i++)
               send_font(fonts[i]);
}
void send_font(char xx)
{
      PTM &= ~BIT5;       /* enable SPI transfer */
      putcspi0(0x05);     /* specify font address pointer */
      putcspi0(xx);       /* send a font value */
      PTM | = BIT5;       /* load data in shift register to destination */
}
void openspi0(void)
{
      SPI0BR  = 0x00;              /* set baud rate to 12 MHz */
      SPI0CR1 = 0x50;             /* disable interrupt, set master mode, shift data on
                                     rising edge, clock idle low */
      SPI0CR2 = 0x02;             /* disable mode fault, disable SPI in wait mode */
      WOMS    = 0;                /* enable Port S pull-up */
}
```

To display user-defined fonts, one needs to send in the appropriate addresses in the range from 0x00 to 0x17. In Example 10.13, the addresses 0x00~0x02 should be used to access those three special characters. The following C statements will display those three Chinese characters followed by letters A, B, C, D, and E from left to right on the matrix displays shown in Figure 10.30:

```
sendtomax(0x20, 0x42, 0x20, 0x00); // 0x00 is the address of the first character font
sendtomax(0x21, 0x43, 0x21, 0x01); // 0x01 is the address of the second character font
sendtomax(0x22, 0x44, 0x22, 0x02); // 0x02 is the address of the third character font
sendtomax(0x23, 0x45, 0x23, 0x41);
```

10.13 Summary

When high-speed data transfer is not needed, using serial data transfer enables us to make the most of the limited number of I/O pins available on the MCU device. Serial data transfer can be performed asynchronously or synchronously. The SPI is a synchronous protocol created by Freescale for serial data exchange between peripheral chips and microcontrollers.

In the SPI format, a device (must be a microcontroller) is responsible for initiating the data transfer and generating the clock pulses for synchronizing data transfer. This device is referred to as the *SPI master*. All other devices in the same system are referred to as *SPI slaves*. The master device needs three signals to carry out the data transfer:

- SCK: a clock signal for synchronizing data transfer
- MOSI: serial data output from the master
- MISO: serial data input to the master

To transfer data to one or more SPI slaves, the MCU writes data into the SPI data register, and eight clock pulses will be generated to shift out the data in the SPI data register from the MOSI pin. If the MISO pin of the MCU is also connected (to the slave), then eight data bits will also be shifted into the SPI data register. To read data from the slave, the MCU also needs to write data into the SPI data register to trigger clock pulses to be sent out from the SCK pin. However, the value written into the SPI data register is unimportant in this case.

When configured as a slave device, the HCS12 MCU also needs the fourth signal called Slave Select ($\overline{SS}$). The $\overline{SS}$ signal enables the HCS12 slave to respond to an SPI data transfer. Most slave peripheral devices have signals called CE (chip enable) or CS (chip select) to enable/disable the SPI data transfer.

Multiple peripheral devices with an SPI interface can be interfaced with a single MCU simultaneously. There are many different methods for interfacing multiple peripheral devices (with SPI interface) to the MCU. Two popular connection methods are

1. *Parallel connection.* In this method, the MISO, MOSI, and SCK signals of all the peripheral devices are connected to the same signals of the HCS12 device. The HCS12 MCU also needs to use certain unused I/O pins to control the CS (or CE) inputs of each individual peripheral device. Using this method, the HCS12 MCU can exchange data with any selected peripheral device without affecting other peripheral devices.

2. *Serial connection.* In this method, the MOSI input of a peripheral device is connected to the MISO pin of its predecessor, and the MISO output of a peripheral device is connected to the MOSI input of its successor. The MOSI input of the peripheral device that is closest (in terms of connection) to the MCU is connected to the MOSI output of the MCU. The MISO output of the last peripheral device (in the loop) is connected to the MISO input of the MCU. The SCK inputs of all peripheral devices are tied to the SCK pin of the MCU. Using this method, the data to be sent to the last device in the loop will need to go through all other peripheral devices. The CE (or CS) signals of all peripheral devices are controlled by the same signal.

In all of the peripheral devices with the SPI interface, shift registers such as 74HC595 and 74HC589 can be used to add parallel output and input ports to the MCU. The widely used digital temperature sensors are quite useful for displaying the ambient temperature. The TC72 and TC77 from Microchip and LM74 from National Semiconductor are digital temperature sensors with the SPI interface. The SPI interface is also often added to LED and LCD display drivers. The MAX7221 from Dallas-Maxim can drive up to eight seven-segment displays. The MC14489 from Freescale can drive up to five seven-segment displays. A tutorial on using the MAX7221 is given in Appendix G. Both driver chips can be cascaded to drive more displays. Many A/D and D/A converters—for example, the 12-bit D/A converters TLV5616 and LTC2621—also have an SPI interface. Matrix displays have been widely used in recent years. They use the time-multiplexing method to display multiple digits. Due to their high demand on CPU time in performing time multiplexing, driver chips such as MAX6952 and MAX6953 have been designed to offload the CPU from time-multiplexing matrix displays.

10.14 Exercises

E10.1 Configure the SPI0 module to operate with the following setting:

- Master mode with all interrupt enabled
- SCK0 idle high and data shifted on the falling edge
- Disable mode fault, disable bidirectional mode, stop SPI in wait mode
- Shift data least significant bit first
- Use $\overline{SS0}$ pin and enable $\overline{SS0}$ output
- Baud rate set to 4 MHz

E10.2 Configure the SPI0 module to operate with the following setting:

- Master mode with all interrupt disabled
- SCK0 idle high and data shifted on the rising edge
- Disable mode fault, disable bidirectional mode, SPI0 stops in wait mode
- Data shifted most significant bit first
- Use $\overline{SS0}$ pin and enable $\overline{SS0}$ output
- Baud rate set to 3 MHz

E10.3 Assume that there is an SPI-compatible peripheral output device that has the following characteristics:

- Has a CLK input pin that is used as the data-shifting clock signal. This signal is idle low and data is shifted in on the rising edge.
- Has an SI pin to shift in data on the falling edge of the CLK input.
- Has a $\overline{CE}$ pin, which enables the chip to shift in data when it is low.
- The highest data-shifting rate is 1 MHz.
- Most significant bit shifted in first.
- Disable all interrupts.
- Stop the SPI0 module in wait mode and freeze mode.
- Use $\overline{SS0}$ pin as a general I/O pin.

Describe how to connect the SPI0 pins for the HCS12 and this peripheral device and write an instruction sequence to configure the SPI subsystem properly for data transfer. Assume that the E clock frequency is 24 MHz.

E10.4 The 74HC165 is another SPI-compatible shift register. This chip has both serial and parallel inputs and is often used to expand the number of parallel input ports. The block diagram of the 74HC165 is shown in Figure E10.4. The operation of this chip is illustrated in the function table of Table E10.4.

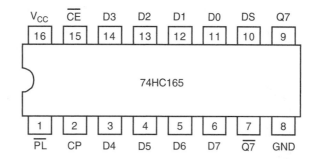

Figure E10.4 ■ The 74HC165 pin assignment

Operating Mode	Inputs					Qn Flip-Flops			Outputs	
	$\overline{PL}$	$\overline{CE}$	CP	D_S	D0–D7	Q0	Q1–Q6	Q7	Q7	$\overline{Q7}$
Parallel load	L	X	X	X	L	L	L-L	L	H
	L	X	X	X	H	H	H-H	H	L
Serial shift	H	L	↑	l	X	L	q0–q5	q6	$\overline{q6}$
	H	L	↑	h	X	H	q0–q5	q6	$\overline{q6}$
Hold	H	H	X	X	X	q0	q1–q6	q7	$\overline{q7}$

Table E10.4 ■ Function table of the 74HC165

The 74HC165 can be cascaded. To cascade, the Q7 output is connected to the DS input of its adjacent 74HC165. All 74HC165s should share the same shift clock SC. Suppose we want to use two 74HC165s to interface with two DIP switches so that four hex digits can be read in by the HCS12. Describe the circuit connection and write an instruction sequence to read the data from these two DIP switches.

E10.5 Suppose you are going to use four 74HC595s to drive four seven-segment displays using the circuit shown in Figure 10.10. Write an instruction sequence to display the number 1982 on these four seven-segment displays.

E10.6 Write a program to display the number 1, 2, . . ., 8 on those eight seven-segment displays in Figure 10.12. Display one digit at a time and turn off the other seven digits. Each digit is displayed for 1 second. Perform this operation continuously.

E10.7 Write the C language version of the program for Example 10.9.

E10.8 The Microchip TC77 is a small-footprint digital temperature sensor with a 13-bit resolution. This device uses a single pin for temperature I/O, which one would need to use the SPI bidirectional mode to read the temperature. The data sheet of this device is included on the complementary CD (it can also be downloaded from the Microchip website, http://www.microchip.com). Describe the circuit connection of this chip and the HCS12. Write a program to configure the TC77 to continuous conversion mode, read the temperature once every second, and display the temperature on the LCD. The block diagram and the pin assignment of the TC77 are shown in Figure E10.8.

Figure E10.8 ■ Pin assignment and block diagram of the TC77

E10.9 Write a C program to generate a sine waveform using the circuit shown in Figure 10.20. Divide one period of the sine wave into 120 points and use these 120 points to represent the waveform. Every two adjacent points are separated by 1.5°.

E10.10 Use the OUT pin of the TLV5616 shown in Figure 10.20 to drive a speaker. Write a C program to generate a two-tone siren. The two frequencies of the siren are 1 kHz and 500 Hz.

E10.11 The Freescale MC14489 is a five-digit, seven-segment display driver chip. The data sheet is on the CD for this text (also available on Freescale's website). Describe how to interface one MC14489 with the HCS12 using the SPI subsystem. Write a program to display the value 12345 on the seven-segment displays driven by the MC14489.

E10.12 Define a new font for each of the characters shown in Figure E10.12, and write a program to display these four characters on the four matrix displays driven by one MAX6952. The PM5 signal drives the $\overline{CS}$ signal.

Figure E10.12 ■ User-definable font example

10.15 Lab Exercises and Assignment

L10.1 Use the Freescale seven-segment display driver MC14489 (with SPI interface) to drive five seven-segment displays, and write a program to display the following pattern forever:

```
0
10
210
3210
43210
54321
65432
76543
87654
98765
09876
10987
21098
32109
```
go to the 5th pattern (43210) and repeat.

L10.2 Use the TC72 to measure the ambient temperature and the LCD on the demo board to display the current temperature:

1. Write a program to read the temperature once every 200 ms and display the new temperature reading on the LCD in two rows:

   Temperature is

   XXX.X°C

2. Change the ambient temperature using your hand, hot water in a plastic bag, ice in a plastic bag, and so on to touch the TC72 and record the temperature reading.

L10.3 Use one MAX6952 and four matrix displays (e.g., common-cathode-row Fairchild GMC7175C) to perform the operation described in Exercise E10.12.

11

Inter-Integrated Circuit (I²C) Interface

11.1 Objectives

After completing this chapter, you should be able to

- Understand the characteristics of the I²C protocol
- Describe the I²C protocol in general
- Explain the I²C signal components
- Explain the I²C bus arbitration method
- Explain the I²C data transfer format
- Interface with the real-time-clock chip DS1307
- Use the DS1631A to measure the ambient temperature
- Store and retrieve data in/from the serial EEPROM chip 24LC08B

11.2 The I²C Protocol

The inter-integrated circuit (I²C) serial interface protocol was developed by Philips in the late 1980s. Version 1.0 was published in 1992. This version supports the following:

- Both the 100 Kbps (*standard mode*) and the 400 Kbps (*fast mode*) data rate
- 7-bit and 10-bit addressing
- Slope control to improve electromagnetic compatibility (EMC) behavior

After the publication of version 1.0, I²C became well received by the embedded application developers.

By 1998, the I²C protocol had become an industrial standard—it has been licensed to more than 50 companies and implemented in over 1000 different integrated circuits. However, many applications require a higher data rate than that provided by the I²C protocol. Version 2.0 incorporates the *high-speed mode* (with a data rate of 3.4 Mbps) to address this requirement. Since the I²C module of the HCS12 does not support the high-speed mode, it will not be discussed in this chapter.

11.2.1 Characteristics of I²C Protocol

The I²C protocol has the following characteristics:

- *Synchronous in nature.* A data transfer is always initiated by a master device. A clock signal (SCL) synchronizes the data transfer. The clock rate can vary without disrupting the data. The data rate will simply change along with the changes in the clock rate.
- *Master/slave model.* The master device controls the clock line (SCL). This line dictates the timing of all data transfers on the I²C bus. Other devices can manipulate this line, but they can only force the line low. By forcing the line low, it is possible to clock more data into any device. This is known as "clock stretching."
- *Bidirectional data transfer.* Data can flow in any direction on the I²C bus.
- *Serial interface method.* I²C uses only signals SCL and SDA. The SCL signal is the serial clock signal; the SDA signal is known as serial data. In reality, the SDA signal can carry both the address and data.

11.2.2 I²C Signal Levels

I²C can have only two possible electrical states: *float high* and *driven low*. A master or a slave device drives the I²C bus using an open-drain (or open-collector) driver. As shown in Figure 11.1, both the SDA and SCL lines are pulled up to V_{DD} via pull-up resistors. Because the driver circuit is open drain, it can pull the I²C bus only to low. When the clock or data output is low, the NMOS transistor is turned off. In this situation, no current flows from (to) the bus to (from) the NMOS transistor, and hence the bus line will be pulled to high by the pull-up resistor. Otherwise, the NMOS transistor is turned on and pulls the bus line to low.

The designer is free to use any resistor value for various speeds. But the calculation of what value to use will depend on the capacitance of the driven line and the speed of the I²C communication. In general, the recommended values for the pull-up resistors are 2.2 kΩ and 1 kΩ for standard mode and fast mode, respectively. These values are found to work frequently. When the data rate is below 100 kbps, the pull-up resistor should be set to 4.7 kΩ.

Figure 11.1 ■ Connecting standard- and fast-mode devices to the I²C bus

11.2.3 I²C Data Transfer Signal Components

An I²C data transfer consists of the following fundamental signal components:

- Start (**S**)
- Stop (**P**)
- Repeated start (**R**)
- Data
- Acknowledge (**A**)

START (S) CONDITION

A start condition indicates that a device would like to transfer data on the I²C bus. As shown in Figure 11.2, a start condition is represented by the SDA line going low when the clock (SCL) signal is high. The start condition will initialize the I²C bus. The timing details for the start condition will be taken care of by the microcontroller that implements the I²C bus. Whenever a data transfer using the I²C bus is to be initiated, the designer must tell the micro-controller that a start condition is wanted.

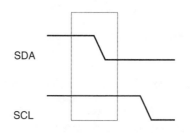

Figure 11.2 ■ I²C start condition

STOP (P) CONDITION

A stop condition indicates that a device wants to release the I²C bus. Once released (the driver is turned off), other devices may use the bus to transmit data. As shown in Figure 11.3, a stop condition is represented by the SDA signal going high when the clock (SCL) signal is high.

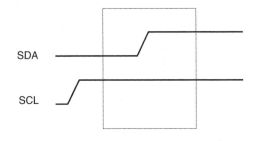

SDA

SCL

Figure 11.3 ■ Stop (P) condition

Once the stop condition completes, both the SCL and the SDA signals will be high. This is considered to be an *idle bus*. After the bus is idle, a start condition can be used to send more data.

REPEATED START (R) CONDITION

A repeated start signal is a start signal generated without first generating a stop signal to terminate the communication. This is used by the master to communicate with another slave or with the same slave in different mode (transmit/receive mode) without releasing the bus. A repeated start condition indicates that a device would like to send more data instead of releasing the line. This is done when a start must be sent but a stop has not occurred. It prevents other devices from grabbing the bus between transfers. The timing diagram of a repeated start condition is shown in Figure 11.4. The repeated start condition is also called *restart* condition. In the figure, there is no stop condition occurring between the start condition and the restart condition.

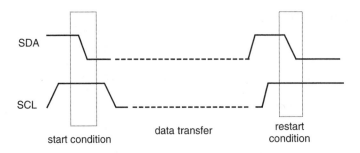

SDA

SCL

start condition   data transfer   restart condition

Figure 11.4 ■ Restart condition

DATA

The data block represents the transfer of 8 bits of information. The data is sent on the SDA line, whereas clock pulses are carried on the SCL line. The clock can be aligned with the data to indicate whether each bit is a 1 or a 0.

Data on the SDA line is considered valid only when the SCL signal is high. When SCL is not high, the data is permitted to change. This is how the timing of each bit works. Data bytes are used to transfer all kinds of information. When communicating with another I²C device, the 8 bits of data may be a control code, an address, or data. An example of 8-bit data is shown in Figure 11.5.

Note. Data bit is always stable when clock (SCL) is high

Figure 11.5 ■ I²C bus data elements

ACKNOWLEDGE (ACK) CONDITION

Data transfer in the I²C protocol needs to be acknowledged either positively (A) or negatively (NACK). As shown in Figure 11.6, a device can acknowledge (A) the transfer of each byte by bringing the SDA line low during the ninth clock pulse of SCL.

Figure 11.6 ■ ACK condition   **Figure 11.7** ■ NACK condition

If the device does not pull the SDA line to low and instead allows the SDA line to float high, it is transmitting a negative acknowledge (NACK). This situation is shown in Figure 11.7.

11.2.4 Synchronization

All masters generate their own clocks on the SCL line to transfer messages on the I²C bus. Data is valid only during the high period of the clock. A defined clock is therefore needed for the bit-by-bit arbitration procedure to take place. For most microcontrollers (including the HCS12 devices), the SCL clock is generated by counting down a programmable reload value using the instruction clock signal.

Clock synchronization is performed using the wired-AND connection of I²C interfaces to the SCL line. This means that a high-to-low transition on the SCL line will cause the devices concerned to start counting off their low period, and once a device clock has gone low, it will hold the SCL line in that state until the high state is reached (CLK1 in Figure 11.8). However, the transition from low to high of this clock may not change the state of the SCL line if another clock (CLK2) is still within its low period. The SCL line will therefore be held low by the device with the longest low period. Devices with shorter low periods enter a high *wait state* during this time.

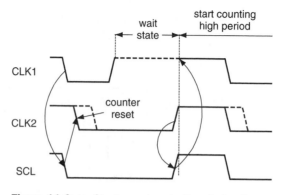

Figure 11.8 ■ Clock synchronization during the arbitration procedure

When all devices concerned have counted off their low period, the clock line will be released and go high. There will then be no difference between the device clocks and the state of the SCL line, and all the devices will start counting their high periods. In this way, a synchronized SCL clock is generated with its low period determined by the device with the longest clock low period and its high period determined by the one with the shortest clock high period.

Clock synchronization occurs when multiple masters attempt to drive the I²C bus and before the arbitration scheme can decide which master is the winner. The I²C bus arbitration process will be discussed in the next section.

HANDSHAKING

The clock synchronization mechanism can be used as a handshake in data transfer. Slave devices may hold the SCL low after completion of a 1-byte transfer (9 bits). In such a case, it halts the bus clock and forces the master clock into wait states until the slave releases the SCL line.

CLOCK STRETCHING

The clock synchronization mechanism can be used by slaves to slow down the bit rate of a transfer. After the master has driven SCL low, the slave can drive SCL low for the required period and then release it. If the slave SCL low period is greater than the master SCL low period, then the resulting SCL bus signal low period is stretched.

11.2.5 Arbitration

I²C allows multiple master devices to coexist in the system. In the event that two or more master devices attempt to begin a transfer at the same time, an arbitration scheme is employed to force one or more masters to give up the bus. The master devices continue transmitting until one attempts a high while the other transmits a low. Since the bus driver has open drain, the bus will be pulled low. The master attempting to transfer a high signal will detect a low on the SDA line and give up the bus by switching off its data output stage. The winning master continues its transmission without interruption; the losing master becomes a slave and receives the rest of the transfer. This arbitration scheme is nondestructive: One device always wins, and no data is lost.

An example of the arbitration procedure is shown in Figure 11.9, where Data1 and Data2 are data driven by device 1 and device 2 and SDA is the resultant data on the SDA line. The moment there is a difference between the internal data level of the master generating Data1 and the actual level on the SDA line, its data output is switched off, which means that a high output level is then connected to the bus. This will not affect the data transfer initiated by the winning master.

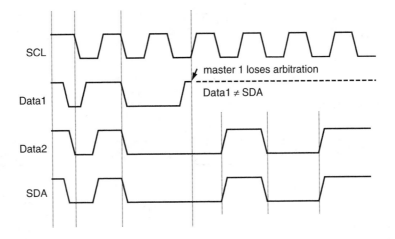

Figure 11.9 ■ Arbitration procedure of two masters

11.2.6 Data Transfer Format

I²C allows a master device to use either the 7-bit or the 10-bit address to specify a slave device for data transfer. The following are the possible I²C data transfer formats:

- *Master transmitter to slave receiver.* The transfer direction is not changed. An example of this format using the 7-bit addressing is shown in Figure 11.10.

- *Master reads slave immediately after the first byte (address byte).* At the moment of the first acknowledgement, the master transmitter becomes a master receiver and the slave receiver becomes a slave transmitter. The first acknowledgement is still generated by the slave. The stop condition is generated by the master, which has previously sent a negative acknowledgement ($\overline{A}$). An example of this format using the 7-bit addressing is shown in Figure 11.11.

- *Combined format.* During a change of direction within a transfer, both the start condition and the slave address are repeated, but with the R/$\overline{W}$ bit reversed. If a master receiver sends a repeated start condition, it has previously sent a negative acknowledgement. An example of this format in the 7-bit addressing is shown in Figure 11.12.

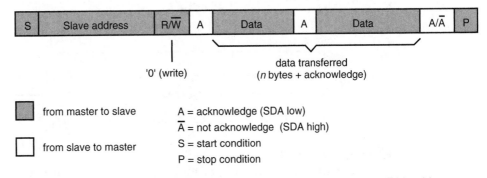

Figure 11.10 ■ A master transmitter addressing a slave receiver with a 7-bit address. The transfer direction is not changed.

Figure 11.11 ■ A master reads a slave immediately after the first byte

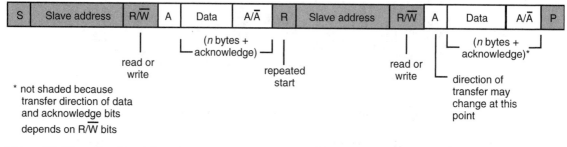

Figure 11.12 ■ Combined format

11.2.7 7-bit Addressing

The addressing procedure for the I²C bus is such that the first byte after the start condition usually determines which slave will be selected by the master. The exception is the *general call address*, which can address all devices. When this address is used, all devices should, in theory, respond with an acknowledgement. However, devices can be made to ignore this address. The second byte of the general call address then defines the action to be taken.

The first byte after the start condition carries the 7-bit address and the direction of the message. The meaning of this byte is shown in Figure 11.13. When the least significant bit is 1, the master device will read information from the selected slave. Otherwise, the master will write information to a selected slave. When an address is sent, each device in a system compares the first seven bits after the start condition with its address. If they match, the device considers itself addressed by the master as a slave receiver or slave transmitter, depending on the R/$\overline{\text{W}}$ bit. The I²C bus committee coordinates the allocation of I²C addresses. Two groups of eight addresses (0000xxx and 1111xxx) are reserved for the purpose, as shown in Table 11.1. The bit combination 11110xx of the slave address is reserved for 10-bit addressing.

Figure 11.13 ■ The first byte after the start condition

GENERAL CALL ADDRESS

The general call address is for addressing every device connected to the I²C bus. However, if a device does not need any of the data supplied within the general call structure, it can ignore this address by not issuing an acknowledgment. If a device does require data from a general call

Slave Address	R̄/W̄ Bit	Description
0000 000	0	General call address
0000 000	1	Start byte [1]
0000 001	x	CBUS address [2]
0000 010	x	Reserved for different bus format [3]
0000 011	x	Reserved for future purpose
0000 1xx	x	Hs-mode master code
1111 1xx	x	Reserved for future purposes
1111 0xx	x	10-bit slave addressing

[1] No device is allowed to acknowledge at the reception of the start byte.

[2] The CBUS address has been reserved to enable the intermixing of CBUS-compatible and I²C-bus compatible devices in the same system. I²C-bus compatible devices are not allowed to respond on reception of this address.

[3] The address reserved for a different bus format is included to enable I²C and other protocols to be mixed. Only I²C-bus compatible devices that can work with such formats and protocols are allowed to respond to this address.

Table 11.1 ■ Definition of bits in the first byte

address, it will acknowledge this address and behave as a slave receiver. The second and following bytes will be acknowledged by every slave receiver capable of handling this data. A slave that cannot process one of these bytes must ignore it by not acknowledging. The meaning of the general call address is always specified in the second byte, as shown in Figure 11.14.

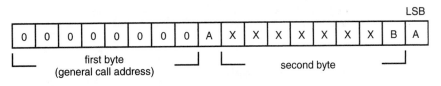

Figure 11.14 ■ General call address format

When bit B in Figure 11.14 is 0, the second byte has the following definitions:

- 00000110 (0x06)—*reset and write programmable part of slave address by hardware.* On receiving this 2-byte sequence, all devices designed to respond to the general call address will reset and take in the programmable part of their address. Precautions have to be taken to ensure that a device is not pulling down the SDA or SCL line after applying the supply voltage, since these low levels would block the bus.

- 00000100 (0x04)—*write programmable part of slave address by hardware.* All devices that define the programmable part of their addresses by hardware will latch this programmable part at the reception of this 2-byte sequence. The device will not reset.

- 00000000 (0x00). This code is not allowed to be used as the second byte.

Sequences of programming procedure are published in the appropriate device data sheets. The remaining codes have not been fixed, and devices must ignore them.

When bit B is 1, the 2-byte sequence is a *hardware general call.* This means that the sequence is transmitted by a hardware master device, such as a keyboard scanner, which cannot be programmed to transmit a desired slave address. Since a hardware master does not know in advance to which device the message has to be transferred, it can only generate this hardware general call and present its own address to identify itself to the system. The seven bits remaining in the second byte contain the address of the hardware master. This address is recognized by an intelligent device (e.g., a microcontroller) connected to the bus, which will then direct the information from the hardware master. If the hardware master can also act as a slave, the slave address is identical to the master address. In some systems, an alternative could be that the hardware master transmitter is set in the slave receiver mode after the system reset. In this way, a system-configuring master can tell the hardware master transmitter (which is now in slave receiver mode) to which address data must be sent (see Figure 11.15). After this programming procedure, the hardware master remains in the master transmitter mode.

Figure 11.15 ■ Data transfer by a hardware transmitter capable of dumping data directly to slave devices. (a) Configuring master sends dump address to hardware master. (b) Hardware master dumps data to selected slave.

11.2.8 10-bit Addressing

Ten-bit addressing is compatible with, and can be combined with, 7-bit addressing. Using 10-bit addressing exploits the reserved combination 1111xxx for the first seven bits of the first byte following a start (S) or repeated start (R) condition. The 10-bit addressing does not affect the existing 7-bit addressing. Devices with 7-bit and 10-bit addresses can be connected to the same I²C bus, and both 7-bit and 10-bit addressing can be used in standard- and fast-mode systems. Although there are eight possible combinations of the reserved address bits 1111xxx, only the four combinations 11110xx are used for 10-bit addressing. The remaining four combinations 11111xx are reserved for future I²C bus enhancements.

DEFINITIONS OF BITS IN THE FIRST TWO BYTES

The 10-bit slave address is formed from the first two bytes following a start (S) condition or repeated start (R) condition. The first seven bits of the first byte are the combination 11110xx, of which the last two bits (xx) are the two most significant bits of the 10-bit address. The eighth bit of the first byte is the $R/\overline{W}$ bit that determines the direction of the message. A 0 in the least significant bit of the first byte means that the master will write information to the selected slave. A 1 in this position indicates that the master will read information from the selected slave.

FORMATS WITH 10-BIT ADDRESS

The following data formats are possible in 10-bit addressing:

- *Master transmitter transmits to slave receiver with a 10-bit slave address.* The transfer direction is not changed (see Figure 11.16). When a 10-bit address follows a start condition, each slave compares the first seven bits of the first byte of the slave address (11110xx) with its own address and tests if the eighth bit is 0. It is possible that more than one device will find a match and generate an acknowledgement (A1). All slaves that find a match will compare the eight bits of the second byte of the slave address with their own addresses, but only one slave will find a match and generate an acknowledgement (A2). The matching slave will remain addressed by the master until it receives a stop condition or a repeated start condition followed by a different slave address.

(write)

Figure 11.16 ■ A master transmitter addresses a slave receiver with a 10-bit address

- *Master receiver reads slave transmitter with a 10-bit address.* The transfer direction is changed after the second $R/\overline{W}$ bit (see Figure 11.17). Up to and including the acknowledgment bit A2, the procedure is the same as that described for a master transmitter addressing a slave receiver. After the repeated start condition, a matching slave remembers that it was addressed before. This slave then checks if the first seven bits of the first byte of the slave address following repeated start (R) are the same as they were after the start condition (S), and tests if the eighth ($R/\overline{W}$) bit is 1. If there is a match, the slave considers that it has been addressed as a transmitter and generates acknowledgment bit A3. The slave transmitter remains addressed until it receives a stop condition (P) or until it

(write)   (read)

Figure 11.17 ■ A master receiver addresses a slave transmitter with a 10-bit address receives another repeated start (R) condition followed by a different slave address. After a repeated start (R) condition, all other slave devices will also compare the first seven bits of the first byte of the slave address (11110xx) with their own addresses and test the eighth bit. However, none of them will be addressed because $R/\overline{W} = 1$ (for 10-bit address) or the 11110xx slave address (for the 7-bit address) does not match.

■ *Combined format*. A master transmits data to a slave and then reads data from the same slave (see Figure 11.18). The same master occupies the bus all the time. The transfer direction is changed after the second $R/\overline{W}$ bit.

Figure 11.18 ■ **Combined format.** A master addresses a slave with a 10-bit address, then transmits data to this slave and reads data from this slave.

■ *Combined format*. A master transmits data to one slave and then transmits data to another slave (see Figure 11.19). The same master occupies the bus all the time.

■ *Combined format*. Ten-bit and 7-bit addressing combined in one serial transfer (see Figure 11.20). After each start (S) condition, or each repeated start condition, a 10-bit or 7-bit slave address can be transmitted. Figure 11.20 shows how a master transmits data to a slave with 7-bit address and then transmits data to a second slave with a 10-bit address. The same master occupies the bus all the time.

Figure 11.19 ■ **Combined format.** A master transmits data to two slaves, both with 10-bit addresses

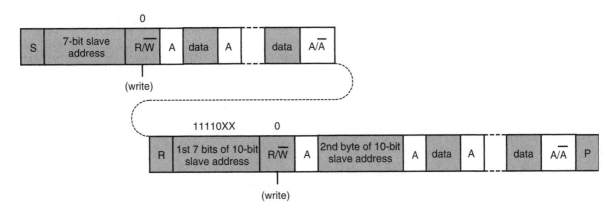

Figure 11.20 ■ **Combined format.** A master transmits data to two slaves, one with a 7-bit address, and one with a 10-bit address.

11.3 An Overview of the HCS12 I²C Module

The I²C module of the HCS12 implements a subset of the I²C standard. It supports most of the master and slave functions and provides interrupts on start and stop bits in hardware to determine if a bus is free (multimaster function). Neither the 10-bit addressing nor the general call addressing is supported by the HCS12 I²C module. The I²C module of the HCS12 may operate at baud rates of up to 100 kbps with maximum capacitive bus loading. With a reduced bus slew rate, the device is capable of operating at a baud rate of up to 400 kbps, provided that the I²C bus slew rate is less than 100 ns. The maximum communications interconnect length and the number of devices that can be connected to the bus are limited by a maximum bus capacitance of 400 pF in all instances.

Two pins are used for data transfer:

- Serial clock (SCL)—PJ7/SCL
- Serial data (SDA)—PJ6/SDA

The I²C module works the same in normal, special, and emulation modes of the HCS12. It has two low power modes: wait and stop modes. The I²C module has five registers to support its operation:

- I²C control register (IBCR)
- I²C status register (IBSR)
- I²C data I/O register (IBDR)
- I²C frequency divider register (IBFD)
- I²C address register (IBAD)

The block diagram of the I²C module is shown in Figure 11.21.

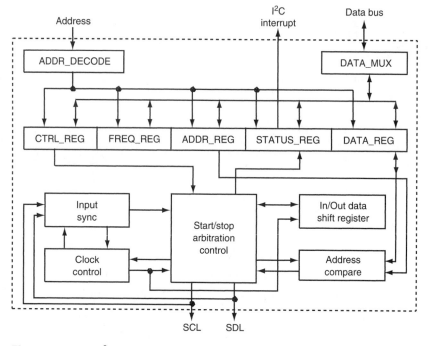

Figure 11.21 ■ I²C block diagram

11.4 Registers for I²C Operation

11.4.1 The I²C Address Register (IBAD)

The IBAD register contains the address to which the I²C module will respond when it is addressed as a slave. The contents of this register are shown in Figure 11.22.

7	6	5	4	3	2	1	0
ADR7	ADR6	ADR5	ADR4	ADR3	ADR2	ADR1	0

Figure 11.22 ■ I²C address register (IBAD)

11.4.2 The I²C Control Register (IBCR)

This register controls all the operation parameters except the baud rate of the I²C module. The contents of this register are shown in Figure 11.23.

When the MS/$\overline{\text{SL}}$ bit is changed from 0 to 1, a start signal is generated on the bus and the master mode is selected. When this bit is changed from 1 to 0, a stop signal is generated and the operation mode changes from master to slave. A stop signal should only be generated if the IBIF flag is set.

The Tx/$\overline{\text{Rx}}$ bit selects the direction of master and slave transfers. When addressed as a slave, this bit should be set by software according to the SRW bit in the status register. In the master mode, this bit should be set according to the type of transfer required. For address cycles, this bit will always be high.

7	6	5	4	3	2	1	0
IBEN	IBIE	MS/$\overline{\text{SL}}$	Tx/$\overline{\text{Rx}}$	TxAK	RSTA	0	IBSWAI

reset: 0   0   0   0   0   0   0   0

IBEN: I²C bus enable

    0 = I²C module is reset and disabled.

    1 = I²C module is enabled. This bit must be set before any other IBCR bits have
        any effect.

IBIE: I²C bus interrupt enable

    0 = interrupts from the I²C module are disabled.

    1 = interrupts from the I²C module enabled.

MS/$\overline{\text{SL}}$: master/slave mode select

    0 = slave mode

    1 = master mode

Tx/$\overline{\text{Rx}}$: transmit/receive mode select

    0 = receive

    1 = transmit

TxAK: transmit acknowledge

    0 = an acknowledge signal will be sent out to the I²C bus on the 9th clock bit
        after receiving 1 byte of data.

    1 = no acknowledge signal response is sent.

RSTA: repeat start

    0 = no action

    1 = generate a repeat start cycle

IBSWAI: I²C bus stop in wait mode

    0 = I²C module clock operates normally.

    1 = stop generating I²C module clock in wait mode.

Figure 11.23 ■ I²C control register (IBCR)

The TxAK bit specifies the value driven onto SDA during data acknowledge cycles for both master and slave receivers. The I²C module will always acknowledge address matches, provided it is enabled, regardless of the value of TxAK. Values written to this bit are only used when the I²C module is a receiver, not a transmitter.

Writing a 1 to the RSTA bit will generate a repeated start condition on the I²C bus, provided it is the current master. This bit will always read as a low.

If the IBSWAI bit is set and the WAI instruction is executed, all clock signals to the I²C module will be stopped and any transmission currently in progress will halt. If the CPU were woken up by a source other than the I²C module, then clocks would restart and the I²C would continue where it left off in the previous transmission. If the IBSWAI bit is cleared when the WAI instruction is executed, the I²C internal clocks and interface would remain active, continuing the operation that is currently underway. It is also possible to configure the I²C such that it will wake up the CPU via an interrupt at the conclusion of the current operation.

11.4.3 The I²C Status Register (IBSR)

This register records the status of all I²C data transmission/reception activities. The contents of this register are shown in Figure 11.24.

7	6	5	4	3	2	1	0
TCF	IAAS	IBB	IBAL	0	SRW	IBIF	RXAK

reset: 1   0   0   0   0   0   0   0

TCF: data transferring bit
  0 = I²C transfer in progress
  1 = I²C transfer complete
IAAS: addressed as a slave
  0 = not addressed
  1 = addressed as a slave
IBB: bus busy bit
  0 = the bus enters idle state.
  1 = I²C bus is busy.
IBAL: arbitration lost
  0 = arbitration is not lost.
  1 = arbitration is lost.
SRW: slave read/write
  0 = slave receive, master writing to slave
  1 = slave transmit, master reading from slave
IBIF: I²C bus interrupt
  0 = no bus interrupt
  1 = bus interrupt
RXAK: receive acknowledge
  This bit reflects the value of SDA during the acknowledge bit of a cycle.
  0 = acknowledge received
  1 = no acknowledge received

Figure 11.24 ■ I²C status register (IBSR)

When a byte is being transferred, the TCF bit is cleared. It is set by the falling edge of the 9th clock of a byte transfer.

The IAAS bit is set (when the I²C module is configured as a slave) when its own address is matched with the calling address. The CPU may be interrupted if the IBIE bit is also set.

The IBB bit indicates the status of the I²C bus. When a start condition is detected, the IBB bit will be set. When a stop condition is detected, this bit is cleared.

The IBAL bit indicates the recent loss in I²C bus arbitration. The arbitration is lost in the following circumstances:

- The SDA signal is sampled low when the master drives a high during an address or data transmit cycle.
- The SDA signal is sampled low when the master drives a high during the acknowledge bit of a data receive cycle.
- A start cycle is attempted when the I²C bus is busy.
- A repeated start cycle is requested in slave mode.
- A stop condition is detected when the master did not request it.

When the IAAS bit is set, the SRW bit indicates the value of the $R/\overline{W}$ command bit of the calling address sent from the master. The SRW bit is valid when the I²C module is in the slave mode, a complete address transfer has occurred with an address match, and no other transfers have been initiated.

The IBIF bit is set when one of the following conditions occurs:

- Arbitration lost (IBAL bit is set)
- Byte transfer complete (TCF bit is set)
- Addressed as slave (IAAS bit set)

This bit is cleared by writing a 1 to it.

11.4.4 I²C Data Register (IBDR)

In master transmit mode (the TxRx bit of the IBCR register set to 1), when data is written into the IBDR register a data transfer is initiated. The most significant bit is sent out first. In master receive mode, reading this register initiates the reception of the next byte. Nine clock pulses will be sent out on the SCL pin to shift in eight data bits and send out the acknowledge bit. In slave mode, the same functions are available after an address match has occurred.

11.4.5 I²C Frequency Divider Register (IBFD)

The most important design consideration of the I²C module is to meet the timing requirements for the start and stop conditions so that data can be correctly transmitted over the bus line. As illustrated in Figure 11.25, there are four timing requirements to be met:

- SCL divider
- SDA hold time
- SCL hold time for start condition
- SCL hold time for stop condition

Figure 11.25 ■ SCL divider and SDA hold

The requirements of these four parameters are listed in Table 11.2. The SCL divider is equal to the bus frequency of the MCU divided by the SCL clock frequency.

Symbol	Parameter	Standard Mode		Fast Mode		Unit
		Min.	Max.	Min.	Max.	
f_{SCL}	SCL clock frequency	0	100	0	400	KHz
$t_{HD;STA}$	SCL hold (start)	4.0	—	0.6	—	μs
$t_{SU;STO}$	SCL hold (stop)	4.0	—	0.6	—	μs
$t_{HD;DAT}$	SDA hold	0	3.45	0	0.9	μs

Table 11.2 ■ I²C bus timing requirements

The I²C timing parameters are set by programming the I²C frequency divider register (IBFD). The contents of this register are shown in Figure 11.26. The contents of this register are used to prescale the bus clock for bit rate selection.

7	6	5	4	3	2	1	0
IBC7	IBC6	IBC5	IBC4	IBC3	IBC2	IBC1	IBC0

Figure 11.26 ■ I²C frequency divider register (IBFD)

The use of these eight bits is as follows:
- IBC7~IBC6: multiply factor (shown in Table 11.3)
- IBC5~IBC3: prescaler divider (shown in Table 11.4)
- IBC2~IBC0: shift register tap points (shown in Table 11.5)

IBC7~IBC6	Multiply Factor
00	01
01	02
10	04
11	Reserved

Table 11.3 ■ Multiply factor

IBC5~IBC3	scl2start (clocks)	scl2stop (clocks)	scl2tap (clocks)	tap2tap (clocks)
000	2	7	4	1
001	2	7	4	2
010	2	9	6	4
011	6	9	6	8
100	14	17	14	16
101	30	33	30	32
110	62	65	62	64
111	126	129	126	128

Table 11.4 ■ Prescaler divider

IBC2~IBC0	SCL Tap (clocks)	SDA Tap (clocks)
000	5	1
001	6	1
010	7	2
011	8	2
100	9	3
101	10	3
110	12	4
111	15	4

Table 11.5 ■ I²C bus tap and prescale values

The circuit for controlling the I²C timing parameters is very complicated and Freescale does not disclose its details. Instead, Freescale provides the equations for computing these four timing parameters and a lookup table for them. A subset of the I²C divider and hold values is shown in Table 11.6; the complete lookup table can be found in the *I²C Block Guide* published by Freescale.

The equation for generating the divider value from the IBFD bits is

$$\text{SCL divider} = \text{MUL} \times \{2 \times (\text{scl2tap} + [(\text{SCL\_tap} - 1) \times \text{tap2tap}] + 2)\} \quad (11.1)$$

The SDA hold delay is equal to the bus clock period multiplied by the SDA hold value shown in Table 11.6. The equation used to generate the SDA hold value from the IBFD bits is

$$\text{SDA hold} = \text{MUL} \times \{\text{scl2tap} + [(\text{SDA\_tap} - 1) \times \text{tap2tap}] + 3)\} \quad (11.2)$$

The equations for HCL hold values to generate the start and stop conditions from IBFD bits are as follows:

$$\text{SCL hold (start)} = \text{MUL} \times [\text{scl2start} + (\text{SCL\_tap} - 1) \times \text{tap2tap}] \quad (11.3)$$
$$\text{SCL hold (stop)} = \text{MUL} \times [\text{scl2stop} + (\text{SCL\_tap} - 1) \times \text{tap2tap}] \quad (11.4)$$

IBC[7:0] (hex)	SCL Divider (clocks)	SDA Hold (clocks)	SCL Hold (start)	SCL Hold (stop)
MUL = 1				
00	20	7	6	11
07	40	10	16	21
0B	40	9	16	21
14	80	17	34	41
18	80	9	38	41
1D	160	25	78	81
1F	240	33	118	121
20	160	17	78	81
25	320	49	158	161
27	480	65	238	241
28	320	33	158	161
MUL = 2				
40	40	14	12	22
45	60	18	22	32
47	80	20	32	42
4B	80	18	32	42
54	160	34	68	82
58	160	18	76	82
5D	320	50	156	162
5F	480	66	236	242
60	320	34	156	162
MUL = 4				
80	80	28	24	44
85	120	36	44	64
87	160	40	64	84
8B	160	36	64	84
94	320	68	136	164
98	320	36	152	164

Table 11.6 ■ I²C divider and hold values

Example 11.1

Assuming that the HCS12 is running with a 24-MHz bus clock, compute the values to be written into the IBFD register to set the baud rate to 100 KHz and 400 KHz.

Solution:
Case 1: Baud rate = 100 KHz
The SCL divider = 24 MHz ÷ 100 KHz = 240

From Table 11.6, there is only one IBFD value ($1F) associated with this SCL divider value. Using this value to look up Tables 11.3, 11.4, and 11.5, we obtain the following parameter values:

- MUL = 1
- scl2tap = 6
- tap2tap = 8

- scl2start = 6
- scl2stop = 9
- SDA_tap = 4
- SCL_tap = 15

SDA hold time	= MUL × {scl2tap + [(SDA_tap − 1) × tap2tap] + 3)}
	= 1 × {6 + [(4 − 1) × 8] + 3)}
	= 33 E cycles = 1.375 μs ($<$ 3.45 μs)
SCL hold time (start)	= MUL × [scl2start + (SCL_tap − 1) × tap2tap]
	= 1 × [6 + (15 − 1) × 8] = 118 E cycles = 4.92 μs ($>$ 4.0 μs)
SCL hold time (stop)	= MUL × [scl2stop + (SCL_tap − 1) × tap2tap]
	= 1 × [9 + (15 − 1) × 8] = 121 E cycles = 5.04 μs ($>$ 4.0 μs)

The computed values satisfy the requirements.

Case 2: Baud rate = 400 KHz
The SCL divider = 24 MHz ÷ 400 KHz = 60

From Table 11.6, there is only one corresponding IBFD value ($45) for this SCL divider value. Using this value to look up Tables 11.3, 11.4, and 11.5, we obtain the following parameter values:

- MUL   = 2
- scl2tap   = 4
- tap2tap   = 1
- scl2start = 2
- scl2stop = 7
- SDA_tap   = 3
- SCL_tap = 10

SDA hold time	= 2 × {scl2tap + [(SDA_tap − 1) × tap2tap] + 3)}
	= 2 × {4 + [(3 − 1) × 1] + 3}
	= 18 E cycles = 0.75 μs ($<$ 0.9 μs)
SCL hold time (start)	= MUL × [scl2start + (SCL_tap − 1) × tap2tap]
	= 2 × [2 + (10 − 1) × 1] = 22 E cycles = 0.917 μs ($>$ 0.6 μs)
SCL hold time (stop)	= MUL × [scl2stop + (SCL_tap − 1) × tap2tap]
	= 2 × [7 + (10 − 1) × 1] = 32 E cycles = 1.33 μs ($>$ 0.6 μs)

The computed values satisfy the requirements. A quicker way of timing verification is to use the numbers listed in Table 11.6 instead of using Equations 11.2 to 11.4.

11.5 Programming the I²C Module

Before the I²C module can transmit and receive data correctly, it must be initialized properly. The initialization procedure is as follows:

- Compute the value that can obtain the SCL frequency from the bus clock and use it to update the IBFD register.
- Optionally load the IBAD register to define its slave address.

- Set the IBEN bit of the IBCR register to enable the I²C system.
- Modify the bits of the IBCR register to select master/slave mode, transmit/receive mode, and interrupt enable mode

The initialization of the I²C module can be performed using a macro or a subroutine. The subroutine that performs the I²C initialization is as follows:

```
;*************************** *************************** ***************************
;
; The following function has two incoming parameters passed in accumulator A and B
; to set up baud rate and slave address, respectively.
;*************************** *************************** ***************************

openI2C   bset   IBCR,IBEN       ; enable I²C module
          staa   IBFD            ; establish SCL frequency
          stab   IBAD            ; establish I²C module slave address
          bclr   IBCR,IBIE       ; disable I²C interrupt
          bset   IBCR,IBSWAI     ; disable I²C in wait mode
          rts
```

The C language version of the subroutine is as follows:

```
void openI²C(char ibc, char i2c_ID)
{
      IBCR |= IBEN;          /* enable I²C module */
      IBFD = ibc;            /* set up I²C baud rate */
      IBAD = i2c_ID;         /* set up slave address */
      IBCR &= ~IBIE;         /* disable I²C interrupt */
      IBCR |= IBSWAI;        /* disable I²C in wait mode */
}
```

A successful data exchange on the I²C bus requires the user to generate the I²C signal components in a proper order. The generation of these signal components is described in the following sections.

11.5.1 Generation of the Start Condition

All data transmissions in the I²C bus begin with the start condition. The start condition can only be sent when the bus is idle. If the MCU is connected to a multimaster bus system, the state of the IBB bit of the IBSR register must be tested to check whether the serial bus is busy.

If the bus is idle (IBB = 0), the start condition and the first byte can be sent. The first byte consists of the slave address and the R/$\overline{W}$ bit. The bus free time (i.e., the time between a *stop* condition and the following *start* condition) is built into the hardware that generates the start cycle.

Example 11.2

▼

Write a subroutine that generates a start condition and also sends out the address of the slave with which the MCU intends to exchange data. The address and the R/$\overline{W}$ bit value is passed in accumulator A.

Solution: The subroutine is as follows:

```
sendSlaveID   brset   IBSR,IBB,*        ; wait until I²C bus is free
              bset    IBCR,TXRX+MSSL   ; generate a start condition
              staa    IBDR             ; send out the slave address
```

```
          brclr    IBSR,IBIF,*        ; wait for address transmission to complete
          movb     #IBIF,IBSR         ; clear the IBIF flag
          rts
```

The C language version of the function is as follows:

```
void sendSlaveID (char cx)
{
    while (IBSR&IBB);          /* wait until I²C bus is idle */
    IBCR | = TXRX+MSSL;        /* generate a start condition */
    IBDR = cx;                 /* send out the slave address with R/W bit */
    while(!(IBSR & IBIF));     /* wait for address transmission to complete */
    IBSR = IBIF;               /* clear IBIF flag */
}
```

11.5.2 I²C Data Transfer in Master Mode

When the HCS12 I²C module is configured as a master, it is responsible for initiating the data transmission and reception. By writing a byte into the IBDR register, nine clock pulses are generated to shift out eight bits of data and shift in the acknowledgment bit. By reading from the IBDR register, nine clock pulses are triggered to shift in eight bits of data and shift out the acknowledgment (positive or negative) bit.

After sending the slave ID and receiving the acknowledgment from the slave, the I²C module can send out a byte by performing the following three steps:

Step 1
Store the data byte in the IBDR register. This step will trigger nine clock pulses on the SCL pin to shift out eight data bits and shift in the acknowledgment bit from the slave.

Step 2
Wait until the IBIF flag is set to 1, which indicates the byte has been shifted out.

Step 3
Clear the IBIF flag by writing a 1 to it.

If the data byte is stored in accumulator A, then the following instruction sequence can send a byte to a slave via the I²C bus:

```
staa     IBDR
brclr    IBSR,IBIF,*       ; wait until IBIF flag is set to 1
movb     #IBIF,IBSR        ; clear the IBIF flag
```

The following C statements send out the variable cx via the I²C bus:

```
IBDR = cx;                  /* send out the value cx */
while (!(IBSR & IBIF));     /* wait until the byte is shifted out */
IBSR = IBIF;                /* clear the IBIF flag */
```

To send multiple bytes to the slave, one can place the previous three instructions in a loop and use an index register to point to the data to be sent. The loop will be executed as many iterations as needed to send out all the data bytes.

After sending out the slave ID and receiving the acknowledgment from the slave, the I²C module can read a data byte by performing the following steps:

Step 1
Clear the Tx/Rx bit of the IBCR register to 0 for data reception. Clear the TxAK bit of the IBCR register to 0 if the user wants to acknowledge the data byte. This should be done

when the user wants to receive multiple bytes from the same slave. If the user wants to read only 1 byte or when the byte to be read is the last byte of a read sequence, then the TxAK bit should be set to 1. Without receiving acknowledgement, the slave will stop driving the SDA line at the end of the eight clock pulses.

Step 2

Perform a dummy read. This action will trigger nine clock pulses to be sent out on the SCL pin to shift in eight data bits and send out an acknowledgment.

Step 3

Wait until the IBIF flag is set to 1.

Step 4

Clear the IBIF flag by writing a 1 to it.

The following instruction sequence reads a byte and acknowledges it:

```
bclr    IBCR,TXRX+TXAK   ; prepare to receive and acknowledge
ldaa    IBDR             ; a dummy read to trigger nine clock pulses
brclr   IBSR,IBIF,*      ; wait until the data byte is shifted in
movb    #IBIF,IBSR       ; clear the IBIF flag
ldaa    IBDR             ; place the received byte in A and also initiate the
                         ; next read sequence
```

The same read operation can be performed by the following C statements:

```
IBCR &= ~(TXRX + TXAK);   /* prepare to receive and acknowledge */
dummy = IBDR;             /* a dummy read */
while(!(IBSR & IBIF));    /* wait for the byte to shift in */
IBSR = IBIF;             /* clear the IBIF flag */
buf = IBDR;              /* place the received byte in buf and also initiate
                            the next read sequence */
```

The following instruction sequence reads a byte, sends out negative acknowledgment, and also generates a stop condition:

```
bclr    IBCR,TXRX        ; prepare to receive
bset    IBCR,TXAK        ; to send negative acknowledgment
ldaa    IBDR             ; dummy read to trigger clock pulses
brclr   IBSR,IBIF,*      ; wait until the byte is shifted in
movb    #IBIF,IBSR       ; clear the IBIF flag
bclr    IBCR,MSSL        ; generate a stop condition
ldaa    IBDR             ; place the received byte in A
```

The same operation can be performed by the following C statements:

```
IBCR    &= ~TXRX;        /* prepare to receive */
IBCR    |= TXAK;         /* prepare not to acknowledge */
dummy   = IBDR;          /* a dummy read to trigger nine clock pulses */
while(!(IBSR & IBIF));   /* wait for a byte to shift in */
IBSR    = IBIF;          /* clear the IBIF flag */
IBCR    &= ~MSSL;        /* generate a stop condition */
buf     = IBDR;          /* place the received byte in buf */
```

11.5.3 I²C Data Transfer in Slave Mode

In the slave mode, the I²C module cannot initiate any data transfer. After reset and a stop condition, the I²C module is in the slave mode. Once the I²C module is enabled in the slave mode, it waits for a start condition to occur. Following the start condition, eight bits are shifted into the IBDR register. The value of the upper seven bits of the received byte is compared with the IBAD register. If the address matches, the following events will occur:

- The bit 0 of the address byte is copied into the SRW bit of the IBSR register.
- The IAAS bit is set to indicate address match.
- An ACK pulse is generated regardless of the value of the TxAK bit.
- The IBIF flag is set.

After the address match, the I²C module in slave mode should acknowledge every data byte received.

When operating in slave mode, the user program needs to make sure that the address match has occurred before it prepares to receive or transmit. The following instruction sequence can detect the address match and take appropriate actions:

```
            brset  IBSR,IAAS,addr_match      ; is address matched?
            . . .
addr_match  brclr  IBSR,SRW,slave_rd
            bset   IBCR,TXRX                 ; prepare to transmit data
            movb   tx_buf,IBDR               ; place data in IBDR to wait for SCL to shift it out
            brclr  IBSR,IBIF,*               ; wait for data to be shifted out
            . . .
slave_rd    bclr   IBCR,TXAK+TXRX            ; prepare to receive and send ACK
            brclr  IBSR,IBIF,*               ; wait for data byte to shift in
            movb   #IBIF,IBSR                ; clear the IBIF flag
            movb   IBDR,rcv_buf              ; save the received data
            . . .
```

Data transmission and reception in I²C protocol can also be made interrupt-driven. Using the interrupt mechanism to control data transmission and reception allows the MCU to perform other tasks during the waiting period.

11.6 The Serial Real-Time Clock DS1307

The DS1307 is a real-time clock (RTC) chip that uses BCD format to represent clock and calendar information. It has 56 bytes of nonvolatile SRAM for storing critical information. The address and data are transferred via the I²C bus. The clock and calendar provide seconds, minutes, hours, day, date, month, and year information. The end-of-the-month data is automatically adjusted for months with fewer than 31 days, including corrections for leap year. The clock operates in either the 24-hour or 12-hour format with an AM/FM indicator. The DS1307 has a built-in power-sense circuit that detects power failure and automatically switches to the battery supply. The pin assignment and the block diagram of the DS1307 are shown in Figure 11.27.

11.6.1 Signal Functions

- V_{CC}, GND: *DC power input.* V_{CC} is the +5 V input. When 5 V is applied within normal limits, the device is fully accessible and data can be written and read. When a 3-V battery is connected to the device and V_{CC} is below $1.25 \times V_{BAT}$, reads and

Figure 11.27 ■ DS1307 pin assignment and block diagram

writes are inhibited. However, the timekeeping function continues unaffected by the lower input voltage. As V_{CC} falls below V_{BAT}, the RAM and timekeeper are switched over to the external power supply at V_{BAT}.

- V_{BAT}: *battery input* for any standard 3-V lithium cell or other energy source. Battery voltage must be held between 2.0 V and 3.5 V for proper operation. The nominal write protect trip point voltage at which access to the RTC and user RAM is denied is set by the internal circuitry as $1.25 \times V_{BAT}$ nominal. A lithium battery with 48-mA-hr or greater will back up DS1307 for more than 10 years in the absence of power at 25°C.

- SCL: *serial clock input.*

- SDA: *serial data input/output.*

- SQWOUT: *square wave output driver.* When enabled (by setting the SQWE bit of the control byte to 1), the SQWOUT pin output may be programmed to one of the four square wave frequencies (1 Hz, 4 kHz, 8 kHz, 32 kHz). The SQWOUT pin is open drain and requires an external pull-up resistor. SQWOUT will operate with either V_{CC} or V_{BAT} applied.

- X1, X2: *crystal connection.* These two pins are used for connections to a standard 32,768-Hz quartz crystal.

11.6.2 RTC and RAM Address Map

The address map for the RTC and RAM registers of the DS1307 is shown in Figure 11.28. The RTC registers are located at address locations $00 to $07. The RAM registers are located at address locations $08 to $3F. During a multibyte access, when the address pointer reaches $3F, it wraps around to location $00, the beginning of the clock space.

$00	seconds
$01	minutes
$02	hours
$03	day
$04	date
$05	month
$06	year
$07	control
$08 ⋮ $3F	RAM 56 x 8

Figure 11.28 ■ DS1307 address map

11.6.3 Clock and Calendar

The time and calendar information is obtained by reading the appropriate register bytes. The contents of the RTC registers are illustrated in Figure 11.29. The time and calendar are set or initialized by writing the appropriate register bytes, and the contents of the registers are in the BCD format. Bit 7 of register 0 is the clock halt (CH) bit. When this bit is set to 1, the oscillator is disabled. When this bit is a 0, the oscillator is enabled. This bit should be cleared to 0 after reset. The initial power-on states of all registers are not defined. It is important that the user remembers enabling the oscillator during the initial configuration.

Bit 7							Bit 0
CH	10 seconds			seconds			
0	10 minutes			minutes			
0	12/24	10 HR / A/P	10 HR	hours			
0	0	0	0	0	day		
0	0	10 date		date			
0	0	0	10 month	month			
10 year				year			
out	0	0	sqwe	0	0	RS1	RS0

Figure 11.29 ■ Contents of RTC registers

The DS1307 can be run in either 12-hour or 24-hour mode. Bit 6 of the hours register is defined as the 12- or 24-hour mode-select bit. When high, the 12-hour mode is selected. In the 12-hour mode, bit 5 is the AM/PM bit, with logic high being PM. In the 24-hour mode, bit 5 is the second 10-hour bit (20~23 hours).

11.6.4 The DS1307 Control Register

The contents of the control register are shown in the bottom row of Figure 11.29. Bit 7 controls the output level of the SQWOUT pin when the square output is disabled (i.e., the SQWE bit is 0). When bit 7 is 1, the SQWOUT output level is 1 when bit 4 (the SQWE bit) is set to 0. Otherwise, the SQWOUT output level is 0. The SQWE bit enables the SQWOUT pin (oscillator) output. The frequency of the SQWOUT output depends on the values of the RS1~RS0 bits. With the square wave output set to 1 Hz, the clock registers are updated on the falling edge of the square wave.

The frequency of the square wave output is set by programming the RS1 and RS0 bits with appropriate values. Table 11.7 lists the square wave frequencies that can be selected with the RS bits.

RS1	RS0	SQW Output Frequency
0	0	1 Hz
0	1	4.096 KHz
1	0	8.192 KHz
1	1	32.768 KHz

Table 11.7 ■ Square wave output frequency

11.6.5 Data Transfer

The DS1307 supports the standard mode (100 Kbps baud rate) of the I²C bus. The device address of the DS1307 is % 1101000. There are two types of data transfer between the DS1307 and an MCU: *slave receiver mode* and *slave transmitter mode*.

SLAVE RECEIVER MODE

In this mode, the MCU sends the device address of the DS1307, the address of the register to be accessed, and 1 or multiple data bytes to the DS1307. The following events occur:

- The MCU generates a start condition.
- The MCU sends a device address byte to the DS1307 with the direction bit (R/$\overline{\text{W}}$) set to 0.
- DS1307 acknowledges the address byte.
- The MCU sends the address of the register to be accessed to the DS1307. This value sets the register pointer. Only the address of the first register needs to be sent to the DS1307 during a multiple-byte transfer.
- The MCU sends 1 or multiple bytes of data to the DS1307 and the DS1307 acknowledges each byte received.
- The MCU generates the stop condition to terminate the data write.

SLAVE TRANSMITTER MODE

In this mode, the MCU sends the device address of the DS1307 and the address of the register to be read to the DS1307, and reads back one or multiple data bytes from the DS1307. The following events occur:

- The MCU generates a start condition.
- The MCU sends a device address byte to the DS1307 with the direction bit (R/$\overline{W}$) set to 0.
- DS1307 acknowledges the address byte.
- The MCU sends the address of the register to be accessed to the DS1307. This value sets the register pointer. The register pointer is incremented by 1 after each register transfer.
- The DS1307 acknowledges the register address byte.
- The MCU generates a restart condition.
- The MCU sends the device address to the DS1307 with the direction bit set to 1.
- The DS1307 acknowledges the device address byte.
- The DS1307 sends 1 or multiple bytes to the MCU and the MCU acknowledges each byte received except the last byte.
- The MCU generates the stop condition to terminate the data read.

11.6.6 Circuit Connection

The circuit connection between the DS1307 and the HCS12 is shown in Figure 11.30. The HCS12 may also connect an LCD or LED circuit to display the current time and calendar. After the HCS12 initializes the current time and calendar information, the DS1307 will interrupt the HCS12 periodically (most commonly, once per second) so that the HCS12 can update the time display periodically.

Figure 11.30 ■ Typical circuit connection between the HCS12 and the DS1307

Example 11.3

▼

Write a function to configure the DS1307 to operate with the following setting:

- SQWOUT output enabled
- SQWOUT output set to 1 Hz
- SQWOUT idle high when it is disabled

Solution: The user needs to send the following bytes to the DS1307:

- Address byte $D0 (bit 0 is 0 to select write operation)
- Register address $07
- Control byte $90 (passed in accumulator B)

The following function will configure the DS1307 accordingly:

```
openDS1307   ldaa    #$D0                      ; place DS1307 ID in A
             jsr     sendSlaveID
             brclr   IBSR,RXAK,sndRegAdr       ; did DS1307 acknowledge?
             ldab    #$FF                      ; return error code −1
             rts
sndRegAdr    movb    #$07,IBDR                 ; send out the control register address
             brclr   IBSR,IBIF,*               ; wait until the register address is shifted out
             movb    #IBIF,IBSR                ; clear the IBIF flag
             brclr   IBSR,RXAK,sndok           ; did DS1307 acknowledge?
             ldab    #$FF
             rts
sndok        stab    IBDR                      ; send out control byte
             brclr   IBSR,IBIF,*               ; wait until the control byte is shifted out
             movb    #IBIF,IBSR
             bclr    IBCR,MSSL                 ; generate stop condition
             rts
```

The same configuration can be performed by the following C function:

```c
char openDS1307(char ctrl)
{
    sendSlaveID(0xD0);      /* send out DS1307's ID */
    if (IBSR & RXAK)        /* if DS1307 did not acknowledge, send error code */
        return −1;
    IBDR = 0x07;            /* send out control register address */
    while(!(IBSR & IBIF));
    IBSR = IBIF;            /* clear IBIF flag */
    if (IBSR & RXAK)        /* if DS1307 did not acknowledge, send error code */
        return −1;
    IBDR = ctrl;            /* send out control byte */
    while(!(IBSR & IBIF));
    IBSR = IBIF;
    if (IBSR & RXAK)        /* if DS1307 did not acknowledge, send error code */
        return −1;
    IBCR &= ~MSSL;          /* generate a stop condition */
    return 0;
}
```

▲

Example 11.4

Write a function to read the time and calendar information from the DIP switches and store them in a buffer to be sent to the DS1307. The DIP switches are driven by Port AD1.

Solution: The user must be informed to enter the time and calendar. Since there are 7 bytes of time and calendar information, the user will need to use the DIP switches to enter data seven times. The procedure for entering a byte of information is as follows:

1. The function outputs a message to ask the user to enter a value. The prompt message is as follows:

     Enter xxxx:

2. The user sets up a value using the DIP switches and presses the button connected to the PJ0 pin to remind the MCU to read the value.

3. The MCU reads the value of the DIP switches and sends it to the DS1307.

The function that reads the time and calendar information and stores them in a buffer is as follows:

```
tready    ds.b    1                    ; a flag to indicate that data is ready
getTime   pshx
          pshy
          ldy     #buf                 ; Y is the pointer to the buffer
          movb    #$FF,ATD1DIEN        ; enable Port AD1 for digital inputs
          bclr    DDRJ,BIT0            ; enable PJ0 pin for input
          bset    PERJ,BIT0            ; enable pull-up or pull-down on PJ0 pin
          bclr    PPSJ,BIT0            ; enable pull-down so that interrupt is rising-edge triggered
          bset    PIEJ,BIT0            ; enable PJ0 interrupt
          cli                         ;          "
          movb    #0,tready            ; clear the data ready flag to 0
          ldaa    #$80                 ; set LCD cursor to the upper left corner
          jsr     cmd2lcd             ;          "
          ldx     #prompty            ; output the prompt "Enter year:"
          jsr     puts2lcd            ;          "
waity     tst     tready              ; is new year info. ready?
          beq     waity              ;          "
          movb    PTAD1,1,y+          ; save year info. in buffer
          movb    #0,tready
          ldaa    #$80                 ; set LCD cursor to the upper left corner
          jsr     cmd2lcd             ;          "
          ldx     #promptm            ; output the prompt "Enter month:"
          jsr     puts2lcd            ;          "
waitm     tst     tready              ; is new month info. ready?
          beq     waitm              ;          "
          movb    PTAD1,1,y+          ; save month info. in buffer
          movb    #0,tready            ; clear the ready flag
          ldaa    #$80                 ; set LCD cursor to the upper left corner
          jsr     cmd2lcd             ;          "
          ldx     #prompte            ; output the prompt "Enter date:"
          jsr     puts2lcd            ;          "
```

```
waite      tst     tready          ; is new date info. ready?
           beq     waite           ;         "
           movb    PTAD1,1,y+      ; save date info. in buffer
           movb    #0,tready       ; clear the ready flag
           ldaa    #$80            ; set LCD cursor to the upper left corner
           jsr     cmd2lcd         ;         "
           ldx     #promptd        ; output the prompt "Enter day:"
           jsr     puts2lcd        ;         "
waitd      tst     tready          ; is new day info. ready?
           beq     waitd           ;         "
           movb    PTAD1,1,y+      ; save day info. in buffer
           movb    #0,tready
           ldaa    #$80            ; set LCD cursor to the upper left corner
           jsr     cmd2lcd         ;         "
           ldx     #prompth        ; output the prompt "Enter hours:"
           jsr     puts2lcd        ;         "
waith      tst     tready          ; is new hour info. ready?
           beq     waith           ;         "
           movb    PTAD1,1,y+      ; save hour info. in buffer
           movb    #0,tready
           ldaa    #$80            ; set LCD cursor to the upper left corner
           jsr     cmd2lcd         ;         "
           ldx     #promptmi       ; output the prompt "Enter minutes:"
           jsr     puts2lcd        ;         "
waitmi     tst     tready          ; is new minute info. ready?
           beq     waitmi          ;         "
           movb    PTAD1,1,y+      ; save hour info. in buffer
           movb    #0,tready
           ldaa    #$80            ; set LCD cursor to the upper left corner
           jsr     cmd2lcd         ;         "
           ldx     #prompts        ; output the prompt "Enter seconds:"
           jsr     puts2lcd        ;         "
waits      tst     tready          ; is new second info. ready?
           beq     waits           ;         "
           movb    PTAD1,1,y+      ; save second info. in buffer
           movb    #0,tready
           puly
           pulx
           rts
#include    "c:\miniIDE\lcd_util_ SSE256.asm"
#include    "c:\miniIDE\delay.asm"
prompts     fcc     "Enter seconds:"
            dc.b    0
promptmi    fcc     "Enter minutes:"
            dc.b    0
prompth     fcc     "Enter hours:"
            dc.b    0
promptd     fcc     "Enter day:"
            dc.b    0
prompte     fcc     "Enter date:"
            dc.b    0
```

```
promptm    fcc      "Enter month:"
           dc.b     0
prompty    fcc      "Enter year:"
           dc.b     0
```

▲

This function is interrupt-driven. It clears the flag tready to 0 after outputting the prompt message. When setting the DIP switches, the designer presses the button connected to the PJ0 pin. Pressing the PJ0 button causes an interrupt to be requested to the MCU. The interrupt service routine simply sets the flag tready to 1 and returns. Since the getTime function is polling this flag, it will detect the value change and read the value of the DIP switches. The PJ0 pin interrupt service routine is very simple:

```
PJ_ISR    movb    #1,tready
          movb    #1,PIFJ      ; clear the PIFJ0 flag
          rti
```

The C language version of the function will be left as an exercise problem.

Example 11.5

▼

Write a function to send the time and calendar information to the DS1307. The time and calendar information is pointed to by X. The device ID and the starting register address are passed in A and B, respectively. X points to the value of year and the second's value is located at [X] + 6.

Solution: The function that sends time and calendar information to the DS1307 is as follows:

```
sendTime      jsr      sendSlaveID               ; send out device ID of the DS1307
              brclr    IBSR,RXAK,sndTimeOK1      ; did DS1307 acknowledge?
              ldab     #$FF                      ; return error code −1 if not acknowledged
              rts
sndTimeOK1    stab     IBDR                      ; send out register address for seconds
              brclr    IBSR,IBIF,*               ; wait until seconds' address has been shifted out
              movb     #IBIF,IBSR                ; clear the IBIF flag
              brclr    IBSR,RXAK,sndTimeOK2      ; did 1307 acknowledge?
              ldab     #$FF                      ; return error code −1 if not acknowledged
              rts
sndTimeOK2    ldy      #7                        ; byte count
              tfr      X,D                       ; set X to point to second's value
              addd     #6                        ;        "
              tfr      D,X                       ;        "
sndloop       movb     1,x-,IBDR                 ; send out 1 byte
              brclr    IBSR,IBIF,*
              movb     #IBIF,IBSR
              brclr    IBSR,RXAK,sndTimeOK3      ; did DS1307 acknowledge?
              ldab     #$FF                      ; return error code −1 if not acknowledged
              rts
sndTimeOK3    dbne     y,sndloop                 ; continue until all bytes have been sent out
              bclr     IBCR,MSSL                 ; generate a stop condition.
              ldab     #0                        ; return normal return code 0
              rts
```

The same operation can be performed by the following C function:

```
char sendTime(char *ptr, char ID)
{
    char i;
    sendSlaveID(0xD0);          /* send ID to DS1307 */
    if(IBSR & RXAK)             /* did DS1307 acknowledge? */
        return −1;
    IBDR = 0x00;                /* send out seconds register address */
    while(!(IBSR & IBIF));
    IBSR = IBIF;
    if(IBSR & RXAK)
        return −1;
    for(i = 6; i >= 0; i−−) {
        IBDR = *(ptr+i);
        while(!(IBSR&IBIF));
        IBSR = IBIF;
        if(IBSR & RXAK)
        return −1;
    }
    return 0;
}
```

Example 11.6

Write a C function to read the time of day from the DS1307 and save it in the array cur_time[0. . .6].

Solution: The following C function will read the time of day from the DS1307:

```
char readTime(char cx)
{
    char i, temp;
    sendSlaveID(0xD0);          /* generate a start condition and send DS1307's ID */
    if (IBSR & RXAK)
            return −1;          /* if DS1307 did not respond, return error code *
    IBDR = cx;                  /* send address of seconds register */
    while(!(IBSR & IBIF));
    IBSR = IBIF;                /* clear the IBIF flag */
    if (IBSR & RXAK)
            return −1;          /* if DS1307 did not respond, return error code */
    IBCR | = RSTA;              /* generate a restart condition */
    IBDR = 0xD1;                /* send ID and set R/W flag to read */
    while(!(IBSR & IBIF));
    IBSR = IBIF;
    if (IBSR & RXAK)
            return −1;          /* if DS1307 did not respond, return error code */
    IBCR &= ~(TXRX + TXAK);     /* prepare to receive and acknowledge */
    temp = IBDR;                /* a dummy read to trigger nine clock pulses */
```

```
for (i = 0; i < 5; i++) {
        while(!(IBSR & IBIF));        /* wait for a byte to shift in */
        IBSR = IBIF;                  /* clear the IBIF flag */
        cur_time[i] = IBDR;           /* save the current time in buffer */
}                                     /* also initiate the next read */
while (!(IBSR & IBIF));               /* wait for the receipt of cur_time[5] */
        IBSR = IBIF;                  /* clear IBIF flag */
IBCR | = TXAK;                        /* not to acknowledge cur_time[6] */
cur_time[5] = IBDR;                   /* save cur_time[5] and initiate next read */
while (!(IBSR & IBIF));
IBSR = IBIF;
IBCR &= ~MSSL;                        /* generate stop condition */
cur_time[6] = IBDR;
return 0;
}
```

▲

Example 11.7

▼

Write a function to format the time information stored in the array cur_time[0..6] so that it can be displayed on the LCD.

Solution: The time and calendar information are converted and stored in two character arrays: **hms[0. . .11]** and **mdy[0. . .11]**. The array hms[0. . .11] holds the hours, minutes, and seconds, whereas the array dmy[0. . .11] holds month, date, and year. The format of display is as follows:
In 24-hour mode,

hh:mm:ss:xx

mm:dd:yy

where xx stands for day of week (can be "SU," "MO," "TU," "WE," "TH," "FR," and "SA").
In 12-hour mode,

hh:mm:ss:ZM

xx:mm:dd:yy

where Z can be "A" or "P" to indicate AM or PM and xx stands for day of week.
The following C function will format the current time of day into two strings:

```
void formatTime(void)
{
    char temp3;
    temp3 = cur_time[3] & 0x07;                      /* extract day-of-week */
    if (cur_time[2] & 0x40) {                        /* if 12-hour mode is used */
        hms[0] = 0x30 + ((cur_time[2] & 0x10) >> 4); /* tens hour digit */
        hms[1] = 0x30 + (cur_time[2] & 0x0F);        /* ones hour digit */
        hms[2] = ':';
        hms[3] = 0x30 + (cur_time[1] >> 4);          /* tens minute digit */
        hms[4] = 0x30 + (cur_time[1] & 0x0F);        /* ones minute digit */
        hms[5] = ':';
        hms[6] = 0x30 + ((cur_time[0] & 0x70) >> 4); /* tens second digit */
```

```
        hms[7] = 0x30 + (cur_time[0] & 0x0F);        /* ones second digit */
        hms[8] = ':';
        if (cur_time[2] & 0x20)
            hms[9] = 'P';
        else
            hms[9] = 'A';
            hms[10] = 'M';
            hms[11] = 0;                              /* terminate the string with a NULL */
        switch(temp3) {                              /* convert to day of week */
            case 1:    mdy[0] = 'S';
                       mdy[1] = 'U';
                       break;
            case 2:    mdy[0] = 'M';
                       mdy[1] = 'O';
                       break;
            case 3:    mdy[0] = 'T';
                       mdy[1] = 'U';
                       break;
            case 4:    mdy[0] = 'W';
                       mdy[1] = 'E';
                       break;
            case 5:    mdy[0] = 'T';
                       mdy[1] = 'H';
                       break;
            case 6:    mdy[0] = 'F';
                       mdy[1] = 'R';
                       break;
            case 7:    mdy[0] = 'S';
                       mdy[1] = 'A';
                       break;
            default:   mdy[0] = 0x20;                 /* space */
                       mdy[1] = 0x20;
                       break;
        }
        mdy[2] = ':';
        mdy[3] = 0x30 + (cur_time[5] >> 4);          /* month */
        mdy[4] = 0x30 + (cur_time[5] & 0x0F);
        mdy[5] = ':';
        mdy[6] = 0x30 + (cur_time[4] >> 4);          /* date */
        mdy[7] = 0x30 + (cur_time[4] & 0x0F);
        mdy[8] = ':';
        mdy[9] = 0x30 + (cur_time[6] >> 4);          /* year */
        mdy[10] = 0x30 + (cur_time[6] & 0x0F);
        mdy[11] = 0;                                 /* NULL character */
    }
    else {                                           /* 24-hour mode */
        hms[0] = 0x30 + ((cur_time[2] & 0x30)>>4);   /* hours */
        hms[1] = 0x30 + (cur_time[2] & 0x0F);
```

```
            hms[2] = ':';
            hms[3] = 0x30 + (cur_time[1] >> 4);         /* minutes */
            hms[4] = 0x30 + (cur_time[1] & 0x0F);
            hms[5] = ':';
            hms[6] = 0x30 + ((cur_time[0] & 0x70)>>4);  /* seconds */
            hms[7] = 0x30 + (cur_time[0] & 0x0F);
            hms[8] = ':';
            switch(temp3) {                             /* convert to day of week */
                case 1:   hms[9] = 'S';
                          hms[10] = 'U';
                          break;
                case 2:   hms[9] = 'M';
                          hms[10] = 'O';
                          break;
                case 3:   hms[9] = 'T';
                          hms[10] = 'U';
                          break;
                case 4:   hms[9] = 'W';
                          hms[10] = 'E';
                          break;
                case 5:   hms[9] = 'T';
                          hms[10] = 'H';
                          break;
                case 6:   hms[9] = 'F';
                          hms[10] = 'R';
                          break;
                case 7:   hms[9] = 'S';
                          hms[10] = 'A';
                          break;
                default:  hms[9] = 0x20;                /* space */
                          hms[10] = 0x20;
                          break;
            }
            hms[11] = 0;                                /* NULL character */
            mdy[0] = 0x30 + (cur_time[5] >> 4);         /* month */
            mdy[1] = 0x30 + (cur_time[5] & 0x0F);
            mdy[2] = ':';
            mdy[3] = 0x30 + (cur_time[4] >> 4);         /* date */
            mdy[4] = 0x30 + (cur_time[4] & 0x0F);
            mdy[5] = ':';
            mdy[6] = 0x30 + (cur_time[6] >> 4);         /* year */
            mdy[7] = 0x30 + (cur_time[6] & 0x0F);
            mdy[8] = 0;                                 /* NULL character */
        }
    }
```

Example 11.8

▼

Write a function to display the current time and calendar information on the LCD.

Solution: This function is as follows:

```
void displayTime(void)
{
        cmd2lcd(0x83);        /* set cursor to row 1 column 3 */
        puts2lcd(hms);        /* output hours, minutes, and seconds */
        cmd2lcd(0xC3);        /* set cursor to row 2 column 3 */
        puts2lcd(mdy);        /* output month, date, and year */
}
```

▲

Example 11.9

▼

Write the $\overline{\text{IRQ}}$ interrupt service routine that reads the time from the DS1307, format the time and calendar information, and display them on the LCD.

Solution: The $\overline{\text{IRQ}}$ service routine simply calls the previous functions that have been written. The C language version of the service routine is as follows:

```
void INTERRUPT irqISR(void)
{
        readTime(0x00);       /* read all time registers starting from seconds */
        formatTime();         /* format time info into two strings */
  ,     displayTime();        /* display the time on LCD */
}
```

▲

11.7 The Digital Thermometer and Thermostat DS1631A

Many embedded products, such as network routers and switches, are used in larger systems, and their failures due to overheating could severely damage the functioning or even cause the total failure of the larger system. Using a thermostat to warn of potential overheating is indispensable for the proper functioning of many embedded systems.

The digital thermostat device DS1631A from Dallas Semiconductor is one such product. The DS1631A will assert a signal (T_{OUT}) whenever the ambient temperature exceeds the *trip point* preestablished by the user. The DS1626 performs the same function but has an SPI interface.

11.7.1 Pin Assignment

As shown in Figure 11.31, the DS1631A is an 8-pin package. The SDA and SCL pins are used as the data and clock lines so that the DS1631A can be connected to an I²C bus. Pins A2~A0 are address inputs to the DS1631A. The T_{OUT} pin is the thermostat output, which is asserted whenever the ambient temperature is above the trip point set by the designer.

Figure 11.31 ■ Pin assignment of DS1631A

11.7.2 Functional Description

The block diagram of the DS1631A is shown in Figure 11.32. The DS1631A converts the ambient temperature into 9-, 10-, 11-, or 12-bit readings over a range of $-55°C$ to $+125°C$. The thermometer accuracy is $\pm0.5°C$ from $0°C$ to $+70°C$ with $3.0\ V \le V_{DD} \le 5.5\ V$.

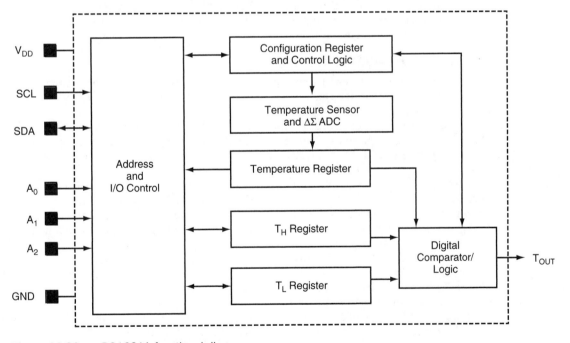

Figure 11.32 ■ DS1631A functional diagram

The thermostat output T_{OUT} is asserted whenever the converted ambient temperature is equal to or higher than the value stored in the T_H register. The DS1631A automatically begins taking temperature measurements at power-up, which allows it to function as a stand-alone thermostat. The DS1631A conforms to the I²C bus specification.

11.7.3 DS1631A Registers

Four registers support the operation of the DS1631A: Config, T_H, T_L, and Temperature registers. The SRAM-based Temperature register is a 2-byte register that holds the converted temperature value in the two's complement format. The converted temperature is stored in the upper bits of the temperature register with lower bits filled with 0s. When the most significant bit of this register is 1, the temperature is negative.

Both the T_H and T_L are EEPROM-based 2-byte registers. T_H holds the upper alarm temperature value in two's complement format. Whenever the converted temperature value is equal to or higher than the value in T_H, the T_{OUT} signal will be asserted. Once asserted, T_{OUT} can be deasserted only when the converted temperature is lower than the value in the T_L register. The value in the T_L register is also represented in the two's complement format.

The contents of the Config register are shown in Figure 11.33. The lower two bits of the configuration register are EEPROM-based, whereas the upper six bits are SRAM-based. The meaning of each bit is well defined. The Config register allows the designer to program various DS1631A options, such as conversion resolution, T_{OUT} polarity, and operation mode. It also provides information about conversion status, EEPROM activity, and thermostat activity. This register can be read from and written into using the Access Config command. When writing to the Config register, conversions should first be stopped using the Stop Convert T command if the device is in continuous conversion mode. Since the POL and 1SHOT bits are stored in EEPROM, they can be programmed prior to installation if desired. All other bits are in SRAM and are powered up in the state shown in Figure 11.33.

11.7.4 The DS1631A Operation

The DS1631A begins conversions automatically at power-up. It can be configured to perform continuous consecutive conversions (continuous conversion mode) or single conversions on command (one-shot mode).

The default resolution of the DS1631A is 12-bit. However, it can also be configured to 9-, 10-, or 11-bit. The resolution is changed via the R1:R0 bits of the Config register. A few samples of temperatures and their converted values are shown in Table 11.8. The lowest 4 bits are always zeros in the table because the resolution is 12-bit.

Both the T_H and T_L registers are in EEPROM. Their resolutions match the output temperature resolution and are determined by the R1:R0 bits. Writing to and reading from these two registers are achieved by using the Access TH and Access TL commands. When making changes to the T_H and T_L registers, conversions should first be stopped using the Stop Convert T command if the device is in continuous conversion mode.

Since the DS1631A automatically begins taking temperature measurements at power-up, it can function as a stand-alone thermostat. For stand-alone operation, the nonvolatile T_H and T_L registers and the POL and 1SHOT bits in the Config register should be programmed to the desired values prior to installation.

Table 11.8 shows that every 16 units (after ignoring the lowest four bits) of conversion result correspond to 1°C. The conversion result cannot be higher than 0x7D00 and lower than 0xC900 because the range of temperature that can be handled by DS1631A cannot be higher than 125°C and lower than −55°C. An easy way for finding the corresponding temperature of a certain conversion result is as follows:

7	6	5	4	3	2	1	0
DONE	THF	TLF	NVB	R1	R0	POL*	1SHOT*

*NV (EEPROM)

power-up value

1	0	0	0	1	1	X	X

Done: temperature conversion done (read-only)

    0 = temperature conversion is in progress.

    1 = temperature conversion is complete. Will be cleared when the Temperature register is read.

THF: temperature high flag (read/write)

    0 = The measured temperature has not exceeded the value in T_H register.

    1 = The measured temperature has exceeded the value in T_H register. THF remains at 1 until it is overwritten with a 0 by the user, the power is recycled, or a software POR command is issued.

TLF: temperature low flag (read/write)

    0 = The measured temperature has not been lower than the value in T_L register.

    1 = At some point after power-up, the measured temperature is lower than the value stored in the T_L register. TLF remains at 1 until it is overwritten with a 0 by the user, the power is recycled, or a software POR command is issued.

NVB: nonvolatile memory busy (read only)

    0 = NV memory is not busy.

    1 = A write to EEPROM memory is in progress.

R1:R0 : resolution bits (read/write)

    00 = 9-bit resolution (conversion time is 93.75 ms)

    01 = 10-bit resolution (conversion time is 187.5 ms)

    10 = 11-bit resolution (conversion time is 375 ms)

    11 = 12-bit resolution (conversion time is 750 ms)

POL: T_{OUT} polarity (read/write)

    0 = T_{OUT} active low

    1 = T_{OUT} active high

1SHOT: conversion mode (read/write)

    0 = continuous conversion mode. The Start Convert T command initiates continuous temperature conversions.

    1 = one-shot mode. The Start Convert T command initiates a single temperature conversion and then the device enters a low-power standby mode.

Figure 11.33 ■ DS1631A Configuration register

POSITIVE CONVERSION RESULT

Step 1

Truncate the lowest four bits.

Step 2

Divide the upper 12 bits by 16.

For example, the conversion result 0x7000 corresponds to 0x700/16 = 112°C. The conversion result 0x6040 corresponds to 0x604/16 = 96.25°C.

Temperature (°C)	Digital Output (binary)	Digital Output (hex)
+125	0111 1101 0000 0000	0x7D00
+25.0625	0001 1001 0001 0000	0x1910
+10.125	0000 1010 0010 0000	0x0A20
+0.5	0000 0000 1000 0000	0x0080
0	0000 0000 0000 0000	0x0000
−0.5	1111 1111 1000 0000	0xFF80
−10.125	1111 0101 1110 0000	0xF5E0
−25.0625	1110 0110 1111 0000	0xE6F0
−55	1100 1001 0000 0000	0xC900

Table 11.8 ■ 12-bit resolution temperature/data relationship

NEGATIVE CONVERSION RESULT

Step 1
Compute the two's complement of the conversion result.

Step 2
Truncate the lowest four bits.

Step 3
Divide the upper 12 bits of the two's complement of the conversion result by 16.

For example, the conversion result 0xE280 corresponds to $-0x1D8/16 = -29.5°C$.

11.7.5 DS1631A Command Set

The DS1631A supports the following commands:

- *Start Convert T [0x51]*—This command initiates temperature conversions. If the part is in one-shot mode, only one conversion is performed. In continuous mode, continuous temperature conversions are performed until a **Stop Convert T** command is issued.

- *Stop Convert T [0x22]*—This command stops temperature conversions when the device is in continuous conversion mode.

- *Read Temperature [0xAA]*—This command reads the last converted temperature value from the 2-byte Temperature register.

- *Access TH [0xA1]*—This command reads or writes the 2-byte T_H register.

- *Access TL [0xA2]*—This command reads or writes the 2-byte T_L register.

- *Access Config [0xAC]*—This command reads or writes the 1-byte Configuration register.

- *Software POR [0x54]*—This command initiates a software power-on-reset operation, which stops temperature conversions and resets all registers and logic to their power-up states. The software POR allows the designer to simulate cycling the power without actually powering-down the device.

11.7.6 I²C Communication with DS1631A

A typical circuit connection between the HCS12 microcontroller and a DS1631A is shown in Figure 11.34. The address input of the DS1631A is arbitrarily set to %001 in Figure 11.34. To initiate I²C communication, the HCS12 MCU asserts a start condition followed by a control byte containing the DS1631A device ID. The R/$\overline{\text{W}}$ bit of the control byte must be a 0 since the HCS12 MCU next will write a command byte to the DS1631A. The format for the control byte is shown in Figure 11.35. The DS1631A responds with an ACK after receiving the control byte. This must be followed by a command byte from the master, which indicates what type of operation is to be performed. The DS1631A again responds with an ACK after receiving the command byte. If the command byte is **Start Convert T** or **Stop Convert T** command, the transaction is finished, and the master must issue a stop condition to signal the end of the communication sequence. If the command byte indicates a write or read operation, additional actions must occur.

Figure 11.34 ■ Typical circuit connection between the HCS12 MCU and DS1631A

Figure 11.35 ■ Control byte for DS1631A

WRITE DATA TO DS1631A

The master can write data to the DS1631A by issuing an **Access Config**, **Access TH**, or **Access TL** command following the control byte. Since the R/$\overline{\text{W}}$ bit in the control byte was a 0, the DS1631A is already prepared to receive data. Therefore, after receiving an ACK in response to the command byte, the master device can immediately begin transmitting data. When writing to the configuration register, the master must send 1 byte of data, and when writing to the T_H or T_L register, the master must send 2 bytes of data. The upper byte is written first followed by the lower byte. After receiving each data byte, the DS1631A responds with an ACK, and the transaction is finished with a stop from the master.

READ DATA FROM DS1631A

The master can read data from the DS1631A by issuing an **Access Config, Access TH, Access TL,** or **Read Temperature** command following the control byte. After receiving an ACK in response to the command, the master must generate a restart condition followed by a control byte with the same slave ID as the first control byte. However, this time the R/$\overline{W}$ bit must be set to 1 to inform the DS1631A that a "read" is being performed. After sending an ACK in response to this control byte, the DS1631A begins transmitting the requested data on the next clock cycle. One byte of data will be transmitted when reading from the configuration register, after which the master must respond with a NACK followed by a stop condition. For 2-byte reads (i.e., from the Temperature, T_H, or T_L register), the master must respond to the first data byte with an ACK and to the second byte with a NACK followed by a stop condition. The upper byte is read first followed by the lower byte. If only the most significant byte of data is needed, the master can issue a NACK followed by a stop condition after reading the first data byte.

Example 11.10

▼

Write a function to configure the DS1631A in Figure 11.34 to operate in continuous conversion mode and set the T_{OUT} polarity to active high. Assume that the I²C has only one master and there is no possibility in getting bus collision.

Solution: The procedure for configuring the DS1631A is as follows:

Step 1
Assert a start condition and send a control byte (0x92) to the DS1631A with $\overline{W}$ = 0.

Step 2
Wait until I²C transfer is completed and check to see if the DS1631A acknowledges. If not, exit.

Step 3
Send an Access Config command to the DS1631A.

Step 4
Wait until the I²C bus is idle and check to see if the DS1631A acknowledges. If not, exit.

Step 5
Write a command byte to configure the DS1631A to operate in continuous conversion mode and set the T_{OUT} polarity to active high.

Step 6
Wait until the I²C bus is idle and check to see if the DS1631A acknowledges. If not, exit.

Step 7
Assert a stop condition to complete the whole process.

The assembly function that performs the desired configuration is as follows:

```
openDS1631   ldaa    #$92
             jsr     sendSlaveID
             brclr   IBSR,RXAK,openOK0    ; did DS1631A acknowledge?
             ldab    #$FF                 ; return error code −1
             rts
openOK0      movb    #$AC,IBDR            ; send the "Access Config" command
             brclr   IBSR,IBIF,*
```

```
                    movb    #IBIF,IBSR              ; clear the IBIF flag
                    brclr   IBSR,RXAK,openOK1       ; did DS1316A acknowledge?
                    ldab    #$FF
                    rts
        openOK1     stab    IBDR                    ; sends configuration data
                    brclr   IBSR,IBIF,*             ; wait until the byte has been shifted out
                    movb    #IBIF,IBSR              ; clear the IBIF flag
                    brclr   IBSR,RXAK,openOK2       ; did DS1316A acknowledge?
                    ldab    #$FF
                    rts
        openOK2     bclr    IBCR,MSSL               ; generate a stop condition
                    ldab    #0                      ; normal return code
                    rts
```

To configure the DS1631A with the desired setting, use the following instruction sequence to call the previous function:

```
        ldab        #$E0
        jsr         openDS1631
```

The C language version of the function is as follows:

```
char openDS1631(char cy)
{
    sendSlaveID(0x92);          /* generate a start condition and send ID */
    if (IBSR & RXAK)
            return −1;          /* error code when DS1631A did not acknowledge */
    IBDR = 0xAC;                /* send command "Access Config" */
    while(!(IBSR & IBIF));
    IBSR = IBIF;                /* clear the IBIF flag */
    if (IBSR & RXAK)
            return −1;          /* error code when DS1631A did not acknowledge */
    IBDR = cy;                  /* send configuration byte */
    while(!(IBSR & IBIF));
    IBSR = IBIF;
    if (IBSR & RXAK)
            return −1;          /* error code when DS1631A did not acknowledge */
    IBCR &= ~MSSL;              /* generate a stop condition */
    return 0;                   /* normal return code */
}
```

Example 11.11

▼

Write a function to start temperature conversion.

Solution: The procedure for starting temperature conversion using the DS1631A is as follows:

Step 1
Assert a start condition and send a control byte (0x92) to the DS1631A with R/$\overline{W}$ = 0.

Step 2

Wait until the I²C bus is idle and check to see if the DS1631A acknowledges. If not, exit.

Step 3

Send a Start Convert T command to the DS1631A.

Step 4

Wait until the I²C transfer is completed and check to see if the DS1631A acknowledges. If not, exit.

Step 5

Assert a stop condition to complete the whole process.

The assembly program that implements this procedure is as follows:

```
startConv   ldaa    #$92
            jsr     sendSlaveID           ; generate a start condition and send DS1631A's ID
            brclr   IBSR,RXAK,startOK0    ; did DS1631A acknowledge?
            ldab    #$FF                  ; return error code −1
            rts
startOK0    movb    #$51,IBDR            ; send "Start Convert T" command
            brclr   IBSR,IBIF,*          ; wait until the byte is shifted out
            movb    #IBIF,IBSR           ; clear the IBIF flag
            brclr   IBSR,RXAK,startOK1   ; did DS1631A acknowledge?
            ldab    #$FF
            rts
startOK1    bclr    IBCR,MSSL            ; generate a stop condition
            ldab    #0                   ; normal return code
            rts
```

The C language version of the function is as follows:

```c
char startConv (void)
{
    sendSlaveID(0x92);              /* generate a start condition and send slave ID */
    if (IBSR & RXAK)
        return −1;                  /* if not acknowledged, return error code −1 */
    IBDR = 0x51;                    /* send command to start a conversion */
    while(!(IBSR & IBIF));
    IBSR = IBIF;                    /* clear the IBIF flag */
    if (IBSR & RXAK)
        return −1;
    IBCR &= ~MSSL;                  /* generate a stop condition */
    return 0;
}
```

Example 11.12

Write a function to set the high thermostat temperature. The upper and lower bytes of the high thermostat temperatures are passed in stack.

Solution: The procedure for setting up the high thermostat temperature is as follows:

Step 1
Assert the start condition and send control byte 0x92 with R/$\overline{\text{W}}$ bit = 0.

Step 2
Wait until I²C transfer is completed and check to see if the DS1631A acknowledges. If not, exit.

Step 3
Send an Access TH command to DS1631A.

Step 4
Wait until the I²C transfer is completed and check to see if the DS1631A acknowledges. If not, exit.

Step 5
Send the upper byte of the thermostat temperature to the DS1631A.

Step 6
Wait until the I²C transfer is completed and check to see if the DS1631A acknowledges. If not, exit.

Step 7
Send the lower byte of the thermostat temperature to the DS1631A.

Step 8
Wait until the I²C transfer is completed and check to see if the DS1631A acknowledges. If not, exit.

Step 9
Assert the stop condition.

The assembly program that implements this procedure is as follows:

```
THhi     equ     2
THlo     equ     3
setTH    ldaa    #$92
         jsr     sendSlaveID
         brclr   IBSR,RXAK,setTHok1     ; did DS1631A acknowledge?
         ldab    #$FF                   ; return error code −1
         rts
setTHok1 movb    #$A1,IBDR             ; send out access TH command */
         brclr   IBSR,IBIF,*           ; wait until command is shifted out
         movb    #IBIF,IBSR            ; clear IBIF flag
         brclr   IBSR,RXAK,setTHok2    ; did DS1631A acknowledge?
         ldab    #$FF
         rts
setTHok2 ldaa    THhi,sp              ; get the upper byte of TH from stack
         staa    IBDR                 ; send out the low byte of TH
         brclr   IBSR,IBIF,*
         movb    #IBIF,IBSR           ; clear the IBIF flag
         brclr   IBSR,RXAK,setTHok3   ; did DS1631A acknowledge?
```

```
                    ldab      #$FF
                    rts
setTHok3            ldaa      THlo,sp                    ; get the lower byte of TH from stack
                    staa      IBDR
                    brclr     IBSR,IBIF,*
                    movb      #IBIF,IBSR                 ; clear the IBIF flag
                    brclr     IBSR,RXAK,setTHok4         ; did DS1631A acknowledge?
                    ldab      #$FF
                    rts
setTHok4            bclr      IBCR,MSSL                  ; generate the stop condition
                    ldab      #0                         ; normal return code
                    rts
```

The C language version of the function is as follows:

```
char setTH(char hibyte, char lobyte)
{
    sendSlaveID(0x92);
    if(IBSR & RXAK)              /* did DS1631A acknowledge? */
            return −1;
    IBDR = 0xA1;                 /* send command to access TH */
    while(!(IBSR & IBIF));
    IBSR = IBIF;                 /* clear the IBIF flag */
    if (IBSR & RXAK)
            return −1;
    IBDR = hibyte;              /* send out the high byte of TH */
    while(!(IBSR & IBIF));
    IBSR = IBIF;
    if (IBSR & RXAK)
            return −1;
    IBDR = lobyte;             /* send out the low byte of TH */
    while(!(IBSR & IBIF));
    IBSR = IBIF;
    if (IBSR & RXAK)
            return −1;
    IBCR &= ~MSSL;             /* generate a stop condition */
    return 0;
}
```

Example 11.13

Write a function to read the Config register from the DS1631A and return its value in accumulator B.

Solution: The procedure for reading the Config register is as follows:

Step 1
Assert the start condition and send the control byte 0x92 with $R/\overline{W} = 0$.

Step 2
Wait until the control byte is shifted out and check to see if the DS1631A acknowledges. If not, exit.

Step 3
Send an Access Config command (0xAC) to the DS1631A.

Step 4
Wait until the command is shifted out and check to see if the DS1631A acknowledges. If not, exit.

Step 5
Assert a restart condition.

Step 6
Send the control byte 0x93 with R/$\overline{\text{W}}$ = 1, which tells the DS1631A that a read is being performed.

Step 7
Wait until the control byte is shifted out and check to see if the DS1631A acknowledges. If not, exit.

Step 8
Read the Config register.

Step 9
Make sure that the bus is idle and send back an NACK to the DS1631A.

Step 10
Assert the stop condition.

The assembly program that implements this procedure is as follows:

```
readConf    ldaa    #$92                    ; generate a start condition and send out DS1631A
            jsr     sendSlaveID             ; ID
            brclr   IBSR,RXAK,rdConfok1     ; did DS1631A acknowledge?
            ldab    #$FF                    ; return error code −1
            rts
rdConfok1   movb    #$AC,IBDR               ; send out the Access Conf command
            brclr   IBSR,IBIF,*             ; wait until the command is shifted out
            movb    #IBIF,IBSR              ; clear the IBIF flag
            brclr   IBSR,RXAK,rdConfok2     ; did DS1631A acknowledge?
            ldab    #$FF                    ; return error code −1
            rts
rdConfok2   bset    IBCR,RSTA               ; generate a restart condition
            movb    #$93,IBDR               ; send out slave ID and set R/W bit to 1
            brclr   IBSR,IBIF,*             ; wait until the command is shifted out
            movb    #IBIF,IBSR              ; clear the IBIF flag
            brclr   IBSR,RXAK,rdConfok3     ; did DS1631A acknowledge?
            ldab    #$FF                    ; return error code −1
            rts
rdConfok3   bclr    IBCR,TXRX               ; prepare to read
            bset    IBCR,TXAK               ; prepare to send NACK
            ldaa    IBDR                    ; perform a dummy read to trigger nine clock pulses
            brclr   IBSR,IBIF,*             ; wait until a byte has been shifted in
            movb    #IBIF,IBSR              ; clear the IBIF flag
            bclr    IBCR,MSSL               ; generate a stop condition
            ldab    IBDR                    ; place the Conf register value in B
            rts
```

The C language version of the function is as follows:

```c
char readConf(void)
{
    char temp;
    sendSlaveID(0x92);
    if(IBSR & RXAK)              /* did DS1631A acknowledge? */
            return -1;
    IBDR = 0xAC;                 /* send command to access Conf register */
    while(!(IBSR & IBIF));
    IBSR = IBIF;
    if (IBSR & RXAK)
            return -1;
    IBCR |= RSTA;               /* generate a restart condition */
    IBDR = 0x93;                /* send out ID and set R/W to 1 */
    while(!(IBSR & IBIF));
    IBSR = IBIF;
    if (IBSR & RXAK)
            return -1;
    IBCR &= ~TXRX;             /* prepare to receive */
    IBCR |= TXAK;             /* prepare to send NACK */
    temp = IBDR;               /* a dummy read to trigger nine clock pulses */
    while(!(IBSR & IBIF));
    IBSR = IBIF;
    IBCR &= ~MSSL;             /* generate a stop condition */
    return IBDR;
}
```

Example 11.14

Write a subroutine to read the converted temperature and return the upper and lower bytes in double accumulator D. Assume that the temperature conversion has been started but this function needs to make sure that the converted temperature value is resulted from the most recent "Start Convert T" command.

Solution: This subroutine will call the previous subroutine to make sure that the DONE bit is set to 1 and reads the converted temperature from the DS1631A. The assembly subroutine is as follows:

```
readTemp    ldx     #0                      ; initialize return error code
rdLoop      jsr     readConf                ; is temperature conversion done yet?
            cmpb    #-1                     ; is there any error?
            beq     rdErr                   ;     "
            andb    #$80                    ; check DONE bit
            bpl     rdLoop                  ; conversion not done yet?
            ldaa    #$92                    ; generate a start condition and send out
            jsr     sendSlaveID             ; the DS1631A ID
            brclr   IBSR,RXAK,rdTempok1     ; did DS1631A acknowledge?
            ldx     #-1
            rts
```

```
rdTempok1   movb    #$AA,IBDR               ; sends "Read Temperature command"
            brclr   IBSR,IBIF,*             ;
            movb    #IBIF,IBSR
            brclr   IBSR,RXAK,rdTempok2     ; did DS1631A acknowledge?
            ldx     #-1
            rts
rdTempok2   bset    IBCR,RSTA              ; generate a restart condition
            movb    #$93,IBDR              ; send DS1631A's ID with R/W set to 1
            brclr   IBSR,IBIF,*
            movb    #IBIF,IBSR
            brclr   IBSR,RXAK,rdTempok3    ; did DS1631A acknowledge?
            ldx     #-1
            rts
rdTempok3   bclr    IBCR,TXRX+TXAK         ; prepare to receive and ACK
            ldaa    IBDR                  ; perform a dummy read
            brclr   IBSR,IBIF,*           ; wait for high byte of temperature to shift in
            movb    #IBIF,IBSR
            bset    IBCR,TXAK             ; prepare to send NACK for the last read
            ldaa    IBDR                 ; place the high byte of temperature in A
            brclr   IBSR,IBIF,*          ; wait for the low byte read to complete
            movb    #IBIF,IBSR           ; clear the IBIF flag
            ldab    IBDR                ; place the low byte of temperature in B
            ldx     #0                  ; correct return code
            rts
rdErr       ldx     #-1
            rts
```

▲

The C language version of the function is straightforward and hence will be left as an exercise problem.

11.8 Interfacing the Serial EEPROM 24LC08B with I²C

Some applications require the use of a large amount of nonvolatile memory because these applications are powered by batteries and may be used in the field for an extended period of time. Many semiconductor manufacturers produce serial EEPROMs with serial interface. Both serial EEPROMs with SPI and I²C interfaces are available.

The 24LC08B is a serial EEPROM from Microchip with the I²C interface. This device is an 8-Kbit EEPROM organized as four blocks of 256×8-bit memory. Low-voltage design permits operation down to 2.5 V with standby and active currents of only 1 μA and 1 mA, respectively. The 24LC08B also has a page write capability for up to 16 bytes of data.

11.8.1 Pin Assignment and Block Diagram

The pin assignment and the block diagram of the 24LC08B are shown in Figure 11.36 and 11.37, respectively. Pins A2, A1, and A0 are not used. The SCL and SDA pins are for I²C bus communications. The frequency of the SCL input can be as high as 400 kHz. The WP pin is used as the write protection input. When this pin is high, the 24LC08B cannot be written into. The pins A2~A0 are not used and can be left floating, grounded, or pulled to high.

Figure 11.36 ■ 24LC08B PDIP package pin assignment

Figure 11.37 ■ Block diagram of 24LC08B

11.8.2 Device Addressing

Like any other I²C slave, the first byte sent to the 24LC08B after the start condition is the control byte. The contents of the control byte for the 24LC08B are shown in Figure 11.38. The upper four bits are the device ID of the 24LC08B, and the value represented by bits B1~B0 is the block address of the memory location to be accessed. For any access to the 24LC08B, the master must also send an 8-bit byte address after the control byte. There is an address pointer inside the 24LC08B. After the access of each byte, the address pointer is incremented by 1.

Figure 11.38 ■ 24LC08B Control byte contents

11.8.3 Write Operation

The 24LC08B supports byte write and page write operations. In a *byte write* operation, the following operations need to be performed:

1. The master asserts the start condition and sends the control byte to the 24LC08B.
2. The 24LC08B acknowledges the data transmission.
3. The master sends the byte address to the 24LC08B.
4. The 24LC08B acknowledges the data transmission.
5. The master sends the data byte to the 24LC08B.
6. The 24LC08B acknowledges the data transmission.
7. The master asserts the stop condition.

In a *page write* operation, the master can send up to 16 bytes of data to the 24LC08B. The write control byte, byte address, and the first data byte are transmitted to the 24LC08B in the same way as in a byte write operation. But instead of asserting a stop condition, the master transmits up to 16 data bytes to the 24LC08B that are temporarily stored in the on-chip page buffer. These 16 data bytes will be written into the memory after the master has asserted a stop condition. After the receipt of each byte, the four lower address pointer bits are internally incremented by 1. The highest six bits of the byte address remain constant. Should the master transmit more than 16 bytes prior to generating the stop condition, the address counter will roll over and the previously received data will be overwritten.

11.8.4 Acknowledge Polling

When the 24LC08B is writing the data held in the write buffer into the EEPROM array, it will not acknowledge any further write operation. This fact can be used to determine when the cycle is completed. Once the stop condition for a write command has been issued from the master, the device initiates the internal write cycle. The ACK polling can be initiated immediately. This involves the master's sending a start condition followed by the control byte for a write command $(R/\overline{W} = 0)$. If the 24LC08B is still busy, then no ACK will be returned. If the cycle is complete, then the device will return the ACK, and the master can then proceed with the next read or write command. The polling process is illustrated in Figure 11.39.

11.8.5 Read Operation

24LC08B supports three types of read operations: *current address read*, *random read*, and *sequential read*.

CURRENT ADDRESS READ

As explained earlier, the internal address counter is incremented by 1 after each access (read or write). The current address read allows the master to read the byte immediately following the location accessed by the previous read or write operation. On receipt of the slave address with the $R/\overline{W}$ bit set to 1, the 24LC08B issues an acknowledgement and transmits an 8-bit data byte. The master will not acknowledge the transfer but asserts a stop condition and the 24LC08B discontinues transmission.

Figure 11.39 ■ Acknowledge polling flow

RANDOM READ

Random read operations allow the master to access any memory location in a random manner. To perform this type of read operation, the address of the memory location to be read must be sent. The procedure for performing a random read is as follows:

Step 1
The master asserts a start condition and sends the control byte with the $R/\overline{W}$ bit set to 0 to the 24LC08B.

Step 2
The 24LC08B acknowledges the control byte.

Step 3
The master sends the address of the byte to be read to the 24LC08B.

Step 4
The 24LC08B acknowledges the byte address.

Step 5
The master asserts a repeated start condition.

Step 6
The master sends the control byte with $R/\overline{W} = 1$.

Step 7
The 24LC08B acknowledges the control byte and sends data to the master.

Step 8
The master asserts NACK to the 24LC08B.

Step 9
The master asserts the stop condition.

SEQUENTIAL READ
If the master acknowledges the data byte returned by the random read operation, the 24LC08B will transmit the next sequentially addressed byte as long as the master provides the clock signal on the SCL line. The master can read the whole chip using sequential read.

11.8.6 Circuit Connection Between the I²C Master and the 24LC08B
The circuit connection of a 24LC08B and the HCS12 MCU is similar to that in Figure 11.34. The diagram of a system with one DS1631A and one 24LC08B and the HCS12 MCU is shown in Figure 11.40.

Figure 11.40 ■ Circuit connection of the HCS12 MCU, 24LC08B, and DS1631A

Example 11.15
▼

Write a function to read a byte from the 24LC08B. Pass the control byte and address in accumulators A and B to this subroutine. Return the data byte and error code in A and B, respectively. This function should check the error that the EEPROM did not acknowledge (implies that 24LC08B may have failed).

Solution: The following assembly subroutine implements the procedure for random read described earlier:

```
EErandomRead
            jsr     sendSlaveID         ; send out 24LC08B ID and block address
            brclr   IBSR,RXAK,ranRdok0  ; does EEPROM acknowledge?
            ldab    #$FF                ; return −1, if EEPROM does not ack
            rts
ranRdok0    stab    IBDR                ; send out EEPROM memory address
            brclr   IBSR,IBIF,*         ; wait until the address is shifted out
```

```
                    movb    #IBIF,IBSR              ; clear IBIF flag
                    brclr   IBSR,RXAK,ranRdok1      ; does EEPROM acknowledge?
                    ldab    #$FF                    ; return −1, if EEPROM does not ack
                    rts
ranRdok1            bset    IBCR,RSTA               ; generate restart condition
                    oraa    #$01                    ; set R/W bit for read
                    staa    IBDR                    ; resend the device ID
                    brclr   IBSR,IBIF,*             ; wait until the EEPROM ID is sent out
                    movb    #IBIF,IBSR              ; clear the IBIF flag
                    brclr   IBSR,RXAK,ranRdok2      ; does EEPROM acknowledge?
                    ldab    #$FF                    ; return −1, if EEPROM does not ack
                    rts
ranRdok2            bset    IBCR,TXAK               ; prepare to send NACK
                    bclr    IBCR,TXRX               ; perform reception
                    ldaa    IBDR                    ; dummy read to initiate reception
                    brclr   IBSR,IBIF,*             ; wait for a byte to shift in
                    movb    #IBIF,IBSR              ; clear the IBIF flag
                    bclr    IBCR,MSSL               ; generate a stop condition
                    ldaa    IBDR                    ; get the data byte
                    ldab    #0                      ; normal read status
                    rts
```

The C language version of the function is as follows:

```c
char EErandomRead(char ID, char addr)
{
    char dummy;
    SendSlaveID(ID);
    if (IBSR & RXAK)
        return −1;
    IBDR = addr;                    /* send out EEPROM address */
    while(!(IBSR & IBIF));          /* wait until the address is shifted out */
    IBSR = IBIF;                    /* clear IBIF flag */
    if (IBSR & RXAK)
        return −1;
    IBCR |= RSTA;                   /* generate restart condition */
    IBDR = ID | 0x01;              /* prepare to read */
    while (!(IBSR & IBIF));
    IBSR = IBIF;
    if (IBSR & RXAK)
        return −1;
    IBCR |= TXAK;                   /* prepare to send NACK */
    IBCR &= ~TXRX;                  /* perform reception */
    dummy = IBDR;                   /* dummy read to trigger nine clock pulses */
    while(!(IBSR & IBIF));          /* wait for data to shift in */
    IBSR = IBIF;
    IBCR &= ~MSSL;                  /* generate a stop condition */
    return IBDR;
}
```

Sequential read is performed in a block read operation. When the MCU saves the contents of the IBDR register, it also initiates a sequential read if the MCU acknowledges the previously received byte.

Example 11.16
▼

Write a subroutine that writes a byte into the 24LC08B. The device ID, memory address, and data to be written are passed to this routine in the stack.

Solution: The assembly subroutine that writes a byte into the 24LC08B to the specified address is as follows:

```
EE_ID        equ     5                           ; stack offset for EE_ID
EE_addr      equ     4                           ; stack offset for EE_addr
EE_dat       equ     3                           ; stack offset for EE_data
EEbyteWrite  psha
             ldaa    EE_ID,sp                    ; get the EEPROM ID from stack
             jsr     sendSlaveID                 ; generate start condition, send EEPROM ID
             brclr   IBSR,RXAK,bywriteok1        ; does EEPROM acknowledge?
             ldab    #$FF                        ; return −1 as the error code
             pula
             rts
bywriteok1   ldaa    EE_addr,sp                  ; get the address to be accessed from the stack
             staa    IBDR                        ; send address to I2C bus
             brclr   IBSR,IBIF,*                 ; wait until address is shifted out
             movb    #IBIF,IBSR                  ; clear the IBIF flag
             brclr   IBSR,RXAK,bywriteok2        ; does EEPROM acknowledge?
             ldab    #$FF                        ; return −1 as the error code
             pula
             rts
bywriteok2   ldaa    EE_dat,sp                   ; get the data byte to be written from stack
             staa    IBDR                        ; send out the data byte
             brclr   IBSR,IBIF,*                 ; wait until data byte is shifted out
             movb    #IBIF,IBSR                  ; clear the IBIF flag
             brclr   IBSR,RXAK,bywriteok3        ; does EEPROM acknowledge?
             ldab    #$FF                        ; return −1 as the error code
             pula
             rts
bywriteok3   bclr    IBCR,MSSL                   ; generate stop condition
             ldab    #0                          ; return error code 0
             pula
             rts
```

The C language version of the function is as follows:

```c
char EEbyteWrite(char ID, char addr, char data)
{
    SendSlaveID(ID);
    if (IBSR & RXAK)         /* error if EEPROM does not respond */
        return −1;
    IBDR = addr;             /* send out address of the location to be written */
    while(!(IBSR & IBIF));
```

```
        IBSR = IBIF;              /* clear the IBIF flag */
        if (IBSR & RXAK)         /* error if EEPROM does not respond */
            return −1;
        IBDR = data;             /* send out the data byte */
        while(!(IBSR&IBIF));
        IBSR = IBIF;             /* clear the IBIF flag */
        if (IBSR & RXAK)         /* error if EEPROM does not respond */
            return −1;
        IBCR &= ~MSSL;           /* generate a stop condition */
        return 0;                /* normal write code */
    }
```

Example 11.17

Write a function that performs a pagewrite operation. The control byte, the starting address of the destination, and the pointer to the data in RAM to be written are passed to this function.

Solution: The following function writes a block of up to 16 bytes to the EEPROM. The block of data to be output is pointed to by X. The EEPROM ID and starting address to be output are passed in A and B, respectively. The number of bytes to be output is passed in Y. The error code −1 will be returned if the EEPROM does not respond to its ID.

```
EEpageWrite   jsr     sendSlaveID            ; generate start condition and send out slave ID
              brclr   IBSR,RXAK,pwriteok1    ; does the EEPROM acknowledge?
              ldab    #$FF                   ; return error code −1
              rts
pwriteok1     stab    IBDR                   ; send out the starting address to be written
              brclr   IBSR,IBIF,*            ; wait until the byte is shifted out
              movb    #IBIF,IBSR             ; clear the IBIF flag
              brclr   IBSR,RXAK,w_loop       ; does the EEPROM acknowledge?
              ldab    #$FF                   ; return error code −1
              rts
w_loop        cpy     #0
              beq     done_EEwrite           ; byte count is 0, done
              movb    1,x+,IBDR              ; send out 1 byte
              brclr   IBSR,IBIF,*            ; wait until the byte is shifted out
              movb    #IBIF,IBSR             ; clear the IBIF flag
              brclr   IBSR,RXAK,okNxt        ; receive ACK?
              ldab    #$FF
              rts
okNxt         dey                            ; decrement byte count
              bra     w_loop
done_EEwrite
              bclr    IBCR,MSSL              ; generate a stop condition
              ldab    #0                     ; return error code 0
              rts
```

The C function that performs the pagewrite operation is as follows:

```
char EEpageWrite(char ID, char addr, char ByteCnt, char *ptr)
{
    SendSlaveID(ID);              /* send out EEPROM ID */
    if (IBSR & RXAK)
        return −1;                /* return −1 if EEPROM did not respond */
    IBDR = addr;                  /* send out starting address of pagewrite */
    while(!(IBSR & IBIF));        /* wait until the address is shifted out */
    IBSR = IBIF;                  /* clear IBIF flag */
    if (IBSR & RXAK)
        return −1;                /* return −1 if EEPROM did not respond */
    while(ByteCnt) {
        IBDR = *ptr++;            /* send out 1 byte of data */
        while(!(IBSR & IBIF));
        IBSR = IBIF;
        if (IBSR & RXAK)
            return −1;            /* return −1 if EEPROM did not respond */
        ByteCnt−;
    }
    IBCR &= ~MSSL;                /* generate a stop condition */
    return 0;
}
```

In some applications, the program may need to write several blocks of data to different locations. An internal write cycle takes about 5 ms to complete. The user can either call a delay function to wait for 5 ms or use acknowledge polling to find out whether the internal write has completed.

Example 11.18

Write a C function to implement the algorithm described in Figure 11.38.

Solution: The function eeAckPoll() will poll the EEPROM until it is not busy.

```
void eeAckPoll(char ID)
{
    SendSlaveID(ID);
    while(IBSR & RXAK){
        IBCR |= RSTA;            /* generate a restart condition */
        IBDR = ID;               /* send out EEPROM ID */
        while(!(IBSR & IBIF));
        IBSR = IBIF;             /* clear the IBIF flag */
    } ;                          /* continue if EEPROM did not acknowledge */
    IBCR &= ~MSSL;               /* generate a stop condition—indispensable */
}
```

11.9 Summary

The I²C protocol is an alternative to the SPI serial interface protocol. Compared with the SPI protocol, the I²C bus offers the following advantages:

- No chip-enable or chip-select signal for selecting slave devices
- Allows multiple master devices to coexist in a system because it provides easy bus arbitration
- Allows many more devices in the same I²C bus
- Allows resources to be shared by multiple master (microcontrollers) devices

However, the SPI has the following advantages over the I²C interface:

- Higher data rates (no longer true for I²C high-speed mode)
- Much lower software overhead to carry out data transmission

Data transfer over the I²C bus requires the user to generate the following signal components:

1. Start condition
2. Stop condition
3. ACK
4. Restart condition
5. Data

Whenever there are multiple master devices attempting to send data over the I²C bus, bus arbitration is carried out automatically. The loser is decided whenever it attempts to drive the data line to high, whereas another master device drives the same data line to low.

To select the slave device without using the chip select (or chip enable) signal, address information is used. Both 7-bit and 10-bit addresses are supported in the same I²C bus. Ten-bit addressing will be used in a system that consists of many slave devices. The I²C bus supports three speed rates:

- 100 kHz
- 400 kHz
- 3.4 MHz

Currently, the HCS12 devices support only 7-bit addressing and a 100-kHz clock rate.

Each data transfer starts with a start condition and ends with the stop condition. One or two control bytes will follow the start condition, which specifies the slave device to receive or send data. For each data byte, the receiver must assert either the ACK or NACK condition to acknowledge or unacknowledge, respectively, the data transfer.

The DS1307 is a real-time clock that can keep track of the current time and calendar information. After the current time and calendar information have been set up, the DS1307 will update it once per clock period. As long as the clock frequency input is accurate, the DS1307 can keep track of the time very accurately. The DS1307 can also optionally interrupt the MCU once per second so that the MCU can update the time display.

The DS1631A is a digital thermometer that allows the ambient temperature to be read using the I²C bus. It has an output signal (T_{OUT}) that will become active whenever the ambient temperature reaches the value preset by the user. This feature is useful to provide early warning about potential overheating in a system. The DS1631A can be configured to operate in one-shot mode or continuous conversion mode. The default temperature conversion resolution is

12-bit. However, it can also be reduced to 9-, or 10-, or 11-bit if high resolution is not so important compared to the required conversion time.

The 24LC08B is an EEPROM with an I²C interface. The capacity of the 24LC08B is 8K bits. This chip has an internal address pointer that will increment automatically after each access. This feature can increase the access efficiency when the access patterns are sequential.

Functions for performing basic read and write access to the I²C slave devices have been provided in this chapter. Users can add their own processing functions to provide further processing or data formatting to make the data more user-friendly.

11.10 Exercises

E11.1 Does the I²C clock frequency need to be exactly equal to 100 kHz or 400 kHz? Why?

E11.2 Suppose that the 7-bit address of an I²C slave is B'10101 A1 A0' with A1 tied to low and A0 pulled to high. What is the 8-bit hex write address for this device? What is the 8-bit hex read address?

E11.3 Assuming that the HCS12 is running with a 16-MHz bus clock, compute the values to be written into the IBFD register to set the baud rate to 100 kHz and 200 kHz. Use Table 11.6.

E11.4 Assuming that the HCS12 is running with a 32-MHz bus clock, compute the values to be written into the IBFD register to set the baud rate to 200 kHz and 400 kHz. Use Table 11.6.

E11.5 Write a function to set the TL value to the DS1631A in the assembly and C languages.

E11.6 Write a instruction sequence to call the openDS1631 function to configure the DS1631A to operate in one-shot mode, 10-bit resolution, and low active polarity for T_{OUT} output.

E11.7 Assuming that the DS1631A has been configured to operate in one-shot mode with 12-bit resolution, write a program to display the converted temperature in the format of three integer digits and one fractional digit in LCD.

E11.8 For the circuit shown in Figure 11.34, add an alarm speaker to the PWM1 output and set the high-temperature trip point to 50°C. Whenever temperature reaches 50°C or higher, turn on the alarm until the temperature drops down to 23°C.

E11.9 Write an assembly routine to implement the algorithm described in Figure 11.39.

E11.10 The MAX5812 is a DAC with 12-bit resolution and I²C interface. The MAX5821 data sheet can be downloaded from the website http://www.maxim-ic.com. What is the 7-bit address of this device? What is the highest operating frequency of this chip? How many commands are available to this chip? How is this device connected to the I²C bus?

E11.11 For the MAX5812 mentioned in E11.10, write a program to generate a sine wave from the OUT pin.

E11.12 The MCP23016 is an I/O expander from Microchip. This chip has an I²C interface and can add 16 I/O pins to the microcontroller. Download the data sheet of this device from Microchip's website. Show the circuit connection of this chip to the microcontroller, and write a sequence of instructions to configure the GP1.0. . .GP1.7 pins for input and configure the GP0.0. . .GP0.7 pins for output. Configure input polarity to active low.

E11.13 What are the corresponding temperatures for the conversion results 0x6800, 0x7200, 0x4800, 0xEE60, and 0xF280 output by the DS1631A? Assume that the 12-bit resolution is used.

E11.14 What will be the conversion results sent out by the DS1631A for the temperature values 40°C, 50°C, 80.5°C, −10.25°C, and −20.5°C? Assume that the 12-bit resolution is used.

E11.15 Write a function that performs a block read from the 24LC08B. The control byte, starting address to be read, and the number of bytes to be read are passed in A, B, and Y respectively. Return the error code in B and store the returned data in a buffer pointed to by X. This function should check the error of no acknowledgement.

E11.16 Write a function that performs a current address read to the 24LC08B. Pass the control byte in A to this subroutine. Return the data byte and error code in A and B, respectively. This function should check the error that the 24LC08B did not acknowledge.

11.11 Lab Exercises and Assignments

L11.1 Write a program to store 0 to 255 in the first block of the 24LC08B of your demo board and then stay in an infinite loop to read out one value from 24LC08B sequentially every half-second and display the value on eight LEDs of the demo board.

L11.2 Connect the DS1631A to the I²C bus of your HCS12 demo board. Write a program to set up the high- and low-temperature trip points to 40°C and 20°C, respectively. Whenever the temperature goes above 40°C, turn on the alarm (speaker). Turn off the alarm when the temperature drops below 20°C. Display the temperature on the LCD display, and update the display once every second.

L11.3 Use the DS1307 (if any) on your demo board to set up a time-of-day display using the LCD. Enter the current time and calendar information using either the DIP switches or keyboard. Update the current time of day display once a second. In addition, add alarm time to your program. Whenever the current time matches the alarm, generate the alarm for 1 minute using the buzzer on the demo board.

12

Analog-to-Digital Converter

12.1 Objectives

After completing this chapter, you should be able to

- Explain the A/D conversion process

- Describe the resolution, the various channels, and the operation modes of the HCS12 A/D converter

- Interpret the A/D conversion results

- Describe the procedure for using the HCS12 A/D converter

- Configure the A/D converter for the application

- Use the temperature sensor TC1047A

- Use the humidity sensor IH-3605

- Use the barometric pressure sensor ASCX30AN from SenSym

12.2 Basics of A/D Conversion

Many embedded applications deal with nonelectric quantities, such as weight, humidity, pressure, weight, massflow, airflow, temperature, light intensity, and speed. These quantities are *analog* in nature because they have a continuous set of values over a given range, in contrast to the discrete values of digital signals. To enable the microcontroller to process these quantities, they need to be represented in digital form; thus an *analog-to-digital* converter is required.

12.2.1 A Data Acquisition System

An A/D converter can deal only with electrical voltage. A nonelectric quantity must be converted into a voltage before A/D conversion can be performed. The conversion of a nonelectric quantity to a voltage requires the use of a *transducer*. In general, a transducer is a device that converts the quantity from one form to another. For example, a temperature sensor is a transducer that can convert the temperature into a voltage. A *load cell* is the transducer that can convert a weight into a voltage.

A transducer may not generate an output voltage in the range suitable for A/D conversion. A voltage *scaler* (or *amplifier*) is often needed to amplify the transducer output voltage into a range that can be handled by the A/D converter. The circuit that performs the scaling and shifting of the transducer output is called a *signal-conditioning circuit*. The overall A/D process is illustrated in Figure 12.1.

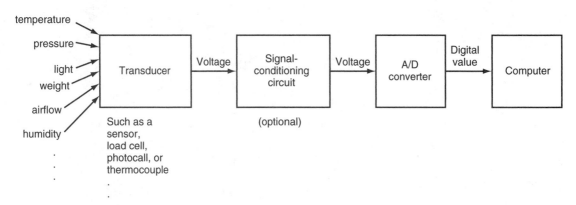

Figure 12.1 ■ The A/D conversion process

12.2.2 Analog Voltage and Digital Code Characteristic

An ideal A/D converter should demonstrate the linear input/output relationship shown in Figure 12.2. However, the output characteristic shown in Figure 12.2 is unrealistic because it requires the A/D converter to use an infinite number of bits to represent the conversion result. The output characteristic of an ideal A/D converter using n bits to represent the conversion result is shown in Figure 12.3. An n-bit A/D converter has 2^n possible output code values. The area between the dotted line and the staircase is called the *quantization error*. The value of $V_{DD}/2^n$ is the resolution of this A/D converter. Using n bits to represent the conversion result, the average *conversion error* is $V_{DD}/2^{n+1}$ if the converter is perfectly linear. For a real A/D converter, the output characteristic may have *nonlinearity* (the staircase may have unequal steps in some values) and *nonmonotonicity* (higher voltage may have smaller code).

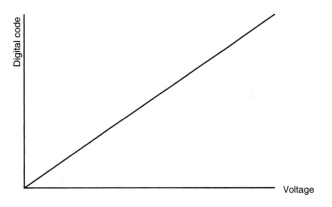

Figure 12.2 ■ An ideal A/D converter output characteristic

Obviously, the more bits used in representing the A/D conversion result, the smaller the conversion error will be. Most microcontrollers use 8 bits, 10 bits, or 12 bits to represent the conversion result. Some microcontrollers (mainly 8051 variants from Silicon Laboratory, TI, and Analog Devices) use 16 bits or even 24 bits to represent conversion results. Whenever the on-chip A/D converter cannot provide the required accuracy, an external A/D converter should be considered.

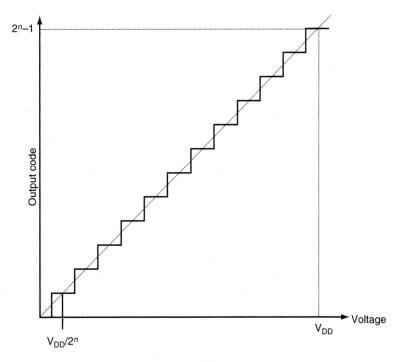

Figure 12.3 ■ Output characteristic of an ideal n-bit A/D converter

12.2.3 A/D Conversion Algorithms

Many A/D conversion algorithms have been introduced in the past. These algorithms can be divided into four categories:

1. Parallel (Flash) A/D converter
2. Slope and double-slope A/D converter
3. Sigma-Delta A/D converter
4. Successive-approximation A/D converter

PARALLEL (FLASH) A/D CONVERTERS

In this type of A/D converter, 2^n comparators are used. One of the inputs to each comparator is the input voltage to be converted; the other input corresponds to the voltage that represents one of the 2^n combinations of n-bit values. The comparator output will be high whenever the analog input (to be converted) is higher than the voltage that represents one of the 2^n combinations of the n-bit value. The largest n-bit value that causes the comparator output to become true is selected as the A/D conversion value through a priority encoder. It is obvious that this type of A/D converter will be very fast. However, they require a lot of hardware resources to implement and therefore are not suitable for implementing high-resolution A/D converter. This type of A/D converter is often used in applications that require high speed but low resolution, such as a video signal. Over the years, several variations to this approach have been proposed to produce high-speed A/D converters. The most commonly used technique is to pipeline a flash A/D converter, which will reduce the amount of hardware required while still achieving high conversion speed.

SLOPE AND DOUBLE-SLOPE A/D CONVERTERS

This type of A/D converter is used in Microchip PIC14000 microcontrollers in which the charging and discharging of a capacitor is used to perform A/D conversion. It requires relatively simple hardware and is popular in low-speed applications, such as digital multimeters. In addition, high resolution (10-bit to 16-bit) can be achieved.

SIGMA-DELTA A/D CONVERTERS

This type of A/D converter uses the *oversampling* technique to perform A/D conversion. It has good noise immunity and can achieve high resolution. Sigma-Delta A/D converters are becoming more and more popular in implementing high-resolution A/D converters. The only disadvantage is the slow conversion speed. However, this weakness is improving because of advancements in CMOS technology.

SUCCESSIVE-APPROXIMATION A/D CONVERTERS

The successive-approximation method approximates the analog signal to n-bit code in n steps. It may be used for low-frequency applications with large DC noise, such as an electrocardiograph. The block diagram of this method is shown in Figure 12.4.

It first initializes the successive-approximation register (SAR) to 0 and then performs a series of guessing, starting with the most significant bit and proceeding toward the least significant bit. The algorithm of the successive-approximation method is illustrated in Figure 12.5. For every bit of the SAR, the algorithm does the following operations:

- Guesses the bit to be a 1
- Converts the value of the SAR register to an analog voltage
- Compares the D/A output with the analog input and clears the bit to 0 if the D/A output is larger (which indicates that the guess is wrong)

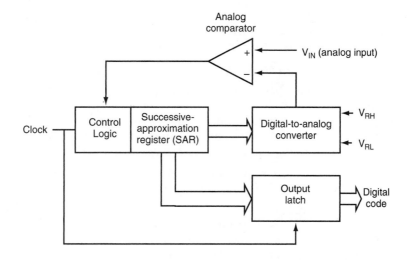

Figure 12.4 ■ Block diagram of a successive approximation A/D converter

Figure 12.5 ■ Successive-approximation A/D conversion method

Because of its balanced speed and precision, this method has become one of the most popular A/D conversion methods. Most microcontrollers use this method to implement the A/D converter. The HCS12 MCUs also use this technique to implement A/D converters.

12.2.4 Optimal Voltage Range for A/D Conversion

An A/D converter needs a *low reference voltage* (V_{RL}) and a *high reference voltage* (V_{RH}) to perform the conversion. The V_{RL} voltage is often set to ground, whereas the V_{RH} voltage is often set to V_{DD}. Some microcontrollers simply tie V_{RL} to the ground voltage and leave only the V_{RH} voltage programmable. Most A/D converters are *ratiometric*, for the following reasons:

- A 0-V (or V_{RL}) analog input is converted to the digital code of n 0s.
- A V_{DD} (or V_{RH}) analog input is converted to the digital code of $2^n - 1$.
- A k-V input will be converted to the digital code of $k \times (2^n - 1) \div V_{DD}$.

Here, n is the number of bits used to represent the A/D conversion result.

The A/D conversion result would be most accurate if the value of the analog signal covers the whole voltage range from V_{RL} to V_{RH}. The A/D conversion result k corresponds to an analog voltage V_K given by the following equation:

$$V_K = V_{RL} + (\text{range} \times k) \div (2^N - 1) \tag{12.1}$$

where range = $V_{RH} - V_{RL}$.

Example 12.1

▼

Suppose that there is a 10-bit A/D converter with $V_{RL} = 1$ V and $V_{RH} = 4$ V. Find the corresponding voltage values for the A/D conversion results of 25, 80, 240, 500, 720, 800, and 900.

Solution: range = $V_{RH} - V_{RL} = 4$ V $- 1$ V $= 3$ V

The voltages corresponding to the A/D conversion results of 25, 80, 240, 500, 720, 800, and 900 are

$$1V + (3 \times 25) \div (2^{10} - 1) = 1.07 \text{ V}$$
$$1V + (3 \times 80) \div (2^{10} - 1) = 1.23 \text{ V}$$
$$1V + (3 \times 240) \div (2^{10} - 1) = 1.70 \text{ V}$$
$$1V + (3 \times 500) \div (2^{10} - 1) = 2.47 \text{ V}$$
$$1V + (3 \times 720) \div (2^{10} - 1) = 3.11 \text{ V}$$
$$1V + (3 \times 800) \div (2^{10} - 1) = 3.35 \text{ V}$$
$$1V + (3 \times 900) \div (2^{10} - 1) = 3.64 \text{ V}$$

▲

12.2.5 Scaling Circuit

Some of the transducer output voltages are in the range of $0 \sim V_Z$, where $V_Z < V_{DD}$. Because V_Z sometimes can be much smaller than V_{DD}, the A/D converter cannot take advantage of the available full dynamic range (0 to $2^n - 1$, where n is the number of bits used to represent the conversion result), and therefore conversion results can be very inaccurate. The voltage scaling (amplifying) circuit can be used to improve the accuracy because it allows the A/D converter to utilize its full dynamic range. The diagram of a voltage scaling circuit is shown in Figure 12.6. Because the OP AMP has an infinite input impedance, the current that flows through the resistor R_2 will be the same as the current that flows through R_1. In addition, the voltage at the inverting input terminal (same as the voltage drop across R_1) would be the same as that at the

Figure 12.6 ■ A voltage scaler

noninverting terminal (V_{IN}). Therefore, the voltage gain of this circuit is given by the following equation:

$$A_V = V_{OUT} \div V_{IN} = (R_1 + R_2) \div R_1 = 1 + R_2/R_1 \qquad (12.2)$$

Example 12.2

▼

Suppose the transducer output voltage ranges from 0 V to 200 mV. Choose the appropriate values for R_1 and R_2 to scale this range to 0 ~ 5 V.

Solution: 5 V ÷ 200 mV = 25 ∴ $R_2/R_1 = 24$

By choosing 240 KΩ for R_2 and 10 KΩ for R_1, we obtain a R_2/R_1 ratio of 24 and achieve the desired scaling goal.

▲

12.2.6 Voltage Translation Circuit

Some transducers have output voltage in the range of V_1 ~ V_2 (V_1 can be negative and V_2 can be unequal to V_{DD}) instead of 0 V ~ V_{DD}. The accuracy of A/D conversion can be improved by using a circuit that shifts and scales the transducer output so that it falls in the full range of 0 V ~ V_{DD}.

An OP AMP circuit that can shift and scale the transducer output is shown in Figure 12.7c. This circuit consists of a summing circuit (Figure 12.7a) and an inverting circuit (Figure 12.7b). The voltage V_{IN} comes from the transducer output; V_1 is an adjusting voltage. By choosing appropriate values for V_1 and resistors R_0, R_1, R_2, and R_f, the desired voltage shifting and scaling can be achieved. Equation 12-5 shows that the resistance R_0 is an independent variable and can be set to a convenient value.

Example 12.3

▼

Choose appropriate resistor values and the adjusting voltage so that the circuit shown in Figure 12.7c can shift the voltage from the range of −1.2 V ~ 3.0 V to the range of 0 V ~ 5 V.

Solution: Applying Equation 12.5:

$$0 = -1.2 \times (R_f/R_1) - (R_f/R_2) \times V_1$$
$$5 = 3.0 \times (R_f/R_1) - (R_f/R_2) \times V_1$$

By choosing $R_0 = R_1 = 10$ KΩ, $R_2 = 50$ KΩ, $R_f = 12$ KΩ, and $V_1 = -5$ V, one can translate and scale the voltage to the desired range. This example tells us that the selection of resistors and the voltage V_1 is a trial-and-error process at best.

▲

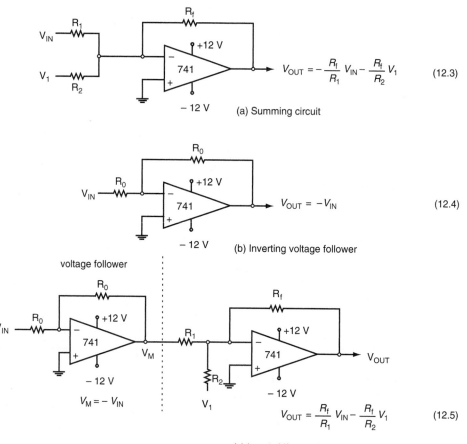

$$V_{OUT} = -\frac{R_f}{R_1}V_{IN} - \frac{R_f}{R_2}V_1 \qquad (12.3)$$

(a) Summing circuit

$$V_{OUT} = -V_{IN} \qquad (12.4)$$

(b) Inverting voltage follower

$$V_M = -V_{IN}$$

$$V_{OUT} = \frac{R_f}{R_1}V_{IN} - \frac{R_f}{R_2}V_1 \qquad (12.5)$$

(c) Level shifter and scaler

Figure 12.7 ■ Level shifting and scaling circuit

12.3 The HCS12 A/D Converter

The block diagram of the HCS12 A/D converter is shown in Figure 12.8. An HCS12 micro-controller may have one or two 8-channel, 10-bit A/D converters. With a 2-MHz conversion clock, the ADC can perform an 8-bit single conversion in 6 μs or a 10-bit single conversion in 7 μs. An A/D conversion sequence can be started by writing into a register or by a valid signal on an external trigger input. The A/D converter uses the successive-approximation method to perform the conversion.

The HCS12 uses two 8-bit registers to hold a conversion result. The result can be stored either right- or left-justified. The A/D conversion is performed in a sequence from one to eight samples. When a sequence consists of multiple samples, the samples may be taken from one channel or multiple channels. An interrupt may be generated upon the completion of an A/D conversion.

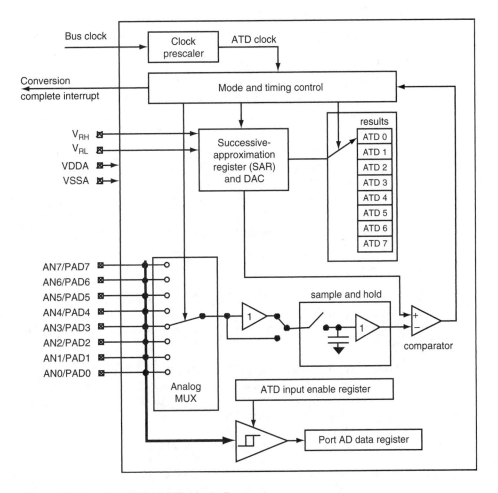

Figure 12.8 ■ The HCS12 ATD block diagram

The analog multiplexer in Figure 12.8 selects one of the three internal or one of the external signal sources for conversion. The sample selected by the multiplexer is amplified before charging the sample capacitor. The comparator indicates whether each successive output of the DAC is higher or lower than the sampled input. The control registers and associated logic select the conversion resolution, multiplexer input, and conversion sequencing mode, sample time, and ADC clock cycle. Each result is available in three formats: right-justified unsigned, left-justified signed, and left-justified unsigned.

12.3.1 Signal Pins Related to A/D Converter

Each of the two A/D modules has eight analog inputs. These input pins are labeled as AN0~AN15. Pins AN0~AN7 are used by the converter AD0; AN8~AN15 are used by the converter AD1. AN7 can optionally be used as the trigger input for AD0, and AN15 can optionally be used as the trigger input for AD1. When a pin is not used as an input to the A/D converter, it can be used as a general-purpose input pin.

The V_{RH} and V_{RL} pins are high reference voltage and low reference voltage inputs, respectively. V_{DDA} and V_{SSA} are power supply inputs for the A/D converters.

12.3.2 Registers Associated with A/D Converter

Each A/D converter has the following registers:

- Six control registers (ATDxCTL0 ~ ATDxCTL5) that control the overall operation of the module. ATDxCTL0 and ATDxCTL1 are intended for factory testing only.
- Two status registers (ATDxSTAT0 ~ ATDxSTAT1) that record the operation status of the converter
- Two test registers (ATDxTEST0 ~ ATDxTEST1) that are dedicated to factory testing purposes
- One input enable register (ATDxDIEN) that enables analog pins to be used as digital input
- One port data register (PORTADx or PTADx)
- Eight 16-bit result registers (ATDxDR0 ~ ATDxDR7)

where x can be 0 or 1.

ATD CONTROL REGISTER 2 (ATD0CTL2, ATD1CTL2)

This register controls power-down, interrupt, and the external trigger. Writes to this register will abort the current conversion sequence but will not start a new conversion sequence. The contents of this register are shown in Figure 12.9.

ATD CONTROL REGISTER 3 (ATD0CTL3, ATD1CTL3)

This register sets the conversion sequence length, enables/disables the FIFO mode for results registers, and controls the ATD behavior in freeze mode. Writes to this register will abort the current conversion but will not start a new conversion. The contents of this register are shown in Figure 12.10.

If the FIFO bit is 0, the A/D conversion results map into result registers based on the conversion sequence; the result of the first conversion appears in the first result register, the second result in the second result register, and so on. If the FIFO bit is 1, the result of a conversion will be stored in the result register specified by the conversion counter. The conversion counter does not reset at the start of a new conversion sequence. The conversion counter value is stored in the lowest three bits of the ATD status register 0.

ATD CONTROL REGISTER 4 (ATD0CTL4, ATD1CTL4)

This register sets the conversion clock frequency, the length of the second phase of the sample time, and the resolution of the A/D conversion. Writes to this register will abort the current conversion sequence but will not start a new sequence. The contents of this register are shown in Figure 12.11.

There are two stages in the analog signal sample time. The first sample stage is fixed at two conversion clock periods. The second stage is selected by SMP1 and SMP0 as shown in Table 12.2.

ATD CONTROL REGISTER 5 (ATD0CTL5, ATD1CTL5)

This register selects the type of conversion sequence and the analog input channels sampled. Writes to this register will abort the current conversion sequence and start a new conversion sequence. The contents of this register are shown in Figure 12.12.

7	6	5	4	3	2	1	0
ADPU	AFFC	AWAI	ETRIGLE	ETRIGP	ETRIGE	ASCIE	ASCIF

reset: 0   0   0   0   0   0   0   0

ADPU: ATD power-down bit
  0 = power-down ATD
  1 = normal ATD operation
AFFC: ATD fast flag clear all bit
  0 = ATD flag is cleared normally, i.e., read the status register before reading the result register.
  1 = any access to a result register will cause the associated CCF flag to clear automatically if it is set at the time.
AWAI: ATD power-down in wait-mode bit
  0 = ATD continues to run when the HCS12 is in wait mode.
  1 = halt conversion and power-down ATD during wait mode.
ETRIGLE: external trigger level/edge control
  This bit controls the sensitivity of the external trigger signal. Details are shown in Table 12.1.
ETRIGP: external trigger polarity
  This bit controls the polarity of the external trigger signal. See Table 12.1 for details.
ETRIGE: external trigger mode enable
  0 = disable external trigger on ATD channel 7
  1 = enable external trigger on ATD channel 7
ASCIE: ATD sequence complete interrupt enable bit
  0 = disables ATD interrupt
  1 = enables ATD interrupt on sequence complete (ASCIF = 1)
ASCIF: ATD sequence complete interrupt flag
  0 = no ATD interrupt occurred.
  1 = ATD sequence complete, interrupt pending.

Figure 12.9 ■ ATD control register 2 (ATDxCTL2, x = 0 or 1)

ETRIGLE	ETRIGP	External Trigger Sensitivity
0	0	Falling edge
0	1	Rising edge
1	0	Low level
1	1	High level

Table 12.1 ■ External trigger configurations

S8C,S4C,S2C,S1C: conversion sequence limit
    0000 = 8 conversions
    0001 = 1 conversion
    0010 = 2 conversions
    0011 = 3 conversions
    0100 = 4 conversions
    0101 = 5 conversions
    0110 = 6 conversions
    0111 = 7 conversions
    1xxx = 8 conversions
FIFO: result register FIFO mode
    0 = conversion results are placed in the corresponding result
       register up to the selected sequence length.
    1 = conversion results are placed in consecutive result registers
       (wrap around at end).
FRZ1 and FRZ0: background debug (freeze) enable bit
    00: continue conversions in active background mode
    01: reserved
    10: finish current conversion, then freeze
    11: freeze immediately when background mode is active

Figure 12.10 ■ ATD control register 3 (ATDxCTL3, x = 0 or 1)

7	6	5	4	3	2	1	0
SRES8	SMP1	SMP0	PRS4	PRS3	PRS2	PRS1	PRS0

reset:
0	0	0	0	0	1	0	1

SRES8: ATD resolution select bit
    0 = 10-bit operation
    1 = 8-bit operation
SMP1 and SMP0: select sample time bits
    These bits are used to select the length of the second phase of the
    sample time in units of ATD conversion clock cycles. See Table
    12.2.
PRS4—PRS0: ATD clock prescaler bits
    These five bits are the binary value prescaler value PRS. The ATD
    conversion clock frequency is calculated as follows:

$$ATDclock = \frac{[bus\ clock]}{PRS + 1} \times 0.5$$

    The ATD conversion frequency must be between 500 kHz and 2
    MHz. The clock prescaler values are shown in Table 12.3.

Figure 12.11 ■ ATD control register 4 (ATDxCTL4, x = 0 or 1)

SMP1	SMP0	Length of 2nd Phase of Sample Time
0	0	2 A/D conversion clock periods
0	1	4 A/D conversion clock periods
1	0	8 A/D conversion clock periods
1	1	16 A/D conversion clock periods

Table 12.2 ■ Sample time select

Prescale Value	Total Divisor Value	Max. Bus Clock[1]	Min. Bus Clock[2]
00000	divide by 2	4 MHz	1 MHz
00001	divide by 4	8 MHz	2 MHz
00010	divide by 6	12 MHz	3 MHz
00011	divide by 8	16 MHz	4 MHz
00100	divide by 10	20 MHz	5 MHz
00101	divide by 12	24 MHz	6 MHz
00110	divide by 14	28 MHz	7 MHz
00111	divide by 16	32 MHz	8 MHz
01000	divide by 18	36 MHz	9 MHz
01001	divide by 20	40 MHz	10 MHz
01010	divide by 22	44 MHz	11 MHz
01011	divide by 24	48 MHz	12 MHz
01100	divide by 26	52 MHz	13 MHz
01101	divide by 28	56 MHz	14 MHz
01110	divide by 30	60 MHz	15 MHz
01111	divide by 32	64 MHz	16 MHz
10000	divide by 34	68 MHz	17 MHz
10001	divide by 36	72 MHz	18 MHz
10010	divide by 38	76 MHz	19 MHz
10011	divide by 40	80 MHz	20 MHz
10100	divide by 42	84 MHz	21 MHz
10101	divide by 44	88 MHz	22 MHz
10110	divide by 46	92 MHz	23 MHz
10111	divide by 48	96 MHz	24 MHz
11000	divide by 50	100 MHz	25 MHz
11001	divide by 52	104 MHz	26 MHz
11010	divide by 54	108 MHz	27 MHz
11011	divide by 56	112 MHz	28 MHz
11100	divide by 58	116 MHz	29 MHz
11101	divide by 60	120 MHz	30 MHz
11110	divide by 62	124 MHz	31 MHz
11111	divide by 64	128 MHz	32 MHz

[1] Maximum ATD conversion clock frequency is 2 MHz. The maximum allowable bus clock frequency is shown in this column.
[2] Minimum ATD conversion clock frequency is 500 kHz. The minimum allowable bus clock frequency is shown in this column.

Table 12.3 ■ Clock prescaler values

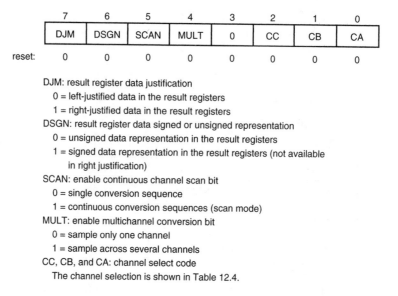

DJM: result register data justification
  0 = left-justified data in the result registers
  1 = right-justified data in the result registers
DSGN: result register data signed or unsigned representation
  0 = unsigned data representation in the result registers
  1 = signed data representation in the result registers (not available
      in right justification)
SCAN: enable continuous channel scan bit
  0 = single conversion sequence
  1 = continuous conversion sequences (scan mode)
MULT: enable multichannel conversion bit
  0 = sample only one channel
  1 = sample across several channels
CC, CB, and CA: channel select code
  The channel selection is shown in Table 12.4.

Figure 12.12 ■ ATD control register 5 (ATDxCTL5, x = 0 or 1)

CC	CB	CA	Analog Input Channel
0	0	0	AN0
0	0	1	AN1
0	1	0	AN2
0	1	1	AN3
1	0	0	AN4
1	0	1	AN5
1	1	0	AN6
1	1	1	AN7

Table 12.4 ■ Analog input channel select code

SRES8	DJM	DSGN	Result Data Formats Description and Bus Bit Mapping
1	0	0	8-bit /left-justified /unsigned—bits 8–15
1	0	1	8-bit /left-justified /signed—bits 8–15
1	1	x	8-bit /right-justified /unsigned—bits 0–7
0	0	0	10-bit /left-justified /unsigned—bits 6–15
0	0	1	10-bit /left-justified /signed—bits 6–15
0	1	x	10-bit /right-justified /unsigned—bits 0–9

Table 12.5 ■ Available result data formats

Table 12.5 summarizes the result data formats available and how they are set up using the control bits. Table 12.6 illustrates the difference between the signed and unsigned, left-justified and right-justified output codes for an input signal range between 0 and 5.12 V.

When single-channel mode is selected, the analog channel to be converted is specified by the channel selection code CC~CA. When multichannel mode is selected, the first analog channel examined is specified by the channel selection code (CC, CB, CA); subsequent channels sampled in the sequence are determined by incrementing the channel selection code.

Input Signal $V_{RL} = 0$ V $V_{RH} = 5.12$ V	Signed 8-bit Codes	Unsigned 8-bit Codes	Signed 10-bit Codes	Unsigned 10-bit Codes
5.120 V	7F	FF	7FC0	FFC0
5.100	7F	FF	7F00	FF00
5.080	7F	FE	7E00	FE00
2.580	01	81	0100	8100
2.560	00	80	0000	8000
2.540	FF	7F	FF00	7F00
0.020	81	01	8100	0100
0.000	80	00	8000	0000

Table 12.6 ■ Left-justified, signed and unsigned ATD output codes

ATD STATUS REGISTER 0 (ATD0STAT0, ATD1STAT1)

This read-only register contains the sequence complete flag, overrun flags for external trigger and FIFO mode, and the conversion counter. The contents of this register are shown in Figure 12.13.

reset:

SCF: sequence complete flag
    0 = conversion sequence not completed.
    1 = conversion sequence has completed.
ETORF: external trigger overrun flag
    0 = no external trigger overrun has occurred.
    1 = external trigger overrun has occurred.
FIFOR: FIFO overrun flag
    0 = no overrun has occurred.
    1 = an overrun has occurred.
CC2, CC1, CC0: conversion counter
    The conversion counter points to the result register that will receive
    the result of the current conversion.
    In non-FIFO mode, this counter is reset to 0 at the begin and end of
    the conversion.
    In FIFO mode, this counter is not reset and will wrap around when
    its maximum value is reached.

Figure 12.13 ■ ATD status register 0 (ATDxSTAT0, x = 0 or 1)

If the AFFC flag is set to 1, the SCF flag can be cleared by reading a result register. Otherwise, it can be cleared by one of the following actions:

- Writing a 1 to SCF
- Writing to the ATDxCTL5 register (a new conversion is started)

While in edge trigger mode, if additional active edges are detected while a conversion sequence is in progress the ETORF flag is set. This flag is cleared when one of the following events occurs:

- Writing a 1 to ETORF
- Writing to ATDxCTL2, ATDxCTL3, or ATDxCTL4 (a conversion sequence is aborted)
- Writing to ATDxCTL5 (a new conversion sequence is started)

The setting of the FIFOR bit indicates that a result register has been written to before its associated conversion complete flag (CCF) has been cleared. This flag is cleared when one of the following events occurs:

- Writing a 1 to FIFOR
- Writing to ATDxCTL5

ATD Test Register 1 (ATD0TEST1, ATD1TEST0)

This register contains the SC bit used to enable special channel conversions. The contents of this register are shown in Figure 12.14.

7	6	5	4	3	2	1	0
0	0	0	0	0	0	0	SC

reset: 0   0   0   0   0   0   0   0

SC: special channel conversion bit
If this bit is set, the special channel conversion can be selected using CC, CB, and CA of the ATDxCTL5 register. Table 12.7 shows the selection.
0 = special channel conversions disabled
1 = special channel conversions enabled

Figure 12.14 ■ ATD test register 1 (ATDxTEST1, x = 0 or 1)

SC	CC	CB	CA	Analog Input Channel
1	0	x	x	Reserved
1	1	0	0	V_{RH}
1	1	0	1	V_{RL}
1	1	1	0	$(V_{RH} + V_{RL})/2$
1	1	1	1	Reserved

Table 12.7 ■ Special channel select code

ATD Status Register 1 (ATD0STAT1, ATD1STAT1)

This read-only register contains the conversion complete flags for all channels. The contents of this register are shown in Figure 12.15.

	7	6	5	4	3	2	1	0
	CCF7	CCF6	CCF5	CCF4	CCF3	CCF2	CCF1	CCF0
reset:	0	0	0	0	0	0	0	0

CCFx: conversion complete flag x (x = 7~0)
  0 = conversion number x not completed
  1 = conversion number x has completed, result in ATDyDRx (y = 0 or 1)

Figure 12.15 ■ ATD status register 1 (ATDxSTAT1, x = 0 or 1)

A CCFx flag can be cleared by one of the following events:

- Writing to ATDxCTL5 (a new conversion is started)
- If AFFC = 0 and read of ATDxSTAT1 followed by read of result register ATDxDRy
- If AFFC = 1 and read of result register ATDxDRy

ATD Input Enable Register (ATD0DIEN, ATD1DIEN)

This register allows the user to enable a Port ADx pin as a digital input. The contents of this register are shown in Figure 12.16.

	7	6	5	4	3	2	1	0
	IEN7	IEN6	IEN5	IEN4	IEN3	IEN2	IEN1	IEN0
reset:	0	0	0	0	0	0	0	0

IENx: ATD digital input enable on channel x (x = 0~7)
  0 = disable digital input buffer to PTADx
  1 = enable digital input buffer to PTADx

Figure 12.16 ■ ATD input enable register (ATDxDIEN, x = 0 or 1)

Port Data Register (PTAD0, PTAD1)

The data port associated with the ATD is input-only. The port pins are shared with the analog inputs AN0~AN7 or AN8~AN15.

ATD Conversion Result Registers (ATDxDRy, x = 0~1, y = 0~7)

Each result register is 16-bit and can be further divided into two 8-bit registers, ATDxDRHy and ATDxDRLy. Depending on the setting, the A/D conversion result can be stored right-justified or left-justified. The conversion result can be signed or unsigned. However, the conversion result is unsigned when it is stored right-justified.

12.4 The Functioning of the ATD Module

The operation of the ATD module is described in detail in this section.

12.4.1 Analog Input Multiplexer

The analog input multiplexer selects one of the external analog inputs to convert. The input analog signals are unipolar and must fall within the potential range of V_{SSA} and V_{DDA}. The input multiplexer includes protection circuitry to prevent crosstalk between channels when the applied input potentials are within specification.

12.4.2 ATD Module Clock

The ATD module needs clock signals to perform the conversion. A clock prescaler allows the ATD module's internal clock to be configured to within the specified frequency range (500 kHz to 2 MHz) for different MCU bus clock frequencies. For maximum accuracy, selecting the 500-kHz clock is recommended where conversion and sample times are not critical. The bus clock is divided by a programmable prescaler value (2, 4, 6, . . . , 64) to generate the ATD clock.

12.4.3 Sample-and-Hold Stage

As shown in Figure 12.8, a sample-and-hold stage accepts analog signals from the input multiplexer and stores them as a charge on the sample capacitor. The sampling process has two stages:

1. Initially, a sample amplifier (of unity gain) is used to buffer the input analog signal for two cycles to charge the sample capacitor almost to the input potential. This stage reduces charging and discharging the sample capacitor by the signal source.

2. The sample buffer is then disconnected and the input signal is directly connected to the storage node for a programmable 2, 4, 8, or 16 cycles.

The conversion time of a sample is given by the following equation:

$$\text{Conversion time} = \frac{(\text{no. of bits in resolution} + \text{no. of programmed sample clocks} + 2)}{\text{ATD clock frequency}}$$

Conversion times for frequencies at 500 kHz and 2 MHz are shown in Table 12.8. For maximum accuracy, selecting 16 programmable sample cycles is recommended whenever the sample time is not critical. For fast-changing signals, sample time and source impedance should be kept low. The actual conversion time is the sum of the converter time and the sample time in one of the rightmost four columns.

ATD Clock Frequency	Resolution	Converter Time	2 + 2 Sample Clocks	2 + 4 Sample Clocks	2 + 8 Sample Clocks	2 + 16 Sample Clocks
2 MHz	8-bit[1]	4 µs	2 µs	3 µs	5 µs	9 µs
2 MHz	10-bit[2]	5 µs				
500 kHz	8-bit	16 µs	8 µs	12 µs	20 µs	36 µs
500 kHz	10-bit	20 µs				

[1] The fastest 8-bit resolution conversion time is 4 µs + 2 µs = 6 µs.
[2] The fastest 10-bit resolution conversion time is 5 µs + 2 µs = 7 µs.

Table 12.8 ■ ATD conversion timings

12.4.4 Input Channel Wraparound

In the case of a multiple channel conversion sequence (MULT bit of the ATDxCTL5 register = 1), when the input selector goes past channel AN7, it wraps around to the channel AN0.

Example 12.4

Assuming that S8C~S1C (ATD0CTL3) are set to 0101 and CC~CA (ATD0CTL5) are set to 110, what is the conversion sequence for this setting?

Solution: The first channel to be converted is AN6 and there are five channels to be converted. The conversion sequence is as follows:

AN6, AN7, AN0, AN1, and AN2

12.4.5 FIFO Mode

All eight 16-bit result registers are organized into a circular ring. The conversion counter value in the ATDxSTAT0 register specifies the result register to hold the current conversion result. In the FIFO mode, the conversion count is not reset to 0 when a new conversion sequence is started. Since one can set the length of a conversion sequence to a value between 1 and 8, the first conversion result in a sequence may not be stored in the ATDxDR0 register.

Example 12.5

Assume that the following setting was programmed before a new conversion is started:

- The conversion counter value in the ATD0STAT0 register is 5.
- The channel select code of the ATD0CTL5 register is 6.
- The conversion sequence limit of the ATD0CTL3 register is set to 5.
- The MULT bit of the ATD0CTL5 register is set to 1.

How would the conversion results be stored when the FIFO mode is selected or not selected?

Solution: The conversion counter specifies the result register to hold the first conversion result. The channel-select code specifies the first analog input channel to be converted, and the conversion sequence limit specifies the number of channels to be converted. The conversion results will be stored as shown in Table 12.9.

Analog Channel	Result Stored in (FIFO mode)	Result Stored in (non-FIFO mode)
AN6	ATD0DR5	ATD0DR0
AN7	ATD0DR6	ATD0DR1
AN0	ATD0DR7	ATD0DR2
AN1	ATD0DR0	ATD0DR3
AN2	ATD0DR1	ATD0DR4

Table 12.9 ■ Conversion results storage

12.4.6 External Trigger Source

An external triggering function allows the user to synchronize the ATD conversion process with external events. The ATD module can be configured to trigger on edges or levels of different polarities. Several bits in the ATDxCTL2 register are used to configure this function. The external triggering capability allows the user to start an A/D conversion when it is desired. The external trigger source is AN7 for AD0 and AN15 for AD1. Connecting the trigger channels together on MCUs with two ATDs allows synchronized conversions on the two converters. Examples of the trigger source could be external logic, software driving an output pin, a timer output-compare output pin, or a PWM channel.

12.4.7 Signed and Unsigned Control

Conversion results can be signed or unsigned. Signed data is represented as two's complement. The ATDxCTL5 register controls this function. A signed conversion sequence treats the value $(V_{RH} - V_{RL})/2$ as zero reference. Input signals greater than this value produce positive results; input signals lower than this value result in negative results.

12.4.8 ATD Operation Modes

The ATD module has three operation modes: power-down mode, idle mode, and run mode.

POWER-DOWN MODE

The ATD module can be powered down to save power consumption in one of three ways:

1. By clearing the ADPU bit of the ATDxCTL2 register
2. By executing the STOP instruction
3. By executing the WAI instruction with the AWAI bit of the ATDxCTL2 register set to 1

The ATD module is in power-down mode out of reset. Once the command to power down has been received, the ATD module aborts any conversion sequence in progress and enters power-down mode. When the module is powered up again, the module requires a recovery time of about 20 μs to stabilize the bias settings in the analog electronics before conversions can be performed. Powering the module up and down does not change the contents of the register file, and in power-down mode, the control and result registers are still accessible.

IDLE MODE

Idle mode for the ATD module is defined as the state in which the ATD module is powered up and ready to perform A/D conversion but not actually performing a conversion at the present time. Access to all control, status, and result registers is available. The module is consuming near the maximum power.

RUN MODE

The run mode for the ATD module is defined as the state in which the ATD module is powered up and currently performing an A/D conversion. Complete access to all control, status, and result registers is available. The module consumes the maximum power.

12.5 Procedure for Performing A/D Conversion

Before the HCS12 can perform A/D conversion, the A/D module must be connected and configured properly. The procedure for performing an A/D conversion is as follows:

Step 1
Connect the hardware properly. The A/D-related pins must be connected as follows:

V_{DDA}: connect to 5 V

V_{SSA}: connect to 0 V

V_{RH}: 5 V or other positive value less than 5 V

V_{RL}: 0 V or other value less than V_{RH} but higher than 0 V

If the transducer output is not in the appropriate range, then a signal-conditioning circuit should be used to shift and scale it to between V_{RL} and V_{RH}.

Step 2
Configure ATD control registers 2 to 4 properly and wait for the ATD to stabilize (need to wait for 20 μs).

Step 3
Select the appropriate channel (s) and operation modes by programming ATD control register 5. Writing into ATD control register 5 starts an A/D conversion sequence.

Step 4
Wait until the SCF flag of the status register ATDxSTAT0 is set, then collect the A/D conversion results and store them in memory.

Example 12.6
▼

Write a subroutine to initialize the AD0 converter for the MC9S12DP256 and start the conversion with the following setting:

- Nonscan mode
- Select channel 7 (single channel mode)
- Fast ATD flag clear all
- Stop ATD in wait mode
- Disable interrupt
- Perform four conversions in a sequence
- Disable FIFO mode
- Finish current conversion then freeze when BDM becomes active
- 10-bit operation and two A/D clock periods of the second-stage sample time
- Choose 2 MHz as the conversion frequency for the 24-MHz bus clock
- Result is unsigned and right-justified

Solution: The settings of ATD control registers 2 to 5 are as follows:
The setting of a *ATD control register 2* is as follows:

- Enable AD0 (set bit 7 to 1)
- Select fast flag clear all (set bit 6 to 1)

- Stop AD0 when in wait mode (set bit 5 to 1)
- Disable external trigger on channel 7 (set bits 4, 3, and 2 to 0)
- Disable AD0 interrupt (set bit 1 to 0)

Write the value $E0 into this control register.

The setting of the *ATD control register 3* is as follows:

- Perform four conversions
- Disable FIFO mode
- When BDM becomes active, complete the current conversion then freeze

Write the value $22 into ATD0CTL3.

The setting of the *ATD control register 4* is as follows:

- Select 10-bit operation (set bit 7 to 0)
- Two A/D clock periods for sample time (set bits 6 and 5 to 00)
- Set the value of PRS4~PRS0 to 00101

Write the value $05 to this control register.

The setting of the *ATD control register 5* is as follows:

- Result register right-justified (set bit 7 to 1)
- Result is unsigned (set bit 6 to 0)
- Nonscan mode (set bit 5 to 0)
- Single-channel mode (set bit 4 to 0)
- Select channel 7 (set bits 2..0 to 111)

Write the value $87 to this control register.

The following subroutine will perform the desired ATD configuration:

```
#include        "c:\miniide\hcs12.inc"

openAD0     movb      #$E0,ATD0CTL2
            jsr       wait20us           ; wait for 20 us
            movb      #$22,ATD0CTL3
            movb      #$05,ATD0CTL4
            rts
wait20us    movb      #$90,TSCR1         ; enable TCNT and fast timer flag clear
            movb      #0,TSCR2           ; set TCNT prescaler to 1
            bset      TIOS,$01           ; enable OC0
            ldd       TCNT               ; start an OC0 operation
            addd      #480               ;      "
            std       TC0                ;      "
            brclr     TFLG1,C0F,*        ; wait for 20 us
            rts
```

This routine does not write into the ATD0CTL5 register because that will start the conversion. One should write into the ATD0CTL5 register only when one wants to perform the conversion.

The C language version of the subroutine is as follows:

```c
void wait20us (void);
void openAD0 (void)
{
        ATD0CTL2 = 0xE0;
        wait20us();
        ATD0CTL3 = 0x22;
        ATD0CTL4 = 0x05;
}
void wait20us (void)
{
        TSCR1    = 0x90;
        TSCR2    = 0;
        TIOS     |= OC0;
        TC0      = TCNT + 480;
        while(!(TFLG1 & COF));
}
```

▲

Example 12.7

▼

Write a C program to perform A/D conversion on the analog signal connected to the AN7 pin. Collect 20 A/D conversion results and store them at memory locations starting from $1000. Use the same configuration as in Example 12.6.

Solution: To collect 20 conversion results, we need to write into the ATD0CTL5 register five times. Each time we will wait for the SCF flag to be set and collect four results. The program is as follows:

```
#include      "c:\miniide\hc12.inc"
              org     $1500
              lds     #$1500
              ldx     #$1000          ; use index register X as a pointer to the buffer
              jsr     openAD0         ; initialize the ATD0 converter
              ldy     #5
loop5         movb    #$87,ATD0CTL5   ; start an A/D conversion sequence
              brclr   ATD0STAT0,SCF,*
              movw    ATD0DR0,2,x+    ; collect and save the conversion results
              movw    ATD0DR1,2,x+    ; postincrement the pointer by 2
              movw    ATD0DR2,2,x+    ;         "
              movw    ATD0DR3,2,x+    ;         "
              dbne    y,loop5
              swi
              end
```

The C language version of the program is as follows:

```c
#include "c:\egnu091\include\hcs12.h"
void openAD0 (void);
int buf[20];
```

```
void main (void)
{
        int i;
        openAD0();
        for (i = 0; i < 5; i++) {
                ATD0CTL5 = 0x87;              /* start an A/D conversion */
                while (!(ATD0STAT0 & SCF));   /* wait for the A/D conversion to complete */
                buf[4*i + 0] = ATD0DR0;       /* save results right-justified */
                buf[4*i + 1] = ATD0DR1;
                buf[4*i + 2] = ATD0DR2;
                buf[4*i + 3] = ATD0DR3;
        }
        asm ("swi");
}
```

Example 12.8

▼

Write a C program to be run on the Dragon12 (or SSE256) demo board to display the voltage (output of a potentiometer) connected to the AN7 pin. Configure the AD0 properly, perform five conversions per second, and display the voltage on the LCD.

Solution: The conversion result 1023 corresponds to 5 V. Therefore, one needs to divide the conversion result by 204.6 to convert the result back to voltage. Since the HCS12 does not support floating-point operation directly, we will multiply the conversion result by 10 and then divide the product by 2046 to obtain the voltage value. The C program is as follows:

```
#include "c:\egnu091\include\hcs12.h"
#include "c:\egnu091\include\delay.c"
#include "c:\egnu091\include\lcd_util_dragon12.c"
void openAD0(void);
void wait20us (void);
main(void)
{
        char buffer[6];          /* used to hold the voltage value */
        int temp;
        char *msg1 = "Voltage =" ;
        openlcd();
        buffer[1] = '.';         /* decimal point */
        buffer[3] = 0x20;        /* space character */
        buffer[4] = 'V';         /* volt character */
        buffer[5] = 0;           /* null character */
        openAD0();
        while(1) {
                ATD0CTL5 = 0x87;                       /* convert AN7, result right-justified */
                while(!(ATD0STAT0 & SCF));             /* wait for conversion to complete */
                buffer[0] = 0x30 + (ATD0DR0 * 10)/2046;
                temp = (ATD0DR0 * 10)%2046;            /* find the remainder */
                buffer[2] = 0x30 + (temp * 10)/2046;   /* compute the fractional digit */
                cmd2lcd(0x80);                         /* set LCD cursor to upper left corner*/
```

```
                    puts2lcd(msg1);              /* output the message "voltage =" */
                    puts2lcd(&buffer[0]);        /* output voltage string */
                    delayby100ms(2);             /* wait for 200 ms */
             }
             return 0;
       }
       void openAD0 (void)
       {
             int i;
             ATD0CTL2 = 0xE0;   /* enable AD0, fast ATD flag clear, disable AD0 in wait mode */
             delayby10us(2);
             ATD0CTL3 = 0x0A;   /* perform one conversion */
             ATD0CTL4 = 0x25;   /* 4 cycles sample time, prescaler set to 12 */
       }
```

When this program is running on the demo board, you can turn the potentiometer and see the voltage value changes on the LCD display.

12.6 Using the Temperature Sensor TC1047A

The TC1047A is a three-pin temperature sensor whose voltage output is directly proportional to the measured temperature. The TC1047A can accurately measure temperatures from −40°C to 125°C with a power supply from 2.7 V to 5.5 V.

The output voltage range for these devices is typically 100 mV at −40°C, 500 mV at 0°C, 750 mV at +25°C, and +1.75 V at +125°C. As shown in Figure 12.17, the TC1047A has a 10 mV/°C voltage slope output response.

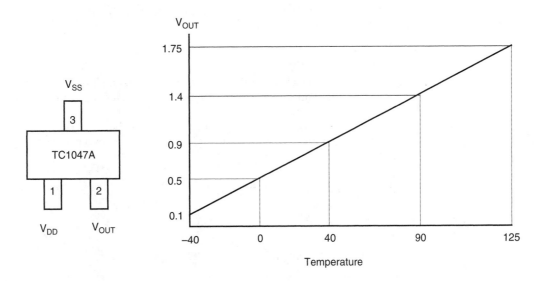

Figure 12.17 ■ TC1047A V_{OUT} vs. temperature characteristic

Example 12.9

Describe a circuit connection and the required program to build a digital thermometer. Display the temperature in three integral and one fractional digits using the LCD. Measure and display the temperature over the whole range of TC1047A, that is, −40°C to +125°C. Update the display data five times per second and assume that the HCS12 operates with a 24-mHz bus clock.

Solution: Since the voltage output of the TC1047A is from 0.1 V to 1.75 V for the temperature range of −40°C through 125°C, we will need to use the circuit shown in Figure 12.7c to perform the translation and scaling for V_{OUT}. The circuit connection is shown in Figure 12.18. This circuit will use V_{DD} and V_{SS} as the reference voltages required in the process of A/D conversion. The range of the voltage connected to the AN7 pin will be between 0 and 5 V for the given temperature range.

Figure 12.18 ■ Circuit connection between the TC1047A and the HCS12

To convert the A/D conversion result to the corresponding temperature, we need to do the following:

1. Divide the conversion result by 6.2
2. Subtract the quotient by 40

Since the HCS12 MCU cannot handle floating-point numbers directly, the operation of "dividing by 6.2" would be implemented by multiplying the conversion result by 10 and then dividing the product by 62.

The procedure for starting an A/D conversion, reading the conversion result, calculating the corresponding temperature, and converting the temperature value into an ASCII string every 200 ms is as follows:

Step 1
Configure the A/D converter properly.

Step 2
Start an A/D conversion.

Step 3
Wait until the A/D conversion is complete.

Step 4
Multiply the conversion result by 10.

Step 5
Divide the product resulting from step 4 by 62 to obtain temperature reading. Use the variables quo and rem to hold the quotient and remainder.

Step 6
Subtract 40 from the variable quo to obtain the actual temperature.

Step 7
If (quo ≥ 0), go to step 9; otherwise, replace quo with its two's complement.

Step 8
If rem ≠ 0, then

1. Decrement quo by 1
2. rem ← 62 − rem

Step 9
Compute the fractional digit by multiplying rem by 10 and then dividing the resulting product by 62.

Step 10
Compute the integral digits by performing repeated division by 10 to quo.

Step 11
Wait for 200 ms and then go to step 2.

The assembly program that implements this algorithm is as follows:

```
#include "c:\miniIDE\hcs12.inc"
period        equ       $2E                  ; ASCII code of period character
degree        equ       223                  ; ASCII code of degree character
              org       $1000
quo           ds.b      1
rem           ds.b      1
sign          ds.b      1
fract         ds.b      1
buf           ds.b      8                    ; to hold string to be output to LCD
              org       $1500
              lds       #$1500               ; set up stack pointer
              ldy       #2                   ; wait for LCD to complete
              jsr       delayby100ms         ; internal configuration
              jsr       openlcd              ; configure LCD
              ldaa      #$80                 ; set LCD cursor to upper
              jsr       cmd2lcd              ; left corner
              ldx       #msg1                ; output "Temperature ="
              jsr       puts2lcd             ;        "
              jsr       openAD0              ; configure ATD0 module
forever       movb      #$20,buf             ; initialize the buffer contents to 0.0°C
              movb      #$20,buf+1           ;        "
              movb      #$30,buf+2           ;        "
              movb      #period,buf+3        ;        "
              movb      #$30,buf+4           ;        "
              movb      #degree,buf+5        ; degree character
              movb      #$43,buf+6           ; letter 'C'
              movb      #0,buf+7             ; null character
```

```
              movb      #$87,ATD0CTL5      ; start an ATD conversion sequence
              movb      #0,sign            ; initialize sign to positive
              movb      #$30,fract         ; initialize fractional digit to 0
              brclr     ATD0STAT0,SCF,*    ; wait for the conversion to complete
              ldd       ATD0DR0            ; read a conversion result
              ldy       #10                ; compute result x 10 / 62
              emul                         ;      "
              ldx       #62                ;      "
              ediv                         ;      "
              stab      rem                ; save the remainder
              tfr       y,d                ; transfer quotient to B
              subb      #40                ; subtract temperature offset
              bhs       save_quo           ; if nonnegative, don't touch remainder
              negb                         ; compute two's complement of quotient
              stab      quo
              movb      #1,sign            ; temperature is negative
              ldab      rem                ; if remainder is 0, skip a few instructions
              beq       convert            ;
              ldab      #62                ; compute 62 - rem
              subb      rem                ;      "
              stab      rem                ;      "
              bra       cal_fract
save_quo      stab      quo                ; save updated quotient
cal_fract     ldab      rem
              beq       convert            ; come here when positive
              ldaa      #10                ; compute fractional digit
              mul                          ;      "
              ldx       #62                ;      "
              idiv                         ;      "
              cmpb      #31                ; round off fractional digit
              blt       no_round           ;      "
              inx                          ;      "
              cpx       #10                ;      "
              bne       no_round           ;      "
              inc       quo                ;      "
              bra       convert            ; prepare to separate integer digits
no_round      tfr       x,d                ; convert fractional digit to ASCII code
              addb      #$30               ;      "
              stab      fract              ;      "
convert       ldab      quo
              clra                         ;      "
              ldx       #10                ; use repeated divide by 10 to separate
              idiv                         ; integral digits
              addb      #$30               ;      "
              stab      buf+2              ; save the one's digit
              tfr       x,d                ; transfer quotient to D
              tstb                         ; is quo zero?
              beq       add_fra            ; if integral part is 0, then add fraction digit
              ldx       #10                ; separate the ten's digit
```

```
                idiv
                addb        #$30                    ; convert and store the ten's digit
                stab        buf+1                   ;      "
                tfr         x,d                     ; test hundred's digit
                tstb                                ; is quotient 0?
                beq         add_fra
                movb        #$31,buf                ; hundred's digit, if any, is 1 only
add_fra         movb        fract,buf+4             ; insert fraction digit
                ldaa        sign                    ; check the sign
                beq         out_it
                movb        #$2D,buf                ; when minus, add minus character
out_it          ldaa        #$C0                    ; set cursor to 2nd row
                jsr         cmd2lcd                 ;      "
                ldx         #spaces                 ; clear the 2nd row of the LCD
                jsr         puts2lcd                ;      "
                ldaa        #$C5                    ; set LCD cursor position
                jsr         cmd2lcd                 ;      "
                ldx         #buf                    ; output the temperature string
                jsr         puts2lcd                ;      "
                ldy         #2                      ; wait for 200 ms
                jsr         delayby100ms            ;      "
                jmp         forever                 ; continue
;****************************************************************************
; The following function perform the AD0 configuration.
;****************************************************************************
openAD0         movb        #$E0,ATD0CTL2           ; enable AD0, fast ATD flag clear, stop in wait mode
                ldy         #2
                jsr         delayby10us             ; wait until AD0 is stabilized
                movb        #$0A,ATD0CTL3           ; perform one A/D conversion
                movb        #$25,ATD0CTL4           ; set four cycles of sample time, set prescaler
                                                    ;   divisor to 12
                rts
#include        "C:\miniIDE\lcd_util_dragon12.asm"
#include        "c:\miniIDE\delay.asm"
msg1            fcc         "Temperature ="
                dc.b        0
spaces          fcc         "          "
                dc.b        0
                end
```

The C language version of the program is as follows:

```
#include        "c:\egnu091\include\hcs12.h"
#include        "c:\egnu091\include\delay.c"
#include        "c:\egnu091\include\convert.c"
#include        "c:\egnu091\include\lcd_util_dragon12.c"
void    openAD0(void);
char    buf[8];
char    *msg1 = "temperature = ";
char    *blanks = "          ";
```

```
void     main (void)
{
         int        temp1,temp2;
         char       sign,fdigit,frem;
         char       *ptr;
         delayby100ms(2);                /* wait for LCD kit to initialize */
         openlcd();                      /* configure LCD kit */
         openAD0();                      /* configure AD0 module */
         cmd2lcd(0x80);                  /* set cursor to upper left corner */
         puts2lcd(msg1);                 /* output the message "temperature = " */
         while(1) {
                  sign = 0;              /* initialize sign to be positive */
                  ATD0CTL5 = 0x87;                       /* start a conversion with result right-justified */
                  while(!(ATD0STAT0 & SCF));             /* wait until conversion is done */
                  temp1 = (ATD0DR0 * 10) / 62;           /* integer part of temperature */
                  temp2 = (ATD0DR0 * 10) % 62;           /* remainder part */
                  temp1 -= 40;          /* subtract the offset from the actual temperature */
                  if (temp1 < 0){       /* temperature is negative */
                           sign = 1;
                           temp1 = ~temp1 + 1;           /* find the magnitude of temperature */
                           if (temp2) { /* remainder not zero */
                                    temp1--;
                                    temp2 = 62 - temp2;
                           }
                  }
                  fdigit = (temp2 * 10) / 62;            /* compute the fractional digit */
                  frem = (temp2 * 10)%62;
                  if (frem > 31) {
                           fdigit ++;
                           if (fdigit == 10) {  /* round off the fractional digit */
                                    fdigit = 0;
                                    temp1 ++;
                           }
                  }
                  if (sign) {
                           ptr = &buf[1];       /* point to the first space to hold ASCII string */
                           buf[0] = 0x2D;       /* store minus sign as the first character */
                  }
                  else
                           ptr = &buf[0];
                  int2alpha(temp1,ptr);         /* convert the integer part to ASCII string */
                  ptr = &buf[0];
                  while(*ptr)                   /* find the end of the integer string */
                           ptr++;
                  *ptr++ = '.';
                  *ptr++ = fdigit + 0x30;
                  *ptr++ = 223;                 /* add a degree character */
                  *ptr++ = 'C';                 /* temperature in Celsius */
                  *ptr = 0;                     /* terminate the temperature string */
                  cmd2lcd(0xC0);                /* move cursor to 2nd row */
                  puts2lcd(blanks);             /* clear the 2nd row */
```

```
                              cmd2lcd(0xC5);              /* set cursor to column 5 row 2 */
                              puts2lcd(&buf[0]);          /* output the temperature */
                              delayby100ms(2);            /* wait for 0.2 seconds */
                    }
            }
            void openAD0(void)
            {
                    ATD0CTL2 = 0xE0;                       /* enable AD0, fast ATD flag clear, power-down on wait */
                    ATD0CTL3 = 0x0A;                       /* perform one ATD conversion */
                    ATD0CTL4 = 0x25;                       /* prescaler set to 12, select four cycles sample time */
                    delayby10us(2);
            }
```

12.7 Using the IH-3606 Humidity Sensor

The IH-3605 is a humidity sensor made by Honeywell. It provides a linear output from 0.8 to 3.9 V in the full range of 0% to 100% relative humidity (RH) when powered by a 5-V power supply. It can operate in a range of 0% to 100% RH, −40° to 185°F. The pins of the IH-3605 are shown in Figure 12.19. The specifications are listed in Table 12.10. The IH-3605 is light sensitive and should be shielded from bright light for best result.

Specification	Description
Total accuracy	± 2% RH, 0%–100% TH @ 25°C
Interchangeability	± 5% RH up to 60% RH, ±8% RH at 90% RH
Operating temperature	−40° to 85°C (−40° to 185°F)
Storage temperature	−51° to 110°C (−60° to 223°F)
Linearity	±0.5% RH typical
Repeatability	±0.5% RH
Humidity stability	±1% RH typical at 50% RH in 5 years
Temp. effect on 0% RH voltage	±0.007% RH/°C (negligible)
Temp. effect on 100% RH voltage	−0.22% RH/°C
Output voltage	$V_{OUT} = (V_S)(0.16$ to $0.78)$ nominal relative to supply voltage for 0%–100% RH; i.e., 1–4.9 V for 6.3-V supply; 0.8–3.9 V for 5-V supply; sink capability 50 μA; drive capability 5 μA typical; low-pass 1-kHz filter required. Turn on time < 0.1 to full output.
VS supply requirement	4 to 9 V, regulated or use output/supply ratio; calibrated at 5 V
Current requirement	200 μA typical @ 5 V, increased to 2 mA at 9 V

Figure 12.19 ■ Honeywell IH-3605 humidity sensor

Table 12.10 ■ Specifications of IH-3605

The IH-3605 can resist contaminant vapors, such as organic solvents, chlorine, and ammonia. It is unaffected by water condensate as well. Because of this capability, the IH-3605 has been used in refrigeration, drying, instrumentation, meteorology and many other applications.

Example 12.10
▼

Construct a humidity measurement system that consists of the HCS12, an IH-3605 humidity sensor, and an LCD. The bus clock frequency of the HCS12 is 24 MHz.

Solution: Since the output of the IH-3605 is between 0.8 V and 3.9 V with a 5-V power supply, it would be beneficial to use a circuit to translate and then scale it to between 0 V and 5 V so that the best accuracy can be achieved. The circuit connection is shown in Figure 12.20. A set of resistor values and V_1 voltage are given in Figure 12.20. A low-pass filter that consists of a 1-kΩ resistor and a 0.16-μF capacitor is added to meet the requirement.

To translate from the conversion result to the relative humidity, divide the conversion result by 10.23. Since the HCS12 MCU does not support floating-point operation directly, one can multiply the conversion result by 100 (or 10) and then divide the product by 1023 (or 102). The C program that configures the A/D module, starts the A/D conversion, translates the conversion result to the relative humidity, converts the humidity into an ASCII string, and displays the string on the LCD is as follows:

Figure 12.20 ■ Relative humidity measurement circuit

```
#include        "c:\egnu091\include\hcs12.h"
#include        "c:\egnu091\include\delay.c"
#include        "c:\egnu091\include\convert.c"
#include        "c:\egnu091\include\lcd_util_dragon12.c"
void openAD0(void);
char buf[10];
char *msg1 = "humidity = ";
char *blanks = "          ";
```

```
void main (void)
{
        unsigned int quo,rem,frem;
        long temp;
        char *ptr,fdigit;
        delayby100ms(2);                /* wait for LCD kit to self-initialize */
        openlcd();                      /* configure LCD kit */
        openAD0();                      /* configure AD0 module */
        cmd2lcd(0x80);                  /* set cursor to upper left corner */
        puts2lcd(msg1);                 /* output the message "humidity = " */
        while(1) {
                ATD0CTL5 = 0x87;                /* start a conversion with result right-justified */
                while(!(ATD0STAT0&SCF));        /* wait until conversion is done */
                temp = (long)ATD0DR0 * 100; /* force compiler to use 32-bit to represent the product */
                quo = temp / 1023;             /* integer part of temperature */
                rem = temp % 1023              /* remainder part */
                if (quo == 100)
                        rem = 0;                /* force highest humidity to 100% only */
                fdigit = (rem * 10)   / 1023;  /* compute the fractional digit */
                frem = (rem * 10) % 1023;
                if (frem > 511)
                        fdigit ++;
                if (fdigit == 10) {            /* round off the fraction digit */
                        fdigit = 0;
                        quo ++;
                }
                ptr = &buf[0];
                int2alpha(quo,ptr);            /* convert the integer part to ASCII string */
                ptr = &buf[0];
                while(*ptr)                     /* find the end of the integer string */
                    ptr++;
                *ptr++ = '.';                  /* decimal point */
                *ptr++ = fdigit + 0x30;        /* fractional digit */
                *ptr++ = '%';
                *ptr = 0;                       /* terminate the humidity string */
                cmd2lcd(0xC0);                  /* move cursor to 2nd row */
                puts2lcd(blanks);              /* clear the 2nd row */
                cmd2lcd(0xC5);                  /* set cursor to column 5 row 2 */
                puts2lcd(&buf[0]);             /* output the humidity */
                delayby100ms(2);               /* wait for 0.2 seconds */
        }
}
void openAD0(void)
{
        ATD0CTL2 = 0xE0;                /* enable AD0, fast ATD flag clear, power-down on wait */
        ATD0CTL3 = 0x0A;                /* perform one ATD conversion */
        ATD0CTL4 = 0x25;                /* prescaler set to 12, select four cycles sample time */
        delayby10us(2);
}
```

12.8 Measuring Barometric Pressure

Barometric pressure refers to the air pressure existing at any point within the earth's atmosphere. This pressure can be measured as an absolute pressure (with reference to absolute vacuum) or can be referenced to some other value or scale. The meteorology and avionics industries traditionally measure the absolute pressure and then reference it to a sea-level pressure value. This complicated process is used in generating maps of weather systems.

Mathematically, atmospheric pressure is exponentially related to altitude. Once the pressure at a particular location and altitude is measured, the pressure at any other altitude can be calculated. Several units have been used to measure the barometric pressure: in-Hg, kPa, mbar, or psi. A comparison of barometric pressure using four different units at sea level up to 15,000 ft is shown in Table 12.11.

Altitude (ft)	Pressure (in-Hg)	Pressure (mbar)	Pressure (kPa)	Pressure (psi)
0	29.92	1013.4	101.4	14.70
500	29.38	995.1	99.5	14.43
1,000	28.85	977.2	97.7	14.17
6,000	23.97	811.9	81.2	11.78
10,000	20.57	696.7	69.7	10.11
15,000	16.86	571.1	57.1	8.28

Table 12.11 ■ Altitude versus pressure data

There are three forms of pressure transducer: *gauge*, *differential*, and *absolute*. Both the gauge pressure (*psig*) and differential (*psid*) transducers measure pressure differentially. The acronym psi stands for "pounds per square inch"; the letters g and d stand for *gauge* and *differential*, respectively. A gauge pressure transducer measures pressure against ambient air, whereas the differential transducer measures against a reference pressure. An absolute pressure transducer measures the pressure against a vacuum (0 psia) and hence it measures the barometric pressure.

The SenSym ASCX30AN is a 0 to 30 psia (psi absolute) pressure transducer. The range of barometric pressure is between 28 to 32 inches mercury (in-Hg) or 13.75 to 15.72 psia or 948 to 1083.8 mbar. The transducer output is about 0.15 V/psi, which would translate to an output voltage from 2.06 V to 2.36 V. The complete specifications of the SenSym ASCX30AN are shown in Table 12.12. Since the range of V_{OUT} is very narrow, the user will need to use a level shifting and scaling circuit in order to take advantage of the available dynamic range.

The pin assignment of the SenSym ASCX30AN is shown in Figure 12.21.

Characteristic	Min.	Typ.	Max.
Pressure range	0 psia	—	30 psia
Zero pressure offset	0.205	0.250	0.295
Full-scale span [2]	4.455	4.500	4.545
Output at FS pressure	4.660	4.750	4.840
Combined pressure nonlinearity and pressure hysteresis [3]	—	±0.1	±0.5
Temperature effect on span [4]	—	±0.2	±1.0
Temperature effect on offset [4]	—	±0.2	±1.0
Response time (10%–90%) [5]	—	0.1	—
Repeatability	—	±0.05	—

[1] Reference conditions: TA = 25°C, supply voltage V_S = 5 V

[2] Full-scale span is the algebraic difference between the output voltage at full-scale pressure and the output at zero pressure. Full-scale span is ratiometric to the supply voltage.

[3] Pressure nonlinearity is based on the best-fit straight line. Pressure hysteresis is the maximum output difference at any point within the operating pressure range for increasing and decreasing pressure.

[4] Maximum error band of the offset voltage or span over the compensated temperature range, relative to the 25°C reading.

[5] Response time for 0 psi to full-scale pressure step response.

[6] If maximum pressure is exceeded, even momentarily, the package may leak or burst or the pressure-sensing die may burst.

Table 12.12 ■ ASCX30AN performance characteristics [1]

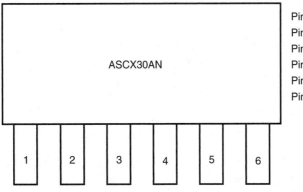

Pin 1: external offset adjust
Pin 2: V_S
Pin 3: V_{OUT}
Pin 4: GND
Pin 5: N/C
Pin 6: N/C

Figure 12.21 ■ ASCX30A pin assignment

Figure 12.22 ■ Barometric pressure sensor output scaling and shifting circuit

Example 12.11

Describe the circuit connection of the ASCX30AN and the voltage level shifting and scaling circuit, and write a program to measure and display the barometric pressure in units of mbar.

Solution: The circuit connection is shown in Figure 12.22. The circuit in Figure 12.22 will shift and scale V_{OUT} to the range of 0 ~ 5 V. According to Table 12.12, the typical value for the offset adjustment is around 0.25 V and can be achieved by using a potentiometer.

The barometric pressure ranges from 948 mbar to 1083.8 mbar. The A/D conversion result will be 0 and 1023 for these two values. Therefore, the A/D conversion result needs to be divided by 7.53 in order to find the corresponding barometric pressure. The following equation shows the conversion:

$$\text{Barometric pressure} = 948 + \text{A/D result} / 7.53$$
$$\approx 948 + (\text{A/D result} \times 100)/753$$

The C program that performs A/D conversion, translates the result into barometric pressure, converts the barometric pressure into an ASCII string, and then displays the pressure on the LCD is as follows:

```
#include        "c:\egnu091\include\hcs12.h"
#include        "c:\egnu091\include\delay.c"
#include        "c:\egnu091\include\convert.c"
#include        "c:\egnu091\include\lcd_util_dragon12.c"
void    openAD0(void);
char    buf[12];
char    *msg1 = "pressure = ";
char    *blanks = "        ";
void    main (void)
{
        unsigned int quo,rem,frem;
        long        temp;
        char        *ptr,fdigit;
```

```
        delayby100ms(2);                        /* wait for LCD kit to self-initialize */
        openlcd();                              /* configure LCD kit */
        openAD0();                              /* configure AD0 module */
        cmd2lcd(0x80);                          /* set cursor to upper left corner */
        puts2lcd(msg1);                         /* output the message "pressure = " */
        while(1) {
                ATD0CTL5 = 0x87;                /* start a conversion with result right-justified */
                while(!(ATD0STAT0&SCF));        /* wait until conversion is done */
                temp = (long)ATD0DR0 * 100;
                quo = temp/753;                 /* integer part of pressure */
                rem = temp%753;                 /* remainder part */
                fdigit = (rem * 10)/753;        /* compute the fractional digit */
                frem = (rem * 10) % 753;
                if (frem > 377)
                        fdigit ++;
                if (fdigit == 10) {             /* round off the fractional digit */
                        fdigit = 0;
                        quo ++;
                }
                ptr = &buf[0];
                quo = quo + 948;
                int2alpha(quo,ptr);             /* convert the integer part to ASCII string */
                ptr = &buf[0];
                while(*ptr)                     /* find the end of the integer string */
                    ptr++;
                *ptr++ = '.';
                *ptr++ = fdigit + 0x30;
                *ptr++ = 0x20;
                *ptr++ = 'm';
                *ptr++ = 'b';
                *ptr++ = 'a';
                *ptr++ = 'r';
                *ptr = 0;                        /* terminate the barometric pressure string */
                cmd2lcd(0xC0);                   /* move cursor to 2nd row */
                puts2lcd(blanks);               /* clear the 2nd row */
                cmd2lcd(0xC5);                   /* set cursor to column 5 row 2 */
                puts2lcd(&buf[0]);              /* output the pressure */
                delayby100ms(2);                /* wait for 0.2 seconds */
        }
}
void openAD0(void)
{
        ATD0CTL2 = 0xE0;                /* enable AD0, fast ATD flag clear, power-down on wait */
        ATD0CTL3 = 0x0A;                /* perform one ATD conversion */
        ATD0CTL4 = 0x25;                /* prescaler set to 12, select four cycles sample time */
        delayby10us(2);
}
```

▲

12.9 Summary

A data acquisition system consists of four major components: a transducer, a signal-conditioning circuit, an A/D converter, and a computer. The transducer converts a nonelectric quantity into a voltage. The transducer output may not be appropriate for processing by the A/D converter. The signal-conditioning circuit shifts and scales the output from a transducer to a range that can take advantage of the full capability of the A/D converter. The A/D converter converts an electric voltage into a digital value that will be further processed by the computer.

Due to the discrete nature of a digital system, the A/D conversion result has a quantization error. The accuracy of an A/D converter is dictated by the number of bits used to represent the analog quantity. The more bits used, the smaller the quantization error will be.

There are four major A/D conversion algorithms:

- Parallel (flash) A/D converter
- Slope and double-slope A/D converters
- Sigma-Delta A/D converters
- Successive-approximation A/D converters

The HCS12 microcontroller uses the successive-approximation algorithm to perform the A/D conversion. All A/D conversion parameters are configured via four ATD control registers: ATDxCTL2, ATDxCTL3, ATDxCTL4, and ATDxCTL5.

The TC1047A temperature sensor, the IH-3606 humidity sensor, and the ASCX30AN pressure sensor are used as examples to illustrate the A/D conversion process. The TC1047A can measure a temperature in the range from $-40°C$ to $125°C$. The IH-3606 can measure relative humidity from 0% to 100%. The ASCX30AN can measure a pressure in the range from 0 to 30 psi absolute. These three examples demonstrate the need for a good voltage shifting and scaling circuit.

There are applications that require A/D accuracy higher than that provided by the HCS12 microcontrollers. In this situation, the user has the option of using an external A/D converter with higher precision or selecting a different microcontroller with higher A/D resolution. Many 8051 variants from Silicon Laboratory, TI, and Analog Devices have much higher A/D resolutions.

12.10 Exercises

E12.1 Design a circuit that can scale the voltage from the range of 0 mV~ 100 mV to the range of 0 V~ 5 V.

E12.2 Design a circuit that can shift and scale the voltage from the range of -80 mV $\sim$ 160 mV to the range of 0 V $\sim$ 5 V.

E12.3 Design a circuit that can shift and scale the voltage from the range of -50 mV $\sim$ 75 mV to the range of 0 V $\sim$ 5 V.

E12.4 Design a circuit that can shift and scale the voltage from the range of 2 V $\sim$ 2.5 V to the range of 0 V $\sim$ 5 V.

E12.5 Suppose that there is a 10-bit A/D converter with $V_{RL} = 2$ V and $V_{RH} = 4$ V. Find the corresponding voltage values for the A/D conversion results of 40, 100, 240, 500, 720, 800, and 1000.

E12.6 Suppose that there is a 12-bit A/D converter with $V_{RL} = 1$ V and $V_{RH} = 4$ V. Find the corresponding voltage values for the A/D conversion results of 80, 180, 480, 640, 960, 1600, 2048, 3200, and 4000.

E12.7 Write a few instructions to configure the HCS12 AD0 converter with the following parameters:

- Bus clock frequency = 16 MHz
- Channel AN3
- Nonscan mode
- Unsigned result representation
- A/D result left-justified, single conversion sequence
- Perform eight conversions in a sequence
- Enable fast flag clear, non-FIFO mode
- Disable the ATD interrupt
- Power down ATD in wait mode
- In background debug mode, finish current conversion, then freeze
- 10-bit resolution, second-stage sample time set to eight clock periods, prescaler set to 8

E12.8 Write an instruction sequence to configure the HCS12DP256 A/D converter with the following characteristics:

- f_{OSC} = 8 MHz
- Channel AN0~AN3
- A/D result right-justified
- 8-bit resolution, unsigned result, continuous conversion
- Four conversions in a sequence
- Prescaler set to 2
- Four cycles second stage sample time, non-FIFO mode
- Enable fast ATD flag clear, enable ATD interrupt
- External falling edge triggered (at ATD7 pin)
- Finish current conversion then freeze in background debug mode

E12.9 Assuming that S8C~S1C (ATD0CTL3) are set to 0111 and CC~CA (ATD0CTL5) are set to 101, what is the conversion sequence for this setting?

E12.10 Assume that the following setting was programmed before a new conversion is started:

- The conversion counter value in the ATD0STAT0 register is 4.
- The channel-select code of the ATD0CTL5 is 7.
- The conversion sequence limit of the ATD0CTL3 register is set to 6.
- The MULT bit of the ATD0CTL5 register is set to 1.

How would the conversion results be stored when the FIFO mode is selected or not selected?

E12.11 Assume that the following setting was programmed before a new conversion is started:

- The conversion counter value in the ATD0STAT0 register is 3.
- The channel-select code of the ATD0CTL5 is 5.
- The conversion sequence limit of the ATD0CTL3 register is set to 4.
- The MULT bit of the ATD0CTL5 register is set to 1.

How would the conversion results be stored when the FIFO mode is selected or not selected?

E12.12 At 8-bit resolution with the second-stage sampling time set to eight ATD clock cycles, how long does it take to complete the conversion of one sample at 500 kHz and at 2 MHz ATD clock frequency, respectively?

E12.13 At 10-bit resolution with the second-stage sampling time set to four ATD clock cycles, how long does it take to complete the conversion of one sample at 500 kHz and at 2 MHz ATD clock frequency, respectively?

E12.14 The LM35 from National Semiconductor is a Centigrade temperature sensor with three external connection pins. The pin assignment and circuit connection for converting temperature are shown in Figure E12.14 Use this device to construct a circuit to display the room temperature in Celsius. The temperature range to be converted is from $-27°C$ to $100°C$. Describe the circuit connection for a digital thermometer comprised of the HCS12 and the LM35, and write a program to display temperature on the LCD.

Figure E12.14 ■ Circuit connection for the LM35

E12.15 The LM34 from National Semiconductor is a Fahrenheit temperature sensor with three external connection pins. The pin assignment and circuit connection for converting temperature are shown in Figure E12.15. Use this device to construct a circuit to display the room temperature in Fahrenheit. Assume that the temperature range of interest is from $-40°F$ to $215°F$. Use an LCD to display the temperature. Describe the circuit connection for a digital thermometer that is comprised of the HCS12 and the LM34, and write a program to display temperature on the LCD.

Figure E12.15 ■ Circuit connection for the LM34

E12.16 The Microbridge AWM3300V is a mass airflow sensor manufactured by Honeywell. The block diagram of the AWM3300V is shown in Figure E12.16. It is designed to measure the airflow. Its applications include air-conditioning, medical ventilation/anesthesia control, gas analyzers, gas metering, fume cabinets, and process control. The AWM3300V operates on a single 10-V ± 10-mV power supply. The sensor output (from V_{OUT}) corresponding to the airflow rate of 0 ~ 1.0 L/min is 1.0 V to 5.0 V. The AWM3300V can operate in the temperature range of −25° to 85°C. It takes 3 ms for the output voltage to settle after power-up. Design a circuit to measure and display the mass airflow using the AWM3300V. Write a program to configure the HCS12 A/D module, start the A/D conversion, and display the mass airflow in an LCD display. Update the display five times per second.

Figure E12.16 ■ Microbridge AWM3300V

12.11 Lab Exercises and Assignments

L12.1 A/D converter testing. Perform the following steps.

Step 1
Set the function generator output (in the lab) to between 0 and 5 V.

Step 2
Connect the AN1 pin to the functional generator output. Set the frequency to about 10 kHz.

Step 3
Write a program to perform the following operations:
- Configure the A/D module properly
- Start the A/D conversion.
- Take 64 samples, convert them, and store the conversion results at appropriate SRAM locations.
- Compute the average value and store it in D.

If your demo board has a square waveform output, then select an output with a frequency close to 16 KHz. Connect it to the AN1 pin.

L12.2 Digital thermometer. Use the LM34 temperature sensor, the 741 OP AMP, and the required resistors to construct a digital thermometer. The temperature should be displayed in three integral and one fractional digits. Use the LCD on your demo board to display the temperature. Update the temperature once every 200 ms.

L12.3 Barometric pressure measurement. The Motorola MPX4115A is a pressure sensor that can measure a pressure ranging from 15 to 115 kPa (2.2 to 16.7 psi) and has a corresponding voltage output from 0.2 V to 4.8 V. The small outline package of this device has eight pins. Among these eight pins, only three pins carry useful signals:

Pin 2. V_S

Pin 3. GND

Pin 4. V_{OUT}

This pressure sensor is often used in aviation altimeters, industrial controls, engine control, and weather stations and weather reporting devices.

Connect an MPX4115AC6U (or MPX4115A6U, case 482) device to your HCS12 demo board to measure the current ambient barometric pressure once every second and display the pressure on the LCD display. The data sheet of the MPX4115A can be found on the complementary CD. Since the barometric pressure is in a narrow range (use the range mentioned in Example 12.11), you may need to construct a signal-conditioning circuit to translate and scale the voltage output of the pressure sensor to the range of 0 V to 5 V.

13

Controller Area Network (CAN)

13.1 Objectives

After completing this chapter, you should be able to:

- Describe the layers of the CAN protocol
- Describe CAN's error detection capability
- Describe the formats of CAN messages
- Describe CAN message handling
- Explain CAN error handling
- Describe CAN fault confinement
- Describe CAN bit timing
- Explain CAN synchronization issue and methods
- Describe the CAN message structures
- Compute timing parameters to meet the requirements of your application
- Write programs to configure the HCS12 CAN module
- Write programs to transfer data over the CAN bus
- Design a CAN-based remote sensing system

13.2 Overview of Controller Area Network

The controller area network (CAN) was initially created by German automotive system supplier Robert Bosch in the mid-1980s for automotive applications as a method for enabling robust serial communication. The goal was to make automobiles more reliable, safe, and fuel-efficient while at the same time decreasing wiring harness weight and complexity. Since its inception, the CAN protocol has gained widespread use in industrial automation and automotive/truck applications. The description of CAN in this chapter is based on the CAN Specification 2.0 published in September 1991 by Bosch.

13.2.1 Layered Approach in CAN

The CAN protocol specified the lowest two layers of the ISO seven-layer model: *data link* and *physical* layers. The data link layer is further divided into two sublayers: logical link control (LLC) layer and medium access control (MAC) layer.

- The LLC sublayer deals with message acceptance filtering, overload notification, and error recovery management.

- The MAC sublayer presents incoming messages to the LLC sublayer and accepts messages to be transmitted that are forwarded by the LLC sublayer. The MAC sublayer is responsible for message framing, arbitration, acknowledgment, error detection, and signaling. The MAC sublayer is supervised by a self-checking mechanism, called *fault confinement*, which distinguishes short disturbances from permanent failures.

The physical layer defines how signals are actually transmitted and deals with the description of bit timing, bit encoding, and synchronization. CAN bus driver/receiver characteristics and the wiring and connectors are not specified in the CAN protocol. These two aspects are not specified so that the implementers can choose the most appropriate transmission medium and hence optimize signal-level implementations for their applications. The system designer can choose from multiple available media technologies including twisted pair, single wire, optical fiber, radio frequency (RF), infrared (IR), and so on. The layered CAN protocol is shown in Figure 13.1.

13.2.2 General Characteristics of CAN

The CAN protocol was optimized for systems that need to transmit and receive relatively small amounts of information (as compared to ethernet or USB, which are designed to move much larger blocks of data). The CAN protocol has the following features.

CARRIER SENSE MULTIPLE ACCESS WITH COLLISION DETECTION (CSMA/CD)

The CAN protocol is a CSMA/CD protocol. Every node on the network must monitor the bus (carrier sense) for a period of no activity before trying to send a message on the bus. Once this period of no activity occurs, every node on the bus has an equal opportunity to transmit a message (multiple access). If two nodes happen to transmit at the same time, the nodes will detect the *collision* and take the appropriate action. In the CAN protocol, a nondestructive bitwise arbitration method is utilized. Messages remain intact after arbitration is completed even if collisions are detected. Message arbitration will not delay higher priority messages. To facilitate bus arbitration, the CAN protocol defines two bus states: *dominant* and *recessive*. The dominant state is represented by logic 0 (low voltage), whereas the recessive state is represented by logic 1 (high voltage). The dominant state will win over the recessive state.

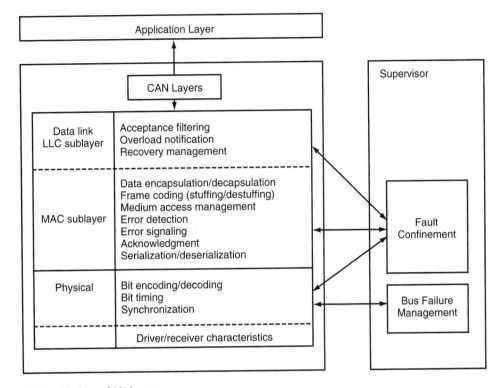

Figure 13.1 ■ CAN layers

MESSAGE-BASED COMMUNICATION

CAN is a message-based protocol, not an address-based protocol. Embedded in each message is an *identifier*. This identifier allows messages to arbitrate the use of the CAN bus and also allows each node to decide whether to work on the message. The value of the identifier is used as the priority of the message. The lower the value, the higher the priority. Each node in the CAN system uses one or more filters to compare the identifier of the incoming message. Once the identifier passes the filter, the message will be worked on by the node. The CAN protocol also provides the mechanism for a node to request data transmission from another node. Since an address is not used in the CAN system, there is no need to reconfigure the system whenever a node is added to or deleted from a system. This capability allows the system to perform node-to-node or multicast communications.

ERROR DETECTION AND FAULT CONFINEMENT

The CAN protocol requires each node to monitor the CAN bus to find out if the bus value and the transmitted bit value are identical. For every message, cyclic redundancy check is calculated and the checksum is appended to the message. CAN is an asynchronous protocol and hence clock information is embedded in the message rather than transmitted as a separate signal. A message with long sequence of identical bits could cause a synchronization problem. To resolve this problem, the CAN protocol requires the physical layer to use bit stuffing to avoid a long sequence of identical bit values. With these measures implemented, the residual probability for undetected corrupted messages in a CAN system is as low as

Message error rate $\times$ 4.7 $\times$ 10^{-11}

CAN nodes are able to distinguish short disturbances from permanent failures. Defective nodes are switched off from the CAN bus.

13.3 CAN Messages

The CAN protocol defines four different types of messages:

- *Data frame.* A data frame carries data from a transmitter to the receivers.
- *Remote frame.* A remote frame is transmitted by a node to request the transmission of the data frame with the same identifier.
- *Error frame.* An error frame is transmitted by a node on detecting a bus error.
- *Overload frame.* An overload frame is used to provide for an extra delay between the preceding and the succeeding data or remote frames.

Data frames and remote frames are separated from preceding frames by an interframe space. Applications do not need to send or handle error and overload frames.

13.3.1 Data Frame

As shown in Figure 13.2, a data frame consists of seven different bit fields: start-of-frame, arbitration, control, data, CRC, ACK, and end-of-frame.

Figure 13.2 ■ CAN Data frame

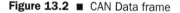

Start-of-Frame Field

This field is a single dominant bit that marks the beginning of a data frame. A node is allowed to start transmission only when the bus is idle. All nodes have to synchronize to the leading edge of the field of the node that starts transmission first.

Arbitration Field

The format of the arbitration field is different for *standard format* and *extended format* frames, as illustrated in Figure 13.3. The identifier's length is 11 bits for the standard format and 29 bits for the extended format.

The identifier of the standard format corresponds to the base ID in the extended format. These bits are transmitted most significant bit first. The most significant 7 bits cannot all be recessive.

The identifier of the extended format comprises two sections: an 11-bit base ID and an 18-bit extended ID. Both the base ID and the extended ID are transmitted most significant bit first. The base ID defines the base priority of the extended frame.

(a) Standard format

(b) Extended format

Figure 13.3 ■ Arbitration field

The *remote transmission request* (RTR) bit in data frames must be dominant. Within a remote frame, the RTR bit has to be recessive.

The *substitute remote request* (SRR) bit is a recessive bit. The SRR bit of an extended frame is transmitted at the position of the RTR bit in the standard frame and therefore substitutes for the RTR bit in the standard frame. As a consequence, collisions between a standard frame and an extended frame, where the base IDs of both frames are identical, are resolved in such a way that the standard frame prevails over the extended frame.

The *identifier extension* (IDE) bit belongs to the arbitration field for the extended format and the control field for the standard format. The IDE bit in the standard format is transmitted dominant, whereas in the extended format the IDE bit is recessive.

CONTROL FIELD

The contents of this field are shown in Figure 13.4. The format of the control field is different for the standard format and the extended format. Frames in standard format include the data length code; the IDE bit, which is transmitted dominant; and the reserved bit r0. Frames in extended format include the data length code and two reserved bits, r0 and r1. The reserved bits must be sent dominant, but the receivers accept dominant and recessive bits in all combinations. The data length code specifies the number of bytes contained in the data field. Data length can be 0 to 8, as encoded in Table 13.1.

Figure 13.4 ■ Control field

DLC3	DLC2	DLC1	DLC0	Data Byte Count
d	d	d	d	0
d	d	d	r	1
d	d	r	d	2
d	d	r	r	3
d	r	d	d	4
d	r	d	r	5
d	r	r	d	6
d	r	r	r	7
r	d	d	d	8

d = dominant, r = recessive

Table 13.1 ■ CAN data length coding

DATA FIELD

The data field consists of the data to be transmitted within a data frame. It may contain from 0 to 8 bytes, each of which contains 8 bits that are transferred most significant bit first.

CRC FIELD

The CRC field contains the CRC sequence followed by a CRC delimiter, as shown in Figure 13.5. The frame-check sequence is derived from a cyclic redundancy code best suited to frames with bit counts less than 127. The CRC sequence is calculated by performing a polynomial division. The coefficients of the polynomial are given by the destuffed bitstream, consisting of the start of frame, arbitration field, control field, data field (if present), and 15 0s. This polynomial is divided (the coefficients are calculated using modulo-2 arithmetic) by the generator polynomial:

$$X^{15} + X^{14} + X^{10} + X^8 + X^7 + X^4 + X^3 + 1$$

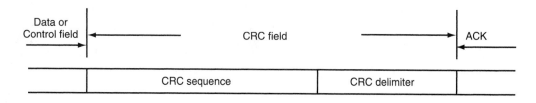

Figure 13.5 ■ CRC field

The remainder of this polynomial division is the CRC sequence. In order to implement this function, a 15-bit shift register CRC_RG (14:0) is used. If nxtbit denotes the next bit of the bit stream, given by the destuffed bit sequence from the start of frame until the end of the data field, the CRC sequence is calculated as follows:

```
CRC_RG = 0;                              /* initialize shift register */
do {
        crcnxt = nxtbit   CRC_RG(14);    /* exclusive OR */
        CRC_RG(14:1) = CRC_RG(13:0);     /* shift left by one bit */
        CRC_RG(0) = 0;
```

```
          if crcnxt
                  CRC_RG(14:0) = CRC_RG(14:0) ^ 0x4599;
      } while (!(CRC SEQUENCE starts or there is an error condition));
```

After the transmission/reception of the last bit of the data field, CRC_RG(14:0) contains the CRC sequence. The *CRC delimiter* is a single recessive bit.

ACK FIELD

As shown in Figure 13.6, the ACK field is 2 bits long and contains the ACK slot and the ACK delimiter. A transmitting node sends two recessive bits in the ACK field. A receiver that has received a valid message reports this to the transmitter by sending a dominant bit in the ACK slot (i.e., it sends ACK). A node that has received the matching CRC sequence overwrites the recessive bit in the ACK slot with a dominant bit. This bit will be received by the data frame transmitter and learn that the previously transmitted data frame has been correctly received. The *ACK delimiter* has to be a recessive bit. As a consequence, the ACK slot is surrounded by two recessive bits (the CRC delimiter and the ACK delimiter).

Figure 13.6 ■ ACK field

END-OF-FRAME FIELD

Each data frame and remote frame is delimited by a flag sequence consisting of 7 recessive bits. This 7-bit sequence is the *end-of-frame* sequence.

13.3.2 Remote Frame

A node that needs certain data can request the relevant source node to transmit the data by sending a remote frame. The format of a remote frame is shown in Figure 13.7. A remote frame consists of six fields: start-of-frame, arbitration, control, CRC, ACK, and end-of-frame. The polarity of the RTR bit in the arbitration field indicates whether a transmitted frame is a *data frame* (RTR bit dominant) or a *remote frame* (RTR bit recessive).

Figure 13.7 ■ Remote frame

13.3.3 Error Frame

The error frame consists of two distinct fields. The first field is given by the superposition of error flags contributed from different nodes. The second field is the error delimiter. The format of the error frame is shown in Figure 13.8. In order to terminate an error frame correctly, an *error-passive node* may need the bus to be idle for at least three bit times (if there is a local error at an error-passive receiver). Therefore, the bus should not be loaded to 100%. An error-passive node has an error count greater than 127 but no more than 255. An *error-active node* has an error count less than 127. There are two forms of error flags:

- *Active-error flag.* This flag consists of six consecutive dominant bits.
- *Passive-error flag.* This flag consists of six consecutive recessive bits unless it is overwritten by dominant bits from other nodes.

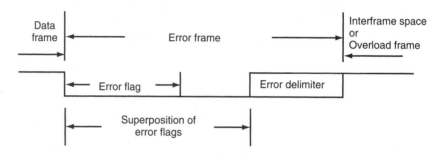

Figure 13.8 ■ Error frame

An error-active node signals an error condition by transmitting an *active-error* flag. The error flag's form violates the law of bit stuffing (to be discussed shortly) and applies to all fields from start-of-frame to CRC delimiter or destroys the fixed-form ACK field or end-of-frame field. As a consequence, all other nodes detect an error condition and each starts to transmit an error flag. Therefore, the sequence of dominant bits, which actually can be monitored on the bus, results from a superposition of different error flags transmitted by individual nodes. The total length of this sequence varies between a minimum of 6 and a maximum of 12 bits.

An error-passive node signals an error condition by transmitting a passive-error flag. The error-passive node waits for six consecutive bits of equal polarity, beginning at the start of the passive-error flag. The passive-error flag is complete when these equal bits have been detected.

The *error delimiter* consists of eight recessive bits. After transmission of an error flag, each node sends recessive bits and monitors the bus until it detects a recessive bit. Afterwards, it starts transmitting seven more recessive bits.

13.3.4 Overload Frame

The *overload frame* contains two bit fields: *overload flag* and *overload delimiter*. There are three different overload conditions that lead to the transmission of an overload frame:

1. The internal conditions of a receiver require a delay of the next data frame or remote frame.

2. At least one node detects a dominant bit during intermission.

3. A CAN node samples a dominant bit at the eighth bit (i.e., the last bit) of an error delimiter or overload delimiter. The error counters will not be incremented.

The format of an overload frame is shown in Figure 13.9. An overload frame resulting from condition 1 is only allowed to start at the first bit time of an expected intermission, whereas an overload frame resulting from overload conditions 2 and 3 starts one bit after detecting the dominant.

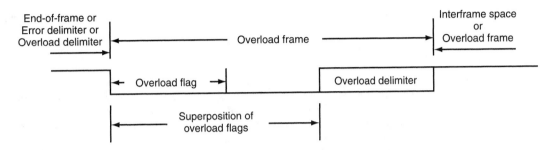

End-of-frame or
Error delimiter or
Overload delimiter

Overload frame

Interframe space
or
Overload frame

Overload flag

Overload delimiter

Superposition of
overload flags

Figure 13.9 ■ Overload frame

No more than two overload frames may be generated to delay the next data frame or remote frame. The overload flag consists of six dominant bits. The format of an overload frame is similar to that of the active-error flag. The overload flag's form destroys the fixed form of the *intermission field*. As a consequence, all other nodes also detect an overload condition and each starts to transmit an overload flag. In the event that there is a dominant bit detected during the third bit of *intermission* locally at some node, it will interpret this bit as the start-of-frame.

The overload delimiter consists of eight recessive bits. The overload delimiter has the same form as the error delimiter. After the transmission of an overload flag, the node monitors the bus until it detects a transition from a dominant to a recessive bit. At this point of time every bus node has finished sending its overload flag and all nodes start transmission of seven more recessive bits in coincidence.

13.3.5 Interframe Space

Data frames and remote frames are separated from preceding frames by a field called *interframe space*. In contrast, overload frames and error frames are not preceded by an interframe space, and multiple overload frames are not separated by an interframe space. For nodes that are not error-passive or have been receivers of the previous message, the interframe space contains the bit fields of *intermission* and *bus idle*, as shown in Figure 13.10. The interframe space of an error-passive node consists of three subfields: *intermission, suspend transmission*, and *bus idle*, as shown in Figure 13.11.

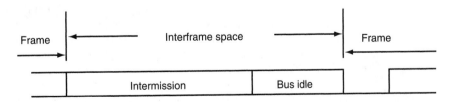

Frame

Interframe space

Frame

Intermission

Bus idle

Figure 13.10 ■ Interframe space for non error-passive nodes or receiver of previous message

Figure 13.11 ■ Interframe space for error-passive nodes

The intermission subfield consists of three recessive bits. During intermission, no node is allowed to start transmission of the data frame or remote frame. The only action permitted is signaling of an overload condition. The period of bus idle may be of arbitrary length. The bus is recognized to be free, and any node having something to transmit can access the bus. A message, pending during the transmission of another message, is started in the first bit following intermission. When the bus is idle, the detection of a dominant bit on the bus is interpreted as a start-of-frame. After an error-passive node has transmitted a frame, it sends eight recessive bits following intermission, before starting to transmit a new message or recognizing the bus as idle. If, meanwhile, a transmission (caused by another node) starts, the node will become the receiver of this message.

13.3.6 Message Filtering

A node uses filter(s) to decide whether to work on a specific message. Message filtering is applied to the whole identifier. A node can optionally implement mask registers that specify which bits in the identifier are examined with the filter. If mask registers are implemented, every bit of the mask registers must be programmable; in other words, they can be enabled or disabled for message filtering. The length of the mask register can comprise the whole identifier or only part of it.

13.3.7 Message Validation

The point in time at which a message is taken to be valid is different for the transmitters and receivers of the message. The message is valid for the transmitter if there is no error until the end-of-frame. If a message is corrupted, retransmission will follow automatically and according to the rules of prioritization. In order to be able to compete for bus access with other messages, retransmission has to start as soon as the bus is idle. The message is valid for the receiver if there is no error until the last but one bit of the end-of-frame.

13.3.8 Bitstream Encoding

The frame segments including start-of-frame, arbitration field, control field, data field, and CRC sequence are encoded by *bit stuffing*. Whenever a transmitter detects five consecutive bits of identical value in the bitstream to be transmitted, it automatically inserts a complementary bit in the actual transmitted bitstream. The remaining bit fields of the data frame or remote frame (CRC delimiter, ACK field, and end-of-frame) are of fixed form and not stuffed. The error frame and overload frame are also of fixed form and are not encoded by the method of bit stuffing.

The bitstream in a message is encoded using the *non-return-to-zero* (NRZ) method. This means that during the total bit time the generated bit level is either dominant or recessive.

13.4 Error Handling

There are five types of errors. These errors are not mutually exclusive.

13.4.1 Bit Error

A node that is sending a bit on the bus also monitors the bus. When the bit value monitored is different from the bit value being sent, the node interprets the situation as an error. There are two exceptions to this rule:

- A node that sends a recessive bit during the stuffed bitstream of the arbitration field or during the ACK slot detects a dominant bit.
- A transmitter that sends a passive-error flag detects a dominant bit.

13.4.2 Stuff Error

A stuff error is detected whenever six consecutive dominant or six consecutive recessive levels occur in a message field.

13.4.3 CRC Error

The CRC sequence consists of the result of the CRC calculation by the transmitter. The receiver calculates the CRC in the same way as the transmitter. A CRC error is detected if the calculated result is not the same as that received in the CRC sequence.

13.4.4 Form Error

A form error is detected when a fixed-form bit field contains one or more illegal bits. For a receiver, a dominant bit during the last bit of end-of-frame is not treated as a form error.

13.4.5 Acknowledgment Error

An acknowledgment error is detected whenever the transmitter does not monitor a dominant bit in the ACK slot.

13.4.6 Error Signaling

A node that detects an error condition signals the error by transmitting an error flag. An error-active node will transmit an *active-error* flag; an error-passive node will transmit a *passive-error* flag. Whenever a node detects a bit error, a stuff error, a form error, or an acknowledgment error, it will start transmission of an error flag at the next bit time. Whenever a CRC error is detected, transmission of an error flag will start at the bit following the ACK delimiter, unless an error flag for another error condition has already been started.

13.5 Fault Confinement

13.5.1 CAN Node Status

A node in error may be in one of three states: error-active, error-passive, or bus-off. An error-active node can normally take part in bus communication and sends an active-error flag when an error has been detected. An error-passive node must not send an active-error flag. It takes part in bus communication, but when an error has been detected only a passive-error flag is sent.

After a transmission, an error-passive node will wait before initiating further transmission. A bus-off node is not allowed to have any influence on the bus.

13.5.2 Error Counts

The CAN protocol requires each node to implement *transmit error count* and *receive error count* to facilitate fault confinement. These two counts are updated according to 12 rules. These 12 rules can be found in the CAN specification. An error count value greater than roughly 96 indicates a heavily disturbed bus. It may be advantageous to provide the means to test for this condition. If during system start-up only one node is on-line and if this node transmits some message, it will get no acknowledgment, detect an error, and repeat the message. It can become error-passive but not bus-off for this reason.

13.6 CAN Message Bit Timing

The setting of a bit time in a CAN system must allow a bit sent out by the transmitter to reach the far end of the CAN bus and allow the receiver to send back acknowledgment and reach the transmitter. In a CAN environment, the *nominal bit rate* is defined to be the number of bits transmitted per second in the absence of resynchronization by an ideal transmitter.

13.6.1 Nominal Bit Time

The inverse of the nominal bit rate is the *nominal bit time*. A nominal bit time can be divided into four nonoverlapping time segments, as shown in Figure 13.12.

Figure 13.12 ■ Nominal bit time

The **sync_seg** segment is used to synchronize the various nodes on the bus. An edge is expected to lie within this segment. The **prop_seg** segment is used to compensate for the physical delay times within the network. It is twice the sum of the signal's propagation time on the bus line, the input comparator delay, and the output driver delay. The **phase_seg1** and **phase_seg2** segments are used to compensate for edge phase errors. These segments can be lengthened or shortened by synchronization. The *sample point* is the point in time in which the bus level is read and interpreted as the value of that respective bit. The sample point is located at the end of phase_seg1. A CAN controller may implement the three-samples-per-bit option in which the majority function is used to determine the bit value. Each sample is separated from the next sample by half a time quanta (CAN clock cycle). The *information processing time* is the time segment starting with the sample point reserved for calculation of the sample bit(s) level. The segments contained in a nominal bit time are represented in the unit of *time quantum*. The time quantum (t_Q) is a fixed unit of time that can be derived from the oscillator period (T_{OSC}). t_Q is expressed as a multiple of a *minimum time quantum*. This multiple is a programmable prescale factor. Thus, the time quantum can have the length of

Time quantum = $m \times$ minimum time quantum

where m is the value of the prescaler.

13.6.2 Length of Time Segments

The segments of a nominal bit time can be expressed in the unit of time quantum as follows:

- **sync_seg** is 1 time quantum long.
- **prop_seg** is programmable to be 1, 2, . . ., 8 time quanta long.
- **phase_seg1** is programmable to be 1, 2, . . ., 8 time quanta long.
- **phase_seg2** is the maximum of **phase_seg1** and information processing time and hence will be programmable from 2 to 8 t_Qs.
- The information processing time is equal to or less than 2 t_Q and is fixed at 2 t_Q for the HCS12 CAN module.

The total number of time quanta in a bit time must be programmable over a range of at least 8 to 25.

13.7 Synchronization Issue

All CAN nodes must be synchronized while receiving a transmission, i.e., the beginning of each received bit must occur during each node's **sync_seg** segment. This is achieved by synchronization. Synchronization is required because of phase errors between nodes, which may arise because of nodes having slightly different oscillator frequencies or because of changes in propagation delay when a different node starts transmitting.

Two types of synchronization are defined: *hard synchronization* and *resynchronization*. Hard synchronization is performed only at the beginning of a message frame, when each CAN node aligns the **sync_seg** of its current bit time to the recessive to dominant edge of the transmitted start-of-frame. After a hard synchronization, the internal bit time is restarted with **sync_seg**. Resynchronization is subsequently performed during the remainder of the message frame whenever a change of bit value from recessive to dominant occurs outside of the expected *sync_seg* segment. Resynchronization is achieved by implementing a digital phase-lock loop (DPLL) function that compares the actual position of a recessive-to-dominant edge on the bus to the position of the expected edge.

13.7.1 Resynchronization Jump Width

There are three possibilities for the occurrence of the incoming recessive-to-dominant edge:

1. *After the sync_seg segment but before the sample point.* This situation is interpreted as a *late edge*. The node will attempt to resynchronize to the bitstream by increasing the duration of its **phase_seg1** segment of the current bit by the number of time quanta by which the edge was late, up to the resynchronization jump width limit.

2. *After the sample point but before the sync_seg segment of the next bit.* This situation is interpreted as an *early bit*. The node will now attempt to resynchronize to the bitstream by decreasing the duration of its **phase_seg2** segment of the current bit by the

number of time quanta by which the edge was early, by up to the resynchronization jump width limit. Effectively, the **sync_seg** segment of the next bit begins immediately.

3. *Within the **sync_seg** segment of the current bit time.* This is interpreted as no synchronization error.

As a result of resynchronization, **phase_seg1** may be lengthened or **phase_seg2** may be shortened. The amount by which the phase buffer segments may be altered may not be greater than the *resynchronization jump width*, which is programmable to be between 1 and the smaller of 4 and **phase_seg1** time quanta.

Clocking information may be derived from transitions from one bit value to the other. The property that only a fixed maximum number of successive bits have the same value provides the possibility of resynchronizing a bus node to the bitstream during a frame.

The maximum length between two transitions that can be used for resynchronization is 29 bit times.

13.7.2 Phase Error of an Edge

The *phase error* of an edge is given by the position of the edge relative to **sync_seg**, measured in time quanta. The sign of phase error is defined as follows:

$e < 0$ if the edge lies after the sample point of the previous bit

$e = 0$ if the edge lies within **sync_seg**

$e > 0$ if the edge lies before the sample point

13.8 Overview of the HCS12 CAN Module

The Motorola implementation of the CAN 2.0A/B protocol for the HCS12 microcontroller family is referred to as MSCAN12. The MSCAN12 provides the following features:

- Full implementation of the CAN 2.0A/B protocol
- Five receive buffers with FIFO storage scheme
- Three transmit buffers with internal prioritization using a "local priority" concept
- Maskable identifier filter supporting two full-size extended identifier filters (two 32-bit) or four 16-bit filters or eight 8-bit filters
- Programmable wakeup functionality with integrated low-pass filter
- Programmable loopback mode supporting self-test operation
- Programmable listen-only mode for monitoring of CAN bus
- Separate signaling and interrupt capabilities for all CAN receiver and transmitter error states (warning, error passive, bus off)
- Clock source coming from either bus clock or oscillator clock
- Internal time for time-stamping of received and transmitted message
- Three low power modes: sleep, power-down, and MSCAN12 enable
- Global initialization of configuration registers

The block diagram of the MSCAN12 module is shown in Figure 13.13. As shown in the figure, each MSCAN module uses two signal pins: TxCAN and RxCAN.

All members of the automotive subfamily in the HCS12 family have implemented one or more CAN modules. The HCS12DP256 implements five CAN modules. For other HCS12 devices without an on-chip CAN module, a CAN controller chip such as the MCP2510 from

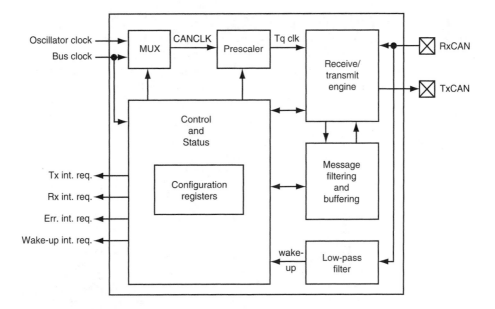

Figure 13.13 ■ MSCAN12 block diagram

Microchip or the PCA82C200 from Philips can be used to interface with the CAN bus. Most CAN controllers do not have enough driving capability and hence require a dedicated CAN bus transceiver chip such as the MCP2551 (from Microchip) or the PCA82C250 (from Philips) to interface with the CAN bus. In addition to providing large driving current, a CAN transceiver chip has current protection against defected CAN or defected nodes. A typical CAN system that includes one or more HCS12 devices is shown in Figure 13.14.

Figure 13.14 ■ A typical CAN system

13.9 MSCAN Module Memory Map

A HCS12 member may have up to five CAN modules and each CAN module occupies 64 bytes of memory space. The MSCAN register organization is shown in Figure 13.15. The detailed MSCAN register memory map is shown in Figure 13.16.

Figure 13.15 ■ MSCAN register organization

Each receive buffer and each transmit buffer occupies 16 bytes of space. Even though each MSCAN module has three transmit buffers, only one of them is accessible to the programmer at a time. Similarly, only one of the five receive buffers is accessible to the programmer at a time. Only the offset of each register relative to the start of MSCAN register block is shown in Figure 13.16. The address of each register is the sum of the offset and the base address of the MSCAN register block.

13.9.1 MSCAN Control Registers

The function of each MSCAN control register is explained in this section. Since the HCS12 may have up to five identical CAN modules having the same set of control registers, Freescale names each CAN register as CANxYYYY, where x indicates the CAN module and YYYY specifies the register.

MSCAN CONTROL REGISTER 0 (CANxCTL0)

The contents of this register are shown in Figure 13.17. The RXFRM bit is set whenever a valid frame is received by the MSCAN module. This flag bit can be cleared by writing a 1 to it. The RXACT flag is a read-only flag and indicates whether the MSCAN is receiving a message. Setting the CSWAI bit will disable all clock signals to the MSCAN during the wait mode and further reduce power consumption.

Address	Register Name	Access
$_00	MSCAN control register 0 (CANCTL0)	R/W
$_01	MSCAN control register 1 (CANCTL1)	R/W
$_02	MSCAN bus timing register 0 (CANBTR0)	R/W
$_03	MSCAN bus timing register 1 (CANBTR1)	R/W
$_04	MSCAN receiver flag register (CANRFLG)	R/W
$_05	MSCAN receiver interrupt enable register (CANRIER)	R/W
$_06	MSCAN transmitter flag register (CANTFLG)	R/W
$_07	MSCAN transmitter interrupt enable register (CANTIER)	R/W
$_08	MSCAN transmitter message abort request (CANTARQ)	R/W
$_09	MSCAN transmitter message abort acknowledge (CANTAAK)	R
$_0A	MSCAN transmit buffer selection (CANTBSEL)	R/W
$_0B	MSCAN identifier acceptance control register (CANIDAC)	R/W
$_0C	reserved	
$_0D	reserved	
$_0E	MSCAN receive error counter register (CANRXERR)	R
$_0F	MSCAN transmit error counter register (CANTXERR)	R
$_10	MSCAN identifier acceptance register 0 (CANIDAR0)	R/W
$_11	MSCAN identifier acceptance register 1 (CANIDAR1)	R/W
$_12	MSCAN identifier acceptance register 2 (CANIDAR2)	R/W
$_13	MSCAN identifier acceptance register 3 (CANIDAR3)	R/W
$_14	MSCAN identifier mask register 0 (CANIDMR0)	R/W
$_15	MSCAN identifier mask register 1 (CANIDMR1)	R/W
$_16	MSCAN identifier mask register 2 (CANIDMR2)	R/W
$_17	MSCAN identifier mask register 3 (CANIDMR3)	R/W
$_18	MSCAN identifier acceptance register 4 (CANIDAR4)	R/W
$_19	MSCAN identifier acceptance register 5 (CANIDAR5)	R/W
$_1A	MSCAN identifier acceptance register 6 (CANIDAR6)	R/W
$_1B	MSCAN identifier acceptance register 7 (CANIDAR7)	R/W
$_1C	MSCAN identifier mask register 4 (CANIDMR4)	R/W
$_1D	MSCAN identifier mask register 5 (CANIDMR5)	R/W
$_1E	MSCAN identifier mask register 6 (CANIDMR6)	R/W
$_1F	MSCAN identifier mask register 7 (CANIDMR7)	R/W
$_20 $_2F	Foreground receive buffer (CANRXFG)	R
$_30 $_3F	Foreground transmit buffer (CANTXFG)	R/W

Figure 13.16 ■ CAN module memory map

The SYNCH bit indicates whether the MSCAN module is synchronized to the CAN bus. The MSCAN module has a 16-bit free-running timer. When the TIME bit is set to 1, the timer value will be assigned to each transmitted and received message within the transmit and receive buffers. The timestamp is assigned as soon as a message is acknowledged. The WUPE bit allows the MSCAN module to be woken up by the CAN bus activity. Setting the SLPRQ bit requests the MSCAN to enter the sleep mode. The sleep mode request is serviced only when the CAN bus is idle.

To make sure that the CAN protocol is not violated, the MSCAN can only be configured when the MSCAN is in the initialization mode. The initialization mode is entered by setting the INITRQ bit. The registers CANxCTL1, CANxBTR0, CANxBTR1, CANxIDAC, CANxIDAR0-7, CANxIDMR0-7 can only be written by the CPU when the MSCAN is in the

7	6	5	4	3	2	1	0
RXFRM	RXACT	CSWAI	SYNCH	TIME	WUPE	SLPRQ	INITRQ

reset: 0   0   0   0   0   0   0   1

RXFRM: received frame flag
  0 = no valid message was received.
  1 = a valid message was received since last clearing of this flag.
RXACT: receiver active status
  0 = MSCAN is transmitting or idle.
  1 = MSCAN is receiving a message (including when arbitration is lost).
CSWAI: CAN stops in wait mode
  0 = the module is not affected during wait mode.
  1 = the module ceases to be clocked during wait mode.
SYNCH: synchronization status
  0 = MSCAN is not synchronized to the CAN bus.
  1 = MSCAN is synchronized to the CAN bus.
TIME: timer enable
  0 = disable internal MSCAN timer.
  1 = enable internal MSCAN timer and hence enable time stamp.
WUPE: wake-up enable
  0 = wake-up disabled (MSCAN ignores traffic on CAN bus)
  1 = wake-up enabled (MSCAN is able to restart)
SLPRQ: sleep mode request
  0 = running—the MSCAN functions normally.
  1 = sleep mode request—the MSCAN enters sleep mode when CAN is idle.
INITRQ: initialization mode request
  0 = normal operation
  1 = MSCAN in initialization mode

Figure 13.17 ■ MSCAN control register 0 (CANxCTL0, x = 0, 1, 2, 3, or 4)

initialization mode. When the INITRQ bit is cleared by the CPU, the MSCAN restarts and then tries to synchronize to the CAN bus.

MSCAN Control Register 1 (CANxCTL1)

This register provides for various control and handshake status information of the MSCAN module. The contents of this register are shown in Figure 13.18. The MSCAN module must be enabled before it can operate. Setting the CANE bit enables the MSCAN module. The clock source of the CAN module can be either the bus clock or the oscillator clock. When the CAN bus is operating at a high data rate and the bus clock is generated by the PLL circuit, the user should select the oscillator clock as the clock source because the PLL circuit has jitter.

The loopback mode can be used to test the software. The loopback mode is entered by setting the LOOPB bit. The MSCAN module can be placed in the listen-only mode to monitor the CAN bus traffic without participating in the CAN data exchange. The listen-only mode is entered by setting the LISTEN bit. When operating in a noisy environment, one can choose to enable the filter to filter out the noise on the CAN bus so that the CAN module won't be woken up by the noise.

7	6	5	4	3	2	1	0
CANE	CLKSRC	LOOPB	LISTEN	0	WUPM	SLPAK	INITAK

reset: 0   0   0   1   0   0   0   1

CANE: MSCAN enable
  0 = the MSCAN module is disabled.
  1 = the MSCAN module is enabled.
CLKSRC: MSCAN clock source
  0 = the MSCAN clock source is the oscillator clock.
  1 = the MSCAN clock source is the bus clock.
LOOPB: loopback self-test mode
  0 = loopback self-test disabled
  1 = loopback self-test enabled
LISTEN: listen-only mode
  0 = normal operation
  1 = listen-only mode activated
WUPM: wake-up mode
  0 = MSCAN wakes up the CPU after any recessive to dominant edge on the
     CAN bus and WUPE bit of the CANCTL0 register is set to 1.
  1 = MSCAN wakes up the CPU only in case of a dominant pulse on the CAN
     bus that has a length of T_{WUP} and the WUPE bit is set to 1.
SLPAK: sleep mode acknowledge
  0 = running—the MSCAN functions normally.
  1 = sleep mode active—the MSCAN has entered sleep mode.
INITAK: initialization mode acknowledge
  0 = normal operation—the MSCAN operates normally.
  1 = initialization mode active—the MSCAN is in initialization mode.

Figure 13.18 ■ MSCAN control register 1 (CANxCTL1, x = 0, 1, 2, 3, or 4)

There is a time delay from requesting to enter the sleep mode or initialization mode until the desired mode is entered. To make sure the requested mode is entered, one should check the SLPAK and INITAK bits.

MSCAN BUS TIMING REGISTER 0 (CANxBTR0)

The CANBTR0 register allows the user to select the synchronization jump width and the divide factor to the selected clock source to derive the time quantum. The contents of the CANBTR0 register are shown in Figure 13.19. The MSCAN uses the time quantum (t_Q) as the minimum unit for timing control. The synchronization jump width defines the maximum number of time quantum (t_Q) clock cycles a bit time can be shortened or lengthened to achieve resynchronization to data transitions on the bus. The time quantum (t_Q) is derived by dividing the prescale factor into the selected clock source. The value of the prescale factor is defined by bits BRP5~BRP0.

MSCAN BUS TIMING REGISTER 1 (CANxBTR1)

This register provides for control on phase_seg1 and phase_seg2 in Figure 13.12. The contents of the CANxBTR1 register are shown in Figure 13.20. When three samples are taken, the bit value is determined by the majority function of these samples. Two adjacent samples are separated by one time quantum.

7	6	5	4	3	2	1	0
SJW1	SJW0	BRP5	BRP4	BRP3	BRP2	BRP1	BRP0

reset: 0 0 0 0 0 0 0 0

SJW1, SJW0: synchronization jump width
00 = 1 t_Q clock cycle
01 = 2 t_Q clock cycle
10 = 3 t_Q clock cycle
11 = 4 t_Q clock cycle
BRP5~BRP0: baud rate prescaler
000000 = 1
000001 = 2
000010 = 3
...
111110 = 63
111111 = 64

Figure 13.19 ■ MSCAN bus timing register 0 (CANxBTR0, x = 0, 1, 2, 3, or 4)

The bit time is determined by the oscillator frequency, the baud rate prescaler, and the number of time quanta (t_Q) clock cycles per bit. The bit time is given by the following expression:

$$\text{Bit time} = \frac{\text{prescaler value}}{f_{CANCLK}} \times (1 + \text{TimeSegment1} + \text{TimeSegment2})$$

TimeSegment1 consists of **prop_seg** and **phase_seg1**. TimeSegment2 is the same as phase_seg2.

7	6	5	4	3	2	1	0
SAMP	TSEG22	TSEG21	TSEG20	TSEG13	TSEG12	TSEG11	TSEG10

reset: 0 0 0 0 0 0 0 0

SAMP: sampling
0 = one sample per bit
1 = three samples per bit
TSEG22~TSEG20: time segment 2
000 = 1 t_Q clock cycle
001 = 2 t_Q clock cycles
...
110 = 7 t_Q clock cycles
111 = 8 t_Q clock cycles
TSEG13~TSEG10: time segment 1
0000 = 1 t_Q clock cycle
0001 = 2 t_Q clock cycles
...
1110 = 15 t_Q clock cycles
1111 = 16 t_Q clock cycles

Figure 13.20 ■ MSCAN bus timing register 1 (CANxBTR1, x = 0, 1, 2, 3, or 4)

MSCAN Receiver Flag Register (CANxRFLG)

This register contains status flags related to CAN reception. The contents of this register are shown in Figure 13.21. The WUPIF flag will be set to 1 if the MSCAN detects the CAN bus activity and the WUPE bit of the CANxCTL0 register is set to 1. The CSCIF flag will be set to 1 when the MSCAN changes its current bus status due to the actual value of the transmit error counter and the receive error counter. If a valid message is received when all five receive buffers are full, then the OVRIF flag will be set to 1 to indicate this overflow condition. When the receive FIFO is not empty, the RXF flag is set to 1. The flag bits WUPIF, CSCIF, OVRIF, RXF are cleared by writing a 1 to them.

	7	6	5	4	3	2	1	0
	WUPIF	CSCIF	RSTAT1	RSTAT0	TSTAT1	TSTAT0	OVRIF	RXF
reset:	0	0	0	0	0	0	0	0

WUPIF: wake-up interrupt flag
  0 = no wake-up activity observed while in sleep mode.
  1 = MSCAN detected activity on the bus and requested wake-up.
CSCIF: CAN status change interrupt flag
  0 = no change in bus status occurred since last interrupt.
  1 = MSCAN changed current bus status.
RSTAT1~RSTAT0: receiver status bits
  00 = RxOK: $0 \leq$ Receive error counter ≤ 96
  01 = RxWRN:96 < Receive error counter ≤ 127
  10 = RxERR: 127 < Receive error counter
  11 = Bus-off[1]:Transmit error counter > 255
TSTAT1~TSTAT0: transmitter status bits
  00 = TxOK: $0 \leq$ Transmit error counter ≤ 96
  01 = TxWRN:96 < Transmit error counter ≤ 127
  10 = TxERR: 127 < Transmit error counter
  11 = Bus-off: Transmit error counter > 255
OVRIF: overrun interrupt flag
  0 = no data overrun occurred.
  1 = a data overrun detected.
RXF: receive buffer full flag
  0 = no new message available within the RxFG.
  1 = the receive FIFO is not empty. A new message is available in the RxFG.
Note 1. This information is redundant. As soon as the transmitter leaves its bus-off state, the receiver state skips to RxOK too.

Figure 13.21 ■ MSCAN receiver flag register (CANxRFLG, x = 0, 1, 2, 3, or 4)

MSCAN Receiver Interrupt Enable Register (CANxRIER)

This register enables and disables all interrupts related to the CAN receiver. The contents of this register are shown in Figure 13.22.

MSCAN Transmitter Flag Register (CANxTFLG)

The contents of this register are shown in Figure 13.23. Each of the three flags indicates whether its associated transmit buffer is empty, and thus not scheduled for transmission. These flags can be cleared by writing a 1 to them.

7	6	5	4	3	2	1	0
WUPIE	CSCIE	RSTATE1	RSTATE0	TSTATE1	TSTATE0	OVRIE	RXFIE

reset: 0   0   0   0   0   0   0   0

WUPIE: wake-up interrupt enable
  0 = no interrupt request is generated from this event.
  1 = a wake-up event causes a wake-up interrupt request.
CSCIE: CAN status change interrupt enable
  0 = no interrupt request is generated from this event.
  1 = a CAN status change event causes an error interrupt request.
RSTATE1~RSTATE0: receiver status change interrupt enable
  00 = do not generate any CSCIF interrupt caused by receiver state changes.
  01 = generate CSCIF interrupt only if the receiver enters or leaves "bus-off"
      state.
  10 = generate CSCIF interrupt only if the receiver enters or leaves "RxErr" or
      "bus-off" state.
  11 = generate CSCIF interrupt on all state changes.
TSTATE1~TSTATE0: transmitter status change interrupt enable
  00 = do not generate any CSCIF interrupt caused by transmitter state changes.
  01 = generate CSCIF interrupt only if the transmitter enters or leaves "bus-
      off" state.
  10 = generate CSCIF interrupt only if the transmitter enters or leaves "bus-
      off" or "TxErr" state.
  11 = generate CSCIF interrupt on all state changes.
OVRIE: overrun interrupt enable
  0 = no interrupt request is generated from this event.
  1 = an overrun event causes an error interrupt request.
RXFIE: receive buffer interrupt enable
  0 = no interrupt request is generated from this event.
  1 = a receive buffer full event causes a receiver interrupt request.

Figure 13.22 ■ MSCAN receiver interrupt enable register (CANxRIER, x = 1, 2, 3, or 4)

7	6	5	4	3	2	1	0
0	0	0	0	0	TXE2	TXE1	TXE0

reset: 0   0   0   0   0   1   1   1

TXE2~TXE0: transmitter buffer empty
  0 = the associated message buffer is full (loaded with a message due for
      transmission).
  1 = the associated message buffer is empty.

Figure 13.23 ■ MSCAN transmitter flag register (CANxTFL, x = 0, 1, 2, 3, or 4)

MSCAN TRANSMITTER INTERRUPT ENABLE REGISTER (CANxTIER)

The contents of this register are shown in Figure 13.24.

7	6	5	4	3	2	1	0
0	0	0	0	0	TXEIE2	TXEIE1	TXEIE0

reset: 0 0 0 0 0 0 0 0

TXEIE2~TXEIE0: transmitter empty interrupt enable
0 = disable interrupt from this buffer.
1 = a transmitter empty event causes a transmitter empty interrupt request.

Figure 13.24 ■ MSCAN transmitter interrupt enable register (CANxTIE, x = 0, 1, 2, 3, or 4)

MSCAN TRANSMITTER MESSAGE ABORT REQUEST REGISTER (CANxTARQ)

When the application has a high-priority message to be sent but cannot find any empty transmit buffer to use, it can request to abort the previous messages that have been scheduled for transmission. The application can request to abort the message in any one of the transmit buffers. This is done via the CANxTARQ register. The contents of this register are shown in Figure 13.25.

7	6	5	4	3	2	1	0
0	0	0	0	0	ABTRQ2	ABTRQ1	ABTRQ0

reset: 0 0 0 0 0 0 0 0

ABTRQ2~ABTRQ0: abort request
0 = no abort request
1 = abort request pending

Figure 13.25 ■ MSCAN transmitter message abort request register (CANxTARQ, x = 0, 1, 2, 3, or 4)

MSCAN TRANSMIT MESSAGE ABORT ACKNOWLEDGE REGISTER (CANxTAAK)

A message that is being transmitted cannot be aborted. Only those messages that have not been transmitted can be aborted. MSCAN answers the abort request by setting or clearing the associated bits in this register. The contents of this register are shown in Figure 13.26.

7	6	5	4	3	2	1	0
0	0	0	0	0	ABTAK2	ABTAK1	ABTAK0

reset: 0 0 0 0 0 0 0 0

ABTAK2~ABTAK0: abort acknowledge
0 = the message was not aborted.
1 = the message was aborted.

Figure 13.26 ■ MSCAN transmitter message abort acknowledge register (CANxTAAK, x = 0, 1, 2, 3, or 4)

MSCAN Transmit Buffer Selection (CANxTBSEL)

This register allows the selection of the actual transmit message buffer, which will then be accessible in the CANxTxFG register space. The contents of this register are shown in Figure 13.27. The lowest-numbered bit that is set makes the respective transmit buffer accessible to the user. For example, the combination TX1 = 1 and TX0 = 1 selects transmit buffer 0, and the combination TX1 = 1 and TX0 = 0 selects transmit buffer TX1.

7	6	5	4	3	2	1	0
0	0	0	0	0	TX2	TX1	TX0

reset: 0 0 0 0 0 0 0 0

TX2~TX0: transmit buffer select bits

0 = the associated message buffer is deselected.

1 = the associated message buffer is selected, if it is the lowest-numbered bit.

Figure 13.27 ■ MSCAN transmitter buffer select register (CANxTBSEL, x = 0, 1, 2, 3, or 4)

When one wants to get the next available transmit buffer, the CANxTFLG register is read and this value is written into the CANxTBSEL register. Suppose that all three transmit buffers are available. The value read from CANxTFLG is therefore %00000111. When writing this value back to CANxTSEL, the transmit buffer 0 is selected as the foreground transmit buffer because the lowest-numbered bit set to 1 is at bit position 0. Reading back this value of CANxTBSEL results in %00000001, because only the lowest-numbered bit position set to 1 is presented. This mechanism eases the application software selection of the next available transmit buffer. This situation is shown in the following instruction sequence:

```
ldaa   CAN1TFLG     ; value read is %00000111
staa   CAN1TBSEL    ; value written is %00000111
ldaa   CAN1TBSEL    ; value read is %00000001
```

MSCAN Identifier Acceptance Control Register (CANxIDAC)

This register provides for identifier acceptance control, as shown in Figure 13.28. The IDHITx indicators are always related to the message in the foreground receive buffer (RxFG). When a message gets shifted into the foreground buffer of the receiver FIFO, the indicators are updated as well.

MSCAN Receive Error Counter Register (CANxRXERR)

This register reflects the status of the MSCAN receive error counter. This register can be read only in sleep or initialization mode. Reading this register in any other mode may return an incorrect value.

MSCAN Transmit Error Counter Register (CANxTXERR)

This register reflects the status of the MSCAN transmit error counter. Like CANxRXERR, this register can be read correctly only in sleep or initialization mode.

7	6	5	4	3	2	1	0
0	0	IDAM1	IDAM0	0	IDHIT2	IDHIT1	IDHIT0

reset: 0   0   0   0   0   0   0   0

IDAM1~IDAM0: identifier acceptance mode
  00 = two 32-bit acceptance filters
  01 = four 16-bit acceptance filters
  10 = eight 8-bit acceptance filters
  11 = filter closed
IDHIT2~IDHIT0: identifier acceptance hit indicator (read only)
  000 = filter 0 hit
  001 = filter 1 hit
  010 = filter 2 hit
  011 = filter 3 hit
  100 = filter 4 hit
  101 = filter 5 hit
  110 = filter 6 hit
  111 = filter 7 hit

Figure 13.28 ■ MSCAN identifier acceptance control register (CANxIDAC, x = 0, 1, 2, 3, or 4)

MSCAN IDENTIFIER ACCEPTANCE REGISTERS (CANxIDAR0~7)

On reception, each message is written into the background receive buffer. The CPU is only signaled to read the message if it passes the criteria in the identifier acceptance and identifier mask registers (accepted); otherwise, the message is overwritten by the next message (dropped).

The acceptance registers of the MSCAN are applied on the IDAR0 to IDAR3 registers of the incoming messages in a bit-by-bit manner. For extended identifiers, all four acceptance and mask registers are applied. For standard identifiers, only the first two are applied. The contents of the first bank and second bank acceptance registers are shown in Figures 13.29 and 13.30,

7	6	5	4	3	2	1	0	
AC7	AC6	AC5	AC4	AC3	AC2	AC1	AC0	CANxIDAR0

reset: 0   0   0   0   0   0   0   0

7	6	5	4	3	2	1	0	
AC7	AC6	AC5	AC4	AC3	AC2	AC1	AC0	CANxIDAR1

reset: 0   0   0   0   0   0   0   0

7	6	5	4	3	2	1	0	
AC7	AC6	AC5	AC4	AC3	AC2	AC1	AC0	CANxIDAR2

reset: 0   0   0   0   0   0   0   0

7	6	5	4	3	2	1	0	
AC7	AC6	AC5	AC4	AC3	AC2	AC1	AC0	CANxIDAR3

reset: 0   0   0   0   0   0   0   0

Figure 13.29 ■ MSCAN identifier acceptance registers (first bank) (x = 0, 1, 2, 3, or 4)

	7	6	5	4	3	2	1	0	
	AC7	AC6	AC5	AC4	AC3	AC2	AC1	AC0	CANxIDAR4
reset:	0	0	0	0	0	0	0	0	

	7	6	5	4	3	2	1	0	
	AC7	AC6	AC5	AC4	AC3	AC2	AC1	AC0	CANxIDAR5
reset:	0	0	0	0	0	0	0	0	

	7	6	5	4	3	2	1	0	
	AC7	AC6	AC5	AC4	AC3	AC2	AC1	AC0	CANxIDAR6
reset:	0	0	0	0	0	0	0	0	

	7	6	5	4	3	2	1	0	
	AC7	AC6	AC5	AC4	AC3	AC2	AC1	AC0	CANxIDAR7
reset:	0	0	0	0	0	0	0	0	

Figure 13.30 ■ MSCAN identifier acceptance registers (second bank) (x = 0, 1, 2, 3, or 4)

respectively. The actual application of acceptance and mask registers is controlled by the CANxIDAC register. AC7~AC0 comprise a user-defined sequence of bits with which the corresponding bits of the related identifier register (IDRn) of the receive message buffer are compared. The result of this comparison is then masked with the corresponding identifier mask register to determine if there is a hit.

MSCAN IDENTIFIER MASK REGISTERS (CANxIDMR0~7)

The identifier mask registers specify which of the corresponding bits in the identifier acceptance register are relevant for acceptance filtering. The contents of the first and second bank of the identifier mask registers are shown in Figures 13.31 and 13.32, respectively. If a mask bit is 1, its corresponding acceptance bit will be ignored.

13.9.2 MSCAN Message Buffers

The receive message and transmit message buffers have the same outline. Each message buffer allocates 16 bytes in the memory map containing a 13-byte data structure (from Identifier register 0 to Data length register). The message buffer organization is illustrated in Figure 13.33.

IDENTIFIER REGISTERS (IDR0~IDR3)

All four identifier registers are compared when a message with an extended identifier is received. The contents of these four identifier registers are shown in Figure 13.34. When a message with the standard identifier is received, only the first two identifier registers are compared. The meaning of the standard identifier is illustrated in Figure 13.35.

The identifier consists of 29 bits (ID28~ID0) for the extended format. The ID28 bit is the most significant bit and is transmitted first. The CAN protocol uses the identifier to arbitrate simultaneous transmissions. The priority of an identifier is defined to be highest for the smallest binary number. The identifier consists of 11 bits (ID10~ID0) for the standard format. The ID10 bit is the most significant bit and is transmitted first. Similar to the extended identifier, the priority of a standard identifier is defined to be highest for the smallest binary number.

	7	6	5	4	3	2	1	0	
	AM7	AM6	AM5	AM4	AM3	AM2	AM1	AM0	CANxIDMR0
reset:	0	0	0	0	0	0	0	0	

	7	6	5	4	3	2	1	0	
	AM7	AM6	AM5	AM4	AM3	AM2	AM1	AM0	CANxIDMR1
reset:	0	0	0	0	0	0	0	0	

	7	6	5	4	3	2	1	0	
	AM7	AM6	AM5	AM4	AM3	AM2	AM1	AM0	CANxIDMR2
reset:	0	0	0	0	0	0	0	0	

	7	6	5	4	3	2	1	0	
	AM7	AM6	AM5	AM4	AM3	AM2	AM1	AM0	CANxIDMR3
reset:	0	0	0	0	0	0	0	0	

Figure 13.31 ■ MSCAN identifier mask registers (first bank) (x = 0, 1, 2, 3, or 4)

	7	6	5	4	3	2	1	0	
	AM7	AM6	AM5	AM4	AM3	AM2	AM1	AM0	CANxIDMR4
reset:	0	0	0	0	0	0	0	0	

	7	6	5	4	3	2	1	0	
	AM7	AM6	AM5	AM4	AM3	AM2	AM1	AM0	CANxIDMR5
reset:	0	0	0	0	0	0	0	0	

	7	6	5	4	3	2	1	0	
	AM7	AM6	AM5	AM4	AM3	AM2	AM1	AM0	CANxIDMR6
reset:	0	0	0	0	0	0	0	0	

	7	6	5	4	3	2	1	0	
	AM7	AM6	AM5	AM4	AM3	AM2	AM1	AM0	CANxIDMR7
reset:	0	0	0	0	0	0	0	0	

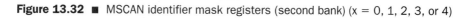

Figure 13.32 ■ MSCAN identifier mask registers (second bank) (x = 0, 1, 2, 3, or 4)

Address	Register Name
$_x0	Identifier register 0
$_x1	Identifier register 1
$_x2	Identifier register 2
$_x3	Identifier register 3
$_x4	Data segment register 0
$_x5	Data segment register 1
$_x6	Data segment register 2
$_x7	Data segment register 3
$_x8	Data segment register 4
$_x9	Data segment register 5
$_xA	Data segment register 6
$_xB	Data segment register 7
$_xC	Data length register
$_xD	Transmit buffer priority register[1]
$_xE	Time stamp register high byte[2]
$_xF	Time stamp register low byte[2]

[1] Not applicable for receive buffer
[2] Read only for CPU

Figure 13.33 ■ MSCAN message buffer organization

	7	6	5	4	3	2	1	0
IDR0	ID28	ID27	ID26	ID25	ID24	ID23	ID22	ID21
IDR1	ID20	ID19	ID18	SRR(=1)	IDE(=1)	ID17	ID16	ID15
IDR2	ID14	ID13	ID12	ID11	ID10	ID9	ID8	ID7
IDR3	ID6	ID5	ID4	ID3	ID2	ID1	ID0	RTR

Figure 13.34 ■ Receive/transmit message buffer extended identifier

	7	6	5	4	3	2	1	0
IDR0	ID10	ID9	ID8	ID7	ID6	ID5	ID4	ID3
IDR1	ID2	ID1	ID0	RTR	IDE(=0)			
IDR2								
IDR3								

Figure 13.35 ■ Receive/transmit message buffer standard identifier

The SRR bit is used only in the extended format and must be set to 1. The IDE bit is used to indicate whether the identifier is an extended identifier. When set to 1, the identifier is used as an extended identifier.

The RTR bit is used to solicit transmission from other CAN nodes. In the case of a receive buffer, it indicates the status of the received frame and supports the transmission of an answering frame in software.

DATA SEGMENT REGISTERS (DSR0~DSR7)

The eight data segment registers, each with its DB7~DB0, contain the data to be transmitted or received. The number of bytes to be transmitted or received is determined by the data length code in the corresponding DLR register.

DATA LENGTH REGISTER (DLR)

The data length register has only the lowest four bits implemented and is used to indicate the number of data bytes contained in the message.

TRANSMIT BUFFER PRIORITY REGISTER (TBPR)

This register defines the local priority of the associated message buffer. The local priority is used for the internal prioritization process of the MSCAN and is defined to be highest for the smallest binary number. The MSCAN implements the following internal prioritization mechanism:

- All transmission buffers with a cleared TXEx flag participate in the prioritization immediately before the SOF (start-of-frame) is sent.
- The transmission buffer with the lowest local priority field wins the prioritization.

In case of more than one buffer having the same lowest priority, the message buffer with the lowest index number wins.

TIMESTAMP REGISTER (TSRH, TSRL)

If the TIME bit of the CANxCTL0 register is set to 1, the MSCAN will write a special timestamp to the respective registers in the active transmit or receive buffer as soon as a message has been acknowledged on the CAN bus. The timestamp is written on the bit sample point for the recessive bit of the ACK delimiter in the CAN frame. In the case of a transmission, the CPU can only read the timestamp after the respective transmit buffer has been flagged empty.

The timer value used for stamping is taken from a free-running internal CAN bit clock. A timer overrun is not indicated by the MSCAN. The contents of the timestamp registers are shown in Figure 11.36. The contents of timestamp registers are not sent out to the CAN bus.

	7	6	5	4	3	2	1	0	
	TSR15	TSR14	TSR13	TSR12	TSR11	TSR10	TSR9	TSR8	TSRH
reset:	x	x	x	x	x	x	x	x	

	7	6	5	4	3	2	1	0	
	TSR7	TSR6	TSR5	TSR4	TSR3	TSR2	TSR1	TSR0	TSRL
reset:	x	x	x	x	x	x	x	x	

Figure 13.36 ■ MSCAN timestamp registers

To facilitate access to the CAN buffers, the variable names for the transmit buffer and receive buffers are added to the **hcs12.inc** file for the assembly language and the **hcs12.h** file for the C language. These variable names are listed in Tables 13.2a and 13.2b.

Name	Address	Description
CANxRIDR0	$_0	CAN foreground receive buffer x identifier register 0
CANxRIDR1	$_1	CAN foreground receive buffer x identifier register 1
CANxRIDR2	$_2	CAN foreground receive buffer x identifier register 2
CANxRIDR3	$_3	CAN foreground receive buffer x identifier register 3
CANxRDSR0	$_4	CAN foreground receive buffer x data segment register 0
CANxRDSR1	$_5	CAN foreground receive buffer x data segment register 1
CANxRDSR2	$_6	CAN foreground receive buffer x data segment register 2
CANxRDSR3	$_7	CAN foreground receive buffer x data segment register 3
CANxRDSR4	$_8	CAN foreground receive buffer x data segment register 4
CANxRDSR5	$_9	CAN foreground receive buffer x data segment register 5
CANxRDSR6	$_A	CAN foreground receive buffer x data segment register 6
CANxRDSR7	$_B	CAN foreground receive buffer x data segment register 7
CANxRDLR	$_C	CAN foreground receive buffer x data length register

Note 1. x can be 0, 1, 2, 3 or 4
    2. The absolute address of each register is equal to the sum of the base address of the CAN foreground receive buffer base x address and the address field of the corresponding register.

Table 13.2a ■ CAN foreground receive buffer x variable names

Name	Address	Description
CANxTIDR0	$_0	CAN foreground transmit buffer x identifier register 0
CANxTIDR1	$_1	CAN foreground transmit buffer x identifier register 1
CANxTIDR2	$_2	CAN foreground transmit buffer x identifier register 2
CANxTIDR3	$_3	CAN foreground transmit buffer x identifier register 3
CANxTDSR0	$_4	CAN foreground transmit buffer x data segment register 0
CANxTDSR1	$_5	CAN foreground transmit buffer x data segment register 1
CANxTDSR2	$_6	CAN foreground transmit buffer x data segment register 2
CANxTDSR3	$_7	CAN foreground transmit buffer x data segment register 3
CANxTDSR4	$_8	CAN foreground transmit buffer x data segment register 4
CANxTDSR5	$_9	CAN foreground transmit buffer x data segment register 5
CANxTDSR6	$_A	CAN foreground transmit buffer x data segment register 6
CANxTDSR7	$_B	CAN foreground transmit buffer x data segment register 7
CANxTDLR	$_C	CAN foreground transmit buffer x data length register
CANxTBPR	$_D	CAN foreground transmit buffer x priority register
CANxTSRH	$_E	CAN foreground transmit buffer x timestamp register high
CANxTSRL	$_F	CAN foreground transmit buffer x timestamp register low

Note 1. x can be 0, 1, 2, 3 or 4
    2. The absolute address of each register is equal to the sum of the base address of the CAN foreground transmit buffer x and the address field of the corresponding register.

Table 13.2b ■ CAN foreground transmit buffer x variable names

13.9.3 Transmit Storage Structure

The design of the MSCAN transmit structure achieves two goals:

- Providing the capability to send out a stream of scheduled messages without releasing the bus between the two messages
- Prioritizing messages so that the message with the highest priority is sent out first

As shown in Figure 13.37, the MSCAN has a triple transmit buffer scheme that allows multiple messages to be set up in advance and achieve a real-time performance. Only one of the three transmit buffers is accessible to the programmer at a time. A transmit buffer is made accessible to the user by writing an appropriate value into the CANxTBSEL register.

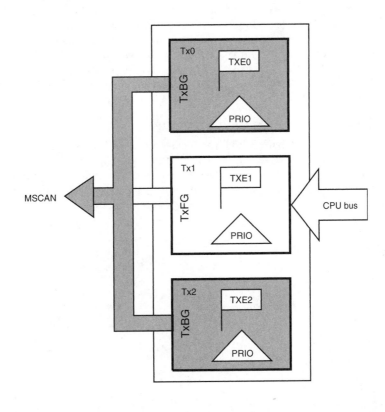

Figure 13.37 ■ User model for transmit buffer organization

The procedure for transmitting a message includes the following steps:

1. Identifying an available transmit buffer by checking the TXEx flag associated with the transmit buffer.
2. Setting a pointer to the empty transmit buffer by writing the CANxTFLG register to the CANxTBSEL register. This makes the transmit buffer accessible to the user.

3. Storing the identifier, the control bits, and the data contents into one of the transmit buffers.

4. Flagging the buffer as ready by clearing the associated TXE flag.

After step 4, the MSCAN schedules the message for transmission and signals the successful transmission of the buffer by setting the associated TXE flag.

If there is more than one buffer scheduled for transmission when the CAN bus becomes available for arbitration, the MSCAN uses the local priority setting to choose the buffer with the highest priority and sends it out. The buffer having the smallest priority field has the highest priority and will be scheduled for transmission first. The internal scheduling process takes place whenever the MSCAN arbitrates for the bus.

When a high-priority message is scheduled by the application, it may be necessary to abort a lower-priority message in one of the three transmit buffers. A message that is being transmitted cannot be aborted. One can make an abort request by setting an appropriate ABTRQ bit of the CANxTARQ register. The MSCAN then grants the request, if possible, by: (1) setting the corresponding Abort Acknowledge flag (ABTAK) in the CANxTAAK register, (2) setting the associated TXE flag to release the buffer, and (3) generating a transmit interrupt. The transmit interrupt handler software can tell from the setting of the ABTAK flag whether the message was aborted (ABTAK = 1) or sent (ABTAK = 0).

13.9.4 Receive Storage Structure

As shown in Figure 13.38, the received messages are stored in a five-stage input FIFO data structure. The message buffers are alternately mapped into a single memory area, which is referred to as the *foreground receive buffer*. The application software reads the foreground receive buffer to access the received message. The background receive buffer is solely used to hold incoming CAN messages and is not accessible to the user.

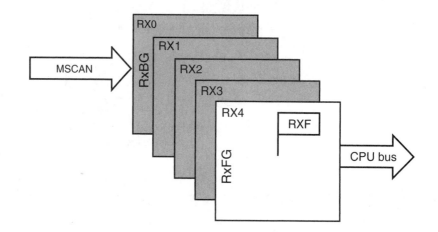

Figure 13.38 ■ User model for receive buffer organization

Whenever a valid message is received at the background receive buffer, it will be transferred to the foreground receive buffer and the RXF flag will be set to 1. The user's receive handler program has to read the received message from the RxFG and then reset the RXF flag to acknowledge the interrupt and to release the foreground buffer.

When the MSCAN module is transmitting, the MSCAN receives its own transmitted messages into the background receive buffer but does not shift it into the receives FIFO nor generate a receive interrupt. An overrun condition occurs when all receive message buffers in the FIFO are filled with correctly received messages with accepted identifiers and another message is correctly received from the bus with an accepted identifier. The latter message is discarded and an error interrupt with overrun indication is generated if enabled. The MSCAN is still able to transmit messages while the receiver FIFO is being filled, but all incoming messages are discarded. As soon as a receive buffer in the FIFO is available again, new valid messages will be accepted.

13.9.5 Identifier Acceptance Filter

The MSCAN identifier acceptance registers define the acceptance patterns of the standard or extended identifier. Any of these bits can be marked "don't care" in the MSCAN identifier mask registers.

A message is accepted only if its associated identifier matches one of the identifier filters. A filter hit is indicated to the application software by a RXF flag setting to 1 and the three hit bits in the CANxIDAC register. These hit bits identify the filter section that caused the acceptance. In case more than one hit occurs, the lower hit has priority. The identifier acceptance filter is programmable to operate in four different modes:

1. Two identifier acceptance filters with each filter applied to (a) the full 29 bits of the extended identifier, the SRR bit, the IDE bit, and the RTR bit or (b) the 11 bits of the standard identifier plus the RTR and IDE bits. This mode may cause up to two hits.

2. Four identifier acceptance filters with each filter applied to (a) the 14 most significant bits of the extended identifier plus the SRR and IDE bits or (b) the 11 bits of the standard identifier plus the RTR and IDE bits. This mode may cause up to four hits.

3. Eight identifier acceptance filters with each filter applied to the first eight bits of the identifier. This mode implements eight independent filters for the first eight bits of the extended or standard identifier. This may cause up to eight hits.

4. Closed filter. No CAN message is copied into the foreground buffer RxFG, and the RXF flag is never set.

13.9.6 MSCAN Clock System

The MSCAN clock generation circuitry is shown in Figure 13.39. This clock circuitry allows the MSCAN to handle CAN bus rates ranging from 10 Kbps up to 1 Mbps. The CLKSRC bit in the CANxCTL1 register defines whether the internal CANCLK is connected to the

Figure 13.39 ■ MSCAN clocking scheme

output of a crystal oscillator or to the bus clock. The clock source has to be chosen such that the tight oscillator tolerance requirements (up to 0.4%) of the CAN protocol are met. Additionally, for a high CAN bus rate (1 Mbps), a 45% to 55% duty cycle of the clock is required.

If the bus clock is generated from a PLL, it is recommended to select the oscillator clock rather than the bus clock due to jitter considerations, especially at the faster CAN bus rate. For microcontrollers without a clock and reset generator (CRG), CANCLK is driven from the crystal oscillator.

A programmable prescaler generates the time quanta (t_Q) clock from CANCLK. A time quantum is the atomic unit of time handled by the MSCAN. The frequency of the time quantum is derived from the CANCLK using the following expression:

$$ft_Q = f_{CANCLK} \div (\text{prescaler value})$$

Slightly deviated from Figure 13.12, the MSCAN divides a bit time into three segments (shown in Figure 13.40) rather than four segments:

- Sync_seg: This segment is fixed at 1 time quantum. Signal edges are expected to happen within this segment.

- Time segment 1: This segment includes the **prop_seg** and the **phase_seg1** of the CAN standard. It can be programmed by setting the TSEG1 parameter of the CANxBTR1 register to consist of 4 to 16 time quanta.

- Time segment 2: This segment represents the **phase_seg2** of the CAN standard. It can be programmed by setting the TSEG2 parameter of the CANxBTR1 register to be 2 to 8 time quanta long.

Figure 13.40 ■ Segments within the bit time

The bit rate of the CAN bus can be derived from the frequency of the time quantum using the following expression:

$$\text{Bit rate} = f_{t_Q} \div (\text{number of time quanta})$$

By computing the number of time quanta contained in a bit time, one can figure out the appropriate values for time segment 1 and time segment 2. This will be elaborated in Section 13.12.

13.9.7 MSCAN Interrupt Operation

The MSCAN supports four interrupt vectors, any of which can be individually masked.

TRANSMIT INTERRUPT

At least one of the three transmit buffers is empty and can be loaded to schedule a message for transmission. The TXEx flag of the empty message buffer is set.

RECEIVE INTERRUPT

A receive interrupt may be requested when a message is successfully received and shifted into the foreground buffer of the receive FIFO. This interrupt is generated immediately after receiving the EOF symbol. The RXF flag is set. If there are multiple messages in the receiver FIFO, the RXF flag is set as soon as the next message is shifted to the foreground buffer.

WAKE-UP INTERRUPT

Activity on the CAN bus occurring during the MSCAN internal sleep mode may generate the wake-up interrupt.

ERROR INTERRUPT

An overrun of the receiver FIFO, error, warning, or bus-off condition may generate an error interrupt.

INTERRUPT ACKNOWLEDGE

Interrupts are directly associated with one or more status flags in either the CANxRFLG or the CANxTFLG register. Interrupts are pending as long as one of the corresponding flags is set. The flags in these registers must be cleared within the interrupt handler to avoid repeated interrupts. The flags are cleared by writing a 1 to the corresponding bit position. A flag cannot be cleared if the respective condition still prevails.

13.9.8 MSCAN Initialization

The MSCAN needs to be configured before it can operate properly. The procedures for configuring the MSCAN out of reset and in normal mode are different.

MSCAN INITIALIZATION OUT OF RESET

The procedure to initialize the MSCAN after reset is as follows:

- Assert the CANE bit.
- Write to the configuration registers (CANxCTL1, CANxBTR0, CANxBTR1, CANxIDAC, CANxIDAR0-7, CANxIDMR0-7) in initialization mode (both the INITRQ and INITAK bits are set).
- Clear the INITRQ bit to leave initialization mode and enter normal mode.

MSCAN INITIALIZATION IN NORMAL MODE

- Make sure that the MSCAN transmission queue gets empty and bring the module into sleep mode by asserting the SLPRQ bit and waiting for the SLPAK bit to be set.
- Enter the initialization mode.
- Write to the configuration registers in initialization mode.
- Clear the INITRQ bit to leave the initialization mode and continue in normal mode.

13.10 Physical CAN Bus Connection

The CAN protocol is designed for data communication over a short distance. It does not specify what medium to use for data transmission. The user can choose optical fiber, shielded cable, or unshielded cable as the transmission medium. Using a shielded or unshielded cable is recommended for a short-distance communication.

A typical CAN bus system setup using a cable is illustrated in Figure 13.41. The resistor R_T is the terminating resistor. Each node uses a transceiver to connect to the CAN bus. The CAN bus transceiver is connected to the bus via two bus terminals, CAN_H and CAN_L, which provide differential receive and transmit capabilities. The nominal CAN bus levels are shown in Figure 13.42.

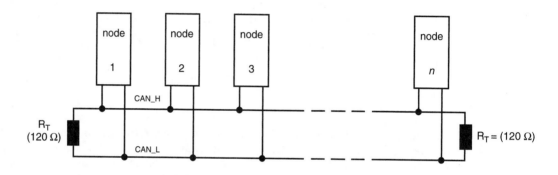

Figure 13.41 ■ A typical CAN bus setup using cable

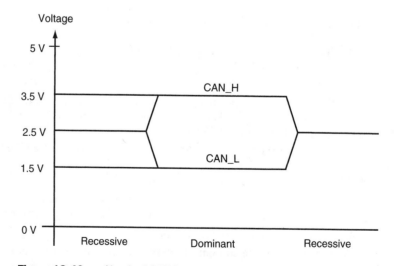

Figure 13.42 ■ Nominal CAN bus levels

Many semiconductor companies produce CAN bus transceivers. The Microchip MCP2551, Philips PCA82C250/251, Texas Instruments SN65HVD251, and MAXIM MAX3050/3057 are among the most popular. All of these chips are compatible. TI SN65HVD251, Philips PCA82C250, and Microchip MCP2551 are drop-in replaceable with each other.

13.10.1 The MCP2551 CAN Transceiver

The MCP2551 provides differential transmit and receive capability for the CAN protocol controller (or module). It operates at speeds up to 1 Mbps. The MCP2551 converts the digital signals generated by a CAN controller to signals suitable for transmission over the CAN bus cabling. It also provides a buffer between the CAN controller and the high-voltage spikes that can be generated on the CAN bus by outside sources, such as EMI, ESD, electrical transients, and so on. The block diagram of the MCP2551 is shown in Figure 13.43.

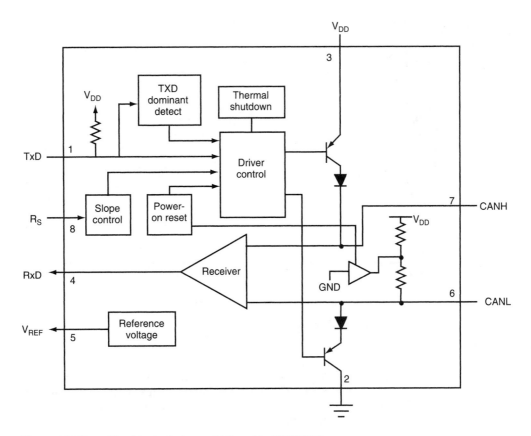

Figure 13.43 ■ The block diagram of Microchip MCP2551

The MCP2551 transceiver outputs will drive a minimum load of 45 Ω, allowing a maximum of 112 nodes to be connected to the same CAN bus. The RxD pin reflects the differential bus voltage between CAN_H and CAN_L. The low and high states of the RxD output pin correspond to the dominant and recessive states of the CAN bus, respectively. The R_S input allows the user to select one of the three operation modes:

- High speed
- Slope control
- Standby

The high-speed mode is selected by grounding the R_S pin. In this mode, the transmission output drivers have fast output rise and fall times to support the high-speed CAN bus. The slope-control mode further reduces the electromagnetic interference (EMI) by limiting the rise and fall times of the CAN_H and CAN_L signals. The slope, or *slew rate*, is controlled by connecting an external resistor (R_{EXT}) between the R_S pin and the ground pin. Figure 13.44 illustrates typical slew rate values as a function of the slope-control resistance value.

Figure 13.44 ■ Slew rate vs. slope control resistance value

The MCP2551 can be placed in the standby or sleep mode by applying a high voltage at the R_S pin. In sleep mode, the transmitter is switched off, and the receiver operates at a lower current. The receive pin on the controller side (RxD) is still functioning but will operate at a slower rate. The attached microcontroller can monitor the RxD pin for CAN bus activity and place the transceiver into normal operation via the R_S pin.

13.10.2 Interfacing the MCP2551 to the HCS12 CAN Devices

A typical method of interfacing the MCP2551 transceiver to the HCS12 CAN module is shown in Figure 13.45. The maximum achievable bus length in a CAN bus network is determined by the following physical factors:

1. The loop delay of the connected bus nodes (CAN controller and transceiver) and the delay of the bus line

2. The differences in bit time quantum length due to the relative oscillator tolerance between nodes

3. The signal amplitude drop due to the series resistance of the bus cable and the input resistance of bus node

Figure 13.45 ■ Interfacing the MCP2551 with the HCS12

The resultant equation after taking these three factors into account would be very complicated. The bus length that can be achieved as a function of the bit rate in the high-speed mode and with CAN bit timing parameters being optimized for maximum propagation delay is shown in Table 13.3. The types and the cross sections suitable for the CAN bus trunk cable that has more than 32 nodes connected and spans more than 100 meters are listed in Table 13.4.

Bit Rate (Kbps)	Bus Length
1000	40
500	100
250	250
125	500
62.5	1000

Table 13.3 ■ CAN bus bit rate/bus length relation

Bus Length/Number of Nodes	32	64	100
100 m	0.25 mm² or AWG 24	0.25 mm² or AWG 24	0.25 mm² or AWG 24
250 m	0.34 mm² or AWG 22	0.50 mm² or AWG 20	0.50 mm² or AWG 20
500 m	0.75 mm² or AWG 18	0.75 mm² or AWG 18	0.75 mm² or AWG 18

Table 13.4 ■ Minimum recommended bus wire cross section for the trunk cable

13.11 Setting the CAN Timing Parameters

The nominal bit rate of a CAN network is uniform throughout the network and is given by the following equation:

$$f_{NBT} = 1/t_{NBT} \qquad\qquad (13.1)$$

where t_{NBT} is the *nominal bit time*. As described earlier, a bit time is divided into four separate nonoverlapping time segments: *sync_seg, prop_seg, phase_seg1,* and *phase_seg2.*

The purpose of the *sync_seg* segment is to inform all nodes that a bit time is just started. The existence of the propagation delay segment, *prop_seg,* is due to the fact that the CAN protocol allows for nondestructive arbitration between nodes contending for access to the bus and the requirement for *in-frame acknowledgment.* In the case of nondestructive arbitration, more than one node may be transmitting during the arbitration field. Each transmitting node samples data from the bus in order to determine whether it has won or lost the arbitration and also to receive the arbitration field in case it loses arbitration. When each node samples a bit, the value sampled must be the logical superposition of the bit values transmitted by each of the nodes arbitrating for bus access. In the case of the acknowledge field, the transmitting node transmits a recessive bit but expects to receive a dominant bit; in other words, a dominant value must be sampled at the sample time. The length of the *prop_seg* segment must be selected for the earliest possible sample of the bit by a node until the transmitted bit values from all the transmitting nodes have reached all of the nodes. When multiple nodes are arbitrating for the control of the CAN bus, the node that transmitted the earliest must wait until the node that transmitted the latest in order to find out if it won or lost. That is, the value for t_{PROP_SEG} is given by

$$t_{PROP\_SEG} = t_{PROP(A, B)} + t_{PROP(B, A)} \qquad\qquad (13.2)$$

where $t_{PROP(A, B)}$ and $t_{PROP(B,A)}$ are the propagation delays from node A to node B and node B to node A, respectively. In the worst case, node A and node B are at the two ends of the CAN bus. The propagation delay from node A to node B is given by

$$t_{PROP(A, B)} = t_{BUS} + t_{Tx} + t_{Rx} \qquad\qquad (13.3)$$

where t_{BUS}, t_{Tx}, and t_{Rx} are data traveling time on the bus, transmitter propagation delay, and receiver propagation delay, respectively.

Let node A and node B be two nodes at opposite ends of the CAN bus, then the value for t_{PROP_SEG} is

$$t_{PROP\_SEG} = 2 \times (t_{BUS} + t_{Tx} + t_{Rx}) \qquad\qquad (13.4)$$

The minimum number of *time quantum* (t_Q) that must be allocated to the *prop_seg* segment is therefore

$$prop\_seg = round\_up\ (t_{PROP\_SEG} \div t_Q) \qquad\qquad (13.5)$$

where the round_up () function returns a value that equals the argument rounded up to the next integer value.

In the absence of bus errors, bit stuffing guarantees a maximum of 10 bit periods between resynchronization edges (5 dominant bits followed by 5 recessive bits and then followed by a dominant bit). This represents the worst-case condition for the accumulation of phase error during normal communication. The accumulated phase error must be compensated for by resynchronization and therefore must be less than the programmed resynchronization jump width (t_{RJW}). The accumulated phase error is due to the tolerance in the CAN system clock, and this requirement can be expressed as

$$(2 \times \Delta f) \times 10 \times t_{NBT} < t_{RJW} \qquad\qquad (13.6)$$

where Δf is the largest crystal oscillator frequency variation (in percentage) of all CAN nodes in the network.

Real systems must operate in the presence of electrical noise, which may induce errors on the CAN bus. A node transmits an error flag after it detects an error. In the case of a local error, only the node that detects the error will transmit the error flag. All other nodes receive the error flag and transmit their own error flags as an echo. If the error is global, all nodes will detect it within the same bit time and will therefore transmit error flags simultaneously. A node can therefore differentiate between a local error and a global error by detecting whether there is an echo after its error flag. This requires that a node can correctly sample the first bit after transmitting its error flag.

An error flag from an error-active node consists of 6 dominant bits, and there could be up to 6 dominant bits before the error flag, if, for example, the error was a stuff error. A node must therefore correctly sample the 13th bit after the last resynchronization. This can be expressed as

$$(2 \times \Delta f) \times (13 \times t_{NBT} - t_{PHASE\_SEG2}) < \text{MIN} \ (t_{PHASE\_SEG1}, t_{PHASE\_SEG2}) \tag{13.7}$$

where the function MIN(arg1,arg2) returns the smaller of the two arguments.

Thus there are two clock tolerance requirements that must be satisfied. The selection of bit timing values involves consideration of various fundamental system parameters. The requirement of the *prop_seg* value imposes a trade-off between the maximum achievable bit rate and the maximum propagation delay, due to the bus length and the characteristics of the bus driver circuit. The highest bit rate can only be achieved with a short bus length, a fast bus driver circuit, and a high-frequency CAN clock source with high tolerance.

The procedure for determining the optimum bit timing parameters that satisfy the requirements for proper bit sampling is as follows:

Step 1
Determine the minimum permissible time for the *prop_seg* segment using Equation 13.4.

Step 2
Choose the CAN system clock frequency. The CAN system clock will be the CPU oscillator output or the bus clock divided by a prescale factor. The CAN system clock is chosen so that the desired CAN bus nominal bit time is an integer multiple of time quanta (CAN system clock period) from 8 to 25.

Step 3
Calculate *prop_seg* duration. The number of time quanta required for the *prop_seg* can be calculated by using Equation 13.5. If the result is greater than 8, go back to step 2 and choose a lower CAN system clock frequency.

Step 4
Determine *phase_seg1* and *phase_seg2*. Subtract the *prop_seg* value and 1 (for *sync_seg*) from the time quanta contained in a bit time. If the difference is less than 3, then go back to step 2 and select a higher CAN system clock frequency. If the difference is an odd number greater than 3, then add 1 to the *prop_seg* value and recalculate. If the difference is equal to 3, then *phase_seg1* = 1 and *phase_seg2* = 2 and only one sample per bit may be chosen. Otherwise, divide the remaining number by 2 and assign the result to *phase_seg1* and *phase_seg2*.

Step 5
Determine the resynchronization jump width (RJW). RJW is the smaller of 4 and *phase_seg1*.

Step 6
Calculate the required oscillator tolerance from Equations 13.6 and 13.7. If *phase_seg1* > 4, it is recommended to repeat steps 2 to 6 with a larger value for the prescaler.

Conversely, if *phase_seg1* < 4, it is recommended to repeat step 2 to 6 with a smaller value for the prescaler, as long as *prop_seg* < 8, as this may result in a reduced oscillator tolerance requirement. If the prescaler is already equal to 1 and a reduced oscillator tolerance is still required, the only option is to consider using a clock source with higher frequency.

Example 13.1

▼

Calculate the CAN bit segments for the following system constraints:

Bit rate = 1 Mbps

Bus length = 25 m

Bus propagation delay = 5×10^{-9} s/m

MCP2551 transceiver plus receiver propagation delay = 150 ns at 85°C

CPU oscillator frequency = bus clock frequency = 24 MHz

Solution: Let's follow the procedure described earlier:

Step 1
Physical delay of bus = 125 ns

$t_{PROP\_SEG} = 2 \times (125 \text{ ns} + 150 \text{ ns}) = 550 \text{ ns}$

Step 2
A prescaler of 1 for a CAN system clock of 24 MHz gives a time quantum of 41.67 ns. One bit time is 1 μs. This gives 1000/41.67 = 24 time quanta per bit.

Step 3
Prop_seg = round_up (550 ns ÷ 41.67 ns) = round_up (13.2) = 14 > 8.
Set prescaler to 2. Then one time quantum is 83.33 ns and one bit time has 12 time quanta. The new prop_seg = round_up(550 ns ÷ 83.33 ns) = 7.

Step 4
Subtracting 7 for prop_seg and 1 for sync_seg from 12 time quanta per bit gives 4. Since the result is equal to 4 and is even, divide it by 2 (quotient is 2) and assign it to phase_seg1 and phase_seg2.

Step 5
RJW is the smaller one of 4 and phase_seg1 and is 2.

Step 6
From Equation 13.6:

$\Delta f < RJW \div (20 \times NBT) = 2 \div (20 \times 12) = 0.83\%$

From Equation 13.7:

$\Delta f < MIN \text{ (phase\_seg1, phase\_seg2)} \div 2(13 \times NBT - \text{phase\_seg2}) = 2 \div 308 = 0.65\%$

The desired oscillator tolerance is the smaller of these values, i.e., 0.65% over a period of 12.83 μs (12.83 bit period).

In summary:

Prescaler	= 2
Nominal bit time	= 12
prop_seg	= 7

sync_seg	= 1
phase_seg1	= 2
phase_seg2	= 2
RJW	= 2
Oscillator tolerance	= 0.65%

Example 13.2

Calculate the bit segments for the following system constraints:

Bit rate = 500 Kbps

Bus length = 50 m

Bus propagation delay = 5×10^{-9} s/m

MCP2551 transmitter plus receiver propagation delay = 150 ns at 85°C

CPU oscillator frequency = bus clock frequency = 16 MHz

Solution: Follow the standard procedure to perform the calculation:

Step 1
Physical delay of bus = 250 ns

$t_{PROP\_SEG} = 2 \times (250 \text{ ns} + 150 \text{ ns}) = 800 \text{ ns}$

Step 2
A prescaler of 2 for a CAN system clock of 16 MHz gives a time quantum of 125 ns. The nominal bit time is 2 μs. This gives 2000/125 = 16 time quanta per bit.

Step 3
Prop_seg = round_up (800 ns ÷ 125 ns) = round_up (6.4) = 7.

Step 4
Subtracting 7 for prop_seg and 1 for sync_seg from 16 time quanta per bit gives 8.
Since the result is greater than 4 and is even, divide it by 2 (quotient is 4) and assign it to phase_seg1 and phase_seg2.

Step 5
RJW (= 4) is the smaller of 4 and phase_seg1.

Step 6
From Equation 13.6:

$\Delta f < RJW \div (20 \times NBT) = 4 \div (20 \times 16) = 1.25\%$

From Equation 13.7:

$\Delta f < MIN (\text{phase\_seg1, phase\_seg2}) \div 2(13 \times NBT - \text{phase\_seg2}) = 4 \div 408 = 0.98\%$

The required oscillator tolerance is the smaller of these values, i.e., 0.98%.

Since phase_seg2 = 4, there is no need to try other prescale values.
In summary:

Prescaler	= 2
Nominal bit time	= 16
prop_seg	= 7

sync_seg	= 1
phase_seg1	= 4
phase_seg2	= 4
RJW	= 4
Oscillator tolerance	= 0.98%

▲

13.12 MSCAN Configuration

The timing parameters for the MSCAN module need be done only once and should be done immediately after reset. Other parameters such as acceptance filters may be changed after reset configuration.

Example 13.3

▼

Write a program to configure the MSCAN module 1 after reset to operate with the timing parameters computed in Example 13.1 and the following setting:

- Enable wake-up
- Disable time stamping
- Select oscillator as the clock source to the MSCAN
- Disable loopback mode, disable listen-only mode
- One sample per bit
- Accept messages with extended identifiers that start with T1 and P1 (use two 32-bit filters)

Solution: The first step is to enter the initialization mode and also make sure the initialization is entered. This can be done as follows:

```
bset   CAN1CTL0,INITRQ
brclr  CAN1CTL1,INITAK,*
```

The assembly program that performs the desired configuration is as follows:

```
openCan1  bset   CAN1CTL1,CANE        ; required after reset
          bset   CAN1CTL0,INITRQ      ; request to enter initialization mode
w1        brclr  CAN1CTL1,INITAK,w1   ; make sure initialization mode is entered
          movb   #$84,CAN1CTL1        ; enable CAN1,select oscillator as clock source
                                      ; enable wake-up filter
          movb   #$41,CAN1BTR0        ; set jump width to 2 tQ, prescaler set to 2
          movb   #$18,CAN1BTR1        ; set phase_seg2 to 2 tQ, phase_seg1 to 2 tQ,
                                      ; set prop_seg to 7 tQ
          movb   #$54,CAN1IDAR0       ; acceptance identifier 'T'
          movb   #$3C,CAN1IDAR1       ; acceptance identifier '1'
          movb   #$40,CAN1IDAR2       ;    "
          movb   #$00,CAN1IDAR3       ;    "
          movb   #$00,CAN1IDMR0       ; acceptance mask for extended identifier "T1"
          movb   #$00,CAN1IDMR1       ;    "
```

```
        movb    #$3F,CAN1IDMR2      ;      "
        movb    #$FF,CAN1IDMR3      ;      "
        movb    #$50,CAN1IDAR4      ; acceptance identifier 'P'
        movb    #$3C,CAN1IDAR5      ; acceptance identifier '1'
        movb    #$40,CAN1IDAR6      ;      "
        movb    #$00,CAN1IDAR7      ;      "
        movb    #$00,CAN1IDMR4      ; acceptance mask for extended identifier "P1"
        movb    #$00,CAN1IDMR5      ;      "
        movb    #$3F,CAN1IDMR6      ;      "
        movb    #$FF,CAN1IDMR7      ;      "
        clr     CAN1IDAC            ; set two 32-bit filter mode
        movb    #$25,CAN1CTL0       ; stop clock on wait mode, enable wake-up
        bclr    CAN1CTL0,INITRQ     ; leave initialization mode
        rts
```

The C language version of the function is as follows:

```
void openCan1(void)
{
    CAN1CTL1  | = CANE;          /* enable CAN, required after reset */
    CAN1CTL0  | = INITRQ;        /* request to enter initialization mode */
    while(!(CAN1CTL1&INITAK));   /* wait until initialization mode is entered */
    CAN1CTL1  = 0x84;            /* enable CAN1, select oscillator as MSCAN clock
                                   source, enable wake-up filter */
    CAN1BTR0  = 0x41;            /* set SJW to 2, set prescaler to 2 */
    CAN1BTR1  = 0x18;            /* set phase_seg2 to 2 t_Q, phase_seg1 to 2 t_Q,
                                   prop_seg to 7 t_Q */
    CAN1IDAR0 = 0x54;            /* set acceptance identifier "T1" */
    CAN1IDAR1 = 0x3C;            /* " */
    CAN1IDAR2 = 0x40;            /* " */
    CAN1IDAR3 = 0x00;            /* " */
    CAN1IDMR0 = 0x00;            /* acceptance mask for "T1" */
    CAN1IDMR1 = 0x00;            /* " */
    CAN1IDMR2 = 0x3F;            /* " */
    CAN1IDMR3 = 0xFF;            /* " */
    CAN1IDAR4 = 0x50;            /* set acceptance identifier "P1" */
    CAN1IDAR5 = 0x3C;            /* " */
    CAN1IDAR6 = 0x40;            /* " */
    CAN1IDAR7 = 0x00;            /* " */
    CAN1IDMR4 = 0x00;            /* acceptance mask for "P1" */
    CAN1IDMR5 = 0x00;            /* " */
    CAN1IDMR6 = 0x3F;            /* " */
    CAN1IDMR7 = 0xFF;            /* " */
    CAN1IDAC  = 0x00;            /* select two 32-bit filter mode */
    CAN1CTL0  = 0x25;            /* stop clock on wait mode, enable wake-up */
    CAN1CTL0  &= ~INITRQ;        /* exit initialization mode *
}
```

The MSCAN application can be very complicated; the initial configuration of the MSCAN may need to be changed to satisfy the changing environment. The following example illustrates the change of acceptance filters.

Example 13.4

▼

Write an instruction sequence to change the configuration of the CAN1 module so that it will accept messages with a standard identifier starting with letter 'T' or 'P'.

Solution: The procedure described in Section 13.10.8 should be followed to reconfigure the CAN1 module. The instruction sequence that performs the reconfiguration is as follows:

```
ct1        brset   CAN1TFLG,$07,tb_empty   ; wait until all transmit buffers are empty
           bra     ct1
tb_empty   bset    CAN1CTL0,SLPRQ          ; request to enter sleep mode
ct2        brclr   CAN1CTL1,SLPAK,ct2      ; wait until sleep mode is entered
           bset    CAN1CTL0,INITRQ         ; request to enter initialization mode
ct3        brclr   CAN1CTL1,INITAK,ct3     ; wait until initialization mode is entered
           movb    #$10,CAN1IDAC           ; select four 16-bit acceptance mode
           movb    #$54,CAN1IDAR0          ; set up filter for letter 'T' for standard
           movb    #0,CAN1IDAR1            ; identifier
           movb    #$50,CAN1IDAR2          ; set up filter for letter 'P' for standard
           clr     CAN1IDAR3              ; identifier
           clr     CAN1IDMR0             ; acceptance mask for 'T'
           movb    #$F7,CAN1IDMR1         ; check IDE bit only (0 for standard identifier)
           clr     CAN1IDMR2             ; acceptance mask for 'P'
           movb    #$F7,CAN1IDMR3         ; check IDE bit only (must be 0)
           movb    #$54,CAN1IDAR4          ; set up filter for letter 'T' for standard
           movb    #0,CAN1IDAR5            ; identifier
           movb    #$50,CAN1IDAR6          ; set up filter for letter 'P' for standard
           clr     CAN1IDAR7             ; identifier
           clr     CAN1IDMR4             ; acceptance mask for 'T'
           movb    #$F7,CAN1IDMR5         ; check IDE bit only (0 for standard identifier)
           clr     CAN1IDMR6             ; acceptance mask for 'P'
           movb    #$F7,CAN1IDMR7         ; check IDE bit only (must be 0)
           bclr    CAN1CTL0,SLPRQ         ; exit sleep mode
           bclr    CAN1CTL0,INITRQ        ; exit initialization mode
```

▲

13.13 Data Transmission and Reception in MSCAN

When the application has only a small amount of data to be transmitted every now and then, it would be easy to find empty transmit buffers to send the message. There is no need to use interrupt in this case. However, if the application has many messages to be sent out at one time, then it may not be able to find an empty transmit buffer to hold the message to be transmitted at some point. In this situation, the designer should consider using the interrupt-driven approach to control the transmission of messages to allow the CPU to perform other operations.

For a received message, the application would take action based on the identifier of the message. If the message arrival time is unpredictable, then it is more convenient to use the interrupt-driven approach. If the arrival times of messages are more predictable, then the designer can consider using either the polling or the interrupt-driven approach.

13.13.1 MSCAN Data Transmission Programming

To prepare data for transmission, one has two options:

1. Identify an empty transmit buffer and place data directly on the available transmit buffer.
2. Store data in a buffer and copy the buffer to the first available transmit buffer that is identified.

The first method can be used in the polling method. The second method involves more overhead. However, it is suitable for both the polling and interrupt-driven approaches.

Example 13.5

▼

Write a function to send out the message stored at a buffer pointed to by index register X from the CAN1 module. The function should find an available buffer to hold the message to be sent out.

Solution:

```
tbuf       equ     0                          ; tbuf offset from top of stack
snd2can1   pshy
           pshb
           leas    -1,sp                      ; allocate one byte for local variable
sloop1     brset   CAN1TFLG,$01,tb0           ; is transmit buffer 0 empty?
           brset   CAN1TFLG,$02,tb1           ; is transmit buffer 1 empty?
           brset   CAN1TFLG,$04,tb2           ; is transmit buffer 2 empty?
           bra     sloop1                     ; if necessary wait until one buffer is empty
tb0        movb    #0,tbuf,sp                 ; mark transmit buffer 0 empty
           bra     tcopy
tb1        movb    #1,tbuf,sp                 ; mark transmit buffer 1 empty
           bra     tcopy
tb2        movb    #2,tbuf,sp                 ; mark transmit buffer 2 empty
tcopy      movb    CAN1TFLG,CAN1TBSEL         ; make the empty transmit buffer accessible
           ldy     #CAN1TIDR0                 ; set y to point to the start of the transmit buffer
           ldab    #7                         ; always copy seven words (place word count in B)
cploop     movw    2,x+,2,y+
           dbne    b,cploop
           ldab    tbuf,sp
           cmpb    #0
           beq     istb0
           cmpb    #1
           beq     istb1
           movb    #$04,CAN1TFLG              ; mark buffer 2 ready for transmission
           bra     dcopy
istb0      movb    #$01,CAN1TFLG              ; mark buffer 0 ready for transmission
           bra     dcopy
```

```
istb1       movb     #$02,CAN1TFLG              ; mark buffer 1 ready for transmission
dcopy       leas     1,sp                       ; deallocate local variables
            pulb
            puly
            rts
```

The C language version of the function is as follows:

```
void snd2can1(char *ptr)
{
    int    tb,i,*pt1,*pt2;
    pt1  = (int *)ptr;                /* convert to integer pointer */
    while(1) {                        /* find an empty transmit buffer */
        if(CAN1TFLG & 0x01){
            tb = 0;
            break;
        }
        if(CAN1TFLG & 0x02){
            tb = 1;
            break;
        }
        if(CAN1TFLG & 0x04){
            tb = 2;
            break;
        }
    }
    CAN1TBSEL = CAN1TFLG;             /* make empty transmit buffer accessible */
    pt2 = (int *)&CAN1TIDR0;          /* pt2 points to the IDR0 of TXFG */
    for (i = 0; i < 7; i++)           /* copy the whole transmit buffer */
        *pt2++ = *pt1++;
    if (tb == 0)
        CAN1TFLG = 0x01;             /* mark buffer 0 ready for transmission */
    else if (tb == 1)
        CAN1TFLG = 0x02;             /* mark buffer 1 ready for transmission */
    else
        CAN1TFLG = 0x04;             /* mark buffer 2 ready for transmission */
}
```

▲

Example 13.6

▼

Write a program to send out the string "3.5 V" from CAN1 and use "V1" as its identifier. Set transmit buffer priority to the highest.

Solution: The assembly program that performs the desired function is as follows:

```
            org      $1000
tbuf0       ds       16
            org      $2000
```

```
        movb    #$56,tbuf0        ; identifier V1
        movb    #$3C,tbuf0+1      ;      "
        movb    #$40,tbuf0+2      ;      "
        movb    #0,tbuf0+3        ;      "
        movb    #$33,tbuf0+4      ; data "3"
        movb    #$2E,tbuf0+5      ; data "."
        movb    #$35,tbuf0+6      ; data "5"
        movb    #$20,tbuf0+7      ; data " "
        movb    #$56,tbuf0+8      ; data "V"
        movb    #5,tbuf0+12       ; data length (=5)
        movb    #0,tbuf0+13       ; set transmit buffer priority to highest
        ldx     #tbuf0
        jsr     snd2can1         ; call subroutine to perform the actual transmission
        swi
```

The C language version of the program is as follows:

```
#include "c:\egnu091\include\hcs12.h"
void snd2can1(char *ptr);
int main(void)
{
    char tbuf0[16];
    tbuf0[0] = 'V';          /* identifier V1 */
    tbuf0[1] = 0x3C;         /* " */
    tbuf0[2] = 0x40;         /* " */
    tbuf0[3] = 0;            /* " */
    tbuf0[4] = '3';          /* letter 3 */
    tbuf0[5] = '.';          /* character. */
    tbuf0[6] = '5';          /* letter 5 */
    tbuf0[7] = 0x20;         /* space */
    tbuf0[8] = 'V';          /* letter V */
    tbuf0[12]= 5;            /* data length */
    tbuf0[13]= 0;            /* tbuf0 priority */
    snd2can1(tbuf0);
    return 0;
}
```

13.13.2 MSCAN Data Reception Programming

Incoming CAN bus messages are sent by other nodes and their arrival times are unpredictable. Therefore, it would be more efficient to use the interrupt-driven approach to handle data reception on the CAN bus. After setting up acceptance filters, one can enable the reception interrupt and wait for the incoming messages to arrive while performing other operations.

Example 13.7

Assuming that the CAN1 receiver has been set up to accept messages with extended identifiers "T1" and "V1". Filter 0 is set up to accept the identifier started with "T1", whereas filter 1 is set up to accept the identifier started with "V1". Write the interrupt handling routine for the RXF interrupt. If the acceptance is caused by filter 0, the RXF service routine would display the following message on a 20 × 2 LCD:

Temperature is
      xxx.y°F

If the acceptance of the message is caused by filter 1, the RXF interrupt service routine would display the following message:

Voltage is
      x.y V

Solution: The service routine will check the RXF bit of the CAN1RFLG register to make sure that the interrupt is caused by the RXF flag. If not, it returns. This routine also ignores the remote transfer request. The assembly language version of the service routine is as follows:

```
; ****************************************************************
; The following function is the service routine for the RXF interrupt. It only services hit 0 and
; hit 1. It ignores the RTR.
; ****************************************************************
;
can1Rx_ISR  brset   CAN1RFLG,RXF,RxfSet     ; is the RXF flag set to 1?
            rti                             ; if not, do nothing
RxfSet      ldab    CAN1IDAC                ; check into IDHIT bits
            andb    #$07                    ; mask out higher five bits
            beq     hit0                    ; filter 0 hit?
            cmpb    #1                      ; filter 1 hit?
            beq     hit1
            rti                             ; not hit 0 nor hit 1, do thing
hit0        ldab    CAN1RDLR                ; get the byte count of incoming data
            beq     rxfDone                 ; byte count 0, return
            ldx     #t1_line1               ; output "Temperature is"
            jsr     puts2lcd                ;       "
            ldx     #CAN1RDSR0
outLoop1    ldaa    1,x+                    ; output one byte at a time
            jsr     putc2lcd                ;       "
            dbne    b,outLoop1              ;       "
            rti
hit1        ldab    CAN1RDLR                ; get the byte count of incoming data
            beq     rxfDone                 ; byte count 0, return
            ldx     #v1_line1               ; output "Voltage is"
            jsr     puts2lcd                ;       "
            ldx     #CAN1RDSR0              ; x points to data segment register 0
outLoop2    ldaa    1,x+
            jsr     putc2lcd
            dbne    b,outLoop2
rxfDone     rti
t1_line1    fcc     "Temperature is"
            dc.b    0
v1_line1    fcc     "Voltage is"
            dc.b    0
```

The C language version of the service routine is as follows:

```c
#include        "c:\egnu091\include\hcs12.h"
#include        "c:\egnu091\include\vectors12.h"
#include        "c:\egnu091\include\delay.c"
#include        "c:\egnu091\include\lcd_util_dragon12.c"
#define         INTERRUPT __attribute__((interrupt))
void            INTERRUPT RxISR(void);
void            openCan1(void);
char            *t1Msg = "Temperature is";
char            *v1Msg = "Voltage is";
int main (void)
{
    UserMSCAN1Rx = (unsigned short) &RxISR;
    openCan1();
    openlcd();
    CAN1RIER = 0x01;                        /* enable CAN1 RXF interrupt only */
    asm("cli");
    while(1);                               /* wait for RXF interrupt */
    return 0;
}
/********************************************************/
/* The following function handles the RXF interrupt. It checks if the RXF flag */
/* is set. If not, return. It also ignores the RTR request.                    */
/********************************************************/
void    INTERRUPT RxISR (void)
{
    char tmp,i,*ptr;
    if (!(CAN1RFLG & RXF))                  /* interrupt not caused by RXF, return */
         return;
    CAN1RFLG = rxf;                         /* vlrst RXF flag */
    tmp = CAN1IDAC & 0x07;                  /* extract filter hit info */
    if (tmp == 0) {                         /* filter 0 hit */
        if (CAN1RDLR==0)                    /* data length 0, do nothing */
             return;
        cmd2lcd(0x80);                      /* set LCD cursor to first row */
        puts2lcd(t1Msg);                    /* output "Temperature is" on LCD */
        cmd2lcd(0xC0);                      /* set LCD cursor to second row */
        ptr = (char *)&CAN1RDSR0;           /* ptr points to the first data byte */
        for (i = 0; i < CAN1RDLR; i++)
             putc2lcd(*ptr++);              /* output temperature value on the LCD 2nd row */
    }
    else if (tmp == 1) {                    /* filter 1 hit */
        if(CAN1RDLR == 0)                   /* data length 0, do nothing */
             return;
        cmd2lcd(0x80);                      /* set LCD cursor to first row */
        puts2lcd(v1Msg);                    /* output "Voltage is" on the 1st row of LCD */
        cmd2lcd(0xC0);                      /* set LCD cursor to second row */
        ptr = (char *)&CAN1RDSR0;           /* ptr points to the first data byte */
```

```
            for(i = 0; i < CAN1RDLR; i++)
                  putc2lcd(*ptr++);                   /* output voltage value on the 2nd row of LCD */
      }
      else asm("nop");                                /* other hit, do nothing */
}
```

▲

This service routine can be modified to handle the remote transmission request (when the RTR bit of the received message is set to 1). The message to be transmitted in response to this request should be stored in a buffer and kept up-to-date. The modification to this service routine is straightforward, and hence will be left as an exercise.

13.13.3 Putting It All Together

The program pieces needed for CAN data communications are already in place. This section will use one example to combine them.

Example 13.8

▼

Write a C program to be run in a CAN environment using the same timing parameters as computed in Example 13.1. Each CAN node measures the voltage (in the range from 0 to 5 V) and sends it out from the CAN bus and also receives the voltage message sent over the CAN bus by other nodes. Configure the CAN1 module to receive messages having an extended identifier starting with "V1". The transmission and reception are to proceed as follows:

- The program measures the voltage connected at the AN7 pin every 200 ms and sends out the value with identifier "V1". The voltage is represented in the format of "x.y V". After sending out a message, the program outputs the following message on the first row of the LCD:

    Sent: x.y V

- Message reception is interrupt-driven. Whenever a new message is accepted, the program outputs the following message on the second row of the LCD:

    Received x.y V

Solution: The C program that performs the desired operation is as follows:

```
#include "c:\egnu091\include\hcs12.h"
#include "c:\egnu091\include\delay.c"
#include "c:\egnu091\include\vectors12.h"
#include "c:\egnu091\include\lcd_util_dragon12.c"
#define INTERRUPT __attribute__((interrupt))
char *msg1 = "Sent: ";
char *msg2 = "Received: ";

void INTERRUPT RxISR(void);
void openAD0(void);
void OpenCan1(void);
void MakeBuf(char *pt1, char *pt2);
void snd2can1(char *ptr);
```

```
int main(void)
{
    char buffer[6];                                    /* to hold measured voltage */
    char buf[16];                                      /* transmit data buffer */
    int temp;
    UserMSCAN1Rx = (unsigned short)&RxISR;             /* set up interrupt vector */
    openlcd();                                         /* configure LCD kit */
    OpenCan1();                                        /* configure CAN1 module */
    buffer[1] = '.';                                   /* decimal point */
    buffer[3] = 0x20;                                  /* space character */
    buffer[4] = 'V';                                   /* volt character */
    buffer[5] = 0;                                     /* null character */
    openAD0();                                         /* configure AD0 module */
    CAN1RIER = 0x01;                                   /* enable RXF interrupt only */
    asm("cli");                                        /* enable interrupt globally */

    while(1) {
        ATD0CTL5 = 0x87;                               /* convert AN7, result right justified */
        while(!(ATD0STAT0 & SCF));                     /* wait for conversion to complete */
        buffer[0] = 0x30 + (ATD0DR0*10)/2046;          /* integral digit of voltage */
        temp = (ATD0DR0 * 10)%2046;                    /* find the remainder */
        buffer[2] = 0x30 + (temp * 10)/2046;           /* compute the fractional digit */
        MakeBuf(&buf[0],&buffer[0]);                   /* format data for transmission */
        snd2can1(&buf[0]);                             /* send out voltage on CAN bus */
        cmd2lcd(0x80);                                 /* set LCD cursor to first row */
        puts2lcd(msg1);                                /* output the message "sent: x.y V" */
        puts2lcd(&buffer[0]);                          /* " " */
        delayby100ms(2);                               /* wait for messages to arrive for .2 seconds */
    }
    return 0;
}
/****************************************************************/
/* The following function formats a buffer into the structure of a CAN transmit buffer so */
/* that it can be copied into any empty transmit buffer for transmission.                */
/****************************************************************/
void MakeBuf(char *pt1, char *pt2)
{
    char i;
    *pt1 = 'V';                      /* set "V1" as the receive identifier
*/
    *(pt1+1) = 0x3C;                 /* " */
    *(pt1+2) = 0x40;                 /* " */
    *(pt1+3) = 0;                    /* " */
    for(i = 4; i < 9; i++)           /* copy voltage data */
        *(pt1 + i) = *(pt2 + i - 4);
    *(pt1 + 12) = 5;                 /* set data to length 5 */
}
```

```
/****************************************************/
/* The following function handles the RXF interrupt. It checks if the RXF   */
/* is set. If not, return. It also ignores the RTR request.                 */
/****************************************************/
void INTERRUPT RxISR (void)
{
    char tmp,i,*ptr;
    if (!(CAN1RFLG & RXF))                        /* interrupt not caused by RXF, return */
        return;
    CAN1FLG = RXF;                                /* clear RXF flag */
    tmp = CAN1IDAC & 0x07;                        /* extract filter hit info */
    if (tmp == 0){                               /* filter 0 hit */
        if (CAN1RDLR==0)                         /* if data length is 0, do nothing */
            return;
        cmd2lcd(0xC0);                           /* set LCD cursor to second row */
        puts2lcd(msg2);                          /* output "received: " */
        ptr = (char *)&CAN1RDSR0;                /* ptr points to the first data byte */
        for (i = 0; i < CAN1RDLR; i++)
            putc2lcd(*ptr++);                    /* output "x.y V" */
        return;
    }
    else return;                                 /* other hit, do nothing */
}
void openAD0 (void)
{
    int i;
    ATD0CTL2 = 0xE0;
    delayby10us(2);
    ATD0CTL3 = 0x0A;                             /* perform one conversion */
    ATD0CTL4 = 0x25;                             /* 4 cycles sample time, prescaler set to 12 */
}
void OpenCan1(void)
{
    CAN1CTL1   |= CANE;                          /* enable CAN, required after reset */
    CAN1CTL0   |= INITRQ;                        /* request to enter initialization mode */
    while(!(CAN1CTL1&INITAK));                   /* wait until initialization mode is entered */
    CAN1CTL1   = 0x84;                           /* enable CAN1, select oscillator as MSCAN clock source,
                                                    enable wake-up filter */
    CAN1BTR0   = 0x41;                           /* set SJW to 2, set prescaler to 2 */
    CAN1BTR1   = 0x18;                           /* set phase_seg2 to 2 tQ, phase_seg1 to 2 tQ,
                                                    prop_seg to 7 tQ */
    CAN1IDAR0  = 0x56;                           /* set acceptance identifier "V1" */
    CAN1IDAR1  = 0x3C;                           /* " " */
    CAN1IDAR2  = 0x40;                           /* " " */
    CAN1IDAR3  = 0x00;                           /* " " */
    CAN1IDMR0  = 0x00;                           /* acceptance mask for "V1" */
    CAN1IDMR1  = 0x00;                           /* " " */
    CAN1IDMR2  = 0x3F;                           /* " " */
    CAN1IDMR3  = 0xFF;                           /* " " */
```

```
        CAN1IDAR4  = 0x56;                /* set acceptance identifier to 'V1'/
        CAN1IDAR5  = 0x3C;                /* duplicate the */
        CAN1IDAR6  = 0x40;                /* acceptance identifier V1 */
        CAN1IDAR7  = 0x00;                /* " */
        CAN1IDMR4  = 0x00;                /* acceptance mask for NULL */
        CAN1IDMR5  = 0x00;                /* " */
        CAN1IDMR6  = 0x3F;                /* " */
        CAN1IDMR7  = 0xFF;                /* " */
        CAN1IDAC   = 0x00;                /* select two 32-bit filter mode */
        CAN1CTL0   = 0x25;                /* stop clock on wait mode, enable wake-up */
        CAN1CTL0 & = ~INITRQ;            /* exit initialization mode */
}
void snd2can1(char *ptr)
{
        int tb,i,*pt1,*pt2;
        pt1 = (int *)ptr;                /* convert to integer pointer */
        while(1) {
            if(CAN1TFLG & TXE0){
                tb = 0;
                break;
            }
            if(CAN1TFLG & TXE1){
                tb = 1;
                break;
            }
            if(CAN1TFLG & TXE2){
                tb = 2;
                break;
            }
        }
        CAN1TBSEL = CAN1TFLG;            /* make empty transmit buffer accessible */
        pt2 = (int *)&CAN1TIDR0;         /* pt2 points to the IDR0 of TXFG */
        for (i = 0; i < 7; i++)          /* copy the whole transmit buffer */
            *pt2++ = *pt1++;
        if (tb == 0)
            CAN1TFLG = TXE0;            /* mark buffer 0 ready for transmission */
        else if (tb == 1)
            CAN1TFLG = TXE1;            /* mark buffer 1 ready for transmission */
        else
            CAN1TFLG = TXE2;            /* mark buffer 2 ready for transmission */
}
```

This example can be extended to construct a distributed sensor system. There are applications that need to monitor temperature, pressure, humidity, and any other parameter in a small area that can be spanned by a CAN network. The purpose might be to monitor these environmental parameters and take appropriate actions in case abnormal values appear. In this type of application, each CAN node sends out the environmental parameters periodically to the CAN bus. A controlling CAN node keeps track of values sent from each node and does some processing. If the controlling CAN node discovers any abnormal situation, it may remind the operator to take appropriate action. To make sure each CAN node works normally, the controlling CAN node may set a counter for each distributed node. If any CAN node did not send in environmental parameters within the time-out period, it will remind the operator to take some corrective actions.

To make the CAN network more fault-tolerant, the designer may consider adding the second CAN bus to each node. The multiple CAN modules of many HCS12 devices serve this purpose well.

13.14 Summary

The CAN bus specification was initially proposed as a data communication protocol for automotive applications. However, it can also fulfill the data communication needs of a wide range of applications, from high-speed networks to low-cost multiplexed wiring. The CAN protocol has gone through several revisions. The latest revision is 2.0A/B. The CAN 2.0A uses standard identifiers, whereas the CAN 2.0B specification accepts extended identifiers. The HCS12 MSCAN module supports the CAN 2.0A and 2.0B specifications.

The CAN protocol supports four types of messages: data frame, remote frame, error frame, and overload frame. Users need to transfer only data frames and remote frames. The other two types of frames are only used by the CAN controller to control data transmission and reception. Data frames are used to carry normal data; remote frames are used to request other nodes to send messages with the same identifier as in the remote frame.

The CAN protocol allows all nodes on the bus to transmit simultaneously. When there are multiple transmitters on the bus, they arbitrate, and the CAN node transmitting a message with the highest priority wins. The simultaneous transmission of multiple nodes will not cause any damage to the CAN bus. CAN data frames are acknowledged in-frame; that is, a receiving node sets a bit only in the acknowledge field of the incoming frame. There is no need to send a separate acknowledge frame.

The CAN bus has two states: *dominant* and *recessive*. The dominant state represents logic 0, and the recessive state represents logic 1 for most CAN implementations. When one node drives the dominant voltage to the bus while other nodes drive the recessive level to the bus, the resultant CAN bus state will be the dominant state.

Synchronization is a critical issue in the CAN bus. Each bit time is divided into four segments: sync_seg, prop_seg, phase_seg1, and phase_seg2. The sync_seg segment signals the start of a bit time. The sample point is between phase_seg1 and phase_seg2. At the beginning of each frame, every node performs a hard synchronization to align its sync_seg segment of its current bit time to the recessive to dominant edge of the transmitted start-of-frame. Resynchronization is performed during the remainder of the frame, whenever a change of bit value from recessive to dominant occurs outside of the expected sync_seg segment.

A CAN module (also called CAN controller) requires a CAN bus transceiver, such as the Philips PCA82C250 or the Microchip MCP2551, to interface with the CAN bus. Most CAN controllers allow more than 100 nodes to connect to the CAN bus. The CAN trunk cable could be a shielded cable, unshielded twisted pair, or simply a pair of insulated wires. It is recommended that shielded cable be used for high-speed transfer when there is a radio frequency interference problem. Up to a 1-Mbps data rate is achievable over a distance of 40 meters.

The CAN module of the HCS12 MCU has six major modes of operations:

- Initialization mode
- Disable mode

- Normal operation mode
- Listen-only mode
- Loopback mode
- Sleep mode
- Stop mode

All configuration operations can only be performed in the initialization mode. Users must make sure that the initialization mode is actually entered before performing the configuration. Most CAN parameters are configured immediately after reset. However, applications may need to change the configuration during the normal operation. Under this situation, the user will need to put the CAN into sleep mode before performing reconfiguration. The first step in CAN configuration is to initialize the timing parameters. The procedure for doing this is described in Section 13.12. All data transfer and reception with other nodes are performed in the normal operation mode.

The CAN module has three transmit buffers and five receive buffer. The first 13 bytes of a transmit buffer and a receive buffer are identical. In addition, the transmit buffer also has a transmit buffer priority register and two transmit timestamp registers. One of the three transmit buffers is mapped to the foreground transmit buffer space and is made accessible to the user. Five receive buffers are organized as a FIFO and only the head of the FIFO is mapped to the receive foreground space and becomes accessible to the user.

Each transmit and receive buffer has eight data registers for holding data, four identifier registers for identification and arbitration purposes, and one length-count register for indicating the number of data segment registers that contain valid data. Data transmission involves copying data into identifier registers, and the data segment registers, and the data length register, setting the priority of the transmit buffer, and clearing the transmit buffer empty bit to mark the buffer ready for transmission. Data reception in the CAN bus is often interrupt-driven due to the unpredictability of message arrival times.

13.15 Exercises

E13.1 Calculate the bit segments for the following system constraints assuming that the Microchip MCP2551 transceiver is used:

Bit rate = 400 Kbps

Bus length = 50 m

Bus propagation delay = 5×10^{-9} s/m

Microchip MCP2551 transceiver and receiver propagation delay = 150 ns at 85°C

Oscillator frequency = bus frequency = 24 MHz

E13.2 Calculate the bit segments for the following system constraints assuming that the Microchip MCP2551 transceiver is used:

Bit rate = 200 Kbps

Bus length = 100 m

Bus propagation delay = 5×10^{-9} s/m

Microchip MCP2551 transceiver and receiver propagation delay = 150 ns at 85°C

Oscillator frequency = bus frequency = 24 MHz

E13.3 Calculate the bit segments for the following system constraints assuming that the Microchip MCP2551 transceiver is used:

> Bit rate = 1 Mbps
>
> Bus length = 20 m
>
> Bus propagation delay = 5×10^{-9} s/m
>
> MCP2551 transceiver and receiver propagation delay = 150 ns at 85°C
>
> Oscillator frequency = bus frequency = 32 MHz

E13.4 Calculate the bit segments for the following system constraints assuming that the Microchip MCP2551 transceiver is used:

> Bit rate = 800 Kbps
>
> Bus length = 20 m
>
> Bus propagation delay = 5×10^{-9} s/m
>
> MCP2551 transceiver and receiver propagation delay = 150 ns at 85°C
>
> Oscillator frequency = 16 MHz (PLL is used to convert to 24-MHz bus frequency)

E13.5 Write a subroutine to configure the CAN1 module of the HCS12DP256 with the bit timing parameters computed in E13.2. Enable receive interrupt but disable transmit interrupt. Configure the CAN so that it accepts only messages with the extended or standard identifiers starting with 'H', 'P', or 'T'. Use an 8-bit filter to control the acceptance.

E13.6 Write a C function to perform the same setting as was done in E13.4.

E13.7 Provide three different ways of accepting messages that have the letter 'K' as the first letter of the identifier.

E13.8 Write an assembly program to send out the string "Monday" using "WD" as the first two characters of the extended identifier. The remaining bits of the identifier are cleared to 0.

E13.9 Write a C program to send out the string "too high" using "ans" as the first three letters of the extended identifier. The remaining bits of the extended identifier are cleared to 0.

13.16 Lab Exercises and Assignments

L13.1 Practice data transfer over the CAN bus using the following procedures:

Step 1

Use a pair of insulated wires about 20 m long. Connect the CAN_H pins of two demo boards (e.g., the SSE256 or Dragon12 demo boards) with one wire and connect the CAN_L pins of both demo boards with another wire. Connect both ends of these two wires together with 120-Ω resistors (SSE256 has terminating resistors on board). The circuit connection is shown in Figure L13.1.

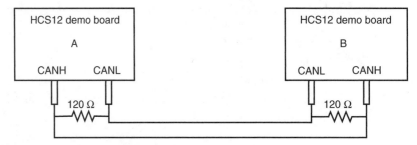

Figure L13.1 ■ CAN circuit connection for L13.1

Step 2

Write a program to be downloaded onto one of the demo board (called board A) that performs A/D conversion 10 times every second. Use the on-board potentiometer to generate a voltage to be converted. Send out the A/D conversion result over the CAN bus every 100 ms. Use the letter 'W' as the identifier of the data frame. (Use the timing parameters computed in E13.4 for the SSE256 demo board.)

Step 3

Write a program to be downloaded onto another demo board (called board B). This program will send out the number of data frames received so far over the CAN bus. This program will use the letter 'R' as the identifier. After the number reaches 99, the program will reset the number to 0 and start over again. Each digit of the decimal number is encoded in ASCII code.

Step 4

Board A will display the number received over the CAN bus on eight LEDS in BCD format.

Step 5

Board B will display the received A/D result in an LCD display.

L13.2 This assignment requires three students to wire three HCS12 demo boards together using the CAN bus. Assign each of these three demo boards a number 0, 1, and 2, respectively. Write a program to be run on each demo board, and perform the following procedure:

Step 1

Use two potentiometers (one is already on board) to emulate the temperature sensor and barometric sensor outputs (0 to 5 V). Connect the demo boards as shown in Figure L13.2. The temperature sensor represents the temperature range from −40° to 125°C. The barometric sensor represents the pressure range from 948 to 1083.8 mbar.

Figure L13.2 ■ CAN circuit connection for L13.2

Step 2

Each demo board performs temperature and barometric pressure measurements once per second and stores the result in separate buffers formatted to have the same structure as a transmit buffer. When storing the measurement data, use Ti (for node i) as the identifier for temperature value and use Pi (for node i) as the identifier for pressure value.

Step 3

After performing temperature and pressure measurements and storing data in buffers, each demo board uses a random number generator to generate a number k in the range of 0 to 2 and waits for that amount of time (seconds) to send out a remote transmission request to request node k to send in temperature or pressure data and display them on the LCD. The received data should be displayed as follows:

> received Tk = xxx.y

or

> received Pk = xxxx.y

The demo board that transmits the data should display the following message:

> send Tk = xxx.y

or

> send Pk = xxxx.y

14

Internal Memory
Configuration and
External Expansion

14.1 Objectives

After completing this chapter, you should be able to:

- Understand the overall HCS12 memory system organization and mapping
- Erase and program the on-chip flash memory
- Control the operation, programming, and protection of the on-chip flash memory
- Erase the on-chip EEPROM
- Understand the external memory expansion issue
- Make memory space assignment
- Design address decoders and the memory control circuit
- Perform timing analysis for the memory system

14.2 Introduction

Each of the HCS12 members has a certain amount of on-chip SRAM, EEPROM, and flash memory. I/O-related registers are implemented in SRAM. These memories share the same memory space and are assigned to a certain default memory space after reset. To meet the requirements of different applications, these memories can be remapped to other locations. This can be achieved by programming the appropriate initialization register associated with each of these memories. Both the EEPROM and flash memory must be erased before they can be programmed correctly. The erasure and programming methods for both the flash memory and EEPROM are similar.

Most of the HCS12-based embedded systems operate in the single-chip mode and do not need external memory. A HCS12 member may have from a minimum of 32 kB of internal memory up to a maximum of 512 kB of on-chip memory. The memory requirements of a wide range of applications can be satisfied by different members of the HCS12 microcontroller family. In case an application needs more than 512 kB of memory, the designer has the option to add external memory to the HCS12 device. After adding external memory, an HCS-based embedded system may have up to 1 MB of system memory. The designer should also consider using a 32-bit microcontroller, because a 32-bit microcontroller may prove to be more cost-effective and has several additional advantages:

- *No paging overhead.* The HCS12 has a 16-bit program counter and hence can only address up to 64 kB of program memory directly. Accessing more than 64 kB of memory would require a certain type of paging or banking scheme, which has been proved to be very inefficient. A 32-bit microcontroller does not have such a problem.

- *Much larger memory space.* A 32-bit microcontroller supports 4 GB of memory space, which can satisfy the memory requirements of many demanding applications.

- *Much higher performance for the same clock frequency.* A 32-bit microcontroller can process 32-bit data in one operation, whereas the HCS12 can handle only 16 bits of data at a time.

Many microcontrollers multiplex the address and data signals on the same set of signal pins in order to economize on the use of signal pins. The HCS12 is no exception. In the HCS12, the address and data signals are time-multiplexed on Port A and Port B pins. However, external memory devices require address signals to be stable throughout the whole access cycle. This can be achieved by using latches to make a copy of the address signals so that the address signals remain valid for the whole access cycle.

When adding external memory to the HCS12, one must make sure that the timing requirements are satisfied. Each bus cycle takes exactly one E clock cycle to complete. However, the bus cycle can be stretched to support slower memory devices. A bus cycle can be stretched by one to three E clock cycles.

14.3 Internal Resource Remapping

The internal register block, SRAM, EEPROM, and flash memory have default locations within the 64-kB standard address space but may be reassigned to other locations during program execution by setting bits in mapping registers INITRG, INITRM, and INITEE. These registers can be written only once in normal operation modes. It is advisable to explicitly establish these resource locations during the initialization phase of program execution, even if default values are chosen, in order to protect the registers from inadvertent modification later. Writes to the mapping registers require one cycle to take effect. To ensure that there are no unintended operations, a write to one of these registers should be followed by a NOP instruction.

If conflict occurs when mapping resources, the register block will take precedence over other resources; RAM or EEPROM addresses occupied by the register block will not be available for storage. When active, BDM ROM takes precedence over other resources, although a conflict between BDM ROM and register space is not possible. Table 14.1 shows mapping precedence. Only one module will be selected at a time. In the case of more than one module sharing a space, only the module with the highest precedence will be selected. Mapping more than one module to the same location won't damage the system. However, it is not wise to map two or more modules to the same space because a significant amount of memory could become unusable.

In expanded modes, all address space not used by internal resources is by default external memory.

| 0 | REG14 | REG13 | REG12 | REG11 | 0 | 0 | 0 | base + $11 |

value after reset:

| 0 | 0 | 0 | 0 | 0 | 0 | 0 | 0 |

REG14~REG11: internal register map position

These four bits along with bit 7 specify the upper five bits of the 16-bit register address. These four bits can be written only once in normal modes and can be written many times in special modes. There is no restriction on the reading of this register.

Table 14.1 ■ Mapping precedence

14.3.1 Register Block Mapping

After reset, the register block (1 kB or 2 kB) can be remapped to any 2-kB space within the first 32 kB of the system address space. Mapping of internal registers is controlled by four bits in the INITRG register. The contents of this register are shown in Figure 14.1. Only bit 6 to bit 3 are implemented. The INITRG register can be written into once in normal mode but many times in special modes.

7	6	5	4	3	2	1	0	
0	REG14	REG13	REG12	REG11	0	0	0	base + $11

value after reset:

| 0 | 0 | 0 | 0 | 0 | 0 | 0 | 0 |

REG14~REG11: internal register map position

These four bits along with bit 7 specify the upper five bits of the 16-bit register address. These four bits can be written only once in normal modes and can be written many times in special modes. There is no restriction on the reading of this register.

Figure 14.1 ■ Contents of the INITRG register

Example 14.1

Write an instruction sequence to remap the register block to $4000.

Solution: The upper five bits of the INITRG register should be 01000. The following instruction sequence will achieve the desired assignment:

```
movb    #$40,INITRG
nop                      ; wait for the remap to take effect
```

14.3.2 SRAM Mapping

SRAM can be remapped to any 2-kB boundary within the 64-kB memory space. The remapping is controlled by the INITRM register. The contents of the INITRM are shown in Figure 14.2. Bits RAM15 to RAM11 determine the upper five bits of the base address for the system's internal SRAM array. The INITRM register can be written into once in normal modes but many times in special modes.

RAM15~RAM11: internal RAM map position
    These bits determine the upper five bits of the base address for
    the system's internal RAM array.
RAMHAL: RAM high-align
    0 = aligns the RAM to the lowest address ($0000) of the
        mappable space
    1 = aligns the RAM to the highest address ($FFFF) of the
        mapping space.

Figure 14.2 ■ RAM initialization register (INITRM)

14.3.3 EEPROM Mapping

The EEPROM block can be remapped to any 2-kB boundary. The remapping of EEPROM is controlled by the INITEE register. The contents of this register are shown in Figure 14.3. The EEPROM is activated by bit 0 of this register. Bits EE15 to EE11 determine the upper five bits of the base address for the system's internal EEPROM array. The INITEE register can be written any time.

14.3.4 Miscellaneous Memory Mapping Control

A few registers are involved in the mapping control of more than one memory module. The functions of these registers are described in this section.

MISCELLANEOUS SYSTEM CONTROL REGISTER (MISC)

This register enables/disables the on-chip ROM (including flash memory and EEPROM) and allows one to stretch the length of the external bus cycle. The contents of this register are

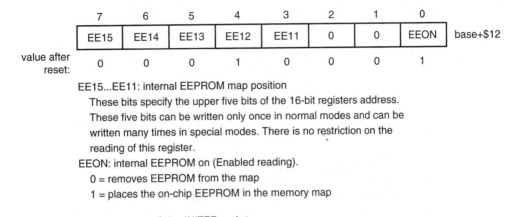

	7	6	5	4	3	2	1	0	
	EE15	EE14	EE13	EE12	EE11	0	0	EEON	base+$12
value after reset:	0	0	0	1	0	0	0	1	

EE15...EE11: internal EEPROM map position

    These bits specify the upper five bits of the 16-bit registers address.
These five bits can be written only once in normal modes and can be
written many times in special modes. There is no restriction on the
reading of this register.

EEON: internal EEPROM on (Enabled reading).

    0 = removes EEPROM from the map

    1 = places the on-chip EEPROM in the memory map

Figure 14.3 ■ Contents of the INITEE register

shown in Figure 14.4. The value of the ROMONE pin is latched to the ROMON bit of the MISC
register on the rising edge of the $\overline{\text{RESET}}$ signal. If the ROMONE signal is high when it is
latched, then the on-chip flash memory is enabled and the reset startup routine will be executed
from the on-chip flash memory rather than the external ROM. After reset, this pin is used as
the $\overline{\text{ECS}}$ signal. This register can only be written once in normal and emulation modes but can
be written any time in special modes.

	7	6	5	4	3	2	1	0
reset value:	0	0	0	0	EXSTR1	EXSTR0	ROMHM	ROMON
expanded or emulation	0	0	0	0	1	1	0	$-^1$
peripheral or single chip	0	0	0	0	1	1	0	1
special test	0	0	0	0	1	1	0	0

EXSTR1...EXSTR0: external access stretch bits 1 and 0

    00 = no stretch to external bus cycle

    01 = stretch the external bus cycle by one E cycle

    10 = stretch the external bus cycle by two E cycles

    11 = stretch the external bus cycle by three E cycles

ROMHM: flash EEPROM or ROM only in second half of memory map

    0 = The fixed page (s) of flash EEPROM or ROM in the lower half of
the lower half of the memory map can be accessed.

    1 = Disable direct access to the flash EEPROM or ROM in the lower
half of the memory map. These physical locations of flash memory
can still be accessed through the program page window.

ROMON: enable flash memory or ROM

    0 = disable the flash memory or ROM in the memory map

    1 = enable the flash memory or ROM in the memory map

Note: 1. The reset state of this bit is determined at the chip integration level.

Figure 14.4 ■ Contents of the MISC register

MEMORY SIZE REGISTER ZERO (MEMSIZ0)

This register is read only. It reflects the state of the on-chip I/O register, EEPROM, and SRAM memory space configuration switches at the core boundary, which are configured at system integration. This register allows read visibility to the state of these switches. The contents of this register are shown in Figure 14.5.

	7	6	5	4	3	2	1	0	
	REG_SW0	0	EEP_SW1	EEP_SW0	0	RAM_SW2	RAM_SW1	RAM_SW0	base+$1C
reset:	--	--	--	--	--	--	--	--	

REG_SW0: allocated system register space
  0 = allocated system register space size is 1 kB.
  1 = allocated system register space size is 2 kB.
EEP_SW1~EEP_SW0: allocated system EEPROM memory space
  00 = 0 kB
  01 = 2 kB
  10 = 4 kB
  11 = 8 kB
RAM_SW2...RAM_SW0: allocated system RAM memory space
  The allocated system RAM space size is as given in Table 14.2.

Figure 14.5 ■ Memory size register zero (MEMSIZ0)

	7	6	5	4	3	2	1	0	
	RAM 15	RAM 14	RAM 13	RAM 12	RAM 11	0	0	RAMHAL	base+
value after reset:	0	0	0	0	1	0	0	0	

RAM15~RAM11: internal RAM map position
  These bits determine the upper five bits of the base address for
  the system's internal RAM array.
RAMHAL: RAM high-align
  0 = aligns the RAM to the lowest address ($0000) of the
      mappable space
  1 = aligns the RAM to the highest address ($FFFF) of the
      mapping space.

Table 14.2 ■ Allocated RAM memory space

MEMORY SIZE REGISTER ONE (MEMSIZ1)

This register is read only and reflects the state of the flash or ROM physical memory and paging switches at the core boundary, which are configured at system integration. This register allows read visibility to the state of these switches. The contents of this register are shown in Figure 14.6. Table 14.3 shows that the HCS12 supports 1 MB of memory space.

PROGRAM PAGE INDEX REGISTER (PPAGE)

This register allows for integrating up to 1 MB of flash or ROM into the system by using the six page index bits to page 16-kB blocks into the program page window located from $8000 to $BFFF. CALL and RTC instructions have special access to read and write this register without using the address bus. The contents of this register are shown in Figure 14.7.

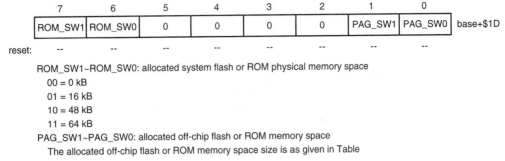

7	6	5	4	3	2	1	0	
ROM_SW1	ROM_SW0	0	0	0	0	PAG_SW1	PAG_SW0	base+$1D

reset: -- -- -- -- -- -- -- --

ROM_SW1~ROM_SW0: allocated system flash or ROM physical memory space
  00 = 0 kB
  01 = 16 kB
  10 = 48 kB
  11 = 64 kB
PAG_SW1~PAG_SW0: allocated off-chip flash or ROM memory space
  The allocated off-chip flash or ROM memory space size is as given in Table 14.3.

Figure 14.6 ■ Memory size register one (MEMSIZ1)

PAG_SW1~PAG_SW0	Off-chip Space	On-chip Space
00	876 kB	128 kB
01	768 kB	256 kB
10	512 kB	512 kB
11	0 kB	1 MB

Table 14.3 ■ Allocated off-chip memory options

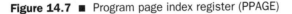

7	6	5	4	3	2	1	0	
0	0	PIX5	PIX4	PIX3	PIX2	PIX1	PIX0	base+$30

reset: -- -- -- -- -- -- -- --

PIX5...PIX0: Program page index bits 5~0

Figure 14.7 ■ Program page index register (PPAGE)

14.4 Expanded Memory Mapping

The HCS12 uses the PPAGE register to determine which of the 64 possible 16-kB pages is active in the program window. Only one page at a time can occupy the window, and the value in the associated register must be changed to access a different page of memory.

Since all memory locations above 64 kB must be accessed via this page window, the direct concatenation of the page number and the page offset does not result in linear addresses. This also causes a problem in encoding machine code using S records because S records use linear addresses in addressing machine code. Freescale uses the following scheme to solve this problem:

1. Use $00 to $3F as the page numbers for the 64 16-kB pages.

2. Use higher page numbers to select on-chip flash memory. For example, for the HCS12 members with 256-kB on-chip flash memory, assign $30~$3F as the page numbers of the 16 on-chip 16-kB pages. Use $00 to $2F as page numbers for external memory. The mapping from PPAGE to S records is illustrated in Table 14.4.

3. Use linear addresses to address external memory and also use the linear method to address machine codes for S records.

4. Compute the page number and page address within the $8000~$BFFF page window using the following equations:

PageNum = SRecAddr / PPAGEWinSize   (14.1)

PageWinAddr = (SRecAddr % PPAGEWinSize) + PPAGEWinStart   (14.2)

where

PageNum is from $00 to $3F.

SRecAddr is the linear address used in S records.

PPAGEWinSize = 16 kB (or 16384)

PageWinAddr is a number between $8000 and $BFFF.

PPAGEWinStart = $8000

5. Compute SRecAddr from PageNum and PageWinAddr as follows:

SRecAddr = PageNum × PPAGEWinSize + PageWinAddr − PPAGEWinStart   (14.3)

6. When addressing external memory, the highest 6 address bits appear on XADDR19~XADDR14 and the lowest 14 address bits appear on A13~A0.

PPAGE value	S-Record Address Range	A15~A0	Page + A13~A0	Memory Type
$00~$2F	$00000~$BFFFF	1	$00000~$BFFFF	Off-chip memory
$30	$C0000~$C3FFF	$0000~$3FFF	$C0000~$C3FFF	On-chip flash
$31	$C4000~$C7FFF	$4000~$7FFF	$C4000~$C7FFF	On-chip flash
$32	$C8000~$CBFFF	$8000~$BFFF	$C8000~$CBFFF	On-chip flash
$33	$CC000~$CFFFF	$C000~$FFFF	$CC000~$CFFFF	On-chip flash
$34	$D0000~$D3FFF	$0000~$3FFF	$D0000~$D3FFF	On-chip flash
$35	$D4000~$D7FFF	$4000~$7FFF	$D4000~$D7FFF	On-chip flash
$36	$D8000~$DBFFF	$8000~$BFFF	$D8000~$DBFFF	On-chip flash
$37	$DC000~$DFFFF	$C000~$FFFF	$DC000~$DFFFF	On-chip flash
$38	$E0000~$E3FFF	$0000~$3FFF	$E0000~$E3FFF	On-chip flash
$39	$E4000~$E7FFF	$4000~$7FFF	$E4000~$E7FFF	On-chip flash
$3A	$E8000~$EBFFF	$8000~$BFFF	$E8000~$EBFFF	On-chip flash
$3B	$EC000~$EFFFF	$C000~$FFFF	$EC000~$EFFFF	On-chip flash
$3C	$F0000~$F3FFF	$0000~$3FFF	$F0000~$F3FFF	On-chip flash
$3D	$F4000~$F7FFF	$4000~$7FFF	$F4000~$F7FFF	On-chip flash
$3E	$F8000~$FBFFF	$8000~$BFFF	$F8000~$FBFFF	On-chip flash
$3F	$FC000~$FFFFF	$C000~$FFFF	$FC000~$FFFFF	On-chip flash

Note: 1. Repetition of $0000~$3FFF, $4000~$3FFF, $8000~$BFFF, and $C000~$FFFF for 12 times.

Table 14.4 ■ MC9S12DP256 PPAGE to S-Record address mapping

Example 14.2

What are the PageNum and PageWinAddr for the SRecAddr of $E1003?

Solution: Apply Equations 14.1 and 14.2 as follows:

PageNum   = $E1003 / $4000 = $38

PageWinAddr = ($E1003 % $4000) + $8000 = $1003 + $8000 = $9003

Example 14.3

▼

What is the corresponding SRecAddr for the pair of (PageNum, PageWinAddr) equal to ($20, $A003)?

Solution: Apply Equation 14.3 as follows:

SRecAddr = $20 × $4000 (16K) + $A003 − $8000 = $82003

▲

14.5 On-Chip Flash Memory

All HCS12 members have a certain amount of on-chip flash memory. The amount of flash memory may be 32 kB, 64 kB, 128 kB, 256 kB, or 512 kB. A flash memory larger than 64 kB is divided into two or more 64-kB blocks. These 64-kB flash memory blocks can be read, programmed, or erased concurrently.

The algorithms for erasing and programming flash memory require very little user involvement. Programming and erasure are performed by sending commands to the command register. Interrupts may be requested when a command is completed or when the command buffer is empty. A two-stage command pipeline is implemented to accelerate the command execution. The flash memory includes a flexible protection scheme against accidental programming and erasure. The flash memory also implements security measures to prevent the application code from being pirated.

The discussion of flash memory will be based on the 256-kB flash memory of several HCS12 members.

14.5.1 Flash Memory Map

The memory map of the 256-kB flash memory is shown in Figure 14.8. The entire flash memory is divided into pages of 16 kB in size. One page is always accessible at $4000 to $7FFF and another fixed page is always accessible at $C000 to $FFFF. The memory space from $0000 to $3FFF has been occupied by registers, EEPROM, and SRAM and hence is not available for flash memory. As shown in Figure 14.8, the first 64 kB of the flash memory is referred to as block 0 and is assigned page numbers $3C to $3F. The pages with addresses from $4000 to $7FFF, $8000 to $BFFF, and $C000 to $FFFF are assigned the page numbers $3E, $3D, and $3F, respectively. Even though it is unnecessary, one can access the pages in address ranges of $4000 to $7FFF and $C000 to $FFFF via the page window.

14.5.2 Flash Memory Protection

Each flash block may be protected against accidental erasure or programming. Flash protection is controlled by a flash protection register (FPROT). For microcontrollers that have multiple flash blocks, there is a separate flash protection register for each flash block. These flash protection registers share a common address, with the active register selected by the bank select bits of the flash configuration register (FCNFG). During the HCS12 reset sequence (execution of reset startup routine), the flash protection registers for each flash block are loaded from the programmed bytes within a flash block. For example, for the MC9S12DP256, the locations $FF0A, $FF0B, $FF0C, and $FF0D store the protection information of blocks three, two, one, and zero, respectively.

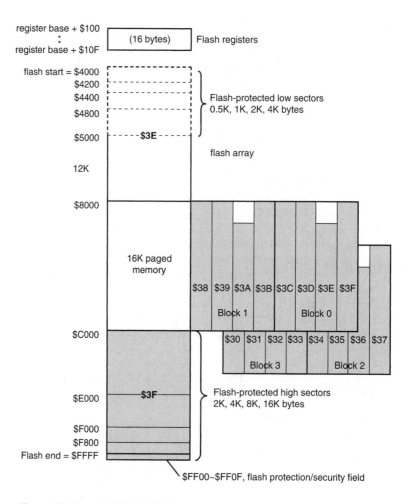

Figure 14.8 ■ MC9S12DP256 memory map

The contents of each FPROT register determine whether the entire block or just subsections are protected from being accidentally erased or programmed. Each flash block can be entirely protected or can have one or two separate protected areas. One of these areas, known as the *lower protected block,* starts at a point 32 kB (the page with the second-largest page number in the same block) below the maximum flash block address and is extendable toward higher addresses. The other area, known as the *upper protected block,* ends at the top of the flash block and is extended toward lower addresses. The upper and lower protected blocks do not meet up. In general, the upper protected area of flash block 0 is used to hold bootloader code since it contains the reset and interrupt vectors. The lower protected area of block 0 and the protected areas of the other flash blocks can be used for critical parameters that would not change when the application program is updated.

The contents of the FPROT register are shown in Figure 14.9. The FPROT register is readable in normal and special modes. The FPOPEN, FPHDIS, and FPLDIS bits in the FPROT register can only be written into the protected state (i.e., 0). The FPLS1~FPLS0 bits can be written

7	6	5	4	3	2	1	0	
FPOPEN	NV6	FPHDIS	FPHS1	FPHS0	FPLDIS	FPLS1	FPLS0	base+$104

reset: F   F   F   F   F   F   F   F

FPOPEN: opens the flash for program or erase
  0 = the whole flash block is protected. In this case, bits 5 to 0 have no effect.
  1 = the flash sectors not protected are enabled for program or erase.
NV6: not volatile flag bit.
  This bit is available for nonvolatile flag usage.
FPHDIS: flash protection higher address range disable
  0 = higher address range protection enabled
  1 = higher address range protection disabled
FPHS1~FPHS0: flash protection higher address size
  00 = 2 kB
  01 = 4 kB
  10 = 8 kB
  11 = 16 kB
FPLDIS: flash protection lower address range disable
  0 = lower address range protection enabled
  1 = lower address range protection disabled
FPLS1~FPLS0: flash protection lower address size
  00 = 512 bytes
  01 = 1 kB
  10 = 2 kB
  11 = 4 kB

Figure 14.9 ■ FPROT register

anytime until the FPLDIS bit is cleared. The FPHS1~FPHS0 bits can be written anytime until the FPHDIS bit is cleared.

To change the flash protection that will be loaded on reset, the upper sector of the flash memory must be unprotected, then the flash protect/security byte located as described in Table 14.5 must be written into.

Address	Size (bytes)	Description
$FF00~$FF07	8	Backdoor comparison keys
$FF08~$FF09	5	Reserved
$FF0A	1	Block 3 flash protection byte
$FF0B	1	Block 2 flash protection byte
$FF0C	1	Block 1 flash protection byte
$FF0D	1	Block 0 flash protection byte
$FF0E	1	Reserved
$FF0F	1	Flash options/security byte

Table 14.5 ■ Flash protection/options field

Trying to program or erase any of the protected areas will result in a protection violation error and the bit PVIOL of the FSTAT register will be set to 1. A bulk erase is possible only if the protection for that block is fully disabled, i.e., FOPEN = 1, FPHDIS = 1, and FPLDIS = 1. The contents are loaded from one byte of four locations, $FF0A~$FF0D. This is indicated by the letter F in Figure 14.9.

14.5.3 Flash Memory Related Registers

In addition to the FPROT registers, the following registers are related to the configuration and operation control of the flash memory.

FLASH CLOCK DIVIDER REGISTER (FLCKDIV)

This register is not banked and is used to control the timed events in programming and erasure algorithms. The contents of this register are shown in Figure 14.10. Bit 7 is read only; other bits are readable and writable.

	7	6	5	4	3	2	1	0	
	FDIVLD	PRDIV8	FDIV5	FDIV4	FDIV3	FDIV2	FDIV1	FDIV0	base+$100
reset:	0	0	0	0	0	0	0	0	

FDIVLD: clock divider loaded
  0 = register has not been written.
  1 = register has been written to since the last reset.
PRDIV8: enable prescaler by 8
  0 = the input oscillator clock is directly fed into the FCLKDIV divider.
  1 = enables a divide-by-8 prescaler, to divide the flash module input oscillator
      clock before feeding into the CLKDIV divider.
FDIV[5:0]: clock divider bits
  The combination of PRDIV8 and FDIV[5:0] effectively divides the flash module
  input oscillator clock down to a frequency of 150 kHz~200 kHz. The maximum
  divide ratio is 512.

Figure 14.10 ■ Flash clock divider register (FCLKDIV)

In order to guarantee that the on-chip flash memory and EEPROM can be programmed and erased properly, the clock signal that controls the timing of programming and erasure should be configured properly. The only clock configuration to be done is to set the prescaler of the clock by programming this register. The startup routine of the HCS12 needs to check the FDIVLD bit to make sure that the prescaler has been configured. Depending on the oscillator frequency, the designer may choose to insert or bypass a divide-by-8 prescaler to the clock signal for controlling the programming and erasure of the flash memory and EEPROM. The procedure for configuring the clock prescaler is discussed in Section 14.5.6.

FLASH SECURITY REGISTER (FSEC)

This register holds all the bits associated with the device security. The contents of this register are shown in Figure 14.11. All the bits of this register are readable but not writable. This register is loaded from the flash protection/options field byte at $FF0F during the reset sequence, indicated by F in Figure 14.11. This register is not banked. The SEC1 and SEC0 bits

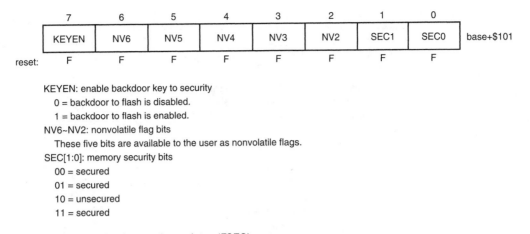

7	6	5	4	3	2	1	0	
KEYEN	NV6	NV5	NV4	NV3	NV2	SEC1	SEC0	base+$101

reset: F   F   F   F   F   F   F   F

KEYEN: enable backdoor key to security
  0 = backdoor to flash is disabled.
  1 = backdoor to flash is enabled.
NV6~NV2: nonvolatile flag bits
  These five bits are available to the user as nonvolatile flags.
SEC[1:0]: memory security bits
  00 = secured
  01 = secured
  10 = unsecured
  11 = secured

Figure 14.11 ■ Flash security register (FSEC)

define the security state of the device. If the flash is unsecured using the Backdoor Key Access, the SEC bits are forced to 10. The HCS12 security function is discussed in Section 14.5.4.

FLASH TEST MODE REGISTER (FTSTMOD)

This register is not banked and is used primarily to control the nonvolatile memory test modes. The contents of this register are shown in Figure 14.12. The WRALL bit is writable only in special modes. The purpose of this bit is to launch a command on all blocks in parallel. This can be useful for bulk erase and blank check operations.

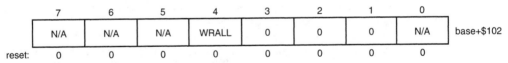

7	6	5	4	3	2	1	0	
N/A	N/A	N/A	WRALL	0	0	0	N/A	base+$102

reset: 0   0   0   0   0   0   0   0

WRALL: write to all register banks
  0 = write only to the bank selected via BKSEL
  1 = write to all register banks

Figure 14.12 ■ Flash test mode register (FTSTMOD)

FLASH CONFIGURATION REGISTER (FCNFG)

This register enables the flash interrupts, gates the security backdoor writes, and selects the register bank to be operated on. The contents of this register are shown in Figure 14.13. This register is not banked.

For a HCS12 device with 256 kB of flash memory, there are four 64-kB blocks. Each block has a set of control registers. Only one set of them is available at a time. The BKSEL1 and BKSEL0 bits select one bank to be accessed.

	7	6	5	4	3	2	1	0	
	CBEIE	CCIE	KEYACC	0	0	0	BKSEL1	BKSEL0	base+$103
reset:	0	0	0	0	0	0	0	0	

CBEIE: command buffer empty interrupt enable

   0 = command buffer empty interrupts disabled.

   1 = an interrupt will be requested whenever the CBEIF flag is set.

CCIE: command completion interrupt enable

   0 = command complete interrupts disabled.

   1 = an interrupt will be requested whenever the CCIF flag is set.

KEYACC: enable security key writing

   0 = flash writes are interpreted as the start of a program or erase sequence.

   1 = writes to flash array are interpreted as keys to open the backdoor. Reads
       of the flash array return invalid data.

BKSEL[1:0]: register bank select

   00 = flash 0

   01 = flash 1

   10 = flash 2

   11 = flash 3

Figure 14.13 ■ Flash configuration register (FCNFG)

FLASH STATUS REGISTER (FSTAT)

The programming and erase of flash memory is controlled by a finite state machine. The FSTAT register defines the flash state machine command status and flash array access, protection, and bank verify status. This register is banked. The contents of this register are shown in Figure 14.14.

The bits CBEIF, PVIOL, and ACCERR are readable and writable, whereas bits CCIF and BLANK are readable but not writable.

The CBEIF flag indicates that the address, data, and command buffers are empty so that a new command sequence can be started. The CBEIF flag is cleared by writing a 1 to it. Writing a 0 to the CBEIF flag has no effect on CBEIF but sets ACCERR, which can be used to abort a command sequence.

The CCIF flag indicates that there are no more commands pending. The CCIF flag is cleared when CBEIF is cleared and sets automatically upon completion of all active and pending commands. Writes to CCIF flag have no effect.

The PVIOL flag indicates an attempt was made to program or erase an address in a protected flash memory area. The PVIOL flag is cleared by writing a 1 to it. Writing a 0 has no effect.

The ACCERR flag indicates an illegal access to the selected flash array. This can be either a violation of the command sequence, issuing an illegal command, or the execution of a CPU STOP instruction while a command is executing. This flag is cleared by writing a 1 to it. Writing a 0 to the ACCERR flag has no effect.

FLASH COMMAND REGISTER (FCMD)

This register defines the flash commands. This register is banked; its contents are shown in Figure 14.15. To perform these commands, one must make sure that the following prerequisites are satisfied:

   1. The FCLKDIV register must be configured correctly.

7	6	5	4	3	2	1	0	
CBEIF	CCIF	PVOL	ACCERR	0	BLANK	0	0	base+$105

reset: 1   1   0   0   0   0   0   0

CBEIF: command buffer empty interrupt flag
  0 = command buffers are full.
  1 = command buffers are ready to accept a new command.
CCIF: command completion interrupt flag
  0 = command in progress.
  1 = all commands are completed.
PVOL: protection violation
  0 = no protection violation has occurred.
  1 = a protection violation has occurred.
ACCERR: flash access error
  0 = no failure.
  1 = access error has occurred.
BLANK: array has been verified as erased.
  0 = if an erase and verify command has been requested, and the CCIF flag is set,
    then a zero in BLANK indicates that the block is not erased.
  1 = flash block verifies as erased.

Figure 14.14 ■ Flash status register (FSTAT)

7	6	5	4	3	2	1	0	
0	CMDB6	CMDB5	0	0	CMDB2	0	CMDB0	base+$106

reset: 1   1   0   0   0   0   0   0

CMDB6, CMDB5, CMDB2, and CMDB0: command bits
  Valid commands include the following:
  $05 = erase and verify
  $20 = program a word (two bytes)
  $40 = sector erase
  $41 = bulk erase

Figure 14.15 ■ Flash command register (FCMD)

2. The sector to be erased must not be protected.

3. The flash word to be programmed must be erased and not be protected.

4. The first flash address must be word-aligned (A0 = 0).

5. If the flash address is in the range of $8000 to $BFFF, then the PPAGE register must be written to select the desired page.

6. If the MCU has multiple flash blocks, the ACCERR and PVIOL flags in all blocks must be cleared.

7. The BSEL bits in the FCNFG register must be written to select the desired block to be programmed or erased.

Flash 16-Bit Address Register (FADDRHI and FADDRLO)

Bit 15 of the address register is tied to 0. In normal modes, the FADDR register reads zero and is not writable. The FADDRHI and FADDRLO registers can be written in special modes by writing to address Base+$108 and Base+$109 in the register space.

Flash 16-bit Data Buffer and Register (FDATA)

In normal modes, all FDATA bits read zero and are not writable. In special modes, all FDATA bits are readable and writable when writing to an address within the flash address range.

14.5.4 Securing the Flash Memory

The flash security feature is designed to prevent unauthorized access to the nonvolatile memory. The memory contents are secured by programming the security bits within the flash options/security byte at the address $FF0F. On devices that have a memory page window, the flash options/security byte is also available at the address $BF0F by setting the value of the PPAGE register to $3F. The contents of this byte are copied into the flash security register (FSEC) during the reset sequence.

The flash sector $FE00 to $FFFF must be erased before the flash options/security byte is programmed. The flash is programmed by aligned word only, so the address $FF0E must be written as the word address to be programmed to program the flash options/security byte. The flash options/security byte can be erased or programmed only when this sector is not protected. Secured operation takes effect on the next reset after programming the security bits of the FSEC register to a secure value. The effects that the secured operation have on the HCS12 microcontroller are listed in Table 14.6.

Operation Mode	Effects
Normal single-chip mode	1. Background debug module operation is completely disabled. 2. Flash and EEPROM commands PROG, bulk erase, sector erase, erase and verify, and sector modify remain enabled.
Special single-chip mode	1. BDM firmware commands are disabled. 2. BDM hardware commands are restricted to register space. 3. Flash and EEPROM commands are limited to bulk erase only.
Expanded modes	1. BDM operation is completely disabled. 2. External access to internal flash and EEPROM is disabled. 3. Internal visibility (IVIS) and CPU pipe (IPIPE) information is disabled. 4. Flash and EEPROM commands cannot be executed from external memory in normal expanded mode.

Table 14.6 ■ Effects of secured operations on the HCS12 operation modes

14.5.5 Unsecuring the Microcontroller

When in secure mode, the HCS12 microcontroller can be unsecured by one of the following methods.

BACKDOOR KEY ACCESS

In normal modes (single chip and expanded), security can be temporarily disabled by means of the backdoor key access method. This method requires the following actions to be taken:

- Program the backdoor key at $FF00~$FF07 to a valid value.
- Set the KEYEN1 and KEYEN0 bits of the flash option/security byte to 10 (enabled).
- In single-chip mode, design applications to have the capability to write to the backdoor key locations.

The backdoor key access method is useful because it allows debugging of a secured microcontroller without having to erase the flash memory. This is useful for failure analysis. The backdoor key is not allowed to have the value of $0000 or $FFFF.

The backdoor key access sequence includes

1. Setting the KEYACC bit in the flash configuration register (FCNFG)
2. Writing the first 16-bit word of the backdoor key to $FF00
3. Writing the second 16-bit word of the backdoor key to $FF02
4. Writing the third 16-bit word of the backdoor key to $FF04
5. Writing the fourth 16-bit word of the backdoor key to $FF06
6. Clearing the KEYACC bit in the flash configuration register FCNFG

Since the flash memory cannot be read when the KEYACC bit is set, the code for backdoor key access sequence must be executed from SRAM.

If all four 16-bit words match the flash contents at $FF00 to $FF07, the microcontroller will be unsecured and the security bits SEC1 and SEC0 of the FSEC register will be forced to unsecured state 10. The contents of the flash options/security byte are not changed by this procedure, and so the HCS12 will revert to the secure state after the next reset, unless further action is taken. If any one of the four 16-bit words does not match the flash contents at $FF00 to $FF07, the HCS12 microcontroller will remain secured.

REPROGRAMMING THE SECURITY BITS

In normal single-chip mode, security can also be disabled by means of erasing and reprogramming the security bits within flash options/security byte to the unsecured value. Since the erase operation will erase the entire sector from $FE00 to $FFFF, the backdoor key and the interrupt vectors will also be erased. Therefore, this method is not recommended for normal single-chip mode. The application software can erase and program the flash options/security byte only if the flash sector containing the flash options/security byte is not protected. Thus, flash protection is a useful means of preventing this method. The microcontroller will enter the unsecured state after the next reset following the programming of the security bits to the unsecured value. One of the following conditions must be satisfied in order for this method to work:

- The application software previously programmed into the MCU has been designed to have the capability to erase and program the flash options/security byte, and the flash section containing the flash options/security byte is not protected.
- Security is first disabled using the backdoor key method allowing the BDM to be used to issue commands to erase and program the flash options/security byte, and the flash section containing the flash options/security byte is not protected.

COMPLETE MEMORY ERASE

The microcontroller can be unsecured in special single-chip modes by erasing the entire EEPROM and flash contents. When a secure microcontroller is reset into special single-chip mode, the BDM firmware verifies whether the EEPROM and flash are erased. If any EEPROM or flash location is not erased, only BDM hardware commands are enabled. BDM hardware commands can then be used to write to the EEPROM and flash registers and also to bulk erase the EEPROM and all flash blocks. When next reset to the special single-chip mode, the BDM firmware will again verify whether all EEPROM and flash are erased, and this being the case, will enable all BDM commands, allowing the flash options/security byte to be programmed to the unsecured value. The security bits SEC1 and SEC0 in the FSEC register will indicate the unsecure state following the next reset.

14.5.6 Configuring the FCLKDIV Register

It is critical to configure the FCLKDIV register properly because this register controls the timing of the programming and erasure operations for the flash memory. It is necessary to divide the oscillator down to within the 150-kHz and 200-kHz range.

Let:

FCLK be the clock of the flash timing control block

T_{bus} be the period of the bus clock

INT(x) take the integer part of x (e.g., INT(3.5) = 3)

The FCLKDIV register bits PRIDIV8 and FDIV5~FDIV0 are to be set according to the algorithm illustrated in Figure 14.16.

Example 14.4

▼

Assume that f_{bus} = 24 MHz and f_{osc} = 16 MHz, respectively. Determine an appropriate value to be written into the FCLKDIV register to set the timing of programming and erasure properly for the flash memory and EEPROM.

Solution: Follow the logic flow illustrated in Figure 14.16:

1. T_{bus} = 41.7 ns (< 1 μs)
2. Initialize the PRDIV8 bit to 0.
3. f_{osc} = 16 MHz (> 12.8 MHz)
4. Set PRDIV8 to 1 and set PRDCLK to (f_{osc} / 8) = 2 MHz.
5. PRDCLK · (5 + 0.0417 μs) = 10.08, not an integer
6. Set FDIV[5:0] to INT(PRDCLK[MHz] · (5 + T_{bus}[μs])) = 10
7. FCLK = PRDCLK / (1 + FDIV[5:0]) = 2 MHz ÷ 11 = 181.81 kHz
8. 1/FCLK[MHz] + T_{bus}[μs] = 5.5 (> 5) and FCLK > 150 kHz, so stop.
9. Write the value of $4A into the FCLKDIV register.

▲

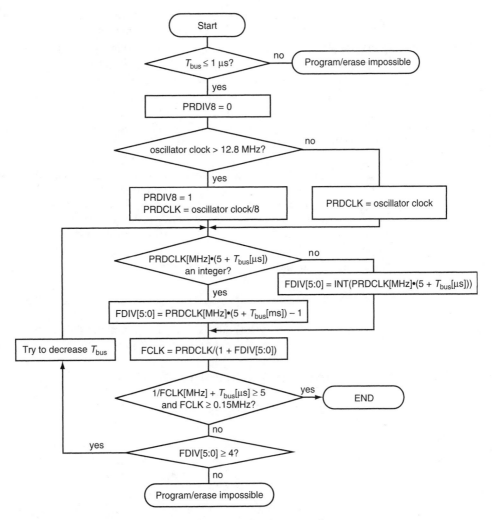

Figure 14.16 ■ PRDIV8 and FDIV bits determination procedure

14.5.7 Flash Memory Programming and Erasure Algorithms

The HCS12 uses a command state machine to supervise the write sequencing for the flash memory programming and erasing. Before starting a command sequence, it is necessary to verify that there is no pending access error or protection violation in any flash blocks. It is then required to set the PPAGE register and the FCNFG register. The procedure for this initial setup is as follows:

1. Verify that all ACCERR and PVIOL flags in the FSTAT register are cleared in all banks. This requires checking the contents of the FSTAT register for all combinations of the BKSEL bits in the FCNFG register.

2. Write to bits BKSEL in the FCNFG register to select the bank of registers corresponding to the flash block to be programmed or erased.

3. Write to the PPAGE register to select one of the pages to be programmed if programming is to be done in the $8000~$BFFF address range. There is no need to set PPAGE when programming in the $4000~$7FFF or $C000~$FFFF address ranges.

After this initialization procedure, the CBEIF flag should be tested to ensure that the address, data, and command buffers are empty. If the CBEIF flag is set to 1, the program/erase command write sequence can be started. The following three-step procedure should be strictly adhered to when programming or erasing the flash memory:

Step 1
Write the aligned data word to be programmed to the valid flash address space. The address and data will be stored in internal buffers. For programming, all address bits are valid. For flash memory erasure, the value of the data bytes is don't-care. For bulk erasure, the address can be anywhere in the available address space of the block to be erased. For sector erasure, the address bits 7 to 0 are ignored for the flash.

Step 2
Write the program or erase command to the command buffer.

Step 3
Clear the CBEIF flag by writing a 1 to it to launch the command. The clearing of the CBEIF flag indicates that the command was successfully launched. The CBEIF flag will be set again, indicating that the address, data, and command buffers are ready for a new command sequence to begin. A summary of the programming algorithm is shown in Figure 14.17.

If a HCS12 device has multiple flash memory blocks, one needs to clear the ACCERR and PVIOL flags in all blocks and also set the BKSEL bits in the FCNFG register properly to select the desired block for erasure or programming.

Example 14.5

▼

Write a function to clear the ACCERR and PVIOL flags in all four blocks in the HCS12 devices with 256 kB of on-chip flash memory.

Solution: The following function clears the ACCERR and PVIOL flags in all four blocks in the HCS12 devices with 256 kB of on-chip flash memory:

```
clearflags    bclr    FCNFG,$03              ; select bank 0
              movb    #ACCERR+PVIOL,FSTAT    ; clear the ACCERR and PVIOL flags
              bset    FCNFG,$01              ; select bank 1
              movb    #ACCERR+PVIOL,FSTAT    ; clear the ACCERR and PVIOL flags
              bset    FCNFG,$03              ; select bank 3
              movb    #ACCERR+PVIOL,FSTAT    ; clear the ACCERR and PVIOL flags
              bclr    FCNFG,$01              ; select bank 2
              movb    #ACCERR+PVIOL,FSTAT    ; clear the ACCERR and PVIOL flags
              rts
```

Figure 14.17 ■ HCS12 flash memory program algorithm

The C language version of the function is as follows:

```
void clearflags (void)
{
    FCNFG    &= ~0x03;            /* select bank 0 */
    FSTAT    = ACCERR+PVIOL;      /* clear the ACCERR and PVIOL flags */
    FCNFG    |= 0x01;             /* select bank 1 */
    FSTAT    = ACCERR+PVIOL;      /* clear the ACCERR and PVIOL flags */
    FCNFG    |= 0x03;             /* select bank 3 */
    FSTAT    = ACCERR+PVIOL;      /* clear the ACCERR and PVIOL flags */
    FCNFG    &= 0xFE;             /* select bank 2 */
    FSTAT    = ACCERR+PVIOL;      /* clear the ACCERR and PVIOL flags */
}
```

Example 14.6

Write a function that erases a sector of flash memory. Index register X contains a word-aligned address within the sector to be erased. Return a 1 in B if the command buffer is not empty.

Solution: The assembly routine that erases the sector pointed to by index register X is as follows:

```
EraseFSector  brclr  FSTAT,CBEIF,err2ES      ; erase prohibited if command buffer not empty
              std    0,X                      ; write any data to sector address
              movb   #SectorErase,FCMD        ; write sector erase command
              movb   #CBEIF,FSTAT             ; launch the erase command
              brclr  FSTAT,ACCERR+PVIOL,OK2ER ; no error flag?
err2ES        ldab   #1                       ; return error code of "1"
              rts
OK2ER         brclr  FSTAT,CCIF,OK2ER         ; wait until command is completed
              ldab   #0                       ; erase successfully, return code "0"
              rts
```

The C language version of the function is as follows:

```c
int EraseFSector(int *pt)
{
    if (!(FSTAT & CBEIF))
            return 1;                /* command buffer not empty, erase prohibited */
    *pt = 0x00;                      /* write any data to the sector */
    FCMD = SectorErase;              /* write sector erase command */
    FSTAT = CBEIF;                   /* launch the erase command */
    if (FSTAT & (ACCERR+PVIOL))
            return 1;                /* return error code 1 */
    while(!(FSTAT & CCIF));          /* wait until erase command is completed */
    return 0;                        /* return normal code */
}
```

Example 14.7

Write a function that performs a bulk erasure operation to the flash memory. The index register contains a word-aligned address of a word inside the sector to be bulk erased. Return a 1 in accumulator B if bulk erasure is not allowed.

Solution: The assembly language function that performs the bulk erasure operation to the flash memory is as follows:

```
BulkEraseF  brset  FPROT,FPOPEN+FPHDIS+FPLDIS,doBL ; Is bulk erasure allowed?
            ldab   #1                               ; return error code 1
            rts
doBL        brclr  FSTAT,CBEIF,errBL                ; bulk erase prohibited if CBEIF == 0
            std    0,X                              ; write any data to sector address
            movb   #BulkErase,FCMD                  ; write bulk erase command
            movb   #CBEIF,FSTAT                     ; launch the erase command
            brclr  FSTAT,ACCERR+PVIOL,OK2BL         ; no error flag?
```

```
errBL       ldab  #1                          ; return error code of "1"
            rts
OK2BL       brclr FSTAT,CCIF,OK2BL            ; wait until command is completed
            ldab  #0                          ; erase successfully, return code "0"
            rts
```

The C language version of the function is as follows:

```
int bulkeraseF(int *ptr)
{
    if(FPROT&(FPOPEN | FPHDIS | FPLDIS)! = 0xA4)
        return 1;                  /* can't bulk erase */
    if(!(FSTAT & CBEIF))
        return 1;                  /* command buffer isn't empty, bulk erase not allowed */
    else {
        *ptr = 0x00;               /* write anything to flash block location */
        FCMD = BulkErase;          /* write bulk erase command */
        FSTAT = CBEIF;             /* launch bulk erase command */
        if (FSTAT & (ACCERR | PVIOL))
            return 1;              /* error flag is set, command failed */
        while(!(FSTAT & CCIF));    /* wait until command completion */
        return 0;
    }
}
```

Example 14.8

Write a function that programs a block of words to the flash memory. The number of words to be programmed, the starting address of the flash memory to be programmed, and the starting address of data are passed to this function in B, X, and Y, respectively.

Solution: The assembly function that performs the flash programming is as follows:

```
feProgBlok  tstb                              ; check word count
            bne   doFLprog                    ; word count is valid
            rts                               ; return if word count is zero
doFLprog    pshb                              ; save the word count in stack
fepwait1    brclr FSTAT,CBEIF,fepwait1        ; wait until command buffer is empty
            movw  2,y+,2,x+                   ; write data word to flash address
            movb  #Program,FCMD               ; write program command
            movb  #CBEIF,FSTAT                ; launch the command
            brclr FSTAT,ACCERR+PVIOL,progK    ; is there any error?
            pulb
            ldab  #1                          ; return error code 1
            rts
progOK      dec   0,SP                        ; one word less to be programmed
            bne   fepwait1                    ; more words to be programmed?
            pulb
            clrb                              ; return error code 0
            rts
```

The C language version of the function is as follows:

```
int feProgBlok(char cnt, int *destptr, int *srcptr)
{
    if(cnt == 0)
        return 0;                        /* if word count is 0, do nothing */
    while(cnt){
        if(FSTAT & CBEIF){               /* if command buffer is not empty, do nothing */
            *destptr++ = *srcptr++;      /*write data word to flash location*/
            FCMD = Program;              /* write program command */
            FSTAT = CBEIF;               /* launch program command */
            if(FSTAT & (ACCERR+PVIOL))
                return 1;                /* program error? */
            cnt--;
        }
    }
    while(!(FSTAT&CCIF));                 /* wait for the last command to complete*/
    return 0;
}
```

Example 14.9

Write a function that performs the erase-and-verify command to the flash memory. The index register contains a word-aligned address to the flash block to be erased and verified.

Solution: The assembly routine that performs the erase-and-verify command is as follows:

```
feraseverify   brclr  FSTAT,CBEIF,cantE    ; command buffer not empty
               std    0,x                  ; write any data to flash sector address
               movb   #EraseVerify,FCMD    ; write the command
               movb   #CBEIF,FSTAT         ; launch the erase and verify command
               brclr  FSTAT,ACCERR+PVIOL,EVNoErr
               ldab   #1                   ; return error code 1
               rts
EVNoErr        brclr  FSTAT,CCIF,EVNoErr   ; wait until command is done
               brset  FSTAT,BLANK,EVFOK    ; successful erase and verify?
cantE          ldab   #1                   ; flash is not blank
               rts
EVFOK          clrb                        ; erase and verify OK
               rts
```

The C function that performs the erase-and-verify operation is as follows:

```
int feraseverify(int *ptr)
{
    if(!(FSTAT & CBEIF))
        return 1;              /* command buffer not empty, returns */
    *ptr = 0x00;               /* write data to flash sector address */
    FCMD = EraseVerify;        /* write erase verify command */
    FSTAT = CBEIF;             /* launch the command */
```

```
        if(FSTAT&(ACCERR | PVIOL))
                return 1;              /* errors have occurred */
        while(!(FSTAT & CCIF));        /* wait until command is completed */
        if(FSTAT & BLANK)
                return 0;              /* command completed successfully */
        else
                return 1;
}
```

▲

14.6 The On-Chip EEPROM Memory

An HCS12 device may have 1 kB, 2 kB, or 4 kB of on-chip EEPROM to store nonvolatile data or programs to be executed out of reset. The EEPROM is organized as an array of 2-byte words. For example, the 4-kB EEPROM is organized as 2048 rows of 2 bytes (1 word). The EEPROM block's erase sector size is two rows or two words.

The programming and erasure of the EEPROM are similar to those of the flash memory. Both the sector and bulk erasures are supported. An erased bit reads 1 and a programmed bit reads 0. The high voltage required for programming and erasure is generated internally by an on-chip charge pump.

The discussion of the HCS12 on-chip EEPROM will be based on the 4-kB EEPROM available in the HCS12DP256.

14.6.1 EEPROM Memory Map

The EEPROM can be mapped to any 2-kB boundary, as explained in Section 14.3.3. The memory map of the 4-kB EEPROM is illustrated in Figure 14.18. The whole 4 kB of EEPROM can be protected by setting the EPOPEN bit of the EPROT register. A 16-byte field is reserved inside the EEPROM module from address $_FF0 to $_FFF. The byte at $_FFD stores the EEPROM protection information. The EEPROM has hardware interlocks that protect data from accidental corruption. A protected sector is located at the higher address end of the EEPROM block, just below $_FFF. The protected sector can be sized from 64 bytes to 512 bytes.

14.6.2 EEPROM Associated Registers

The following registers control the operation of the EEPROM block.

EEPROM Clock Divider Register (ECLKDIV)

The contents of this register are shown in Figure 14.19. The EDIVLD bit is read only. This register controls the timing of the EEPROM programming and erasure operation. The contents of this register are identical to those of the FCLKDIV register. Bits 0–6 can be written once after reset.

EEPROM Configuration Register (ECNFG)

This register enables the EEPROM interrupts. The contents of this register are shown in Figure 14.20. This register is readable and writable any time.

EEPROM_start = $_000

(base+$110)   EEPROM control
(base+$11B)   registers (12 bytes)

3.5 kB

$_E00

$_E40

$_E80

$_EC0   EEPROM-protected high sectors
$_F00   64, 128, 192, 256, 320, 384,
                      448, 512 bytes
$_F40

$_F80

$_FC0
                      $_FF0~$_FFF, EEPROM
EEPROM_END = $_FFF   protection/reserved field

Figure 14.18 ■ HCS12 4KB EEPROM memory map

7	6	5	4	3	2	1	0	
EDIVLD	PRDIV8	EDIV5	EDIV4	EDIV3	EDIV2	EDIV1	EDIV0	base+$100

reset: 0   0   0   0   0   0   0   0

EDIVLD: clock divider loaded
    0 = register has not been written.
    1 = register has been written to since the last reset.
PRDIV8: enable prescaler by 8
    0 = the input oscillator clock is directly fed into the FCLKDIV divider.
    1 = enables a divide-by-8 prescaler, to divide the flash module input oscillator
       clock before feeding into the CLKDIV divider.
EDIV[5:0]: clock divider bits
    The combination of PRDIV8 and FDIV[5:0] effectively divides the flash module
    input oscillator clock down to a frequency of 150 kHz~200 kHz. The maximum
    divide ratio is 512.

Figure 14.19 ■ EEPROM clock divider register (ECLKDIV)

7	6	5	4	3	2	1	0	
CBEIE	CCIE	0	0	0	0	0	0	base+$113

reset: 0   0   0   0   0   0   0   0

CBEIE: command buffer empty interrupt enable
  0 = command buffer empty interrupts disabled.
  1 = an interrupt will be requested whenever the CBEIF flag is set.
CCIE: command completion interrupt enable
  0 = command complete interrupts disabled.
  1 = an interrupt will be requested whenever the CCIF flag is set.

Figure 14.20 ■ EEPROM configuration register (ECNFG)

EEPROM PROTECTION REGISTER (EPROT)

This register defines which EEPROM sectors are protected against program or erase. The contents of this register are shown in Figure 14.21. This register is loaded from EEPROM array at $_FFD in the reset sequence as indicated by the F letter. All bits are readable; bits NV[6:4] are not writable. The EPOPEN and EPDIS bits in the EPROT register can only be written to the protected state (i.e., 0). The EP[2:0] bits can be written any time until the EPDIS bit is cleared. If the EPOPEN bit is cleared, then the state of the EPDIS and EP[2:0] bits is irrelevant. Trying to alter any of the protected areas will result in a protection violation error and the PVIOL bit of the ESTAT register will be set. A bulk erasure is only possible when protection is fully disabled.

7	6	5	4	3	2	1	0	
EPOPEN	NV6	NV5	NV4	EPDIS	EP2	EP1	EP0	base+$114

reset: F   F   F   F   F   F   F   F

EPOPEN: opens the EEPROM for programming or erasure
  0 = the whole EEPROM array is protected. In this case, EPDIS and EP bits within
      the protection register have no effect.
  1 = the EEPROM sectors not protected are enabled for program or erase.
NV6~NV4: not volatile flag bit
  These bits are available for nonvolatile flag usage.
EPDIS: EEPROM protection address range disable
  0 = protection enabled
  1 = protection disabled
EP2~EP0: EEPROM protection address size
  000 = 64 bytes ($_FC0~$_FFF)
  001 = 128 bytes ($_F80~$_FFF)
  010 = 192 bytes ($_F40~$_FFF)
  011 = 256 bytes ($_F00~$_FFF)
  100 = 320 bytes ($_EC0~$_FFF)
  101 = 384 bytes ($_E80~$_FFF)
  110 = 448 bytes ($_E40~$_FFF)
  111 = 512 bytes ($_E00~$_FFF)

Figure 14.21 ■ EPROT register

EEPROM Status Register (ESTAT)

The contents of this register are shown in Figure 14.22. The bits CBEIF, PVIOL, and ACCERR are readable and writable; bits CCIF and BLANK are readable but not writable.

EEPROM Command Register (ECMD)

The ECMD register defines the EEPROM commands. The contents of this register are shown in Figure 14.23.

	7	6	5	4	3	2	1	0	
	CBEIF	CCIF	PVIOL	ACCERR	0	BLANK	0	0	base+$115
reset:	1	1	0	0	0	0	0	0	

CBEIF: command buffer empty interrupt flag
    0 = command buffers are full.
    1 = command buffers are ready to accept a new command.
CCIF: command completion interrupt flag
    0 = command in progress.
    1 = all commands are completed.
PVIOL: protection violation
    0 = no failure.
    1 = a protection violation has occurred.
ACCERR: flash access error
    0 = no failure.
    1 = access error has occurred.
BLANK: array has been verified as erased
    0 = if an erase and verify command has been requested, and the CCIF flag is set,
        then a zero in BLANK indicates that the block is not erased.
    1 = flash block verifies as erased.

Figure 14.22 ■ EEPROM status register (ESTAT)

	7	6	5	4	3	2	1	0	
	0	CMDB6	CMDB5	0	0	CMDB2	0	CMDB0	base+$116
reset:	1	1	0	0	0	0	0	0	

CMDB6, CMDB5, CMDB2, and CMDB0: command bits
    Valid commands include the following:
    $05 = erase and verify
    $20 = program a word (2 bytes)
    $40 = sector erase
    $41 = bulk erase
    $60 = sector modify

Figure 14.23 ■ EEPROM command buffer and register

14.6.3 EEPROM Protection

The EEPROM can be protected against accidental erasure and programming. EEPROM protection is controlled by the EPROT register. During the HCS12 reset sequence, the EPROT register is loaded from the EEPROM protection byte located within the EEPROM. The EPROT register is located within the smallest EEPROM protected area, so protecting EEPROM always protects the EEPROM protection byte, thus guaranteeing the reset state of EEPROM protection. The value of the EPROT register determines whether the entire EEPROM or just subsections are protected from being accidentally erased or programmed. Trying to program or erase any of the protected areas will result in a protection violation error and the PVIOL bit in the ESTAT register will be set. A bulk erasure is possible only if protection is fully disabled.

14.6.4 Configuring the ECLKDIV Register

The timing for EEPROM programming and erasure is controlled by the ECLKDIV register. The bit definitions and configuration are identical to those of the FCLKDIV register for the flash memory. One needs to write a value to divide the oscillator frequency down to a range from 150 kHz to 200 kHz.

14.6.5 Programming and Erasure of EEPROM

The algorithm for programming the EEPROM is illustrated in Figure 14.24. It is almost identical to the algorithm for programming the flash memory.

The execution of the first four EEPROM commands as listed at the end of Section 14.6.2 follows the same three-step sequence as their counterparts in flash memory:

1. Write the aligned data word to be programmed to the valid EEPROM address space. The address and data will be stored in internal buffers. For programming, all address bits are valid. For erasure, the value of the data bytes is don't-care. For bulk erasure, the address can be anywhere in the available address space of the block to be erased. For sector erasure, the address bits 1 and 0 are ignored for the EEPROM.

2. Write the program or erase command to the command register.

3. Clear the CBEIF flag to launch the command. The clearing of the CBEIF bit indicates that the command was successfully launched.

The completion of the command is indicated by the setting of the CCIF flag. An erroneous command write sequence will abort and set the appropriate flag. If set, the designer must clear the ACCERR or PVIOL flags before commencing another command write sequence. By writing a 0 to the CBEIF flag, the command sequence can be aborted after writing a word to the EEPROM address space or after writing a command to the ECMD register and before the command is launched. Writing a 0 to the CBEIF flag in this way will set the ACCERR flag.

The sector-modify command ($60) is a two-step command, which first erases a sector (two words) of EEPROM and then reprograms one of the words in that sector. The EEPROM sector that is erased by the command is the sector containing the address of the aligned data write that starts the valid command sequence. That same address is reprogrammed with the data that is written. By launching a sector-modify command and then pipelining a program command, it is possible to completely replace the contents of an EEPROM sector.

Figure 14.24 ■ HCS12 EEPROM program algorithm

Example 14.10

▼

Write a function that erases a sector (4 bytes) of EEPROM. Index register X contains a word-aligned EEPROM address within the sector to be erased.

Solution: The following assembly function erases the specified EEPROM sector by following the three-step procedure:

```
eraseEEsector
        movb    #ACCERR+PVIOL,ESTAT         ; clear error flags
        brclr   ESTAT,CBEIF,EERErr          ; command buffer not empty, return
        std     0,X                         ; write any data to EEPROM sector
        movb    #SectorErase,ECMD           ; write sector-erase command
        movb    #CBEIF,ESTAT                ; launch erase command
        brclr   ESTAT,ACCERR+PVIOL,EEROK    ; no error?
```

```
EERErr      ldab    #1                              ; error code set to 1
            rts
EEROK       brclr   ESTAT,CCIF,EEROK                ; wait until command completion
            clrb
            rts
```

The C language version of the function is as follows:

```
int eraseEEsector (int *ptr)
{
    ESTAT = ACCERR | PVIOL;   /* clear error flags */
    if(!(ESTAT & CBEIF))
          return 1;                    /* command buffer not empty, can't issue new command*/
    *ptr = 0x00;                       /* write any data to EEPROM sector location */
    ECMD = SectorErase;                /* write sector-erase command */
    ESTAT = CBEIF;                     /* launch the command */
    if(ESTAT & (ACCERR | PVIOL))
          return 1;                    /* error occurred */
    while(!(ESTAT&CCIF));              /* wait for command completion */
    return 0;                          /* command completed correctly */
}
```

▲

Example 14.11

▼

Write a function that performs bulk erasure to the EEPROM. Index register X contains a word-aligned EEPROM address.

Solution: The assembly function that performs EEPROM bulk erasure is as follows:

```
bulkeraseEE  movb    #ACCERR+PVIOL,ESTAT       ; clear error flags
             brclr   ESTAT,CBEIF,EEBEErr       ; command buffer not empty, return
             std     0,X                       ; write any data to EEPROM
             movb    #BulkErase,ECMD           ; write bulk-erase command
             movb    #CBEIF,ESTAT              ; launch bulk-erase command
             brclr   ESTAT,ACCERR+PVIOL,EEBROK ; no error?
EEBEErr      ldab    #1                        ; error code set to 1
             rts
EEBROK       brclr   ESTAT,CCIF,EEBROK         ; wait until command completion
             clrb
             rts
```

▲

The C language version of the function is left as an exercise problem.

Example 14.12

▼

Write a function that bulk erases the EEPROM and verifies if the whole EEPROM is blank. The index register X contains a valid word-aligned address to the EEPROM.

Solution: The assembly function that performs the specified operation is as follows:

```
eeraseverify  movb   #ACCERR+PVIOL,ESTAT   ; clear error flags
              brclr  ESTAT,CBEIF,cantEE     ; command buffer not empty
              std    0,x                    ; write any data to EEPROM address
              movb   #EraseVerify,ECMD      ; write the command
              movb   #CBEIF,ESTAT           ; launch the erase and check command
              brclr  ESTAT,ACCERR+PVIOL,EEEVOK
              ldab   #1                     ; return error code 1
              rts
EEEVOK        brclr  ESTAT,CCIF,EEEVOK      ; wait until command is done
              brset  ESTAT,BLANK,EVEOK      ; successful erase and verify?
cantEE        ldab   #1                     ; EEPROM is not blank
              rts
EVEOK         clrb
              rts
```

The C language version of the function is straightforward and hence will be left as an exercise problem.

Example 14.13

Write a function that programs a word to the EEPROM. X contains the address of the EEPROM location to be programmed and double accumulator D holds the data to be programmed. A 0 is returned in accumulator B if the command is performed correctly; otherwise, a 1 is returned.

Solution: The function that performs the specified function is as follows:

```
eeprogram    movb   #ACCERR+PVIOL,ESTAT   ; clear error flags
             brclr  ESTAT,CBEIF,cantPRE    ; command buffer not empty
             std    0,x                    ; write data to EEPROM address
             movb   #Program,ECMD          ; write the command
             movb   #CBEIF,ESTAT           ; launch the erase and check command
             brclr  ESTAT,ACCERR+PVIOL,EEPROK
cantPRE      ldab   #1                     ; return error code 1
             rts
EEPROK       brclr  ESTAT,CCIF,EEPROK      ; wait until command is done
             clrb                          ; successful program code is 0
             rts
```

The C language version of the function is as follows:

```
int eeprogram (unsigned int data, unsigned int *ptr)
{
    ESTAT = ACCERR+PVIOL;   /* clear error flag */
    if(!(ESTAT&CBEIF))
         return 1;
    *ptr = data;            /* write data to word-aligned address */
    ECMD = Program;         /* write program command */
    ESTAT = CBEIF;          /* launch command */
```

```
        if(ESTAT &(ACCERR | PVIOL))
                return 1;
        while(!(ESTAT&CCIF));          /* wait for command to complete */
        return 0;

}
```

Example 14.14

Write a function that executes the sector-modify command to change the contents of an EEPROM sector. The index register X holds the word-aligned address of the first EEPROM word to be modified. The index register Y holds the word-aligned address of the first word of data to be programmed.

Solution: The assembly function that performs the desired function is as follows:

```
eesectormodify
                movb    #ACCERR+PVIOL,ESTAT   ; clear error flags
                brclr   ESTAT,CBEIF,cantmod   ; command buffer not empty
                movw    0,Y,0,X               ; write data to EEPROM address
                movb    #SectorModify,ECMD    ; write sector-modify command
                movb    #CBEIF,ESTAT          ; launch the erase and check command
                brclr   ESTAT,ACCERR+PVIOL,EEModOK
cantmod         ldab    #1                    ; return error code 1
                rts
EEModOK         brclr   ESTAT,CBEIF,EEModOK   ; wait for command buffer to empty
                movw    2,Y,2,X               ; write second data word to second word of EEPROM
                movb    #Program,ECMD         ; write program command
                movb    #CBEIF,ESTAT          ; launch the program command
                brclr   ESTAT,ACCERR+PVIOL,EEPR2OK
                ldab    #1
                rts
EEPR2Ok         brclr   ESTAT,CCIF,EEPR2OK    ; wait for program command completion
                clrb                          ; successful program code is 0
                rts
```

The C language version of the function is as follows:

```
int EESectorModify(unsigned int *src, unsigned int *dest)
{
    ESTAT = ACCERR | PVIOL;     /* clear error flags */
    if(!(ESTAT&CBEIF))
            return 1;                /* command buffer not empty is error */
    *dest = *src;               /* write first data word */
    ECMD = SectorModify;        /* write sector-modify command */
    ESTAT = CBEIF;              /* launch the sector-modify command */
    if(ESTAT&(ACCERR | PVIOL))
            return 1;                /* command failed */
    while(!(ESTAT&CBEIF));      /* wait for command buffer becomes empty */
    *(dest+1) = *(src+1);       /* write second data word */
    ECMD = Program;             /* write the program command */
```

```
        ESTAT = CBEIF;              /* launch the program command */
        if(ESTAT&(ACCERR | PVIOL))
              return 1;             /* command failed */
        while(!(ESTAT&CCIF));       /* wait for command buffer becomes empty */
              return 0;
   }
```

14.7 HCS12 External Memory Interface

The HCS12 can access external memory when it is reset into one of the expanded modes. Address and data signals are multiplexed onto the Port A and Port B pins; the Port E and Port K supply control signals are required for accessing external memories.

14.7.1 HCS12 Pins for External Memory Interfacing

The HCS12 external memory interface uses 27 pins and is implemented across four I/O ports (A, B, E, and K). The signal pins used in external memory interface are listed in Table 14.7. The signals ADDR0...ADDR15 and XADDR14...XADDR19 are used to select a memory location to access and are referred to as the *address bus*. The signals DATA0...DATA15 are used to carry data and hence are referred to as the *data bus*. The remaining signals are referred to as the *control bus*. When the external memory is not paged, only the lower 16 address/data pins

Signal Name	Function
ADDR0/DATA0	EMI address bit 0 or data bit 0
ADDR1/DATA1	EMI address bit 1 or data bit 1
ADDR2/DATA2	EMI address bit 2 or data bit 2
ADDT3/DATA3	EMI address bit 3 or data bit 3
ADDT4/DATA4	EMI address bit 4 or data bit 4
ADDT5/DATA5	EMI address bit 5 or data bit 5
ADDT6/DATA6	EMI address bit 6 or data bit 6
ADDT7/DATA7	EMI address bit 7 or data bit 7
ADDR8/DATA8	EMI address bit 8 or data bit 8
ADDR9/DATA9	EMI address bit 9 or data bit 9
ADDR10/DATA10	EMI address bit 10 or data bit 10
ADDT11/DATA11	EMI address bit 11 or data bit 11
ADDT12/DATA12	EMI address bit 12 or data bit 12
ADDT13/DATA13	EMI address bit 13 or data bit 13
ADDT14/DATA14	EMI address bit 14 or data bit 14
ADDT15/DATA15	EMI address bit 15 or data bit 15
XADDR14	EMI extended address bit 14
XADDR15	EMI extended address bit 15
XADDR16	EMI extended address bit 16
XADDR17	EMI extended address bit 17
XADDR18	EMI extended address bit 18
XADDR19	EMI extended address bit 19
R/W̄	Read/write
LSTRB	Lower byte strobe
ECLK	E clock
ECS/ROMONE	Emulated chip select/on-chip ROM enable
XCS	External data chip select

Note: EMI stands for external memory interface

Table 14.7 ■ HCS12 external memory interface signal pins are used. However, if expanded memory is enabled, then ADDR14 and ADDR15 are not used. Instead, signals XADDR19. . .XADDR14 are used as expanded address signals. In the following discussion, we will use A0~A15, D0~D15, and XA14~XA19 to refer to ADDR0~ADDR15, DATA0~DATA15, and XADDR14~XADDR19, respectively.

The R/$\overline{\text{W}}$ signal is used to indicate the direction of data transfer. When this signal is high, the MCU reads data from external memory chips. When this signal is low, the MCU writes data to the external memory. The $\overline{\text{LSTRB}}$ signal is used to indicate whether the lower data bus (D7. . .DA0) carries valid data. There are two situations in which the lower data bus does not carry valid data:

1. In expanded narrow mode, external memory data pins are connected to the D15. . .D8 pins. D7. . .D0 are not used to carry data.

2. In expanded wide mode, the MCU may execute instructions to write byte data to memory locations at even addresses.

The signals $\overline{\text{LSTRB}}$, R/$\overline{\text{W}}$, and A0 indicate the type of bus access that is taking place. Accesses to the internal RAM module are the only type of access that would produce $\overline{\text{LSTRB}}$ = A0 = 1, because the internal RAM is specifically designed to allow misaligned 16-bit accesses in a single cycle. In these cases, the data at the given address is on the lower half of the data bus, and the data at address +1 is on the upper half of the data bus. These are summarized in Table 14.8.

$\overline{\text{LSTRB}}$	A0	R/$\overline{\text{W}}$	Type of Access
1	0	1	8-bit read of an even address
0	1	1	8-bit read of an odd address
1	0	0	8-bit write of an even address
0	1	0	8-bit write of an odd address
0	0	1	16-bit read of an even address
1	1	1	16-bit read of an odd address (low/high bytes swapped)
0	0	0	16-bit write of an even address
1	1	0	16-bit write of an odd address (low/high bytes swapped)

Table 14.8 ■ Access type versus bus control signals

The ECLK signal is provided as a timing reference for external memory accesses. This signal is controlled by three bits:

- The NECLK bit of the PEAR register
- The IVIS bit of the MODE register
- The ESTR bit of the EBICTL register

The NECLK bit enables/disable the E clock signal. The IVIS bit allows the E clock to be visible when the MCU is performing an internal access cycle. The ESTR bit allows the E clock to be stretched.

The $\overline{\text{ECS}}$ signal is used as the chip select signal for external memory chips. Because it is asserted immediately after the address signals become valid, the $\overline{\text{ECS}}$ signal is often used to latch address signals so that they can remain valid throughout a complete bus cycle. The rising edge of the E clock can also be used to latch address signals. The $\overline{\text{XCS}}$ signal, available in some HCS12 devices only, is also used as the external chip select signal for external chips. It can also be used to latch address signals because of its timing.

14.7.2 Waveforms of Bus Signals

The waveform of a typical digital signal is shown in Figure 14.25. A bus signal cannot rise from low to high or drop from high to low instantaneously. The time needed for a signal to rise from 10% of the power supply voltage to 90% of the power supply voltage is referred to as the *rise time* (t_R). The time needed for a signal to drop from 90% of the power supply voltage to 10% of the power supply voltage is referred to as the *fall time* (t_F).

Figure 14.25 ■ A typical digital waveform

A single bus signal is often represented as a set of line segments (see Figure 14.26). The horizontal axis and vertical axis represent the time and the magnitude (in volts), respectively, of the signal. Multiple signals of the same nature, such as address and data, are often grouped together and represented as parallel lines with crossovers, as illustrated in Figure 14.27. A crossover in the waveform represents the point at which one or multiple signals change values. The HCS12 literature shows that Freescale uses 0.65 V_{CC} and 0.35 V_{CC} as high input voltage (V_{IH}) and low input voltage (V_{IL}). All timing indications use these two values as the reference points.

Figure 14.26 ■ Single-signal waveform

Figure 14.27 ■ Multiple-signal waveform

Sometimes a signal value is unknown because the signal is changing. Hatched areas in the timing diagram, shown in Figure 14.28, represent single and multiple unknown signals. Sometimes one or multiple signals are not driven (because their drivers are not enabled) and hence cannot be received. An undriven signal is said to be *floating* or in *high-impedance* state. Single and multiple floating signals are represented by a value between the high and low levels, as shown in Figure 14.29.

In a microcontroller system, a bus signal falls into one of the three categories: *address*, *data*, or *control*.

Figure 14.28 ■ Unknown signals: (a) single signal, (b) multiple signal

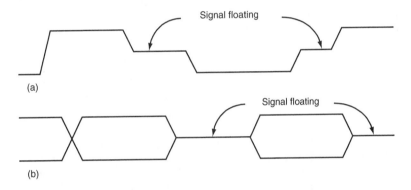

Figure 14.29 ■ Floating signals: (a) single signal, (b) multiple signal

14.7.3 Bus Transactions

A bus transaction includes sending the address and receiving or sending the data. A *read* transaction (also called a *read bus cycle*) transfers data from memory to either the CPU or the I/O device, and a *write* transaction (also called a *write bus cycle*) writes data to the memory. In a read transaction, the address is first sent down the bus to the memory, together with the appropriate control signals indicating a read. In Figure 14.30, this means pulling the read signal to high. The memory responds by placing the data on the bus and driving the $\overline{\text{Ready}}$ signal to low. The $\overline{\text{Ready}}$ signal (asserted low) indicates that the data on the data bus is valid.

In Figure 14.30, a read bus cycle takes one clock cycle to complete. For some microcontrollers, the $\overline{\text{Ready}}$ signal is used to extend the bus cycle to more than one clock cycle to accommodate slower memory components. When data from external memory is not ready, the control circuitry does not pull the $\overline{\text{Ready}}$ signal to low. When the $\overline{\text{Ready}}$ signal is sampled to be low, the microcontroller copies the data into the CPU. The HCS12 microcontroller does not use the $\overline{\text{Ready}}$ signal to accommodate slower memory. In a write bus cycle, the CPU sends both the address and the data and requires no return of data.

In a bus transaction, there must be a device that can initiate a read or write transaction. The device that can initiate a bus transaction is called a *bus master*. A microcontroller is always a

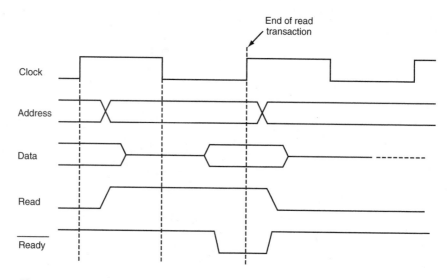

Figure 14.30 ■ A typical bus read transaction

bus master. A device such as a memory chip that cannot initiate a bus transaction is called a *bus slave*.

In a bus transaction, there must be a signal to synchronize the data transfer. The signal that is used most often is a clock signal. The bus is *synchronous* when a clock signal is used to synchronize the data transfer. In a synchronous bus, the timing parameters of all signals use the clock signal as the reference. As long as all timing requirements are satisfied, the bus transaction will be successful. An *asynchronous bus*, on the other hand, is not clocked. Instead, self-timed, handshaking protocols are used between the bus sender and receiver. Figure 14.31 shows the steps of a master performing a read on an asynchronous bus. A synchronous bus is often used between the CPU and the memory system, whereas an asynchronous bus is often used to handle different types of I/O devices. A synchronous bus is usually faster because it avoids the overhead of synchronizing the bus for each transaction.

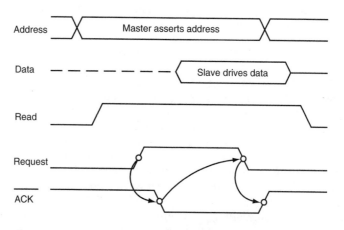

Figure 14.31 ■ Asynchronous read bus transaction

14.7.4 Bus Multiplexing

Designers of microcontrollers prefer to minimize the number of signal pins because it will make the chip less expensive to manufacture and use. By multiplexing the address bus and the data bus, some pins can be used for other purposes than address data. The drawback of multiplexing bus signals is that the achievable bus transaction performance is compromised. Most 8-bit and many 16-bit microcontrollers multiplex their address and data buses.

For any bus transaction, the address signal input to the memory chips must be stable throughout the whole bus transaction. The memory system will need to use a circuit to latch the address signals so that they stay valid throughout the bus cycle. In a microcontroller that multiplexes the address and data buses, the address signals are placed on the multiplexed bus first, along with certain control signals to indicate that the address signals are valid. After the address signals are on the bus long enough so that the external logic has time to latch them, the microcontroller stops driving the address signals and either waits for the memory devices to place data on the multiplexed bus (in a read bus cycle) or places data on the multiplexed bus (in a write bus cycle).

14.7.5 The HCS12 Bus Cycles

When adding external memory chips to the HCS12, it is important to make sure that all the timing requirements of the MCU and memory chips are satisfied. The timing requirements of the HCS12 are specified using a diagram as shown in Figure 14.32. The value for each timing

Figure 14.32 ■ The HCS12 read/write bus cycle timing diagram

requirement is listed in Table 14.9. The interval when the ECLK signal is high can be stretched by one to three minimum ECLK cycles (t_{cyc}) to accommodate slower memory devices.

The timing diagram in Figure 14.32 does not clearly describe how a read bus cycle and a write bus cycle proceed. A read bus cycle starts from the HCS12 driving the address signals onto the address bus to select a location to read. The sequence of events that occurs when the HCS12 family MCU performs a read from an external memory is illustrated in Figure 14.33. In a read bus cycle, the R/$\overline{\text{W}}$ signal stays high and the external memory drives the data onto the data bus.

Num	Parameter Name	Symbol	Min	Typ	Max	Unit
1	Frequency of operation (E-clock)	f_o	0	—	25.0	MHz
2	Cycle time	t_{cyc}	40	—	—	ns
3	Pulse width, E low	PW_{EL}	19	—	—	ns
4	Pulse width, E high	PW_{EH}	19	—	—	ns
5	Address delay time	t_{AD}	—	—	8	ns
6	Address valid time to E rise ($PW_{EL} - t_{AD}$)	t_{AV}	11	—	—	ns
7	Muxed address hold time	t_{MAH}	2	—	—	ns
8	Address hold to data valid	t_{AHDS}	7	—	—	ns
9	Data hold to address	t_{DHA}	2	—	—	ns
10	Read data setup time	t_{DSR}	13	—	—	ns
11	Read data hold time	t_{DHR}	0	—	—	ns
12	Write data delay time	t_{DDW}	—	—	7	ns
13	Write data hold time	t_{DHW}	2	—	—	ns
14	Write data setup time[1] ($PW_{EH} - t_{DDW}$)	t_{DSW}	12	—	—	ns
15	Address access time[1] ($t_{cyc} - t_{AD} - t_{DSR}$)	t_{ACCA}	19	—	—	ns
16	E high access time[1] ($PW_{EH} - t_{DSR}$)	t_{ACCE}	6	—	—	ns
17	Nonmultiplexed address delay time	t_{NAD}	—	—	6	ns
18	Nonmuxed address valid to E rise ($PW_{EL} - t_{NAD}$)	t_{NAV}	15	—	—	ns
19	Nonmultiplexed address hold time	t_{NAH}	2	—	—	ns
20	Chip select delay time	t_{CSD}	—	—	16	ns
21	Chip select access time[1] ($t_{cyc} - t_{CSD} - t_{DSR}$)	t_{ACCS}	11	—	—	ns
22	Chip select hold time	t_{CSH}	2	—	—	ns
23	Chip select negated time	t_{CSN}	8	—	—	ns
24	Read/write delay time	t_{RWD}	—	—	7	ns
25	Read/write valid time to E rise ($PW_{EL} - t_{RWD}$)	t_{RWV}	14	—	—	ns
26	Read/write hold time	t_{RWH}	2	—	—	ns
27	Low strobe delay time	t_{LSD}	—	—	7	ns
28	Low strobe valid time to E rise ($PW_{EL} - t_{LSD}$)	t_{LSV}	14	—	—	ns
29	Low strobe hold time	t_{LSH}	2	—	—	ns
30	NOACC strobe delay time	t_{NOD}	—	—	7	ns
31	NOACC valid time to E rise ($PW_{EL} - t_{NOD}$)	t_{NOV}	14	—	—	ns
32	NOACC hold time	t_{NOH}	2	—	—	ns
33	IPIPO[1:0] delay time	t_{POD}	2	—	7	ns
34	IPIPO[1:0] valid time to E rise ($PW_{EL} - t_{POD}$)	t_{POV}	11	—	—	ns
35	IPIPO[1:0] delay time[1] ($PW_{EL} - t_{P1V}$)	t_{P1D}	2	—	25	ns
36	IPIPO[1:0] valid time to E fall	t_{P1V}	11	—	—	ns

[1] Affected by clock stretch: add $N \times t_{cyc}$, where N = 0, 2, or 3, depending on the number of clock stretches.

Table 14.9 ■ HCS12 expanded bus timing characteristics

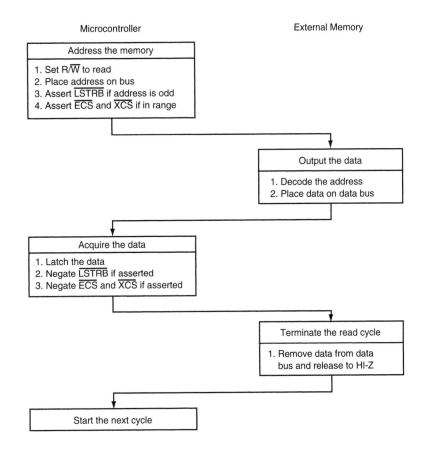

Figure 14.33 ■ Sequence of events occurring in an HCS12 read bus cycle

A write cycle also starts with the HCS12 driving the address signals onto the address bus to select a memory location to receive the data. The sequence of events that occurs when the HCS12 family MCU performs a write access to the external memory is shown in Figure 14.34. In a write bus cycle, the $R/\overline{W}$ signal goes low and the MCU drives the data onto the data bus.

14.8 Issues Related to Adding External Memory

Adding external memory becomes necessary when the application gets larger than the on-chip memory. When external memory is added, the designer should also consider treating external peripheral chips (when possible) as memory devices because it will make the programming of the peripheral chips easier. When adding external memory, there are three issues that need to be considered:

- Memory space assignment
- Address decoder and control circuitry design
- Timing verification

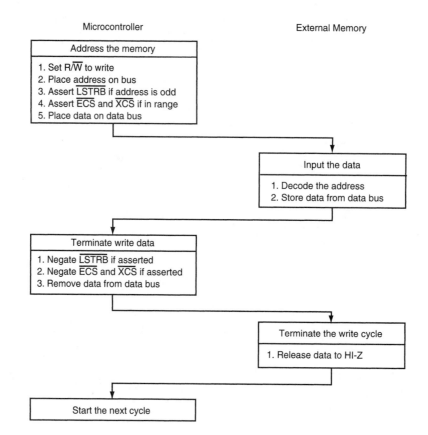

Figure 14.34 ■ Sequence of events occurring in an HCS12 write bus cycle

14.8.1 Memory Space Assignment

Any space unoccupied by the on-chip flash memory can be assigned to external memory devices. When making memory space assignment, the designer has two options to choose from:

- *Equal size assignment*. In this method, the available memory space is divided into blocks of equal size and then each block is assigned to a memory device without regard for the actual size of each memory-mapped device. A memory-mapped device could be a memory chip or a peripheral device. Memory space tends to be wasted using this approach because most memory-mapped peripheral chips need only a few bytes to be assigned to their internal registers.

- *Demand assignment*. In this approach, the designer assigns the memory space according to the size of memory devices.

Example 14.15

▼

Suppose that a designer is to design a HCS12DP256-based embedded product that requires 256 kB of external 16-bit SRAM, 256 kB of 16-bit EEPROM, and a parallel peripheral interface (PPI) that requires only 4 bytes of address space. The only SRAM available to this designer is the 128K × 8 SRAM chips (This chip has 128K locations with each location containing 8 bits). The only available EEPROM is the 128K × 8 EEPROM chips. Suggest a workable memory space assignment.

Solution: The designer is to design a 16-bit-wide memory system using the 8-bit-wide SRAM and EEPROM chips. Two 8-bit-wide memory chips are needed to construct a 16-bit memory module. One 16-bit-wide SRAM module can provide the 256-kB capacity. One 16-bit-wide EEP-ROM module is needed to offer the 256-kB capacity.

The on-chip FLASH memory occupies the space from $C0000 to $FFFFF. The address space from 0x00000 to 0xBFFFF is available for assignment. The following memory space assignment will be appropriate for this project:

SRAM:	$00000~$3FFFF	; 256 kB
EEPROM:	$40000~$7FFFF	; 256 kB
PPI:	$BFFFC~$BFFFF	; 4 bytes

▲

14.8.2 Address Decoder Design

The function of an address decoder is to make sure that there is no more than one memory device enabled to drive the data bus at a time. If there are two or more memory devices driving the same bus lines, *bus contention* will occur and could cause severe damage to the system. All memory devices or peripheral devices have control signals such as *chip-enable* (CE), *chip-select* (CS), or *output-enable* (OE) to control their read and write operations. These signals are often asserted low. The address decoder outputs will be used as the chip-select or chip-enable signals of external memory devices.

Two address-decoding schemes have been used: *full decoding* and *partial decoding*. A memory device is said to be *fully decoded* when each of its addressable locations responds to only a single address on the system bus. A memory component is said to be *partially decoded* when each of its addressable locations responds to more than one address on the system bus. Memory components such as DRAM, SRAM, EPROM, EEPROM, and flash-memory chips use the full address-decoding scheme more often, whereas peripheral chips or devices use the partial address-decoding scheme more often.

Address decoder design is closely related to memory space assignment. For the address space assignment made in Example 14.15, the higher address signals are used as inputs to the decoder, and the lower address signals are applied to the address inputs of memory devices. Before *programmable logic devices* (PLDs) became popular and inexpensive, designers used transistor-transistor logic (TTL) chips such as the 74LS138 as address decoders. However, the off-the-shelf TTL decoders force designers to use equal-size memory space assignment. When PLDs became popular and inexpensive, designers started to use them to implement address decoders. PLDs allow the designer to implement demand assignment.

One of the methods for implementing the address decoder is to use one of the hardware description languages (HDL), such as ABEL, CUPL, VHDL, or VERILOG. Low-density PLDs, such as GAL18V10, GAL20V8, GAL20V8, SPLD16V8, SPLD20V8, and SPLD20V8, are often used to implement address decoders for their ability to implement product terms of many variables. The *Generic Array Logic* (GAL) devices are produced by Lattice Semiconductor. The *Simple Programmable Logic Devices* (SPLDs) are produced by Atmel. Both ABEL and CUPL are simple hardware description languages that are very suitable for describing circuit behaviors for address decoders. ABEL is supported by Lattice Semiconductor, and CUPL is supported by Atmel.

14.8.3 Timing Verification

When designing a memory system, the designer needs to make sure that timing requirements for both the microcontroller and the memory system are satisfied. In a read cycle, the most critical timing requirements are the *data setup time* (t_{DSR}, parameter #10) and *data hold time* (t_{DHR}, parameter #11) required by the HCS12 microcontroller. In addition, the designer must make sure that the address setup-time and hold-time requirements for the memory devices are met. The control signals needed by memory devices during a read cycle must be asserted at the appropriate times.

In a write cycle, the most critical timing requirements are the *write data setup time* and *write data hold time* required by the memory devices. As in a read cycle, the address setup time and address hold time must also be satisfied. Control signals required during a write cycle must also be generated at proper times.

14.9 Memory Devices

The control circuit designs for interfacing the SRAM, the EPROM, the EEPROM, and the flash memory to the HCS12 MCU are quite similar. The following sections illustrate how to add SRAM and EEPROM chips with the 128K × 8 organization to the HCS12 microcontroller.

14.9.1 The K6R1008C1D

The K6R1008C1D is a 128K × 8 bit asynchronous SRAM from Samsung that operates with a 5-V power supply. The K6R1008C1D has a short access time (10 ns) and three-state outputs. The pin assignment of the K6R1008C1D is shown in Figure 14.35. The address signals A16–A0 select one of the 128K locations within the chip to be read or written. Pins I/O8–I/O1 carry the data to be transferred between the chip and the microcontroller. The chip-select ($\overline{CS}$) input allows/disallows the read/write access request to the K6R1008C1D. The $\overline{OE}$ signal is the output enable signal. When the $\overline{OE}$ signal is high, all eight I/O pins will be in high-impedance state.

Depending on the assertion times of control signals, there are two timing diagrams for the read cycle and three timing diagrams for the write cycle (shown in Figures 14.36 and 14.37). The values of the related timing parameters for the read and write cycles are listed in Table 14.10.

In Figures 14.36a and 14.36b, the $\overline{CS}$ signal must be asserted for at least t_{RC} ns during a read cycle. The signal that is asserted the latest determines the time that data will become available. For example, in Figure 14.36b, the $\overline{OE}$ signal is asserted the latest; therefore, data becomes valid t_{OE} ns later. Data pins will go to high-impedance state t_{OHZ} ns after the $\overline{OE}$ signal goes to high or t_{HZ} ns after the $\overline{CS}$ signal goes to high.

In Figure 14.37a, the $\overline{OE}$ is controlled by a signal. Whenever the MCU wants to write data into the SRAM, the $\overline{OE}$ signal is pulled to high. In Figure 14.37b and 14.37c, the $\overline{OE}$ signal is tied to low. One can see that the Data Out gets out of high-impedance state t_{OW} ns after the $\overline{WE}$ and $\overline{CS}$ signals go to high. The write cycle in Figure 14.37a and 14.37b are $\overline{WE}$-controlled because the $\overline{WE}$ signal is asserted later than the $\overline{CS}$ signal. The write cycle in Figure 14.37c is $\overline{CS}$-controlled because the $\overline{CS}$ signal is asserted later than the $\overline{WE}$ signal.

Figure 14.35 ■ The K6R1008C1D pin assignment

14.9.2 The AT28C010 EEPROM

The AT28C010 is a 128K × 8 electrically erasable, programmable, read-only memory. It needs only a 5-V power supply to operate and achieves access times ranging from 120 ns to 200 ns.

The AT28C010 supports a page-write operation that can write from 1 to 128 bytes. The device contains a 128-byte page register to allow writing up to 128 bytes simultaneously. During a write cycle, the address and 1 to 128 bytes of data are internally latched, freeing the address and data bus for other operations. Following the initiation of a write cycle, the device will automatically write the latched data using an internal control timer. The end of an internal write cycle can be detected by polling the I/O7 pin or checking whether the I/O6 pin has stopped toggling.

The device utilizes internal error correction for extended endurance and improved data retention. To prevent an unintended write operation, an optional software data protection mechanism is available. The device also includes an additional 128 bytes of EEPROM for device identification or tracking. The pin assignment for the AT28C010 is shown in Figure 14.38.

(a) Read cycle 1 ($\overline{OE}$ and $\overline{CE}$ are asserted in the whole cycle)

Notes: (b) Read cycle 2 ($\overline{OE}$ controlled)
1. $\overline{WE}$ is high for read cycle.
2. All read cycle timing is referenced from the last valid address to the first transition address.
3. t_{HZ} and t_{OHZ} are defined as the time at which the output achieves the open circuit condition and are not referenced to the V_{OH} or V_{OL} level.
4. At any given temperature and voltage condition, $t_{HZ}(max)$ is less than $t_{LZ}(min)$ both for a given device and from device to device.
5. Transition is measured ±200 mV from steady-state voltage with load. This parameter is sampled and not 100% tested.
6. Device is continuously selected with $\overline{CS} = V_{IL}$.
7. For common I/O applications, minimization or elimination of bus contention conditions is necessary during the read and write cycle.

Figure 14.36 ▪ K6R1008C1D read cycle timing diagram

DEVICE OPERATION

The AT28C010 is accessed like a SRAM. When the $\overline{CE}$ and $\overline{OE}$ signals are low and the $\overline{WE}$ signal is high, the data stored at the location determined by the address pins is driven on the I/O pins. The I/O pins are put in the high-impedance state when either the $\overline{CE}$ or $\overline{OE}$ signal is high.

A byte-write operation is started by a low pulse on the $\overline{WE}$ or the $\overline{CE}$ input with the $\overline{CE}$ or the $\overline{WE}$ input low (respectively) and the $\overline{OE}$ pin high. The address inputs are latched on the falling edge of the $\overline{CE}$ or the $\overline{WE}$ signal, whichever occurs last. The data is latched by the first rising edge of the $\overline{CE}$ or the $\overline{WE}$ signal.

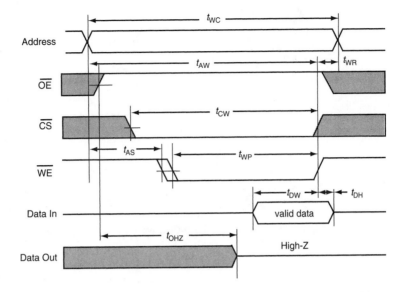

Figure 14.37a ■ K6R1008C1D write cycle timing diagram

Figure 14.37b ■ K6R1008C1D write cycle timing diagram ($\overline{OE}$ tied to low)

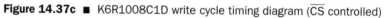

Figure 14.37c ■ K6R1008C1D write cycle timing diagram ($\overline{\text{CS}}$ controlled)

Parameter	Description	K6R1008C1D-10		Unit
		Min	Max	
Read Cycle				
t_{RC}	Read cycle time	10	—	ns
t_{AA}	Address to data valid	—	10	ns
t_{OH}	Data hold from address change	3	—	ns
t_{CO}	$\overline{\text{CS}}$ low to data valid	—	10	ns
t_{OE}	$\overline{\text{OE}}$ low to data valid	—	5	ns
t_{OLZ}	$\overline{\text{OE}}$ low to low Z	0	—	ns
t_{OHZ}	$\overline{\text{OE}}$ high to high Z	—	5	ns
t_{LZ}	$\overline{\text{CS}}$ low to low Z	3	—	ns
t_{HZ}	$\overline{\text{CS}}$ high to high Z	—	5	ns
t_{PU}	$\overline{\text{CS}}$ low to power up	0	—	ns
t_{PD}	$\overline{\text{CS}}$ high to power down	—	10	ns
Write Cycle				
t_{WC}	Write cycle time	10	—	ns
t_{CW}	$\overline{\text{CS}}$ low to write end	7	—	ns
t_{AW}	Address valid to end of write	7	—	ns
t_{AS}	Address setup to write start	0	—	ns
t_{WP}	$\overline{\text{WE}}$ pulse width ($\overline{\text{OE}}$ high)	7	—	ns
t_{WP1}	$\overline{\text{WE}}$ pulse width ($\overline{\text{OE}}$ low)	10	—	ns
t_{DW}	Data setup to write end	5	—	ns
t_{DH}	Data hold from write end	0	—	ns
t_{OW}	$\overline{\text{WE}}$ high to low Z	3	—	ns
t_{WHZ}	$\overline{\text{WE}}$ low to high Z	0	5	ns
t_{WR}	Write recovery time	0	—	ns

Table 14.10 ■ K6R1008C1D read and write timing parameters

Figure 14.38 ■ The AT28C010 pin assignment

A page-write operation is initiated in the same manner as a byte-write operation; after the first byte is written, it can then be followed by 1 to 127 additional bytes. Each successive byte must be loaded within 150 μs (t_{BLC}) of the previous byte. If the t_{BLC} limit is exceeded, the AT28C010 will cease accepting data and begin the internal programming operation. All bytes involved in a page-write operation must reside on the same page as defined by the state of the A16–A7 inputs. For each high-to-low transition of the $\overline{WE}$ signal during the page-write operation, the address signals A16–A7 must be the same. The inputs A6–A0 are used to specify which bytes within the page are to be written. The bytes may be loaded in any order and may be altered within the same load period. Only bytes that are specified for writing will be written.

The AT28C010 allows the designer to poll the I/O7 pin to find out if an internal write operation has completed. Before an internal write operation has completed, a read of the last byte written will result in the complement of the written data being presented on the I/O7 pin. Once the write cycle has been completed, true data is valid on all outputs, and the next write cycle can begin.

In addition to data polling, the AT28C010 provides another method for determining the end of a write cycle. During the internal write operation, successive attempts to read data from the device will result in I/O6 toggling between 1 and 0. Once the write operation has completed, the I/O6 pin will stop toggling, and valid data will be read.

DATA PROTECTION

Atmel has incorporated both hardware and software features to protect the memory against inadvertent write operations. The hardware protection method works as follows:

V_{CC} *sense.* If V_{CC} is below 3.8 V, the write function is inhibited.

V_{CC} *power-on delay.* Once V_{CC} has reached 3.8 V, the device will automatically time out for 5 ms before allowing a write operation.

Write inhibit. Holding the $\overline{OE}$ signal low, or the $\overline{CE}$ signal high, or the $\overline{WE}$ signal high inhibits write cycles.

Noise filter. Pulses shorter than 15 ns on the $\overline{WE}$ or the $\overline{CE}$ input will not initiate a write cycle.

A software data protection (SDP) feature is included that can be enabled to prevent inadvertent write operations. The SDP is enabled by the host system issuing a series of three write commands; three specific bytes of data are written to three specific addresses. After writing the 3-byte command sequence and after the t_{WC} delay, the entire AT28C010 will be protected against inadvertent write operations. It should be noted that once protected, the host may still perform a byte write or page write to the AT28C010. This is done by preceding the data to be written by the same 3-byte command sequence used to enable the SDP. Once set, the SDP will remain active unless the disable command sequence is issued. Power transitions do not disable the SDP, and the SDP will protect the AT28C010 during the power-up and power-down conditions. After setting the SDP, any attempt to write to the device without the 3-byte command sequence will start the internal write timer. No data will be written into the device.

The algorithm for enabling software data protection is as follows:

Step 1
Write the value of 0xAA to the memory location at 0x5555.

Step 2
Write the value of 0x55 to the memory location at 0x2AAA.

Step 3
Write the value of 0xA0 to the memory location at 0x5555. At the end of write, the write-protect state will be activated. After this step, write operation is also enabled.

Step 4
Write any value to any location (1–128 bytes of data are written).

Step 5
Write last byte to last address.

Software data protection can be disabled any time when it is undesirable. The algorithm for disabling software data protection is as follows:

Step 1
Write the value of 0xAA to the memory location at 0x5555.

Step 2
Write the value of 0x55 to the memory location at 0x2AAA.

Step 3
Write the value of 0x80 to the memory location at 0x5555.

Step 4
Write the value of 0xAA to the memory location at 0x5555.

Step 5
Write the value of 0x55 to the memory location at 0x2AAA.

Step 6
Write the value of 0x20 to the memory location at 0x5555. After this step, software data protection is exited.

Step 7
Write any value(s) to any location(s).

Step 8
Write the last byte to last address.

Device Identification

An extra 128 bytes of EEPROM memory are available to the user for device identification. By raising the voltage at the A9 pin to 12 V ± 0.5 V and using the address locations 0x1FF80 to 0x1FFFF, the bytes may be written to or read from in the same manner as the regular memory array.

Read and Write Timing

The read cycle timing diagram is shown in Figure 14.39. The AT28C010 has three read access times:

- Address access time t_{ACC}
- $\overline{CE}$ access time t_{CE}
- $\overline{OE}$ access time t_{OE}

Figure 14.39 ■ AT28C010 read timing diagram

Each of the read access times is measured by assuming that the other control signals and/or the address have been valid. For example, the address access time is the time from the moment that the address inputs to the AT28C010 become valid until data is driven out of data pins assuming that the $\overline{CE}$ and $\overline{OE}$ signals have been valid (low) before the required moment (t_{CE} or t_{OE} ns before data become valid). After address change, the data value will change immediately (data hold time is 0). The values of read timing parameters are listed in Table 14.11.

Symbol	Parameter	AT28C010-12		AT28C010-15		AT28C010-20		Unit
		Min	Max	Min	Max	Min	Max	
t_{ACC_1}	Address to output delay		120		150		200	ns
t_{CE_2}	CE to output delay		120		150		200	ns
$t_{OE_{3,4}}$	OE to output delay	0	50	0	55	0	55	ns
t_{DF}	OE to output float	0	50	0	55	0	55	ns
t_{DH}	Output hold from OE, CE, or address, which occurred first	0		0		0		ns

[1] $\overline{CE}$ may be delayed up to $t_{ACC} - t_{CE}$ after the address transition without impact on t_{ACC}.
[2] $\overline{OE}$ may be delayed up to $t_{CE} - t_{OE}$ after the falling edge of $\overline{CE}$ without impact on t_{CE} or by $t_{ACC} - t_{OE}$ after an address change without impact on t_{ACC}.
[3] t_{DF} is specified from $\overline{OE}$ or $\overline{CE}$, whichever occurs first (CL = 5 pF).
[4] This parameter is characterized and is not 100% tested.

Table 14.11 ■ AT28C010 read characteristics

The write-cycle timing diagram is shown in Figure 14.40. There are two write timing diagrams: $\overline{WE}$-controlled and $\overline{CE}$-controlled diagrams, depending on which signal is asserted the latest. In a $\overline{WE}$-controlled write timing waveform, the $\overline{CE}$ signal is asserted (goes low) earlier than the $\overline{WE}$ signal and becomes inactive (goes high) after the $\overline{WE}$ signal goes high. In a $\overline{CE}$-controlled write cycle, the $\overline{WE}$ signal goes low earlier and returns to high later than the $\overline{CE}$ signal. In Figure 14.40, the control signal that is asserted the latest must have a minimal pulse width of 100 ns. The address input must not be valid after the assertion (going low) of the latest control signal and must remain valid for at least 50 ns after the assertion of the same signal. The values of write cycle timing parameters are shown in Table 14.12. The write data must be valid at least 50 ns before the latest control signal starts to rise. The write data need not be stable after the latest control signal (data hold time) rises.

The page-mode write-cycle timing diagram is shown in Figure 14.41, and the values of the timing parameters are shown in Table 14.13.

The write timing waveform illustrates only how the CPU writes data into the EEPROM. The EEPROM still needs to initiate an internal programming process to actually write data into the specified location. The CPU can find out whether the internal programming process has been completed by using the data polling or the toggle bit polling method.

14.10 Example of External Memory Expansion for the HCS12

One can choose to use or not to use expanded memory (paging) when adding external memory chips to the HCS12, but being limited to adding less than 64 kB of external memory without using the expanded memory doesn't make much sense. When expanded memory is used, one can add up to 1 MB of external memory if the on-chip flash memory is disabled. The external memory can be 8-bit wide or 16-bit wide. In an 8-bit wide external memory, the HCS12 performs two read bus cycles when handling 16-bit data. This reduces the external memory performance to half but may lower the system cost slightly. When adding 8-bit-wide external memory, the data output of memory is connected to Port A.

(a) $\overline{\text{WE}}$-controlled write timing waveform

(b) $\overline{\text{CE}}$-controlled write timing waveform

Figure 14.40 ■ AT28C010 write-cycle timing waveform

Symbol	Parameter	Min	Max	Unit
t_{AS}, t_{OES}	Address, $\overline{\text{OE}}$ setup time	0		ns
t_{AH}	Address hold time	50		ns
t_{CS}	Chip-select setup time	0		ns
t_{CH}	Chip-select hold time	0		ns
t_{WP}	Write pulse width ($\overline{\text{WE}}$ or $\overline{\text{CE}}$)	100		ns
t_{DS}	Data setup time	50		ns
t_{DH}, t_{OEH}	Data, $\overline{\text{OE}}$ hold time	0		ns

Table 14.12 ■ AT28C010 write characteristics

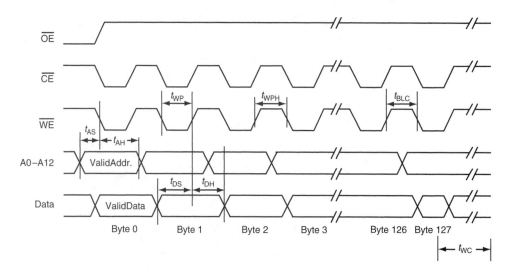

Figure 14.41 ■ AT28C010 page-mode write waveform

Symbol	Parameter	Min	Max	Unit
t_{WC}	Write cycle time		10	ms
t_{AS}	Address setup time	0		ns
t_{AH}	Address hold time	50		ns
t_{DS}	Data setup time	50		ns
t_{DH}	Data hold time	0		ns
t_{WP}	Write pulse width	100		ns
t_{BLC}	Byte load cycle time		150	μs
t_{WPH}	Write pulse width high	50		ns

Table 14.13 ■ AT28C010 page-write characteristics

This section illustrates how to add 256 kB of SRAM and 256 kB of EEPROM to the HCS12 by using the paging technique. The application program needs to go through the paging window located from $8000 to $BFFF in order to access these two memory modules.

14.10.1 Memory Space Assignment

Because this design will be using two 128-kB SRAM chips (K6R1008C1D) to construct 256 kB of 16-bit-wide memory, the lowest 13 address signals (A13. . .A1) and the expanded address signals XA14. . .XA17 will be used as address inputs for both SRAM chips. The A0 signal will not be needed. Because these two chips must be selected by the same chip-select signal, the address space must be divided into 256-kB blocks when making memory space assignment to the SRAM module. The following assignment will be appropriate:

- SRAM module: $00000–$3FFFF
- EEPROM module: $40000–$7FFFF

14.10.2 Address Latch

Since the address signals A15 to A0 and data signals D15 to D0 are time-multiplexed, they need to be latched into a memory device such as the dual octal latch 74ABT16373B (made by Philips) and held valid throughout the whole bus cycle. This device uses the falling edge of its enable inputs (1E and 2E) to latch the data inputs. The $\overline{ECS}$ signal (or $\overline{XCS}$ for some devices) of the HCS12 can be used to drive this input and latch the address signals A15 to A0 into the 74ABT16373B. The pin assignment of the 74ABT16373B is shown in Figure 14.42.

Figure 14.42 ■ Pin assignment of 74ABT16373B

The values of 1D0~1D7 and 2D0~2D7 are latched into 1Q0~1Q7 and 2Q0~2Q7 on the falling edges of the 1E and 2E signals, respectively. The propagation delay from the falling edge of 1E (or 2E) to 1Q0~1Q7 (or 2Q0~2Q7) is 4.4 ns. The propagation delay from 1Dx (x = 0. . .7) to 1Qx is also 4.4 ns in the worst case. When the 1E (or 2E) signal is high, the latch is transparent, i.e., the value on 1Dx (2Dx) will appear on 1Qx (2Qx) after the propagation delay. The required data setup and hold times for 1Dx (and 2Dx) are 1 ns and 0.5 ns, respectively. The outputs 1Q0~1Q7 (2Q0~2Q7) will be in high impedance if the $\overline{1OE}$ ($\overline{2OE}$) signal is high.

At 25 MHz, address signals A15~A0 (driven by the HCS12) become valid 8 ns after the falling edge of the E clock or up to 8 ns before the falling edge of the $\overline{ECS}$ signal. Since A15~A0 stay valid until 2 ns after the rising edge of the E clock, A15~A0 have a hold time of at least 6 ns. Therefore, the input setup and hold-time requirements of the 74ABT16373B are satisfied. Since the propagation delay relative to the 1E and 2E signals is 4.4 ns, A15~A0 will become valid (to memory chips) 0.4 ns after the rising edge of the E clock or 20.4 ns after the start of a bus cycle. The address latching circuit is shown in Figure 14.43.

Figure 14.43 ■ Address latch circuit for the HCS12

The latched outputs A0, A14, and A15 are not used in a paged memory module. A0 is not used because the memory module is 16-bit.

14.10.3 Address Decoder Design

In this simple memory design example, the 1-MB memory is divided into four 256-kB modules. Therefore, the highest two address bits, XA19 and XA18, should be used as decoder inputs. A 3-to-8 decoder (e.g., 74F138) or a dual 2-to-4 decoder (e.g., 74F139) can be used as the address decoder. The chip-select signal equations for the SRAM and EEPROM are

$$\overline{SRAM\_CS} = !(!XA19 * !XA18 * ECS)$$
$$\overline{EEPROM\_CS} = !(!XA19 * XA18 * ECS)$$

The decoder circuit is shown in Figure 14.44.

Figure 14.44 ■ HCS12 external memory decoder circuit

14.10.4 Write-Control Signals

Because the HCS12 may sometimes write to a byte rather than a word of the external memory, the $\overline{WE}$ signal for the upper byte (even address) and lower byte (odd address) must be generated separately. Let $\overline{WE1}$ and $\overline{WE0}$ be the write enable signals of the high and low bytes, then

$$\overline{WE1} = !(LSTRB * !A0 * !R/\overline{W} + \overline{LSTRB} * !A0 * !R/\overline{W}) = !(!A0 * !R/\overline{W})$$
$$\overline{WE0} = !(\overline{LSTRB} * A0 * !R/\overline{W} + LSTRB * A0 * !R/\overline{W}) = !(A0 * !R/\overline{W})$$

These two control signals can be generated by using an inverter chip and a NAND gate chip. A simpler solution is to generate the chip-select signals and write-enable signals using a single GAL device, such as the GAL16V8. The GAL16V8 has a propagation delay of 3.5 ns. By using a GAL16V8 to generate the chip-select and write-enable signals as shown in Figure 14.45, the decoder circuit shown in Figure 14.44 can be eliminated. In Figure 14.45, the output-enable signal ($\overline{OE}$) required to control the output of SRAM and EEPROM can be generated without additional cost. The $\overline{OE}$ signal is simply the complement of the R/$\overline{W}$ signal output of the HCS12.

Figure 14.45 ■ Address decoder and write-enable signal-generating circuit

The design of the circuit shown in Figure 14.45 can be done by using the ispLEVER software from Lattice Semiconductor.

14.10.5 Example HCS12 External Memory Design

As shown in Figure 14.46, the circuit required to interface with external SRAMs and EEPROMs is minimal. A 16-bit latch is needed to keep address signals A15~A0 valid throughout the whole bus cycle. A GAL16V8 chip is used to generate the chip-select signals for selecting the SRAM and EEPROM modules. The $\overline{ECS}$ signal is asserted (goes low) only when the address outputs are valid. It must be included as one of the inputs to the address decoder circuit to make sure that the decoder asserts one of its outputs to low when its address inputs XA18 and XA19 are valid.

Example 14.16
▼

For the circuit shown in Figure 14.46, can the SRAM be accessed without stretching the E clock? Assume that the E clock is 25 MHz.

Solution: The timing verification consists of the read cycle and the write cycle.

Address Timing
The address inputs A12~A0 to the memory chips become valid 20.5 ns after the start of a bus cycle; the address inputs A16~A13 become valid 6 ns after the start of a bus cycle. Therefore, the complete address inputs A16~A0 to the memory chips become valid 20.5 ns after the start of a bus cycle.

Figure 14.46 ■ HCS12DP256B paged external memory example

Chip-Select Signals Timing

The $\overline{ECS}$ signal becomes valid 16 ns after the start of a bus cycle. XA19~XA18 are valid 6 ns after the start of a bus cycle. The GAL16V8 has a propagation delay of 3.5 ns. Therefore, the chip-select signals $\overline{SRAM\_CS}$ and $\overline{EEPROM\_CS}$ become valid 19.5 ns after the start of a bus cycle. Because the HCS12 stops driving the XA19, XA18, and $\overline{ECS}$ signals 2 ns after the end of a bus cycle, these two chip-select signals will become invalid 5.5 ns after the end of a bus cycle.

Write-Enable Signals Timing

Since $\overline{WE1}$ and $\overline{WE0}$ are a function of A0 and R/$\overline{W}$ and A0 is valid later than R/$\overline{W}$, the valid times of $\overline{WE1}$ and $\overline{WE0}$ are determined by A0. Since A0 is valid 20.5 ns after the start of a bus cycle and the propagation delay of the GAL16V8 is 3.5 ns, $\overline{WE1}$ and $\overline{WE0}$ will become valid 24 ns after the start of a bus cycle. The HCS12 de-asserts the R/$\overline{W}$ signal 2 ns after the end of a bus cycle; both the $\overline{WE1}$ and $\overline{WE0}$ signals become invalid 5.5 ns after the end of a write cycle.

Output-Enable Signal Timing

The $\overline{OE}$ signal is the complement of the R/$\overline{W}$ signal and is generated by the GAL16V8. The R/$\overline{W}$ signal is valid 7 ns after the start of a bus cycle. Therefore, the $\overline{OE}$ signal will become valid 10.5 ns after the start of a bus cycle and become invalid 5.5 ns after the end of a bus read cycle.

Read Cycle Timing Verification

For the K6R1008C1D SRAM, address inputs A16~A0 become valid later than $\overline{OE}$ and $\overline{CS}$. Therefore, read data is valid 30.5 ns after the start of a read cycle or 9.5 ns before the end of a read cycle. The HCS12DP256 requires a read data setup time of 13 ns. Therefore, the read data does not provide enough setup time for the HCS12DP256 to correctly read it. One needs to stretch the E clock by one period.

Write Cycle Timing Verification

The address inputs to the K6R1008C1D will stay valid until the new address is latched in the next bus cycle. Therefore, A16~A0 stay valid for a whole E clock period (40 ns at 25 MHz) and satisfy the minimum requirement 10 ns (t_{WC}).

The write enable signals $\overline{WE1}$ and $\overline{WE0}$ have a low pulse width of 21.5 ns (40 ns − 24 ns + 5.5 ns) and satisfy the requirement of the SRAM.

The HCS12 drives the write data 28 ns after the start of a write cycle (or 12 ns before the end of a write cycle). Both the $\overline{WE}$ and $\overline{CS}$ signal inputs to the K5R1008C1D are de-asserted 5.5 ns after the end of a write cycle. This provides 17.5 ns of write data setup time for the K5R1008C1D and exceeds the minimum requirement of 5 ns. The K5R1008C1D requires a write data hold time of 0 ns. Since the $\overline{CS}$ and $\overline{WE}$ signal become invalid 5.5 ns after the end of a write cycle. The write data must remain valid for at least 5.5 ns after a write cycle. The HCS12 stops driving the write data 2 ns after the end of a write bus cycle. This is shorter than the requirement. However, the HCS12 does not drive the multiplexed address/data bus for 8 ns after the start of the next bus cycle. During this period, the capacitance of the printed circuit board can keep the write data for a little while longer and satisfies the write data hold-time requirement. Because both the write data setup- and hold-time requirements are satisfied, the E clock needs not be stretched for the write cycle.

Because the E clock stretching applies to all the bus cycles, both the read and write cycles need to be stretched for one E clock period for the K5R1008C1D.

▲

Example 14.17

▼

Provide an analytical proof that the capacitance of the printed circuit board can hold the data voltage for enough time to satisfy the data hold time requirement of the SRAM.

Solution: The voltage of the data input to the K5R1008C1D is degraded by the following leakage currents:

1. Input current into the HCS12 data pin (typically on the order of 2.5 μA for Freescale microcontroller products)
2. Input current into the EEPROM chips (on the order of 10 μA)
3. Other leakage paths (assumed to be as large as 10 μA).

The capacitance of the printed circuit board is estimated to be 20 pF per foot. Let C, I, ΔV, and Δt be the printed circuit board capacitance of one data line, total leakage current, voltage change due to leakage current, and the time it takes for the voltage to degrade by ΔV, respectively.

The elapsed time before the data bus signal degrades to an invalid level can be estimated by the following equation:

$$\Delta t \approx C\Delta V \div I$$

The voltage degradation of 2.5 V (for a 5-V V_{CC}) is considered enough to cause data input to the SRAM to change the logic value from 1 to 0. The data bus line on the printed circuit board is normally not longer than 1 foot for a single-board computer, so 1 foot will be used as its length. The elapsed time before the data bus signal degrades to an invalid level is

$$\Delta t \approx 20 \text{ pF} \times 2.5 \text{ V} \div 22.5 \text{ } \mu A \approx 2.2 \text{ } \mu s$$

Although this equation is oversimplified, it does give some idea about the order of the time over which the charge across the data bus capacitor will hold after the HCS12 stops driving the data bus. Even if the leakage current is 10 times larger, the Δt value will be 225 ns and is still longer than the minimum hold-time requirement of the K5R1008C1D.

Example 14.18

▼

For the circuit shown in Figure 14.46, can the EEPROM be accessed without stretching the E clock, assuming that the 120-ns access time version of the AT28C010 is used? If not, how many cycles should the E clock be stretched?

Solution: The timing analysis of this problem differs from that of the SRAM mainly in the read/write data setup times. According to the analysis done in Example 14.16, A16~A0 become valid the latest (30.5 ns after the start of a read cycle). The data output from the AT28C010 becomes valid 150.5 ns after the start of a read cycle. For a 25-MHz E clock, this is 110.5 ns (150.5 ns − 40 ns) too late. Therefore, the read cycle for the AT28C010 needs to be stretched for at least three E clock periods at this frequency. After stretching the E clock by three cycles, the read data is valid 9.5 ns before the end of the read cycle. This doesn't provide enough data setup time for the HCS12. The only solution is to lower the frequency of the E clock. By slowing the E clock frequency to 24 MHz, the read data setup time is increased to 16.17 ns and satisfies the requirement. The AT28C010 provides 0 ns of data hold time, which is acceptable for the HCS12.

According to Example 14.16, the HCS12 provides 17.5 ns of write data setup time, which is shorter than the requirement (50 ns minimum) of the AT28C010. Stretching the E clock by one E clock period can satisfy this requirement. However, the write pulse width (both the $\overline{EEPROM\_CS}$ and $\overline{WE0}$) must be at least 100 ns. This would require the write bus cycle to be stretched by at least two E clock periods. Since the read cycle needs to be stretched by three E clock periods, the EBICTL and MISC registers need to be programmed to provide this requirement.

Example 14.19

▼

Write an instruction sequence to configure the EBICTL and MISC registers to stretch the external bus cycles by three E clock periods.

Solution: The value to be written into the EBICTL is $01. This value enables the E clock stretch. The value to be written into the MISC register is $0D, which will stretch the bus cycles by three E clock periods. The following instruction sequence will achieve the required stretching:

```
movb #$01,EBICTL    ; enable E stretch
movb #$0D,MISC      ; stretch E by 3 periods
```

The control signals that are multiplexed on Port E pins must be programmed properly in order for external bus cycles to proceed properly. The designer needs to program the PEAR register to enable the required signals, including R/$\overline{\text{W}}$, E clock, $\overline{\text{LSTRB}}$, and others.

The signals NOACC, IPIPE1, and IPIPE0 need not be enabled if the HCS12 is not designed to be debugged by the in-circuit emulator.

Example 14.20

▼

Give a value to be written into the PEAR register to achieve the following setting:

- Enable E clock output
- Enable the R/$\overline{\text{W}}$ output
- Enable the $\overline{\text{LSTRB}}$ output
- Disable the NOACC, IPIPE1 and IPEP0 outputs

Solution: The value to be written into the PEAR register to achieve this setting is $0C.

▲

14.11 Summary

The HCS12 has on-chip memory resources, including I/O registers, SRAM, EEPROM, and flash memory. The I/O registers, SRAM, and EEPROM can be remapped to other memory space when the application requires it. It is possible to have two or more memory resources overlapped. The HCS12 assigns precedence to each memory resource. The memory resource with the highest precedence is selected when two or more memory resources overlap in their spaces. The remapping of memory resources is performed by programming their associated initialization registers.

Limited by its 16-bit architecture, the HCS12 uses the paging technique to allow programs to access memory space larger than the 64-kB limit. The paging window is located at $8000 to $BFFF. An additional 6-bit PPAGE register is used to facilitate the mapping from the paging window to the actual memory block to be accessed. The HCS12 uses this technique to allow the application program to access up to 1 MB of memory space. Whenever the program wants to access expanded memory, it sets the address to be in the range of this window and also sets the appropriate value to the PPAGE register. After this, the hardware will map to the corresponding location to perform the actual access. Writing programs to take advantage of the expanded memory incurs quite a bit of overhead in assembly language. Neither the asmIDE nor the MiniIDE supports expanded memory. Many C compilers (including the ICC12 from Imagecraft and the Code Warrior from Metrowerks) support the expanded memory. It is recommended that C language be used to write application programs whenever the expanded memory is needed.

Both the flash memory and the EEPROM need to be erased before they can be programmed with new values. The algorithms of erasure and programming for these two memory technologies are quite similar. Before erasing and programming the flash memory and EEPROM, the designer needs to make sure that they are not protected. The erasure and programming operations are controlled by a clock signal that is derived by dividing the bus clock by a prescaler. The following four operations are common in flash-memory and EEPROM erasure and programming:

- Clear the error flags in ESTAT (or FSTAT).
- Write a word to a location in the sector to be erased or programmed.

■ Write a command to the command register (ECMD or FCMD).

■ Clear the CBEIF flag to launch the command.

Both the flash memory and EEPROM can be protected from inadvertent erasure. The software stored in the flash memory can be secured from being pirated.

Certain applications require more memory than the amount available in any of the HCS12 members. Adding external memory is one solution to this problem. External memory can be added only in the expanded modes. There are two expanded modes: narrow and wide. The expanded narrow mode allows the designer to add 8-bit wide external memory; the expanded wide mode allows the user to add 16-bit-wide memory to the HCS12.

Most 8-bit and 16-bit microcontrollers multiplex their address and data buses onto the same set of pins in order to lower the pin count and device cost. All of the read and write cycles take exactly one E clock cycle to complete. However, the HCS12 allows the user to stretch the E clock so that the slower but less expensive memory chips can be used in the product. A read or write cycle starts with the E clock going low. When the E clock is low, the HCS12 drives the address onto the multiplexed address/data bus. The control circuitry outside the MCU latches the address to keep it valid throughout the whole bus cycle. When the E clock goes high, the HCS12 stops driving the address onto the bus and either waits for the read data to be returned from the memory devices or drives data to be written into external memory on the multiplexed bus.

There are three major issues in the external memory expansion:

■ Memory space assignment

■ Address decoder design and certain additional signals generation

■ Timing verification

14.12 Exercises

E14.1 Write a program to erase the first hundred words of the on-chip EEPROM and then to write the values 1, 2, . . ., 100 to these 100 words, read them back, and store them in the first 100 words of the on-chip SRAM.

E14.2 Write an instruction sequence to remap the register block so that it starts at $1000.

E14.3 Write an instruction sequence to remap the SRAM block so that it starts at $2000.

E14.4 Write an instruction sequence to disable E clock stretch and disable on-chip ROM for expanded mode.

E14.5 Write an instruction sequence to map the EEPROM block to start from $0000.

E14.6 Give an instruction to protect the highest 64 bytes of the EEPROM.

E14.7 Write an instruction sequence to protect the 1 kB in the lower address range and the 4 kB in the higher address range of the flash 3 of the HCS12DP256.

E14.8 Write a C function that erases the EEPROM in bulk.

E14.9 Write a C function that erases the EEPROM in bulk and verifies it.

E14.10 What is the slowest SRAM read access time that does not require the designer to stretch the E clock of the HCS12? Assume the E clock frequency is 25 MHz.

14.13 Lab Exercises and Assignments

L14.1 Write a program to store the first one hundred prime numbers (starting from 2 and upward) in the on-chip EEPROM starting from address $400. Read out the prime numbers and display them to the terminal monitor with each line displaying 5 numbers.

Appendix A: Instruction Set Reference

Freescale Semiconductor, Inc.

Appendix A. Instruction Set Reference

A.1 Contents

A.2 Introduction

This appendix provides quick references for the instruction set, opcode map, and encoding.

Freescale Semiconductor, Inc.

Instruction Set Reference

Figure A-1. Programming Model

For More Information On This Product,
Go to: www.freescale.com

Reprinted with permission of Freescale Semiconductor, Inc.

Freescale Semiconductor, Inc.

A.3 Stack and Memory Layout

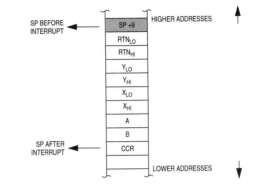

A.4 Interrupt Vector Locations

$FFFE, $FFFF	Power-On (POR) or External Reset
$FFFC, $FFFD	Clock Monitor Reset
$FFFA, $FFFB	Computer Operating Properly (COP Watchdog Reset
$FFF8, $FFF9	Unimplemented Opcode Trap
$FFF6, $FFF7	Software Interrupt Instruction (SWI)
$FFF4, $FFF5	XIRQ
$FFF2, $FFF3	IRQ
$FFC0–$FFF1 (M68HC12)	Device-Specific Interrupt Sources
$FF00–$FFF1 (HCS12)	Device-Specific Interrupt Sources

Freescale Semiconductor, Inc.

Instruction Set Reference

A.5 Notation Used in Instruction Set Summary

CPU Register Notation

Accumulator A — A or a	Index Register Y — Y or y
Accumulator B — B or b	Stack Pointer — SP, sp, or s
Accumulator D — D or d	Program Counter — PC, pc, or p
Index Register X — X or x	Condition Code Register — CCR or c

Explanation of Italic Expressions in Source Form Column

abc — A or B or CCR

abcdxys — A or B or CCR or D or X or Y or SP. Some assemblers also allow T2 or T3.

abd — A or B or D

abdxys — A or B or D or X or Y or SP

dxys — D or X or Y or SP

msk8 — 8-bit mask, some assemblers require # symbol before value

opr8i — 8-bit immediate value

opr16i — 16-bit immediate value

opr8a — 8-bit address used with direct address mode

opr16a — 16-bit address value

oprx0_xysp — Indexed addressing postbyte code:

    oprx3,–xys Predecrement X or Y or SP by 1 . . . 8

    oprx3,+xys Preincrement X or Y or SP by 1 . . . 8

    oprx3,xys– Postdecrement X or Y or SP by 1 . . . 8

    oprx3,xys+ Postincrement X or Y or SP by 1 . . . 8

    oprx5,xysp 5-bit constant offset from X or Y or SP or PC

    abd,xysp Accumulator A or B or D offset from X or Y or SP or PC

oprx3 — Any positive integer 1 . . . 8 for pre/post increment/decrement

oprx5 — Any integer in the range –16 . . . +15

oprx9 — Any integer in the range –256 . . . +255

oprx16 — Any integer in the range –32,768 . . . 65,535

page — 8-bit value for PPAGE, some assemblers require # symbol before this value

rel8 — Label of branch destination within –256 to +255 locations

rel9 — Label of branch destination within –512 to +511 locations

rel16 — Any label within 64K memory space

trapnum — Any 8-bit integer in the range $30-$39 or $40-$FF

xys — X or Y or SP

xysp — X or Y or SP or PC

Operators

+ — Addition

– — Subtraction

• — Logical AND

+ — Logical OR (inclusive)

Continued on next page

For More Information On This Product,
Go to: www.freescale.com

Reprinted with permission of Freescale Semiconductor, Inc.

Freescale Semiconductor, Inc.

Operators (continued)

$\oplus$ — Logical exclusive OR

$\times$ — Multiplication

$\div$ — Division

$\overline{M}$ — Negation. One's complement (invert each bit of M)

: — Concatenate
Example: A : B means the 16-bit value formed by concatenating 8-bit accumulator A with 8-bit accumulator B.
A is in the high-order position.

$\Rightarrow$ — Transfer
Example: (A) $\Rightarrow$ M means the content of accumulator A is transferred to memory location M.

$\Leftrightarrow$ — Exchange
Example: D $\Leftrightarrow$ X means exchange the contents of D with those of X.

Address Mode Notation

INH — Inherent; no operands in object code

IMM — Immediate; operand in object code

DIR — Direct; operand is the lower byte of an address from $0000 to $00FF

EXT — Operand is a 16-bit address

REL — Two's complement relative offset; for branch instructions

IDX — Indexed (no extension bytes); includes:
    5-bit constant offset from X, Y, SP, or PC
    Pre/post increment/decrement by 1 . . . 8
    Accumulator A, B, or D offset

IDX1 — 9-bit signed offset from X, Y, SP, or PC; 1 extension byte

IDX2 — 16-bit signed offset from X, Y, SP, or PC; 2 extension bytes

[IDX2] — Indexed-indirect; 16-bit offset from X, Y, SP, or PC

[D, IDX] — Indexed-indirect; accumulator D offset from X, Y, SP, or PC

Machine Coding

dd — 8-bit direct address $0000 to $00FF. (High byte assumed to be $00).

ee — High-order byte of a 16-bit constant offset for indexed addressing.

eb — Exchange/Transfer post-byte. See **Table A-5** on page 436.

ff — Low-order eight bits of a 9-bit signed constant offset for indexed addressing, or low-order byte of a 16-bit constant offset for indexed addressing.

hh — High-order byte of a 16-bit extended address.

ii — 8-bit immediate data value.

jj — High-order byte of a 16-bit immediate data value.

kk — Low-order byte of a 16-bit immediate data value.

lb — Loop primitive (DBNE) post-byte. See **Table A-6** on page 437.

ll — Low-order byte of a 16-bit extended address.

mm — 8-bit immediate mask value for bit manipulation instructions.
Set bits indicate bits to be affected.

Freescale Semiconductor, Inc.

Instruction Set Reference

pg — Program page (bank) number used in CALL instruction.

qq — High-order byte of a 16-bit relative offset for long branches.

tn — Trap number $30–$39 or $40–$FF.

rr — Signed relative offset $80 (–128) to $7F (+127).
Offset relative to the byte following the relative offset byte, or low-order byte of a 16-bit relative offset for long branches.

xb — Indexed addressing post-byte. See **Table A-3** on page 434 and **Table A-4** on page 435.

Access Detail

Each code letter except (,), and comma equals one CPU cycle. Uppercase = 16-bit operation and lowercase = 8-bit operation. For complex sequences see the *CPU12 Reference Manual* (CPU12RM/AD) for more detailed information.

f — Free cycle, CPU doesn't use bus

g — Read PPAGE internally

I — Read indirect pointer (indexed indirect)

i — Read indirect PPAGE value (CALL indirect only)

n — Write PPAGE internally

O — Optional program word fetch (P) if instruction is misaligned and has an odd number of bytes of object code — otherwise, appears as a free cycle (f); Page 2 prebyte treated as a separate 1-byte instruction

P — Program word fetch (always an aligned-word read)

r — 8-bit data read

R — 16-bit data read

s — 8-bit stack write

S — 16-bit stack write

w — 8-bit data write

W — 16-bit data write

u — 8-bit stack read

U — 16-bit stack read

V — 16-bit vector fetch (always an aligned-word read)

t — 8-bit conditional read (or free cycle)

T — 16-bit conditional read (or free cycle)

x — 8-bit conditional write (or free cycle)

( ) — Indicate a microcode loop

, — Indicates where an interrupt could be honored

Special Cases

PPP/P — Short branch, PPP if branch taken, P if not

OPPP/OPO — Long branch, OPPP if branch taken, OPO if not

Reprinted with permission of Freescale Semiconductor, Inc.

Freescale Semiconductor, Inc.

Condition Codes Columns

   – — Status bit not affected by operation.

   0 — Status bit cleared by operation.

   1 — Status bit set by operation.

   Δ — Status bit affected by operation.

   fl — Status bit may be cleared or remain set, but is not set by operation.

   ⇑ — Status bit may be set or remain cleared, but is not cleared by operation.

   ? — Status bit may be changed by operation but the final state is not defined.

   ! — Status bit used for a special purpose.

For More Information On This Product,
Go to: www.freescale.com

Freescale Semiconductor, Inc.

Instruction Set Reference

Table A-1. Instruction Set Summary (Sheet 1 of 14)

Source Form	Operation	Addr. Mode	Machine Coding (hex)	Access Detail HCS12	Access Detail M68HC12	S X H I	N Z V C
ABA	(A) + (B) ⇒ A Add Accumulators A and B	INH	18 06	OO	OO	– – Δ –	Δ Δ Δ Δ
ABX	(B) + (X) ⇒ X *Translates to LEAX B,X*	IDX	1A E5	Pf	PP[1]	– – – –	– – – –
ABY	(B) + (Y) ⇒ Y *Translates to LEAY B,Y*	IDX	19 ED	Pf	PP[1]	– – – –	– – – –
ADCA #opr8i ADCA opr8a ADCA opr16a ADCA oprx0_xysp ADCA oprx9,xysp ADCA oprx16,xysp ADCA [D,xysp] ADCA [oprx16,xysp]	(A) + (M) + C ⇒ A Add with Carry to A	IMM DIR EXT IDX IDX1 IDX2 [D,IDX] [IDX2]	89 ii 99 dd B9 hh ll A9 xb A9 xb ff A9 xb ee ff A9 xb A9 xb ee ff	P rPf rPO rPf rPO frPP fIfrPf fIPrPf	P rfP rOP rfP rPO frPP fIfrfP fIPrfP	– – Δ –	Δ Δ Δ Δ
ADCB #opr8i ADCB opr8a ADCB opr16a ADCB oprx0_xysp ADCB oprx9,xysp ADCB oprx16,xysp ADCB [D,xysp] ADCB [oprx16,xysp]	(B) + (M) + C ⇒ B Add with Carry to B	IMM DIR EXT IDX IDX1 IDX2 [D,IDX] [IDX2]	C9 ii D9 dd F9 hh ll E9 xb E9 xb ff E9 xb ee ff E9 xb E9 xb ee ff	P rPf rPO rPf rPO frPP fIfrPf fIPrPf	P rfP rOP rfP rPO frPP fIfrfP fIPrfP	– – Δ –	Δ Δ Δ Δ

ADDA #opr8i
ADDA opr8a
ADDA opr16a
ADDA oprx0_xysp
ADDA oprx9,xysp
ADDA oprx16,xysp
ADDA [D,xysp]
ADDA [oprx16,xysp]
	(A) + (M) ⇒ A
	Add without Carry to A
		IMM
		DIR
		EXT
		IDX
		IDX1
		IDX2
		[D,IDX]
		[IDX2]
			8B ii
			9B dd
			BB hh ll
			AB xb
			AB xb ff
			AB xb ee ff
			AB xb
			AB xb ee ff
				P
				rPf
				rPO
				rPf
				rPO
				frPP
				fIfrPf
				fIPrPf
					P
					rfP
					rOP
					rfP
					rPO
					frPP
					fIfrfP
					fIPrfP
						– – Δ –	Δ Δ Δ Δ

ADDB #opr8i
ADDB opr8a
ADDB opr16a
ADDB oprx0_xysp
ADDB oprx9,xysp
ADDB oprx16,xysp
ADDB [D,xysp]
ADDB [oprx16,xysp]
	(B) + (M) ⇒ B
	Add without Carry to B
		IMM
		DIR
		EXT
		IDX
		IDX1
		IDX2
		[D,IDX]
		[IDX2]
			CB ii
			DB dd
			FB hh ll
			EB xb
			EB xb ff
			EB xb ee ff
			EB xb
			EB xb ee ff
				P
				rPf
				rPO
				rPf
				rPO
				frPP
				fIfrPf
				fIPrPf
					P
					rfP
					rOP
					rfP
					rPO
					frPP
					fIfrfP
					fIPrfP
						– – Δ –	Δ Δ Δ Δ

ADDD #opr16i
ADDD opr8a
ADDD opr16a
ADDD oprx0_xysp
ADDD oprx9,xysp
ADDD oprx16,xysp
ADDD [D,xysp]
ADDD [oprx16,xysp]
	(A:B) + (M:M+1) ⇒ A:B
	Add 16-Bit to D (A:B)
		IMM
		DIR
		EXT
		IDX
		IDX1
		IDX2
		[D,IDX]
		[IDX2]
			C3 jj kk
			D3 dd
			F3 hh ll
			E3 xb
			E3 xb ff
			E3 xb ee ff
			E3 xb
			E3 xb ee ff
				PO
				RPf
				RPO
				RPf
				RPO
				fRPP
				fIfRPf
				fIPRPf
					OP
					RfP
					ROP
					RfP
					RPO
					fRPP
					fIfRfP
					fIPRfP
						– – – –	Δ Δ Δ Δ

ANDA #opr8i ANDA opr8a ANDA opr16a ANDA oprx0_xysp ANDA oprx9,xysp ANDA oprx16,xysp ANDA [D,xysp] ANDA [oprx16,xysp]	(A) • (M) ⇒ A Logical AND A with Memory	IMM DIR EXT IDX IDX1 IDX2 [D,IDX] [IDX2]	84 ii 94 dd B4 hh ll A4 xb A4 xb ff A4 xb ee ff A4 xb A4 xb ee ff	P rPf rPO rPf rPO frPP fIfrPf fIPrPf	P rfP rOP rfP rPO frPP fIfrfP fIPrfP	– – – –	Δ Δ 0 –
ANDB #opr8i ANDB opr8a ANDB opr16a ANDB oprx0_xysp ANDB oprx9,xysp ANDB oprx16,xysp ANDB [D,xysp] ANDB [oprx16,xysp]	(B) • (M) ⇒ B Logical AND B with Memory	IMM DIR EXT IDX IDX1 IDX2 [D,IDX] [IDX2]	C4 ii D4 dd F4 hh ll E4 xb E4 xb ff E4 xb ee ff E4 xb E4 xb ee ff	P rPf rPO rPf rPO frPP fIfrPf fIPrPf	P rfP rOP rfP rPO frPP fIfrfP fIPrfP	– – – –	Δ Δ 0 –
ANDCC #opr8i	(CCR) • (M) ⇒ CCR Logical AND CCR with Memory	IMM	10 ii	P	P	⇓⇓⇓⇓	⇓⇓⇓⇓

Note 1. Due to internal CPU requirements, the program word fetch is performed twice to the same address during this instruction.

Freescale Semiconductor, Inc.

Table A-1. Instruction Set Summary (Sheet 2 of 14)

Source Form	Operation	Addr. Mode	Machine Coding (hex)	Access Detail HCS12	Access Detail M68HC12	S X H I	N Z V C
ASL opr16a ASL oprx0_xysp ASL oprx9,xysp ASL oprx16,xysp ASL [D,xysp] ASL [oprx16,xysp] ASLA ASLB	Arithmetic Shift Left Arithmetic Shift Left Accumulator A Arithmetic Shift Left Accumulator B	EXT IDX IDX1 IDX2 [D,IDX] [IDX2] INH INH	78 hh 11 68 xb 68 xb ff 68 xb ee ff 68 xb 68 xb ee ff 48 58	rPwO rPw rPwO frPwP fIfrPw fIPrPw O O	rOPw rPw rPOw frPPw fIfrPw fIPrPw O O	- - - -	Δ Δ Δ Δ
ASLD	Arithmetic Shift Left Double	INH	59	O	O	- - - -	Δ Δ Δ Δ
ASR opr16a ASR oprx0_xysp ASR oprx9,xysp ASR oprx16,xysp ASR [D,xysp] ASR [oprx16,xysp] ASRA ASRB	Arithmetic Shift Right Arithmetic Shift Right Accumulator A Arithmetic Shift Right Accumulator B	EXT IDX IDX1 IDX2 [D,IDX] [IDX2] INH INH	77 hh 11 67 xb 67 xb ff 67 xb ee ff 67 xb 67 xb ee ff 47 57	rPwO rPw rPwO frPwP fIfrPw fIPrPw O O	rOPw rPw rPOw frPPw fIfrPw fIPrPw O O	- - - -	Δ Δ Δ Δ
BCC rel8	Branch if Carry Clear (if C = 0)	REL	24 rr	PPP/P[1]	PPP/P[1]	- - - -	- - - -
BCLR opr8a, msk8 BCLR opr16a, msk8 BCLR oprx0_xysp, msk8 BCLR oprx9,xysp, msk8 BCLR oprx16,xysp, msk8	(M) • (mm) ⇒ M Clear Bit(s) in Memory	DIR EXT IDX IDX1 IDX2	4D dd mm 1D hh 11 mm 0D xb mm 0D xb ff mm 0D xb ee ff mm	rPwO rPwP rPwO rPwP frPwPO	rPOw rPPPw rPOw rPwP frPwOP	- - - -	Δ Δ 0 -
BCS rel8	Branch if Carry Set (if C = 1)	REL	25 rr	PPP/P[1]	PPP/P[1]	- - - -	- - - -
BEQ rel8	Branch if Equal (if Z = 1)	REL	27 rr	PPP/P[1]	PPP/P[1]	- - - -	- - - -
BGE rel8	Branch if Greater Than or Equal (if N ⊕ V = 0) (signed)	REL	2C rr	PPP/P[1]	PPP/P[1]	- - - -	- - - -
BGND	Place CPU in Background Mode see CPU12 Reference Manual	INH	00	VfPPP	VfPPP	- - - -	- - - -
BGT rel8	Branch if Greater Than (if Z + (N ⊕ V) = 0) (signed)	REL	2E rr	PPP/P[1]	PPP/P[1]	- - - -	- - - -
BHI rel8	Branch if Higher (if C + Z = 0) (unsigned)	REL	22 rr	PPP/P[1]	PPP/P[1]	- - - -	- - - -
BHS rel8	Branch if Higher or Same (if C = 0) (unsigned) same function as BCC	REL	24 rr	PPP/P[1]	PPP/P[1]	- - - -	- - - -
BITA #opr8i BITA opr8a BITA opr16a BITA oprx0_xysp BITA oprx9,xysp BITA oprx16,xysp BITA [D,xysp] BITA [oprx16,xysp]	(A) • (M) Logical AND A with Memory Does not change Accumulator or Memory	IMM DIR EXT IDX IDX1 IDX2 [D,IDX] [IDX2]	85 ii 95 dd B5 hh 11 A5 xb A5 xb ff A5 xb ee ff A5 xb A5 xb ee ff	P rPf rPO rPf rPO frPP fIfrPf fIPrPf	P rfP rOP rfP rPO frPP fIfrfP fIPrfP	- - - -	Δ Δ 0 -
BITB #opr8i BITB opr8a BITB opr16a BITB oprx0_xysp BITB oprx9,xysp BITB oprx16,xysp BITB [D,xysp] BITB [oprx16,xysp]	(B) • (M) Logical AND B with Memory Does not change Accumulator or Memory	IMM DIR EXT IDX IDX1 IDX2 [D,IDX] [IDX2]	C5 ii D5 dd F5 hh 11 E5 xb E5 xb ff E5 xb ee ff E5 xb E5 xb ee ff	P rPf rPO rPf rPO frPP fIfrPf fIPrPf	P rfP rOP rfP rPO frPP fIfrfP fIPrfP	- - - -	Δ Δ 0 -
BLE rel8	Branch if Less Than or Equal (if Z + (N ⊕ V) = 1) (signed)	REL	2F rr	PPP/P[1]	PPP/P[1]	- - - -	- - - -
BLO rel8	Branch if Lower (if C = 1) (unsigned) same function as BCS	REL	25 rr	PPP/P[1]	PPP/P[1]	- - - -	- - - -

Note 1. PPP/P indicates this instruction takes three cycles to refill the instruction queue if the branch is taken and one program fetch cycle if the branch is not taken.

For More Information On This Product,
Go to: www.freescale.com

Reprinted with permission of Freescale Semiconductor, Inc.

Freescale Semiconductor, Inc.

Instruction Set Reference

Table A-1. Instruction Set Summary (Sheet 3 of 14)

Source Form	Operation	Addr. Mode	Machine Coding (hex)	Access Detail HCS12	Access Detail M68HC12	S X H I	N Z V C
BLS rel8	Branch if Lower or Same (if C + Z = 1) (unsigned)	REL	23 rr	PPP/P[1]	PPP/P[1]	- - - -	- - - -
BLT rel8	Branch if Less Than (if N ⊕ V = 1) (signed)	REL	2D rr	PPP/P[1]	PPP/P[1]	- - - -	- - - -
BMI rel8	Branch if Minus (if N = 1)	REL	2B rr	PPP/P[1]	PPP/P[1]	- - - -	- - - -
BNE rel8	Branch if Not Equal (if Z = 0)	REL	26 rr	PPP/P[1]	PPP/P[1]	- - - -	- - - -
BPL rel8	Branch if Plus (if N = 0)	REL	2A rr	PPP/P[1]	PPP/P[1]	- - - -	- - - -
BRA rel8	Branch Always (if 1 = 1)	REL	20 rr	PPP	PPP	- - - -	- - - -
BRCLR opr8a, msk8, rel8 BRCLR opr16a, msk8, rel8 BRCLR oprx0_xysp, msk8, rel8 BRCLR oprx9,xysp, msk8, rel8 BRCLR oprx16,xysp, msk8, rel8	Branch if (M) • (mm) = 0 (if All Selected Bit(s) Clear)	DIR EXT IDX IDX1 IDX2	4F dd mm rr 1F hh ll mm rr 0F xb mm rr 0F xb ff mm rr 0F xb ee ff mm rr	rPPP rfPPP rPPP rfPPP PrfPPP	rPPP rfPPP rPPP rffPPP frPffPPP	- - - -	- - - -
BRN rel8	Branch Never (if 1 = 0)	REL	21 rr	P	P	- - - -	- - - -
BRSET opr8, msk8, rel8 BRSET opr16a, msk8, rel8 BRSET oprx0_xysp, msk8, rel8 BRSET oprx9,xysp, msk8, rel8 BRSET oprx16,xysp, msk8, rel8	Branch if (M̄) • (mm) = 0 (if All Selected Bit(s) Set)	DIR EXT IDX IDX1 IDX2	4E dd mm rr 1E hh ll mm rr 0E xb mm rr 0E xb ff mm rr 0E xb ee ff mm rr	rPPP rfPPP rPPP rfPPP PrfPPP	rPPP rfPPP rPPP rffPPP frPffPPP	- - - -	- - - -
BSET opr8, msk8 BSET opr16a, msk8 BSET oprx0_xysp, msk8 BSET oprx9,xysp, msk8 BSET oprx16,xysp, msk8	(M) + (mm) ⇒ M Set Bit(s) in Memory	DIR EXT IDX IDX1 IDX2	4C dd mm 1C hh ll mm 0C xb mm 0C xb ff mm 0C xb ee ff mm	rPwO rPwP rPwO rPwP frPwPO	rPOw rPPw rPOw rPwP frPwOP	- - - -	Δ Δ 0 -
BSR rel8	(SP) – 2 ⇒ SP; RTN_H:RTN_L ⇒ M_(SP):M_(SP+1) Subroutine address ⇒ PC Branch to Subroutine	REL	07 rr	SPPP	PPPS	- - - -	- - - -
BVC rel8	Branch if Overflow Bit Clear (if V = 0)	REL	28 rr	PPP/P[1]	PPP/P[1]	- - - -	- - - -
BVS rel8	Branch if Overflow Bit Set (if V = 1)	REL	29 rr	PPP/P[1]	PPP/P[1]	- - - -	- - - -
CALL opr16a, page CALL oprx0_xysp, page CALL oprx9,xysp, page CALL oprx16,xysp, page CALL [D,xysp] CALL [oprx16, xysp]	(SP) – 2 ⇒ SP; RTN_H:RTN_L ⇒ M_(SP):M_(SP+1) (SP) – 1 ⇒ SP; (PPG) ⇒ M_(SP); pg ⇒ PPAGE register; Program address ⇒ PC Call subroutine in extended memory (Program may be located on another expansion memory page.) Indirect modes get program address and new pg value based on pointer.	EXT IDX IDX1 IDX2 [D,IDX] [IDX2]	4A hh ll pg 4B xb pg 4B xb ff pg 4B xb ee ff pg 4B xb 4B xb ee ff	gnSsPPP gnSsPPP gnSsPPP fgnSsPPP fIignSsPPP fIignSsPPP	gnfSsPPP gnSsPPP gnfSsPPP fgnSsPPP fIignSsPPP fIignSsPPP	- - - -	- - - -
CBA	(A) – (B) Compare 8-Bit Accumulators	INH	18 17	OO	OO	- - - -	Δ Δ Δ Δ
CLC	0 ⇒ C Translates to ANDCC #$FE	IMM	10 FE	P	P	- - - -	- - - 0
CLI	0 ⇒ I Translates to ANDCC #$EF (enables I-bit interrupts)	IMM	10 EF	P	P	- - - 0	- - - -
CLR opr16a CLR oprx0_xysp CLR oprx9,xysp CLR oprx16,xysp CLR [D,xysp] CLR [oprx16,xysp] CLRA CLRB	0 ⇒ M Clear Memory Location 0 ⇒ A Clear Accumulator A 0 ⇒ B Clear Accumulator B	EXT IDX IDX1 IDX2 [D,IDX] [IDX2] INH INH	79 hh ll 69 xb 69 xb ff 69 xb ee ff 69 xb 69 xb ee ff 87 C7	PwO Pw PwO PwP PIfw PIPw O O	wOP Pw PwO PwP PIfPw PIPPw O O	- - - -	0 1 0 0
CLV	0 ⇒ V Translates to ANDCC #$FD	IMM	10 FD	P	P	- - - -	- - 0 -

Note 1. PPP/P indicates this instruction takes three cycles to refill the instruction queue if the branch is taken and one program fetch cycle if the branch is not taken.

Freescale Semiconductor, Inc.

Table A-1. Instruction Set Summary (Sheet 4 of 14)

Source Form	Operation	Addr. Mode	Machine Coding (hex)	Access Detail HCS12	Access Detail M68HC12	S X H I	N Z V C

CMPA #opr8i
CMPA opr8a
CMPA opr16a
CMPA oprx0_xysp
CMPA oprx9,xysp
CMPA oprx16,xysp
CMPA [D,xysp]
CMPA [oprx16,xysp]
	(A) – (M)
	Compare Accumulator A with Memory
		IMM
		DIR
		EXT
		IDX
		IDX1
		IDX2
		[D,IDX]
		[IDX2]
			81 ii
			91 dd
			B1 hh ll
			A1 xb
			A1 xb ff
			A1 xb ee ff
			A1 xb
			A1 xb ee ff
				P
				rPf
				rPO
				rPf
				rPO
				frPP
				fIfrPf
				fIPrPf
					P
					rfP
					rOP
					rfP
					rPO
					frPP
					fIfrfP
					fIPrfP
						– – – –	Δ Δ Δ Δ

CMPB #opr8i
CMPB opr8a
CMPB opr16a
CMPB oprx0_xysp
CMPB oprx9,xysp
CMPB oprx16,xysp
CMPB [D,IDX]
CMPB [oprx16,xysp]
	(B) – (M)
	Compare Accumulator B with Memory
		IMM
		DIR
		EXT
		IDX
		IDX1
		IDX2
		[D,IDX]
		[IDX2]
			C1 ii
			D1 dd
			F1 hh ll
			E1 xb
			E1 xb ff
			E1 xb ee ff
			E1 xb
			E1 xb ee ff
				P
				rPf
				rPO
				rPf
				rPO
				frPP
				fIfrPf
				fIPrPf
					P
					rfP
					rOP
					rfP
					rPO
					frPP
					fIfrfP
					fIPrfP
						– – – –	Δ Δ Δ Δ

COM opr16a COM oprx0_xysp COM oprx9,xysp COM oprx16,xysp COM [D,xysp] COM [oprx16,xysp] COMA COMB	(M̄) ⇒ M equivalent to $FF – (M) ⇒ M 1's Complement Memory Location (Ā) ⇒ A Complement Accumulator A (B̄) ⇒ B Complement Accumulator B	EXT IDX IDX1 IDX2 [D,IDX] [IDX2] INH INH	71 hh ll 61 xb 61 xb ff 61 xb ee ff 61 xb 61 xb ee ff 41 51	rPwO rPw rPwO frPwP fIfrPw fIPrPw O O	rOPw rPw rPOw frPPw fIfrPw fIPrPw O O	– – – –	Δ Δ 0 1
CPD #opr16i CPD opr8a CPD opr16a CPD oprx0_xysp CPD oprx9,xysp CPD oprx16,xysp CPD [D,xysp] CPD [oprx16,xysp]	(A:B) – (M:M+1) Compare D to Memory (16-Bit)	IMM DIR EXT IDX IDX1 IDX2 [D,IDX] [IDX2]	8C jj kk 9C dd BC hh ll AC xb AC xb ff AC xb ee ff AC xb AC xb ee ff	PO RPf RPO RPf RPO fRPP fIfRPf fIPRPf	OP RfP ROP RfP RPO fRPP fIfRfP fIPRfP	– – – –	Δ Δ Δ Δ
CPS #opr16i CPS opr8a CPS opr16a CPS oprx0_xysp CPS oprx9,xysp CPS oprx16,xysp CPS [D,xysp] CPS [oprx16,xysp]	(SP) – (M:M+1) Compare SP to Memory (16-Bit)	IMM DIR EXT IDX IDX1 IDX2 [D,IDX] [IDX2]	8F jj kk 9F dd BF hh ll AF xb AF xb ff AF xb ee ff AF xb AF xb ee ff	PO RPf RPO RPf RPO fRPP fIfRPf fIPRPf	OP RfP ROP RfP RPO fRPP fIfRfP fIPRfP	– – – –	Δ Δ Δ Δ
CPX #opr16i CPX opr8a CPX opr16a CPX oprx0_xysp CPX oprx9,xysp CPX oprx16,xysp CPX [D,xysp] CPX [oprx16,xysp]	(X) – (M:M+1) Compare X to Memory (16-Bit)	IMM DIR EXT IDX IDX1 IDX2 [D,IDX] [IDX2]	8E jj kk 9E dd BE hh ll AE xb AE xb ff AE xb ee ff AE xb AE xb ee ff	PO RPf RPO RPf RPO fRPP fIfRPf fIPRPf	OP RfP ROP RfP RPO fRPP fIfRfP fIPRfP	– – – –	Δ Δ Δ Δ
CPY #opr16i CPY opr8a CPY opr16a CPY oprx0_xysp CPY oprx9,xysp CPY oprx16,xysp CPY [D,xysp] CPY [oprx16,xysp]	(Y) – (M:M+1) Compare Y to Memory (16-Bit)	IMM DIR EXT IDX IDX1 IDX2 [D,IDX] [IDX2]	8D jj kk 9D dd BD hh ll AD xb AD xb ff AD xb ee ff AD xb AD xb ee ff	PO RPf RPO RPf RPO fRPP fIfRPf fIPRPf	OP RfP ROP RfP RPO fRPP fIfRfP fIPRfP	– – – –	Δ Δ Δ Δ
DAA	Adjust Sum to BCD Decimal Adjust Accumulator A	INH	18 07	OfO	OfO	– – – –	Δ Δ ? Δ
DBEQ abdxys, rel9	(cntr) – 1 ⇒ cntr if (cntr) = 0, then Branch else Continue to next instruction Decrement Counter and Branch if = 0 (cntr = A, B, D, X, Y, or SP)	REL (9-bit)	04 lb rr	PPP (branch) PPO (no branch)	PPP	– – – –	– – – –

**For More Information On This Product,
Go to: www.freescale.com**

Reprinted with permission of Freescale Semiconductor, Inc.

Freescale Semiconductor, Inc.

Instruction Set Reference

Table A-1. Instruction Set Summary (Sheet 5 of 14)

Source Form	Operation	Addr. Mode	Machine Coding (hex)	Access Detail HCS12	M68HC12	S X H I	N Z V C
DBNE abdxys, rel9	(cntr) − 1 ⇒ cntr If (cntr) not = 0, then Branch; else Continue to next instruction Decrement Counter and Branch if ≠ 0 (cntr = A, B, D, X, Y, or SP)	REL (9-bit)	04 lb rr	PPP (branch) PPO (no branch)	PPP	– – – –	– – – –
DEC opr16a DEC oprx0_xysp DEC oprx9,xysp DEC oprx16,xysp DEC [D,xysp] DEC [oprx16,xysp] DECA DECB	(M) − $01 ⇒ M Decrement Memory Location (A) − $01 ⇒ A Decrement A (B) − $01 ⇒ B Decrement B	EXT IDX IDX1 IDX2 [D,IDX] [IDX2] INH INH	73 hh 11 63 xb 63 xb ff 63 xb ee ff 63 xb 63 xb ee ff 43 53	rPwO rPw rPwO frPwP fIfrPw fIPrPw O O	rOPw rPw rPOw frPPw fIfrPw fIPrPw O O	– – – –	Δ Δ Δ –
DES	(SP) − $0001 ⇒ SP Translates to LEAS −1,SP	IDX	1B 9F	Pf	PP[1]	– – – –	– – – –
DEX	(X) − $0001 ⇒ X Decrement Index Register X	INH	09	O	O	– – – –	– Δ – –
DEY	(Y) − $0001 ⇒ Y Decrement Index Register Y	INH	03	O	O	– – – –	– Δ – –
EDIV	(Y:D) ÷ (X) ⇒ Y Remainder ⇒ D 32 by 16 Bit ⇒ 16 Bit Divide (unsigned)	INH	11	ffffffffffO	ffffffffffO	– – – –	Δ Δ Δ Δ
EDIVS	(Y:D) ÷ (X) ⇒ Y Remainder ⇒ D 32 by 16 Bit ⇒ 16 Bit Divide (signed)	INH	18 14	OffffffffffO	OffffffffffO	– – – –	Δ Δ Δ Δ
EMACS opr16a [2]	(M(X):M(X+1)) × (M(Y):M(Y+1)) + (M~M+3) ⇒ M~M+3 16 by 16 Bit ⇒ 32 Bit Multiply and Accumulate (signed)	Special	18 12 hh 11	ORROfffRRfWWP	ORROfffRRfWWP	– – – –	Δ Δ Δ Δ
EMAXD oprx0_xysp EMAXD oprx9,xysp EMAXD oprx16,xysp EMAXD [D,xysp] EMAXD [oprx16,xysp]	MAX((D), (M:M+1)) ⇒ D MAX of 2 Unsigned 16-Bit Values N, Z, V and C status bits reflect result of internal compare ((D) − (M:M+1))	IDX IDX1 IDX2 [D,IDX] [IDX2]	18 1A xb 18 1A xb ff 18 1A xb ee ff 18 1A xb 18 1A xb ee ff	ORPf ORPO OfRPP OfIfRPf OfIPRPf	ORfP ORPO OfRPP OfIfRfP OfIPRfP	– – – –	Δ Δ Δ Δ
EMAXM oprx0_xysp EMAXM oprx9,xysp EMAXM oprx16,xysp EMAXM [D,xysp] EMAXM [oprx16,xysp]	MAX((D), (M:M+1)) ⇒ M:M+1 MAX of 2 Unsigned 16-Bit Values N, Z, V and C status bits reflect result of internal compare ((D) − (M:M+1))	IDX IDX1 IDX2 [D,IDX] [IDX2]	18 1E xb 18 1E xb ff 18 1E xb ee ff 18 1E xb 18 1E xb ee ff	ORPW ORPWO OfRPWP OfIfRPW OfIPRPW	ORPW ORPWO OfRPWP OfIfRPW OfIPRPW	– – – –	Δ Δ Δ Δ
EMIND oprx0_xysp EMIND oprx9,xysp EMIND oprx16,xysp EMIND [D,xysp] EMIND [oprx16,xysp]	MIN((D), (M:M+1)) ⇒ D MIN of 2 Unsigned 16-Bit Values N, Z, V and C status bits reflect result of internal compare ((D) − (M:M+1))	IDX IDX1 IDX2 [D,IDX] [IDX2]	18 1B xb 18 1B xb ff 18 1B xb ee ff 18 1B xb 18 1B xb ee ff	ORPf ORPO OfRPP OfIfRPf OfIPRPf	ORfP ORPO OfRPP OfIfRfP OfIPRfP	– – – –	Δ Δ Δ Δ
EMINM oprx0_xysp EMINM oprx9,xysp EMINM oprx16,xysp EMINM [D,xysp] EMINM [oprx16,xysp]	MIN((D), (M:M+1)) ⇒ M:M+1 MIN of 2 Unsigned 16-Bit Values N, Z, V and C status bits reflect result of internal compare ((D) − (M:M+1))	IDX IDX1 IDX2 [D,IDX] [IDX2]	18 1F xb 18 1F xb ff 18 1F xb ee ff 18 1F xb 18 1F xb ee ff	ORPW ORPWO OfRPWP OfIfRPW OfIPRPW	ORPW ORPWO OfRPWP OfIfRPW OfIPRPW	– – – –	Δ Δ Δ Δ
EMUL	(D) × (Y) ⇒ Y:D 16 by 16 Bit Multiply (unsigned)	INH	13	ffO	ffO	– – – –	Δ Δ – Δ
EMULS	(D) × (Y) ⇒ Y:D 16 by 16 Bit Multiply (signed)	INH	18 13	OfO (if followed by page 2 instruction) OffO	OfO OffO	– – – –	Δ Δ – Δ
EORA #opr8i EORA opr8a EORA opr16a EORA oprx0_xysp EORA oprx9,xysp EORA oprx16,xysp EORA [D,xysp] EORA [oprx16,xysp]	(A) ⊕ (M) ⇒ A Exclusive-OR A with Memory	IMM DIR EXT IDX IDX1 IDX2 [D,IDX] [IDX2]	88 ii 98 dd B8 hh 11 A8 xb A8 xb ff A8 xb ee ff A8 xb A8 xb ee ff	P rPf rPO rPf rPO frPP fIfrPf fIPrPf	P rfP rOP rfP rOP frPP fIfrfP fIPrfP	– – – –	Δ Δ 0 –

Notes:
1. Due to internal CPU requirements, the program word fetch is performed twice to the same address during this instruction.
2. opr16a is an extended address specification. Both X and Y point to source operands.

For More Information On This Product,
Go to: www.freescale.com

Freescale Semiconductor, Inc.

Table A-1. Instruction Set Summary (Sheet 6 of 14)

Source Form	Operation	Addr. Mode	Machine Coding (hex)	Access Detail HCS12	M68HC12	S X H I	N Z V C
EORB #opr8i EORB opr8a EORB opr16a EORB oprx0_xysp EORB oprx9,xysp EORB oprx16,xysp EORB [D,xysp] EORB [oprx16,xysp]	(B) ⊕ (M) ⇒ B Exclusive-OR B with Memory	IMM DIR EXT IDX IDX1 IDX2 [D,IDX] [IDX2]	C8 ii D8 dd F8 hh ll E8 xb E8 xb ff E8 xb ee ff E8 xb E8 xb ee ff	P rPf rPO rPf rPO frPP fIfrPf fIPrPf	P rfP rOP rfP rPO frPP fIfrfP fIPrfP	– – – –	Δ Δ 0 –
ETBL oprx0_xysp	(M:M+1)+ [(B)×((M+2:M+3) – (M:M+1))] ⇒ D 16-Bit Table Lookup and Interpolate Initialize B, and index before ETBL. <ea> points at first table entry (M:M+1) and B is fractional part of lookup value (no indirect addr. modes or extensions allowed)	IDX	18 3F xb	ORRfffffP	ORRfffffP	– – – –	Δ Δ – Δ ? C Bit is undefined in HC12
EXG abcdxys,abcdxys	(r1) ⇔ (r2) (if r1 and r2 same size) or $00:(r1) ⇒ r2 (if r1=8-bit; r2=16-bit) or (r1low) ⇔ (r2) (if r1=16-bit; r2=8-bit) r1 and r2 may be A, B, CCR, D, X, Y, or SP	INH	B7 eb	P	P	– – – –	– – – –
FDIV	(D) ÷ (X) ⇒ X; Remainder ⇒ D 16 by 16 Bit Fractional Divide	INH	18 11	OfffffffffffO	OfffffffffffO	– – – –	– Δ Δ Δ
IBEQ abdxys, rel9	(cntr) + 1⇒ cntr If (cntr) = 0, then Branch else Continue to next instruction Increment Counter and Branch if = 0 (cntr = A, B, D, X, Y, or SP)	REL (9-bit)	04 lb rr	PPP (branch) PPO (no branch)	PPP	– – – –	– – – –
IBNE abdxys, rel9	(cntr) + 1⇒ cntr if (cntr) not = 0, then Branch; else Continue to next instruction Increment Counter and Branch if ≠ 0 (cntr = A, B, D, X, Y, or SP)	REL (9-bit)	04 lb rr	PPP (branch) PPO (no branch)	PPP	– – – –	– – – –
IDIV	(D) ÷ (X) ⇒ X; Remainder ⇒ D 16 by 16 Bit Integer Divide (unsigned)	INH	18 10	OfffffffffffO	OfffffffffffO	– – – –	– Δ 0 Δ
IDIVS	(D) ÷ (X) ⇒ X; Remainder ⇒ D 16 by 16 Bit Integer Divide (signed)	INH	18 15	OfffffffffffO	OfffffffffffO	– – – –	Δ Δ Δ Δ
INC opr16a INC oprx0_xysp INC oprx9,xysp INC oprx16,xysp INC [D,xysp] INC [oprx16,xysp] INCA INCB	(M) + $01 ⇒ M Increment Memory Byte (A) + $01 ⇒ A Increment Acc. A (B) + $01 ⇒ B Increment Acc. B	EXT IDX IDX1 IDX2 [D,IDX] [IDX2] INH INH	72 hh ll 62 xb 62 xb ff 62 xb ee ff 62 xb 62 xb ee ff 42 52	rPwO rPw rPwO frPwP fIfrPw fIPrPw O O	rOPw rPw rPOw frPPw fIfrPw fIPrPw O O	– – – –	Δ Δ Δ –
INS	(SP) + $0001 ⇒ SP Translates to LEAS 1,SP	IDX	1B 81	Pf	PP[1]	– – – –	– – – –
INX	(X) + $0001 ⇒ X Increment Index Register X	INH	08	O	O	– – – –	– Δ – –
INY	(Y) + $0001 ⇒ Y Increment Index Register Y	INH	02	O	O	– – – –	– Δ – –
JMP opr16a JMP oprx0_xysp JMP oprx9,xysp JMP oprx16,xysp JMP [D,xysp] JMP [oprx16,xysp]	Routine address ⇒ PC Jump	EXT IDX IDX1 IDX2 [D,IDX] [IDX2]	06 hh ll 05 xb 05 xb ff 05 xb ee ff 05 xb 05 xb ee ff	PPP PPP PPP fPPP fIfPPP fIfPPP	PPP PPP PPP fPPP fIfPPP fIfPPP	– – – –	– – – –

Note 1. Due to internal CPU requirements, the program word fetch is performed twice to the same address during this instruction.

For More Information On This Product,
Go to: www.freescale.com

Reprinted with permission of Freescale Semiconductor, Inc.

Freescale Semiconductor, Inc.

Instruction Set Reference

Table A-1. Instruction Set Summary (Sheet 7 of 14)

Source Form	Operation	Addr. Mode	Machine Coding (hex)	Access Detail HCS12	Access Detail M68HC12	S X H I	N Z V C
JSR opr8a JSR opr16a JSR oprx0_xysp JSR oprx9,xysp JSR oprx16,xysp JSR [D,xysp] JSR [oprx16,xysp]	(SP) − 2 ⇒ SP; RTN$_H$:RTN$_L$ ⇒ M$_{(SP)}$:M$_{(SP+1)}$; Subroutine address ⇒ PC Jump to Subroutine	DIR EXT IDX IDX1 IDX2 [D,IDX] [IDX2]	17 dd 16 hh ll 15 xb 15 xb ff 15 xb ee ff 15 xb 15 xb ee ff	SPPP SPPP PPPS PPPS fPPPS fIfPPPS fIfPPPS	PPPS PPPS PPPS PPPS fPPPS fIfPPPS fIfPPPS	– – – –	– – – –
LBCC rel16	Long Branch if Carry Clear (if C = 0)	REL	18 24 qq rr	OPPP/OPO[1]	OPPP/OPO[1]	– – – –	– – – –
LBCS rel16	Long Branch if Carry Set (if C = 1)	REL	18 25 qq rr	OPPP/OPO[1]	OPPP/OPO[1]	– – – –	– – – –
LBEQ rel16	Long Branch if Equal (if Z = 1)	REL	18 27 qq rr	OPPP/OPO[1]	OPPP/OPO[1]	– – – –	– – – –
LBGE rel16	Long Branch Greater Than or Equal (if N ⊕ V = 0) (signed)	REL	18 2C qq rr	OPPP/OPO[1]	OPPP/OPO[1]	– – – –	– – – –
LBGT rel16	Long Branch if Greater Than (if Z + (N ⊕ V) = 0) (signed)	REL	18 2E qq rr	OPPP/OPO[1]	OPPP/OPO[1]	– – – –	– – – –
LBHI rel16	Long Branch if Higher (if C + Z = 0) (unsigned)	REL	18 22 qq rr	OPPP/OPO[1]	OPPP/OPO[1]	– – – –	– – – –
LBHS rel16	Long Branch if Higher or Same (if C = 0) (unsigned) same function as LBCC	REL	18 24 qq rr	OPPP/OPO[1]	OPPP/OPO[1]	– – – –	– – – –
LBLE rel16	Long Branch if Less Than or Equal (if Z + (N ⊕ V) = 1) (signed)	REL	18 2F qq rr	OPPP/OPO[1]	OPPP/OPO[1]	– – – –	– – – –
LBLO rel16	Long Branch if Lower (if C = 1) (unsigned) same function as LBCS	REL	18 25 qq rr	OPPP/OPO[1]	OPPP/OPO[1]	– – – –	– – – –
LBLS rel16	Long Branch if Lower or Same (if C + Z = 1) (unsigned)	REL	18 23 qq rr	OPPP/OPO[1]	OPPP/OPO[1]	– – – –	– – – –
LBLT rel16	Long Branch if Less Than (if N ⊕ V = 1) (signed)	REL	18 2D qq rr	OPPP/OPO[1]	OPPP/OPO[1]	– – – –	– – – –
LBMI rel16	Long Branch if Minus (if N = 1)	REL	18 2B qq rr	OPPP/OPO[1]	OPPP/OPO[1]	– – – –	– – – –
LBNE rel16	Long Branch if Not Equal (if Z = 0)	REL	18 26 qq rr	OPPP/OPO[1]	OPPP/OPO[1]	– – – –	– – – –
LBPL rel16	Long Branch if Plus (if N = 0)	REL	18 2A qq rr	OPPP/OPO[1]	OPPP/OPO[1]	– – – –	– – – –
LBRA rel16	Long Branch Always (if 1=1)	REL	18 20 qq rr	OPPP	OPPP	– – – –	– – – –
LBRN rel16	Long Branch Never (if 1 = 0)	REL	18 21 qq rr	OPO	OPO	– – – –	– – – –
LBVC rel16	Long Branch if Overflow Bit Clear (if V=0)	REL	18 28 qq rr	OPPP/OPO[1]	OPPP/OPO[1]	– – – –	– – – –
LBVS rel16	Long Branch if Overflow Bit Set (if V = 1)	REL	18 29 qq rr	OPPP/OPO[1]	OPPP/OPO[1]	– – – –	– – – –

LDAA #opr8i
LDAA opr8a
LDAA opr16a
LDAA oprx0_xysp
LDAA oprx9,xysp
LDAA oprx16,xysp
LDAA [D,xysp]
LDAA [oprx16,xysp]
	(M) ⇒ A
	Load Accumulator A
		IMM
		DIR
		EXT
		IDX
		IDX1
		IDX2
		[D,IDX]
		[IDX2]
			86 ii
			96 dd
			B6 hh ll
			A6 xb
			A6 xb ff
			A6 xb ee ff
			A6 xb
			A6 xb ee ff
				P
				rPf
				rPO
				rPf
				rPO
				frPP
				fIfrPf
				fIPrPf
					P
					rfP
					rOP
					rfP
					rPO
					frPP
					fIfrfP
					fIPrfP
						– – – –	Δ Δ 0 –

LDAB #opr8i
LDAB opr8a
LDAB opr16a
LDAB oprx0_xysp
LDAB oprx9,xysp
LDAB oprx16,xysp
LDAB [D,xysp]
LDAB [oprx16,xysp]
	(M) ⇒ B
	Load Accumulator B
		IMM
		DIR
		EXT
		IDX
		IDX1
		IDX2
		[D,IDX]
		[IDX2]
			C6 ii
			D6 dd
			F6 hh ll
			E6 xb
			E6 xb ff
			E6 xb ee ff
			E6 xb
			E6 xb ee ff
				P
				rPf
				rPO
				rPf
				rPO
				frPP
				fIfrPf
				fIPrPf
					P
					rfP
					rOP
					rfP
					rPO
					frPP
					fIfrfP
					fIPrfP
						– – – –	Δ Δ 0 –

LDD #opr16i
LDD opr8a
LDD opr16a
LDD oprx0_xysp
LDD oprx9,xysp
LDD oprx16,xysp
LDD [D,xysp]
LDD [oprx16,xysp]
	(M:M+1) ⇒ A:B
	Load Double Accumulator D (A:B)
		IMM
		DIR
		EXT
		IDX
		IDX1
		IDX2
		[D,IDX]
		[IDX2]
			CC jj kk
			DC dd
			FC hh ll
			EC xb
			EC xb ff
			EC xb ee ff
			EC xb
			EC xb ee ff
				PO
				RPf
				RPO
				RPf
				RPO
				fRPP
				fIfRPf
				fIPRPf
					OP
					RfP
					ROP
					RfP
					RPO
					fRPP
					fIfRfP
					fIPRfP
						– – – –	Δ Δ 0 –

Note 1. OPPP/OPO indicates this instruction takes four cycles to refill the instruction queue if the branch is taken and three cycles if the branch is not taken.

For More Information On This Product,
Go to: www.freescale.com

Reprinted with permission of Freescale Semiconductor, Inc.

Freescale Semiconductor, Inc.

Instruction Set Reference

Table A-1. Instruction Set Summary (Sheet 8 of 14)

Source Form	Operation	Addr. Mode	Machine Coding (hex)	Access Detail HCS12	Access Detail M68HC12	S X H I	N Z V C

LDS #opr16i
LDS opr8a
LDS opr16a
LDS oprx0_xysp
LDS oprx9,xysp
LDS oprx16,xysp
LDS [D,xysp]
LDS [oprx16,xysp]
	(M:M+1) ⇒ SP
	Load Stack Pointer
		IMM
		DIR
		EXT
		IDX
		IDX1
		IDX2
		[D,IDX]
		[IDX2]
			CF jj kk
			DF dd
			FF hh ll
			EF xb
			EF xb ff
			EF xb ee ff
			EF xb
			EF xb ee ff
				PO
				RPf
				RPO
				RPf
				RPO
				fRPP
				fIfRPf
				fIPRPf
					OP
					RfP
					ROP
					RfP
					RPO
					fRPP
					fIfRfP
					fIPRfP
						- - - -	Δ Δ 0 -

LDX #opr16i
LDX opr8a
LDX opr16a
LDX oprx0_xysp
LDX oprx9,xysp
LDX oprx16,xysp
LDX [D,IDX]
LDX [oprx16,xysp]
	(M:M+1) ⇒ X
	Load Index Register X
		IMM
		DIR
		EXT
		IDX
		IDX1
		IDX2
		[D,IDX]
		[IDX2]
			CE jj kk
			DE dd
			FE hh ll
			EE xb
			EE xb ff
			EE xb ee ff
			EE xb
			EE xb ee ff
				PO
				RPf
				RPO
				RPf
				RPO
				fRPP
				fIfRPf
				fIPRPf
					OP
					RfP
					ROP
					RfP
					RPO
					fRPP
					fIfRfP
					fIPRfP
						- - - -	Δ Δ 0 -

LDY #opr16i
LDY opr8a
LDY opr16a
LDY oprx0_xysp
LDY oprx9,xysp
LDY oprx16,xysp
LDY [D,xysp]
LDY [oprx16,xysp]
	(M:M+1) ⇒ Y
	Load Index Register Y
		IMM
		DIR
		EXT
		IDX
		IDX1
		IDX2
		[D,IDX]
		[IDX2]
			CD jj kk
			DD dd
			FD hh ll
			ED xb
			ED xb ff
			ED xb ee ff
			ED xb
			ED xb ee ff
				PO
				RPf
				RPO
				RPf
				RPO
				fRPP
				fIfRPf
				fIPRPf
					OP
					RfP
					ROP
					RfP
					RPO
					fRPP
					fIfRfP
					fIPRfP
						- - - -	Δ Δ 0 -

LEAS oprx0_xysp LEAS oprx9,xysp LEAS oprx16,xysp	Effective Address ⇒ SP Load Effective Address into SP	IDX IDX1 IDX2	1B xb 1B xb ff 1B xb ee ff	Pf PO PP	PP[1] PO PP	- - - -	- - - -
LEAX oprx0_xysp LEAX oprx9,xysp LEAX oprx16,xysp	Effective Address ⇒ X Load Effective Address into X	IDX IDX1 IDX2	1A xb 1A xb ff 1A xb ee ff	Pf PO PP	PP[1] PO PP	- - - -	- - - -
LEAY oprx0_xysp LEAY oprx9,xysp LEAY oprx16,xysp	Effective Address ⇒ Y Load Effective Address into Y	IDX IDX1 IDX2	19 xb 19 xb ff 19 xb ee ff	Pf PO PP	PP[1] PO PP	- - - -	- - - -
LSL opr16a LSL oprx0_xysp LSL oprx9,xysp LSL oprx16,xysp LSL [D,xysp] LSL [oprx16,xysp] LSLA LSLB	C b7 ← 0 Logical Shift Left same function as ASL Logical Shift Accumulator A to Left Logical Shift Accumulator B to Left	EXT IDX IDX1 IDX2 [D,IDX] [IDX2] INH INH	78 hh ll 68 xb 68 xb ff 68 xb ee ff 68 xb 68 xb ee ff 48 58	rPwO rPw rPwO frPPw fIfrPw fIPrPw O O	rOPw rPw rPOw frPPw fIfrPw fIPrPw O O	- - - -	Δ Δ Δ Δ
LSLD	C b7 A b0 b7 B b0 ← 0 Logical Shift Left D Accumulator same function as ASLD	INH	59	O	O	- - - -	Δ Δ Δ Δ
LSR opr16a LSR oprx0_xysp LSR oprx9,xysp LSR oprx16,xysp LSR [D,xysp] LSR [oprx16,xysp] LSRA LSRB	0 → b7 b0 C Logical Shift Right Logical Shift Accumulator A to Right Logical Shift Accumulator B to Right	EXT IDX IDX1 IDX2 [D,IDX] [IDX2] INH INH	74 hh ll 64 xb 64 xb ff 64 xb ee ff 64 xb 64 xb ee ff 44 54	rPwO rPw rPwO frPPw fIfrPw fIPrPw O O	rOPw rPw rPOw frPPw fIfrPw fIPrPw O O	- - - -	0 Δ Δ Δ
LSRD	0 → b7 A b0 b7 B b0 C Logical Shift Right D Accumulator	INH	49	O	O	- - - -	0 Δ Δ Δ
MAXA oprx0_xysp MAXA oprx9,xysp MAXA oprx16,xysp MAXA [D,xysp] MAXA [oprx16,xysp]	MAX((A), (M)) ⇒ A MAX of 2 Unsigned 8-Bit Values N, Z, V and C status bits reflect result of internal compare ((A) − (M)).	IDX IDX1 IDX2 [D,IDX] [IDX2]	18 18 xb 18 18 xb ff 18 18 xb ee ff 18 18 xb 18 18 xb ee ff	OrPf OrPO OfrPP OfIfrPf OfIPrPf	OrfP OrPO OfrPP OfIfrfP OfIPrfP	- - - -	Δ Δ Δ Δ

Note 1. Due to internal CPU requirements, the program word fetch is performed twice to the same address during this instruction.

Freescale Semiconductor, Inc.

Reprinted with permission of Freescale Semiconductor, Inc.

Freescale Semiconductor, Inc.

Instruction Set Reference

Table A-1. Instruction Set Summary (Sheet 9 of 14)

Source Form	Operation	Addr. Mode	Machine Coding (hex)	Access Detail HCS12	M68HC12	S X H I	N Z V C
MAXM oprx0_xysp MAXM oprx9,xysp MAXM oprx16,xysp MAXM [D,xysp] MAXM [oprx16,xysp]	MAX((A), (M)) ⇒ M MAX of 2 Unsigned 8-Bit Values N, Z, V and C status bits reflect result of internal compare ((A) – (M)).	IDX IDX1 IDX2 [D,IDX] [IDX2]	18 1C xb 18 1C xb ff 18 1C xb ee ff 18 1C xb 18 1C xb ee ff	OrPw OrPwO OfrPwP OfIfrPw OfIPrPw	OrPw OrPwO OfrPwP OfIfrPw OfIPrPw	– – – –	Δ Δ Δ Δ
MEM	μ (grade) ⇒ M(Y); (X) + 4 ⇒ X; (Y) + 1 ⇒ Y; A unchanged if (A) < P1 or (A) > P2 then μ = 0, else μ = MIN[((A) – P1)×S1, (P2 – (A))×S2, $FF] where: A = current crisp input value; X points at 4-byte data structure that describes a trapezoidal membership function (P1, P2, S1, S2); Y points at fuzzy input (RAM location). See CPU12 Reference Manual for special cases.	Special	01	RRfOw	RRfOw	– – ? –	? ? ? ?
MINA oprx0_xysp MINA oprx9,xysp MINA oprx16,xysp MINA [D,xysp] MINA [oprx16,xysp]	MIN((A), (M)) ⇒ A MIN of 2 Unsigned 8-Bit Values N, Z, V and C status bits reflect result of internal compare ((A) – (M)).	IDX IDX1 IDX2 [D,IDX] [IDX2]	18 19 xb 18 19 xb ff 18 19 xb ee ff 18 19 xb 18 19 xb ee ff	OrPf OrPO OfrPP OfIfrPf OfIPrPf	OrfP OrPO OfrPP OfIfrP OfIPrPf	– – – –	Δ Δ Δ Δ
MINM oprx0_xysp MINM oprx9,xysp MINM oprx16,xysp MINM [D,xysp] MINM [oprx16,xysp]	MIN((A), (M)) ⇒ M MIN of 2 Unsigned 8-Bit Values N, Z, V and C status bits reflect result of internal compare ((A) – (M)).	IDX IDX1 IDX2 [D,IDX] [IDX2]	18 1D xb 18 1D xb ff 18 1D xb ee ff 18 1D xb 18 1D xb ee ff	OrPw OrPwO OfrPwP OfIfrPw OfIPrPw	OrPw OrPwO OfrPwP OfIfrPw OfIPrPw	– – – –	Δ Δ Δ Δ

MOVB #opr8, opr16a[1]
MOVB #opr8i, oprx0_xysp[1]
MOVB opr16a, opr16a[1]
MOVB opr16a, oprx0_xysp[1]
MOVB oprx0_xysp, opr16a[1]
MOVB oprx0_xysp, oprx0_xysp[1]
	(M₁) ⇒ M₂
	Memory to Memory Byte-Move (8-Bit)
		IMM-EXT
		IMM-IDX
		EXT-EXT
		EXT-IDX
		IDX-EXT
		IDX-IDX
			18 0B ii hh ll
			18 08 xb ii
			18 0C hh ll hh ll
			18 09 xb hh ll
			18 0D xb hh ll
			18 0A xb xb
				OPwP
				OPwO
				OrPwPO
				OrPrPw
				OrPwP
				OrPwO
					OPwP
					OPwO
					OrPwPO
					OPrPw
					OrPwP
					OrPwO
						– – – –	– – – –

MOVW #oprx16, opr16a[1]
MOVW #opr16i, oprx0_xysp[1]
MOVW opr16a, opr16a[1]
MOVW opr16a, oprx0_xysp[1]
MOVW oprx0_xysp, opr16a[1]
MOVW oprx0_xysp, oprx0_xysp[1]
	(M:M+1₁) ⇒ M:M+1₂
	Memory to Memory Word-Move (16-Bit)
		IMM-EXT
		IMM-IDX
		EXT-EXT
		EXT-IDX
		IDX-EXT
		IDX-IDX
			18 03 jj kk hh ll
			18 00 xb jj kk
			18 04 hh ll hh ll
			18 01 xb hh ll
			18 05 xb hh ll
			18 02 xb xb
				OPWPO
				OPPW
				ORPWPO
				OPRPW
				ORPWP
				ORPWO
					OPWPO
					OPPW
					ORPWPO
					OPRPW
					ORPWP
					ORPWO
						– – – –	– – – –

MUL	(A) × (B) ⇒ A:B 8 by 8 Unsigned Multiply	INH	12	O	ffO	– – – –	– – – Δ
NEG opr16a NEG oprx0_xysp NEG oprx9,xysp NEG oprx16,xysp NEG [D,xysp] NEG [oprx16,xysp] NEGA	0 – (M) ⇒ M equivalent to (M̄) + 1 ⇒ M Two's Complement Negate 0 – (A) ⇒ A equivalent to (Ā) + 1 ⇒ A Negate Accumulator A	EXT IDX IDX1 IDX2 [D,IDX] [IDX2] INH	70 hh ll 60 xb 60 xb ff 60 xb ee ff 60 xb 60 xb ee ff 40	rPwO rPw rPwO frPwP fIfrPw fIPrPw O	rOPw rPw rPOw frPPw fIfrPw fIPrPw O	– – – –	Δ Δ Δ Δ
NEGB	0 – (B) ⇒ B equivalent to (B̄) + 1 ⇒ B Negate Accumulator B	INH	50	O	O		
NOP	No Operation	INH	A7	O	O	– – – –	– – – –
ORAA #opr8i ORAA opr8a ORAA opr16a ORAA oprx0_xysp ORAA oprx9,xysp ORAA oprx16,xysp ORAA [D,xysp] ORAA [oprx16,xysp]	(A) + (M) ⇒ A Logical OR A with Memory	IMM DIR EXT IDX IDX1 IDX2 [D,IDX] [IDX2]	8A ii 9A dd BA hh ll AA xb AA xb ff AA xb ee ff AA xb AA xb ee ff	P rPf rPO rPf rPO frPP fIfrPf fIPrPf	P rfP rOP rfP rPO frPP fIfrfP fIPrfP	– – – –	Δ Δ 0 –

Note 1. The first operand in the source code statement specifies the source for the move.

Freescale Semiconductor, Inc.

Table A-1. Instruction Set Summary (Sheet 10 of 14)

Source Form	Operation	Addr. Mode	Machine Coding (hex)	Access Detail		S X H I	N Z V C
				HCS12	**M68HC12**		
ORAB #opr8i ORAB opr8a ORAB opr16a ORAB oprx0_xysp ORAB oprx9,xysp ORAB oprx16,xysp ORAB [D,xysp] ORAB [oprx16,xysp]	(B) + (M) ⇒ B Logical OR B with Memory	IMM DIR EXT IDX IDX1 IDX2 [D,IDX] [IDX2]	CA ii DA dd FA hh ll EA xb EA xb ff EA xb ee ff EA xb EA xb ee ff	P rPf rPO rPf rPO frPP fIfrPf fIPrPf	P rfP rOP rfP rPO frPP fIfrfP fIPrfP	– – – –	Δ Δ 0 –
ORCC #opr8i	(CCR) + M ⇒ CCR Logical OR CCR with Memory	IMM	14 ii	P	P	⇑ – ⇑ ⇑	⇑ ⇑ ⇑ ⇑
PSHA	(SP) – 1 ⇒ SP; (A) ⇒ M$_{(SP)}$ Push Accumulator A onto Stack	INH	36	Os	Os	– – – –	– – – –
PSHB	(SP) – 1 ⇒ SP; (B) ⇒ M$_{(SP)}$ Push Accumulator B onto Stack	INH	37	Os	Os	– – – –	– – – –
PSHC	(SP) – 1 ⇒ SP; (CCR) ⇒ M$_{(SP)}$ Push CCR onto Stack	INH	39	Os	Os	– – – –	– – – –
PSHD	(SP) – 2 ⇒ SP; (A:B) ⇒ M$_{(SP)}$:M$_{(SP+1)}$ Push D Accumulator onto Stack	INH	3B	OS	OS	– – – –	– – – –
PSHX	(SP) – 2 ⇒ SP; (X$_H$:X$_L$) ⇒ M$_{(SP)}$:M$_{(SP+1)}$ Push Index Register X onto Stack	INH	34	OS	OS	– – – –	– – – –
PSHY	(SP) – 2 ⇒ SP; (Y$_H$:Y$_L$) ⇒ M$_{(SP)}$:M$_{(SP+1)}$ Push Index Register Y onto Stack	INH	35	OS	OS	– – – –	– – – –
PULA	(M$_{(SP)}$) ⇒ A; (SP) + 1 ⇒ SP Pull Accumulator A from Stack	INH	32	ufO	ufO	– – – –	– – – –
PULB	(M$_{(SP)}$) ⇒ B; (SP) + 1 ⇒ SP Pull Accumulator B from Stack	INH	33	ufO	ufO	– – – –	– – – –
PULC	(M$_{(SP)}$) ⇒ CCR; (SP) + 1 ⇒ SP Pull CCR from Stack	INH	38	ufO	ufO	Δ ⇓ Δ Δ	Δ Δ Δ Δ
PULD	(M$_{(SP)}$:M$_{(SP+1)}$) ⇒ A:B; (SP) + 2 ⇒ SP Pull D from Stack	INH	3A	UfO	UfO	– – – –	– – – –
PULX	(M$_{(SP)}$:M$_{(SP+1)}$) ⇒ X$_H$:X$_L$; (SP) + 2 ⇒ SP Pull Index Register X from Stack	INH	30	UfO	UfO	– – – –	– – – –
PULY	(M$_{(SP)}$:M$_{(SP+1)}$) ⇒ Y$_H$:Y$_L$; (SP) + 2 ⇒ SP Pull Index Register Y from Stack	INH	31	UfO	UfO	– – – –	– – – –
REV	MIN-MAX rule evaluation Find smallest rule input (MIN). Store to rule outputs unless fuzzy output is already larger (MAX). For rule weights see REVW. Each rule input is an 8-bit offset from the base address in Y. Each rule output is an 8-bit offset from the base address in Y. $FE separates rule inputs from rule outputs. $FF terminates the rule list. REV may be interrupted.	Special	18 3A	Orf(t,tx)O (exit + re-entry replaces comma above if interrupted) ff + Orf(t,	Orf(t,tx)O ff + Orf(t,	– – ? –	? ? Δ ?
REVW	MIN-MAX rule evaluation Find smallest rule input (MIN), Store to rule outputs unless fuzzy output is already larger (MAX). Rule weights supported, optional. Each rule input is the 16-bit address of a fuzzy input. Each rule output is the 16-bit address of a fuzzy output. The value $FFFE separates rule inputs from rule outputs. $FFFF terminates the rule list. REVW may be interrupted.	Special	18 3B	ORf(t,Tx)O (loop to read weight if enabled) (r,RfRf) (exit + re-entry replaces comma above if interrupted) ffff + ORf(t,	ORf(t,Tx)O (r,RfRf) fff + ORf(t,	– – ? –	? ? Δ !

Reprinted with permission of Freescale Semiconductor, Inc.

Freescale Semiconductor, Inc.

Instruction Set Reference

Table A-1. Instruction Set Summary (Sheet 11 of 14)

Source Form	Operation	Addr. Mode	Machine Coding (hex)	Access Detail HCS12	Access Detail M68HC12	S X H I	N Z V C
ROL opr16a ROL oprx0_xysp ROL oprx9,xysp ROL oprx16,xysp ROL [D,xysp] ROL [oprx16,xysp] ROLA ROLB	Rotate Memory Left through Carry Rotate A Left through Carry Rotate B Left through Carry	EXT IDX IDX1 IDX2 [D,IDX] [IDX2] INH INH	75 hh ll 65 xb 65 xb ff 65 xb ee ff 65 xb 65 xb ee ff 45 55	rPwO rPw rPwO frPwP fIfrPw fIPrPw O O	rOPw rPw rPOw frPPw fIfrPw fIPrPw O O	– – – –	Δ Δ Δ Δ
ROR opr16a ROR oprx0_xysp ROR oprx9,xysp ROR oprx16,xysp ROR [D,xysp] ROR [oprx16,xysp] RORA RORB	Rotate Memory Right through Carry Rotate A Right through Carry Rotate B Right through Carry	EXT IDX IDX1 IDX2 [D,IDX] [IDX2] INH INH	76 hh ll 66 xb 66 xb ff 66 xb ee ff 66 xb 66 xb ee ff 46 56	rPwO rPw rPwO frPwP fIfrPw fIPrPw O O	rOPw rPw rPOw frPPw fIfrPw fIPrPw O O	– – – –	Δ Δ Δ Δ
RTC	$(M_{(SP)}) \Rightarrow$ PPAGE; (SP) + 1 $\Rightarrow$ SP; $(M_{(SP)}:M_{(SP+1)}) \Rightarrow PC_H:PC_L$; (SP) + 2 $\Rightarrow$ SP Return from Call	INH	0A	uUnfPPP	uUnPPP	– – – –	– – – –
RTI	$(M_{(SP)}) \Rightarrow$ CCR; (SP) + 1 $\Rightarrow$ SP $(M_{(SP)}:M_{(SP+1)}) \Rightarrow$ B:A; (SP) + 2 $\Rightarrow$ SP $(M_{(SP)}:M_{(SP+1)}) \Rightarrow X_H:X_L$; (SP) + 4 $\Rightarrow$ SP $(M_{(SP)}:M_{(SP+1)}) \Rightarrow PC_H:PC_L$; (SP) – 2 $\Rightarrow$ SP $(M_{(SP)}:M_{(SP+1)}) \Rightarrow Y_H:Y_L$; (SP) + 4 $\Rightarrow$ SP Return from Interrupt	INH	0B	uUUUUPPP (with interrupt pending) uUUUUVfPPP	uUUUUPPP uUUUUfVfPPP	Δ ⇓ Δ Δ	Δ Δ Δ Δ
RTS	$(M_{(SP)}:M_{(SP+1)}) \Rightarrow PC_H:PC_L$; (SP) + 2 $\Rightarrow$ SP Return from Subroutine	INH	3D	UfPPP	UfPPP	– – – –	– – – –
SBA	(A) – (B) $\Rightarrow$ A Subtract B from A	INH	18 16	OO	OO	– – – –	Δ Δ Δ Δ
SBCA #opr8i SBCA opr8a SBCA opr16a SBCA oprx0_xysp SBCA oprx9,xysp SBCA oprx16,xysp SBCA [D,xysp] SBCA [oprx16,xysp]	(A) – (M) – C $\Rightarrow$ A Subtract with Borrow from A	IMM DIR EXT IDX IDX1 IDX2 [D,IDX] [IDX2]	82 ii 92 dd B2 hh ll A2 xb A2 xb ff A2 xb ee ff A2 xb A2 xb ee ff	P rPf rPO rPf rPO frPP fIfrPf fIPrPf	P rfP rOP rfP rPO frPP fIfrPf fIPrPf	– – – –	Δ Δ Δ Δ
SBCB #opr8i SBCB opr8a SBCB opr16a SBCB oprx0_xysp SBCB oprx9,xysp SBCB oprx16,xysp SBCB [D,xysp] SBCB [oprx16,xysp]	(B) – (M) – C $\Rightarrow$ B Subtract with Borrow from B	IMM DIR EXT IDX IDX1 IDX2 [D,IDX] [IDX2]	C2 ii D2 dd F2 hh ll E2 xb E2 xb ff E2 xb ee ff E2 xb E2 xb ee ff	P rPf rPO rPf rPO frPP fIfrPf fIPrPf	P rfP rOP rfP rPO frPP fIfrPf fIPrPf	– – – –	Δ Δ Δ Δ
SEC	1 $\Rightarrow$ C *Translates to ORCC #$01*	IMM	14 01	P	P	– – – –	– – – 1
SEI	1 $\Rightarrow$ I; (inhibit I interrupts) *Translates to ORCC #$10*	IMM	14 10	P	P	– – – 1	– – – –
SEV	1 $\Rightarrow$ V *Translates to ORCC #$02*	IMM	14 02	P	P	– – – –	– – 1 –
SEX abc,dxys	$00:(r1) \Rightarrow$ r2 if r1, bit 7 is 0 or $FF:(r1) \Rightarrow$ r2 if r1, bit 7 is 1 Sign Extend 8-bit r1 to 16-bit r2 r1 may be A, B, or CCR r2 may be D, X, Y, or SP *Alternate mnemonic for TFR r1, r2*	INH	B7 eb	P	P	– – – –	– – – –

**For More Information On This Product,
Go to: www.freescale.com**

Reprinted with permission of Freescale Semiconductor, Inc.

Freescale Semiconductor, Inc.

Table A-1. Instruction Set Summary (Sheet 12 of 14)

Source Form	Operation	Addr. Mode	Machine Coding (hex)	Access Detail HCS12	Access Detail M68HC12	S X H I	N Z V C
STAA *opr8a* STAA *opr16a* STAA *oprx0_xysp* STAA *oprx9,xysp* STAA *oprx16,xysp* STAA [D,*xysp*] STAA [*oprx16,xysp*]	(A) ⇒ M Store Accumulator A to Memory	DIR EXT IDX IDX1 IDX2 [D,IDX] [IDX2]	5A dd 7A hh ll 6A xb 6A xb ff 6A xb ee ff 6A xb 6A xb ee ff	Pw PwO Pw PwO PwP PIfw PIPw	Pw wOP Pw PwO PwP PIfPw PIPPw	– – – –	Δ Δ 0 –
STAB *opr8a* STAB *opr16a* STAB *oprx0_xysp* STAB *oprx9,xysp* STAB *oprx16,xysp* STAB [D,*xysp*] STAB [*oprx16,xysp*]	(B) ⇒ M Store Accumulator B to Memory	DIR EXT IDX IDX1 IDX2 [D,IDX] [IDX2]	5B dd 7B hh ll 6B xb 6B xb ff 6B xb ee ff 6B xb 6B xb ee ff	Pw PwO Pw PwO PwP PIfw PIPw	Pw wOP Pw PwO PwP PIfPw PIPPw	– – – –	Δ Δ 0 –
STD *opr8a* STD *opr16a* STD *oprx0_xysp* STD *oprx9,xysp* STD *oprx16,xysp* STD [D,*xysp*] STD [*oprx16,xysp*]	(A) ⇒ M, (B) ⇒ M+1 Store Double Accumulator	DIR EXT IDX IDX1 IDX2 [D,IDX] [IDX2]	5C dd 7C hh ll 6C xb 6C xb ff 6C xb ee ff 6C xb 6C xb ee ff	PW PWO PW PWO PWP PIfW PIPW	PW WOP PW PWO PWP PIfPW PIPPW	– – – –	Δ Δ 0 –
STOP	(SP) – 2 ⇒ SP; RTN$_H$:RTN$_L$ ⇒ M$_{(SP)}$:M$_{(SP+1)}$; (SP) – 2 ⇒ SP; (Y$_H$:Y$_L$) ⇒ M$_{(SP)}$:M$_{(SP+1)}$; (SP) – 2 ⇒ SP; (X$_H$:X$_L$) ⇒ M$_{(SP)}$:M$_{(SP+1)}$; (SP) – 2 ⇒ SP; (B:A) ⇒ M$_{(SP)}$:M$_{(SP+1)}$; (SP) – 1 ⇒ SP; (CCR) ⇒ M$_{(SP)}$; STOP All Clocks Registers stacked to allow quicker recovery by interrupt. If S control bit = 1, the STOP instruction is disabled and acts like a two-cycle NOP.	INH	18 3E	(entering STOP) OOSSSSsf (exiting STOP) fVfPPP (continue) ff (if STOP disabled) OO	(entering STOP) OOSSSfSs (exiting STOP) fVfPPP (continue) fO (if STOP disabled) OO	– – – –	– – – –
STS *opr8a* STS *opr16a* STS *oprx0_xysp* STS *oprx9,xysp* STS *oprx16,xysp* STS [D,*xysp*] STS [*oprx16,xysp*]	(SP$_H$:SP$_L$) ⇒ M:M+1 Store Stack Pointer	DIR EXT IDX IDX1 IDX2 [D,IDX] [IDX2]	5F dd 7F hh ll 6F xb 6F xb ff 6F xb ee ff 6F xb 6F xb ee ff	PW PWO PW PWO PWP PIfW PIPW	PW WOP PW PWO PWP PIfPW PIPPW	– – – –	Δ Δ 0 –
STX *opr8a* STX *opr16a* STX *oprx0_xysp* STX *oprx9,xysp* STX *oprx16,xysp* STX [D,*xysp*] STX [*oprx16,xysp*]	(X$_H$:X$_L$) ⇒ M:M+1 Store Index Register X	DIR EXT IDX IDX1 IDX2 [D,IDX] [IDX2]	5E dd 7E hh ll 6E xb 6E xb ff 6E xb ee ff 6E xb 6E xb ee ff	PW PWO PW PWO PWP PIfW PIPW	PW WOP PW PWO PWP PIfPW PIPPW	– – – –	Δ Δ 0 –
STY *opr8a* STY *opr16a* STY *oprx0_xysp* STY *oprx9,xysp* STY *oprx16,xysp* STY [D,*xysp*] STY [*oprx16,xysp*]	(Y$_H$:Y$_L$) ⇒ M:M+1 Store Index Register Y	DIR EXT IDX IDX1 IDX2 [D,IDX] [IDX2]	5D dd 7D hh ll 6D xb 6D xb ff 6D xb ee ff 6D xb 6D xb ee ff	PW PWO PW PWO PWP PIfW PIPW	PW WOP PW PWO PWP PIfPW PIPPW	– – – –	Δ Δ 0 –

SUBA #*opr8i*
SUBA *opr8a*
SUBA *opr16a*
SUBA *oprx0_xysp*
SUBA *oprx9,xysp*
SUBA *oprx16,xysp*
SUBA [D,*xysp*]
SUBA [*oprx16,xysp*]
	(A) – (M) ⇒ A
	Subtract Memory from Accumulator A
		IMM
		DIR
		EXT
		IDX
		IDX1
		IDX2
		[D,IDX]
		[IDX2]
			80 ii
			90 dd
			B0 hh ll
			A0 xb
			A0 xb ff
			A0 xb ee ff
			A0 xb
			A0 xb ee ff
				P
				rPf
				rPO
				rPf
				rPO
				frPP
				fIfrPf
				fIPrPf
					P
					rfP
					rOP
					rfP
					rPO
					frPP
					fIfrfP
					fIPrfP
						– – – –	Δ Δ Δ Δ

Reprinted with permission of Freescale Semiconductor, Inc.

Freescale Semiconductor, Inc.

Instruction Set Reference

Table A-1. Instruction Set Summary (Sheet 13 of 14)

Source Form	Operation	Addr. Mode	Machine Coding (hex)	Access Detail HCS12	M68HC12	S X H I	N Z V C

SUBB #opr8i
SUBB opr8a
SUBB opr16a
SUBB oprx0_xysp
SUBB oprx9,xysp
SUBB oprx16,xysp
SUBB [D,xysp]
SUBB [oprx16,xysp]
	(B) − (M) ⇒ B
	Subtract Memory from Accumulator B
		IMM
		DIR
		EXT
		IDX
		IDX1
		IDX2
		[D,IDX]
		[IDX2]
			C0 ii
			D0 dd
			F0 hh ll
			E0 xb
			E0 xb ff
			E0 xb ee ff
			E0 xb
			E0 xb ee ff
				P
				rPf
				rPO
				rPf
				rPO
				frPP
				fIfrPf
				fIPrPf
					P
					rfP
					rOP
					rfP
					rPO
					frPP
					fIfrPf
					fIPrfP
						– – – –	Δ Δ Δ Δ

SUBD #opr16i
SUBD opr8a
SUBD opr16a
SUBD oprx0_xysp
SUBD oprx9,xysp
SUBD oprx16,xysp
SUBD [D,xysp]
SUBD [oprx16,xysp]
	(D) − (M:M+1) ⇒ D
	Subtract Memory from D (A:B)
		IMM
		DIR
		EXT
		IDX
		IDX1
		IDX2
		[D,IDX]
		[IDX2]
			83 jj kk
			93 dd
			B3 hh ll
			A3 xb
			A3 xb ff
			A3 xb ee ff
			A3 xb
			A3 xb ee ff
				PO
				RPf
				RPO
				RPf
				RPO
				fRPP
				fIfRPf
				fIPRPf
					OP
					RfP
					ROP
					RfP
					RPO
					fRPP
					fIfRPf
					fIPRfP
						– – – –	Δ Δ Δ Δ

SWI	(SP) − 2 ⇒ SP; RTN$_H$:RTN$_L$ ⇒ M$_{(SP)}$:M$_{(SP+1)}$; (SP) − 2 ⇒ SP; (Y$_H$:Y$_L$) ⇒ M$_{(SP)}$:M$_{(SP+1)}$; (SP) − 2 ⇒ SP; (X$_H$:X$_L$) ⇒ M$_{(SP)}$:M$_{(SP+1)}$; (SP) − 2 ⇒ SP; (B:A) ⇒ M$_{(SP)}$:M$_{(SP+1)}$; (SP) − 1 ⇒ SP; (CCR) ⇒ M$_{(SP)}$ 1 ⇒ I; (SWI Vector) ⇒ PC Software Interrupt	INH	3F	VSPSSPSsP* (for Reset) VfPPP	VSPSSPSsP* VfPPP	– – – 1 1 1 – 1	– – – – – – – –

*The CPU also uses the SWI microcode sequence for hardware interrupts and unimplemented opcode traps. Reset uses the VfPPP variation of this sequence.

Source Form	Operation	Addr. Mode	Machine Coding (hex)	Access Detail HCS12	M68HC12	S X H I	N Z V C
TAB	(A) ⇒ B Transfer A to B	INH	18 0E	OO	OO	– – – –	Δ Δ 0 –
TAP	(A) ⇒ CCR *Translates to* TFR A , CCR	INH	B7 02	P	P	Δ ⇓ Δ Δ	Δ Δ Δ Δ
TBA	(B) ⇒ A Transfer B to A	INH	18 0F	OO	OO	– – – –	Δ Δ 0 –
TBEQ abdxys,rel9	If (cntr) = 0, then Branch; else Continue to next instruction Test Counter and Branch if Zero (cntr = A, B, D, X, Y, or SP)	REL (9-bit)	04 lb rr	PPP (branch) PPO (no branch)	PPP	– – – –	– – – –
TBL oprx0_xysp	(M) + [(B) × ((M+1) − (M))] ⇒ A 8-Bit Table Lookup and Interpolate Initialize B, and index before TBL. <ea> points at first 8-bit table entry (M) and B is fractional part of lookup value. (no indirect addressing modes or extensions allowed)	IDX	18 3D xb	ORfffP	OrrffffP	– – – – C Bit is undefined in HC12	Δ Δ – Δ ?
TBNE abdxys,rel9	If (cntr) not = 0, then Branch; else Continue to next instruction Test Counter and Branch if Not Zero (cntr = A, B, D, X, Y, or SP)	REL (9-bit)	04 lb rr	PPP (branch) PPO (no branch)	PPP	– – – –	– – – –
TFR abcdxys,abcdxys	(r1) ⇒ r2 or $00:(r1) ⇒ r2 or (r1[7:0]) ⇒ r2 Transfer Register to Register r1 and r2 may be A, B, CCR, D, X, Y, or SP	INH	B7 eb	P	P	– – – – or Δ ⇓ Δ Δ	– – – – Δ Δ Δ Δ
TPA	(CCR) ⇒ A *Translates to* TFR CCR ,A	INH	B7 20	P	P	– – – –	– – – –

**For More Information On This Product,
Go to: www.freescale.com**

Reprinted with permission of Freescale Semiconductor, Inc.

Freescale Semiconductor, Inc.

Table A-1. Instruction Set Summary (Sheet 14 of 14)

Source Form	Operation	Addr. Mode	Machine Coding (hex)	Access Detail HCS12	Access Detail M68HC12	S X H I	N Z V C
TRAP *trapnum*	$(SP) - 2 \Rightarrow SP;$ $RTN_H:RTN_L \Rightarrow M_{(SP)}:M_{(SP+1)};$ $(SP) - 2 \Rightarrow SP; (Y_H:Y_L) \Rightarrow M_{(SP)}:M_{(SP+1)};$ $(SP) - 2 \Rightarrow SP; (X_H:X_L) \Rightarrow M_{(SP)}:M_{(SP+1)};$ $(SP) - 2 \Rightarrow SP; (B:A) \Rightarrow M_{(SP)}:M_{(SP+1)};$ $(SP) - 1 \Rightarrow SP; (CCR) \Rightarrow M_{(SP)}$ $1 \Rightarrow I; (TRAP\ Vector) \Rightarrow PC$ Unimplemented opcode trap	INH	18 tn tn = $30-$39 or $40-$FF	OVSPSSPSsP	OfVSPSSPSsP	- - - 1	- - - -
TST *opr16a* TST *oprx0_xysp* TST *oprx9,xysp* TST *oprx16,xysp* TST [D,*xysp*] TST [*oprx16,xysp*] TSTA TSTB	$(M) - 0$ Test Memory for Zero or Minus $(A) - 0$ Test A for Zero or Minus $(B) - 0$ Test B for Zero or Minus	EXT IDX IDX1 IDX2 [D,IDX] [IDX2] INH INH	F7 hh ll E7 xb E7 xb ff E7 xb ee ff E7 xb E7 xb ee ff 97 D7	rPO rPf rPO frPP fIfrPf fIPrPf O O	rOP rfP rPO frPP fIfrfP fIPrfP O O	- - - -	Δ Δ 0 0
TSX	$(SP) \Rightarrow X$ *Translates to TFR SP,X*	INH	B7 75	P	P	- - - -	- - - -
TSY	$(SP) \Rightarrow Y$ *Translates to TFR SP,Y*	INH	B7 76	P	P	- - - -	- - - -
TXS	$(X) \Rightarrow SP$ *Translates to TFR X,SP*	INH	B7 57	P	P	- - - -	- - - -
TYS	$(Y) \Rightarrow SP$ *Translates to TFR Y,SP*	INH	B7 67	P	P	- - - -	- - - -
WAI	$(SP) - 2 \Rightarrow SP;$ $RTN_H:RTN_L \Rightarrow M_{(SP)}:M_{(SP+1)};$ $(SP) - 2 \Rightarrow SP; (Y_H:Y_L) \Rightarrow M_{(SP)}:M_{(SP+1)};$ $(SP) - 2 \Rightarrow SP; (X_H:X_L) \Rightarrow M_{(SP)}:M_{(SP+1)};$ $(SP) - 2 \Rightarrow SP; (B:A) \Rightarrow M_{(SP)}:M_{(SP+1)};$ $(SP) - 1 \Rightarrow SP; (CCR) \Rightarrow M_{(SP)};$ WAIT for interrupt	INH	3E	OSSSSsf (after interrupt) fVfPPP	OSSSfSsf VfPPP	- - - - or - - - 1 or - 1 - 1	- - - - or - - - - or - - - -
WAV	$\sum_{i=1}^{B} S_i F_i \Rightarrow Y{:}D$ and $\sum_{i=1}^{B} F_i \Rightarrow X$ Calculate Sum of Products and Sum of Weights for Weighted Average Calculation Initialize B, X, and Y before WAV. B specifies number of elements. X points at first element in S$_i$ list. Y points at first element in F$_i$ list. All S$_i$ and F$_i$ elements are 8-bits. If interrupted, six extra bytes of stack used for intermediate values	Special	18 3C	Of(frr,fffff)O (add if interrupt) SSS + UUUrr,	Off(frr,fffff)O SSSf + UUUrr	- - ? -	? Δ ? ?
wavr pseudo-instruction	*see* WAV Resume executing an interrupted WAV instruction (recover intermediate results from stack rather than initializing them to zero)	Special	3C	UUUrr,ffff UUUrrfffff (frr,ffff)O (frr,fffff)O (exit + re-entry replaces comma above if interrupted) SSS + UUUrr, SSSf + UUUrr		- - ? -	? Δ ? ?
XGDX	$(D) \Leftrightarrow (X)$ *Translates to EXG D, X*	INH	B7 C5	P	P	- - - -	- - - -
XGDY	$(D) \Leftrightarrow (Y)$ *Translates to EXG D, Y*	INH	B7 C6	P	P	- - - -	- - - -

Reprinted with permission of Freescale Semiconductor, Inc.

Appendix B: Summary of Features of HCS12 Devices

Device	RAM (bytes)	Flash (bytes)	EEPROM (bytes)	Timer	I/O	Serial	MUX	A/D	PWM	Operating Frequency (MHz)	Operating Voltage (V)	Packaging
MC9S12A128	4K	128K	2K	8-ch 16-bit IC, OC, PA	up to 91	2 SCI 2 SPI IIC	—	16 ch 10-bit	8-ch 8-bit or 4-ch 16-bit	25.0	5.0	80-pin QFP (FU) 112-pin LQFP (PV)
MC9S12A32	2K	32K	1K	8-ch 16-bit ECT	up to 59	2 SCI 1 SPI	—	8-ch 10-bit	7-ch 8-bit or 3-ch 16-bit	25.0	3.0, 5.0	80-pin QFP (FU)
MC9S12A512	14K	512K	4K	8-ch 16-bit ECT	up to 91	2 SCI 3 SPI I2C	—	16 ch 10-bit	8-ch 8-bit or 4-ch 16-bit	25, 33	5.0	112-pin LQFP (PV)
MC9S12A64	4K	64K	1K	8-ch 16-bit IC, OC, PA	up to 91	2 SCI 1 SPI IIC	—	8 ch 10 bit	up to 8-ch 8 bit or 4-ch 16-bit	25.0	5.0	80-pin QFP (FU) 112-pin LQFP (PV)
MC9S12C32	2K	32K	0K	8 ch 16-bit IC, OC, or PWM	up to 60	SCI SPI	CAN	8 ch 10-bit	See timer	16, 25	3.15 - 5.5	48-pin QFP(FA) 52-pin QFP(PB) 80-pin QFP(FU)
MC9S12D32	2K	32K	1K	8-ch 16-bit ECT	up to 59	2 SCI 1 SPI	CAN	8 ch 10-bit	7-ch 8-bit or 3-ch 16-bit	25.0	3.0, 5.0	80-pin QFP (FU)
MC9S12D64	4K	64K	1K	8-ch 16-bit IC, OC, PA	up to 91	2 SCI 1 SPI IIC	CAN	16 ch 10 bit	8 ch 8 bit or 4-ch 16-bit	25.0	5.0	80-pin QFP (FU) 112-pin LQFP (PV)
MC9S12DJ64	4K	64K	1K	8-ch 16-bit IC, OC, PA	up to 91	2 SCI 1 SPI IIC	1 CAN 1 J1850	16 ch 10 bit	8 ch 8 bit or 4-ch 16-bit	25.0	5.0	80-pin QFP (FU) 112-pin LQFP (PV)
MC9S12DB128	8K	128K	2K	8-ch 16-bit IC, OC, PA	up to 91	2 SCI 2 SPI	1 CAN byte flight	16 ch 10 bit	8 ch 8 bit or 4-ch 16-bit	25.0	5.0	80-pin QFP (FU) 112-pin LQFP (PV)
MC9S12DG128	8K	128K	2K	8-ch 16-bit IC, OC, PA	up to 91	2 SCI 2 SPI IIC	2 CAN	16 ch 10 bit	8 ch 8 bit or 4-ch 16-bit	25.0	5.0	80-pin QFP (FU) 112-pin LQFP (PV)
MC9S12DJ128	8K	128K	2K	8-ch 16-bit IC, OC, PA	up to 91	2 SCI 2 SPI IIC	2 CAN 1 J1850	16 ch 10 bit	8 ch 8 bit or 4-ch 16-bit	25.0	5.0	80-pin QFP (FU) 112-pin LQFP (PV)

Table B.1 ■ Features of the HCS12 family devices (1 of 3)

Device	RAM (bytes)	Flash (bytes)	EEPROM (bytes)	Timer	I/O	Serial	MUX	A/D	PWM	Operating Frequency (MHz)	Operating Voltage (V)	Packaging
MC9S12DP512	14K	512K	4K	8-ch 16-bit ECT	up to 91	2 SCI 3 SPI I2C	5 CAN	16 ch 10-bit	8-ch 8 bit or 4-ch 16-bit	25, 33	5.0	112-pin LQFP (PV)
MC9S12DT128	8K	128K	2K	8-ch 16-bit IC, OC, PA	up to 91	2 SCI 3 SPI I2C	3 CAN	16 ch 10-bit	8-ch 8-bit or 4-ch 16-bit	25.0	5.0	112-pin LQFP (PV)
MC9S12A256	12K	256K	4K	8-ch 16-bit IC, OC, PA	up to 91	Up to 2 SCI 3 SPI I2C	–	16 ch 10-bit	8-ch 8 bit or 4-ch 16-bit	25	5.0	80-pin QFP (FU) 112-pin LQFP (PV)
MC9S12DG256	12K	256K	4K	8-ch 16-bit IC, OC, PA	up to 91	2 SCI 3 SPI I2C	2 CAN	16 ch 10-bit	8-ch 8-bit or 4-ch 16-bit	25.0	5.0	112-pin LQFP (PV)
MC9S12DJ256	12K	256K	4K	8-ch 16-bit IC, OC, PA	up to 91	2 SCI 3 SPI I2C	2 CAN 1 J1850	up to 16 ch 10-bit	8-ch 8-bit or 4-ch 16-bit	25	5.0	80-pin QFP (FU) 112-pin LQFP (PV)
MC9S12DP256	12K	256K	4K	8-ch 16-bit IC, OC, PA	up to 91	2 SCI 3 SPI I2C	5 CAN	16 ch 10-bit	8-ch 8-bit or 4-ch 16-bit	25.0	5.0	112-pin LQFP (PV)
MC9S12DT256	12K	256K	4K	8-ch 16-bit IC, OC, PA	up to 91	2 SCI 3 SPI I2C	3 CAN	16 ch 10-bit	8-ch 8 bit or 4-ch 16-bit	25.0	5.0	112-pin LQFP (PV)
MC9S12E64	4K	64K	0	3 4-ch 16-bit IC, OC, or PWM	up to 90	1 SCI 1 SPI 1 I2C	–	16-ch 10-bit	see timer	16.0, 25.0	3.3-5.0	80-pin QFP (FU) 112-pin LQFP (PV)
MC9S12E128	8K	128K	0	3 4-ch 16-bit IC, OC, or PWM	up to 90	1 SCI 1 SPI 1 I2C	–	16-ch 10-bit	see timer	16.0, 25.0	3.3-5.0	80-pin QFP (FU) 112-pin LQFP (PV)
MC9S12H128B	12K	128K	4K	8-ch 16-bit IC, OC, PA	99 plus 18 in	1 SCI 1 SPI 1 I2C	2 CAN	16-ch 10-bit	6-ch 8 bit or 3-ch 16-bit	16.0	5.0	112-pin LQFP (PV)
MC9S12H256B	12K	256K	4K	8-ch 16-bit IC, OC, PA	99 plus 18 in	1 SCI 1 SPI 1 I2C	2 CAN	16-ch 10-bit	6-ch 8 bit or 3-ch 16-bit	16.0	5.0	112-pin LQFP (PV) 144-pin LQFP (FV)

Table B.1 ■ Features of the HCS12 family devices (2 of 3)

Device	RAM (bytes)	Flash (bytes)	EEPROM (bytes)	Timer	I/O	Serial	MUX	A/D	PWM	Operating Frequency (MHz)	Operating Voltage (V)	Packaging
MC9S12NE64	8K	64K	0	4-ch 16-bit IC,OC,or PWM	up to 70	2 SCI SPI	–	8 ch 10-bit	see timer	16, 25	3.3-5.0	80-pin TQFP-EP(TU) 112-pin LQFP (PV)
MC9S12UF32	3.5K	32K	0	8-ch 16-bit IC, OC, or PWM	up to 75	SCI USB 2.0	–	–	see timer	30	5.0	100-pin LQFP (PU) 64-pin LQFP
MC9S12T64	2K + 2K CALRAM	64K	–	8-ch 16-bit IC, OC, PA	25	2 SCI 1 SPI	–	8 ch 10-bit	8-ch 8 bit or 4-ch 16-bit	16.0	5.0	80-pin QFP (PK)

Note: 1. The MC9S12NE64 has an on-chip integrated media access controller (EMAC), 10/100 Ethernet PHY (EPHY) and was designed for interfacing with Ethernet.
2. The MC9S12UF32 was designed for interfacing with USB 2.0 interface.

Table B.1 ■ Features of the HCS12 family devices (3 of 3)

Appendix C: Tutorial for Using MiniIDE

Step 1 *Invoke MiniIDE by double-clicking on the icon of MiniIDE.* A window similar to that in Figure C.1 appears. The lower half is the status window. The gray area is to be used for entering a new program or editing an existing file.

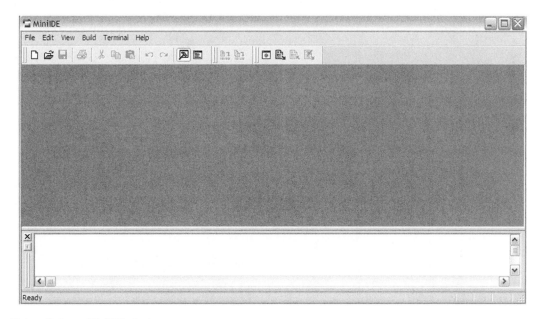

Figure C.1 ■ MiniIDE startup screen

Step 2 *Communicating with the Demo Board.* Press the **Terminal** menu (as shown in Figure C.2) and select **Show Terminal Window**. The terminal window will appear. Press the reset button on the demo board and the screen will change to Figure C.3. There are three sections in Figure C.3. The middle section of the IDE window is the status window. In order to communicate with the demo board, both the items **Show Terminal Window** and **Connected** must be selected, as demonstrated in Figure C.3.

Figure C.2 ■ Press the Terminal menu and select Show Terminal window

Figure C.3 ■ Terminal window after pressing reset button on the demo board

Step 3 *Setting options.* There are two places for setting options. One can select either **Options. . .** under the Build menu (as shown in Figure C.4) or **Options. . .** under the Terminal menu (as shown in Figure C.5). Either place allows the user to set options. Choose one of them to set options in four categories.

Figure C.4 ■ Options under the Build menu

Figure C.5 ■ Options under the Terminal menu

The options in the general category (shown in Figure C.6) allow one to select the font for the editor, terminal, and assembler output and the size for each of them. The tab stop is also set in this category. The options in the terminal category are shown in Figure C.7. One can select the COM port, baud rate, number of bits per character, and flow control in this dialog box. The settings in Figure C.7 are appropriate. In order to select the COM port, one must disconnect the terminal.

Figure C.6 ■ Setting the General options

Figure C.7 ■ Options for the Terminal category

The tool option allows the user to select the appropriate assembler to assemble the program. The correct setting should be as shown in Figure C.8.

The last category of options is related to the selected assembler. The settings in Figure C.9 are acceptable.

Figure C.8 ■ Select the right tool (asm12.exe for HCS12)

Figure C.9 ■ Options for the assembler

Step 4 *Open a new file for entering an assembler program.* Press the **File** menu and select **New** and the screen will change to that in Figure C.10. Enter the assembly program in Example 8.7 (Figure C.11). This example program generates a two-tone siren. Save the file after the whole program is entered. Press the **File** menu and select **Save** as shown in Figure C.12 and a dialog box will appear that asks the user to enter the filename. Type **siren.asm** in the rectangle to the right of **File Name** and click on **Save**.

Figure C.10 ■ Open a new file for entering the program

```
                #include        "c:\miniide\hcs12.inc"
hi_freq  equ    1250            ; delay count for 1200 Hz (with 1:8 prescaler)
lo_freq  equ    7500            ; delay count for 300 Hz (with 1:8 prescaler)
toggle   equ    $04             ; value to toggle the TC5 pin
         org    $1000
delay    ds.w   1               ; store the delay for output compare operation
         org    $2000
         lds    #$2000
         movb   #$90,tscr1      ; enable TCNT, fast timer flag clear
         movb   #$03,tscr2      ; set main timer prescaler to 8
         bset   tios,$20        ; enable OC5
         movb   #toggle,tctl1   ; select toggle for OC5 pin action
         ldd    #lo_freq
         std    delay           ; use low frequency delay count first
         ldd    tcnt            ; start the low frequency sound
         addd   delay           ;          "
         std    tc5             ;          "
         bset   tie,BIT5        ; enable OC5 interrupt
         cli                    ;          "
forever  ldy    #50             ; wait for half a second
         jsr    delayby10ms     ;          "
         movw   #hi_freq,delay  ; switch to high frequency delay count
         ldy    #50
         jsr    delayby10ms
         movw   #lo_freq,delay  ; switch to low frequency delay count
         bra    forever
oc5_isr  ldd    tc5
         addd   delay
         std    tc5
         rti
```

```
>load
*
>g 2000

D-Bug12 4.0.0b24
Copyright 1996 - 2002 Motorola Semiconductor
For Commands type "Help"
```

Ready   Ln 18, Col 47

Figure C.11 ■ Enter the siren program in Example 8.7 for this tutorial

Step 5 *Assemble the program.* Press the **Build** menu and select **Build siren.asm,** as shown in Figure C.13. If the program is assembled successfully, the status window will display the corresponding message, as shown in Figure C.14.

Figure C.12 ■ Dialog box for entering filename to save

Figure C.13 ■ Prepare to assemble the siren.asm program

Figure C.14 ■ Status window message shows that the previous assembling is successful

Step 6 *Download the program onto the demo board.* The previous assembly generated the output **siren.s19**. This file contains a set of S-records that represent the machine code of the **siren.asm** program. One needs to type the **load** command followed by the **enter** key in the terminal window and then press the **Terminal** menu and select **Download File** (shown in Figure C.15), and a dialog box as shown in Figure C.16 appears. Enter the name of the file to be downloaded, click on **Open**, and the file will be transferred to the demo board. The resultant screen after a successful download is shown in Figure C.17.

Figure C.15 ■ Select Download File command

Figure C.16 ■ Select an S-record file to download

Figure C.17 ■ Screen for a successful file download

Step 7 *Running and debugging the program.* To run the siren program, type **g 2000** followed by **enter** key. After this, you should hear the two-tone siren. This program never stops. Therefore, you won't see the D-Bug12 prompt reappear. To return control to the D-Bug12, press either the reset button or the abort button.

This concludes the tutorial for using the MiniIDE.

Appendix D: MC9S12DP256 Registers

1.6 Detailed Register Map

The following tables show the detailed register map of the MC9S12DP256B.

$0000 - $000F　　　**MEBI map 1 of 3 (Core User Guide)**

Address	Name		Bit 7	Bit 6	Bit 5	Bit 4	Bit 3	Bit 2	Bit 1	Bit 0
$0000	PORTA	Read: Write:	Bit 7	6	5	4	3	2	1	Bit 0
$0001	PORTB	Read: Write:	Bit 7	6	5	4	3	2	1	Bit 0
$0002	DDRA	Read: Write:	Bit 7	6	5	4	3	2	1	Bit 0
$0003	DDRB	Read: Write:	Bit 7	6	5	4	3	2	1	Bit 0
$0004	Reserved	Read: Write:	0	0	0	0	0	0	0	0
$0005	Reserved	Read: Write:	0	0	0	0	0	0	0	0
$0006	Reserved	Read: Write:	0	0	0	0	0	0	0	0
$0007	Reserved	Read: Write:	0	0	0	0	0	0	0	0
$0008	PORTE	Read: Write:	Bit 7	6	5	4	3	2	Bit 1	Bit 0
$0009	DDRE	Read: Write:	Bit 7	6	5	4	3	Bit 2	0	0
$000A	PEAR	Read: Write:	NOACCE	0	PIPOE	NECLK	LSTRE	RDWE	0	0
$000B	MODE	Read: Write:	MODC	MODB	MODA	0	IVIS	0	EMK	EME
$000C	PUCR	Read: Write:	PUPKE	0	0	PUPEE	0	0	PUPBE	PUPAE
$000D	RDRIV	Read: Write:	RDPK	0	0	RDPE	0	0	RDPB	RDPA
$000E	EBICTL	Read: Write:	0	0	0	0	0	0	0	ESTR
$000F	Reserved	Read: Write:	0	0	0	0	0	0	0	0

$0010 - $0014　　　**MMC map 1 of 4 (Core User Guide)**

Address	Name		Bit 7	Bit 6	Bit 5	Bit 4	Bit 3	Bit 2	Bit 1	Bit 0
$0010	INITRM	Read: Write:	RAM15	RAM14	RAM13	RAM12	RAM11	0	0	RAMHAL
$0011	INITRG	Read: Write:	0	REG14	REG13	REG12	REG11	0	0	0

MC9S12DP256B Device User Guide — V02.14

$0010 - $0014   MMC map 1 of 4 (Core User Guide)

Address	Name		Bit 7	Bit 6	Bit 5	Bit 4	Bit 3	Bit 2	Bit 1	Bit 0
$0012	INITEE	Read: Write:	EE15	EE14	EE13	EE12	0	0	0	EEON
$0013	MISC	Read: Write:	0	0	0	0	EXSTR1	EXSTR0	ROMHM	ROMON
$0014	MTST0	Read: Write:	Bit 7	6	5	4	3	2	1	Bit 0

$0015 - $0016   INT map 1 of 2 (Core User Guide)

Address	Name		Bit 7	Bit 6	Bit 5	Bit 4	Bit 3	Bit 2	Bit 1	Bit 0
$0015	ITCR	Read: Write:	0	0	0	WRINT	ADR3	ADR2	ADR1	ADR0
$0016	ITEST	Read: Write:	INTE	INTC	INTA	INT8	INT6	INT4	INT2	INT0

$0017 - $0017   MMC map 2 of 4 (Core User Guide)

Address	Name		Bit 7	Bit 6	Bit 5	Bit 4	Bit 3	Bit 2	Bit 1	Bit 0
$0017	MTST1	Read: Write:	Bit 7	6	5	4	3	2	1	Bit 0

$0018 - $001B   Miscellaneous Peripherals (Device User Guide, Table 1-3)

Address	Name		Bit 7	Bit 6	Bit 5	Bit 4	Bit 3	Bit 2	Bit 1	Bit 0
$0018	Reserved	Read: Write:	0	0	0	0	0	0	0	0
$0019	Reserved	Read: Write:	0	0	0	0	0	0	0	0
$001A	PARTIDH	Read: Write:	ID15	ID14	ID13	ID12	ID11	ID10	ID9	ID8
$001B	PARTIDL	Read: Write:	ID7	ID6	ID5	ID4	ID3	ID2	ID1	ID0

$001C - $001D   MMC map 3 of 4 (Core and Device User Guide, Table 1-4)

Address	Name		Bit 7	Bit 6	Bit 5	Bit 4	Bit 3	Bit 2	Bit 1	Bit 0
$001C	MEMSIZ0	Read: Write:	reg_sw0	0	eep_sw1	eep_sw0	0	ram_sw2	ram_sw1	ram_sw0
$001D	MEMSIZ1	Read: Write:	rom_sw1	rom_sw0	0	0	0	0	pag_sw1	pag_sw0

MC9S12DP256B Device User Guide — V02.14

$001E - $001E MEBI map 2 of 3 (Core User Guide)

Address	Name		Bit 7	Bit 6	Bit 5	Bit 4	Bit 3	Bit 2	Bit 1	Bit 0
$001E	INTCR	Read:	IRQE	IRQEN	0	0	0	0	0	0
		Write:								

$001F - $001F INT map 2 of 2 (Core User Guide)

Address	Name		Bit 7	Bit 6	Bit 5	Bit 4	Bit 3	Bit 2	Bit 1	Bit 0
$001F	HPRIO	Read:	PSEL7	PSEL6	PSEL5	PSEL4	PSEL3	PSEL2	PSEL1	0
		Write:								

$0020 - $0027 Reserved

Address	Name		Bit 7	Bit 6	Bit 5	Bit 4	Bit 3	Bit 2	Bit 1	Bit 0
$0020	Reserved	Read:	0	0	0	0	0	0	0	0
		Write:								
$0021	Reserved	Read:	0	0	0	0	0	0	0	0
		Write:								
$0022	Reserved	Read:	0	0	0	0	0	0	0	0
		Write:								
$0023	Reserved	Read:	0	0	0	0	0	0	0	0
		Write:								
$0024	Reserved	Read:	0	0	0	0	0	0	0	0
		Write:								
$0025	Reserved	Read:	0	0	0	0	0	0	0	0
		Write:								
$0026	Reserved	Read:	0	0	0	0	0	0	0	0
		Write:								
$0027	Reserved	Read	0	0	0	0	0	0	0	0
		Write:								

$0028 - $002F BKP (Core User Guide)

Address	Name		Bit 7	Bit 6	Bit 5	Bit 4	Bit 3	Bit 2	Bit 1	Bit 0
$0028	BKPCT0	Read:	BKEN	BKFULL	BKBDM	BKTAG	0	0	0	0
		Write:								
$0029	BKPCT1	Read:	BK0MBH	BK0MBL	BK1MBH	BK1MBL	BK0RWE	BK0RW	BK1RWE	BK1RW
		Write:								
$002A	BKP0X	Read:	0	0	BK0V5	BK0V4	BK0V3	BK0V2	BK0V1	BK0V0
		Write:								
$002B	BKP0H	Read:	Bit 15	14	13	12	11	10	9	Bit 8
		Write:								
$002C	BKP0L	Read:	Bit 7	6	5	4	3	2	1	Bit 0
		Write:								

MC9S12DP256B Device User Guide — V02.14

$0028 - $002F   BKP (Core User Guide)

Address	Name		Bit 7	Bit 6	Bit 5	Bit 4	Bit 3	Bit 2	Bit 1	Bit 0
$002D	BKP1X	Read: Write:	0	0	BK1V5	BK1V4	BK1V3	BK1V2	BK1V1	BK1V0
$002E	BKP1H	Read: Write:	Bit 15	14	13	12	11	10	9	Bit 8
$002F	BKP1L	Read: Write:	Bit 7	6	5	4	3	2	1	Bit 0

$0030 - $0031   MMC map 4 of 4 (Core User Guide)

Address	Name		Bit 7	Bit 6	Bit 5	Bit 4	Bit 3	Bit 2	Bit 1	Bit 0
$0030	PPAGE	Read: Write:	0	0	PIX5	PIX4	PIX3	PIX2	PIX1	PIX0
$0031	Reserved	Read: Write:	0	0	0	0	0	0	0	0

$0032 - $0033   MEBI map 3 of 3 (Core User Guide)

Address	Name		Bit 7	Bit 6	Bit 5	Bit 4	Bit 3	Bit 2	Bit 1	Bit 0
$0032	PORTK	Read: Write:	Bit 7	6	5	4	3	2	1	Bit 0
$0033	DDRK	Read: Write:	Bit 7	6	5	4	3	2	1	Bit 0

$0034 - $003F   CRG (Clock and Reset Generator)

Address	Name		Bit 7	Bit 6	Bit 5	Bit 4	Bit 3	Bit 2	Bit 1	Bit 0
$0034	SYNR	Read: Write:	0	0	SYN5	SYN4	SYN3	SYN2	SYN1	SYN0
$0035	REFDV	Read: Write:	0	0	0	0	REFDV3	REFDV2	REFDV1	REFDV0
$0036	CTFLG TEST ONLY	Read: Write:	TOUT7	TOUT6	TOUT5	TOUT4	TOUT3	TOUT2	TOUT1	TOUT0
$0037	CRGFLG	Read: Write:	RTIF	PROF	0	LOCKIF	LOCK	TRACK	SCMIF	SCM
$0038	CRGINT	Read: Write:	RTIE	0	0	LOCKIE	0	0	SCMIE	0
$0039	CLKSEL	Read: Write:	PLLSEL	PSTP	SYSWAI	ROAWAI	PLLWAI	CWAI	RTIWAI	COPWAI
$003A	PLLCTL	Read: Write:	CME	PLLON	AUTO	ACQ	0	PRE	PCE	SCME
$003B	RTICTL	Read: Write:	0	RTR6	RTR5	RTR4	RTR3	RTR2	RTR1	RTR0
$003C	COPCTL	Read: Write:	WCOP	RSBCK	0	0	0	CR2	CR1	CR0

$0034 - $003F   CRG (Clock and Reset Generator)

Address	Name		Bit 7	Bit 6	Bit 5	Bit 4	Bit 3	Bit 2	Bit 1	Bit 0
$003D	FORBYP TEST ONLY	Read:	RTIBYP	COPBYP	0	PLLBYP	0	0	FCM	0
		Write:	RTIBYP	COPBYP		PLLBYP			FCM	
$003E	CTCTL TEST ONLY	Read:	TCTL7	TCTL6	TCTL5	TCTL4	TCLT3	TCTL2	TCTL1	TCTL0
		Write:								
$003F	ARMCOP	Read:	0	0	0	0	0	0	0	0
		Write:	Bit 7	6	5	4	3	2	1	Bit 0

$0040 - $007F   ECT (Enhanced Capture Timer 16 Bit 8 Channels)

Address	Name		Bit 7	Bit 6	Bit 5	Bit 4	Bit 3	Bit 2	Bit 1	Bit 0
$0040	TIOS	Read: Write:	IOS7	IOS6	IOS5	IOS4	IOS3	IOS2	IOS1	IOS0
$0041	CFORC	Read:	0	0	0	0	0	0	0	0
		Write:	FOC7	FOC6	FOC5	FOC4	FOC3	FOC2	FOC1	FOC0
$0042	OC7M	Read: Write:	OC7M7	OC7M6	OC7M5	OC7M4	OC7M3	OC7M2	OC7M1	OC7M0
$0043	OC7D	Read: Write:	OC7D7	OC7D6	OC7D5	OC7D4	OC7D3	OC7D2	OC7D1	OC7D0
$0044	TCNT (hi)	Read:	Bit 15	14	13	12	11	10	9	Bit 8
		Write:								
$0045	TCNT (lo)	Read:	Bit 7	6	5	4	3	2	1	Bit 0
		Write:								
$0046	TSCR1	Read: Write:	TEN	TSWAI	TSFRZ	TFFCA	0	0	0	0
$0047	TTOV	Read: Write:	TOV7	TOV6	TOV5	TOV4	TOV3	TOV2	TOV1	TOV0
$0048	TCTL1	Read: Write:	OM7	OL7	OM6	OL6	OM5	OL5	OM4	OL4
$0049	TCTL2	Read: Write:	OM3	OL3	OM2	OL2	OM1	OL1	OM0	OL0
$004A	TCTL3	Read: Write:	EDG7B	EDG7A	EDG6B	EDG6A	EDG5B	EDG5A	EDG4B	EDG4A
$004B	TCTL4	Read: Write:	EDG3B	EDG3A	EDG2B	EDG2A	EDG1B	EDG1A	EDG0B	EDG0A
$004C	TIE	Read: Write:	C7I	C6I	C5I	C4I	C3I	C2I	C1I	C0I
$004D	TSCR2	Read:	TOI	0	0	0	TCRE	PR2	PR1	PR0
		Write:	TOI				TCRE	PR2	PR1	PR0
$004E	TFLG1	Read: Write:	C7F	C6F	C5F	C4F	C3F	C2F	C1F	C0F
$004F	TFLG2	Read:	TOF	0	0	0	0	0	0	0
		Write:	TOF							
$0050	TC0 (hi)	Read: Write:	Bit 15	14	13	12	11	10	9	Bit 8
$0051	TC0 (lo)	Read: Write:	Bit 7	6	5	4	3	2	1	Bit 0
$0052	TC1 (hi)	Read: Write:	Bit 15	14	13	12	11	10	9	Bit 8

MC9S12DP256B Device User Guide — V02.14

$0040 - $007F **ECT (Enhanced Capture Timer 16 Bit 8 Channels)**

Address	Name		Bit 7	Bit 6	Bit 5	Bit 4	Bit 3	Bit 2	Bit 1	Bit 0
$0053	TC1 (lo)	Read: Write:	Bit 7	6	5	4	3	2	1	Bit 0
$0054	TC2 (hi)	Read: Write:	Bit 15	14	13	12	11	10	9	Bit 8
$0055	TC2 (lo)	Read: Write:	Bit 7	6	5	4	3	2	1	Bit 0
$0056	TC3 (hi)	Read: Write:	Bit 15	14	13	12	11	10	9	Bit 8
$0057	TC3 (lo)	Read: Write:	Bit 7	6	5	4	3	2	1	Bit 0
$0058	TC4 (hi)	Read: Write:	Bit 15	14	13	12	11	10	9	Bit 8
$0059	TC4 (lo)	Read: Write:	Bit 7	6	5	4	3	2	1	Bit 0
$005A	TC5 (hi)	Read: Write:	Bit 15	14	13	12	11	10	9	Bit 8
$005B	TC5 (lo)	Read: Write:	Bit 7	6	5	4	3	2	1	Bit 0
$005C	TC6 (hi)	Read: Write:	Bit 15	14	13	12	11	10	9	Bit 8
$005D	TC6 (lo)	Read: Write:	Bit 7	6	5	4	3	2	1	Bit 0
$005E	TC7 (hi)	Read: Write:	Bit 15	14	13	12	11	10	9	Bit 8
$005F	TC7 (lo)	Read: Write:	Bit 7	6	5	4	3	2	1	Bit 0
$0060	PACTL	Read: Write:	0	PAEN	PAMOD	PEDGE	CLK1	CLK0	PAOVI	PAI
$0061	PAFLG	Read: Write:	0	0	0	0	0	0	PAOVF	PAIF
$0062	PACN3 (hi)	Read: Write:	Bit 7	6	5	4	3	2	1	Bit 0
$0063	PACN2 (lo)	Read: Write:	Bit 7	6	5	4	3	2	1	Bit 0
$0064	PACN1 (hi)	Read: Write:	Bit 7	6	5	4	3	2	1	Bit 0
$0065	PACN0 (lo)	Read: Write:	Bit 7	6	5	4	3	2	1	Bit 0
$0066	MCCTL	Read: Write:	MCZI	MODMC	RDMCL	0 ICLAT	0 FLMC	MCEN	MCPR1	MCPR0
$0067	MCFLG	Read: Write:	MCZF	0	0	0	POLF3	POLF2	POLF1	POLF0
$0068	ICPAR	Read: Write:	0	0	0	0	PA3EN	PA2EN	PA1EN	PA0EN
$0069	DLYCT	Read: Write:	0	0	0	0	0	0	DLY1	DLY0
$006A	ICOVW	Read: Write:	NOVW7	NOVW6	NOVW5	NOVW4	NOVW3	NOVW2	NOVW1	NOVW0
$006B	ICSYS	Read: Write:	SH37	SH26	SH15	SH04	TFMOD	PACMX	BUFEN	LATQ

MC9S12DP256B Device User Guide — V02.14

$0040 - $007F　　　　ECT (Enhanced Capture Timer 16 Bit 8 Channels)

Address	Name		Bit 7	Bit 6	Bit 5	Bit 4	Bit 3	Bit 2	Bit 1	Bit 0
$006C	Reserved	Read: Write:								
$006D	TIMTST Test Only	Read: Write:	0	0	0	0	0	0	TCBYP	0
$006E	Reserved	Read: Write:								
$006F	Reserved	Read: Write:								
$0070	PBCTL	Read: Write:	0	PBEN	0	0	0	0	PBOVI	0
$0071	PBFLG	Read: Write:	0	0	0	0	0	0	PBOVF	0
$0072	PA3H	Read: Write:	Bit 7	6	5	4	3	2	1	Bit 0
$0073	PA2H	Read: Write:	Bit 7	6	5	4	3	2	1	Bit 0
$0074	PA1H	Read: Write:	Bit 7	6	5	4	3	2	1	Bit 0
$0075	PA0H	Read: Write:	Bit 7	6	5	4	3	2	1	Bit 0
$0076	MCCNT (hi)	Read: Write:	Bit 15	14	13	12	11	10	9	Bit 8
$0077	MCCNT (lo)	Read: Write:	Bit 7	6	5	4	3	2	1	Bit 0
$0078	TC0H (hi)	Read: Write:	Bit 15	14	13	12	11	10	9	Bit 8
$0079	TC0H (lo)	Read: Write:	Bit 7	6	5	4	3	2	1	Bit 0
$007A	TC1H (hi)	Read: Write:	Bit 15	14	13	12	11	10	9	Bit 8
$007B	TC1H (lo)	Read: Write:	Bit 7	6	5	4	3	2	1	Bit 0
$007C	TC2H (hi)	Read: Write:	Bit 15	14	13	12	11	10	9	Bit 8
$007D	TC2H (lo)	Read: Write:	Bit 7	6	5	4	3	2	1	Bit 0
$007E	TC3H (hi)	Read: Write:	Bit 15	14	13	12	11	10	9	Bit 8
$007F	TC3H (lo)	Read: Write:	Bit 7	6	5	4	3	2	1	Bit 0

$0080 - $009F　　　　ATD0 (Analog to Digital Converter 10 Bit 8 Channel)

Address	Name		Bit 7	Bit 6	Bit 5	Bit 4	Bit 3	Bit 2	Bit 1	Bit 0
$0080	ATD0CTL0	Read: Write:	0	0	0	0	0	0	0	0
$0081	ATD0CTL1	Read: Write:	0	0	0	0	0	0	0	0

MC9S12DP256B Device User Guide — V02.14

$0080 - $009F   ATD0 (Analog to Digital Converter 10 Bit 8 Channel)

Address	Name		Bit 7	Bit 6	Bit 5	Bit 4	Bit 3	Bit 2	Bit 1	Bit 0
$0082	ATD0CTL2	Read: Write:	ADPU	AFFC	AWAI	ETRIGLE	ETRIGP	ETRIG	ASCIE	ASCIF
$0083	ATD0CTL3	Read: Write:	0	S8C	S4C	S2C	S1C	FIFO	FRZ1	FRZ0
$0084	ATD0CTL4	Read: Write:	SRES8	SMP1	SMP0	PRS4	PRS3	PRS2	PRS1	PRS0
$0085	ATD0CTL5	Read: Write:	DJM	DSGN	SCAN	MULT	0	CC	CB	CA
$0086	ATD0STAT0	Read: Write:	SCF	0	ETORF	FIFOR	0	CC2	CC1	CC0
$008B	Reserved	Read: Write:	0	0	0	0	0	0	0	0
$0088	ATD0TEST0	Read: Write:	0	0	0	0	0	0	0	0
$0089	ATD0TEST1	Read: Write:	0	0	0	0	0	0	0	SC
$008A	Reserved	Read: Write:	0	0	0	0	0	0	0	0
$008B	ATD0STAT1	Read: Write:	CCF7	CCF6	CCF5	CCF4	CCF3	CCF2	CCF1	CCF0
$008C	Reserved	Read: Write:	0	0	0	0	0	0	0	0
$008D	ATD0DIEN	Read: Write:	Bit 7	6	5	4	3	2	1	Bit 0
$008E	Reserved	Read: Write:	0	0	0	0	0	0	0	0
$008F	PORTAD0	Read: Write:	Bit7	6	5	4	3	2	1	BIT 0
$0090	ATD0DR0H	Read: Write:	Bit15	14	13	12	11	10	9	Bit8
$0091	ATD0DR0L	Read: Write:	Bit7	Bit6	0	0	0	0	0	0
$0092	ATD0DR1H	Read: Write:	Bit15	14	13	12	11	10	9	Bit8
$0093	ATD0DR1L	Read: Write:	Bit7	Bit6	0	0	0	0	0	0
$0094	ATD0DR2H	Read: Write:	Bit15	14	13	12	11	10	9	Bit8
$0095	ATD0DR2L	Read: Write:	Bit7	Bit6	0	0	0	0	0	0
$0096	ATD0DR3H	Read: Write:	Bit15	14	13	12	11	10	9	Bit8
$0097	ATD0DR3L	Read: Write:	Bit7	Bit6	0	0	0	0	0	0
$0098	ATD0DR4H	Read: Write:	Bit15	14	13	12	11	10	9	Bit8
$0099	ATD0DR4L	Read: Write:	Bit7	Bit6	0	0	0	0	0	0
$009A	ATD0DR5H	Read: Write:	Bit15	14	13	12	11	10	9	Bit8

MC9S12DP256B Device User Guide — V02.14

$0080 - $009F   **ATD0 (Analog to Digital Converter 10 Bit 8 Channel)**

Address	Name		Bit 7	Bit 6	Bit 5	Bit 4	Bit 3	Bit 2	Bit 1	Bit 0
$009B	ATD0DR5L	Read:	Bit7	Bit6	0	0	0	0	0	0
		Write:								
$009C	ATD0DR6H	Read:	Bit15	14	13	12	11	10	9	Bit8
		Write:								
$009D	ATD0DR6L	Read:	Bit7	Bit6	0	0	0	0	0	0
		Write:								
$009E	ATD0DR7H	Read:	Bit15	14	13	12	11	10	9	Bit8
		Write:								
$009F	ATD0DR7L	Read:	Bit7	Bit6	0	0	0	0	0	0
		Write:								

$00A0 - $00C7   **PWM (Pulse Width Modulator 8 Bit 8 Channel)**

Address	Name		Bit 7	Bit 6	Bit 5	Bit 4	Bit 3	Bit 2	Bit 1	Bit 0
$00A0	PWME	Read:/Write:	PWME7	PWME6	PWME5	PWME4	PWME3	PWME2	PWME1	PWME0
$00A1	PWMPOL	Read:/Write:	PPOL7	PPOL6	PPOL5	PPOL4	PPOL3	PPOL2	PPOL1	PPOL0
$00A2	PWMCLK	Read:/Write:	PCLK7	PCLK6	PCLK5	PCLK4	PCLK3	PCLK2	PCLK1	PCLK0
$00A3	PWMPRCLK	Read:/Write:	0	PCKB2	PCKB1	PCKB0	0	PCKA2	PCKA1	PCKA0
$00A4	PWMCAE	Read:/Write:	CAE7	CAE6	CAE5	CAE4	CAE3	CAE2	CAE1	CAE0
$00A5	PWMCTL	Read:/Write:	CON67	CON45	CON23	CON01	PSWAI	PFRZ	0	0
$00A6	PWMTST Test Only	Read:	0	0	0	0	0	0	0	0
		Write:								
$00A7	PWMPRSC	Read:	0	0	0	0	0	0	0	0
		Write:								
$00A8	PWMSCLA	Read:/Write:	Bit 7	6	5	4	3	2	1	Bit 0
$00A9	PWMSCLB	Read:/Write:	Bit 7	6	5	4	3	2	1	Bit 0
$00AA	PWMSCNTA	Read:	0	0	0	0	0	0	0	0
		Write:								
$00AB	PWMSCNTB	Read:	0	0	0	0	0	0	0	0
		Write:								
$00AC	PWMCNT0	Read:	Bit 7	6	5	4	3	2	1	Bit 0
		Write:	0	0	0	0	0	0	0	0
$00AD	PWMCNT1	Read:	Bit 7	6	5	4	3	2	1	Bit 0
		Write:	0	0	0	0	0	0	0	0
$00AE	PWMCNT2	Read:	Bit 7	6	5	4	3	2	1	Bit 0
		Write:	0	0	0	0	0	0	0	0
$00AF	PWMCNT3	Read:	Bit 7	6	5	4	3	2	1	Bit 0
		Write:	0	0	0	0	0	0	0	0
$00B0	PWMCNT4	Read:	Bit 7	6	5	4	3	2	1	Bit 0
		Write:	0	0	0	0	0	0	0	0

MC9S12DP256B Device User Guide — V02.14

$00A0 - $00C7   PWM (Pulse Width Modulator 8 Bit 8 Channel)

Address	Name		Bit 7	Bit 6	Bit 5	Bit 4	Bit 3	Bit 2	Bit 1	Bit 0
$00B1	PWMCNT5	Read:	Bit 7	6	5	4	3	2	1	Bit 0
		Write:	0	0	0	0	0	0	0	0
$00B2	PWMCNT6	Read:	Bit 7	6	5	4	3	2	1	Bit 0
		Write:	0	0	0	0	0	0	0	0
$00B3	PWMCNT7	Read:	Bit 7	6	5	4	3	2	1	Bit 0
		Write:	0	0	0	0	0	0	0	0
$00B4	PWMPER0	Read: Write:	Bit 7	6	5	4	3	2	1	Bit 0
$00B5	PWMPER1	Read: Write:	Bit 7	6	5	4	3	2	1	Bit 0
$00B6	PWMPER2	Read: Write:	Bit 7	6	5	4	3	2	1	Bit 0
$00B7	PWMPER3	Read: Write:	Bit 7	6	5	4	3	2	1	Bit 0
$00B8	PWMPER4	Read: Write:	Bit 7	6	5	4	3	2	1	Bit 0
$00B9	PWMPER5	Read: Write:	Bit 7	6	5	4	3	2	1	Bit 0
$00BA	PWMPER6	Read: Write:	Bit 7	6	5	4	3	2	1	Bit 0
$00BB	PWMPER7	Read: Write:	Bit 7	6	5	4	3	2	1	Bit 0
$00BC	PWMDTY0	Read: Write:	Bit 7	6	5	4	3	2	1	Bit 0
$00BD	PWMDTY1	Read: Write:	Bit 7	6	5	4	3	2	1	Bit 0
$00BE	PWMDTY2	Read: Write:	Bit 7	6	5	4	3	2	1	Bit 0
$00BF	PWMDTY3	Read: Write:	Bit 7	6	5	4	3	2	1	Bit 0
$00C0	PWMDTY4	Read: Write:	Bit 7	6	5	4	3	2	1	Bit 0
$00C1	PWMDTY5	Read: Write:	Bit 7	6	5	4	3	2	1	Bit 0
$00C2	PWMDTY6	Read: Write:	Bit 7	6	5	4	3	2	1	Bit 0
$00C3	PWMDTY7	Read: Write:	Bit 7	6	5	4	3	2	1	Bit 0
$00C4	PWMSDN	Read: Write:	PWMIF	PWMIE	PWMRS TRT	PWMLVL	0	PWM7IN	PWM7IN L	PWM7E NA
$00C5	Reserved	Read:	0	0	0	0	0	0	0	0
		Write:								
$00C6	Reserved	Read:	0	0	0	0	0	0	0	0
		Write:								
$00C7	Reserved	Read:	0	0	0	0	0	0	0	0
		Write:								

MC9S12DP256B Device User Guide — V02.14

$00C8 - $00CF   SCI0 (Asynchronous Serial Interface)

Address	Name		Bit 7	Bit 6	Bit 5	Bit 4	Bit 3	Bit 2	Bit 1	Bit 0
$00C8	SCI0BDH	Read:	0	0	0	SBR12	SBR11	SBR10	SBR9	SBR8
		Write:								
$00C9	SCI0BDL	Read:	SBR7	SBR6	SBR5	SBR4	SBR3	SBR2	SBR1	SBR0
		Write:								
$00CA	SCI0CR1	Read:	LOOPS	SCISWAI	RSRC	M	WAKE	ILT	PE	PT
		Write:								
$00CB	SCI0CR2	Read:	TIE	TCIE	RIE	ILIE	TE	RE	RWU	SBK
		Write:								
$00CC	SCI0SR1	Read:	TDRE	TC	RDRF	IDLE	OR	NF	FE	PF
		Write:								
$00CD	SCI0SR2	Read:	0	0	0	0	0	BRK13	TXDIR	RAF
		Write:								
$00CE	SCI0DRH	Read:	R8	T8	0	0	0	0	0	0
		Write:								
$00CF	SCI0DRL	Read:	R7	R6	R5	R4	R3	R2	R1	R0
		Write:	T7	T6	T5	T4	T3	T2	T1	T0

$00D0 - $00D7   SCI1 (Asynchronous Serial Interface)

Address	Name		Bit 7	Bit 6	Bit 5	Bit 4	Bit 3	Bit 2	Bit 1	Bit 0
$00D0	SCI1BDH	Read:	0	0	0	SBR12	SBR11	SBR10	SBR9	SBR8
		Write:								
$00D1	SCI1BDL	Read:	SBR7	SBR6	SBR5	SBR4	SBR3	SBR2	SBR1	SBR0
		Write:								
$00D2	SCI1CR1	Read:	LOOPS	SCISWAI	RSRC	M	WAKE	ILT	PE	PT
		Write:								
$00D3	SCI1CR2	Read:	TIE	TCIE	RIE	ILIE	TE	RE	RWU	SBK
		Write:								
$00D4	SCI1SR1	Read:	TDRE	TC	RDRF	IDLE	OR	NF	FE	PF
		Write:								
$00D5	SCI1SR2	Read:	0	0	0	0	0	BRK13	TXDIR	RAF
		Write:								
$00D6	SCI1DRH	Read:	R8	T8	0	0	0	0	0	0
		Write:								
$00D7	SCI1DRL	Read:	R7	R6	R5	R4	R3	R2	R1	R0
		Write:	T7	T6	T5	T4	T3	T2	T1	T0

$00D8 - $00DF   SPI0 (Serial Peripheral Interface)

Address	Name		Bit 7	Bit 6	Bit 5	Bit 4	Bit 3	Bit 2	Bit 1	Bit 0
$00D8	SPI0CR1	Read:	SPIE	SPE	SPTIE	MSTR	CPOL	CPHA	SSOE	LSBFE
		Write:								
$00D9	SPI0CR2	Read:	0	0	0	MODFEN	BIDIROE	0	SPISWAI	SPC0
		Write:								
$00DA	SPI0BR	Read:	0	SPPR2	SPPR1	SPPR0	0	SPR2	SPR1	SPR0
		Write:								
$00DB	SPI0SR	Read:	SPIF	0	SPTEF	MODF	0	0	0	0
		Write:								

MC9S12DP256B Device User Guide — V02.14

$00D8 - $00DF   SPI0 (Serial Peripheral Interface)

Address	Name		Bit 7	Bit 6	Bit 5	Bit 4	Bit 3	Bit 2	Bit 1	Bit 0
$00DC	Reserved	Read:	0	0	0	0	0	0	0	0
		Write:								
$00DD	SPI0DR	Read:	Bit7	6	5	4	3	2	1	Bit0
		Write:								
$00DE	Reserved	Read:	0	0	0	0	0	0	0	0
		Write:								
$00DF	Reserved	Read:	0	0	0	0	0	0	0	0
		Write:								

$00E0 - $00E7   IIC (Inter IC Bus)

Address	Name		Bit 7	Bit 6	Bit 5	Bit 4	Bit 3	Bit 2	Bit 1	Bit 0
$00E0	IBAD	Read:	ADR7	ADR6	ADR5	ADR4	ADR3	ADR2	ADR1	0
		Write:								
$00E1	IBFD	Read:	IBC7	IBC6	IBC5	IBC4	IBC3	IBC2	IBC1	IBC0
		Write:								
$00E2	IBCR	Read:	IBEN	IBIE	MS/$\overline{SL}$	TX/$\overline{RX}$	TXAK	0		IBSWAI
		Write:						RSTA		
$00E3	IBSR	Read:	TCF	IAAS	IBB	IBAL	0	SRW	IBIF	RXAK
		Write:								
$00E4	IBDR	Read:	D7	D6	D5	D4	D3	D2	D1	D0
		Write:								
$00E5	Reserved	Read:	0	0	0	0	0	0	0	0
		Write:								
$00E6	Reserved	Read:	0	0	0	0	0	0	0	0
		Write:								
$00E7	Reserved	Read:	0	0	0	0	0	0	0	0
		Write:								

$00E8 - $00EF   BDLC (Bytelevel Data Link Controller J1850)

Address	Name		Bit 7	Bit 6	Bit 5	Bit 4	Bit 3	Bit 2	Bit 1	Bit 0
$00E8	DLCBCR1	Read:	IMSG	CLKS	0	0	0	0	IE	WCM
		Write:								
$00E9	DLCBSVR	Read:	0	0	I3	I2	I1	I0	0	0
		Write:								
$00EA	DLCBCR2	Read:	SMRST	DLOOP	RX4XE	NBFS	TEOD	TSIFR	TMIFR1	TMIFR0
		Write:								
$00EB	DLCBDR	Read:	D7	D6	D5	D4	D3	D2	D1	D0
		Write:								
$00EC	DLCBARD	Read:	0	RXPOL	0	0	BO3	BO2	BO1	BO0
		Write:								
$00ED	DLCBRSR	Read:	0	0	R5	R4	R3	R2	R1	R0
		Write:								
$00EE	DLCSCR	Read:	0	0	0	BDLCE	0	0	0	0
		Write:								
$00EF	DLCBSTAT	Read:	0	0	0	0	0	0	0	IDLE
		Write:								

MC9S12DP256B Device User Guide — V02.14

$00F0 - $00F7   SPI1 (Serial Peripheral Interface)

Address	Name		Bit 7	Bit 6	Bit 5	Bit 4	Bit 3	Bit 2	Bit 1	Bit 0
$00F0	SPI1CR1	Read: Write:	SPIE	SPE	SPTIE	MSTR	CPOL	CPHA	SSOE	LSBFE
$00F1	SPI1CR2	Read: Write:	0	0	0	MODFEN	BIDIROE	0	SPISWAI	SPC0
$00F2	SPI1BR	Read: Write:	0	SPPR2	SPPR1	SPPR0	0	SPR2	SPR1	SPR0
$00F3	SPI1SR	Read: Write:	SPIF	0	SPTEF	MODF	0	0	0	0
$00F4	Reserved	Read: Write:	0	0	0	0	0	0	0	0
$00F5	SPI1DR	Read: Write:	Bit7	6	5	4	3	2	1	Bit0
$00F6	Reserved	Read: Write:	0	0	0	0	0	0	0	0
$00F7	Reserved	Read: Write:	0	0	0	0	0	0	0	0

$00F8 - $00FF   SPI2 (Serial Peripheral Interface)

Address	Name		Bit 7	Bit 6	Bit 5	Bit 4	Bit 3	Bit 2	Bit 1	Bit 0
$00F8	SPI2CR1	Read: Write:	SPIE	SPE	SPTIE	MSTR	CPOL	CPHA	SSOE	LSBFE
$00F9	SPI2CR2	Read: Write:	0	0	0	MODFEN	BIDIROE	0	SPISWAI	SPC0
$00FA	SPI2BR	Read: Write:	0	SPPR2	SPPR1	SPPR0	0	SPR2	SPR1	SPR0
$00FB	SPI2SR	Read: Write:	SPIF	0	SPTEF	MODF	0	0	0	0
$00FC	Reserved	Read: Write:	0	0	0	0	0	0	0	0
$00FD	SPI2DR	Read: Write:	Bit7	6	5	4	3	2	1	Bit0
$00FE	Reserved	Read: Write:	0	0	0	0	0	0	0	0
$00FF	Reserved	Read: Write:	0	0	0	0	0	0	0	0

$0100 - $010F   Flash Control Register (fts512k4)

Address	Name		Bit 7	Bit 6	Bit 5	Bit 4	Bit 3	Bit 2	Bit 1	Bit 0
$0100	FCLKDIV	Read: Write:	FDIVLD	PRDIV8	FDIV5	FDIV4	FDIV3	FDIV2	FDIV1	FDIV0
$0101	FSEC	Read: Write:	KEYEN	NV6	NV5	NV4	NV3	NV2	SEC1	SEC0
$0102	FTSTMOD	Read: Write:	0	0	0	WRALL	0	0	0	0
$0103	FCNFG	Read: Write:	CBEIE	CCIE	KEYACC	0	0	0	BKSEL1	BKSEL0

MC9S12DP256B Device User Guide — V02.14

$0100 - $010F   Flash Control Register (fts512k4)

Address	Name	R/W	Bit 7	Bit 6	Bit 5	Bit 4	Bit 3	Bit 2	Bit 1	Bit 0
$0104	FPROT	Read:/Write:	FPOPEN	NV6	FPHDIS	FPHS1	FPHS0	FPLDIS	FPLS1	FPLS0
$0105	FSTAT	Read:/Write:	CBEIF	CCIF	PVIOL	ACCERR	0	BLANK	0	0
$0106	FCMD	Read:/Write:	0	CMDB6	CMDB5	0	0	CMDB2	0	CMDB0
$0107	Reserved for Factory Test	Read:/Write:	0	0	0	0	0	0	0	0
$0108	FADDRHI	Read:/Write:	0	Bit 14	13	12	11	10	9	Bit 8
$0109	FADDRLO	Read:/Write:	Bit 7	6	5	4	3	2	1	Bit 0
$010A	FDATAHI	Read:/Write:	Bit 15	14	13	12	11	10	9	Bit 8
$010B	FDATALO	Read:/Write:	Bit 7	6	5	4	3	2	1	Bit 0
$010C	Reserved	Read:/Write:	0	0	0	0	0	0	0	0
$010D	Reserved	Read:/Write:	0	0	0	0	0	0	0	0
$010E	Reserved	Read:/Write:	0	0	0	0	0	0	0	0
$010F	Reserved	Read:/Write:	0	0	0	0	0	0	0	0

$0110 - $011B   EEPROM Control Register (eets4k)

Address	Name	R/W	Bit 7	Bit 6	Bit 5	Bit 4	Bit 3	Bit 2	Bit 1	Bit 0
$0110	ECLKDIV	Read:/Write:	EDIVLD	PRDIV8	EDIV5	EDIV4	EDIV3	EDIV2	EDIV1	EDIV0
$0111	Reserved	Read:/Write:	0	0	0	0	0	0	0	0
$0112	Reserved for Factory Test	Read:/Write:	0	0	0	0	0	0	0	0
$0113	ECNFG	Read:/Write:	CBEIE	CCIE	0	0	0	0	0	0
$0114	EPROT	Read:/Write:	EPOPEN	NV6	NV5	NV4	EPDIS	EP2	EP1	EP0
$0115	ESTAT	Read:/Write:	CBEIF	CCIF	PVIOL	ACCERR	0	BLANK	0	0
$0116	ECMD	Read:/Write:	0	CMDB6	CMDB5	0	0	CMDB2	0	CMDB0
$0117	Reserved for Factory Test	Read:/Write:	0	0	0	0	0	0	0	0
$0118	EADDRHI	Read:/Write:	0	0	0	0	0	10	9	Bit 8

MC9S12DP256B Device User Guide — V02.14

$0110 - $011B   **EEPROM Control Register (eets4k)**

Address	Name		Bit 7	Bit 6	Bit 5	Bit 4	Bit 3	Bit 2	Bit 1	Bit 0
$0119	EADDRLO	Read: Write:	Bit 7	6	5	4	3	2	1	Bit 0
$011A	EDATAHI	Read: Write:	Bit 15	14	13	12	11	10	9	Bit 8
$011B	EDATALO	Read: Write:	Bit 7	6	5	4	3	2	1	Bit 0

$011C - $011F   **Reserved for RAM Control Register**

Address	Name		Bit 7	Bit 6	Bit 5	Bit 4	Bit 3	Bit 2	Bit 1	Bit 0
$011C	Reserved	Read: Write:	0	0	0	0	0	0	0	0
$011D	Reserved	Read: Write:	0	0	0	0	0	0	0	0
$011E	Reserved	Read: Write:	0	0	0	0	0	0	0	0
$011F	Reserved	Read: Write:	0	0	0	0	0	0	0	0

$0120 - $013F   **ATD1 (Analog to Digital Converter 10 Bit 8 Channel)**

Address	Name		Bit 7	Bit 6	Bit 5	Bit 4	Bit 3	Bit 2	Bit 1	Bit 0
$0120	ATD1CTL0	Read: Write:	0	0	0	0	0	0	0	0
$0121	ATD1CTL1	Read: Write:	0	0	0	0	0	0	0	0
$0122	ATD1CTL2	Read: Write:	ADPU	AFFC	AWAI	ETRIGLE	ETRIGP	ETRIG	ASCIE	ASCIF
$0123	ATD1CTL3	Read: Write:	0	S8C	S4C	S2C	S1C	FIFO	FRZ1	FRZ0
$0124	ATD1CTL4	Read: Write:	SRES8	SMP1	SMP0	PRS4	PRS3	PRS2	PRS1	PRS0
$0125	ATD1CTL5	Read: Write:	DJM	DSGN	SCAN	MULT	0	CC	CB	CA
$0126	ATD1STAT0	Read: Write:	SCF	0	ETORF	FIFOR	0	CC2	CC1	CC0
$0127	Reserved	Read: Write:	0	0	0	0	0	0	0	0
$0128	ATD1TEST0	Read: Write:	0	0	0	0	0	0	0	0
$0129	ATD1TEST1	Read: Write:	0	0	0	0	0	0	0	SC
$012A	Reserved	Read: Write:	0	0	0	0	0	0	0	0
$012B	ATD1STAT1	Read: Write:	CCF7	CCF6	CCF5	CCF4	CCF3	CCF2	CCF1	CCF0
$012C	Reserved	Read: Write:	0	0	0	0	0	0	0	0

MC9S12DP256B Device User Guide — V02.14

$0120 - $013F   ATD1 (Analog to Digital Converter 10 Bit 8 Channel)

Address	Name		Bit 7	Bit 6	Bit 5	Bit 4	Bit 3	Bit 2	Bit 1	Bit 0
$012D	ATD1DIEN	Read: Write:	Bit 7	6	5	4	3	2	1	Bit 0
$012E	Reserved	Read: Write:	0	0	0	0	0	0	0	0
$012F	PORTAD1	Read: Write:	Bit7	6	5	4	3	2	1	BIT 0
$0130	ATD1DR0H	Read: Write:	Bit15	14	13	12	11	10	9	Bit8
$0131	ATD1DR0L	Read: Write:	Bit7	Bit6	0	0	0	0	0	0
$0132	ATD1DR1H	Read: Write:	Bit15	14	13	12	11	10	9	Bit8
$0133	ATD1DR1L	Read: Write:	Bit7	Bit6	0	0	0	0	0	0
$0134	ATD1DR2H	Read: Write:	Bit15	14	13	12	11	10	9	Bit8
$0135	ATD1DR2L	Read: Write:	Bit7	Bit6	0	0	0	0	0	0
$0136	ATD1DR3H	Read: Write:	Bit15	14	13	12	11	10	9	Bit8
$0137	ATD1DR3L	Read: Write:	Bit7	Bit6	0	0	0	0	0	0
$0138	ATD1DR4H	Read: Write:	Bit15	14	13	12	11	10	9	Bit8
$0139	ATD1DR4L	Read: Write:	Bit7	Bit6	0	0	0	0	0	0
$013A	ATD1DR5H	Read: Write:	Bit15	14	13	12	11	10	9	Bit8
$013B	ATD1DR5L	Read: Write:	Bit7	Bit6	0	0	0	0	0	0
$013C	ATD1DR6H	Read: Write:	Bit15	14	13	12	11	10	9	Bit8
$013D	ATD1DR6L	Read: Write:	Bit7	Bit6	0	0	0	0	0	0
$013E	ATD1DR7H	Read: Write:	Bit15	14	13	12	11	10	9	Bit8
$013F	ATD1DR7L	Read: Write:	Bit7	Bit6	0	0	0	0	0	0

$0140 - $017F   CAN0 (Motorola Scalable CAN - MSCAN)

Address	Name		Bit 7	Bit 6	Bit 5	Bit 4	Bit 3	Bit 2	Bit 1	Bit 0
$0140	CAN0CTL0	Read: Write:	RXFRM	RXACT	CSWAI	SYNCH	TIME	WUPE	SLPRQ	INITRQ
$0141	CAN0CTL1	Read: Write:	CANE	CLKSRC	LOOPB	LISTEN	0	WUPM	SLPAK	INITAK
$0142	CAN0BTR0	Read: Write:	SJW1	SJW0	BRP5	BRP4	BRP3	BRP2	BRP1	BRP0

MC9S12DP256B Device User Guide — V02.14

$0140 - $017F **CAN0 (Motorola Scalable CAN - MSCAN)**

Address	Name		Bit 7	Bit 6	Bit 5	Bit 4	Bit 3	Bit 2	Bit 1	Bit 0
$0143	CAN0BTR1	Read: Write:	SAMP	TSEG22	TSEG21	TSEG20	TSEG13	TSEG12	TSEG11	TSEG10
$0144	CAN0RFLG	Read: Write:	WUPIF	CSCIF	RSTAT1	RSTAT0	TSTAT1	TSTAT0	OVRIF	RXF
$0145	CAN0RIER	Read: Write:	WUPIE	CSCIE	RSTATE1	RSTATE0	TSTATE1	TSTATE0	OVRIE	RXFIE
$0146	CAN0TFLG	Read: Write:	0	0	0	0	0	TXE2	TXE1	TXE0
$0147	CAN0TIER	Read: Write:	0	0	0	0	0	TXEIE2	TXEIE1	TXEIE0
$0148	CAN0TARQ	Read: Write:	0	0	0	0	0	ABTRQ2	ABTRQ1	ABTRQ0
$0149	CAN0TAAK	Read: Write:	0	0	0	0	0	ABTAK2	ABTAK1	ABTAK0
$014A	CAN0TBSEL	Read: Write:	0	0	0	0	0	TX2	TX1	TX0
$014B	CAN0IDAC	Read: Write:	0	0	IDAM1	IDAM0	0	IDHIT2	IDHIT1	IDHIT0
$014C	Reserved	Read: Write:	0	0	0	0	0	0	0	0
$014D	Reserved	Read: Write:	0	0	0	0	0	0	0	0
$014E	CAN0RXERR	Read: Write:	RXERR7	RXERR6	RXERR5	RXERR4	RXERR3	RXERR2	RXERR1	RXERR0
$014F	CAN0TXERR	Read: Write:	TXERR7	TXERR6	TXERR5	TXERR4	TXERR3	TXERR2	TXERR1	TXERR0
$0150 - $0153	CAN0IDAR0 - CAN0IDAR3	Read: Write:	AC7	AC6	AC5	AC4	AC3	AC2	AC1	AC0
$0154 - $0157	CAN0IDMR0 - CAN0IDMR3	Read: Write:	AM7	AM6	AM5	AM4	AM3	AM2	AM1	AM0
$0158 - $015B	CAN0IDAR4 - CAN0IDAR7	Read: Write:	AC7	AC6	AC5	AC4	AC3	AC2	AC1	AC0
$015C - $015F	CAN0IDMR4 - CAN0IDMR7	Read: Write:	AM7	AM6	AM5	AM4	AM3	AM2	AM1	AM0
$0160 - $016F	CAN0RXFG	Read: Write:	FOREGROUND RECEIVE BUFFER see **Table 1-2**							
$0170 - $017F	CAN0TXFG	Read: Write:	FOREGROUND TRANSMIT BUFFER see **Table 1-2**							

Table 1-2 Detailed MSCAN Foreground Receive and Transmit Buffer Layout

Address	Name		Bit 7	Bit 6	Bit 5	Bit 4	Bit 3	Bit 2	Bit 1	Bit 0
$xxx0	Extended ID	Read:	ID28	ID27	ID26	ID25	ID24	ID23	ID22	ID21
	Standard ID	Read:	ID10	ID9	ID8	ID7	ID6	ID5	ID4	ID3
	CANxRIDR0	Write:								
$xxx1	Extended ID	Read:	ID20	ID19	ID18	SRR=1	IDE=1	ID17	ID16	ID15
	Standard ID	Read:	ID2	ID1	ID0	RTR	IDE=0			
	CANxRIDR1	Write:								

MC9S12DP256B Device User Guide — V02.14

Table 1-2   Detailed MSCAN Foreground Receive and Transmit Buffer Layout

Address	Name		Bit 7	Bit 6	Bit 5	Bit 4	Bit 3	Bit 2	Bit 1	Bit 0
	Extended ID	Read:	ID14	ID13	ID12	ID11	ID10	ID9	ID8	ID7
$xxx2	Standard ID	Read:								
	CANxRIDR2	Write:								
	Extended ID	Read:	ID6	ID5	ID4	ID3	ID2	ID1	ID0	RTR
$xxx3	Standard ID	Read:								
	CANxRIDR3	Write:								
$xxx4-	CANxRDSR0 -	Read:	DB7	DB6	DB5	DB4	DB3	DB2	DB1	DB0
$xxxB	CANxRDSR7	Write:								
$xxxC	CANRxDLR	Read:					DLC3	DLC2	DLC1	DLC0
		Write:								
$xxxD	Reserved	Read:								
		Write:								
$xxxE	CANxRTSRH	Read:	TSR15	TSR14	TSR13	TSR12	TSR11	TSR10	TSR9	TSR8
		Write:								
$xxxF	CANxRTSRL	Read:	TSR7	TSR6	TSR5	TSR4	TSR3	TSR2	TSR1	TSR0
		Write:								
$xx10	Extended ID CANxTIDR0	Read: Write:	ID28	ID27	ID26	ID25	ID24	ID23	ID22	ID21
	Standard ID	Read: Write:	ID10	ID9	ID8	ID7	ID6	ID5	ID4	ID3
$xx10	Extended ID CANxTIDR1	Read: Write:	ID20	ID19	ID18	SRR=1	IDE=1	ID17	ID16	ID15
	Standard ID	Read: Write:	ID2	ID1	ID0	RTR	IDE=0			
$xx12	Extended ID CANxTIDR2	Read: Write:	ID14	ID13	ID12	ID11	ID10	ID9	ID8	ID7
	Standard ID	Read: Write:								
$xx13	Extended ID CANxTIDR3	Read: Write:	ID6	ID5	ID4	ID3	ID2	ID1	ID0	RTR
	Standard ID	Read: Write:								
$xx14-	CANxTDSR0 -	Read:	DB7	DB6	DB5	DB4	DB3	DB2	DB1	DB0
$xx1B	CANxTDSR7	Write:								
$xx1C	CANxTDLR	Read: Write:					DLC3	DLC2	DLC1	DLC0
$xx1D	CONxTTBPR	Read: Write:	PRIO7	PRIO6	PRIO5	PRIO4	PRIO3	PRIO2	PRIO1	PRIO0
$xx1E	CANxTTSRH	Read: Write:	TSR15	TSR14	TSR13	TSR12	TSR11	TSR10	TSR9	TSR8
$xx1F	CANxTTSRL	Read: Write:	TSR7	TSR6	TSR5	TSR4	TSR3	TSR2	TSR1	TSR0

MC9S12DP256B Device User Guide — V02.14

$0180 - $01BF   **CAN1 (Motorola Scalable CAN - MSCAN)**

Address	Name		Bit 7	Bit 6	Bit 5	Bit 4	Bit 3	Bit 2	Bit 1	Bit 0
$0180	CAN1CTL0	Read:	RXFRM	RXACT	CSWAI	SYNCH	TIME	WUPE	SLPRQ	INITRQ
		Write:	RXFRM		CSWAI		TIME	WUPE	SLPRQ	INITRQ
$0181	CAN1CTL1	Read:	CANE	CLKSRC	LOOPB	LISTEN	0	WUPM	SLPAK	INITAK
		Write:	CANE	CLKSRC	LOOPB	LISTEN		WUPM		
$0182	CAN1BTR0	Read:	SJW1	SJW0	BRP5	BRP4	BRP3	BRP2	BRP1	BRP0
		Write:	SJW1	SJW0	BRP5	BRP4	BRP3	BRP2	BRP1	BRP0
$0183	CAN1BTR1	Read:	SAMP	TSEG22	TSEG21	TSEG20	TSEG13	TSEG12	TSEG11	TSEG10
		Write:	SAMP	TSEG22	TSEG21	TSEG20	TSEG13	TSEG12	TSEG11	TSEG10
$0184	CAN1RFLG	Read:	WUPIF	CSCIF	RSTAT1	RSTAT0	TSTAT1	TSTAT0	OVRIF	RXF
		Write:	WUPIF	CSCIF					OVRIF	RXF
$0185	CAN1RIER	Read:	WUPIE	CSCIE	RSTATE1	RSTATE0	TSTATE1	TSTATE0	OVRIE	RXFIE
		Write:	WUPIE	CSCIE	RSTATE1	RSTATE0	TSTATE1	TSTATE0	OVRIE	RXFIE
$0186	CAN1TFLG	Read:	0	0	0	0	0	TXE2	TXE1	TXE0
		Write:						TXE2	TXE1	TXE0
$0187	CAN1TIER	Read:	0	0	0	0	0	TXEIE2	TXEIE1	TXEIE0
		Write:						TXEIE2	TXEIE1	TXEIE0
$0188	CAN1TARQ	Read:	0	0	0	0	0	ABTRQ2	ABTRQ1	ABTRQ0
		Write:						ABTRQ2	ABTRQ1	ABTRQ0
$0189	CAN1TAAK	Read:	0	0	0	0	0	ABTAK2	ABTAK1	ABTAK0
		Write:								
$018A	CAN1TBSEL	Read:	0	0	0	0	0	TX2	TX1	TX0
		Write:						TX2	TX1	TX0
$018B	CAN1IDAC	Read:	0	0	IDAM1	IDAM0	0	IDHIT2	IDHIT1	IDHIT0
		Write:			IDAM1	IDAM0				
$018C	Reserved	Read:	0	0	0	0	0	0	0	0
		Write:								
$018D	Reserved	Read:	0	0	0	0	0	0	0	0
		Write:								
$018E	CAN1RXERR	Read:	RXERR7	RXERR6	RXERR5	RXERR4	RXERR3	RXERR2	RXERR1	RXERR0
		Write:								
$018F	CAN1TXERR	Read:	TXERR7	TXERR6	TXERR5	TXERR4	TXERR3	TXERR2	TXERR1	TXERR0
		Write:								
$0190	CAN1IDAR0	Read:	AC7	AC6	AC5	AC4	AC3	AC2	AC1	AC0
		Write:	AC7	AC6	AC5	AC4	AC3	AC2	AC1	AC0
$0191	CAN1IDAR1	Read:	AC7	AC6	AC5	AC4	AC3	AC2	AC1	AC0
		Write:	AC7	AC6	AC5	AC4	AC3	AC2	AC1	AC0
$0192	CAN1IDAR2	Read:	AC7	AC6	AC5	AC4	AC3	AC2	AC1	AC0
		Write:	AC7	AC6	AC5	AC4	AC3	AC2	AC1	AC0
$0193	CAN1IDAR3	Read:	AC7	AC6	AC5	AC4	AC3	AC2	AC1	AC0
		Write:	AC7	AC6	AC5	AC4	AC3	AC2	AC1	AC0
$0194	CAN1IDMR0	Read:	AM7	AM6	AM5	AM4	AM3	AM2	AM1	AM0
		Write:	AM7	AM6	AM5	AM4	AM3	AM2	AM1	AM0
$0195	CAN1IDMR1	Read:	AM7	AM6	AM5	AM4	AM3	AM2	AM1	AM0
		Write:	AM7	AM6	AM5	AM4	AM3	AM2	AM1	AM0
$0196	CAN1IDMR2	Read:	AM7	AM6	AM5	AM4	AM3	AM2	AM1	AM0
		Write:	AM7	AM6	AM5	AM4	AM3	AM2	AM1	AM0
$0197	CAN1IDMR3	Read:	AM7	AM6	AM5	AM4	AM3	AM2	AM1	AM0
		Write:	AM7	AM6	AM5	AM4	AM3	AM2	AM1	AM0
$0198	CAN1IDAR4	Read:	AC7	AC6	AC5	AC4	AC3	AC2	AC1	AC0
		Write:	AC7	AC6	AC5	AC4	AC3	AC2	AC1	AC0

MC9S12DP256B Device User Guide — V02.14

$0180 - $01BF   CAN1 (Motorola Scalable CAN - MSCAN)

Address	Name		Bit 7	Bit 6	Bit 5	Bit 4	Bit 3	Bit 2	Bit 1	Bit 0
$0199	CAN1IDAR5	Read: Write:	AC7	AC6	AC5	AC4	AC3	AC2	AC1	AC0
$019A	CAN1IDAR6	Read: Write:	AC7	AC6	AC5	AC4	AC3	AC2	AC1	AC0
$019B	CAN1IDAR7	Read: Write:	AC7	AC6	AC5	AC4	AC3	AC2	AC1	AC0
$019C	CAN1IDMR4	Read: Write:	AM7	AM6	AM5	AM4	AM3	AM2	AM1	AM0
$019D	CAN1IDMR5	Read: Write:	AM7	AM6	AM5	AM4	AM3	AM2	AM1	AM0
$019E	CAN1IDMR6	Read: Write:	AM7	AM6	AM5	AM4	AM3	AM2	AM1	AM0
$019F	CAN1IDMR7	Read: Write:	AM7	AM6	AM5	AM4	AM3	AM2	AM1	AM0
$01A0 - $01AF	CAN1RXFG	Read: Write:	FOREGROUND RECEIVE BUFFER see **Table 1-2**							
$01B0 - $01BF	CAN1TXFG	Read: Write:	FOREGROUND TRANSMIT BUFFER see **Table 1-2**							

$01C0 - $01FF   CAN2 (Motorola Scalable CAN - MSCAN)

Address	Name		Bit 7	Bit 6	Bit 5	Bit 4	Bit 3	Bit 2	Bit 1	Bit 0
$01C0	CAN2CTL0	Read: Write:	RXFRM	RXACT	CSWAI	SYNCH	TIME	WUPE	SLPRQ	INITRQ
$01C1	CAN2CTL1	Read: Write:	CANE	CLKSRC	LOOPB	LISTEN	0	WUPM	SLPAK	INITAK
$01C2	CAN2BTR0	Read: Write:	SJW1	SJW0	BRP5	BRP4	BRP3	BRP2	BRP1	BRP0
$01C3	CAN2BTR1	Read: Write:	SAMP	TSEG22	TSEG21	TSEG20	TSEG13	TSEG12	TSEG11	TSEG10
$01C4	CAN2RFLG	Read: Write:	WUPIF	CSCIF	RSTAT1	RSTAT0	TSTAT1	TSTAT0	OVRIF	RXF
$01C5	CAN2RIER	Read: Write:	WUPIE	CSCIE	RSTATE1	RSTATE0	TSTATE1	TSTATE0	OVRIE	RXFIE
$01C6	CAN2TFLG	Read: Write:	0	0	0	0	0	TXE2	TXE1	TXE0
$01C7	CAN2TIER	Read: Write:	0	0	0	0	0	TXEIE2	TXEIE1	TXEIE0
$01C8	CAN2TARQ	Read: Write:	0	0	0	0	0	ABTRQ2	ABTRQ1	ABTRQ0
$01C9	CAN2TAAK	Read: Write:	0	0	0	0	0	ABTAK2	ABTAK1	ABTAK0
$01CA	CAN2TBSEL	Read: Write:	0	0	0	0	0	TX2	TX1	TX0
$01CB	CAN2IDAC	Read: Write:	0	0	IDAM1	IDAM0	0	IDHIT2	IDHIT1	IDHIT0
$01CC	Reserved	Read: Write:	0	0	0	0	0	0	0	0

MC9S12DP256B Device User Guide — V02.14

$01C0 - $01FF   CAN2 (Motorola Scalable CAN - MSCAN)

Address	Name		Bit 7	Bit 6	Bit 5	Bit 4	Bit 3	Bit 2	Bit 1	Bit 0
$01CD	Reserved	Read:	0	0	0	0	0	0	0	0
		Write:								
$01CE	CAN2RXERR	Read:	RXERR7	RXERR6	RXERR5	RXERR4	RXERR3	RXERR2	RXERR1	RXERR0
		Write:								
$01CF	CAN2TXERR	Read:	TXERR7	TXERR6	TXERR5	TXERR4	TXERR3	TXERR2	TXERR1	TXERR0
		Write:								
$01D0	CAN2IDAR0	Read:	AC7	AC6	AC5	AC4	AC3	AC2	AC1	AC0
		Write:								
$01D1	CAN2IDAR1	Read:	AC7	AC6	AC5	AC4	AC3	AC2	AC1	AC0
		Write:								
$01D2	CAN2IDAR2	Read:	AC7	AC6	AC5	AC4	AC3	AC2	AC1	AC0
		Write:								
$01D3	CAN2IDAR3	Read:	AC7	AC6	AC5	AC4	AC3	AC2	AC1	AC0
		Write:								
$01D4	CAN2IDMR0	Read:	AM7	AM6	AM5	AM4	AM3	AM2	AM1	AM0
		Write:								
$01D5	CAN2IDMR1	Read:	AM7	AM6	AM5	AM4	AM3	AM2	AM1	AM0
		Write:								
$01D6	CAN2IDMR2	Read:	AM7	AM6	AM5	AM4	AM3	AM2	AM1	AM0
		Write:								
$01D7	CAN2IDMR3	Read:	AM7	AM6	AM5	AM4	AM3	AM2	AM1	AM0
		Write:								
$01D8	CAN2IDAR4	Read:	AC7	AC6	AC5	AC4	AC3	AC2	AC1	AC0
		Write:								
$01D9	CAN2IDAR5	Read:	AC7	AC6	AC5	AC4	AC3	AC2	AC1	AC0
		Write:								
$01DA	CAN2IDAR6	Read:	AC7	AC6	AC5	AC4	AC3	AC2	AC1	AC0
		Write:								
$01DB	CAN2IDAR7	Read:	AC7	AC6	AC5	AC4	AC3	AC2	AC1	AC0
		Write:								
$01DC	CAN2IDMR4	Read:	AM7	AM6	AM5	AM4	AM3	AM2	AM1	AM0
		Write:								
$01DD	CAN2IDMR5	Read:	AM7	AM6	AM5	AM4	AM3	AM2	AM1	AM0
		Write:								
$01DE	CAN2IDMR6	Read:	AM7	AM6	AM5	AM4	AM3	AM2	AM1	AM0
		Write:								
$01DF	CAN2IDMR7	Read:	AM7	AM6	AM5	AM4	AM3	AM2	AM1	AM0
		Write:								
$01E0 - $01EF	CAN2RXFG	Read:	FOREGROUND RECEIVE BUFFER see **Table 1-2**							
		Write:								
$01F0 - $01FF	CAN2TXFG	Read:	FOREGROUND TRANSMIT BUFFER see **Table 1-2**							
		Write:								

MC9S12DP256B Device User Guide — V02.14

$0200 - $023F   CAN3 (Motorola Scalable CAN - MSCAN)

Address	Name		Bit 7	Bit 6	Bit 5	Bit 4	Bit 3	Bit 2	Bit 1	Bit 0
$0200	CAN3CTL0	Read: Write:	RXFRM	RXACT	CSWAI	SYNCH	TIME	WUPE	SLPRQ	INITRQ
$0201	CAN3CTL1	Read: Write:	CANE	CLKSRC	LOOPB	LISTEN	0	WUPM	SLPAK	INITAK
$0202	CAN3BTR0	Read: Write:	SJW1	SJW0	BRP5	BRP4	BRP3	BRP2	BRP1	BRP0
$0203	CAN3BTR1	Read: Write:	SAMP	TSEG22	TSEG21	TSEG20	TSEG13	TSEG12	TSEG11	TSEG10
$0204	CAN3RFLG	Read: Write:	WUPIF	CSCIF	RSTAT1	RSTAT0	TSTAT1	TSTAT0	OVRIF	RXF
$0205	CAN3RIER	Read: Write:	WUPIE	CSCIE	RSTATE1	RSTATE0	TSTATE1	TSTATE0	OVRIE	RXFIE
$0206	CAN3TFLG	Read: Write:	0	0	0	0	0	TXE2	TXE1	TXE0
$0207	CAN3TIER	Read: Write:	0	0	0	0	0	TXEIE2	TXEIE1	TXEIE0
$0208	CAN3TARQ	Read: Write:	0	0	0	0	0	ABTRQ2	ABTRQ1	ABTRQ0
$0209	CAN3TAAK	Read: Write:	0	0	0	0	0	ABTAK2	ABTAK1	ABTAK0
$020A	CAN3TBSEL	Read: Write:	0	0	0	0	0	TX2	TX1	TX0
$020B	CAN3IDAC	Read: Write:	0	0	IDAM1	IDAM0	0	IDHIT2	IDHIT1	IDHIT0
$020C	Reserved	Read: Write:	0	0	0	0	0	0	0	0
$020D	Reserved	Read: Write:	0	0	0	0	0	0	0	0
$020E	CAN3RXERR	Read: Write:	RXERR7	RXERR6	RXERR5	RXERR4	RXERR3	RXERR2	RXERR1	RXERR0
$020F	CAN3TXERR	Read: Write:	TXERR7	TXERR6	TXERR5	TXERR4	TXERR3	TXERR2	TXERR1	TXERR0
$0210	CAN3IDAR0	Read: Write:	AC7	AC6	AC5	AC4	AC3	AC2	AC1	AC0
$0211	CAN3IDAR1	Read: Write:	AC7	AC6	AC5	AC4	AC3	AC2	AC1	AC0
$0212	CAN3IDAR2	Read: Write:	AC7	AC6	AC5	AC4	AC3	AC2	AC1	AC0
$0213	CAN3IDAR3	Read: Write:	AC7	AC6	AC5	AC4	AC3	AC2	AC1	AC0
$0214	CAN3IDMR0	Read: Write:	AM7	AM6	AM5	AM4	AM3	AM2	AM1	AM0
$0215	CAN3IDMR1	Read: Write:	AM7	AM6	AM5	AM4	AM3	AM2	AM1	AM0
$0216	CAN3IDMR2	Read: Write:	AM7	AM6	AM5	AM4	AM3	AM2	AM1	AM0
$0217	CAN3IDMR3	Read: Write:	AM7	AM6	AM5	AM4	AM3	AM2	AM1	AM0
$0218	CAN3IDAR4	Read: Write:	AC7	AC6	AC5	AC4	AC3	AC2	AC1	AC0

MC9S12DP256B Device User Guide — V02.14

$0200 - $023F   **CAN3 (Motorola Scalable CAN - MSCAN)**

Address	Name		Bit 7	Bit 6	Bit 5	Bit 4	Bit 3	Bit 2	Bit 1	Bit 0
$0219	CAN3IDAR5	Read: Write:	AC7	AC6	AC5	AC4	AC3	AC2	AC1	AC0
$021A	CAN3IDAR6	Read: Write:	AC7	AC6	AC5	AC4	AC3	AC2	AC1	AC0
$021B	CAN3IDAR7	Read: Write:	AC7	AC6	AC5	AC4	AC3	AC2	AC1	AC0
$021C	CAN3IDMR4	Read: Write:	AM7	AM6	AM5	AM4	AM3	AM2	AM1	AM0
$021D	CAN3IDMR5	Read: Write:	AM7	AM6	AM5	AM4	AM3	AM2	AM1	AM0
$021E	CAN3IDMR6	Read: Write:	AM7	AM6	AM5	AM4	AM3	AM2	AM1	AM0
$021F	CAN3IDMR7	Read: Write:	AM7	AM6	AM5	AM4	AM3	AM2	AM1	AM0
$0220 - $022F	CAN3RXFG	Read: Write:	FOREGROUND RECEIVE BUFFER see **Table 1-2**							
$0230 - $023F	CAN3TXFG	Read: Write:	FOREGROUND TRANSMIT BUFFER see **Table 1-2**							

$0240 - $027F   **PIM (Port Integration Module PIM_9DP256)**

Address	Name		Bit 7	Bit 6	Bit 5	Bit 4	Bit 3	Bit 2	Bit 1	Bit 0
$0240	PTT	Read: Write:	PTT7	PTT6	PTT5	PTT4	PTT3	PTT2	PTT1	PTT0
$0241	PTIT	Read: Write:	PTIT7	PTIT6	PTIT5	PTIT4	PTIT3	PTIT2	PTIT1	PTIT0
$0242	DDRT	Read: Write:	DDRT7	DDRT7	DDRT5	DDRT4	DDRT3	DDRT2	DDRT1	DDRT0
$0243	RDRT	Read: Write:	RDRT7	RDRT6	RDRT5	RDRT4	RDRT3	RDRT2	RDRT1	RDRT0
$0244	PERT	Read: Write:	PERT7	PERT6	PERT5	PERT4	PERT3	PERT2	PERT1	PERT0
$0245	PPST	Read: Write:	PPST7	PPST6	PPST5	PPST4	PPST3	PPST2	PPST1	PPST0
$0246	Reserved	Read: Write:	0	0	0	0	0	0	0	0
$0247	Reserved	Read: Write:	0	0	0	0	0	0	0	0
$0248	PTS	Read: Write:	PTS7	PTS6	PTS5	PTS4	PTS3	PTS2	PTS1	PTS0
$0249	PTIS	Read: Write:	PTIS7	PTIS6	PTIS5	PTIS4	PTIS3	PTIS2	PTIS1	PTIS0
$024A	DDRS	Read: Write:	DDRS7	DDRS7	DDRS5	DDRS4	DDRS3	DDRS2	DDRS1	DDRS0
$024B	RDRS	Read: Write:	RDRS7	RDRS6	RDRS5	RDRS4	RDRS3	RDRS2	RDRS1	RDRS0
$024C	PERS	Read: Write:	PERS7	PERS6	PERS5	PERS4	PERS3	PERS2	PERS1	PERS0
$024D	PPSS	Read: Write:	PPSS7	PPSS6	PPSS5	PPSS4	PPSS3	PPSS2	PPSS1	PPSS0

MC9S12DP256B Device User Guide — V02.14

$0240 - $027F   **PIM (Port Integration Module PIM_9DP256)**

Address	Name		Bit 7	Bit 6	Bit 5	Bit 4	Bit 3	Bit 2	Bit 1	Bit 0
$024E	WOMS	Read: Write:	WOMS7	WOMS6	WOMS5	WOMS4	WOMS3	WOMS2	WOMS1	WOMS0
$024F	Reserved	Read: Write:	0	0	0	0	0	0	0	0
$0250	PTM	Read: Write:	PTM7	PTM6	PTM5	PTM4	PTM3	PTM2	PTM1	PTM0
$0251	PTIM	Read: Write:	PTIM7	PTIM6	PTIM5	PTIM4	PTIM3	PTIM2	PTIM1	PTIM0
$0252	DDRM	Read: Write:	DDRM7	DDRM7	DDRM5	DDRM4	DDRM3	DDRM2	DDRM1	DDRM0
$0253	RDRM	Read: Write:	RDRM7	RDRM6	RDRM5	RDRM4	RDRM3	RDRM2	RDRM1	RDRM0
$0254	PERM	Read: Write:	PERM7	PERM6	PERM5	PERM4	PERM3	PERM2	PERM1	PERM0
$0255	PPSM	Read: Write:	PPSM7	PPSM6	PPSM5	PPSM4	PPSM3	PPSM2	PPSM1	PPSM0
$0256	WOMM	Read: Write:	WOMM7	WOMM6	WOMM5	WOMM4	WOMM3	WOMM2	WOMM1	WOMM0
$0257	MODRR	Read: Write:	0	MODRR6	MODRR5	MODRR4	MODRR3	MODRR2	MODRR1	MODRR0
$0258	PTP	Read: Write:	PTP7	PTP6	PTP5	PTP4	PTP3	PTP2	PTP1	PTP0
$0259	PTIP	Read: Write:	PTIP7	PTIP6	PTIP5	PTIP4	PTIP3	PTIP2	PTIP1	PTIP0
$025A	DDRP	Read: Write:	DDRP7	DDRP7	DDRP5	DDRP4	DDRP3	DDRP2	DDRP1	DDRP0
$025B	RDRP	Read: Write:	RDRP7	RDRP6	RDRP5	RDRP4	RDRP3	RDRP2	RDRP1	RDRP0
$025C	PERP	Read: Write:	PERP7	PERP6	PERP5	PERP4	PERP3	PERP2	PERP1	PERP0
$025D	PPSP	Read: Write:	PPSP7	PPSP6	PPSP5	PPSP4	PPSP3	PPSP2	PPSP1	PPSS0
$025E	PIEP	Read: Write:	PIEP7	PIEP6	PIEP5	PIEP4	PIEP3	PIEP2	PIEP1	PIEP0
$025F	PIFP	Read: Write:	PIFP7	PIFP6	PIFP5	PIFP4	PIFP3	PIFP2	PIFP1	PIFP0
$0260	PTH	Read: Write:	PTH7	PTH6	PTH5	PTH4	PTH3	PTH2	PTH1	PTH0
$0261	PTIH	Read: Write:	PTIH7	PTIH6	PTIH5	PTIH4	PTIH3	PTIH2	PTIH1	PTIH0
$0262	DDRH	Read: Write:	DDRH7	DDRH7	DDRH5	DDRH4	DDRH3	DDRH2	DDRH1	DDRH0
$0263	RDRH	Read: Write:	RDRH7	RDRH6	RDRH5	RDRH4	RDRH3	RDRH2	RDRH1	RDRH0
$0264	PERH	Read: Write:	PERH7	PERH6	PERH5	PERH4	PERH3	PERH2	PERH1	PERH0
$0265	PPSH	Read: Write:	PPSH7	PPSH6	PPSH5	PPSH4	PPSH3	PPSH2	PPSH1	PPSH0
$0266	PIEH	Read: Write:	PIEH7	PIEH6	PIEH5	PIEH4	PIEH3	PIEH2	PIEH1	PIEH0

MC9S12DP256B Device User Guide — V02.14

$0240 - $027F PIM (Port Integration Module PIM_9DP256)

Address	Name		Bit 7	Bit 6	Bit 5	Bit 4	Bit 3	Bit 2	Bit 1	Bit 0
$0267	PIFH	Read: Write:	PIFH7	PIFH6	PIFH5	PIFH4	PIFH3	PIFH2	PIFH1	PIFH0
$0268	PTJ	Read: Write:	PTJ7	PTJ6	0	0	0	0	PTJ1	PTJ0
$0269	PTIJ	Read: Write:	PTIJ7	PTIJ6	0	0	0	0	PTIJ1	PTIJ0
$026A	DDRJ	Read: Write:	DDRJ7	DDRJ7	0	0	0	0	DDRJ1	DDRJ0
$026B	RDRJ	Read: Write:	RDRJ7	RDRJ6	0	0	0	0	RDRJ1	RDRJ0
$026C	PERJ	Read: Write:	PERJ7	PERJ6	0	0	0	0	PERJ1	PERJ0
$026D	PPSJ	Read: Write:	PPSJ7	PPSJ6	0	0	0	0	PPSJ1	PPSJ0
$026E	PIEJ	Read: Write:	PIEJ7	PIEJ6	0	0	0	0	PIEJ1	PIEJ0
$026F	PIFJ	Read: Write:	PIFJ7	PIFJ6	0	0	0	0	PIFJ1	PIFJ0
$0270 - $027F	Reserved	Read:								

$0280 - $02BF CAN4 (Motorola Scalable CAN - MSCAN)

Address	Name		Bit 7	Bit 6	Bit 5	Bit 4	Bit 3	Bit 2	Bit 1	Bit 0
$0280	CAN4CTL0	Read: Write:	RXFRM	RXACT	CSWAI	SYNCH	TIME	WUPE	SLPRQ	INITRQ
$0281	CAN4CTL1	Read: Write:	CANE	CLKSRC	LOOPB	LISTEN	0	WUPM	SLPAK	INITAK
$0282	CAN4BTR0	Read: Write:	SJW1	SJW0	BRP5	BRP4	BRP3	BRP2	BRP1	BRP0
$0283	CAN4BTR1	Read: Write:	SAMP	TSEG22	TSEG21	TSEG20	TSEG13	TSEG12	TSEG11	TSEG10
$0284	CAN4RFLG	Read: Write:	WUPIF	CSCIF	RSTAT1	RSTAT0	TSTAT1	TSTAT0	OVRIF	RXF
$0285	CAN4RIER	Read: Write:	WUPIE	CSCIE	RSTATE1	RSTATE0	TSTATE1	TSTATE0	OVRIE	RXFIE
$0286	CAN4TFLG	Read: Write:	0	0	0	0	0	TXE2	TXE1	TXE0
$0287	CAN4TIER	Read: Write:	0	0	0	0	0	TXEIE2	TXEIE1	TXEIE0
$0288	CAN4TARQ	Read: Write:	0	0	0	0	0	ABTRQ2	ABTRQ1	ABTRQ0
$0289	CAN4TAAK	Read: Write:	0	0	0	0	0	ABTAK2	ABTAK1	ABTAK0
$028A	CAN4TBSEL	Read: Write:	0	0	0	0	0	TX2	TX1	TX0
$028B	CAN4IDAC	Read: Write:	0	0	IDAM1	IDAM0	0	IDHIT2	IDHIT1	IDHIT0
$028C	Reserved	Read: Write:	0	0	0	0	0	0	0	0

MC9S12DP256B Device User Guide — V02.14

$0280 - $02BF   **CAN4 (Motorola Scalable CAN - MSCAN)**

Address	Name		Bit 7	Bit 6	Bit 5	Bit 4	Bit 3	Bit 2	Bit 1	Bit 0
$028D	Reserved	Read:	0	0	0	0	0	0	0	0
		Write:								
$028E	CAN4RXERR	Read:	RXERR7	RXERR6	RXERR5	RXERR4	RXERR3	RXERR2	RXERR1	RXERR0
		Write:								
$028F	CAN4TXERR	Read:	TXERR7	TXERR6	TXERR5	TXERR4	TXERR3	TXERR2	TXERR1	TXERR0
		Write:								
$0290	CAN4IDAR0	Read:/Write:	AC7	AC6	AC5	AC4	AC3	AC2	AC1	AC0
$0291	CAN4IDAR1	Read:/Write:	AC7	AC6	AC5	AC4	AC3	AC2	AC1	AC0
$0292	CAN4IDAR2	Read:/Write:	AC7	AC6	AC5	AC4	AC3	AC2	AC1	AC0
$0293	CAN4IDAR3	Read:/Write:	AC7	AC6	AC5	AC4	AC3	AC2	AC1	AC0
$0294	CAN4IDMR0	Read:/Write:	AM7	AM6	AM5	AM4	AM3	AM2	AM1	AM0
$0295	CAN4IDMR1	Read:/Write:	AM7	AM6	AM5	AM4	AM3	AM2	AM1	AM0
$0296	CAN4IDMR2	Read:/Write:	AM7	AM6	AM5	AM4	AM3	AM2	AM1	AM0
$0297	CAN4IDMR3	Read:/Write:	AM7	AM6	AM5	AM4	AM3	AM2	AM1	AM0
$0298	CAN4IDAR4	Read:/Write:	AC7	AC6	AC5	AC4	AC3	AC2	AC1	AC0
$0299	CAN4IDAR5	Read:/Write:	AC7	AC6	AC5	AC4	AC3	AC2	AC1	AC0
$029A	CAN4IDAR6	Read:/Write:	AC7	AC6	AC5	AC4	AC3	AC2	AC1	AC0
$029B	CAN4IDAR7	Read:/Write:	AC7	AC6	AC5	AC4	AC3	AC2	AC1	AC0
$029C	CAN4IDMR4	Read:/Write:	AM7	AM6	AM5	AM4	AM3	AM2	AM1	AM0
$029D	CAN4IDMR5	Read:/Write:	AM7	AM6	AM5	AM4	AM3	AM2	AM1	AM0
$029E	CAN4IDMR6	Read:/Write:	AM7	AM6	AM5	AM4	AM3	AM2	AM1	AM0
$029F	CAN4IDMR7	Read:/Write:	AM7	AM6	AM5	AM4	AM3	AM2	AM1	AM0
$02A0 - $02AF	CAN4RXFG	Read:	FOREGROUND RECEIVE BUFFER see **Table 1-2**							
		Write:								
$02B0 - $02BF	CAN4TXFG	Read:/Write:	FOREGROUND TRANSMIT BUFFER see **Table 1-2**							

$02C0 - $03FF   **Reserved space**

Address	Name		Bit 7	Bit 6	Bit 5	Bit 4	Bit 3	Bit 2	Bit 1	Bit 0
$02C0 - $03FF	Reserved	Read:	0	0	0	0	0	0	0	0
		Write:								

Appendix E: LCD Functions for the SSE256 Demo Board

LCD data width: 8 bits

Port used as control port to LCD kit: Port K

Port pins for data: PH7~PH0 (corresponds to DB7~DB0)

Port pin for E signal: PK7

Port pin for R/W signal: PK5

Port pin for RS signal: PK4

Four functions are provided for the LCD kit of the SSE256 demo board:

1. **cmd2lcd:** This function sends the command contained in accumulator A to the LCD.
2. **openlcd:** This function initializes the LCD kit to operate with the following features:
   (a) 8-bit data
   (b) enable display, cursor, and blinking
   (c) cursor shift right
   (d) clear the display, and return cursor to home position
3. **putc2lcd:** This function outputs the character contained in accumulator A to the LCD.
4. **puts2lcd:** This function outputs a NULL-terminated string pointed to by X to the LCD kit.

Functions in assembly language:

```
lcdPort      equ    PTH                      ; LCD data pins
lcdDIR       equ    DDRH                     ; LCD data direction port
lcdCtl       equ    PTK                      ; LCD control port
lcdCtlDIR    equ    DDRK                     ; LCD control port direction
lcdE         equ    $80                      ; E signal pin (PK7)
lcdRW        equ    $20                      ; R/W signal pin (PK5)
lcdRS        equ    $10                      ; RS signal pin (PK4)
;*******************************************************************************
; This function sends a command in accumulator A to the LCD.
;*******************************************************************************
cmd2lcd      bclr   lcdCtl,lcdRS+lcdRW       ; select instruction register and write
             bset   lcdCtl,lcdE             ; pull the E signal high
             staa   lcdPort                  ; send the command, along with RS, E signals
             nop
             nop
             bclr   lcdCtl,lcdE             ; pull the E signal low
             bset   lcdCtl,lcdRW            ; pull R/W to high
             ldy    #1                       ; adding this delay will complete the internal
             jsr    delayby50us              ; operation for most instructions
             rts
```

```
; ********************************************************************************
; The following function performs initialization to the LCD used in SSE256
; demo board with 8-bit data width, two-line display, turn on display, cursor,
; and blinking. Shift cursor right, clear display and return to home position.
; ********************************************************************************
openlcd   movb  #$FF,lcdDIR          ; configure port H for output
          bset  lcdCtlDir,$B0        ; configure control pins for output
          ldy   #5                   ; wait for LCD to complete internal
          jsr   delayby100ms         ; configuration
          ldaa  #$38                 ; set 8-bit data, 2-line display, 5x8 font
          jsr   cmd2lcd              ;        "
          ldaa  #$0F                 ; turn on display, cursor, and blinking
          jsr   cmd2lcd              ;        "
          ldaa  #$06                 ; move cursor right (entry mode set instruction)
          jsr   cmd2lcd              ;        "
          ldaa  #$01                 ; clear cursor and return to home position
          jsr   cmd2lcd              ;        "
          ldy   #2                   ; wait until "clear display" command is complete
          jsr   delayby1ms           ;        "
          rts
; ********************************************************************************
; The following function outputs the character in accumulator in A to LCD.
; ********************************************************************************
putc2lcd  bset  lcdCtl,lcdRS         ; select LCD data register
          bclr  lcdCtl,lcdRW         ; enable write to LCD
          bset  lcdCtl,lcdE          ; pull E to high
          staa  lcdPort              ; send data to LCD
          nop                        ; provide enough length to E signal
          nop                        ;        "
          bclr  lcdCtl,lcdE          ; pull the E signal low
          bset  lcdCtl,lcdRW         ; pull R/W high to complete the write cycle
          ldy   #1                   ; wait until the write operation is
          jsr   delayby50us          ; complete
          rts
; ********************************************************************************
; The following function outputs a NULL-terminated string pointed to by X.
; ********************************************************************************
puts2lcd  ldaa  1,x+                 ; get one character from the string
          beq   done_puts            ; reach NULL character?
          jsr   putc2lcd
          bra   puts2lcd
done_puts rts
```

A test assembly program for the preceding LCD functions is as follows:

```
#include        "c:\miniIDE\hcs12.inc"
          org   $1500
          lds   #$1500               ; set up stack pointer
          ldy   #2
          jsr   delayby100ms
```

```
            jsr     openlcd                   ; initialize the LCD
            ldx     #msg1lcd
            jsr     puts2lcd
            ldaa    #$C0                      ; move to the second row
            jsr     cmd2lcd                   ;       "
            ldx     #msg2lcd
            jsr     puts2lcd
            swi
msg1lcd     fcc     "hello world!"
            dc.b    0
msg2lcd     fcc     "I am ready!"
            dc.b    0
#include    "c:\miniIDE\delay.asm"            ; include delay routines here
#include    "c:\miniIDE\lcd_util_SSE256.asm"
            end
```

These functions are contained in the **lcd_util_SSE256.asm** file and must be copied into the installation directory of miniIDE or asmIDE before they can be used.

The C language versions of the LCD functions for the SSE256 demo board are as follows:

```
#define    lcdPort    PTH          /* Port H drives LCD data pins */
#define    lcdCtl     PTK          /* Port K drives LCD kit control signals */
#define    lcdDIR     DDRH         /* Direction of LCD port */
#define    lcdCtlDir  DDRK         /* LCD control port direction */
#define    lcdE       0x80         /* E signal (PK7) */
#define    lcdRW      0x20         /* R/W signal (PK5) */
#define    lcdRS      0x10         /* RS signal (PK4) */
void       cmd2lcd(char cmd);
void       openlcd(void);
void       putc2lcd(char cx);
void       puts2lcd (char *ptr);

void       cmd2lcd(char cmd)
{
           char temp;
           char xa, xb;
           lcdCtl    &= ~(lcdRS+lcdRW); /* select instruction register & pull R/W low */
           lcdCtl    |= lcdE;        /* pull E signal to high */
           lcdPort   = cmd ;         /* output command */
           xa        = 1;            /* dummy statements to lengthen E */
           xb        = 2;            /*       "        */
           lcdCtl    &= ~lcdE;       /* pull E signal to low */
           lcdCtl    |= lcdRW;       /* pull R/W to high */
           delayby50us(1);           /* wait until the command is complete */
}
void openlcd(void)
{
           lcdDIR = 0xFF;            /* configure lcdPort port for output */
           lcdCtlDIR = 0xB0;        /* configure LCD control pins for output */
           delayby100ms(5);         /* wait for LCD to become ready */
           cmd2lcd(0x38);           /* set 8-bit data, 2-line display, 5x8 font */
```

```
            cmd2lcd(0x0F);               /* turn on display, cursor, blinking */
            cmd2lcd(0x06);               /* move cursor right */
            cmd2lcd(0x01);               /* clear screen, move cursor to home */
            delayby1ms(2);               /* wait until "clear display" command is complete */
}
void putc2lcd(char cx)
{
            char temp;
            char xa, xb;
            lcdCtl      |= lcdRS;         /* select LCD data register */
            lcdCtl      &= ~lcdRW;        /* pull R/W low */
            lcdCtl      |= lcdE;          /* pull E signal to high */
            lcdPort     = cx;             /* output data byte */
            xa          = 1;             /* create enough width for E */
            xb          = 2;             /* create enough width for E */
            lcdCtl      &= ~lcdE;         /* pull E to low */
            lcdCtl      |= lcdRW;         /* pull R/W signal to high */
            delayby50us(1);
}
void puts2lcd (char *ptr)
{
            while (*ptr) {
            putc2lcd(*ptr);
            ptr++;
            }

}
```

A test program in C language for these LCD functions is as follows:

```
#include "c:\egnu091\include\hcs12.h"
#include "c:\egnu091\include\delay.c"
#include "c:\egnu091\include\lcd_util_SSE256.c"
main (void)
{
            char *msg1 = "hello world!";
            char *msg2 = "I am ready!";
            openlcd();
            cmd2lcd(0x80);               /* make sure cursor is in the 1st column of row 1 */
            puts2lcd(msg1);
            cmd2lcd(0xC0);               /* move cursor to 2nd row, 1st column */
            puts2lcd(msg2);
            return 0;
}
```

Appendix F: Music Note Frequencies

Note	Frequency (Hz)	Wavelength (cm)	Note	Frequency (Hz)	Wavelength (cm)
C_0	16.35	2100	$G^\#_1/A^b_1$	51.91	665
$C^\#_0/D^b_0$	17.32	1990	A_1	55.00	627
D_0	18.35	1870	$A^\#_1/B^b_1$	58.27	592
$D^\#_0/E^b_0$	19.45	1770	B_1	61.74	559
E_0	20.60	1670	C_2	65.41	527
F_0	21.83	1580	$C^\#_2/D^b_2$	69.30	498
$F^\#_0/G^b_0$	23.12	1490	D_2	73.42	470
G_0	24.50	1400	$D^\#_2/E^b_2$	77.78	444
$G^\#_0/A^b_0$	25.96	1320	E_2	82.41	419
A_0	27.50	1250	F_2	87.31	395
$A^\#_0/B^b_0$	29.14	1180	$F^\#_2/G^b_2$	92.50	373
B_0	30.87	1110	G_2	98.00	352
C_1	32.70	1050	$G^\#_2/A^b_2$	103.83	332
$C^\#_1/D^b_1$	34.65	996	A_2	110.00	314
D_1	36.71	940	$A^\#_2/B^b_2$	116.54	296
$D^\#_1/E^b_1$	38.89	887	B_2	123.47	279
E_1	41.20	837	C_3	130.81	264
F_1	43.65	790	$C^\#_3/D^b_3$	138.59	249
$F^\#_1/G^b_1$	46.25	746	D_3	146.83	235
G_1	49.00	704	$D^\#_3/E^b_3$	155.56	222

Table F.1 ■ Complete list of frequencies of music notes (page 1)

A4 = 440 Hz
Speed of sound = 345 m/s
(Middle C is C_4)

Note	Frequency (Hz)	Wavelength (cm)	Note	Frequency (Hz)	Wavelength (cm)
E_3	164.81	209	C_5	523.25	65.9
F_3	174.61	198	$C^\#_5/D^\flat_5$	554.37	62.2
$F^\#_3/G^\flat_3$	185.00	186	D_5	587.33	58.7
G_3	196.00	176	$D^\#_5/E^\flat_5$	622.25	55.4
$G^\#_3/A^\flat_3$	207.65	166	E_5	659.26	52.3
A_3	220.00	157	F_5	698.46	49.4
$A^\#_3/B^\flat_3$	233.08	148	$F^\#_5/G^\flat_5$	739.99	46.6
B_3	246.94	140	G_5	783.99	44.0
C_4	261.63	132	$G^\#_5/A^\flat_5$	830.61	41.5
$C^\#_4/D^\flat_4$	277.18	124	A_5	880.00	39.2
D_4	293.66	117	$A^\#_5/B^\flat_5$	932.33	37.0
$D^\#_4/E^\flat_4$	311.13	111	B_5	987.77	34.9
E_4	329.63	105	C_6	1046.50	33.0
F_4	349.23	98.8	$C^\#_6/D^\flat_6$	1108.73	31.1
$F^\#_4/G^\flat_4$	369.99	93.2	D_6	1174.66	29.4
G_4	392.00	88.0	$D^\#_6/E^\flat_6$	1244.51	27.7
$G^\#_4/A^\flat_4$	415.30	83.1	E_6	1318.51	26.2
A_4	440.00	78.4	F_6	1396.91	24.7
$A^\#_4/B^\flat_4$	466.16	74.0	$F^\#_6/G^\flat_6$	1479.98	23.3
B_4	493.88	69.9	G_6	1567.98	22.0

Table F.1 ■ Complete list of frequencies of music notes (page 2)

Note	Frequency (Hz)	Wavelength (cm)
$G^{\#}_6/A^{b}_6$	1661.22	20.8
A_6	1760.00	19.6
$A^{\#}_6/B^{b}_6$	1864.66	18.5
B_6	1975.53	17.5
C_7	2093.00	16.5
$C^{\#}_7/D^{b}_7$	2217.46	15.6
D_7	2349.32	14.7
$D^{\#}_7/E^{b}_7$	2489.02	13.9
E_7	2637.02	13.1
F_7	2793.83	12.3
$F^{\#}_7/G^{b}_7$	2959.96	11.7
G_7	3135.96	11.0
$G^{\#}_7/A^{b}_7$	3322.44	10.4
A_7	3520.00	9.8
$A^{\#}_7/B^{b}_7$	3729.31	9.3
B_7	3951.07	8.7
C_8	4186.01	8.2
$C^{\#}_8/D^{b}_8$	4434.92	7.8
D_8	4698.64	7.3
$D^{\#}_8/E^{b}_8$	4978.03	6.9

Table F.1 ■ Complete list of frequencies of music notes (page 3)

Appendix G: Using the MAX7221 LED Display Driver

The MAX7221 from MAXIM is an 8-digit, seven-segment display driver with a serial interface that is compatible with SPI and Microwire. In addition to driving seven-segment displays, it can also be used to drive bar-graph displays, industrial controllers, panel meters, and LED matrix displays. When being used to drive seven-segment displays, the MAX7221 can address individual digits and update without rewriting the entire display. Only common-cathode seven-segment displays can be driven directly. The user is allowed to select code-B decoding or no decoding for each digit.

The MAX7221 has a scan limit register that allows the user to display from one to eight digits. It also has a test mode that forces all LEDs on. This feature is useful for finding out whether seven-segment displays are connected properly.

G.1 MAX7221 Pins

The MAX7221 has 23 signals and is housed in a 24-pin DIP package. Its pin assignment is shown in Figure G.1. The function of each pin is as follows:

Figure G.1 ■ Pin assignment of MAX7221

- DIN: *Serial data input.* Data is loaded into the internal 16-bit shift register on CLK's rising edge.

- DIG 0–DIG 7: *Eight-digit drive lines*. These lines sink current from the display common cathode. The MAX7221's digit drivers are in high impedance when they are turned off.
- GND: ground.
- $\overline{CS}$: *Chip-select input*. Serial data is loaded into the shift register while $\overline{CS}$ is low. The last 16 bits of serial data are latched on $\overline{CS}$'s rising edge.
- CLK: *Serial-clock input*. 10–MHz maximum data shift rate. Data is shifted into the internal shift register on CLK's rising edge and shifted out of the DOUT pin on CLK's falling edge.
- SEG A–SEG G, DP: *Seven-segment drive and decimal point drive*. These signals are in a high-impedance state when they are turned off.
- ISET: *Current setting resistor*. This pin is connected to the V_{DD} through a resistor (R_{SET}) to set the peak segment current.
- V_{DD}: Positive power supply.
- DOUT: *Serial-data out*. This pin is used to daisy-chain several MAX7221s and is never in a high-impedance state.

G.2 MAX7221 Functioning

The functional diagram of the MAX7221 is shown in Figure G.2. The function of every block is as follows:

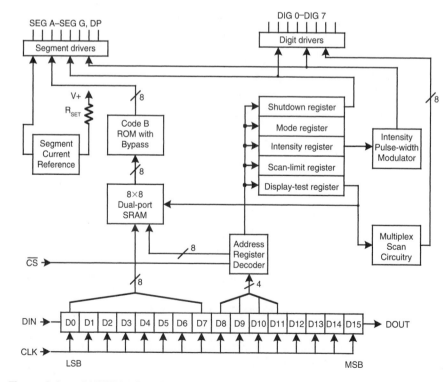

Figure G.2 ■ MAX7221 functional diagram

G.2.1 Segment Drivers

This block provides the current to drive the segment pattern outputs SEG A, ..., SEG G and DP. The source of segment patterns can come from the Code B ROM (in decode mode) or dual-port SRAM (in no-decode mode). The dual-port SRAM consists of eight digit registers. The intensity of the segment pattern is modulated by the *intensity pulse-width modulator*. The magnitude of the current is controlled by the *segment current reference* block.

G.2.2 Code B ROM with Bypass

This ROM stores the segment pattern of decimal digits 0 to 9 and characters -, E, H, L, and P. A 4-bit address presented by the dual-port SRAM is used to select the segment pattern from this ROM. The output of this ROM will be used to drive the segment outputs SEG A, ..., DP. The intensity of the segment pattern is under the control of the output intensity register.

If the designer doesn't like the patterns provided by the Code B ROM, they can be bypassed and the designer's own patterns can be supplied. To do this, one needs to choose the *no-decode mode*. In no-decode mode, one needs to store the desired patterns (corresponding to eight digits) in the dual-port SRAM, and these patterns will be used to drive the segment outputs. One can choose to bypass from one to all eight digits.

The segment pattern stored in the ROM is shown in Table G.1.

7-Segment Character	Register Data (output from SRAM)						On Segment = 1							
	D7*	D6-D4	D3	D2	D1	D0	DP*	A	B	C	D	E	F	G
0		x	0	0	0	0	1	1	1	1	1	1	1	0
1		x	0	0	0	1	0	1	1	0	0	0	0	0
2		x	0	0	1	0	1	1	0	1	1	0	1	
3		x	0	0	1	1	1	1	1	1	0	0	1	
4		x	0	1	0	0	0	1	1	0	0	1	1	
5		x	0	1	0	1	1	0	1	1	0	1	1	
6		x	0	1	1	0	1	0	1	1	1	1	1	
7		x	0	1	1	1	1	1	1	0	0	0	0	
8		x	1	0	0	0	1	1	1	1	1	1	1	
9		x	1	0	0	1	1	1	1	1	0	1	1	
—		x	1	0	1	0	0	0	0	0	0	0	1	
E		x	1	0	1	1	1	0	0	1	1	1	1	
H		x	1	1	0	0	0	1	1	0	1	1	1	
L		x	1	1	0	1	0	0	0	1	1	1	0	
P		x	1	1	1	0	1	1	0	0	1	1	1	
blank		x	1	1	1	1	0	0	0	0	0	0	0	

* The decimal point is set by bit D7 =1.

Table G.1 ■ Code B pattern

G.2.3 Digit Drivers and Multiplex Scan Circuitry

The MAX7221 uses the time-multiplexing technique to display eight digits. When DIG i (i = 0 to 7) is asserted, the corresponding pattern is sent to the segment driver from either ROM or SRAM. Since only one-digit output is asserted, only one digit will be displayed at a time. However, the multiplex scan circuitry will only allow one digit to be lighted for a short period of time (slightly over 1 ms) before it switches to display the next digit. In this way, all eight digits are lighted in turn many times in a second. Due to the effect of persistence of vision, all eight digits appear to be lighted at the same time.

G.2.4 Mode Register

The mode register sets BCD code B (0–9, E, H, L, P, and -) or no-decode operation for each digit. Each bit in the register corresponds to one digit. A logic high selects code B decoding, and low bypasses the decoder. Examples of the decode mode control-register format are shown in Table G.2.

Decode Mode	Register Data								Hex Code
	D7	D6	D5	D4	D3	D2	D1	D0	
No decode for digits 7–0	0	0	0	0	0	0	0	0	00
Code B decode for digit 0 No decode for digits 7–1	0	0	0	0	0	0	0	1	01
Code B decode for digits 3–0 No decode for digits 7–4	0	0	0	0	1	1	1	1	0F
Code B decode for digits 7–0	1	1	1	1	1	1	1	1	FF

Table G.2 ■ Decode-Mode register examples

When the code B decode mode is used, the decoder looks only at the lower nibble of the data in the digit registers (D3–D0), disregarding bits D4–D6. D7, which sets the decimal point (SEG DP), is independent of the decoder and is positive logic (D7 = 1 turns on the decimal point). Digit registers are the dual-port SRAM.

G.2.5 Intensity Control and Interdigit Blanking

The MAX7221 allows display brightness to be controlled with an external resistor (R_{SET}) connected between V_{DD} and ISET. The peak current sourced from the segment drivers is nominally 100 times the current entering ISET. This resistor can either be fixed or variable to allow brightness adjustment from the front panel. Its minimum value should be 9.53 kΩ, which typically sets the segment current at 40 mA. Display brightness can also be controlled digitally by using the intensity register.

Digital control of display brightness is provided by an internal pulse-width modulator, which is controlled by the lower nibble of the intensity register. The modulator scales the average segment current in 16 steps from a maximum of 15/16 down to 1/16 of the peak current set by R_{SET}. Table G.3 lists the intensity register format. The minimum interdigit blanking time is set to 1/32 of a cycle.

Duty Cycle	D7	D6	D5	D4	D3	D2	D1	D0	Hex Code
(min on)									
1/16	x	x	x	x	0	0	0	0	x0
2/16	x	x	x	x	0	0	0	1	x1
3/16	x	x	x	x	0	0	1	0	x2
4/16	x	x	x	x	0	0	1	1	x3
5/16	x	x	x	x	0	1	0	0	x4
6/16	x	x	x	x	0	1	0	1	x5
7/16	x	x	x	x	0	1	1	0	x6
8/16	x	x	x	x	0	1	1	1	x7
9/16	x	x	x	x	1	0	0	0	x8
10/16	x	x	x	x	1	0	0	1	x9
11/16	x	x	x	x	1	0	1	0	xA
12/16	x	x	x	x	1	0	1	1	xB
13/16	x	x	x	x	1	1	0	0	xC
14/16	x	x	x	x	1	1	0	1	xD
15/16	x	x	x	x	1	1	1	0	xE
15/16	x	x	x	x	1	1	1	1	xF
(max on)									

Table G.3 ■ Intensity register format

G.2.6 Scan-Limit Register

The scan-limit register sets how many digits are displayed, from one to eight. They are displayed in a multiplexed manner with a typical scan rate of 800 Hz with 8 digits displayed. If fewer digits are displayed, the scan rate is $8\ f_{OSC}/N$, where N is the number of digits scanned. Since the number of scanned digits affects the display brightness, the scan limit register should not be used to blank portions of the display (such as leading zero suppression). Table G.4 lists the scan-limit register format.

Scan Limit	Register Data								Hex Code
	D7	D6	D5	D4	D3	D2	D1	D0	
Display digit 0 only	x	x	x	x	x	0	0	0	x0
Display digits 0 & 1	x	x	x	x	x	0	0	1	x1
Display digits 0 to 2	x	x	x	x	x	0	1	0	x2
Display digits 0 to 3	x	x	x	x	x	0	1	1	x3
Display digits 0 to 4	x	x	x	x	x	1	0	0	x4
Display digits 0 to 5	x	x	x	x	x	1	0	1	x5
Display digits 0 to 6	x	x	x	x	x	1	1	0	x6
Display digits 0 to 7	x	x	x	x	x	1	1	1	x7

Table G.4 ■ Scan-limit register format

G.2.7 Shutdown Mode Register

When the MAX7221 is in shutdown mode, the segment drivers are in a high-impedance state. Data in the digit and control registers remain unaltered. Shutdown can be used to save power or as an alarm to flash the display by successively entering and leaving shutdown mode. For the minimum supply current in shutdown mode, logic inputs should be at ground or V_{DD} (CMOS-logic levels).

Typically, it takes less than 250 μs for the MAX7221 to leave shutdown mode. The display driver can be programmed while in shutdown mode, and shutdown mode can be overridden by the display test function. When the bit 0 of the shutdown register is zero, the MAX7221 is in shutdown mode. Otherwise, the chip is in normal mode.

G.2.8 Display Test Register

The display test register operates in two modes: normal and display test. The display test mode turns all LEDs on by overriding, but alternating, all controls and digit registers (including the shutdown register). In display test mode, eight digits are scanned and the duty cycle is 15/16. When the bit of the display test register is 0, the MAX7221 is in normal operation mode. Otherwise, it is in display test mode.

G.2.9 Serial Shift Register

At the bottom of Figure G.2 is a 16-bit serial shift register. The meaning of the contents of this register is illustrated in Table G.5. When the $\overline{CS}$ signal is low, data will be shifted in from the DIN pin on the rising edge of the CLK signal. On the rising edge of $\overline{CS}$, the lower eight bits of the shift register will be transferred to a destination specified by the value represented by bits D11 to D8 of the shift register. The mapping of these four address bits is shown in Table G.6. The upper four bits of this register are not used.

D15	D14	D13	D12	D11	D10	D9	D8	D7	D6	D5	D4	D3	D2	D1	D0
X	X	X	X	Address				MSB			Data				LSB

Table G.5 ■ Serial data format

In Table G.6, digit 0 to digit 7 are registers in dual-port SRAM that are used to hold the decimal digits or the patterns (when no-decode mode is selected) to be displayed. Whenever the register **Digit i** (i = 0...7) is selected to supply segment pattern directly (in no decode mode) or indirectly (in decode mode) to the segment drivers, the associated digit driver pin **DIG i** is also asserted.

The register name in the first row of the table is *no op*. This register is used when cascading multiple MAX7221s. To cascade multiple MAX7221s, connect all devices $\overline{CS}$ inputs together and connect DOUT to DIN on adjacent devices. For example, if four MAX7221s are cascaded, then to write to the fourth chip, send the desired 16-bit word, followed by three no-op codes (hex *X0XX*, msb first). When $\overline{CS}$ goes high, data is latched in all devices. The first three chips receive no-op commands, and the fourth receives the intended data.

Register	Register Data					Hex Code
	D15–D12	D11	D10	D9	D8	
No op	x	0	0	0	0	x0
Digit 0	x	0	0	0	1	x1
Digit 1	x	0	0	1	0	x2
Digit 2	x	0	0	1	1	x3
Digit 3	x	0	1	0	0	x4
Digit 4	x	0	1	0	1	x5
Digit 5	x	0	1	1	0	x6
Digit 6	x	0	1	1	1	x7
Digit 7	x	1	0	0	0	x8
Decode Mode	x	1	0	0	1	x9
Intensity	x	1	0	1	0	xA
Scan limit	x	1	0	1	1	xB
Shutdown	x	1	1	0	0	xC
Display test	x	1	1	1	1	xF

Table G.6 ■ Register address map

G.2.10 Choosing the Value for R_{SET}

Appropriate value must be chosen for the R_{SET} resistor in order to supply enough current to the LED segments. Table G.7 shows us how to choose the R_{SET} value for a given LED segment current (I_{SEG}) and the LED forward voltage drop (V_{LED}). Any other combinations of I_{SEG} and V_{LED} that are not in the table can be obtained by using interpolation. For example, suppose V_{LED} is 2.3 V when I_{SEG} is 10 mA, then the R_{SET} value can be calculated as follows:

$$(2.3 - 2.0) / (2.5 - 2.0) = (63.7 - R_{SET}) / (63.7 - 59.3)$$

The value of R_{SET} is solved to be 61 kΩ, which is not a standard resistor. The closest standard resistor to 62 kΩ will be used instead.

I_{SEG} (mA)	V_{LED}(V)				
	1.5	2.0	2.5	3.0	3.5
40	12.2 kΩ	11.8 kΩ	11.0 kΩ	10.6 kΩ	9.69 kΩ
30	17.8 kΩ	17.1 kΩ	15.8 kΩ	15.0 kΩ	14.0 kΩ
20	29.8 kΩ	28.0 kΩ	25.9 kΩ	24.5 kΩ	22.6 kΩ
10	66.7 kΩ	63.7 kΩ	59.3 kΩ	55.4 kΩ	51.2 kΩ

Table G.7 ■ R_{SET} vs. segment current and LED forward voltage drop

G.3 Using the MAX7221

In this section, we will use examples to illustrate the application of the MAX7221.

Example G.1
▼

Use a MAX7221 and seven seven-segment displays to display the value $-126.5°F$. Show the circuit connection, and write a program to display the given value. Assume enough brightness can be obtained when there is 10-mA current flowing through the LED and the V_{LED} at this current is 2.3 V.

Solution: The circuit connection is shown in Figure G.3. The R_{SET} value is chosen to be 62 kΩ to satisfy the V_{LED} and I_{SEG} requirements. In this circuit, we need to send 12 16-bit data to the MAX7221 in order to complete the set up. The values to be written are as follows:

Figure G.3 ■ Circuit connection with one MAX7221

Scan Limit Register
We want to display digits 0, 1, 2, 3, 4, 5, 6. We need to write the value $06 into the scan limit register. To achieve this, one should write the value $0B06 to the MAX7221. The hex value B is the address of the scan limit register.

Decode Mode Register

Among the seven characters (−126.5°F) to be displayed, the degree character and letter F cannot be decoded. We will use *decode mode* for the leading five characters and *no-decode mode* for the degree character and letter F. Therefore, the value to be stored in the decode register is $FC. To achieve this, we need to send the value $09FC to the MAX7221, where 9 is the address of the decode mode register.

Digit 0 Register (to display letter F)

We have chosen no-decode mode for letter F. To display letter F, we need to turn on segments A, E, F, and G and turn off segments B, C, D, and decimal point, DP. The value to be stored in digit 0 is $47. The address of this register is 1. We need to send the value $0147 to the MAX7221.

Digit 1 Register (to display degree character)

The degree character is displayed in no-decode mode. The seven-segment pattern for degree character is %1100011 (from segment a to g). Since a decimal point is not associated with this character, the value to be stored in digit 1 is $63. Therefore, we need to send the value $0263 to the MAX7221.

Digit 2 Register (to display the value 5)

This digit can be displayed in decode mode. Decimal point is not associated with it. The address for this register is 3. Therefore, the value to be stored in digit 2 is $05. To achieve this, we need to send the value $0305 to the MAX7221.

Digit 3 Register (to display the value 6)

This digit can be displayed in decode mode and a decimal point is associated with this digit. The value to be stored at this register is %10000110 (from segment DP, A to G). The address of this register is 4. Therefore, the value to be written to the MAX7221 is $0486.

Digit 4 Register (to display the value 2)

This digit can be displayed in decode mode and the decimal point is not associated with it. The value to be stored at this register is $02. The address of this register is 5. Therefore, the value to be written to the MAX7221 is $0502.

Digit 5 Register (to display the value 1)

This digit can be displayed in decode mode and a decimal point is not associated with it. The value to be stored in this register is $01. The address of this register is 6. Therefore, we need to write the value $0601 to the MAX7221.

Digit 6 Register (to display negative sign −)

This character can be displayed in decode mode and the decimal point is not associated with it. The value to be stored in this register is $0A. The address of this register is 7. Therefore, we need to send the value $070A to the MAX7221.

Intensity Register

We will use the maximum intensity to display the value −126.5°F. Therefore, the value to be stored in this register is $0F. The address of this register is $A. Therefore, we need to send the value $0A0F to the MAX7221.

Shutdown Register

The chip should perform normal operation. Therefore, the value to be stored in this register is $01 (bit 0 must be 1). The address of this register is $C. Therefore, the value to be sent to the MAX7221 is $0C01.

Display Test Register

The chip should perform in normal operation. Therefore, the value to be stored in this register is $00 (bit 0 must be 0). The address of this register is $F. Therefore, the value to be sent to the MAX7221 is $0F00.

The SPI function must be configured properly:

1. *Pin directions*: The SCK and MOSI pins are forced for output. The SS pin must be configured as general output.

2. *SPI0CR1 register*: Disable SPI interrupt, enable SPI, set the SCK pin to be idle low, choose rising edge to shift out data, set SS pin as general output, data transfer MSB first. Write the value of $50 into the SP0CR1 register.

3. *SPI0CR2 register*: Disable bidirectional mode, stop SPI in wait mode, write the value of $02 into SP0CR2.

4. *SPI0BR register*: Select 2 as the E clock divide factor. Write the value $10 into the SPI0BR register. This will set the data shift rate to 6 MHz and will be within the maximum MAX7221 transfer rate.

5. *WOMS register*: Enable Port S pull up. Write the value $00 into this register.

The following subroutine will configure the SPI function accordingly:

```
#include     "c:\miniide\hcs12.inc"
OpenSPI   movb   #$10,SPI0BR     ; set baud rate to 6 MHz for 24-MHz clock
          movb   #$50,SPI0CR1    ; disable interrupt, enable SPI, SCK idle low,
                                 ; data latched on rising edge, data transferred msb first
          movb   #$02,SPI0CR2    ; disable bidirectional mode, stop SPI in wait mode
          movb   #0,WOMS         ; enable Port S pull-up
          bset   DDRS,BIT7       ; configure port SS pin for output
          rts
```

We will use the following assembler directives to define the value to be written into the MAX7221 device:

```
display_dat db   $0B,$06        ; value scan limit register
            db   $09,$FC        ; value for decode mode register
            db   $0A,$0F        ; value for intensity register
            db   $0C,$01        ; value for shutdown register
            db   $0F,$00        ; value for display test register
            db   $01,$47        ; value for Digit 0 register
            db   $02,$63        ; value for Digit 1 register
            db   $03,$05        ; value for Digit 2 register
            db   $04,$86        ; value for Digit 3 register
            db   $05,$02        ; value for Digit 4 register
            db   $06,$01        ; value for Digit 5 register
            db   $07,$0A        ; value for Digit 6 register
```

The following subroutine will transfer the appropriate data to the MAX7221:

```
; ****************************************************************************
; The following subroutine transfers the array of 16-bit data to the MAX7221.
; The starting address of the data is passed in index register X and the number
; of 16-bit words is passed in accumulator B.
; ****************************************************************************
```

```
transfer2   bclr    PTS,$80              ; pull SS pin to low to enable SPI transfer
            ldaa    1,x+                 ; get one byte and also move the pointer
            jsr     putcspi0             ; start an SPI transfer
            ldaa    1,x+                 ; get one byte and also move the pointer
            jsr     putcspi0
            bset    PTS,$80              ; transfer data to appropriate register
            dbne    b,transfer2          ; continue until data is complete
            rts
#include    "c:\miniIDE\spi0util.asm"
```

The C language version of this subroutine will be left as an exercise.

Example G.2

▼

Use the MAX7221 in Figure G.3 to drive five seven-segment displays and display the following patterns with each pattern last for 1 second:

```
0
10
210
3210
43210        ; repeat from this line until the last line
54321
65432
76543
87654
98765
09876
10987
21098
32109        ; last line of the pattern
```

Solution: The setting of the MAX7221 is identical to that in Example G.1. The display pattern is changed by a table lookup method. The addresses of digits 4, 3, 2, 1, and 0 from left to right are 5, 4, 3, 2, and 1. To blank a digit, send $FF as its pattern. The following assembly program does the digit shifting on five seven-segment displays:

```
#include "c:\miniIDE\hcs12.inc"
            org     $1000
pat_cnt     ds.b    1
            org     $1500
            jsr     OpenSPI              ; configure SPI function
            jsr     OpenMax              ; configure MAX7221
            movb    #4,pat_cnt
            ldx     #pattern
loop1       ldd     2,x+
            jsr     send2max
            ldd     2,x+
            jsr     send2max
            ldd     2,x+
```

```
                    jsr      send2max
                    ldd      2,x+
                    jsr      send2max
                    ldd      2,x+
                    jsr      send2max
                    ldy      #3
                    jsr      delayby100ms
                    dec      pat_cnt
                    bne      loop1
          forever   ldx      #forLP
                    movb     #10,pat_cnt
          next_i    ldd      2,x+
                    jsr      send2max
                    ldd      2,x+
                    jsr      send2max
                    ldd      2,x+
                    jsr      send2max
                    ldd      2,x+
                    jsr      send2max
                    ldd      2,x+
                    jsr      send2max
                    ldy      #3
                    jsr      delayby100ms
                    dec      pat_cnt
                    bne      next_i
                    bra      forever
;*********************************************************************************
; The following function initializes the MAX7221 parameters.
;*********************************************************************************
          OpenMax   ldx      #maxConf       ; configure max7221
                    ldd      0,X
                    jsr      send2max
                    ldd      2,X
                    jsr      send2max
                    ldd      4,X
                    jsr      send2max
                    ldd      6,X
                    jsr      send2max
                    ldd      8,X
                    jsr      send2max
                    rts
;*********************************************************************************
; The following function configures the SPI module for data transfer.
;*********************************************************************************
          OpenSPI   movb     #$10,SPI0BR    ; set baud rate to 6 MHz for 24-MHz clock
                    movb     #$50,SPI0CR1   ; disable interrupt, enable SPI, SCK idle low,
                                            ; data latched on rising edge, data transferred msb first
                    movb     #$02,SPI0CR2   ; disable bidirectional mode, stop SPI in wait mode
```

```
                movb    #0,WOMS          ; enable Port S pull-up
                bset    DDRS,BIT7        ;
                movb    #$EA,DDRS        ; configure port S pins
                rts
;***************************************************************************
; The following subroutine sends 2 bytes of data to the MAX7221. The data to be sent
; are passed in accumulator D. Accumulator holds address and B holds data.
;***************************************************************************
send2max  bclr    PTS,bit7         ; enable SPI transfer
          brclr   SPIOSR,SPTEF,*
          staa    SPIODR
          brclr   SPIOSR,SPIF,*
          ldaa    SPIODR           ; clear the SPIF flag
          brclr   SPIOSR,SPTEF,*
          stab    SPIODR
          brclr   SPIOSR,SPIF,*
          ldab    SPIODR           ; clear the SPIF flag
          bset    PTS,bit7
          rts
#include   "c:\miniIDE\delay.asm"   ; include time delay functions
;***************************************************************************
; Values to configure MAX7221 for pattern shifting
;***************************************************************************
maxConf   dc.w    0x0B05           ; display lowest five digits.
          dc.w    0x09FF           ; all five digits use code B decode mode
          dc.w    0x0A0F           ; set to highest intensity
          dc.w    0x0C01           ; normal mode, no shutdown
          dc.w    0x0F00           ; normal mode, no display test;
;***************************************************************************
; values to be transferred to digit 4 through 0 driven by the MAX7221
;***************************************************************************
pattern   dc.w    0x0500, 0x04FF, 0x03FF, 0x02FF, 0x01FF  ; display 0 on digit 4
          dc.w    0x0501, 0x0400, 0x03FF, 0x02FF, 0x01FF  ; display 10 on digits 4 & 3
          dc.w    0x0502, 0x0401, 0x0300, 0x02FF, 0x01FF  ; display 210 on digits 4, 3, 2
          dc.w    0x0503, 0x0402, 0x0301, 0x0200, 0x01FF  ; display 3210 on digits 4,3,2,1
forLP     dc.w    0x0504, 0x0403, 0x0302, 0x0201, 0x0100  ; display 43210 on digits 4 to 0
          dc.w    0x0505, 0x0404, 0x0303, 0x0202, 0x0101  ; display 54321 on digits 4 to 0
          dc.w    0x0506, 0x0405, 0x0304, 0x0203, 0x0102  ; display 65432 on digits 4 to 0
          dc.w    0x0507, 0x0406, 0x0305, 0x0204, 0x0103  ; display 76543 on digits 4 to 0
          dc.w    0x0508, 0x0407, 0x0306, 0x0205, 0x0104  ; display 87654 on digits 4 to 0
          dc.w    0x0509, 0x0408, 0x0307, 0x0206, 0x0105  ; display 98765 on digits 4 to 0
          dc.w    0x0500, 0x0409, 0x0308, 0x0207, 0x0106  ; display 09876 on digits 4 to 0
          dc.w    0x0501, 0x0400, 0x0309, 0x0208, 0x0107  ; display 10987 on digits 4 to 0
          dc.w    0x0502, 0x0401, 0x0300, 0x0209, 0x0108  ; display 21098 on digits 4 to 0
          dc.w    0x0503, 0x0402, 0x0301, 0x0200, 0x0109  ; display 32109 on digits 4 to 0
          end
```

G.4 Cascading Multiple MAX7221s

Cascading multiple MAX7221s is easy. The procedure is as follows:

- Tie the $\overline{\text{CS}}$ signal of all the MAX7221s together and connect them to an output pin of the microcontroller.
- Tie the SCK signal of all the MAX7221s together and connect them to the SCK pin of the microcontroller.
- Wire the DIN pin of the first MAX7221 to the MOSI pin of the HCS12, connect the DOUT pin of the first MAX7221 to the DIN pin of its adjacent MAX7221, and so on.

The circuit for cascading two MAX7221s is shown in Figure G.4. In this example, we use the $\overline{\text{SS}}$ pin as a general-purpose output pin to drive the $\overline{\text{CS}}$ pin. Again, we use the same R_{SET} to achieve the same I_{SEG} and V_{LED} values. When multiple MAX7221s are cascading together, we need to shift data to all devices before transferring data (by pulling the $\overline{\text{CS}}$ signal to high) to the destination register of each MAX7221.

Example G.3

Write a program to configure the SPI subsystem and display the value

38.9°C 02 13 04 20 01

on the 16 seven-segment displays. Assume that the E clock frequency is 24 MHz.

Solution: We can use the same configuration for the SPI subsystem in this circuit. We will divide the digit string into two halves so that each MAX7221 displays one half as follows:

38.9°C 02

13 04 20 01

The first MAX7221 will display seven characters; the second MAX7221 will display eight digits. We will pair these two substrings as shown in Figure G.5 for data transfer purposes.

In this circuit configuration, we need to send thirteen 32-bit data to perform the data setup. The values to be transferred are as follows:

Scan Limit Register
The first MAX7221 will display seven digits and the second MAX7221 will display eight digits. For this setup, we need to send the value $0B06 to the first MAX7221 and send the value $0B07 to the second MAX7221.

Decode Mode Register
For the first MAX7221, we will use no-decode mode for digits 2 and 3 and decode mode for the other digits. For the second MAX7221, all digits will be displayed in decode mode. For this setup, we will send the value $09F3 to the first MAX7221 and send the value $09FF to the second MAX7221.

Intensity Register
For this example, we will use the maximum intensity to display the values. For this setup, we will send the value $0A0F to both MAX7221s.

Shutdown Register
Both chips should perform normally. Therefore, the value to be stored in this register is $01, and so the value to be sent to both MAX7221 is $0C01.

Figure G.4 ■ Cascading two MAX7221 to display 16 BCD digits

digit	7	6	5	4	3	2	1	0	
		3	9	8	∘	C	0	2	← first MAX7221
	1	3	0	4	2	0	0	1	← second MAX7221

Figure G.5 ■ Data arrangement for transfer

Display Test Register   Both chips should perform normal operation. Therefore, the value to be stored in this register is $00 (bit 0 must be 0). We need to send the value $0F00 to both MAX7221s.

Digit 7 Register

The first MAX7221 won't display anything on digit 7, whereas the second MAX7221 will display the value 1 on digit 7. For this value, we can send $08FF to the first MAX7221 and send the value $0801 to the second MAX7221.

Digit 6 Register

Both MAX7221s will display 3 on this digit. Therefore, send the value $0703 to both MAX7221s.

Digit 5 Register

The first MAX7221 will display 9 on digit 5, whereas the second MAX7221 will display 0 on digit 5. Therefore, we will send the value $0609 to the first MAX7221 and send the value $0600 to the second MAX7221.

Digit 4 Register

The first MAX7221 will display 8 on digit 4, whereas the second MAX7221 will display 4 on digit 4. Therefore, we will send the value $0508 to the first MAX7221 and send the value $0504 to the second MAX7221.

Digit 3 Register

The first MAX7221 will display º on digit 3, whereas the second MAX7221 will display 2 on digit 3. Remember the first MAX7221 selected no-decode mode for this digit. Therefore, we will send the value $0463 to the first MAX7221 and send the value $0402 to the second MAX7221.

Digit 2 Register

The first MAX7221s will display C on digit 2, whereas the second MAX7221 will display 0 on digit 2. Remember the first MAX7221 selected no-decode mode for this digit. Therefore, we will send the value $034E to the first MAX7221 and send the value $0300 to the second MAX7221.

Digit 1 Register

Both MAX7221s will display 0 on this digit. Therefore, we should send the value $0200 to both MAX7221s.

Digit 0 Register

The first MAX7221 will display 2 on this digit, whereas the second MAX7221 will display 1 on this digit. Therefore, we will send the value $0102 to the first MAX7221 and send the value $0101 to the second MAX7221.

We can use the same configuration as in Example G.1 for the SPI subsystem in this example. The data to be transferred to the MAX7221s are defined as follows:

```
;*************************************************************************
; The data to be sent to two MAX7221s are paired together. The data for the second
; MAX7221 are listed in the front because they should be sent out first.
;*************************************************************************
max_data    db      $0B,$07,$0B,$06     ; data for scan limit registers
            db      $09,$FF,$09,$F3     ; data for decode mode registers
            db      $0A,$0F,$0A,$0F     ; data for intensity registers
            db      $0C,$01,$0C,$01     ; data for shutdown registers
            db      $0F,$00,$0F,$00     ; data for display test registers
            db      $08,$01,$08,$FF     ; data for digit 7 registers
            db      $07,$03,$07,$03     ; data for digit 6 registers
```

```
                    db      $06,$00,$06,$09      ; data for digit 5 registers
                    db      $05,$04,$05,$08      ; data for digit 4 registers
                    db      $04,$02,$04,$63      ; data for digit 3 registers
                    db      $03,$00,$03,$4E      ; data for digit 2 registers
                    db      $02,$00,$02,$00      ; data for digit 1 registers
                    db      $01,$01,$01,$02      ; data for digit 0 registers
```

The following subroutine will transfer the appropriate data to the two MAX7221s:

```
; **************************************************************************
; The following subroutine transfers the array of 32-bit data to the MAX7221.
; The starting address of the data is passed in index register X and the number
; of 32-bit words is passed in accumulator B.
; **************************************************************************
transfer4     bclr    PORTS,$80     ; enable SPI transfer to two MAX7221s
              ldaa    1,x+
              jsr     putcspi0      ; send the first byte of the second MAX7221
              ldaa    1,x+
              jsr     putcspi0      ; send the second byte of the second MAX7221
              ldaa    1,x+
              jsr     putcspi0      ; send the first byte of the first MAX7221
              ldaa    1,x+
              jsr     putcspi0      ; send the second byte of the first MAX7221
              bset    PTS,$80       ; pull SS pin to high and transfer data to appropriate
;                                   ; register
              dbne    b,transfer    ; continue until data is complete
              rts
#include      "c:\miniIDE\spiOutil.asm"
```

The C language version of the subroutine is left as an exercise.

▲

G.5 Exercises

EG.1 Write a C program to display the value of −126.5°F on the seven-segment displays shown in Figure G.3.

EG.2 Write a C program to display the pattern specified in Example G.2.

EG.3 Write a C program to display the following value on the seven-segment displays in Figure G.4:

    38.9°C 02 13 04 20 01

References 807

1. Analog Devices, *AD7302 Datasheet*, Norwood, Massachusetts: Analog Devices, 1997.
2. Bosch, *CAN Specification, Version 2.0*, Stuggart Germany, 1991.
3. CANopen, *Cabling and Connector Pin Assignment, Version 1.1*, Stuttgart, Germany: Bosch, April 2001.
4. Dallas Semiconductor, *DS1307 Datasheet*, Dallas Texas: Dallas Semiconductor, 2004.
5. Dallas Semiconductor, *DS1631A Datasheet*, Dallas Texas: Dallas Semiconductor, 2004.
6. Honeywell, *IH-3605 Datasheet*, Honeywell, 1999.
7. Honeywell, *ASCX30AN Datasheet*, Honeywell, 2004.
8. Han-Way Huang, *MC68HC11: An Introduction, 2nd ed. Clifton Park, New York: Delmar Thomson Learning, 2001.*
9. Han-Way Huang, *Using the MCS51 Microcontroller*, New York: Oxford University Press, 1998.
10. Han-Way Huang, *MC68HC12: An Introduction*, Clifton Park, New York: Delmar Thomson Learning, 2002.
11. Han-Way Huang, *PIC Microcontroller*, Clifton Park, New York: Delmar Thomson Learning, 2004.
12. Linear Technology, *LTC1661 Datasheet*, Milpitas, California: Linear Technology, 1999.
13. Maxim Semiconductor, *MAX6952 Datasheet*, Sunnyvale, California: Maxim Semiconductor, 2004.
14. Microchip, *MCP2551 Datasheet DS21667D*, Chandler, Arizona: Microchip Technology, 2003.
15. Microchip, *24LC08B Data sheet DS21710A*, Chandler, Arizona: Microchip Technology, 2002.
16. Microchip, *24LC16 Data sheet DS21703D*, Chandler, Arizona: Microchip Technology, 2003.
17. Microchip, *MCP23016 Data sheet DS21710A*, Chandler, Arizona: Microchip Technology, 2002.
18. Microchip, *TC1047A Data sheet DS21498B*, Chandler, Arizona: Microchip Technology, 2002.
19. Microchip, *TC72 Data sheet DS21743A*, Chandler, Arizona: Microchip Technology, 2002.
20. Freescale, *MC9S12DP256B Device User Guide, V02.14*, 9S12DP256BDGV2/D, Austin, Texas: Freescale, 2003.
21. Freescale, *S12ATD10B8CV2/D Block User Guide*, Austin, Texas: Freescale, 2002.
22. Freescale, *S12CPUV2 Reference Manual*, Austin, Texas: Freescale, 2003.
23. Freescale, *Background Debug Module V4, S12BDMV3*, Austin, Texas: Freescale, 2003.
24. Freescale, *Breakpoint Module Block User Guide, S12BKPV1/D*, Austin, Texas: Freescale, 2003.
25. Freescale, *Clock and Reset Generator Block User Guide, S12CRGV2/D*, Austin, Texas: Freescale, 2002.
26. Freescale, *MC9S12DP256 Port Integration Module Block User Guide, V02.07, S12DP256PIMV2*, Austin, Texas: Freescale, 2002.

27. Freescale, *Enhanced Captured Timer Block User Guide, S12ECT16B8CV1/D*, Austin, Texas: Freescale, 2004.

28. Freescale, *EETS4K Block User Guide V02.03, HCS12EETS4KUG/D*, Austin, Texas: Freescale, 2001.

29. Freescale, *FTS256K Block User Guide V02.02, HCS12FTS256KUG/D*, Austin, Texas: Freescale, 2001.

30. Freescale, *HCS12 Inter-Integrated Circuit Block Guide, V02.08, S12IICV2/D*, Austin, Texas: Freescale, 2004.

31. Freescale, *HCS12 Interrupt Module V1, S12INTV1/D*, Austin, Texas: Freescale, 2003.

32. Freescale, *Multiplexed External Bus Interface (MEBI) Module V3 Block User Guide, S12MEBIV3/D, Rev. 3.00*, Austin, Texas: Freescale, 2003.

33. Freescale, *Module Mapping Control V4, S12MMCV4/D, Rev. 4.00*, Austin, Texas: Freescale, 2003.

34. Freescale, *MSCAN Block Guide,V02.15, S12MSCANV2/D*, Austin, Texas: Freescale, 2004.

35. Freescale, *PWM_8B8C Block User Guide, V01.17, S12PWM8B8CV1/D*, Austin, Texas: Freescale, 2004.

36. Freescale, *HCS12 Serial Communications Interface Block Guide, V02.05, S12SCIV2/D*, Austin, Texas: Freescale, 2001.

37. Freescale, *SPI Block User Guide, V02.07, S12SPIV2/D*, Austin, Texas: Freescale, 2000.

38. Freescale, *Reference Guide for D-Bug12 Version 4.x.x*, Austin, Texas: Freescale, 2004.

39. National Semiconductor, *LM34 Precision Fahrenheit Temperature Sensor Datasheet*, Santa Clara, California: National Semiconductor, 2000.

40. National Semiconductor, *LM35 Precision Centigrade Temperature Sensor Datasheet*, Santa Clara, California: National Semiconductor, 2000.

41. Texas Instrument, *TLV5616 Datasheet*, Dallas, Texas: Texas Instrument, 2004.

Glossary

Accumulator   A register in a computer that contains an operand to be used in an arithmetic operation.

Acknowledgment error   In a CAN bus protocol, an acknowledgment error is detected whenever the transmitter does not monitor a dominant bit in the ACK slot.

Activation record   Another term for stack frame.

Address access time   The amount of time it takes for a memory component to send out valid data to the external data pins after address signals have been applied (assuming that all other control signals have been asserted).

Addressing   The application of a unique combination of high and low logic levels to select a corresponding unique memory location.

Address multiplexing   A technique that allows the same address pin to carry different signals at different times; used mainly by DRAM technology. Address multiplexing can dramatically reduce the number of address pins required by DRAM chips and reduce the size of the memory chip package.

Algorithm   A set of procedure steps represented in pseudo code that is designed to solve certain computation issues.

ALU (arithmetic logic unit)   The part of the processor in which all arithmetic and logical operations are performed.

Array   An ordered set of elements of the same type. The elements of the array are arranged so that there is a zeroth, first, second, third, and so forth. An array may be one-, two-, or multidimensional.

ASCII (American Standard Code for Information Interchange) code.   A code that uses 7 bits to encode all printable and control characters.

Assembler   A program that converts a program in assembly language into machine instructions so that it can be executed by a computer.

Assembler directive   A command to the assembler for defining data and symbols, setting assembling conditions, and specifying output format. Assembler directives do not produce machine code.

Assembly instruction   A mnemonic representation of a machine instruction.

Assembly program   A program written in assembly language.

Automatic variable   A variable defined inside a function that comes into existence when the function is entered and disappears when the function returns.

Background receive buffer   The receive buffer where the incoming frame is placed before it is forwarded to the foreground receive buffer.

Barometric pressure   The air pressure existing at any point within the earth's atmosphere.

BDM mode   A mode provided by the HCS12 and some other Freescale microcontrollers to facilitate the design of debugging tools. This mode has been utilized to implement source-level debugger.

Binary Coded Decimal (BCD)   A coding method that uses four binary digits to represent one decimal digit. The binary codes 0000_2–1001_2 correspond to the decimal digits 0 to 9.

Bit error   In the CAN bus protocol, a node that is sending a bit on the bus also monitors the bus. When the bit value monitored is different from the bit value being sent, the node interprets the situation as a bit error.

Branch instruction   An instruction that causes the program flow to change.

Break   The transmission or reception of a low for at least one complete character time.

Breakpoint   A memory location in a program where the user program execution will be stopped and the monitor program will take over the CPU control and display the contents of CPU registers.

Bubble sort   A simple sorting method in which an array or a file to be sorted is gone through sequentially several times. Each iteration consists of comparing each element in the array or file with its successor (x[i] with x[i + 1]) and interchanging the two elements if they are not in proper order (either ascending or descending)

Bus   A set of signal lines through which the processor of a computer communicates with memory and I/O devices.

Bus cycle timing diagram   A diagram that describes the transitions of all the involved signals during a read or write operation.

Bus multiplexing   A technique that allows more than one set of signals to share the same group of bus lines.

Bus off   The situation in which a CAN node has a transmit error count above 256.

CAN transceiver   A chip used to interface a CAN controller to the CAN bus.

Central processing unit (CPU)   The combination of the register file, the ALU, and the control unit.

Charge pump   A circuit technique that can raise a low voltage to a level above the power supply. A charge pump is often used in an A/D converter, in EEPROM and EPROM programming, etc.

Clock monitor reset   A mechanism provided to detect whether the frequency of the system clock signal inside a CPU is lower than a certain value.

Clock stretching   In the I²C protocol, the slave device can drive the SCL low for an interval longer than the master's SCL low period. The resulting SCL bus signal low period is stretched. A slave stretches the clock signal when it needs more time to make data ready.

Column address strobe (CAS)   The signal used by DRAM chips to indicate that column address logic levels are applied to the address input pins.

Comment   A statement that explains the function of a single instruction or directive or a group of instructions or directives. Comments make a program more readable.

Communication program   A program that allows a PC to communicate with another computer.

Computer   A computer consists of hardware and software. The hardware includes four major parts: the central processing unit, the memory unit, the input unit, and the output unit. Software is a sequence of instructions that control the operations of the hardware.

Computer operate properly (COP) watchdog timer   A special timer circuit designed to detect software processing errors. If software is written correctly, then it should complete all operations within a certain amount of time. Software problems can be detected by enabling a watchdog timer so that the software resets the watchdog timer before it times out.

Contact bounce   A phenomenon in which a mechanical key switch will go up and down several times before it settles down when it is pressed.

Control unit   The part of the processor that decodes and monitors the execution of instructions. It arbitrates the use of computer resources and makes sure that all computer operations are performed in proper order.

Controller Area Network (CAN)   A serial communication protocol initially proposed to be used in automotive applications. In this protocol, data is transferred frame by frame. Each frame can carry up to 8 bytes of data. Each data frame is acknowledged in frame.

CRC error   Cyclic redundancy check error. In data communications, the CRC sequence consists of the result of the CRC calculation by the transmitter. The receiver calculates the CRC in the same way as the transmitter. A CRC error is detected if the calculated result is not the same as that received in the CRC sequence.

Cross assembler   An assembler that runs on one computer but generates machine instructions that will be executed by another computer that has a different instruction set.

Cross compiler   A compiler that runs on one computer but generates machine instructions that will be executed by another computer that has a different instruction set.

D/A converter   A circuit that can convert a digital value into an analog voltage.

Data hold time   The length of time over which the data must remain stable after the edge of the control signal that latches the data.

Datapath   The part of the processor that consists of a register file and the ALU.

Data setup time   The amount of time over which the data must become valid before the edge of the control signal that latches the data.

D-Bug12 Monitor   A monitor program designed for the HCS12 microcontroller.

DCE   The acronym for data communication equipment. DCE usually refers to equipment such as a modem, concentrator, router, etc.

Demo board   A single-board computer that contains the target microcontroller as the CPU and a monitor to help the user to perform embedded product development.

Direct mode   An addressing mode that uses an 8-bit value to represent the address of a memory location.

Dominant level   A voltage level in a CAN bus that will prevail when a voltage level at this state and a different level (*recessive level*) are applied to the CAN bus at the same time.

DTE   The acronym for data terminal equipment. DTE usually refers to a computer or terminal.

Dynamic memories   Memory devices that require periodic refreshing of the stored information, even when power is on.

EIA   The acronym for Electronic Industry Association.

Electrically erasable programmable read-only memory (EEPROM)   A type of read-only memory that can be erased and reprogrammed using electrical signals. EEPROM allows each individual location inside the chip to be erased and reprogrammed.

Embedded system   A product that uses a microcontroller as the controller to provide the features. End users are interested in these features rather than the power of the microcontroller. A cell phone, a charge card, a weather station, and a home security system are examples of embedded systems.

Erasable programmable read-only memory (EPROM)   A type of read-only memory that can be erased by subjecting it to strong ultraviolet light. It can be reprogrammed using an EPROM programmer. A quartz window on top of the EPROM chip allows light to be shone directly on the silicon chip inside.

Error active   A CAN node that has both the transmit error count and receive error count lower than 127.

Error passive   A CAN node that has either the transmit error count or receive error count between 128 and 256.

Exception   Software interrupts such as an illegal opcode, an overflow, division by zero, or an underflow.

Expanded mode   The operation modes in which the HCS12 can access external memory components by sending out address signals. A 1-MB memory space is available in this mode.

Extended mode   An addressing mode that uses a 16-bit value to represent the address of a memory location.

Fall time   The amount of time a digital signal takes to go from logic high to logic low.

Floating signal   An undriven signal.

Foreground receive buffer   The CAN receive buffer that is accessible to the programmer.

Form error   An error detected when a fixed-form bit field contains one or more illegal bits (in CAN bus protocol).

Frame pointer   A pointer used to facilitate access to parameters in a stack frame.

Framing error   A data communication error in which a received character is not properly framed by the start and stop bits.

Full-duplex link   A four-wire communication link that allows both transmission and reception to proceed simultaneously.

General call address   A special value (all 0s) used by the master to address every device connected to the I^{2}C bus.

Global memory   Memory that is available to all programs in a computer system.

Half duplex link   A communication link that can be used for either transmission or reception, but only in one direction at a time.

Hard synchronization   The synchronization performed by all CAN nodes at the beginning of a frame.

Hardware breakpoint   A hardware circuit that compares actual address and data values to predetermined data in setup registers. A successful comparison places the CPU in background debug mode or initiates a software interrupt (SWI).

Identifier acceptance filter   A group of registers that can be programmed to select those identifier bits of the incoming frames to be compared for acceptance.

Idle   A continuous logic high on the RxD line for one complete character time.

Illegal opcode   A binary bit pattern of the opcode byte for which an operation is not defined.

Immediate mode   An addressing mode that will be used as the operand of the instruction.

Indexable data structure   A data structure in which each element is associated with an integer that can be used to access it. Arrays and matrices are examples of indexable data structures.

Indexed addressing mode   An addressing mode that uses the sum of the contents of an index register and a value contained in the instruction to specify the address of a memory location. The value to be added to the index register can be a 5-bit, 9-bit, or 16-bit signed value. The contents of an accumulator (A, B, or D) can also be used as the value to be added to the index register to compute the address.

Inline assembly instruction   Assembly instructions that are embedded in a high-level language program.

Input capture   The HCS12 function that captures the value of the 16-bit free-running main timer into a latch when the falling or rising edge of the signal connected to the input capture pin arrives.

Input handshake   A protocol that uses two handshake signals to make sure that the peripheral chip receives data correctly from the input device.

Input port   The part of the microcontroller that consists of input pins, the input data register, and other control circuitry to perform the input function.

Instruction queue   A circuit in the CPU to hold the instruction bytes prefetched by the CPU. The HCS12 and some other microprocessors utilize the bus idle time to perform instruction prefetch in the hope of enhancing the processor throughput.

Instruction tagging   A feature that provides a way of forcing the HCS12 to stop executing an application program when the tagged instruction reaches the CPU. When the tagged instruction is detected, the HCS12 CPU enters background debug mode rather than executing the tagged instruction.

Integrated development environment   A piece of software that combines a text editor, a terminal program, a cross compiler and/or cross assembler, and/or simulator to allow the user to perform program development activities without quitting any one of the programs.

Interframe space   A field in the CAN bus that is used to separate data frames or remote frames from the previous frames.

Inter-Integrated Circuit   A serial communication protocol proposed by Philips in which two wires (SDA and SCL) are used to transmit data. One wire carries the clock signal, and the other wire carries the data or address information. This protocol allows multiple master devices to coexist in the same bus system.

Interrupt   An unusual event that requires the CPU to stop normal program execution and perform some service to the event.

Interrupt overhead   The time spent on handling an interrupt. This time consists of the saving and restoring of registers and the execution of instructions contained in the service routine.

Interrupt priority   The order in which the CPU will service interrupts when all of them occur at the same time.

Interrupt service   The service provided to a pending interrupt by the CPU's execution of a program called a service routine.

Interrupt vector   The starting address of an interrupt service routine.

Interrupt vector table   A table that stores all interrupt vectors.

I/O synchronization   A mechanism that can make sure that CPU and I/O devices exchange data correctly.

ISO   The acronym for the International Standard Organization.

Keyboard debouncing   A process that can eliminate the keyswitch bouncing problem so that the computer can detect correctly whether a key has indeed been pressed.

Keyboard scanning   A process that is performed to detect whether any key has been pressed by the user.

Key wakeup   A mechanism that can generate interrupt requests to wake up a sleeping CPU. The key wakeup ability is associated with I/O ports.

Label field   The field in an assembly program statement that represents a memory location.

Linked list   A data structure that consists of linked nodes. Each node consists of two fields, an information field and a next address field. The information field holds the actual element on the list, and the next address field contains the address of the next node in the list.

Load cell   A transducer that can convert a weight into a voltage.

Local variable   Temporary variables that exist only when a subroutine is called. They are used as loop indices, working buffers, etc. Local variables are often allocated in the system stack.

Low power mode   An operation mode in which less power is consumed. In CMOS technology, the low power mode is implemented by either slowing down the clock frequency or turning off some circuit modules within a chip.

Machine instruction   A set of binary digits that tells the computer what operation to perform.

Mark   A term used to indicate a binary 1.

Maskable interrupts   Interrupts that can be ignored by the CPU. This type of interrupt can be disabled by setting a mask bit or by clearing an enable bit.

Masked ROM (MROM)   A type of ROM that is programmed when the chip is manufactured.

Matrix   A two-dimensional data structure that is organized into rows and columns. The elements of a matrix are of the same length and are accessed using their row and column numbers (i, j), where i is the row number and j is the column number.

Memory   Storage for software and information.

Memory capacity   The total amount of information that a memory device can store; also called *memory density*.

Memory organization   A description of the number of bits that can be read from or written into a memory chip during a read or write operation.

Microcontroller   A computer system implemented on a single, very-large-scale integrated circuit. A microcontroller contains everything that is in a microprocessor and may contain memories, an I/O device interface, a timer circuit, an A/D converter, and so on.

Microprocessor   A CPU packaged in a single integrated circuit.

Mode fault   An SPI error that indicates that there may have been a multimaster conflict for system control. Mode fault is detected when the master SPI device has its $\overline{SS}$ pin pulled low.

Modem   A device that can accept digital bits and change them into a form suitable for analog transmission (modulation) and can also receive a modulated signal and transform it back to its original digital representation (demodulation).

Multi-drop   A data communication scheme in which more than two stations share the same data link. One station is designated as the master, and the other stations are designated as slaves. Each station has its own unique address, with the primary station controlling all data transfers over the link.

Multiprecision arithmetic   Arithmetic operation (add, subtract, multiply, or divide) performed by a computer that deals with operands longer than the computer's word length.

Multitasking   A computing technique in which CPU time is divided into slots that are usually 10 to 20 ms in length. When multiple programs are resident in the main memory waiting for execution, the operating system assigns a program to be executed to one time slot. At the end of a time slot or when a program is waiting for completion of I/O, the operating system takes over and assigns another program to be executed.

Nibble   A group of 4-bit information.

Nonmaskable interrupts   Interrupts that the CPU cannot ignore.

Nonvolatile memory Memory that retains stored information even when power to the memory is removed.

Null modem A circuit connection between two DTEs in which the leads are interconnected in such a way as to fool both DTEs into thinking that they are connected to modems. A null modem is only used for short-distance interconnections.

Object code The sequence of machine instructions that results from the process of assembling and/or compiling a source program.

Output compare An HCS12 timer function that allows the user to make a copy of the value of the 16-bit main timer, add a delay to the copy, and then store the sum in a register. The output-compare function compares the sum with the main timer in each of the following E clock cycles. When these two values are equal, the circuit can trigger a signal change on an output-compare pin and may also generate an interrupt request to the HCS12.

Output handshake A protocol that uses two handshake signals to make sure that an output device correctly receives the data driven by the peripheral chip (sent by the CPU).

Output port The part of the circuit in a microcontroller that consists of output pins, data register, and control circuitry to send data to the output device.

Overflow A condition that occurs when the result of an arithmetic operation cannot be accommodated by the preset number of bits (say, 8 or 16 bits); it occurs fairly often when numbers are represented by fixed numbers of bits.

Parameter passing The process and mechanism of sending parameters from a caller to a subroutine, where they are used in computations; parameters can be sent to a subroutine using CPU registers, the stack, or global memory.

Parity error An error in which an odd number of bits change value; it can be detected by a parity checking circuit.

Phase_seg1 & Phase_seg2 Segments that are used to compensate for edge phase errors. These segments can be lengthened or shortened by synchronization.

Physical layer The lowest layer in the layered network architecture. This layer deals with how signals are actually transmitted, the descriptions of bit timing, bit encoding, and synchronization.

Physical time In the HCS12 timer system, the time represented by the count in the 16-bit main timer counter.

Point-to-point A data communication scheme in which two stations communicate as peers.

Precedence of operators The order in which operators are processed.

Program A set of instructions that the computer hardware can execute.

Program counter (PC) A register that keeps track of the address of the next instruction to be executed.

Program loops A group of instructions or statements that are executed by the processor more than once.

PROM (programmable read-only memory) A type of ROM that allows the end user to program it once and only once using a device called PROM programmer.

Prop_seg The segment within a bit time used to compensate for the physical delay times within the CAN network.

Pseudo code An expressive method that combines the use of plain English and statements similar to certain programming languages to represents an algorithm.

Pull The operation that removes the top element from a stack data structure.

Pulse accumulator   A timer function that uses a counter to count the number of events that occur or measure the duration of a single pulse.

Pulse-width modulation   A timer function that allows the user to specify the frequency and duty cycle of the digital waveform to be generated.

Push   The operation that adds a new element to the top of a stack data structure.

Queue   A data structure to which elements can be added at only one end and removed from only the other end. The end to which new elements can be added is called the *tail* of the queue, and the end from which elements can be removed is called the *head* of the queue.

RAM (random-access memory)   RAM allows read and write access to every location inside the memory chip. Furthermore, read access and write access take the same amount of time for any location within the RAM chip.

Receiver overrun   A data communication error in which a character or a number of characters were received but not read from the buffer before subsequent characters are received.

Refresh   An operation performed on dynamic memories in order to retain the stored information during normal operation.

Refresh period   The time interval within which each location of a DRAM chip must be refreshed at least once in order to retain its stored information.

Register   A storage location in the CPU. It is used to hold data and/or a memory address during the execution of an instruction.

Relative mode   An addressing mode that uses an 8-bit or 16-bit value to specify the branch distance for branch instructions. If the sign of the value is negative, then the branch is a backward branch. Otherwise, the branch is a forward branch.

Remote frame   A frame sent out by a CAN node to request another node to send data frames.

Repeated start condition   This condition is used in I²C protocol. A repeated start signal is a start signal generated without first generating a stop signal to terminate the current communication. This condition is used to change the direction of data transfer or change the partner of data communication.

Reset   A signal or operation that sets the flip-flops and registers of a chip or microprocessor to some predefined values or states so that the circuit or microprocessor can start from a known state.

Reset handling routine   The routine that will be executed when the microcontroller or microprocessor gets out of the reset state.

Reset state   The state in which the voltage level of the RESET pin of the HCS12 is low. In this state, a default value is established for most on-chip registers, including the program counter. The operation mode is established when the HCS12 exits the reset state.

Resynchronization   All CAN nodes perform resynchronization within a frame whenever a change of bit value from recessive to dominant occurs outside of the expected sync_seg segment after the hard synchronization.

Resynchronization jump width   The amount of lengthening in phase_seg1 or shortening in phase_seg2 in order to achieve resynchronization in every bit within a frame. The resynchronization jump width is programmable to between 1 and 4.

Return address   The address of the instruction that immediately follows the subroutine call instruction (either JSR or BSR).

Rise time   The amount of time a digital signal takes to go from logic low to logic high.

ROM (read-only memory)   A type of memory that is nonvolatile in the sense that when power is removed from ROM and then reapplied, the original data are still there. ROM data can only be read—not written—during normal computer operation.

Row address strobe (RAS)   The signal used by DRAM chips to indicate that row address logic levels are applied to the address input pins.

RS232   An interface standard recommended for interfacing between a computer and a modem. This standard was established by EIA in 1960 and has since been revised several times.

Serial Peripheral Interface (SPI)   A protocol proposed by Freescale that uses three wires to perform data communication between a master device and a slave device.

Seven-bit addressing   In the I²C protocol, the master device may use a 7-bit value to specify the device for communication.

Signal conditioning circuit   A circuit added to the output of a transducer to scale and shift the voltage output from the transducer to a range that can take advantage of the whole dynamic range of the A/D converter being used.

Simplex link   A line that is dedicated either for transmission or reception, but not both.

Simulator   A program that allows the user to execute microcontroller programs without having the actual hardware.

Single-chip mode   The operation mode in which the HCS12 functions without external address and data buses.

Source code   A program written in either an assembly language or a high-level language; also called a *source program*.

Source-level debugger   A program that allows the user to find problems in user code at the high-level language (such as C) or assembly-language level.

Space   A term space is used to indicate a binary 0.

Special test mode   The HCS12 operation mode used primarily during Freescale's internal production testing.

Stack   A last-in-first-out data structure whose elements can be accessed only from one end. A stack structure has a top and a bottom. A new item can be added only to the top, and the stack elements can be removed only from the top.

Stack frame   A region in the stack that holds incoming parameters, the subroutine return address, local variables, saved registers, and so on.

Standard timer module   The timer module implemented in some HCS12 members. This module consists of a 16-bit main timer, 8 channels of output-compare/input-capture function, and one 16-bit pulse accumulator.

Start condition   A signal condition that the I²C protocol requires all data transmission to start with. A start condition is a condition that the SDA signal goes from high to low when the SCL signal is high.

Static memories   Memory devices that do not require periodic refreshing in order to retain the stored information as long as power is applied.

Status register   A register located in the CPU that keeps track of the status of instruction execution by noting the presence of carries, zeros, negatives, overflows, and so on.

Stepper motor   A digital motor that rotates certain degrees clockwise or counterclockwise whenever a certain sequence of values is applied to the motor.

Stop condition   In the I²C protocol, all data transfer must be terminated by a stop condition. A stop condition is generated by bringing the SDA line from low to high when the SCL signal is high.

String   A sequence of characters terminated by a NULL character or other special character.

Subroutine   A sequence of instructions that can be called from various places in the program and will return to the caller after its execution. When a subroutine is called, the return address will be saved on the stack.

Subroutine call   The process of invoking the subroutine to perform the desired operations. The HCS12 has BSR, JSR, and CALL instructions for making subroutine calls.

Successive approximation method   A method for performing an A/D conversion that works from the most significant bit toward the least significant bit. For every bit, the algorithm guesses the bit to be 1 and then converts the resultant value into the analog voltage and then compares it with the input voltage. If the converted voltage is smaller than the input voltage, the guess is right. Otherwise, the guess is wrong and the bit is cleared to 0.

Switch statement   A multiway decision based on the value of a control expression.

Sync_seg   In the CAN format, the sync_seg segment is the segment within a bit time used to synchronize all CAN nodes.

Temperature sensor   A transducer that can convert temperature into a voltage.

Ten-bit addressing   In the I²C protocol, a master device can use either a 7-bit or 10-bit value to address a slave device. When a 10-bit value is used, it is called 10-bit addressing.

Text editor   A program that allows the end user to enter and edit text and program files.

Thermocouple   A transducer that converts a high temperature into a voltage.

Transducer   A device that can convert a nonelectric quantity into a voltage.

Trap   A software interrupt; an exception.

UART   The acronym for Universal Asynchronous Receiver and Transmitter, an interface chip that allows the microprocessor to perform asynchronous serial data communication.

Union   A variable that may hold (at different times) objects of different types and sizes, with the compiler keeping track of size and alignment requirements.

Vector   A vector is a one-dimensional data structure in which each element is associated with an index, *i*. The elements of a vector are of the same length.

Volatile memory   Semiconductor memory that loses its stored information when power is removed.

Volatile variable   A variable that has a value that can be changed by something other than user code. A typical example is an input port or a timer register. These variables must be declared as *volatile* so that the compiler makes no assumptions on their values while performing optimizations.

Write collision   The SPI error that occurs when an attempt is made to write to the SPDR register while data transfer is taking place.

Index